use coupon code: gurumurthy
to avail online services
at www.nipaers.com

# Descriptive and Objective Soil Science

AF755079

nipa books for everyone
PAPERBACKS

NIPA GENX ELECTRONIC RESOURCES & SOLUTIONS P. LTD.
New Delhi-110 034

## About the Author

**P. Gurumurthy,** MSc(Ag.) PhD, Professor & Head, Department of Soil Science & Agricultural Chemistry, Agricultural College, Naira, Acharya N.G. Ranga Agricultural University, had 17 years of teaching experience of BSc.(Ag.) and MSc.(Ag.) courses and 2 years of research experience and 5 years experience in the field of extension as Programme Coordinator of University extension unit. He taught and is teaching Soil Science courses of Soil Physics; Soil Chemistry, Fertility and Nutrient management; Manures and fertilizers; Diagnostic techniques for soil, plant and irrigation waters; Soil survey, Soil Taxonomy and Remote sensing techniques for Undergraduate students and Soil pollution; Problem Soils and their management; Remote sensing and GIS techniques for soil, water and crop studies for post graduate students with passion. He published three books, forty six full length research articles in peered review national and international journals, participated in eighteen national/ international symposiums/ seminars/ workshops. He is member of professional societies Indian Society of Soil Science, Indian society of soil survey and land use planning, the Andhra Agricultural union, Indian society of oil seed research. Dr. Murthy also participated in more than dozen summer/ winter schools and updated the knowledge in Soil Science. He is recipient of prestigious Jawaharlal Nehru award for outstanding post graduate research in 2002, Indian Society of Soil Science (ISSS) Best Doctoral research presentation award in 2001, prestigious State level Best University Teacher Award in 2019, prestigious Ugadi puraskaram awarded by the Government of Andhra Pradesh in 2013 and state level Padmisri IV SubbaRao-Rytunestam foundation award in 2014 for his overall contributions in the field of Agricultural sciences, particularly in Soil Science and Agricultural Chemistry.

# Descriptive and Objective Soil Science

*(for B.Sc (Agriculture), B.Sc. (Horticulture) B.Sc. (Environment Science), JRF, SRF, ARS SAUs and other Competitive Examinations)*

**P. Gurumurthy**

Professor & Head
Department of Soil Science & Agricultural Chemistry
Agricultural College, Naira
Acharya N.G. Ranga Agricultural University
Guntur, Andhra Pradesh-522034

nipa books for everyone PAPERBACKS

**NIPA GENX ELECTRONIC RESOURCES & SOLUTIONS P. LTD.**
New Delhi-110 034

nipa books for everyone PAPERBACKS

**NIPA GENX ELECTRONIC RESOURCES & SOLUTIONS P. LTD.**
101,103, Vikas Surya Plaza, CU Block
L.S.C. Market, Pitam Pura, New Delhi-110 034
Ph : +91 11 27341616, 27341717, 27341718
E-mail:newindiapublishingagency@gmail.com
www: www.nipabooks.com

***For customer assistance, please contact***
Phone: + 91-11-27 34 17 17 Fax: + 91-11- 27 34 16 16
E-Mail: feedbacks@nipabooks.com

© 2022, Publisher

ISBN: 978-93-91383-62-6

All rights reserved. No part of this publication may be reproduced, stored in a retrieval system or transmitted in any form or by any means, including electronic, mechanical, photocopying recording or otherwise without the prior written permission of the publisher or the copyright holder.

This book contains information obtained from authentic and highly reliable sources. Reasonable efforts have been made to publish reliable data and information, but the author/s, editor/s and publisher cannot assume responsibility for the validity, accuracy or completeness of all materials or information published herein or the consequences of their use. The work is published with the understanding that the publisher and author/s are not attempting to render any professional services. The author/s, editor/s and publisher have attempted to trace and acknowledge the copyright holders of all material reproduced in this publication and apologize to copyright holders if permission and/or acknowledgements to publish in this form have not been taken. If any copyrighted material has not been acknowledged, please write to us and let us know so that we may rectify the error, in subsequent reprints.

**Trademark Notice:** NIPA, the NIPA logos and their presentations (the way they are written/ presented) in this book are the trademarks of the publisher and hence may not be used without written permission, if copied or used without authorization, the infringer will be prosecuted as per law.

NIPA also publishes books in a variety of electronic formats. Some content that appears in print may not be available in electronic books, and vice versa.

Composed and Designed by NIPA.

**D. Balaguraviah**
Former Dean P G Studies and University Head (Soil Science)
Acharya N.G. Ranga Agricultural Univeristy Andhra Pradesh, India

# Foreword

Soils perform multiple functions viz., (i) recycling organic wastes, (ii) providing physical support to terrestrial plants, (iii) supplying fundamental resources viz., water, nutrients and oxygen required for terrestrial primary production, (iv) providing habitat to a variety of soil organisms, with taxonomic identity and functions of several organisms still unknown/lesser known to the scientific and wider community, (v) regulating hydrological and mineral/nutrient cycling, with significant impacts on global climate, (vi) detoxification of organic and inorganic substances, leading to purification of water resource and (vii) resisting erosion. Soil Science deals with courses such as Soil Physics, Soil Chemistry, Soil Mineralogy, Soil Degradation, Soil Pollution, Soil & Water Conservation etc.

According to 5th Deans Committee recommendations of uniform Course curricular across the 54 State Agricultural Universities (SAUs), 9 ICAR educational units, 4 Central Agricultural Universities (CAUs) in the discipline of Soil Science & Agricultural Chemistry both at Undergraduate and Post graduate level. Many workshops have been held involving subject experts with in State Agricultural Universities and across the Universities to place commonalities in the curriculum which not only meets the present needs and also keeping in view of future needs.

In this contest it was some vacuum found to have suitable text book covering most of the topics in the discipline of Soil Science & Agricultural chemistry, although some authors published some text books covering some of the chapters. It is in this context the present publication 'Descriptive and Objective Soil Science' contained systematic and comprehensive information as per the curriculum recommended by the 5th Deans committee in the discipline of Soil Science & Agricultural Chemistry.

I appreciate the author Dr. P. Gurumurthy compiling the information on the topic and brought out a useful publication entitled 'Descriptive and Objective Soil Science'. I hope students of under graduation in Agricultural sciences and Post graduation (Soil Science) will be highly benefitted from the contents of the publication.

**D. Balaguraviah**
Former Dean P G Studies and University Head (Soil Science)

# Preface

Soil is the most fundamental resource to fulfill basic requirements of food, fiber and shelter of human race. Soil forms basis for all terrestrial life. Soil provides a wide range of ecosystem services that make humankind to survive and progress. Soils are complex mixtures of minerals, water, air, organic matter, and countless organisms that are the decaying remains of once-living things. It forms at the surface of land and is the "skin of the earth". Soil is capable of supporting plant life and is vital to life on earth.

The rate of soil formation is very slow (during every 100 to 400 years, only 1 cm soil is formed) and the enough soil depth is formed in 3000 to 12000 years to have a productive land. The soil covering the earth surface has taken millions of years to develop. Thus, when soil, a non-renewable natural resource is ruined then it will be lost entirely Soil degradation disturbs agricultural, environmental and ecological functions performed by the soil. Degradation of soil results in depletion of soil fertility decreased moisture storage capacity and consequently in decreased crop productivity. Among different soil degradation processes, soil erosion causes land deterioration. Soil erosion is detachment and dislocation of soil due to the action of water or wind. Soil loss due to erosion has great consequences because it leads to loss of its productivity. Soil erosion occurs though out the world but it is a very common feature and more serious problem in dry areas. In addition to loss of soil fertility and crop yields, soil erosion also increases environmental pollution, increasing the sediment load in streams and rivers, thereby disturbing the aquatic life, particularly fish. In the long run, soil erosion affects socio-economic conditions of the society by causing floods, silting up of water reservoirs and disruption of communication systems.

The present publication 'Descriptive and Objective Soil Science' aimed to provide chapter wise brief information on different concepts of soil science viz., Soil genesis, Soil physics, Soil chemistry, Soil mineralogy, Soil heath, Soil fertility, Manures and Fertilisers, Soil degradation, Problematic soils, Soils of India, Irrigation water quality, so on along with objective questions at the end of the each chapter. Fifth deens' committee recommendation was taken in view while compiling the information. This information will be much useful for the learners for overall comprehension aswell as successful attempts of competitive examination such as JRF, SFR, ASRB and competitive examinations of State Agricultural Universities in the disciplines of Soil Science & Agricultural Chemistry, besides it is also useful for the ambitious candidates of completive examinations conducted by UPSC. References were mentioned at the end to facilitate readers for additional information. This text book is useful for students of BSc(Ag.) BSc. (Horti), BSc. Environment Science), MSc. (Soil Science) MSc. (Agronomy), BSc. (Agrl.Eng.).

I, take this opportunity to express my gratitude to University officers of Acharya N.G.Ranga Agricultural University and Associate Dean of Agricultural College, Naira for their moral support in bringing out this publication. I am also grateful to the faculty and other staff of the Department of Soil Science & Agricultural chemistry, Agricultural college, Naira for extending their help during the preparation of the manuscript. It is hoped that the students, faculty, researchers and environmentalists will find this publication useful. Constructive criticism is appreciated for improvements in the revised editions.

Naira, Srikakulam District **P. Gurumurthy**
Andhra Pradesh, India

# Contents

# 1

# Introduction

Soil is the most fundamental resource to fulfill basic requirements of food, fiber and shelter of human race. Although soil is perceived as something of insignificant value and it is considered as dirt but humans cannot survive without soil. Soil forms basis for all terrestrial life. Soil provides a wide range of ecosystem services (Table 1) that make humankind to survive and progress. Soils are complex mixtures of minerals, water, air, organic matter, and countless organisms that are the decaying remains of once-living things. It forms at the surface of land and is the "skin of the earth". Soil is capable of supporting plant life and is vital to life on earth.

**Table 1:** Ecosystem services provided by soil, (*Adopted from Blanco and Lal, 2008).*

| Food security, biodiversity | Water quality control | Climate change mitigation | Energy resource |
|---|---|---|---|
| Fiber<br>Food Housing<br>Recreation<br>Infrastructure<br>Disposal of waste<br>Diversity of microbes<br>Flora and fauna conservation | Pollutants filtration<br>$CH_4$ and $CO_2$ sinks<br>Purification of water<br>Retention of sediments andchemicals<br>Chemicals buffering and transformation | Sequestration of Carbon in soil and biota<br>Nitrification inhibitors<br>Deposition and burial of carbon enriched sediments | Bioenergy crops(warm season grasses and short rotation woody crops Prairie grasses) |

Soil is also an important non-renewable natural resource is subjecting for degradation due to faulty land use practices. Soil degradation disturbs agricultural, environmental and ecological functions performed by the soil. Degradation of soil results in depletion of soil fertility decreased moisture storage capacity and consequently in decreased crop productivity. Among different soil degradation processes, soil erosion causes land deterioration. Soil erosion is detachment and dislocation of soil due to the action of water or wind. Soil loss due to erosion has great consequences because it leads to loss of its productivity. Soil erosion occurs though out the world but it is a very common feature and more serious problem in dry areas. In addition to loss of soil fertility and crop yields, soil erosion also increases environmental pollution, increasing the sediment load in streams and rivers, thereby disturbing the aquatic life, particularly fish. In the long run, soil erosion affects socio-economic conditions of the society by causing floods, silting up of water reservoirs and disruption of communication systems. The rate of soil formation is very slow (on average for every

100 to 400 years, only 1 cm soil is formed) and the enough soil depth is formed in 3000 to 12000 years to have a productive land. The soil covering the earth surface has taken millions of years to develop. Thus, when soil, a non-renewable natural resource is ruined then it will be lost entirely (Lal, 2001).

## 1.1.Earth's evolution

The origin and evolution of earth is best explained by 'Planetesimal Hypothesis' put forth by T.C. Chamberlain and F.R. Moulton in 19th century and Nebular hypothesis propered by Immanuel Kant and Piorre Laplace 17th century. According to Planetesimal hypothesis this when a big star comes closer to sun it creates certain gravitational pull because of which tearing away of the outer matter took place. The teared matter got condensed and started to move around the sun in orbits. These are called planetary nucleoids. According to 'Nebular Hypothesis', the whole solar system at one time is a cloudy (nebula) mass of matter in form of vapor or gas. It was intensely hot and incandescent. It was in perpetual motion. On lapse of time nebula began to cool and the vapors in the outer layer are condensed and thrown out in the form of rings. Further condensation and disruption lead to formation of planets. The central mass which is still incandescent and not yet condensed is Sun. The planet earth consisting of the solid zone (Lithosphere) which covers about three forth with water forming seas and oceans (liquid spheres) is Hydrosphere and the gaseous envelop over the earth surface is Atmosphere.

## 1.2. Spheres of earth

Depending on the state of matter the earth has been geologically divided into three sphere viz. atmosphere, hydrosphere, and lithosphere.

**i) Atmosphere:** The thin mantle of gases surrounding the outer surface of the earth as an envelope is called the atmosphere of the earth. Mass of the atmosphere is about $5.249 \times 10^{18}$ kg, while that of the earth is about $5.973 \times 10^{24}$ kg. Atmosphere thus accounts for less than one millionth of the total mass of the earth. Atmosphere extends up to a height of 500 Km. from the earth's surface and is divided into four distinct regions (Table 1).

**Table 1:** The regions of the atmosphere and important chemical species.

| Region | Attitude range | Important chemical species |
|---|---|---|
| Troposphere | 0 to 11 | Oxygen and nitrogen |
| Stratosphere | 12 to 50 | Ozone |
| Mesosphere | 51 to 85 | Oxygen, nitric oxide |
| Thermosphere | 86 to 500 | Oxygen and nitrate |

The region of 200 kilometers from thermosphere, considered a part of the atmosphere, is referred to as exosphere. An approximate composition of the first few kilometers of the atmosphere with which biosphere is intimately concerned given in table 2:

**Table 2:** Approximate composition of the atmosphere.

| S.No | Category | % by volume |
|---|---|---|
| 1 | **Major constituents** | |
| | i) Nitrogen | 78.09 |
| | ii) Oxygen | 20.94 |
| | iii) Water vapour | 0.10 |
| 2 | **Minor constituents** | |
| | i) Argon | 0.90 |
| | ii) Carbon dioxide | 0.032 |
| 3 | **Trace constituents** | |
| | i) Neon | 0.0018 |
| | ii) Helium | 0.005 |
| | iii) Methane | 0.002 |
| | iv) Others (ammonia, hydrogen and Ozone | Balance |

At sea level atmosphere exerts a pressure of 14.71 psi (1037 cm of water column). As attitude increases from the earth's surface, the density of the atmosphere decreases with a concomitant decline in pressure. Thus, atmospheric pressure drops from one atmosphere at sea level to $3x10^{-7}$ atmospheres at 100 kilometers above sea level. Temperature of the atmosphere ranges from - 92° to 1200°C.

## ii) Hydrosphere

Seas, oceans and fresh water reserve of the earth constitute hydrosphere of the earth. The total surface area of the earth is $5.1007x10^8$ $Km^2$, while that of the hydrosphere is $3.6113x10^8$ $Km^2$. Hydrosphere thus accounts for about 70.80 per cent of the area of the earth. On mass basis however, hydrosphere constitutes about 1/4400th of that of the earth. The total water reserve of the earth under hydrosphere is 1.4 billion $Km^3$. However, as much as 97 per cent of this water reserve is in the seas and oceans is unfit for human consumption and other uses due to high salt concentration, of the remaining 3 per cent water, 2.3 per cent is locked in the form of polar ice caps and out of man's reach. The balance of 0.70 per cent only is available as fresh water; again, as much 0.66 per cent of the 0.70 is in the form of ground water under the rock 0.03 to 0.04 percent is available in the rivers, lakes and ponds. Biological world is primarily dependent on this fraction of the fresh water reserve.

## iii) Lithosphere

Lithosphere is the solid ouster-most part of a rocky planet. If an egg is considered like a planet, then the shell of the egg is like the lithosphere of the planet. Earth's lithosphere accounts for 99.9 per cent of the mass of the planet. Lithosphere covers an area of $1.489x10^8$ $Km^2$. In other words, land mass covers 29.2 per cent of the total surface area of the earth. There are two kinds of lithosphere.

1. Continental lithosphere – 150 Km. thick consisting of about 50 Km of the crust and 100 Km or more the upper most mantles.
2. Oceanic lithosphere – associated with oceanic crust; typically 50 to 100 Km thick. Oceanic lithosphere is much younger than continental lithosphere.

The earth is nearly spherical with an equatorial radius of 6,378 km therefore the depth of the earth is the same as its radius. However, man's distinct geological knowledge is confined to the upper 50 to 60 km of the lithosphere. This upper region of the earth with which man is geologically much familiar is called the earth's crust. The base of the earth's crust is called the Moho. Thus, the region between the Moho of the lithosphere and the ground surface of the earth is in fact the earth's crust.

The earth's crust can be sub-divided in to two distinct regions called Sial and Sima. While the sial is a largely heterogeneous rocky material, Sima is fairly homogenous and forms a concentric layer. As the names imply, the key elements of sial and sima area silicon (about 65 to 72% $SiO_2$) and aluminum. Aluminum is time lighter than magnesium. Sial is therefore lighter than Sima- the precise densities of sial and sima are 2.72 and 2.94 g $cm^{-3}$ respectively. Obviously, therefore, sial remains underlain by sima.

Below the crust, there lies the viscous mantle. The mantle starts where the crust ends and runs up to 2890 kilometers depth. All the crust is thus in the lithosphere; but the lithosphere contains more mantle than the crust.

Below the mantle, there lies a much less viscous liquid called 'liquid outer core' of the earth. It runs up to 5100 Km. from its take off point ie. where the mantle ends. The outer core is followed by the solid inner core of the earth that goes up to the full depth of the earth ie., located between 5100 and 6378 kilometers depth of the earth.

Mention must also be made of the frequently used term Asthenosphere which is between 100 and 700 kilometers depth of the earth – which is the weaker hotter and deeper part of the upper mantle. Ignoring the masses of the atmosphere and hydrosphere (together less than 0.1%), crust, mantle, and core (both outer and inner together) account for 0.4, 68% and 31.5 per cent of the mass of the earth respectively.

## 1.3. Composition of the earth's crust

All the elements found naturally in the earth either in elemental state or chemically combined forms are present in the earth's crust. These elements found in the earth's crusts are chemically distributed in five principal categories, the criterion being their bonding characteristics. These categories are:|

i) Lithophile elements:- Those elements which readily ionize or form oxyanions. They include: O, Si, Ti, Fe, Mn, Al, H, Li, Na, K, Rb, Cs, Be, Mg, Ca, Sr, Ba, B, Ga, Ge, Sn, Sg, Y, F, Cl, Br, I, C, Hf, Th, P, V, Nb, Ta, G, W, U, Zr, Mo, Cu, Zn, Pb, Tl, As, Sb, Bi, S, Sr, Te, Ni, Co and rare earth elements.

ii) Chalcophile elements:- These elements tend to form covalent bonds with sulphides. These elements include: S, Se, Te, Fe, Ni, Co, Cu, n, Pb, Mo, Ag, Sb, Sn, Cd, In, Tl, Pb, As, Bi, Re, Mn, Ga and Ge

iii) Siderophile elements:- These elements form metallic bonds. These elements include: Fe, Ni, Co, Ru, Rh, Pd, Pt, Lr, Os and Au.

iv) Atmosphile elements:- These elements are found in the atmosphere and include: N, O, He, Ne, Ar, Kr and Xn

v) Biophile elements:- These elements are found in the biological systems and include: C, H, O, N, P, S, Cl, I, B, Ca, Mg, K, Na, Mn, Fe, Zn, Cu, Ag, Mo, Co, Ni, Au, Be, Cd, Se, Tl, Sn, Pb, As and V.

This distribution is however, not absolute and many elements are found to be distributed in more than one category. For example, the element, cobalt is in all the categories, except the atmosphile one. Though the earth's crust is the store house of all the elements, lithophilic ones are the most dominant. The top ten or so elements abundant in the earth's crust and their per cent mass as reported by Jackson are as presented in table 3:

**Table 3:** Abundance of elements in the earth's crust

| Element | Percent | Element | Percent | Element | Percent |
|---|---|---|---|---|---|
| O | 46.5 | Na | 2.8 | Mn | 0.09 |
| Si | 27.6 | K | 2.6 | S | 0.06 |
| Al | 8.1 | Mg | 2.1 | Cl | 0.05 |
| Fe | 5.0 | Ti | 0.6 | C | 0.04 |
| Ca | 3.6 | P | 0.12 | Others | 0.74 |

## Some generalizations can be drawn from the elemental composition of the earth's crust

1. Only eight elements viz. O, Na, Mg, Al, Si, K, Ca and Fe surpass 1 per cent of the mass of the earth's crust.
2. The abundance of the eight elements surpassing 1 per cent of the mass of the earth's crust is in the order: O>Si>Al>Fe>Ca>Na>K>Mg.
3. The fourteen elements listed above constitute as much as 99.36 per cent of the mass of the earth's crust, while all the remaining more than hundred elements make up just 0.64 per cent.
4. The most abundant element of the earth's crust is oxygen and alone accounts for nearly half of the total mass of the crust. On volume basis, oxygen accounts for almost 92%.
5. In abundance in the earth's crust, oxygen is followed by silicon accounting for more than 25 per cent of the total mass.
6. Oxygen and silicon together thus constitute nearly three fourth of the mass of the earth's crust. Since both of them are non-metals and added to that there are other non-metals as well, the non-metals greatly surpass the metals in abundance in little amounts.
7. Though, metals are more than the non-metals in number, only six elements viz. Al, Fe, Ca, Mg, Na and K are found in high amounts in the earth's crust. By comparison, the other elements like Cu, Zn, Pb, Cz, Ni, Sn etc., and not to speak of the noble metals are found in little amounts. The most abundant metal element is Al follows by Fe.

8. The third most non-metallic element is Ti followed by P and S. Incidentally, nature's usable reserve of P and S are not that high in the light of their demand and use and thus they are to be consumed with utmost restraint and misery.

The elements of the earth are seldom found to occur in their free elemental state in any one of the three spheres of the earth, except however only three elements. They are nitrogen a relatively non-reactive element, in the atmosphere, and carbon present in the lithosphere as diamond, graphite and coal. The other element on this count is oxygen present in the atmosphere. Oxygen unlike nitrogen is relatively far more chemically reactive and still present in elemental state. The origin of elemental nitrogen in the atmosphere is attributed to the development of modern oxygenated photosynthesis of the plant world, a process believed to have taken place in the last part of the biotic evolution. Once this happened, oxygen largely changed the composition of the earlier existing highly reduced atmosphere to the modern day atmosphere where oxygen is a key atmosphile element.

The rest of the elements are found in the nature-mostly in the lithosphere in chemically combined forms. These chemically combined forms are vastly different types, the common ones of which are oxides, nitrates, phosphates, carbonates, bicarbonates, sulphates, sulphides, halides, silicates, borates molyldates chromates etc. They are in fact the minerals of the nature. Among all these forms, only the following three forms are of prime importance in abundance. Silicates accounting for nearly 72 per cent. Oxides that constitute about 24 per cent. Carbonates making up about 3 per cent. All the other forms viz., sulphates, phosphates, chloride etc. constitute just 1 per cent. Among the minerals, the silicates are the most abundant followed by chlorides and carbonates.

The abundance of three forms is presumably due to:

1. The abundance of Si and O and their tendency to form silicates owing to their appreciable difference in electro negativity values.
2. Oxygen is abundant and highly electro negative and its tendency to form electrovalent bonds with metal elements, which are highly electropositive, thus forming a variety of oxides.
3. Ready availability of $CO_2$ an acidic oxide and its tendency to react with metal oxides, which are alkaline in nature, remitting in the formation of carbonates.

**Lithophilic elements:** The type of igneous rocks that form from magma is a function of three factors

i) Chemical composition of the magma: Accordingly the silica content of igneous rocks ranges from 75 to 45 percent. Further, if the magma was deficient in silica and rich in Na and K, nepheline was formed, while olivine was formed when Fe and Mg contents were very high.

ii) The rapidity of the solidification and crystallization process.

iii) The temperature of cooling determining the types of minerals that are found dominating the composition of such rocks. Rocks that begin their cooling at low temperature tend to be rich in minerals composed of Si, K and Al. Minerals with higher quantities of Ca, Mg, Na and Fe by contrast dominate high temperature cooling igneous rocks. Basalt and gabbros are such rocks and rich in these elements. Granite on the other side is a low temperature cooling rock and rich in quartz

Depending on the cooling process, igneous rocks are again sub classified in to two categories.

i) Volcanic or extrusive rocks-these rocks are formed by the cooling and solidification of lava thrown out mostly by volcanic eruption on the surface of the earth. Cooling thus was more rapid, producing many fractures and fine-grained rocks. Granite, grandiorite and diorite are all volcanic rocks.

ii) Plutonic or intrusive or abyssal rocks. These rocks are formed by the cooling of magma at considerable depth within the earth's surface. The cooling was slow on account of depth and pressure, giving the rocks enough time to form large crystals with well-defined faces.

Soil is the interface of the organic and inorganic chemistry of the terrestrial world, combining nitrogen and carbon from the atmosphere with the various elements of mineral lithosphere via the organisms anchored in the soil, intercepting energy of the sun and moisture from the hydrosphere and atmosphere. *"With out life, there is no soil and with out soil, there is no life on the earth planet"*

## 1.4. Soil components

The word **"Soil"** is derived from Latin word 'Solum' means 'floor' or 'ground'. Mineral soil consists of four major components i.e., inorganic or mineral materials, organic matter, water and air. In a representative loam surface soil, the solid mineral particles comprise about 45% of the soil volume and organic matter 5%. At optimum moisture for plant growth, the pore space is divided roughly in half, 25 %, of volume being water space and 25 % air. The proportions of air and water are subjected to rapid and great fluctuations.

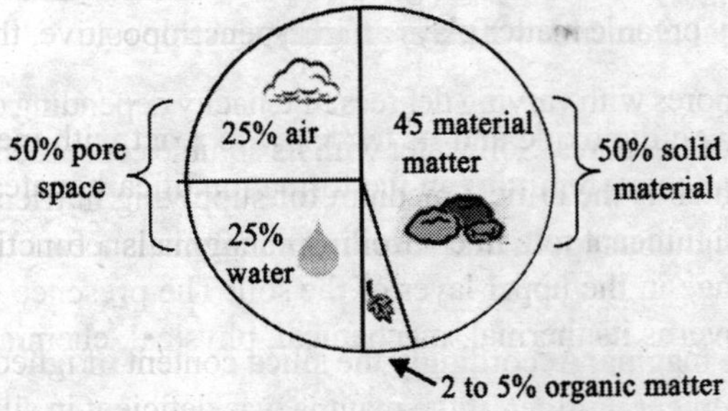


**Fig. 1:** Components of soil

The four soil components occur in a thoroughly mixed condition in soil and this mixture encourages interactions within and between the groups and permits marked variations in the environment for the growth of plants. The proportion of different components in the diagram depicts the good soil condition for plant growth (loamy surface soil). The air and water are extremely variable and their proportions determine in large degree the soil's suitability for the plant growth.

**Mineral matter**: The inorganic portion of soils is quite variable in size and composition. It is composed of small rock fragments and minerals of various kinds.

**Table 4:** Classification of mineral matter: The mineral matter of soil is classified by the size viz., gravel, cobbles, stones, sand, silt and clay.

| Size of the fragment | Name |
|---|---|
| **Rock Fragments (Non soil)** | |
| More than 250.0 mm | Stones or boulders |
| 75.0 - 250.0 mm | Cobbles (round) flags (flat) |
| 2.0 - 75.0 mm | Gravel or pebbles |
| **Soil Particles** | |
| 0.2 - 2.0mm | Coarse sand (gritty) |
| 0.02 - 0.2mm | Fine sand (less gritty) |
| 0.002 - 0.02mm | Silt (powdery) |
| Less than 0.002mm | Clay (sticky) |

The proportion of different sized particles (texture) determines the nutrient supplying power of the soil, considerably. Primary minerals (original) are prominent in sand and silt fractions; whereas the secondary minerals (formed) dominate in clay fraction and in some cases the silt fraction. The inorganic fraction of soil is the original source of most of the mineral elements that are essential for plant growth.

**Organic Matter**: It comprises an accumulation of partially disintegrated and decomposed plant and animal residues and other organic compounds synthesized by soil microbes as the decay occurs. It is a transitory soil constituent as it is continuously broken down by soil organisms and lasts from few hours to several hundred years. It requires maintenance by regular addition to the soil of plant and/ or animal residues. Organic matter content varies from 1.0 to 6.0 % by weight in top soil and very less in sub soil. In respect of soil productivity organic matter plays an indispensable role.

**Soil Water:** Soil water is held in soil pores with varying degrees of tenacity depending on the amount of water present and size of the pores. Soil water with its soluble constituents (nutrients) makes up soil solution, which is the critical medium for supplying nutrients to growing plants. Soil water plays significant role in controlling energy balance of the soil and regulates the gaseous exchange in the upper layer of the soil. The presence of water in different amounts in soil governs its thermal, mechanical, physical, chemical and biological properties.

**Soil Air:** The content and composition of soil air are determined largely by the water content of the soil, since the air occupies those soil pores not filled with water. Soil air always differs from atmosphere air in composition because of moisture content, root and microbial activities. Relative humidity of soil air may approach 100% at optimum soil moisture. The $CO_2$ content of soil air is often several times higher than that of atmospheric air (0.033%). Oxygen content of soil air usually less than 20%, in extreme cases only 5-10%. In cases of poor soil aeration, the diffusion rate of air in to and out of the soil would be slow and this leads to unsatisfactory conditions for optimum plant growth.

## 1.5. Definitions of soil

According to **Dokuchaev (1900)**, "Soil is a natural body composed of inorganic and organic constituents, having a definite genesis and a distinct nature of its own". **Jenny (1941)** defined soil as a natural occurring body that has been evolved owing to combined influence of climate and other organisms, acting on parent material, as conditioned by relief over a period of time. According to **Buckman and Brady (1969)**, Soil is a dynamic natural body on the surface of the earth in which plants grow, composed of mineral and organic materials and living forms.

According to Soil Science Society of America (**SSSA, 1970),** Soil is the unconsolidated mineral matter on the surface of the earth that has been subjected to and influenced by genetic and environmental factors of parent material, climate, macro and micro organisms and topography, all acting over a period of time and producing a product, that is soil, that differs from the material from which it is derived in many physical, chemical, biological and morphological properties and characteristics.

Soil also can be defined as a "natural body synthesized in a profile form from a variable mixture of broken and weathered minerals and decaying organic matter, which covers the earth in a thin layer and which supplies, when containing proper amounts of air and water, mechanical support and in part sustenance to plants"

## 1.6. Concepts of soil

The basic concept of soil, since humankind as an entity worthy of attention first recognized it, seems to have been that of a thin mantle over the land surface. The past conceptions of soil are both the foundations and major building stones for the conceptions prevailing now. The examination of some earlier conceptions of soil and the ways these developments can throw light on the nature of present conceptions. Knowledge about soils comes from two basic sources, first from the farmer experience based on centuries of trial and error, and second from the scientific investigations of soils and their management

### Early scientific investigations on soil science

French agriculturist **J.B. Boussingault** (1834) proved that air and water were the primary sources of C, H, and O in plant tissue. German chemist, **Justus Von Leibig** (1840) disproved the humus theory and postulated 'Crop yields were increased by adding minerals or inorganic elements to the soil'. He proposed that the mineral elements in soil and in added manures and fertilizers are essential for plant growth. Leibig's research led to the concept **Law of Minimum,** that certain factors were essential for plant growth and that if any one of these factors was limiting, plant production would be reduced accordingly. In 1851, Sir **G.G.Stokes** formulated the law governing the rate of settling of spherical particles in a viscous medium. 'the terminal velocity of a spherical particle settling under the influence of gravity in a fluid of a given density and viscosity is proportional to the square of the particle's radius'. **Henry Darcy** stated that the 'velocity of flow of a liquid through a porous medium is proportional to the force causing the flow and to the hydraulic conductivity of the medium'. **R. Warrington** hypothesized nitrogen transformations are biological. **Fouriers**'s law states that the negative gradient of temperature and the time rate of heat transfer is proportional to the area at right angles of that gradient through which the heat flows. **Winogradski** (1890), isolated two groups of bacteria responsible for nitrogen transformations. He proved nitrogen assimilating bacteria grow in nodules of leguminous plants. **C.G. Hopkins, d**eveloped effective soil management system based on lime stone, rock phosphate for legumes. **Milton Whitney** of USDA emphasized field studies and initiated the first National Soil Survey Systems. **F H King** (Wisconsin) studied the movement and storage of water in soils in relation to root penetration and crop growth.

## Pedologica and edaphalogical concepts of soil

According to Pedological concept, the origin of the soil, its classification and its description are examined in Pedology (Greek word pedon, means soil or earth). Pedology is the study of soil as a natural body and does not focus on the soil's immediate practical use. A pedologist studies, examines and classifies soil as they occur in their natural environment. According to Edaphological concept, (Greek word edaphos, means soil or ground) it is the study of soil from the stand point of higher plants. Edaphologists consider the various properties of soil in relation to plant growth. They are practical and have the production of food and fibre as their ultimate goal. They must determine the reasons for variation in the productivity of soils and find means for improvement.

## 1.7. Branches of soil science

**Pedology:-** The Science dealing with the genesis, survey and classification and the laws of geographical distribution of soils as a body in nature. Pedology is the study of soil as a natural body and does not focus primarily on the soil's immediate practical use.

**Edaphology:-** The science that deals with the influence of soils on living things, particularly plants, including man's use of land for plant growth. Edaphology is the study of soil from the stand point of higher plants.

**Soil Fertility: -** The branch of soil science that deals with quality of the soil that enables it to provide essential chemical elements in quantities and proportions for the growth of specified plants.

**Soil Chemistry:-** Deals with the chemical constituents , the chemical properties and the chemical reactions of soil in relation to crop needs.

**Soil Physics:-** Study of various physical processes that are taking place in and through the soils.

**Soil Microbiology:-** Deals with microscopic population of the soil, its role in various transformations and its importance in plant nutrition and crop production.

**Soil Conservation:-** Deals with the protection of soil against physical loss by erosion and against chemical deterioration.

**Soil Genesis:-** The study of the mode of origin of soils, with special reference to the processes responsible for the development of Solum or true soil from the unconsolidated parent material.

**Soil Survey:-** The systematic examination, description, classification and mapping of soils in an area.

## Review questions

1. List out various spheres of earth and describe characters of Lithosphere
2. What are principal categories Elements found in the earth's crust?
3. Explain the components of soil.
4. Differentiate between soil pedology and edaphology

## Objective type questions

1. Soil provide the following ecosystem service

   a) Food security, biodiversity  b) Water quality control
   c) Climate change mitigation  d) all three

2. Soil is

   a) non-renewable natural resource  b) renewable natural resource
   c) non-renewable synthetic resource  d) non-renewable synthetic resource

3. Degradation of soil results in

   a) depletion of soil fertility  b) decreased moisture storage capacity
   c) decreased crop productivity  d) all three

4. Among different soil degradation processes, causes land deterioration.

   a) soil erosion  b) soil acidity
   c) soil pollution  d) soil salinity

5. Enough soil depth is formed to have a productive land in

   a) in 100 to 120 years  b) in 300 to 800 years
   c) in 1000 to 2000 years  d) in 3000 to 12000 years

6. Planetesimal Hypothesis of earth evolution was laid out by

   a) Henry Darcy  b) T.C. Chamberlain
   c) F.R. Moulton  d) GG Stokes

7. Nubular Hypothesis of earth evolution was laid out by
   a) Henry Darcy
   b) T.C. Chamberlain
   c) F.R. Moulton
   d) GG Stokes
8. The thin mantle of gases surrounding the outer surface of the earth as an envelope is
   a) Hydrosphere
   b) Atmosphere
   c) Lithosphere
   d) Troposphere
9. Important chemical species in stratosphere is
   a) Oxygen
   b) Nitric oxide
   c) Ozone
   d) Nitrogen
10. Oxygen content in atmosphere is
    a) 0.032%
    b) 20.94
    c) 78.09
    d) 0.90
11. Nitrogen content of the atmosphere is
    a) 0.032%
    b) 20.94
    c) 78.09
    d) 0.90
12. $CO_2$ content of atmosphere is
    a) 0.032%
    b) 20.94%
    c) 78.09%
    d) 0.90%
13. At sea level atmosphere exerts a pressure of
    a) 14.71 psi
    b) 1037cm water column
    c) 76cm Hg column
    d) all three
14. The range of atmosphere temperature is
    a) 4000 to 6000 °C
    b) - 92° to 1200°C
    c) 500 to 2000 °C
    d) -197 to 600 °C
15. Hydrosphere accounts for about per cent of the area of the earth
    a) 33.3
    b) 66.7
    c) 45.2
    d) 70.80
16. Out of the total hydrosphere ------- % is present as fresh water
    a) 1.7
    b) 7.0
    c) 0.7
    d) 17
17. Out of the total hydrosphere ------- % is present in the rivers, lakes and ponds
    a) 0.3 to 0.4.
    b) 0.03 to 0.04
    c) 0.1 to 0.2
    d) 0.6 to 0.7

18. Earth's lithosphere accounts for ---per cent of the mass of the planet
    a) 99.9 b) 75
    c) 33 d) 20.9
19. The base of the earth's crust is called
    a) Geta b) Ohm
    c) Moho d) Soil
20. The key elements of sial and sima of earths crust are
    a) Ca& Mg b) Cu &Zn
    c)B & Mo d) Si & Al
21. The density of sial if earth crust is
    a) 2.94 b) 2.72
    c) 2.83 d) 2.65
22. The density of sima of earth crust is
    a) 2.94 b) 2.72
    c) 2.83 d) 2.65
23. Earth's crust accounts for ---- % mass of earth
    a) 31.5 b) 68
    c) 0.4 d) 0.1
24. Hydrosphere accounts for ---- % mass of earth
    a) 31.5 b) 68
    c) 0.4 d) 0.1
25. N, O, He, Ne, Ar, Kr and Xn are
    a) Lithophile elements b) Chalcophile elements
    c) Atmosphile elements d) Siderophile elements
26. C, H, O, N, P, S are
    a) Lithophile elements b) Biophile elemetns
    c) Atmosphile elements d) Siderophile elements
27. The abundance of top eight elements (on mass basis) of the earth's crust is in the order of
    a) Si>O> Al> Ca>Fe> Na>K>Mg b) O>Al >Si >Fe>Ca>Na>Mg >K
    c) O>Si>Fe >Al >Ca>K >Na >Mg d) O>Si>Al>Fe>Ca>Na> K>Mg
28. The most abundant element of the earth's crust is
    a) Oxygen b) Silicon
    c) Aluminium d) Iron

29. The most abundant metal element of the earth's crust is

a) oxygen b) Silicon
c) Aluminium d) Iron

30. The elements that constitute nearly three fourth of the mass of the earth's crust are

a) iron and aluminium b) Oxygen and silicon
c) Zinc and Copper d) Calcium and Magnesium

31. The third most non-metallic element in earth crust is

a) P b) S
c) Ti d) Si

32. The key atmosphile element is

a) Si b) Al
c) O d) Ar

33. Silicates of the earth crust constitutes

a) 24 per cent b) 72 per cent
c) 3 per cent d) 1 percent

34. Oxides of earth crust constitutes

a) 24 per cent b) 72 per cent
c) 3 per cent d) 1 percent

35. Sulphates, phosphates, chloride of earth crust constitute

a) 24 per cent b) 72 per cent
c) 3 per cent d) 1 percent

36. The range of silica content in igneous rocks

a) 10 to 25 percent b) 75 to 45 percent
c) 25 to 40 percent d) 40to 50percent

37. The dynamic soil component is

a) mineral matter b) soil air
c) soil water d) B&C

38. the size range of sand particle is

a) 0.002to 0.02mm b) less than 0.002mm
c) 0.02 to 2.0mm d) 2.0 to 75mm

39. Study of soil from the stand point of higher plants is

a) Pedology b) Botany
c) Edaphology d) Agronomy

40. The Science dealing with the genesis ,survey and classification and the laws of geographical distribution of soils as a body in nature is

a) Geology b) Pedology
c) Edaphology d) Agronomy

41. The definition of soil "Soil is a natural body composed of inorganic and organic constituents, having a definite genesis and a distinct nature of its own" given by

a) Jenny b) Marbut
c) Joffe d) Dokuchaev

42. The definition of soil as 'a natural occurring body that has been evolved owing to combined influence of climate and other organisms, acting on parent material, as conditioned by relief over a period of time' was given by

a) Jenny b) Marbut
c) Joffe d) Dokuchaev

43. The scientist proved that air and water were the primary sources of C, H, and O in plant tissue was

a) Justus Von Leibig b) J.B. Boussingault
c) GG Stoke d) Henry Darcy

44. The scientist disproved the humus theory was

a) GG Stokes b) J.B. Boussingault
c)Justus Von Leibig d) Henry Darcy

45. Humus theory (humus was considered the sole and direct source plant nutrients) was formulated by

a) J.B. Boussingault b) Justus Von Leibig
c) Albrecht Thaer d) GG Stoke

46. The concept of Law of Minimum was proposed by

a) J.B. Boussingault b) Justus Von Leibig
c) Albrecht Thaer d) GG Stoke

47. The hypothesis of R. Warrington was

a) nitrogen transformations are chemical b) nitrogen transformations are biological
c)phosphorous transformations are biological d) phosphorous transformations are chemical

48. 'The negative gradient of temperature and the time rate of heat transfer are proportional to the area at right angles of that gradient through which the heat flows', choose the scientist forposed the hypothesis

a) GG Stokes b) J.B. Boussingault
c) Fouriers d) Henry Darcy

49. The scientist proved nitrogen assimilating bacteria grow in nodules of leguminous plants and isolated two groups of bacteria responsible for nitrogen transformations is

 a) J.B. Boussingault  b) Winogradski
 c) Albrecht Thaer  d) GG Stoke

50. The scientist first emphasized field studies and initiated the first National Soil Survey Systems at USDA was

 a) Milton Whitney  b) Justus Von Leibig
 c) Albrecht Thaer  d) GG Stoke

51. The branch of soil science deals with systematic examination, description, classification and mapping of soils in an area is

 a) Soil genesis  b) Soil survey
 d) Soil Classification  d) Soil fertility

52. The branch of soil science that deals with the protection of soil against physical loss by erosion and against chemical deterioration is

 a) Soil genesis  b) Soil survey
 d) Soil Classification  d) Soil conservation

**Answer key**

| | | | | | | | | | | | | | | | | | | | | | |
|---|---|---|---|---|---|---|---|---|---|---|---|---|---|---|---|---|---|---|---|---|---|
| **1** | d | **6** | c | **11** | c | **16** | c | **21** | b | **26** | b | **31** | c | **36** | b | **41** | d | **46** | d | **51** | b |
| **2** | a | **7** | c | **12** | a | **17** | b | **22** | a | **27** | d | **32** | c | **37** | d | **42** | a | **47** | b | **52** | d |
| **3** | d | **8** | b | **13** | d | **18** | a | **23** | c | **28** | a | **33** | b | **38** | c | **43** | b | **48** | c | | |
| **4** | a | **9** | c | **14** | b | **19** | c | **24** | d | **29** | c | **34** | a | **39** | c | **44** | c | **49** | b | | |
| **5** | d | **10** | b | **15** | d | **20** | d | **25** | c | **30** | b | **35** | a | **40** | b | **45** | c | **50** | a | | |

# 2

# Soil Genesis and Classification

Soil genesis, or pedogenesis, is the combined effect of physical, chemical, biological and anthropogenic processes working on soil parent material (rocks & minerals). Soil is said to be formed when organic matter has accumulated and colloids are washed downward, leaving deposits of clay, humus, iron oxide, carbonate, and gypsum, producing a distinct layers. These constituents are moved from one level to another by water and animal activity. As a result, layers (horizons) form in the soil profile. The alteration and movement of materials within a soil causes the formation of distinctive soil horizons. Soil formation starts with weathering of rocks and minerals to smaller fragments and new compounds followed by certain of pedogenic processes makes the soil development. Therefore it is pertinent to study rocks and minerals before understanding the processes of soil genesis.

## 2.1. Rocks and Minerals

The weathering of rocks forms soils and minerals spread over a long time and in concert with other factors. Soils therefore were made exclusively of mineral matter at the time of their origin. This matter was in turn obtained from the once existing entirely rocky mass of the earth—more precisely the lithosphere of the earth. Lithosphere is the solid sphere of the three different spheres of the earth.

### 2.1.1. Rocks

Rocks are hard solid materials present in the earth's crust. Rocks are naturally occurring aggregates of one or more minerals. Cementing agents performs the task of joining the constituents together is performed by cementing agents. The rocks thus can be monomonieralitic made up of only one mineral like limestone in which the single mineral is calcite and sand stone made up of quartz as the only mineral. Polymineralitic rocks include granite made up of the minerals quartz and orthoclase and basalt made up of anorthite, antigite and hornblende.

Rocks form the units on the earth's crust. Since they are mix-ups of different constituents in varying proportions, they do not possess well-demarcated physical chemical and mineralogical properties. Rocks are identified and characterized on the basis of their various proportions like colour, specific gravity, optical properties, clearage, fracture and most importantly their mineralogical composition. However, as they are mere mixtures of different minerals in intimate association in the solid states, these criteria are not adhered to with a high degree of rigidity. The subject of the study of rocks is called petrology.

## Classification Rocks

Rocks are classified based on various criteria, the most important of which is their origin. Depending on the mode of their formation, rocks are classified in to three categories viz., igneous, sedimentary and metamorphic rocks.

### A. Igneous rocks

Igneous rocks also known as fine or crystalline rocks are the earliest formed rocks. The cooling down and subsequent solidification of magma formed them. Magma is the silicate malt of the fractures is thus much less and the rocks are coarse-grained rocks. Basalt and andesite are such coarse grained rocks, while gabbro and diorite mentioned earlier are fine-grained rocks.

Other well-known examples of intrusive rocks are rhyolite, scoria, obsidian (nature's glass) and pumice. Pumice does have plenty of air pockets that were trapped when the lava cooled while it was frothing. It is the only rock that flots on the surface of water. Intrusive rocks by the cooling of magma within the vertical, horizontal and irregular cracks of the existing rocks are called dykes, sills and veins, respectively.

Igneous rocks can also be classified based on their chemical composition such as acid and basic constituents. Silica is the notable acid constituent, while lime, soda and feldspar form the common basic constituents. Depending on silica content, igneous rocks are of four types:

i) Acid igneous or felsic rocks containing more than 65 per cent silica e.g. granite, grandiorite, rhyoite and dacite. They are light coloured due to dominance of quartz and sodium, potassium and plagioclase feldspars.

ii) Neutral or intermediate igneous rocks containing 52 to 65 per cent silica e.g. andesite, diorite, syenite and trachyte. They are composed of mainly plagioclase feldspars, amphiboles and pyroxenes.

iii) Basic igneous or mafic rocks containing 45 to 52 per cent silica e.g. basalt, gabbro, dolerite and diabase. Basalt is more widespread then gabbro. Basic rocks are dark coloured since they contain a high proportion of minerals rich in Fe and Mg (Pyroxenes, amphiboles and olivine). Mafic rocks are thus rich in Fe and Mg while the felsic rocks are comparatively richer in Al, Ca and Na.

iv) Ultra-basic rocks containing less than 45 per silica. These rocks are relatively less abundant and dominated by olivines, calcium rich plagioclase feldspars and pyroxenes. The most common ultra basic rock is periodolite. Other such rocks include dunite, pyroxene and hornblende.

Igneous rocks are made up of almost entirely primary minerals that include feldspars (nearly 60%), pyroxenes and amphiboles (almost 18%), quartz (nearly 12%) and micas (nearly 4%) other minerals constitute the balance of about 6 per cent. The mean chemical composition of igneous rocks is: $SiO_2$ (59.07%), $Al_2O_3$ (15.22%), $Fe_3O_4$(3.10%), FeO (3.71%), MgO (3.95%), CaO (5.10%), $Na_2O$ (3.71%), $K_2O$ (3.11%), $H_2O$ (1.30%), $TiO_2$ (1.03)%, $P_2O_5$ (0.30%), $H_2O$ (1.30%), Others (0.29%)

Igneous rocks account for about 95 per cent of the total rock mass and are thus the most abundant rocks. They are large in size and some times their thickness some times runs up to several kilometers. They are thus referred to as massive rocks too.

## B. Sedimentary Rocks

The principal agency involved in the formation of sedimentary rocks also known as aqueous rocks is water. Sedimentary rocks cover much of the land area of the earth's surface. Igneous and metamorphic rocks and some times some minerals too were subjected to various environmental agents like rains, sun light, heat etc., and were broken down to a loose mass. This loose material was then carried out by flowing water and deposited inside the lakes, rivers, seas and oceans and low lying areas in the form of sedimentary layers depending on the density of the materials. Some times dissolved materials present in the water were also added after evaporation of the water. At times cementing agents like silica or lime or sesquioxides to form rocks consolidated these deposits. These rocks are called sedimentary rocks. The cementation process is referred to as citification-a process that converts the newly deposited sedimentary rocks in to compact sedimentary rocks by the combined effects of pressure, heat, and cementing action. Citification can happen by way of (1) drying and compaction, (2) Oxidation of iron and aluminium and (3) precipitation of silica and calcium.

Sedimentary rocks are far less consolidated than igneous rocks and often contain fossils. Important minerals present in sedimentary rocks are quartz, calcite, dolomite and kaolin. They may be variously coloured due to coloured minerals like hematite (red), limonite (yellow), rutile (black) and malachite (green to blue). Well-known examples of sedimentary rocks are conglomerate and breccia composed of coarse fragments of rounded and angular gravels respectively.

Depending on the origin of the sediment material and mode and mechanism of consolidation, sedimentary rocks are of four types.

i. Residual sedimentary rocks formed due to the accumulation and subsequent lithification of those materials, which were left out as residues in the process of weathering and transportation e.g. Laterite and bauxite.

ii. Mechanical sedimentary rocks due to the accumulation and consolidation of the materials carried by flowing water in a state of suspension such as small gravels, pebbles, sand and silt e.g. shale, sand stone, silt stone, mud stone, shale and conglomerate.

iii. Chemical sedimentary rocks formed by the accumulation and cementation of some soluble materials in the flowing water by produced by precipitation. Precipitation might have taken place because change in the chemical environment of the water due to various reasons, notably microbial activity e.g. limestone and dolomite.

iv. Organic sedimentary rocks formed by the accumulation and consolidation of organic matter of plant and animal origin under high pressure and preserved under suitable conditions for long persons e.g. coal, lignite and chalk.

### C. Metamorphic Rocks

Metamorphic rocks are formed by the action of heat and pressure on igneous and metamorphic rocks. The rocks formed from the igneous and sedimentary rocks are often referred to as ortho and Para metamorphic rocks respectively.

Metamorphism involves the alteration of the existing rocks by either excessive heat or pressure. The alteration can cause chemical changes or structural modification of the minerals making up the rocks. Structural modification may involve simple reorganization of the minerals into layers or the aggregation of the minerals in to specific areas within the rocks. Some important metamorphic rocks and their parent rocks are as follows:

**Table 1:** Important metamorphic rocks developed from igneous and sedimentary rocks.

| Original rocks | Metamorphic rocks |
|---|---|
| **I. Igneous rocks:** | |
| 1. Granite | 1. Gneiss |
| 2. Basalt | 2. Schist |
| **II. Sedimentary rocks:** | |
| 1. Limestone | 1. Marble |
| 2. Shale | 2. Slate. |
| 3. Sandstone | 3. Quartzite |

Marble, a sedimentary rock of high value and aesthelic look is thus formed from limestone, a relatively innocuous rock to a common man. Marble as such is white coloured, but may have different shades or tints of other colours due to the presence of associated minerals. Other examples of metamorphic rocks are greenstone, amphibolites, tactile etc., Metamorphic rocks can be foliated with layers or bands e.g. slate, genesis and schist and non-foliated that are not layers e.g. marble and quartzite.

## 2.1.2. Minerals

Minerals are naturally occurring substances with more or less well defined chemical composition and properties. Great majority of them are crystalline and do possess identifiable crystal systems. Minerals form the constituent of rocks. Ores, which are of great significance in metallurgical industries, are either rocks or minerals.

### Classification of Minerals

Minerals are classified based on criteria of retention of inherent chemical composition, abundance in rocks specific gravity and chemical nature and ststage of weatherful. Based on the retention of the inherent chemical composition, minerals are classified into two groups:

i. **Primary minerals**: These minerals are inherited from the parent rocks and still retain the same chemical composition as in the original rocks, are called primary minerals. Almost all primary minerals are found in igneous and metamorphic rocks.

ii. **Secondary minerals**: These minerals have been formed due to subsequent, change in the chemical composition of the earlier minerals. They thus did not exist at the beginning and have been formed to physical and chemical changes taken place in the rocks and minerals. Clay minerals are secondary minerals. Secondary minerals are relatively more abundant in sedimentary rocks.

Most primary and secondary minerals however, are not absolute in their belongingness. A mineral characterized as a primary mineral in a particular rock or even soil may in fact be a secondary mineral in another rock or soil.

Based on the abundance of minerals in the rocks, minerals can be classified as:

i. **Essential minerals** – those minerals that are found in rocks in high amounts, so much so that they can be used for the characterization of a rock.

ii. **Accessory minerals** – those minerals that are present in the rocks in such small amounts that there presence or absence is not of any relevance for the identification and characterization of rocks.

This classification is therefore in reference to a particular rock and not absolute, since an essential mineral in one rock may as well be a secondary mineral in another rock.

Based on specific gravity/density the mineral are classified to two groups.

**i) Light Minerals :** Are the minerals which have specific gravity below 2.85 g/cc Eg. Quartz (2.60), Feldspar (2.65), Muscovite (2.50-2.75)

**ii) Heavy Minerals :** Having specific gravity **above 2.85 g/cc**

E.g. Haematite (5.30), Pyrite (5.0), Limonite (3.8), Augite (pyroxene) (3.1 – 3.6), Hornblende (amphiboles) (2.9 – 3.8), Olivine (3.5)

Minerals can also be classified depending on their chemical nature as follows,

i. Phosphate minerals e.g. appetites, stengite and variscite.

ii. Sulphate minerals e.g. gypsum, epsomite and barite.

iii. Silicate minerals eg. Orthoclase, muscovite and biolite.

iv. Carbonate minerals e.g. Calcite, magnesite and dolomite.

v. Chloride minerals e.g. Sylvite and sylvinite.

vi. Nitrate minerals e.g. Nitre and chilli salt peter

vii. Oxide minerals e.g., hematite, magnetite, rutile and limonite (hydrated oxide).

viii. Sulphides e.g. pyrites and galena.

Based on stage of weathering / weatherability the minerals are classified as

**Early stage of weathering: Ex.** Gypsum $CaSO_4.2H_2O$; Calcite $CaCO_3$; Dolomite $CaCO_3$. $MgCO_3$

**Intermediate stage of weathering:** Secondary clay minerals like Illite, Vermiculite, Montmorillonite.

**Advanced stage of weathering:** Silicate minerals like Kaolinite and Halloysite. Non- silicate minerals like Gibbisite $Al_2O_3.3H_2O$; Hematite - $Fe_2O_3$; Goethite FeOOH, Rutile and Anatase $TiO_2$; Zircon $ZrSiO_4$

A summarized version of the classification of minerals is depicted in the following in Figure1.

**Identification of minerals:** Minerals are identified by their characteristic properties. Some of these properties are as follows:

**Colour:** Most minerals do have varied colours due to the presence of associated minerals and impurities. Colour thus may not be a very reliable criterion; however, for the identification of a particular mineral, colour may help in narrowing down the range of minerals to zoom on a particular one.

**Specific gravity:** Specific gravity of a substance is a mere number indicating its heaviness in relation to water i.e., how many times heavier or lighter than water. Minerals, though they are not very pure in chemical state do have specific gravity within a narrow range. Specific gravity thus can be used in the identification of a mineral.

**Crystal nature:** The most fundamental distinction between two kinds of solids is that some are crystalline and some are amorphous, crystalline solids are those the atoms, ions, or molecules of which have an ordered arrangement extending over a long range. This order at the atomic or ionic or molecular level is also seen in the visible level because crystalline solids have flat faces and distinct angles. By contrast, those solids the said constituents of which are randomly arranged and have no ordered long-range structure are referred to as amorphous solids.

Crystals of a mineral can accordingly be (a) enhedral with perfectly developed faces, (b) sub-hedral without properly developed faces and (c) anhedral with no developed faces. Crystalline solids can also be variously grouped based on their axes – imaginary lines around which crystals can be rotated when two or more identical views of the crystal can be seen in one rotation. The axis, based on crystalline solids may be:

i. Monoclinic – a crystal structure of three unequal axes, two of which intersect obliquely and are perpendicular to the third one of gypsum, muscovite.

ii. Triclinic – a crystal structure of three unequal axes intersecting at oblique angles e.g. labradorite.

iii. Hexagonal – a crystal structure of four axes – Three equal axes intersect at 60° angles in one plane and the fourth axis of variable length is perpendicular to the other three axes e.g. covellite.

iv. Isometric or cubic – a crystal structure with three mutually perpendicular axes of equal lengths e.g. Magnetite.

v. Tetragonal – a crystal structure with three mutually perpendicular axes of equal lengths e.g. chalcopyrite.

vi. Trigonal – a crystal structure with ex bromide.

vii. Orthorhombic – a crystal structure of three mutually perpendicular axes all of different lengths e.g. epsomite.

**Streak:** The colour of the powder of a mineral may be different from its own colour. The colour of the power of a mineral obtained by rubbing against an unglazed porcelain plate is known as the streak of a mineral. Gray type hematite and galena have blood red and lead gray streak; pyrite is frassy gold, white its streak is black. A streak plate is 6.5 in Moh's hardness scale; a mineral harder than 6.5 shall not have a streak on streak plate, but might scratch the plate leaving a white powder of porcelain and not the mineral. Fortunately, however, most minerals harder than 6.5 do have a white streak.

**Cleavage:** Cleavage is the property of breaking along definite planes. Cleavage can be:

i. Perfect, if the broken surface is very smooth.

ii. Imperfect, if the broken surface is not smooth.

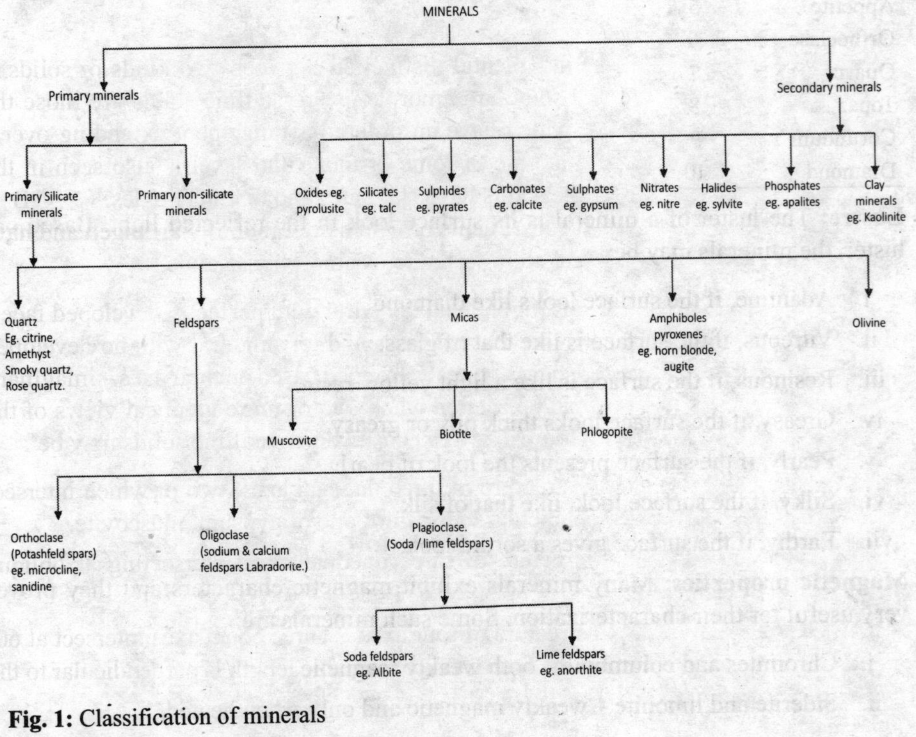


**Fig. 1:** Classification of minerals

**Fracture:** Fracture refers to the nature of surface obtained by breaking a mineral along a plane other than the cleavage plane. Minerals that do not have cleavage are said to have fracture. Based on fracture, the miners may be:

i. Conchoidal – the broken surface is concave.
ii. Even – the broken surface is smooth.
iii. Uneven – the broken surface is irregular.
iv. Filorous – the broken surface exhibits hair like strands.
v. Hackly – the broken surface is extremely irregular.
vi. Splintery – the broken surface exhibits splinter like projections.

**Hardness:** Hardness of minerals is measured by moh's scale suggested by Fried Rich Mohs in 1822. The scale selects ten minerals as standards in order of increasing hardness is given in table 2. Mohs scale of hardness of minerals is of great use in their identification.

**Table 2:** Hardness value of some selected minerals asper Moh's scale

| | |
|---|---|
| Talc | 1 |
| Gypsum | 2 |
| Calcite | 3 |
| Fluorite | 4 |
| Appetite | 5 |
| Orthoclase | 6 |
| Quartz | 7 |
| Topaz | 8 |
| Corundum | 9 |
| Diamond | 10 |

**Lustre:** The luster of a mineral is its surface look in the reflected light. Based on luster, the minerals may be:

i. Adantine, if the surface looks like diamond.
ii. Vitreous, if the surface is like that of glass.
iii. Resinous, if the surface is like a light yellow resin.
iv. Greasy, if the surface looks thick oily or greasy.
v. Pearly, if the surface presents the look of pearl.
vi. Silky, if the surface look, like that of silk.
vii. Earthy, if the surface gives a sordid look.

**Magnetic properties:** Many minerals exhibit magnetic characters and they prove very useful for their characterization. Some such minerals are:

i. Chromites and columbine – both weakly magnetic.
ii. Siderite and limonite – weakly magnetic and only when heated.

iii. Magnetite and magnetite – strongly magnetic.

**Other properties:** Other properties of minerals include florescence, phosphorescence (emitting light in visible range up on absorbing in visible or U.V.range), refractive index tenacity (firmness of the packing of the constituents of a crystal system).

**Characters of commonly observed primary minerals in soil**

**Quartz:** Quartz is the most abundant mineral in the earth's crust. It is abundantly present in igneous rocks like granite, grandiorite and rhyolite; sedimentary rocks like sand stone and silt stone; and metamorphic rocks like gneiss and quartzite. Quartz constitutes 50 to 90 per cent of the sand and silt fractions of soils. Quartz exists in a number of polymorphic forms like citrine, amethyst, rose quartz; milky quartz and smoky quartz etc. quartz is the mineral most resistant to chemical weathering.

Quartz in its chemical nature is the crystalline form of silica, $SiO_2$. Silica is made up of a continuous framework of silicon tetrahedron units. In each silicon tetrahedron, one Si ion occupies the center point between four oxygen ions. The five together forms a tetrahedron called Silicon tetrahedron. It however, keeps 4 negative charges in each unit. The basal oxygen ions of one tetrahedron are shared by Si ions of the neighbouring tetrahedron units. This takes care of the charge factor mentioned and results in the formation of a continuous silicon tetrahedral sheet as seen in quartz.

**Feldspars:** Feldspars are the second most abundant minerals – next only to quartz. Igneous rocks contain as much as 60 percent feldspars. Feldspars are lightweight light coloured. They are easily weather able due to their composition, solubility and size. Plagioclase feldspars weather more easily and sometimes form clay minerals. Feldspars are anhydrous alumino-silicates, principally of potassium, sodium, calcium and less frequently some other cautions. Feldspars are thus of different type:

i. Potash feldspars (KA ($Si_3O_8$). They are found in granite, gneiss and syenite. They include orthoclase, sanidine, microcline and adularia.

ii. Soda feldspars ($NaAlSi_3O_8$) – Found abundantly in granite, syenite and diorite. E.g. albite.

iii. Lime feldspars ($CaAl_2Si_2O_8$) – found in gabbro and basalt e.g. anorthite.

iv. Soda lime feldspars – found in e.g. laboratories.

**Olivines:** A large number of olivine is olive green minerals ranging between the two members forsterite ($Mg_2SiO_4$) and fayalite ($Fe_2SiO_4$). Olivines are neosilicate minerals. The structure of neosilicates consists of individual silicon tetrahedrons [$(SiO_4)^{4-}$] linked with another tetrahedron by a caution lying between them. They can be represented by the formula $R_2SiO_4$, where R is a divalent caution like Fe (II) or Mg. As they are neosilicates, they are ortho-silicates too. Olivines are widely present in igneous rocks like basalt and gabbro.

**Pyroxenes and Amphiboles:** Pyroxenes and amphiboles are ferromagnesian minerals. They are profusely substituted by various cations and thus act as sources of macro, micro and beneficial plant nutrients in soils. Pyroxenes a represent extensively in igneous rocks like granite and syenite and metamorphic rocks like gneiss and schist. These minerals are ion-silicates. Inosilicates exhibit a continuous framework of silicon tetrahedron units in which oxygen gives the connecting bridge between the adjoining tetrahedral. They may be single or double chained. Both pyroxenes and amphiboles weather with ease and form ferrogenous clays. Pyroxenes include dark coloured angite, institute and rhodonite. Amphiboles, which are more complicated and still more weatherable, include horn-blended and action-ophyllites.

**Micas:** Micas are double silicates of potassium and aluminium. In soils micas provide a good source of potassium for plant nutrition. They have a layer lattice crystal structure and are thus phyllo-silicates. In the structure of micas, one aluminium octa hedral layer (made up of aluminium Octahedron units) remains sand witched between two silicon tetrahedral layers. Micas therefore have plate like structures associated with perfect cleavage. There are two major types of micas.

i. Muscovite or white mica : In muscovite, nearly one fourth of the $Si^{4+}$ ions in the tetrahedral sheets gets replaced by $Al^{3+}$ ions. As +4 charge is being replaced by a+3 charge entity, the crystal system acquires a negarive charge. To maintain electro neutrality in the crystal system, K+ ions rush in from the surroundings and occupy the intercrystal space. Thus.

$$Si_8Al_4O_{20}(OH)_4 \longrightarrow K_2(Si_6Al_2)Al_4O_{20}(OH)_4$$

Such mica is also referred to as diocta hedral mica.

Muscovite is somewhat resistant to chemical weathering. It is found mostly in acid rocks like granite and gneiss and is normally not found in basic rocks.

ii. Biotite or black mica : In biotite, all the octahedral aluminium ions of muscovite get replaced and substituted by divalent Fe(2+) or Mg ions. Thus:

$$\underset{\text{Muscovite}}{K_2(Si_6Al_2)Al_4O_{20}(OH)_4} \longrightarrow \underset{\text{biotite}}{K_2(Si_6Al_2)(Mg.Fe)_6O_{20}(OH)_4}$$

Biotite is also referred to as trioctahedral mica. Biotite weathers much faster than muscovite and is found to occur in both acidic and basic rocks. Yet, another but much less common variety of mica is phlogopite. Phologipite is potassium, magnesium aluminosilicate and hydroxide with the approximate formula $K_2Kg_6Al_2\ Si_6O_{20}\ (OH)_4$

**Albite** ($NaAlSi_3O_8$): It is a tectosilicate of feldspar group. Albite, name derived from the root word 'albino'. Usually white or colourless, but can have sheds of blue, yellow, orange and brown. Luster-vitreous to dull crystal systeɱ-triclinic. Perfect cleavage in one direction and good in another direction forming nearly right angled prisms. Fracture conchoidal. Hardness 6.0 to 6.5. Specific gravity 2.6 (average) streak-white.

**Apatite**, $Ca_{10}(PO_4)_6$: Appetites rather than apatite depending on predominance of F, Cl, O, $OH^-$ and$CO_3$. Colour is typically green, but yellow, blue, reddish brown and purple varieties found. Lustre-vitreous to greasy and gumdrop, crystal system-hexagonal system. Indistinct cleavage in one basal direction. Conchoidal fracture. Hardness 5, specific gravity 3.1 to 3.2, streak-white.

**Anorthite,** Ca $Al_2Si_2O_8$: A tectosilicate of feldspar group. White or gray or colourless, vitreous to dull luster crystal system-triclinic, cleavage-perfect in one direction and good in another forming nearly right angled prisms. Conchoidal fracture hardness 6.0 to 6.5, specific gravity 2.76, streak-white.

**Augite,** (Ca, Na)(Mg, Fe, Al)(Al, Si)$_2O_6$: It is an ino-silicate of pyroxene group, colour-dark green, brown and black, vitreous to sub-metallic lustre, crystal system-monoclinic. Cleavage-perfect in two lengthwise directions. Uneven fracture, hardness 5.0 to 6.0, specific gravity 3.2 to 3.6 streak-greenish white.

**Azurite**, $2CuCo_3.Cu(OH)_2$: Deep blue colour, vitreous to dull luster. A crystal system-monoclinic, cleavage good in one direction and fair in another. Fracture conchoidal, hardness 3.0 to 3.5. Specific gravity-4.5, streak-white.

**Biotite**, $K_2$ $Si_6A_{12}(Fe.Mg)_6O_{20}(OH)_4$: A phyllo-silicate of mica group with black to brown colour, vitreou8s to pearly lustre, crystal monoclinic, cleavage perfect in one direction forming sheets or flakes. Fracture not readily observed due to perfect cleavage, but is uneven hardness-2.5. Specific gravity 2.9 to 3.4, white streak.

**Borax**, $Na_3B_4O_7$,$10H_2O$: White coloured with vitreous lustre, monoclinic crystal system, cleavage perfect in one direction, conchoidal fracture, hardness-2.0 to 2.5, specific gravity-1.7. White streak.

**Calcite**, $CaCO_3$: colour is extremely variable, but generally white or colourless. Vitreous to resinous to dull lustre, crystal system-trigonal. Perfect cleavage in three directions forming rhombohedron, conchoidal fracture. Hardness-3.0, specific gravity-2.7, white streak.

**Carnallite**, $K_2Mg_2Cl_6$. $6H_2O$: Colourless or white or yellow with vitreous, greasy, resinous or dull lustre, crystal system – orthorhombic cleavage-absent. Conchoidal fracture, hardness-2.5. Specific gravity 1.6 white streak of potash in fertilizer industry. An important source.

**Chalcocite**, $Cu_2S$: Dark gray to brown colour with metallic lustre, crystal system-orthorhombic. Imperfect cleavage in two directions, conchoidal fracture. Hrdness-2.5 to 3.0, specific gravity-5.5 to 5.8 streak-shiny black to lead gray, high Cu content (80%) makes it an important source of Cu.

**Chalcopyrite**, $CuFeS_2$: Brassy yellow coloured with metallic lustre, crystal system tetragonal cleavage is rather poor in one direction. Conchoidal and brittle fracture hardness-3.5 to 4.0. Specific gravity-4.2 streak-dark green.

**Corundum**, $Al_2O_3$: Colour-highly variable mainly white, blue, red, green, purple or colourless. Vitreous to adamantine lustre. Crystal system trigonal, cleavage is absent, though there is parting in three directions. Conchoidal fracture, hardness-9.0, specific gravity-4.0+. White streak corundum is the second hardest mineral next only to diamond. The red variety of corundum is ruby widely used in gum making, while the non-red variety is sapphire.

**Dolomite**, Ca Mg $(CO_3)_2$: Pinkish, but can also be white, gray or yellow or colourless. Pearly to vitreous luster. Trigonal crystal system. Cleavage is perfect in three directions forming rhombo hedrons. Conchoidal fracture. Hardness-3.5 to 4.0. Specific gravity – 2.86. White streak.

**Epsomite**, $MgSo_4.7H_2O$: Colourless, luster-vitreous in large form, silky to earthy in microforms. Crystal system orthorhombic. Cleavage-perfect in two directions, distinct in two other directions. Conchoidal fracture hardness 2.0 to 2.5 specific gravity 1.7 white of colourless streak.

**Fayalite**, $Fe_2SiO_4$: It is a neosilicate mineral of olivine group. Colour-yellowish green to greenish brown and even black. Vitreous luster and orthorhombic crystal system. Cleavage-distinct in two directionsat 90°. Conchoidal fracture, hardness 6.5 to 7.0. Specific gravity 4.3 streaks – white.

**Forsterite**, $Mg_2SiO_4$: Fayalite is the close relative of fayalite. The properties are more or less the same. It is Mg rich with a specific gravity of 3.2

**Gibbsite**, $Al(OH)_3$ White colour with vitreous to dull luster. Crystal system – monoclinic. Cleavage perfect in one direction, basal. Uneven fracture, Hardness 2.5 to 3.5. Specific gravity 2.4. White streak. Gibbsite is often found as a part of the structure of other minerals like kaolinite, illite and montmorillonite (all clay minerals).

**Goethite**, FeO.OH: Yellow or brown or black coloured with earthy to dull luster in large forms and metallic in small forms. Crystal system – orthorhombic. Cleavage – perfect in one direction. Fracture is uneven to splintery hardness 5.0 to 5.5. Specific gravity 3.3 to 4.3 streak – brown or yellow or orange.

**Gypsum**, $CaSO_4.2H_2O$: Gypsum, a common mineral of sedimentary environment has unusually large crystals. Plaster of Paris known for various uses is made by heating gypsum to 300°F, thus driving 75 per cent of the water. Heating further to 350°F expels the remaining moisture forming a hydrate. Gypsum is white in colour with vitreous to pearly luster and monoclinic crystal system. Good in one direction cleavage and distinct in two others directions in cleavage. Uneven fracture, hardness of 2.0, Specific gravity of 2.3 g/cm$^3$, white in streak. The colourless variety of gypsum with a pearly luster is called salenite.

**Halite** (NaCl): it is usually white coloured with vitreous luster. Crystalline system isometric. Prfect cleavage in three directions forming cubes, conchoidal fracture. Hrdness is2.0, specific gravity 2.1 g/cm$^3$ , streak is white.

**Hematite**, $Fe_2O_3$: Colour – red to brown in earthy forms, luster – metallic or dull. Crystal system trigonal cleavage absent but there is parting on two planes. Uneven fracture hardness 5.0 to 6.0. Specific gravity 5.3 streak blood red to brownish.

**Hornblende**, $Ca_2(Mg, Fe,Al)_5(Al,Si)_8\ O_{22}\ (OH)_2$: An in silicate of amphibole group black to dark green coloured with vitreous to dull luster and monoclinic crystal system. Imperfect cleavage in two directions at 56 and 124 angles uneven fracture. Hardness 5.0 to 6.0. Specific gravity2.9 to 3.4 streaks Brown to gray 5.0 to 6.0.

**Magnesite**, $MgCO_3$: White or gray coloured mineral with vitreous luster, trigonal crystal system and white streak cleavage perfect in three directions forming rhombohedrons. Fracture – conchoidal and uneven. Hardness-4.0 to 4.5. Specific gravity 3.0. It exhibits strong magnetism and is an important ore of iron.

**Magnetite**, $Fe_3O_4$: Black coloured mineral with metallic or dull luster, isometric crystal system, almost no cleavage and black streak. Fracture – Conchoidal, Hardness – 5.5 to 6.5. Specific gravity 5.1.

**Muscovite**: ($K_2$ ($Si_6Al_2$) $Al_4O_{20}(OH)_2$): It is a phyllosilicate of mica group with usually white colour and streak, Crystal system–monoclinic luster– vitreous to pearly. Cleavage perfect in one direction forming, thin sheets or flakes. Fracture – not easily observed due to perfect cleavage hardness – 2.0 to 2.5. Specific gravity 2.8.

**Nitre**: ($KNO_3$): It is not abundant due to high water solubility. White or gray coloured with white streak, vitreous luster, orthorhombic crystal system. Cleavage – good in two directions. Fracture – uneven. Hardness – 2.0 specific gravity 2%.

**Microcline**, K Al $Si_3O_8$. It is one of the polymorphs of potash feldspars. Colour off white, yellowish, fresh pink. Lustre – vitreous to pearly and dull. Crystal system triclinic, cleavage – perfect in one and good in another forming nearly right angled prisms. Conchoidal fracture, hardness – 6.0 to 6.5. Specific gravity 2.5. Streak – white.

**Pyrites**, $FeS_2$. A brassy yellow coloured mineral with metallic luster, isometric crystal system, greenish black streak and conchoidal fracture. Cleavage – indistinct, hardness – 6.0 to 6.5. Specific gravity 5.1+. It forms a good source of S.

**Pyrolusite**, $MnO_2$. Colour – steel gray to a solid black, luster – metallic to dull. Crystal system – tetragonal, cleavage good in two directions, but not normally observed. Fracture – conchoidal to uneven hardness – 6.0 in individual crystals; aggregates are much stronger. Specific gravity – 4.1 to 5% streak – black. It is the main source of Mn.

**Phologopite**, $K_2Mg_6Al_2Si_6O_{20}$ $(OH)_4$. It is a phyllosilicate and relatively less common variety of mica group. A pale brown to brown coloured mineral with white streak and vitreous to pearly luster. Crystals are monoclinic with perfect cleavage in one direction providing thin sheets or flakes. Hardness–2.5 to 3.0. Specific gravity 2.9.

**Serpentine**, $(Mg, Fe)_6$ $Si_4O_{10}$ $(OH)_8$. It is a phyllo-silicate mineral with different varieties. Colour – olive green, yellow, golden brown and black, luster – greasy, waxy or silky. Crystal system and cleavage – variable, fracture – conchoidal and splintery. Hardness – 3.0 to 4.5. Specific gravity 2.2 to 2.6. Streak – white.

**Siderite**, $FeCo_3$. Colour is gray, yellow, yellowish, brown or reddish brown. Streak – white, luster – vitreous pearly and silky. Crystal system – Trigonal cleavage – perfect in three directions forming rhombos. Fracture – conchoidal to even. Hardness – 3.5 to 4.5. Specific gravity – 3.9+.

**Stengite**, $FePO_4.2H_2O$: It has splendid colours ranging from pink to violet red. Lustre – vitreous crystal system – orthorhombic, cleavage – good in one direction, poor in another Fracture – conchoidal, hardness – 3.0 to 4.0 specific gravity – 2.87. Streak – white.

**Sylvite**, KCl: It is a major mineral ore in potash fertilizer industry. A white, sometimes tinted yellow, red, or blue coloured mineral with white streak, vitreous luster and good cleavage in three directions. Cubic crystals with frequent octahedral faces. Uneven fracture. Hardness 2.0 to 2.5. Specific gravity-3.9 to 4%.

**Tourmalines** : Tourmalines are boro silicate minerals with an approximate composition of $Ax_3Y_6(BO_3)_3\ Si_6O_{18}(O, OH, F)_4$, where X = either Al, Fe, Li or Mg, Y = usually Al, but can also be Cr or Fe.

**Variscite**, $Al\ PO_4.2H_2O$: Like stengite, variscite too is a rare phosphate mineral. It is a green to emerald green coloured mineral with white streak, vitreous or waxy luster and orthorhombic crystal system. Cleavage – good in one direction and poor in another. Fracture – conchoidal, splintery and uneven hardness – 3.5 to 5.0. Specific gravity 2.5%.

## 2.2. Weathering

Rocks and minerals on being subjected to the environmental agencies for hundreds and thousands of years eventually form the soils. The process is indeed a very complex one and the result of action and interaction of a number of forces, factors and processes, both physical and chemical.

The rocks and even the mi too in their native state are large unconsolidated masses with a high packing density of their constituents and obviously with a low surface area. For a faster and effective alteration to take place either physical or chemical the contact areas of the participating ingredients must be high enough. Stated otherwise, the massive sized hard enough rocks and minerals must first be broken down into small fragments or pieces or a loose mass so that they can effectively experience varieties of processes operating upon them, the net result of which is a change in their structural arrangement and molecular make-up. Such changes in the rocks and minerals involve both physical and chemical processes and are collectively referred to as Weathering.

Weathering is a slow and continuous process that affects all substances exposed to the atmosphere and even underneath. In the context of the soils, physical dis-integration and chemical decomposition of the rocks and minerals preceding the formation of soils is called weathering. Weathering of the rocks and minerals is thus a prerequisite to soil formation and the fresh rocks and minerals that once occupied the outer most part of the earth reached their present state of decay through a complex interaction of physical, chemical and biological processes called weathering.

### 2.2.1. Types of weathering

Depending on the agents responsible and the processes involved, weathering is divided in to three types viz., physical, chemical and biological weathering.

#### 2.2.1.1. Physical weathering

In physical weathering also referred to as mechanical or disintegration weathering, the large masses of rocks and minerals get broken down to smaller fragments or pieces, without however, being accompanied by an alteration in their chemical composition. Physical weathering is thus just a mechanical process pertaining to a decline in the size of the materials and is the nature's way of breaking down in to smaller sizes.

Physical weathering involves different kinds of stresses experienced by the large rocks and mineral bodies such as those caused by phase changes, hydration, changes in crystal structure, thermal stresses and the "unloading" phenomenon. Physical weathering is of greater significance in very hot and dry and very cold conditions where either moisture is scantly available and/or rate of chemical reactions is very slow due to low temperature. The soils of such areas where only physical weathering is the dominant one among the three kinds are characterized by coarse texture i.e. sandy soils. Principally, there are five agents involved in physical weathering viz. temperature, water, glaciers (ice), wind, and waves.

**a) Temperature:** The processes involved in physical weathering caused by temperature are frost wedging, exfoliation and thermal expansion. Rocks are not very efficient conductors of heat. Therefore, during the day time, the outer regions of the rocks get heated up and expanded more than the inner regions. Again, on the onset of the night, the expanded outer regions get cooled quicker and contracted faster than the inner regions. Diurnal temperature fluctuations thus cause differential expansions and contractions between the outer and inner parts of the rocks. This causes peeling away of the outer-most part of the rocks in the form of a distinct layer. This process is called exfoliation, often pinningly referred to as 'onion skin weathering.' Exfoliation is quite common in dry regions; weathering of rocks due to such diurnal temperature fluctuations is some times called insulation, the reason being the poor heat conducting nature of the rocks.

In addition, most rocks are polymineralitic. Each mineral of such rocks has its own co-efficient of expansion due to temperature rise. For instance, quartz and feldspars in intimate association are the two main minerals of granite, a silica rich igneous rock. Quartz expands twice more than feldspar per unit temperature rise. Differential expansion and contraction of such minerals that remain very close to each other in the same rock system do cause enough stress. Such stresses cause cracks in the rocks that prove great help in breaking a massive rock in to smaller pieces.

**b) Water:** Water in flow is an important agent of physical weathering. Flouring water carries materials of different sizes ranging from very small size silt and clay (diameter less than 0.004mm) to as high as foulders (diameter more than 256mm).

Flowing water thus does have a scouring action (cutting power) and in the process depends and widens its own path. Valleys, ravines and gorges amply demonstrate this property of water. Besides sorting reorganizing and shifting of the materials carried depending on size and density is also performed by flowing water. This helps in the formation of different layers of rocks and minerals bedding to the formation of newer rocks and minerals.

Higher the velocity of flowing water, higher is its capacity to carry materials both with rigid to the size and density. At times of flow or in deluging water flowing heavy incessant down pour the velocity of water reaches high and it can carry rock and materials of the size bounders like floating objects. They collide, if they happen to meet and in the process get broken down to small pieces. As the pieces are now smaller and lighter, water even with much less speed can carry them and bring out the same collision though with a lesser impact.

i. Water thus can cause disintegration of rocks and minerals masses in to smaller sizes.
ii. It can carry and transport, and the materials carried over long distances and deposit them at a new place, loading to the formation of transported parent materials and/or soils.

**c) Wind:** Air in motion is called wind. The role played by wind in the weathering of rocks and minerals is largely like that by water. Wind too remains laden with various materials like gravels, sand, silt, clay and dust. As water does have scouring action on its path, wind does have abrasive action over a surface it strikes. If the surface so hit is that of rocks and minerals, abrasion causes removal of the particles present in them. In course of time, a large rock mass is reduced to a powdery material.

The particles so dislodged are then blown by wind and deposited at a new place where they can experience further physical and chemical weathering at ease due to large surface area. Wind thus acts not only as a disintegrating agent, but also as an agent of transportation. Higher the velocity of wind higher is its capacity to carry and transport. Obviously therefore, wind carries high amounts of suspended materials at times of typhoon, cyclone and hurricane. The role played by wind in the weathering of rocks is of special significance in arid regions and dry weather. Formation of sand dunes in the deserts – some of them as much 1500' deep in the Sahara desert – testify the carrying capacity of wind.

**d) Waves:** Waves refer to the sea or ocean water hitting the beach or the shore. The role of wages in weathering is therefore confined only to the coastline. Seawater carries a considerable amount of dissolved salts and suspended materials. It thus acts as an abrasive agent. The outer surfaces of the rocks when hit by such water thus get scrapped away in the form of fine particles. Such rocks present nearby get completely disintegrated over a period of time.

**e) Glaciers:** A glacier is a large slow moving river of thick ice formed from compacted layers of ice that slowly deforms and flows in response to gravity. Glaciers may also roll due to ice melting at their bottom region. While rolling down, they exert a tremendous pressure over the rocks over which they slide over. Under the impact of such high weight, the rocks get cut, broken and crushed. Glaciers also act as an agent of transportation of rocks, minerals and weathered materials. Glaciers, though not playing an important role in the current day environments, had a decisive to play in the physical weathering of rocks and minerals in the past.

### 2.2.1.2. Chemical weathering

Decomposition of rocks and minerals under natural conditions and accompanied by a change in their chemical composition is referred to as chemical weathering. Chemical weathering involves chemical reactions. Chemical reactions are facilitated by water. The extent of chemical weathering is thus more in upper layers of the earth that remain moist for longer period and in high rainfall areas where water is abundantly available. High temperature along with high rainfall, common in humid tropics and sub-trooping, make chemical weathering all the more intense. Depending on the site of operation, chemical weathering is divided in to:

i. Geochemical weathering referring to chemical weathering taking place below the solum, and
ii. Pedochemical weather embracing decomposition of rocks and minerals of the soils taking place within the solum.

There are principally six forms of chemical reactions involved in chemical weathering viz., solution, hydrolysis, hydration, oxidation, reduction and carbonation. As a result of these chemical reactions, primary and secondary minerals are altered to secondary and newer secondary minerals respectively in the process of chemical weathering.

A brief account of each of the types of chemical reactions shall help understand the process further.

**a) Solution:** Solution in chemical weathering refers to the solvency of water, universal solvent, for the mineral components and cementing agents of the rocks. Water thus dissolves the minerals and the cementing agents that join minerals to form the rocks. The individual minerals are then released and brought in a soluble state. In soluble state, the possibility of experiencing chemical reactions like hydrolysis, hydration, and carbonation is greatly increased.

Solution helps in the formation of new minerals too. The minerals like halite, gypsum and calcite present in rocks get dissolved in water – more so if it is acidic in pH. Rainwater with its normal pH of 5.6 in unpolluted areas is thus more effective in the context. These minerals then get carried away by water as 'dissolved load' and flow to the lakes or seas or low-lying areas. As water evaporates, the dissolved minerals precipitate or crystallize out as newly formed solid minerals (halite, gypsum and calcite). Such minerals that form from the evaporation of water are called evaporites.

**b) Hydrolysis:** Hydrolysis refers to the chemical reaction in which a substance is decomposed by water. Some times, it resembles the double decomposition reaction wherein there is mere exchange of the caution of the substance with that of $H^+$ of water. Thus, water to form alumino silicic acid hydrolyses orthoclase.

$$KHAlSi_3O_8 + H.OH \longrightarrow HAlSi_3O_8 + KOH$$

Orthoclase alumino-silicic acid

Alumino Silicic acid then may get dehydrated to form the oxides or may further be acted upon by water to form silicic acid.

$$2H.AlSi_3O_8 \longrightarrow H_2O + 6SiO_2 + Al_2O_3$$

Alumino silicic acid

$$2H.AlSi_3O_8 + 11H_2O\ 6H_4 \longrightarrow SiO_4 + Al_2O_3$$

Alumino silicic acid Silicic acid alumina.

Alumina may get hydrated to form bauxite ($Al_2O_3.3H_2O$).

$$Al_2O_3 + 3H_2O \longrightarrow Al_2O_3.3H_2O$$

The latter in turn may react with silicic acid to form Kaolinite, a secondary clay mineral common in soils.

$$4H_4SiO_4 + 2Al_2O_3.3H_2O \longrightarrow Si_4Al_4O_{10}(OH)_8 + 10H_2O$$

Silicic acid Kaolinite

The reactions illustrate how a primary mineral like orthoclase gets converted to a secondary clay mineral such as Kaolinite. In humid regions, much of the orthoclase, a feldspar mineral present in rocks such as granite weather to form clays.

Potassium hydroxide (KOH) formed earlier usually reacts with $CO_2$ an acidic oxide and forms the respective carbonate. The carbonate carries potassium in soluble ionic form ($K^+$), permitting its uptake by plants or loss through leaching and run off or reaction with other minerals to undergo other kinds of chemical reactions.

$$2KOH + CO_2 \longrightarrow K_2CO_3 + H_2O$$

Among the different kinds of chemical reactions known to cause chemical weathering of rocks and minerals, hydrolysis is normally the most important one.

**c) Hydration:** Hydration refers to the absorption of water by a substance in such mananer that the absorbed water becomes an integral part of the molecule and must be taken in to account for weighing a given mass of the substance based on its anhydrous molecular water. The absorbed water in such cases survey as a point for other reactants, say $H^+$ or Oxygen or hydrogen, to strike the molecule. Some well-known hydration reactions are listed below:

$$CaSO_4 + 2H_2O \longrightarrow CaSO_4.2H_2O$$

Anhydrite Gypsum

$$2Fe_2O_3 + 3H_2O \longrightarrow 2Fe_2O_3.3H_2O$$

Hematite (red) Limonite (yellow)

$Al_2O_3 + 3H_2O \longrightarrow Al_2O_3.\ 3H_2O$
Alumina (gibbsite) Bauxite

$5Fe_2O_3 + 9\ H_2O \longrightarrow Fe_{10}O_{15}.\ 9H_2O$
Hematite ferrihydrite

Hydration is often accompanied by a large increase in volume. This means that the crystals become porous and susceptible to chemical decomposition, especially for the entry of water, oxygen, hydrogen and other reactive constituents. Besides when rock materials take up water, the increased volume creates physical stresses within the rocks thus helping in their physical disintegration.

**d) Oxidation:** Oxidation refers to the reaction of a substance with oxygen or a reaction with loss of electrons or hydrogen. Oxygen is abundantly avoidable in the earth. Oxygen is highly reactive too and readily combines with many minerals causing a change in their chemical composition. Oxidation is therefore very important in the chemical weathering of rocks and minerals. Oxidation is particularly important in decomposing minerals containing iron as if undergoes *ous* to *ic* (eg. ferrous to ferric) changes i.e. lower to higher valiancy state most frequently. Obviously, pyroxenes, amphiboles olivine and biotitic are common among the minerals weathered by oxidation. Oxidation is important also for weathering of other minerals containing elements that can exist in multi-valence states e.g. Mn, Cu, Sn, Mo, Cr, S etc. Some typical oxidation reactions suffered by minerals are furnished below:

$4FeO + O_2 \longrightarrow 2Fe_2O_3$
Green hematite red-hematite

$4FeSiO_3 + 6H_2O + O_2 \longrightarrow 4Fe(OH)_3 + 4SiO_2$

$6FeSio_3 + O_2 \longrightarrow 2Fe_3O_4 + 6SiO_2$

$FeS + 2O_2 \longrightarrow FeSO_4$
black green

$2FeS_2 + 2H_2O + 7O_2 \longrightarrow 2FeSo_4 + 2H_2SO_4$
Pyrites (golden)

$4FeS + 9O_2 + 10H_2O \longrightarrow 4Fe(OH)_3 + 4H_2SO_4$

In the above reactions, either iron or sulfur, or both are undergoing oxidation. Oxidation is often accompanied by a decline in size. This size decline along with the change in charge creates structural and electrical imbalances in the crystal system, the remit of which is a change in the mineralogy of the minerals. Oxidation takes place under aerated environments. Obviously therefore, the degree of operation of oxidation is more and more in upper layers of the earth and in aerated soils or parent materials.

**e) Reduction:** Reduction is the reverse of oxidation and refers to the reactions in which hydrogen or electrons are added or oxygen is eliminated. Like oxidation, reduction too is important for the weathering of minerals containing elements capable of existing in polyvalent elements e.g. Fe, Mn, Cu, Mo, S etc., In the absence of oxygen, these elements present in their higher valiancy states get reduced to their lower valiancy states forming new minerals. Thus

$$2Fe_2O \longrightarrow 4FeO + O_2$$

$$FeSo_4 \longrightarrow FeS + 2O_2$$

In soils where reduction, processes are very active green or light blue colours due mainly to iron and manganese usually predominate in contrast to red colours of soils where oxidation is active. The reduced forms of the polyvalent elements are more soluble and thus more mobile them their oxidized forms, thus paving the way for their removal by leaching or run off. Reduction takes place in anaerobic environments and is thus favoured in lower depths of the earth ill-drained or water logged or compact conditions and in places where there is intense biological activity exhausting available molecular oxygen required for cellular respiration.

Reduction is accompanied by an increase in size. This factor along with change in valiancy state also contributes to disintegration and decomposition of various minerals. Both oxidation and reduction are very important in the chemical weathering of rocks and minerals; however, alternate oxidation and reduction taking place in short intervals or regular cycles is more active than either of them operating alone.

**f) Carbonation**: Reaction of a substance with dissolved $CO_2$ or carbonic acid – the nature's acid – farming a new substance is called carbonation. Higher the $CO_2$ content higher is the extent of carbonation reaction. Rainwater is for more high in $CO_2$ in comparison to normal water carbonation therefore becomes more significant in high rainfall humid area. Carbonation is particularly important for the dissolution of the insoluble carbonates of the alkaline earth metals i.e. Ca and Mg. The insoluble carbonates get decomposed to their soluble carbonates. Thus,

$$\underset{\text{insoluble}}{CaCO_3} + H_2CO_3 \longrightarrow \underset{\text{soluble}}{Ca(HCO_3)_2}$$

$$\underset{\text{Insoluble}}{MgCO_3} + H_2CO_3 \longrightarrow \underset{\text{soluble}}{Mg(HCO_3)_2}$$

The insoluble carbonates thus can be dislodged from the rocks, transported elsewhere for deposition by water, lost by leaching of run off absorbed by plants, or react with other substances. The reactions of chemical weathering make it amply clear that water is the most important ingredient in chemical weathering as it is directly involved in solution, hydrolysis, hydration and carbonation.

### 2.2.1.3. Biological weathering

Both physical and chemical processes are involved in biological weathering. Biological weathering in essence is therefore either physical or chemical weathering, but carried out by biological agents. The role of biological weathering hitherto largely over shadowed by a biotic weathering is being increasingly understood. There are cases where biological weathering surpasses the role played by both physical and chemical weathering in the overall weathering and decomposition of rocks and minerals. Also of interest to note is the fact that biological weathering complements physical and chemical weathering and takes place mainly in the solemn.

**Higher plants:** Physical weathering of rocks causes cracks. Higher plants start growing inside the cracks. The roots of the plants pry apart the rocks due to the pressure exerted by the roots, as the plants grow older and larger.

**Lower forms of life:** Lower forms of life like fungi, algae, bacteria etc., grow on the exposed rock surface and thus keep the rock surface moist. This in turn facilitates chemical weathering for which water is an essential input. The microbes and the root calls carry out respiration and form CO2. This forms carbonic and as a result of which solvent action of water and the rate of carbonation reaction in chemical weathering are greatly enhanced.

**Lichens:** The role played by lichens in biological weathering is of special importance. It is a crusty rubbery light green and live organic material that grows in patches on rock surfaces. It is an association of fungi and algae living together in symbiotic association. The fungus part of the association forms the acids which help in dissolving the minerals which are then used by algae. Later, water seeps in to the crevices etched by the acids and assist in the breakdown through freezing (frost wedging) and chemical weathering.

**Animals including man:** Animals like rats, rodents; moles etc. are borrowing by nature and constantly excavate othered rock materials from inside to outside and vice versa. The activities of men like tillage, cultivation lining, quarrying and mining, road laying bore rigging etc., also cause similar effects. These activities help the rock materials in getting thoroughly mixed a process being helped further by animals like earthworms. The resulting homogeneity enhances the scope for chemical weathering.

### 2.2.2. Factors affecting weathering

Factors affecting the weathering of rocks and minerals are (i)climatic factors, (ii) physical characteristics of the rocks and minerals, (iii) chemical composition and nature of the rocks and minerals and (iv) stability/weather ability of the minerals.

**Climatic factors:** Climate plays an important role in determining both the kind and the rate of weathering of the rocks and minerals. Among the different parameters of climate, rainfall and temperature are normally the most important in the context.

If the temperature is high or low, but the rainfall is scarce, physical weathering plays a dominant role. Again, if the diurnal and seasonal temperature fluctuations are wide then physical weathering processes like exfoliation freezing and thawing or frost wedging and insolation weathering play a decisive role. In such conditions the soils are characterized by coarse texture rich in coarse sand fractions and the chemical composition of the soils remain largely the same as the original parent material.

If rainfall is high and water does not become a limiting factor for any other reason, then chemical weathering is greatly encouraged. The mineral constituents suffer a number of chemical reactions like hydrolysis, hydration and carbonation all of which contribute to forming secondary minerals from a suite of primary or secondary minerals.

Climate assists in weathering also by supporting luxuriant vegetation and thus making the role of biological weathering all the more important higher biological activity forms more organic matter and makes it a decisive ingredient in chemical weathering. This is principally because of the acid qualities of the organic matter and also its ability to form complexes with metallic ions by collation and thus to promote the removal of the products from the weathering process resulting in a lower ion concentration in the weathering fluid or soil solution.

**Physical characteristics of rocks and minerals:** Physical characteristics of the rocks and minerals that determine the rate of their weathering are their shape, specific gravity, grain sizes, surface area, porosity, permeability, fractures, hardness etc.,

Specific surface or surface area governs the rate of weathering since the process takes place on the surface of the materials. Large minerals therefore are harder to weather than small ones occupying the some volume, size however, is not the only determinant factor and is inconsistently related to the rate of weathering. This, finer grained granites are to be more susceptible to weathering due to the greater surface area to volume ratio. However, coarser grained ones especially those that are in strain and traversed by an array of micro cracks are in fact more weatherable.

Crystal shape and perfection are also important. Platy crystals are more weatherable than enhedral crystals. Perfect crystals on the other hand are less weatherable than those with crystal defects, crystal defects imply loose bonding of the constituents i.e. atoms or ions, within the crystal system and thus making them weatherable case. Igneous and metamorphic rocks are composed of a three dimensional mosaic of interlocking crystals which make them very strong resistant to weathering. Sedimentary rocks by contrast are merely aggregates of minerals joined by cementing agents and are far more prone to weathering.

**Chemical composition and nature of the rocks and minerals:** The chemical nature of the constituents of the rocks and minerals and the atomic arrangement of the constituents are by far the most important factors that determine the rate of their weathering.

The presence of oxidisable elements like iron greatly increases the rate of weathering. The nature of iron to undergo reversible $Fe^{3+}$ to $Fe^{2+}$ change at ease makes iron-bearing minerals more susceptible to chemical weathering. Olivine's like fayalite, pyroxenes, amphiboles and micas containing iron are thus very susceptible to weathering. Biotite and muscovite are both phyllo silicates and micas. Biotite however is much more weatherable to environmental agencies them muscovite due to (i)biotite unlike muscovite contains oxidisable iron and (ii) atoms of biotite are much less tightly packed to form layer lattice sheet structure.

Under high rainfall conditions the insoluble carbonate minerals like calcite and dolomite got dissolved by water carrying dissolved $CO_2$ and are thus removed by water. Chalk and limestone, both sedimentary rocks and rich in calcite are weathered mainly by carbonation and clay rocks like slate by hydration. Igneous rocks on the other hand are weathered principally by hydrolysis yet, another factor determining the weather ability of rocks and minerals is their basic nature. More basic minerals weather at a faster rate. Basalt thus weathers faster granite.

## Weatherability of rocks and minerals

The weatherability of rocks and minerals are not absolute and is determined to a large extent by the climatic conditions. Even, here also, the order of stability valid for humid temperate climate may not be the same under arid tropics.

However, there are certain features and inherent characteristics that largely determine the weatherability of minerals. For example, higher the temperature of formation of a mineral, the move unstable it is at the earth's surface. Accordingly, olivine weathers the most rapidly, while quartz remains at the other extreme. Olivine is followed by biotitic and plagioclase feldspars.

Thus, the order in which minerals chemically weather is essentially the same as their order of crystallization from a silicate melt. Olivine, the first to crystallize is also the first to undergo alteration i.e. weathering. The reverse is the case with quartz. The entire series of the minerals forms Goldrich's stability series Figure 2.

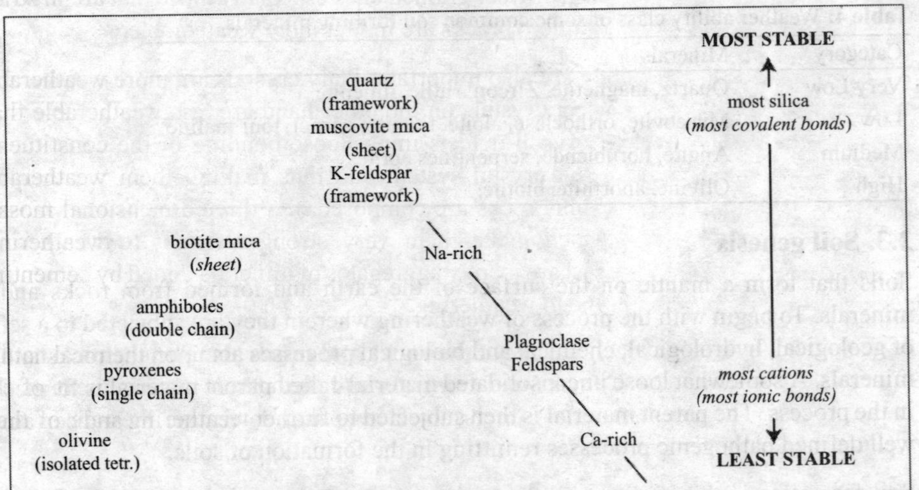


**Fig. 2:** Goldrich's stability series of minerals

A factor contributing to stability is the degree of tetrahedral linkage. Higher the degree of tetrahedral linkage, higher is the stability. Quartz is a tecto silicate where all the oxygen atoms of the tetrahedrons are interlinked with those of the other tetrahedrons, hence more resistant to weathering. Olivine on the other hand is a neosilicate consisting of single tetrahedrons and unbonded to other tetrahedrons. Based on tetrahedral linkage, quartz thus must very resistant, while olivine must be very susceptible.

Higher the basicity of a mineral higher is the degree of weatherability i.e. less stable the mineral is Gypsum is thus more stable than calcite, the reason being the higher basicity of calcite. Yet another factor in this context is the alumina-silica ratio. Higher the ratio, more is the weatherability of the minerals. The following table listing the mean lifetime of 1mm crystals of few minerals at 25°C and pH 5 confirm the Goldrich's stability series.

**Table 3:** Mean life time of some common soil forming minerals at 25°C and pH 5.

| Mineral | Mean life time (years) |
|---|---|
| Anorthite (Ca plagioclase) | 112 |
| Nepheline (feldspathoid) | 211 |
| Forsterite (Olivine) | 2300 |
| Enstatite (Pyroxene) | 10,100 |
| Albite (Na plagioclase) | 5,75,000 |
| Microcline (K feldspar) | 9,21,000 |
| Muscovite (mica) | 2,600,000 |
| Quartz (silica) | 34,000,000 |

The following table illustrates the relative stability of some common soil forming minerals categorized in to weather ability classes.

**Table 4:** Weather ability class of some common soil forming minerals.

| Category | Minerals |
|---|---|
| Very Low | Quartz, magnetite, Zircon, rutile, ilmenite |
| Low | Muscovite, orthoclase, albite, titanite, garnet, tour maline |
| Medium | Angite, hornblende, serpentines aptites |
| High | Olivine, anorthite, biotite. |

## 2.3. Soil genesis

Soils that form a mantle on the surface of the earth and formed from rocks and minerals. To begin with the process of weathering wherein they are subjected to a set of geological, hydrological, chemical and biological processes act upon the rocks and minerals. A somewhat loose unconsolidated material called parent material is formed in the process. The parent material is then subjected to further weathering and certain well defined pathogenic processes remitting in the formation of soils.

Soil formation is therefore a two phase process; the first one that forms the parent materials by the weathering of rocks and minerals is destructive in nature, while the second one that forms soils from the parent materials is largely constructive. Thus,

$$\text{Rocks \& Minerals} \xrightarrow[\text{Destructive phase}]{\text{Weathering}} \text{Parent materials} \xrightarrow[\text{Constructive phase.}]{\text{Soil genesis}} \text{Soils}$$

There is thus considerable distinction between weathering and soil formation. Weathering is geological in its nature while soil formation is pedogenic in the same perspective. Weathering includes disintegration and dissolution of the earthy materials and chemical reactions like hydrolysis, hydration, etc., experienced by the letter. Soil formation involves some additional localized processes and the pedogenic processes described in head 'weathering' continue during soil formation too.

### 2.3.1. Soil forming factors

In 1883, V.V.Dokuchaev, a Russian Soil Scientist undertook a summary field study and made the observation 'soil as independent natural body formed under the influence of five factors'. They included climate, relief, organisms, parent material and time and are now called the *factor of soil formation*. The five factors are often designated by the acronym 'CROPT'. The 'Organism' factor has later been replaced by a wider and more acceptable term called 'biosphere'.

Later H.Jenny (1941) suggested that the properties of soils are the results of interactions of these factors of soil formation acting over the soil forming material. A particular soil characteristic, say pH or texture or clay content, is therefore determined by the integrated influence of these five factors of soil formation. Jenny went ahead to define the relationship between a particular soil property and five factors that are regarded as central variables by a function of the process.

S = f (cl, pm, b, r, t……), where S is any soil property; cl, pm, le, r and t have their usual meanings and the dots indicating additional unspecified factors, if any, that may some times arise under certain conditions.

To solve the equation for changes in a soil property S, when only one of the central variables say climate, varies, the others being constant or nearly so, the equation can be rearranged as

S = f (cl) pm, b, r, t ………………

Incidentally, the relationship is called a climo-function and the range of soils is referred to as climosequence. Likewise, there are ranges of soils known as Biosequence and toposequence. Toposequence of soil is consequently by the same as described by Milne (1935) in reference to catena.

Joffe (1949) categorized the above five factors in to active and passive factors of soil formation *climate* and *biosphere* are the active factors of soil formation. Active factors are the agents that act on the soil forming materials and make happen certain soil forming processes because of which distinct changes in them leading to the soil formation/maturity of the soils formed can be observed. Passive factors on the other hand include the soil forming material and the conditions like elevation of the land, slope etc., under which the soil forming material is being acted upon for soil formation as a function of time. The *parent material*, *relief* and *time* are thus the passive factors of soil formation.

### a) Parent material

Among the five factors of soil formation, parent material is the only inherited factor, while the others are somewhat like acquired in nature. The parent material is defined as the state of the soil system at zero time of soil formation. The loose unconsolidated material, which over a long period of time forms the solum, is referred to as parent material. The parent material contributes largely to the mineralogical make up of the soils, especially in the case of residual or cumulus soils. As a fall not of that the parent materials of the respective soils also influence physical properties like the texture, structure, water holding capacity and the downward movement of water for profile development.

The mineralogical composition of the coarse sand fractions of the soils and immature soils in general are very close to that of their parent materials. The chemical compositions of the soils depend to a large extent on the types of rocks on which they have developed and their mineralogical compositions.

Like the rocks, the parent materials too can be classified in to acidic, intermediate, basic and ultra basic ones based on their silica contents. Acidic parent material is gradually and slowly decomposed by weathering agents to form lighter coloured and textured soils of a relatively poor fertility status because the acidic parent material is rich in sand or silica and poor in basic nutrient elements. Such soils are often acidic in pH. By contrast, basic parent materials richer in basic elements, plant nutrients and poorer in sand or silica are weathered faster to form darker and heavier soils of relatively higher fertility and alkaline in reaction.

Granites and rhyolite are acidic rocks and on weathering form acidic parent materials which over a period of time forms sandy to sandy loam soils that are lighter in colour due to high silica and low iron content. Almost, similar is the case with sandstone and quartzite. Trachyte and syenite are intermediate (neutral) rocks rich in orthoclase and poor in biotite and amphiboles. Orthoclase whether to form clays. These intermediate rocks therefore, whether to form soils usually of medium texture, fertility and almost neutral in reaction.

Basalt, basic andesine and gabbro are all basic rocks, poor in silica and rich in highly weatherable ferro magnesium pyroxenes and amphiboles. These rocks therefore weather rapidly forming fine textured soils rich in basic elements. Such soils are alkaline due to inherently low silica content. Early formed clays are those of smectites group, which on weathering and maturity change to Kaolinite. In terms of weathering, kaolinite is more weathered than smectite.

Shale, a weakly cemented sedimentary rock rich in clay minerals and clay forming minerals like the micas and feldspars on weathering form dark coloured clayey soils that are high in basic cations and are thus alkaline in reaction. Limestone and dolomite rocks are rich in insoluble carbonates of calcium and magnesium. They are alkaline in nature and normally whether somewhat fast to form clayey soils of comparable pH.

Clay minerals are basically the hydrated silicates of aluminium, magnesium and iron. The nature of the parent materials greatly influences the nature of the clay minerals. If silica to Alumina ratio is higher, clay minerals formed are of smectite group, more so if the chemical environment is alkaline, potassium ions because of its unique size get snugly fitted into the crystal lattice of some clay minerals of mica group forming illite. To sum up, acid environment, permeable parent material, good drainage and high leaching are conducive for the formation kaolinites, while the reverse favours the formation of montmorillonite and illite clays.

Clays beget clays. In nature due to the weathering of natural clay sediments, further transformations continue. Thus, illite clay mineral may weather to form active montmorillonite clay mineral due to low pH of sediments. Such slow transformations of one clay mineral to another under suitable environment are called Digenesis.

Parent material thus greatly influences soil properties like texture (clay content) nature of the clays, pH, permeability depth etc. If sufficient time has elapsed the parent materials on weathering forms more stable, secondary minerals like anatase, magnetite, ilmenite and secondary clay minerals like kaolinite are more abundant.

## b) Climate

Climate represents the variability of the attributes of the latter like temperature, rainfall, sunshine etc., over a long period of time. The importance of climate in soil formation can be understood from the fact that the same rock type can form different soils in different climatic regimes. The climate exerts its role in soil formation transfer its four principal attributes viz., temperature, rainfall, wind and evaporation.

The rate of a chemical reaction increases with increase in temperature. For every 10 $^0$C rise in temperature, the reaction rate is doubled. Soil formation is a process in which numerous chemical reactions are involved. Obviously these chemical reactions and therefore soil formation do proceed at a faster rate at a higher temperature. However, there are indirect effects also. For example, decomposition of organic matter is higher at a higher temperatures and less organic matter remains left to contribute to soil formation. Tropical soils are thus low in organic matter, humus and nitrogen in comparison to those of temperate soils under similar rainfall.

As regards rainfall three aspects of it viz., total amount distribution pattern and the net quantum of water received after deduction of evapo-transpiration loss are equally important in soil formation. If the rainfall is high more so its intensity, leaching of soil constituents is also very high and this becomes a determinant affair in soil formation. Besides, more water remains available for chemical reactions to take place. With rise in rainfall there is thus rise in clay content, organic matter nitrogen and biological activity. But high leaching in high rainfall areas causes loss of exchangeable cations like $Ca^{++}$, $Mg^{++}$, $K^+$ and $Na^+$. This causes the formation of acid soils.

The role of wind in soil formation is related to its ability to act as a transporting agent. Wind transports soil particles, especially the lighter ones like clay and silt from one place and deposit them at a new one with such transport, the soils of one area become coarse textured rich in sand, while those of the deposit area become fine textured rich in clay and silt. Wind also transports salt particles and contributes to the formation salt rich soils at certain times. Besides, the rate of evaporation is greatly influenced by the velocity of wind. Wind thus governs or losing soil moisture too.

Under tropical and subtropical climates, both rainfall and temperature are very high. As a result, the weathering is very intense. High rainfall gives luxurious forest vegetation, which makes the role of biological weathering to be played at a higher note. The soils of such area are thus acidic (low in pH), low in per cent base saturation, fine textured rich in clay and dark coloured due to high organic matter and some times reddish due to the presence of oxidized iron as hematite. The soils are deep also unless interfered by erosion. By contrast in semiarid climate both temperature and rainfall are low and the extent of weathering therefore continues at a lower. The soils formed are not basic in reaction, shallow in depth, high in per cent base saturation, lower in clay content.

Kind of clay minerals in soil is also influenced by climate. Certain kinds of clay minerals are more common under certain climates, though all kinds of them are normally found under different climates. In humid tropics, due to intense weathering, kaolinite is more common; in arid climates illite and chlorite are more abundant, while in semiarid climates montmorillonite is the dominant clay mineral.

### c) Relief

Relief, a synonym of topography, includes features of the land like elevation, length and steepness of slope and landscape position. Relief can act as a climate modifier by way of creating a different microclimate within a vastly different macroclimate. Thus, changes in elevation affect the temperature e.g. decreases of approximately 0.5°C per 100m increases in elevation with a concomitant decrease in mean air and soil temperature and evapotranspiration. This helps in forming a range of hydrothermal regimes, from a hilltop to its base, soils formed do differ accordingly.

Relief governs the processes that further the development of soils like leaching, translocation, drainage, run off and erosion. For instance, if elevation is very high as in mountainous or hilly regions, most of the precipitation received goes off as run off and very little water remains available for the process of soil formation. Soil formation is thus slowed down due to inadequate water in spite of high rainfall. To add to this, slope further contributes to soil erosion very relevant. Since large part of the soil formed is lost by erosion, soils formed are very shallow. Some times, the extent of erosion is so high due to steep slope that the entire solum is lost and the parent material gets exposed. In such cases the process of soil formation has to start afresh. On the other side, the soils of the valley receiving eroded soil material become much thicker.

The soils of the upper and lower areas of the slope show altogether different soil profiles with extremes of many soil properties e.g. depth, texture, pH, mineralogy, cations etc., Significance of relief in soil formation was illustrated by Milne (1935) when he introduced the term catena to refer to regularly repeating sequence of soils formed on different slopes in regularly undulating land scope. A catena is group of different soils formed when relief as a soil forming factor surpasses the influence of other factors of soil formation. In other words, a catena is a regular range of soils formed under the same parent material and climate but under different relief associated with varying degrees of slope.

A catenary sequence of soils ranging from red earth on top to black soils on the bottom or valley of the slope on a recurring basis is found in the Chhatishgarh state of India. In between the red earth and black soils, yellow and brown earth soils in the same order from the slope remain interspaced.

**d) Biosphere**

Biosphere in reference to the soil formation denotes the role played by microbes, higher plants and animals including man in the development of soils. Once the parent materials or the early soils can sustain plant growth, plants start growing over them. The channels so created by the inserts of the plant roots provide better drainage and aeration those in turn better conditions for more biological activity. The acidic extracts of the plant leaves and humus solubilise the compounds of iron and aluminium and shifts them down the soil depth. This gives characteristic soil profiles with markedly different A or B horizons or sub horizons within each of them.

Deciduous trees accelerate soil formation in comparison to grasslands over the same parent materials and under largely similar climatic conditions. Differences in the chemical composition of the acid teachates of different accounts give rise to a varying pattern of soil. Thus, the acid litters of pines favours the formation of acid soils with a poor soil structure. By contrast, the litter of deciduous trees favours the formation of soils with more favourable soil structure.

The roots of the higher plants stabilize the soil particles contribute in reducing soil erosion and help in the formation of deep soils. Plants grown add organic matter to the soil. Added organic matter acts as the source of carbon and energy for the heterotrophic micro flora and fauna. Microbial respiration forms $CO_2$, which forms carbonic acid. Microbial activity also produces a site of organic acids during the course of their metabolism. The latter increase in the capacity of water to solubilise rocks, minerals and the cementing agents and the extent of carbonation reaction discussed earlier in weathering chapter. In addition, organic matter also acts as mulching agents, which binds the soil particles together and form deep soils by preventing erosion.

Among the macro fauna involved in soil formation, insects like termites moles, centipedes, earthworms, squirrels and rats and rabbits deserve special mention. Most of them are burrowing by nature and constantly excavate turn down and mix the soil materials. This homogeneity hastens up soil formation but interferes in horizon differentiation. Such soils have therefore though deep had less number of horizons. Earthworms are the most important soil forming fauna in acidic medium and temperate regions. If their population is very high, a stone and gravel free highly productive sub horizon is formed on the soil surface.

Man also contributes to soil formation by his acts like cultivation, manuring and fertilizing soils, liming, crop rotation and the alike. Some of his activities however retard soil formation or even degrade soils formed in the course of thousands of years.

### e) Time

Soil formation takes place as a definite function of time. The effect of time in soil formation can be distinctly watched in the age of a soil. Time exerts its role in soil formation in two ways

(1) the degree of a pedogenic even depends on the length of time for which it has operated.

(2) The nature of a soil-forming factor may change with time.

Under normal conditions, it takes about 6000 years to form one foot of topsoil. However, there are conditions, which lengthen or shorten this time lag. Factors that hasten up soil formation are warm humid climate. Permeable and unconsolidated parent material low in lime content; forest vegetation more so with coniferous trees; and flat topography with perfect drainage. Mohr and Barren recognized five stages of soil formation as follows:

1. Initial stage – Unweathered parent material.
2. Juvenile stage – Weathering just started, but much of the original material is still unweathered.
3. Virile stage – Easily decomposable materials have been weathered.
4. Senile stage – Decomposition has reached its final stage, except the most resistant minerals.
5. Final stage – Soil formation is completed and the parent material is fully weathered.

A dynamic system as soil is cannot be at rest indefinitely because a tendency to reach more stable state by destabilization of the equilibrium is a basic rule of the nature. Soil formation is therefore a continuous process and even the soil supposedly to have reached 'climax' or final stage continues to be subjected to progressive or digressive or both kinds of changes.

## 2.3.2. Pedogenic processes or Soil forming processes

The geological weathering produces weathered rock material i.e. the parent material and the genetic factors set the stage for soil development. The pedogenic processes change the parent material in to soil with varying horizonations. The pedogenic processes are extremely complex and dynamic involving many chemical and biological reactions, and usually operate simultaneously in a given area. One process may counteract another, or two other processes may work simultaneously to achieve the same result. The basic pedogenic processes involved in soil formation, according to Simonson (1959) include:

- Gains or additions of water (mostly as rainfall) organic and mineral matter to the soil.
- Losses of above material from soil.
- Transformation of mineral and organic substances with in the soil.
- Translocation or movement of soil materials from one point to another with in the soil. It is usually divided in to i) movement in solution (leaching) and ii)movement in suspension (eluviation) of clay, organic matter and hydrous oxides.

In contrast, the major changes that retard or offset horizon differentiation are due to:

- Mixing of materials by burrowing animals.
- Removal of surface soil by erosion (water or wind)
- Creep (by shifting old and its replacement by new materials); and
- Accretion of sediments in cultivated flood – plain areas, for instance silting of irrigated areas

### 2.3.2.1. Fundamental pedogenic processes

**a)Humification:** Humification is the process of transformation (decomposition) of raw organic matter in to 'Humus'. It is an extremely complex process involving various organisms such as bacteria, fungi, actinomycetes, earth worms and termites.

The decomposition of organic matter takes place in two phases: (i) mineralization and (ii) humification. Mineralization is a biochemical breakdown of dead plant tissues by soil microorganisms to produce simple structured soluble organic substances, mineral compounds, metal cations and gases ($CO_2$). During humification, soluble organic substances regroup themselves in to large molecules by polymerization and become poorly soluble. They form major part of soil humus and provide site for retention of cations. The other part of humus is the polysaccharides – gummy products of microbial excretions, which help in soil aggregation.

The activities of microorganisms and soil formation are categorized as

*Mor* : It refers to surface soil horizon developed under acid litter and humus from coniferous and heath vegetation, where fungi activity predominates.

*Mull* : Designated as forest soil horizon (A1) is of intimately mixed mineral matter and amorphous humus. It is slightly acid and is best developed under base rich litter, where bacterial activity predominates.

*Sward*: Is a dominantly rhizogenous A1 horizon in grasslands as contrasted with zoogenous mull horizon of forest soils. This includes mollic epipedon or Ap horizon formed by cultivation of forest soils, in general.

*Orterde*: Is a humus rich B horizon in podzols.

**b)Eluviation :** Eluviation means "Washing out". It is the process of removal of constituents in suspension or solution by the percolating water from the upper to lower layers. The eluviation encompasses mobilization and translocation of mobile constituents resulting in textural differences. Mechanical movement of clay and iron oxides from 'A' horizon without undergoing chemical alteration is referred as Lessivage. Leaching refers to the movement and removal of material in solution from the soil. It connotes the removal of the dissolved material from the entire solum.

Elemental mobility during leaching is in the order

$Ca^{2+}$ $Na^{+}$ > $K^{+}$, $Mg^{2+}$ >>>> $Fe^{2+}$ >> $Si^{4+}$ >> $Al^{3+}$

(Most mobile) (Least mobile)

The leaching of an element depends not only on its relative mobility but also on the rate of water percolation through the soil. The effect of leaching is well illustrated with the depth of accumulation of CaCO3 in soils (Jenny, 1941).

Hot arid zone : < 50 cm

Less hot semi arid zone :100 – 150 cm

Sub humid zone : >150 cm

**c)Illuvation:** The process of deposition of soil materials (removed from the eluvial horizon "E") in the lower layer (or horizon of gains having the property of stabilizing translocated clay materials) is termed as "illuviation". The horizons formed by this process are termed as illuvial horizons (B-horizon especially Bt).

All these basic pedogenic processes, combine to result in a number of wide ranging soils that are observed on surface of the earth.

Hot semi arid - Calcisol

Cool humid - Podsol

Hot humid - Ferralsol

## 2.3.2.2. Specific Pedogenic Processes

The basic pedogenic processes provide a frame work for later operation of more specific processes.

**a) Calcification:** The process of precipitation and accumulation of calcium carbonate in some part of the profile is called calcification.This is a common process in arid and semi-arid regions, which are low in rainfall (Rainfall<PET). The illuviated horizon of CaCO 3 is designated as 'calcic horizon'. Whenever high carbon dioxide is produced in soils, it combines with water and forms into carbonic acid. This dissolves the calcium carbonate in soils into soluble calcium bicarbonate, which moves along the percolating water. Again wherever a situation of high temperature and low carbon dioxide prevails, there calcium carbonate precipitates.

$$CO_2 + H_2O \longrightarrow H_2CO_3$$

$H_2CO_3 + Ca \longrightarrow Ca(HCO_3)_2$

$Ca(HCO_3)_2 \longrightarrow 2CaCO3 + 2H_2O + 2CO_2$

The calcium compounds are in solution as long as the $CO_2$ concentration or supply is maintained. The depth of "calcic horizon" depends on percolating rain water, ground water depth, amount of rainfall and the texture of the soil. The depth of calcareous layer In hot arid zone-<50cm, in semi arid zone 100-150cm and in sub-humid zone the calcium carbonate accumulates at a depth of >150cm

**b) Decalcification :** It is the reverse of calcification that is the process of removal of CaCO3 or calcium ions from the soil by leaching.

$CaCO3 + CO2 + H_2O \longrightarrow Ca(HCO_3)_2$ (Soluble)

This occurs mostly in high rainfall or humid regions.

**c) Podzolisation :** (Russian term, Pod = under and zola = ashlike). It is a process of soil formation resulting in the formation of podzols and podzolic soils. It is the process of accumulation of silica and eluviation of sesquioxides It is almost a reverse ofcalcification process due to leaching of all bases including calcium.

The favourable conditions for podzolization are

- A cool and humid climate (Invariably found at high altitudes)
- Siliceous (sandy) or acidic parent material, having poor reserves of weatherable minerals, favor the operation of podsolization, as it helps in easy percolation of water.
- Acid loving vegetation, such as coniferous pines (*Pinus roxburghii*), hemlock (*Tsuga Canadensis*) and heath ( *Calluna vulgaris*) are essential for this process.
- Under calcium free, acidic environment (pH <5.0), fungi plays active role in organic matter decomposition.
- Less microbial activity declines the polysaccharide production and keeps the soluble organic products in soluble form.
- The soluble organic acids react with sesquioxides and the remaining clay minerals, forming organic - sesquioxide and organic - clay complexes, which are soluble and move with the percolating water to the lower horizons .
- Alluminium ions in solution hydrolyse and make the soil solution very acidic.

$2Al + 6H2O \longrightarrow 2Al(OH)3 + 6H^+$

As the materials move out, it gives a bleached appearance (E-horizon) below the surface. The eluviated materials deposit in B horizon as dark coloured Bh (precipitated humus), reddish brown Bs (deposition of sesquioxides) and a yellowishish brown(silicate clay) layer which gradually merges with parent material. Hence a mature podzol has well developed horizonation. Podzols are highly acidic, low in fertility and used for forestry or pastures. Rarely crops like oats. Potato and clover can be cultivated.

**d) Laterisation :** The term laterite is derived from the word "later" means "brick" or "tile". In tropics, certain soils are massively impregnated with sesquioxides to the extent of 70 to 80% of the total mass, and forms a cemented horizon, which when dried becomes very hard like a brick. This soil forming process is called "laterization" or "Lotozation" Eg: Soils of Malabar hills of Kerala. In laterization, unlike podzolisation, silica is removed leaving sesquioxides to remain in solum. The favorable conditions for Laterisation processes are

- Warm and humid (tropical) climate with 2000 to 2500 mm rainfall and continuous high temperature (±25°C) throughout the year. Rapid decomposition of parent material and organic matter, and intensive leaching are very likely in this climate.
- The rain forests of tropical areas are the suitable vegetation for this process. Under this vegetation organic additions are low but organic matter decomposition is at very high rate.
- Basic parent materials, having sufficient ferro-mangnesion minerals (Pyroxenes, amphiboles, biotite and chlorite) are congenial for the development of laterites.

The iron released during weathering is oxidized to form FeO, $Fe_2O_3$ and coats clay, silt or sand particles imparting characteristic red color to soils. The Al-oxides / hydroxides impart grey coatings to the soil particles. The high temperature, intense leaching and basic kind of parent material all favor the removal of silica (de-silication) and accumulation of sesquioxides. The soluble basic cations are quickly released during weathering, moves freely in the soil profile and shoots up the pH to neutrality. Under this basic environment silica liberated from parent material is solubilized and leached. The solubility of quartz and amorphous silica increases with increased temperature. The sesquioxides are left behind as these are more stable under these conditions. As the alkaline bases are removed from the seat o f their formation, the residual soil is acidic in reaction. Though considerable eluviation takes place, there is no marked horizonsiation as the eluviated materials are not re-deposited in the lower layers.

Laterite soils are non-plastic, non cohesive and have granular structure. They are low in cation exchange capacity and fertility. Phosphorus fixation is high in these soils hence phosphorous availability to crop plants is a probelm. Plantation crops are usually grown on these soils.

**e) Gleization:** "Glei" means blue, grey or green clay. The gleization is a process of soil formation resulting in the development of a glei (or gley) horizon in the lower part of the profile above the parent material due to poor drainage conditions or water logged conditions. Such soils are called "hydromorphic soils". This process is not particularly dependant on climate (high rainfall as in humid regions) but often on poor drainage conditions. Poor drainage may be due to lower topographic position, impervious soil, puddle paddy cultivation and lack of aeration.

Under anaerobic conditions, iron compounds are reduced to soluble ferrous forms. The reduction of iron is primarily biological and requires both organic matter and the micro organisms capable of respiring anaerobically. Iron exists as $Fe^{2+}$ organo-compexes in solution or as a mixed precipitate of ferric and ferrous hydroxides, which is responsible for the production of typical bluish to grayish horizon. Due to seasonal fluctuations of ground waters, the gley shows distinct mottling of yellow and rusty brown colors caused by the alternate oxidation and reduction phenomena.

**f) Salinization:** It is the process of accumulation of salts such as sulphates, chlorides of calcium, magnesium, sodium and potassium in soils in the form of salty (salic) horizon. As a result of the accumulation of salts, solonchalks or saline soils develop with an electrical conductivity of > 4 $dSm^{-1}$. The soils are called saline soils, which have ESP less than15 per cent and pH between 7 and 8.5. The factors responsible for salinisation are :

- Climate: Arid or semi-arid climatic conditions, associated with shallow and brackish (high amounts of sulphates and chlorides) ground waters.
- Topography: Lower topographic positions / depression land forms.
- Drainage: Imperfect or poor drainage conditions
- Old lake bottoms
- Alluvial deposits along the sea coasts
- Use of saline irrigation waters.

The ground water containing high salts moves in an upward direction by capillary action. The water on evaporation leaves the salts behind, which accumulate at the surface or at some depth depending upon the capillary fringe. Surface accumulation of salts gives white appearance to soils .Hence the soils are called as white alkali soils. These soils can be managed by leaching of salts followed by provision of sub-surface drainage.

**g) Solonization or Alkakization:** The process involves the accumulation of sodium ions on the exchange complex of the clay to an extent of >15%, resulting in the formation of sodic soils (solonetz) under arid and semi-arid conditions. This occurs when anions like carbonates and bicarbonates predominate in soil.

The calcium and magnesium in soil solution will precipitate as corresponding carbonates or bicarbonate whenever the ionic product of solution exceeds the solubility products of respective carbonates. This reduces the concentration of Ca and Mg in soil solution , there by releasing them from exchange complex. As this process continues, the sodium concentration on exchangeable complex increase. When the ESP in soils exceeds 15 %, the soil is designated as alkali soil with a high pH of >8.5, which results in less nutrient availability.

The high pH in soils results in dissolution of humus, which moves upward along the capillary water giving black colour to soils. Hence the soils are called black alkali soils. High sodium on clay surfaces results in dispersion of soil aggregates leading to physical problems like poor aeration, low infiltration and percolation of water. This process results in a very thin, friable horizon followed by a dark horizon of hard and impermeable heavy soils generally with illuviated clay and having a typical columnar structure, which is characteristic of "solonetz".

**h) Solodization or Dealkalization:** This process refers to the removal of $Na^+$ from exchange sites. The $Na^+$ can be eliminated by increasing the concentration of $Ca^{2+}$ or $Mg^{2+}$ in the water, followed by improved drainage facilities.

$2\ NaX + CaSO_4 \longrightarrow Na_2SO_4 + CaX$

Exch. Sodium soluble

**i) Pedoturbation:** It is the process of mixing of the soils. Mixing to some extent takes place in all soils.

- Faunal pedoturbation : Mixing by animals such as ants, earthworms, moles, rodents and man himself.
- Floral pedoturbation : Mixing by plants, as in tree tipping that forms pits and mounds.
- Argillopedoturbation : Mixing of materials in solum by churning process caused by swell - shrink clays as is observed in deep black cotton soil.

Calcification, podzolization and laterization are zonal soil forming processes, where in the profile characteristics are influenced by prevailing conditions of climate and vegetation. Gleization, salinization, solonization and solodization are the Intra -zonal soil forming processes, wherein, the profile characteristics are more influenced by certain local conditions, such as relief (topography) and / or parent material than the climate and vegetation. Salinization, alkalization and dealkalization processes operate in sequence and advance the soil to a certain point in transition until the zonal soils are formed.

## 2.4. Soil profile

A vertical section of soil through all its horizons and extending in to the parent material. A vertical exposure of the horizon sequence is termed as "soil profile". Theoretical soil profile has following horizons

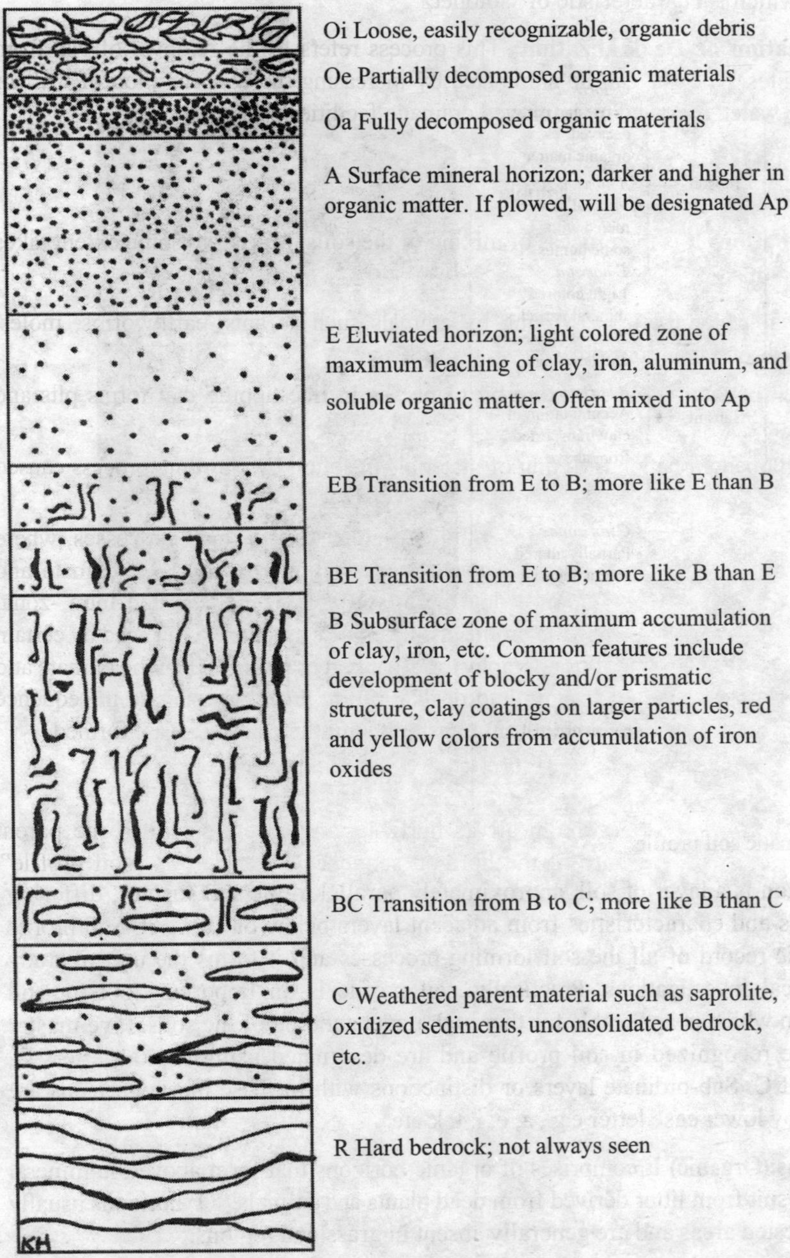


**Fig. 3:** Hypothetical Soil profile

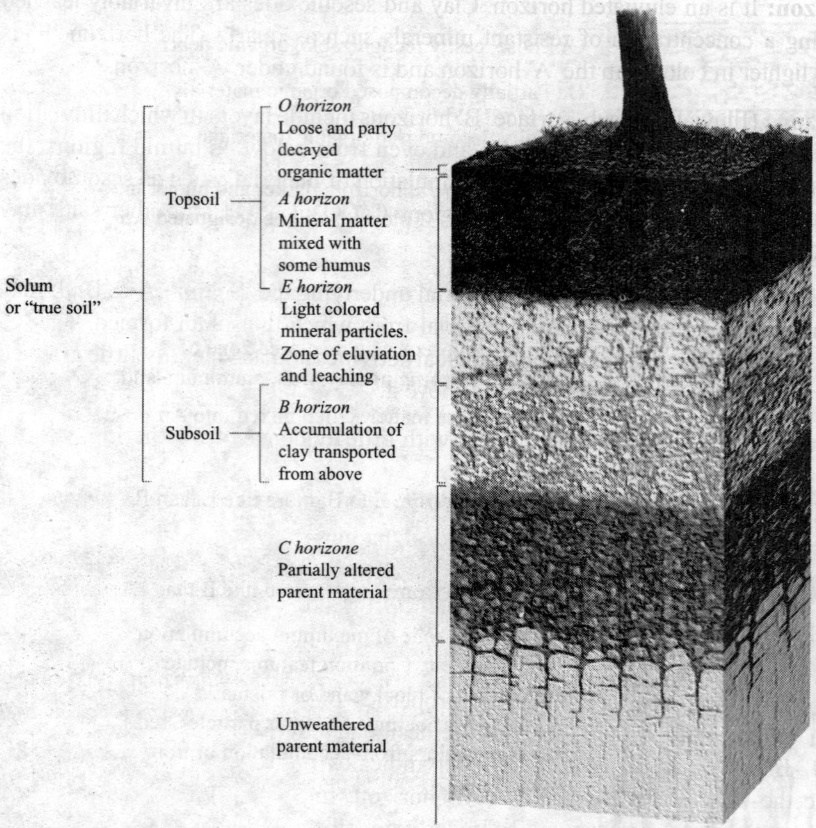


**Fig. 4:** Schematic soil profile

A soil horizon is a layer of soil, approximately parallel to the soil surface, differing in properties and characteristics from adjacent layers below or above it. Soil profile is an historic record of all the soil forming processes and it forms the unit of study in pedological investigations. Practically, soil profile is an important tool for soil classification which is applicable for thorough understanding of the soils. Five master horizons are recognized in soil profile and are designated using capital letters O, A, E, B and C. Sub-ordinate layers or distinctions with in these mater horizons are designated by lower case letter e.g., a, e, i, t, k etc.

**O- Horizons:**(Organic) It comprises of organic horizons that form above the mineral soil. They result from litter derived from dead plants and animals. 'O' horizons usually occur in forested areas and are generally absent in grassland regions.

**A - Horizon:** It is the top most mineral horizon. It contains a strong mixture of decomposed (humified) organic matter, which tends to impart a darker color than that of the lower horizons.

**E - Horizon:** It is an eluviated horizon. Clay and sesquioxides are invariably leached out, leaving a concentration of resistant minerals such as quartz. The horizon 'E' is generally lighter in color than the 'A' horizon and is found under 'A' horizon.

**B – Horizon :** (Illuvial) The sub -surface 'B' horizons include layers in which illuviation of materials has taken place from above and even from below. In humid regions, the B horizons are the layers of maximum accumulation of materials such as sesquioxides and silicate clays. In arid and semi-arid regions CaCO3, CaSO4 and other salts may accumulate in the B horizon.

**C – Horizon:** It is the unconsolidated material underlying the 'Solum' (A & B). It may or may not be the same as the parent material from which the solum formed. The 'C' horizon is out side the zones of major biological activities and is generally little affected by the processes that formed the horizons above it.

**R- Layer :** Underlying consolidated rock, with little evidence of weathering.

## Transitional and Combination Horizons

In some cases a single master horizon designation does not adequately convey information about the layer, such as where the horizon transitions to another layer or where it contains distinct parts from two kinds of master horizons.

## Transitional Horizons

Transitional horizons are dominated by properties of one master horizon but have subordinate properties of another. They are designated by two capital-letter symbols, e.g., AB, EB, BE, or BC. The first letter indicates the horizon whose properties dominate the transitional horizon. An AB horizon, for example, has characteristics of both an overlying A horizon and an underlying B horizon, but it is more like the A horizon than the B.

In some cases, a horizon can be designated as transitional even if one of the master horizons to which it transitions is not present. For example, a BE horizon recognized in a truncated soil has properties similar to those of a BE horizon in a soil from which the overlying E horizon has not been removed by erosion. A BC horizon may be recognized even if no underlying C horizon is present: it transitions to assumed parent materials.

## Combination Horizons

Combination horizons have two distinct parts that have recognizable properties of the two kinds of master horizons. They are designated by two capital-letter symbols (master horizons) separated by a virgule (/), e.g., E/B, B/E, or B/C. Most of the individual parts of one horizon component are surrounded by the other. The designation may be used even when horizons similar to one or both of the components are not present, provided that the separate components can be recognized in the combination horizon. The first letter indicates the horizon with the greater volume.

## Sub-Ordinate Distinctions with in Master Horizons

**Suffix Symbols:** Lowercase letters are used as suffixes to designate specific subordinate distinctions within master horizons and layers. If there is any evidence of accumulation, a suffix (or suffixes) can be used. The suffix symbols and their meanings follow:

***a.*** *Highly decomposed organic material.*

***b.*** *Buried genetic horizon.*

***c.*** *Concretions or nodules.*

***d.*** *Physical root restriction.*

***e.*** *Organic material of intermediate composition.*

***f.*** *Frozen soil.*

***g.*** *Strong gleying.*

***h.*** *Illuvial accumulation of organic matter.*

***i.*** *Slightly decomposed organic matter.*

***k.*** *accumulation of carbonates.*

***m.*** *cementation or induration,*

*km* . Cementation by carbonates;

*qm* . Cementation by silica;

*sm*. Cementation by iron;

*ym*. Cementation by gypsum;

*kqm*. Cementation by lime and silica; and

*zm*. Cementation by salts more soluble than gypsum.

***n.*** *Accumulation of sodium.*

***o.*** *Residual accumulation of sesquioxides*

***p.*** *Tillage or other disturbance.*

***q.*** *Accumulation of silica.*

***r.*** *Weathered or soft bedrock.*

***s.*** *illuvial accumulation of sesquioxides and organic matter.*

***s.*** *Presence of slickensides*

***t.*** *Accumulation of silicate clay*

***v.*** *plinthite.*

***w.*** *development of color or structure.*

***x.*** *fragipan character.*

***y.*** *Accumulation of gypsum*

***z.*** *accumulation of salts more soluble than gypsum*

## 2.5. Soil classification

Soil Classification is grouping of soils in to various categories by orderly and logical manner.

### 2.5.1. Early systems of classification

They are quite simple and practicable.

**Economic classification**: Grouping soils according to their productivity. This system is adopted by the revenue department for the purpose of taxation.

**Physical classification**: based on the physical properties like soil texture & soil structure. Based on texture the soils are grouped to sandy soils, loamy soils, clay soil etc. Based on structure the soils categorised to massive soils, single grain soils, aggregated soils etc.

**Chemical classification**: It is based on the chemical properties of the soils. Calcareous soils, gypsiferous soils, alkaline soils, acidic soils etc.

**Geological classification:** Two broad groups are recognized. i) Residual/sedentary soils – soils developed in-situ from the underlying rocks. ii)Transported soils –soils developed form transported & deposited sediments.

**Physiographic classification**: it is based on the landscape. The soils are termed as Pleatus, Levee soils, Terrace soils, Mountain soils, Hilly soils, Upland soils, Lowland soils etc.

**Other types of classifications:** Based on organic matter content, the soils were grouped to Organic soils and inorganic soils. Based on climate the soils grouped to Arid soils, Humid soils, Subhumid soils etc. Based on vegetation, Prairie soils, grassland soils, forest soils, etc.

**Genetic System of Soil Classification:** Russian scientist V.V. Docuchaev first suggested the genetic system of soil classification, which was further developed by the American Soil Scientist C.F. Marbut. According to this system, each soil possesses certain properties which develop as a consequence of the effects of a particular combination of factors affecting soil formation. The soils inherit their properties from the parent material in the initial stages of soil formation. For example, soils developed from basic parent material are rich in basic elements and plant nutrients and are alkaline in reaction in the initials stages of development. As soils mature, the inherited character gradually disappears and the soils are dominated by the character acquired from the climate and vegetation. For example, as time passes, basic elements are gradually washed down and ultimately soils which had developed from basic parent materials become poor in basic elements, plant nutrients and acidic in reaction.

According to the genetic system, soils of the world have been classified into three orders: Zonal, Intra-zonal and Azonal.

(1) **Zonal Soils:** The characteristics of soils belongings this order are mainly determined by the climate and vegetation. This means that soils that have developed from different parent material under the same climatic and vegetation conditions possess same general properties.

(2) **Intra-zonal Soils:** General characteristics of soils belonging to this order are mainly determined by certain local conditions like poor drainage, soluble salts etc. The soil characteristics due to these conditions, dominate over the characteristics due to climate and vegetation.

(3) **Azonal soils:** Soils belongings to this order to do not possess well developed profiles because they have not been subjected to the influence of the climate and the resultant vegetation long enough for the soil profile to develop.

Zonal soils have been divided into six suborders on the basic of wider ranges of variation in climate and the resultant vegetation. Each of the sub-orders have been subdivided into great groups on the basis of narrower ranges of variation of climate and the resultant vegetation.

Intra-zonal soils have been divided into three sub-orders viz., (i) Hydromorphic soils that contain excessive amounts of water. (ii) Holomorphic soils that contain excessive amounts of soluble salts. (iii) Calciomorphic soils that contain excessive amounts of lime or gypsum. Hydromorphic soils have been subdivided into great groups on the basis of increasing degrees of accumulation of water in soils. Holomorphic soils have been subdivided into the two great groups, Saline and Alkali soils. Saline soils contain excessive amounts of neutral salts i.e. sulphates and chlorides of sodium, calcium and magnesium. Alkali soils contain excessive amounts of sodium saturated clays and sodium carbonate. Calciomorphic soils are divided into two great groups: (i) Brown forest soils that are dark brown at the surface and gradually grade into gray calcareous parent material. (ii) Black earth or Grumosols that are dark brown grassland soils developed from limestone.

The order Azonal doesn't have any suborders, but the three following great groups have been recognized under viz., (i) Lithosols, (ii)Regosols and (iii) Alluvial soils. Lithosols, which are commonly found on steep slopes and consist of very shallow stony soils over bed rock. Regosols have developed from deep soft mineral deposits. They are also very young soils like Lithosols and differ from Lithosols in that they are not stony. Alluvial soils have been derived by the deposition of products of weathering of rocks by the great river systems like the Ganges, the Brahmaputra and the Indus etc. Great groups have been divided into families, each of which includes a certain number of identical soil series possessing certain common properties.

Each family has been divided into a number of series. Soils belonging to a series have originated from the same parent material, under same climate and resultant vegetation and therefore possess the same profile characteristics. Each series has been divided into a number of soil types on the basis of the texture of the surface soil. Soil types have been divided into phases on the basis of some important deviations like slope, erosion (removal of top soil by running water or blowing wind is called soil erosion), wetness, stoniness etc.

### 2.5.2. Modern Systems of soil classification

#### 2.5.2.1. FAO/UNESCO classification system

The FAO/UNESCO system was developed by a panel set up by UNESCO for providing the basic unit for the soil map of the world. This classification was compiled from diverse systems in term of category and nomenclature. The FAO system has two categories, a higher and a lower one. These categories have not been given name. From their definition, the higher category is equivalent to the great group of the USDA taxonomy while the lower category cannot be fitted into any category of the USDA.

The criteria for classification are similar to those of great group and sub group in the USDA taxonomy. The definition of diagnostic horizons in the FAO system is different from that of the USDA system, although there are many equivalent definitions. For example, argillic horizon in the USDA is the same as argic horizon in the FAO, while Albic, calcic, cambic, duric, histic, melanic, gypsic and nitric horizons has definition similar to those horizons bearing the same nomenclature in the USDA system.

There are 32 reference groups in the FAO (WRB, 2006) system:- Acrisols, Albeluvisols, Alisols, Andosols, Anthrosols, Arenosols, Calcisols, Cambisols, Chernozems, Cryosols, Durisols, Ferralsols, Fluvisols, Gleysols, Cypsisols, Histosols, Kastanozems, Leptosols, Lixisols, Luvisols, Nitosols, Phaeozems, Planosols, Plinthosols, Podzols, Regosols, Solonckak, Solonetz, Stagnosols, Technosols, Umbrisols and Vertisols.

Common group in Nigeria soils are Plinthosols, Ferralsols, Stagnosols (mangrove soils), Alisols (ultisols), Acrisols (ultisols), Luvisols, Lixisols (alfisols), Arenosols, Cambisols and Regosols. The names of soils are indicated by adding prefix and suffix adjectives from the qualifier lists to the reference group, for example Gleyic Luvisol oxyaquic.

**Local soil classification Systems**: There were a number of soil classification systems exist within the some countries that is native of the country. These includes Smith and Montgomery (1962) system of soil classification, Moss (1957) system of soil classification, Jungerius (1964) system of soil classification, Klinkenberg and Higgins (1968) system of soil classification.

In 1951 a decision was taken in US to develop a new system of classification because of the following defects in the genetic system:

1) Zonal, intra-zonal and azonal soils were not clearly defined.
2) Classification is based on assumed soil genesis.
3) Definition of certain taxa was based on virgin soils.
4) Too much emphasis on soil colour.
5) Taxa of lower category were defined in terms of comparative and subjective definitions.
6) No suitable classes were provided at some of the category levels.
7) Some families were not clearly defined.

### 2.5.2.2. USDA Soil Taxonomy

C.F. Marbut strongly emphasized that the classification of soils should be based on morphology instead of on theories of soil genesis, because theories are both ephemeral and dynamic. He perhaps overemphasized this point because some scientists assumed that soils had certain characteristics without ever actually examining them. Marbut stressed that examination of the soils themselves was essential in developing a system of soil classification and in making usable soil maps. However, Marbut's work reveals his personal understanding of the contributions of geology to soil science. His soil classification of 1935 relied heavily on the concept of a "normal soil," the product of equilibrium on a landscape where downward erosion keeps pace with soil formation. Continued work in soil classification by the U.S. Department of Agriculture culminated in the release of a new system published in the 1938 Yearbook of Agriculture in the chapter "Soil Classification" (Baldwin et al., 1938).

In both the early classification developed by Marbut and the later 1938 classification developed by USDA, the classes were described mainly in qualitative terms. Because the central concept of each class was described but the limits between classes were not, some soils seemed to be members of more than one class. The classes were not defined in quantitative terms that would permit consistent application of the system by different scientists. Neither system definitely linked the classes of its higher categories, which were largely influenced by the genetic concepts initiated by the Russian soil scientists, to the soil series and their subdivisions that were used in soil mapping in the United States. Both systems reflected the concepts and theories of soil genesis of the time, which were themselves predominantly qualitative in character. Modification of the 1938 system in 1949 corrected some deficiencies but also illustrated the need for a reappraisal of concepts and principles. One continuing problem was that a scientist required knowledge about the genesis of the soil to classify it. This information was often lacking or was disagreed upon by soil surveyors. It was determined that a new classification system was required, one that could be applied consistently by an increasingly large and varied cadre of soil surveyors.

More than 15 years of work under the leadership of Dr. Guy Smith culminated in a new soil classification system. Categories and classes of the new taxonomy were direct consequences of new and revised concepts and theories. This system became the official classification system of the U.S. National Cooperative Soil Survey in 1965 and was published in 1975 as *Soil Taxonomy: A Basic System of Soil Classification for Making and Interpreting Soil Surveys* (Soil Survey Staff, 1975). The system's most significant contribution was the establishment of taxonomic class limits and their quantitative definitions, whereby an individual soil could belong to only one class. Soil genesis was no longer used directly in determining the correct classification. Instead, diagnostic horizons and features that are the morphological expression of major known genetic processes were defined and used. In this way the current understanding of soil genesis, while indirectly incorporated in the taxonomy, is one step removed from the process of classifying a soil (Smith, 1963). The application of quantitative diagnostic horizons and features as criteria to be used in soil classification has been widely adopted in other soil classification systems around the world, perhaps most notably by the World Reference Base (IUSS Working Group WRB, 2014), sponsored by the Food and Agriculture Organization of the United Nations.

The system of soil classification discussed in *Soil Taxonomy* is dynamic and can change as new knowledge is obtained. The theories on which the system is based are tested every time the taxonomy is applied. During the 1980s and 1990s, nine international committees contributed to major revisions of the taxonomy. This work culminated in the printing of the second edition of *Soil Taxonomy* (Soil Survey Staff, 1999). In addition, many individual proposals for change have been incorporated in editions of the *Keys to Soil Taxonomy*, which have been published periodically since the first edition of *Soil Taxonomy* was published in 1975. The work of a 10th international committee, which addressed the impact of human influences on soils, resulted in important changes. These changes are reflected in the 12th edition of the *Keys to Soil Taxonomy* (Soil Survey Staff, 2014). The pioneers of soil pedology are V V Dokuchaev (1846 – 1903) – Founder of Modern Pedology, C F Marbut (1920 – 1940) – Founder of American Pedology

## 2.5.2.2.1. Criteria adopted for soil taxonomy

1.Diagnostic soil horizons

i. Surface diagnostic horizons (Epipedons)

ii. Subsurface diagnostic horizons (Endopedons)

iii. Other diagnostic features

2.Soil moisture regimes (SMRs)

3.Soil temperature regimes (STRs)

**1)Diagnostic soil horizons:**

Master horizons describe a soil profile, while diagnostic horizons are used to classify soils. Whereas master horizons are based on appearance, diagnostic horizons are based on soil formation processes. There are two types of diagnostic horizons. These are surface (epipedon) and subsurface horizon (Endopedon).

**i. Epipedons (Diagnostic surface soil horizons)**: An epipedon is the surface, or uppermost soil diagnostic horizon. They are not synonymous to the A horizon. They may be thinner than the A horizon, or include the E or part or the entire B horizon.

***Histic epipedon*** : This organic horizon is water saturated long enough for reduced conditions to occur unless artificially drained. It is 40 to 60 cm thick and has a low bulk density often less than 1 g $cm^3$.

***Mollic epipedon*** : This epipedon is a soft dark grassland soil. Its organic carbon content is 0.6 percent or more; its base saturation is 50 percent or more; It has a minimum thickness of 18 cm and contains less than 250 ppm $P_2O_5$.

***Anthropic epipedon*** : While similar to the mollic epipedon, the anthropic epipedon contains greater than 250 ppm citric acid soluble $P_2O_5$ with or without a 50 percent base saturation.

***Umbric epipedon*** : Mollic-like in thickness, organic carbon content, color, $P_2O_5$ content, consistence, and structure, this epipedon has less than 50 percent base saturation.

***Ochric epipedon*** Ochric epipedon is light in colour and low in organic matter and too thin to be any of the other five epipedons. Ochric epipedon is the most common epipedon in Nigerian soils.

***Plaggen epipedon*** : This man-made horizon is 50 cm or more thick and has resulted from centuries of accumulation of sod, straw, and manure, for example. It commonly contains artifacts such as pottery and bricks.

**Melanic Epipedon:** This horizon is commonly formed on volcanic ash deposits. It has andic soil properties in top 30cm, a colour value (moist) 2 or less, a melanic index of 1.7 or less and 6 % or more organic carbon

**Folistic (L. *folia, leaf*) Epipedon:** *Saturated for < 30 days & consists of organic* carbon of 8 – 16% or more depeding upon the clay content 0f 0-60%.

**Grossarenic Epipedon**: It has sandy texture of thickness 100cm or more over an argilling horizon.

**ii. Endopedons (Diagnostic subsurface soil horizons)**

Diagnostic subsurface horizons can be categorized as weakly developed horizons, as horizons featuring an accumulation of clay, organic matter, or inorganic salts, as cemented horizons, or as strongly acidic horizons.

***Agric horizon***: The agric horizon is an illuvial horizon that has formed under cultivation and contains significant amounts of illuvial silt, clay, and humus. It is usually formed after prolonged years of cultivation.

***Albic horizon*** : Albic (L. *albus*, white) materials are soil materials with a color that is largely determined by the color of primary sand and silt particles rather than by the color of their coatings.

***Argillic horizon(Bt)***: Argillic horizon is normally a subsurface horizon with a significantly higher percentage of phyllosilicate clay than the overlying soil material. It shows evidence of clay illuviation.

***Calcic horizon***: The calcic horizon is an illuvial horizon in which secondary calcium carbonate or other carbonates have accumulated to a significant extent.

***Cambic horizon***: This horizon shows some evidence of alterations but is very weakly developed between A and C horizons. The cambic horizon has less illuviation evidence than found in the argillic and spodic horizons.

***Gypsic horizon*** **(By)**: Accumulation of $CaSO_4$ ; 15 cm or more thick 5% or more calcium sulfate.

***Glossic horizon***: It consists of an eluvial part (E - albic) and an illuvial part (Bt - argillic, or Bto- kandic or Btn-natric) sometimes referred to as tounging

***Kandic horizon*** **(Bto)**: IT is fine textured subsurface horizon that has evidence of clay translocation and is composed mostly of low activity clays (low CEC and ECEC). The CEC (by the 1N $NH_4OAC$ method at pH=7) is 16 cmol(+) or less per kg of clay and the ECEC (by the basic cations displaced by ammonium plus KCl extractable aluminum methods) is 12 cmol(+) or less per kg of clay.

***Natric horizon*** **(Btn):** The borizon meets all the requirements of the argillic horizon and, in addition, has: a. either prismatic or columnar structure b. either (1) an exchangeable sodium percentage of 15 percent or more or (2) a sodium adsorption ratio (SAR) of 13 or more.

***Oxic horizon*** **(Bo):** It is highly weathered horizon that consists of oxides of iron and aluminum, 1:1 clays and resistant minerals such as quartz. It has a thickness of 30 cm or more; less than 10 percent weatherable minerals; low CEC (≤ 16 cmol(+) per kg of clay

***Salic horizon*** **(Bz)**: It is formed by accumulation of soluble salts. Salts more soluble than gypsum 15 cm or more thick EC > 30dS/m in a 1:1 soil water extract.

***Spodic*** **(Bs or Bhs):** Illuvial accumulations of sesquioxides and/or organic matter composed of spodic materials.

**iii. Other Diagnostic Soil Features**

***Abrupt textural change***: Ochric or Albic overlies an argillic considerable increase in clay over a short distance in the profile usually double the clay content within a vertical distance of 7.5 cm

***Glacic*** **(Bf):** Ice in the form of ice lenses or wedges.

***Lamellae*** **(Bt or Bw):** Number of clay-enriched layers separated by coarser textured layers illuvial horizon less than 7.5 cm thick horizontal bands contain an accumulation of oriented clay bridging sand grains lamellae occur in a vertical series of two or more; each lamellae has an overlying eluvial.

***Lithologic discontinuity***: (Numerical prefix; e.g. $2Bt_1$) change in parent material geologic process rather than pedogenic possibly indicated by abrupt textural change, contrasting particle sizes, stone lines, mineral composition or orientation of rocks.

***Plinthite*** **(Bv):** Humus poor, iron-rich, ironstone hardpan hardens irreversibly after being exposed to repeated wetting and drying in place begins as soft, red iron segregations (mottled appearance) due to oxidation.

***Slickensides*** **(Bss):** Polished or grooved clay surfaces produced by one soil mass sliding against another common in shrink-swell clays indicates vertic.

***Densic contact*** **(Bd or Cd):** Boundary between soil and underlying densic material (plow pan, glacial till, etc).

***Lithic contact*** **(R):** Boundary between soil and consolidated underlying material digging with hand tools is impractical.

***Paralithic contact*** **(Cr):** Boundary between soil and weathered bedrock underlying material can be dug with hand tools.

**2)Soil Moisture Regimes (SMRs)** : The term "soil moisture regime" refers to the presence or absence water held at a tension of less than 1500 kPa in soil moisture control section during periods of the year. Various Classes of Soil Moisture Regimes are described hereunder.

Soil Moisture control section (SMCC) soil moisture control section is to facilitate estimation of soil moisture regimes from climatic data. The upper boundary of this control section is the depth to which a dry (tension of more than 1500 kPa, but not air-dry) soil will be moistened by 2.5 cm of water within 24 hours. The lower boundary is the depth to which a dry soil will be moistened by 7.5 cm of water within 48 hours. These depths do not include the depth of moistening along any cracks or animal burrows that are open to the surface.

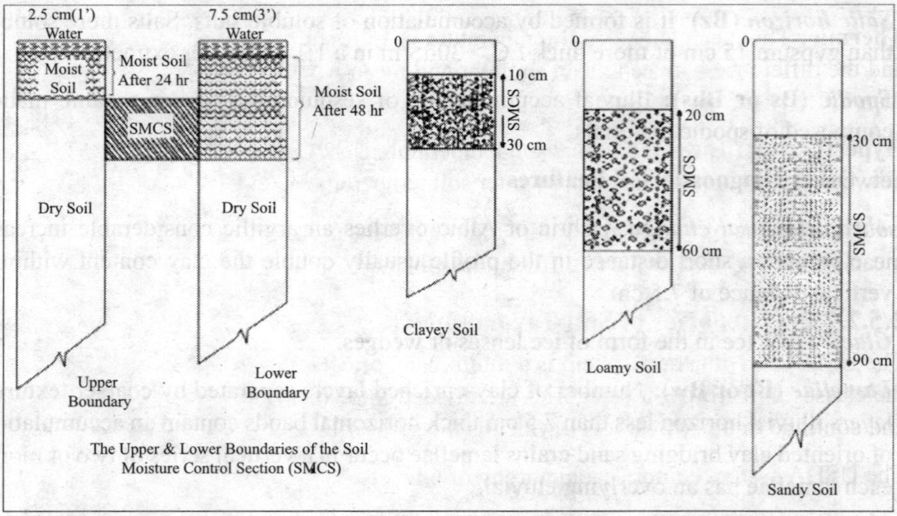


**Fig. 6:** Upper and lower boundaries of SMCC

**Aquic moisture regime:** The aquic moisture regime is a reducing regime in a soil that is virtually free of dissolved oxygen because it is saturated by water.

**Aridic and torric moisture regimes:** These terms are used for the same moisture regime but in different categories of the taxonomy. In the aridic (torric) moisture regime, the moisture control section is dry in all parts for more than half of the cumulative days per year.

**Udic moisture regime:** The udic moisture regime is one in which the soil moisture control section is not dry in any part for as long as 90 cumulative days in normal years.

**Ustic moisture regime:** The ustic moisture regime is intermediate between the aridic regime and the udic regime. Its concept is one of moisture that is limited but is present at a time when conditions are suitable for plant growth.

**Xeric moisture regime:** The xeric moisture regime is the typical moisture regime in areas of Mediterranean climates, where winters are moist and cool and summers are warm and dry.

**3)Soil Temperature Regimes (STRs)** : Various Classes of Soil Moisture Regimes are described hereunder.

**Cryic (Gr. *kryos*, coldness; meaning very cold soils):** Soils in this temperature regime have a mean annual temperature lower than 8 °C but do not have permafrost.

**Frigid:**A soil with a frigid temperature regime is warmer in summer than a soil with a cryic regime, but its mean annual temperature is lower than 8°C and the difference between mean summer and mean winter soil temperatures is more than 6°C

**Mesic:** The mean annual soil temperature is 8°C or higher but lower than 15°C, and the difference between mean summer and mean winter soil temperatures is more than 6°C.

**Thermic:** The mean annual soil temperature is 15 °C or higher but lower than 22°C, and the difference between mean summer and mean winter soil temperatures is more than 6°C.

**Hyperthermic:** The mean annual soil temperature is 22°C or higher, and the difference between mean summer and mean winter soil temperatures is more than 6°C.

***iso* :** If the name of a soil temperature regime has the prefix *iso*, the mean summer and mean winter soil temperatures differ by less than 6°C.

## 2.5.2.2.2. Categories of Soil Taxonomy

The USDA classification system is a multi- categorical and hierarchical system. Thus the classes in the highest categories are divided into smaller classes in the lower one and continue to the lowest level, which is the soil series.

The USDA system comprise a hierarchy of 6 levels

1. Orders (12) (surface and subsurface diagnostic horizons)
2. Suborders (55) (Soil temperature. and moisture regimes)
3. Great group (238) (subsurface diagnostic horizon)
4. Subgroup (1243) (drainage, lithic contact, PM, clay type)
5. Family (7504) (Texture of diagnostic surface horizon)
6. Series (about 19,000) in U.S.

### 2.5.2.2.3. Nomenclature of soil taxonomy

Soil classification is polynomial nomenclature. The highest category Order ends with Sol e.g. arid soil. Sub-order is composed of formative element of order + any differentiating character like temperature, moisture, drainage, diagnostic horizon etc. e.g. Aquolls wet soil-Mollisols (oil), Aquents wet soil- Entisols (ent), Argid aridi soil (id). Great group is composed of order + sub-order + one or more specific characters which modify the sub-order e.g. Argiaqoll Molli soil –order, aqoll - sub-order wetness, Arg -argillic horizon. Sub-group is identified with a binomial nomenclature. e.g. A 'Typic' adjective is used when sub-group represent a 'Central concept' of the great group. e.g. Typic Argiaqoll. There are integrated between different sub-groups or between great groups in the same order. In nomenclature of family, we have textural class, mineralogy, temperature regime e.g. Fine-clay, mixed montmorillonitic, hyperthermic, Typic Chromustert. The name of the 'series' is given on the basis where it was described first e.g. Naira Typic Chromusterts.

**Table 5:** Nomenclature at various categories of soil taxonomy

| Category | Nomenclature | Example |
|---|---|---|
| Order | The names of orders are combination of two parts; first word containing the formative element and sol connected by vowel 'I' or 'O' for Latin and Greek words respectively. Second word is sol. | Mollisol, Alfisol, Ultisol, Entisol, Histosol. Spodosol etc |
| Suborder | The name consists of two syllables. The first is the suggestive of a property of that suborder and the second the name of the order as reflected by the formative element of the order. | Udoll, Cryept, Ustalf etc |
| Great group | the name made of prefixing one additional word (distinguishing character of the great group) to the suborder name. | Tropustalf, Hapludoll, Calciustert etc. |
| Subgroup | The name consists of the name of the great group modified by one or more adjectives (binomial nomenclature) | Vertic Haplustept, Typic Calciustert |
| Family | The name consists particle size class, mineralogy class, and STR class as prefix to subgroup. This category meet the need for making practical prediction for land use planning | Fine silty, kaolinitic, isohyperthermic, Typic Hapludoll. |
| Series | The name of the soil series represent a prominent geographic name (has no pedogenic significance). | Naira red soils |

**Soil Orders**: There are 12 Soil Orders in Soil Taxonomy including two recently formed orders viz., Andisol and Gelisol. Each Order have *formative element* to use in classification. The formative elements of the Soil Orders as follows, Entisol, *ent*; Inceptisol, *ept*; Andisols, *and*; Spodosols, *od*; Mollisols oll, *oll*; Alfisols, *alf*; Ultisols, *ult*; Oxisols, *ox*; Aridisols, *id*; Vertisols, *ert* ; Histosols, *ist* and Gelisols, *el*.

## Simplified KEYs to Soil Orders (a flow diagram)

Permafrost within 100/200 cm of the soil surface → Yes → Gelisol

↓ No

More than 30% organic matter to a depth of 40cm → Yes → Histosol

↓ No

Spodic horizon within 2m from the surface → Yes → Spodosol

↓ No

Andic soil properties of > 35cm thickness soil → Yes → Andisol

↓ No

Oxic horizon within 1.5m but no kandic horizon → Yes → Oxisol

↓ No

30% or more clay up to 50cm, cracks, slickensides → Yes → Vertisol

↓ No

Dry for more than 50% of the year → Yes → Aridisol

↓ No

Argillic/kandic horizon with < 35% BS → Yes → Ultisol

↓ No

Mollic Epipedon / dark coloured soil of grasslands → Yes → Mollisol

↓ No

Argillic / kandic horizon with 35% or more BS → Yes → Alfisol

↓ No

Cambic horizon → Yes → Inceptisol

↓ No

No diagnostic horizon other than ochric/anthropic horizon → Yes → Entisol

To summarize, there are 3 orders with unique parent materials (Andisols, Histisols, Vertisols); 3 orders with unique environments (Aridisols, Gelisols, Oxisols); 3 orders by age of development (Entisols, Inceptisols, Ultisols) and 3 orders by unique vegetative influence (Alfisols, Mollisols, Spodosols).

### 2.5.2.2.4. Important features of soil orders

**Gelisols:** Gelisols (from the Latin *gelare* – to freeze) are soils that are permanently frozen (contain "permafrost") or contain evidence of permafrost near the soil surface. Gelisols are found in the Arctic and Antarctic, as well as at extremely high elevations. Permafrost influences land use through its effect on the downward movement of water and freeze-thaw activity (cryoturbation) such as frost heaves. Permafrost can also restrict the rooting depth of plants. Gelisols make up about 9% of the world's glacier-free land surface.

**Histosols:** Histosols (from the Greek *histos* – tissue) are dominantly composed of organic material in their upper portion. The Histosol order mainly contains soils commonly called bogs, moors, peat lands, muskegs, fens, or peats and mucks. These soils form when organic matter, such as leaves, mosses, or grasses, decomposes more slowly than it accumulates due to a decrease in microbial decay rates. This most often occurs in extremely wet areas or underwater; thus, most of these soils are saturated year-round. Histosols can be highly productive farmland when drained; however, drained Histosols can decompose rapidly and subside dramatically. They are also not stable for foundations or roadways, and may be highly acidic. Histosols make up about 1% of the world's glacier-free land surface.

**Spodosols:** Spodosols (from the Greek *spodos* – wood ash) are among the most attractive soils. They often have a dark surface underlain by an ashy gray layer, which is subsequently underlain by a reddish, rusty, coffee-colored, or black subsoil horizon. These soils form as rainfall interacts with acidic vegetative litter, such as the needles of conifers, to form organic acids. These acids dissolve iron, aluminum, and organic matter in the topsoil and ashy gray (eluvial) horizons. The dissolved materials then move (illuviate) to the colorful subsoil horizons. Spodosols most often develop in coarsely textured soils (sands and loamy sands) under coniferous vegetation in humid regions of the world. They tend to be acidic, and have low fertility and low clay content. Spodosols occupy about 4% of the world's glacier-free land surface.

**Andisols:** Andisols (from the Japanese *ando* – black soil) typically form from the weathering of volcanic materials such as ash, resulting in minerals in the soil with poor crystal structure. These minerals have an unusually high capacity to hold both nutrients and water, making these soils very productive and fertile. Andisols include weakly weathered soils with much volcanic glass, as well as more strongly weathered soils. They typically occur in areas with moderate to high rainfall and cool temperatures. They also tend to be highly erodible when on slopes. These soils make up about 1% of the glacier-free land surface.

**Oxisols:** Oxisols (from the French *oxide* – oxide) are soils of tropical and subtropical regions, which are dominated by iron oxides, quartz, and highly weathered clay minerals such as kaolinite. These soils are typically found on gently sloping land surfaces of great age that have been stable for a long time. For the most part, they are nearly featureless soils without clearly marked layers, or horizons. Because they are highly weathered, they have low natural fertility, but can be made productive through wise use of fertilizers and lime. Oxisols are found over about 8% of the glacier-free land surface.

**Vertisols:** Vertisols (from the Latin *verto* – turn) are clay-rich soils that contain a type of "expansive" clay that shrinks and swells dramatically. These soils therefore shrink as they dry and swell when they become wet. When dry, vertisols form large cracks that may be more than one meter (three feet) deep and several centimeters, or inches, wide. The movement of these soils can crack building foundations and buckle roads. Vertisols are highly fertile due to their high clay content; however, water tends to pool on their surfaces when they become wet. Vertisols are located in areas where the underlying parent materials allow for the formation of expansive clay minerals. They occupy about 2% of the glacier-free land surface.

**Aridisols:** Aridisols (from the Latin *aridus* – dry) are soils that occur in climates that are too dry for "mesophytic" plants—plants adapted to neither too wet nor too dry environments—to survive. The climate in which Aridisols occur also restricts soil weathering processes. Aridisols often contain accumulations of salt, gypsum, or carbonates, and are found in hot and cold deserts worldwide. They occupy about 12% of the Earth's glacier-free land area, including some of the dry valleys of Antarctica.

**Ultisols:** Ultisols (from the Latin *ultimus* – last) are soils that have formed in humid areas and are intensely weathered. They typically contain a subsoil horizon that has an appreciable amount of translocated clay, and are relatively acidic. Most nutrients are held in the upper centimeters of Ultisol soils, and these soils are generally of low fertility although they can become productive with additions of fertilizer and lime. Ultisols make up about 8% of the glacier-free land surface.

**Mollisols:** Mollisols (from the Latin *mollis* – soft) are prairie or grassland soils that have a dark colored surface horizon, are highly fertile, and are rich in chemical "bases" such as calcium and magnesium. The dark surface horizon comes from the yearly addition of organic matter to the soil from the roots of prairie plants. Mollisols are often found in climates with pronounced dry seasons. They make up approximately 7% of the glacier-free land surface.

**Alfisols:** Alfisols (from the soil science term *Pedalfer* – aluminum and iron) are similar to Ultisols but are less intensively weathered and less acidic. They tend to be more inherently fertile than Ultisols and are located in similar climatic regions, typically under forest vegetation. They are also more common than Ultisols, occupying about 10% of the glacier-free land surface.

**Inceptisols:** Inceptisols (from the Latin *inceptum* – beginning) exhibit a moderate degree of soil development, lacking significant clay accumulation in the subsoil. They occur over a wide range of parent materials and climatic conditions, and thus have a wide range of characteristics. They are extensive, occupying approximately 17% of the earth's glacier-free surface.

**Entisols:** Entisols (from *recent* – new) are the last order in soil taxonomy and exhibit little to no soil development other than the presence of an identifiable topsoil horizon. These soils occur in areas of recently deposited sediments, often in places where deposition is faster than the rate of soil development. Some typical landforms where Entisols are located include: active flood plains, dunes, landslide areas, and behind retreating glaciers. They are common in all environments. Entisols make up the second largest group of soils after Inceptisols, occupying about 16% of the Earth's surface.

### 2.5.2.2.5. Salient features of Soil Taxonomy

1) It is based on measurable soil properties
2) It considers all such properties which affect soil genesis
3) The nomenclature is most logical & helps in relating the place of taxon
4) It is truly multi -categorical system with 6 categories
5) It is an orderly scheme without prejudice and
6) It facilitate easy remembering of objects.

### 2.5.2.2.6. Advantages of soil taxonomy over other systems

1) It is the classification of soils rather than soil forming processes
2) It focuses on soils rather than related sciences such as geology and climatology
3) It permits the classification of soils of unknown genesis
4) It permits the greater uniformity of Soil classification than other systems
5) The soil with a different genesis but having identical properties are classified within the same unit
6) There is no particular order for strictly hydromorphic and halomorphic soils as in the other systems.

### 2.5.2.3. Land capability classification

Soil survey reports provide the basic material for the land capability classification system.

The system consists of three categories, viz., Land capability unit, land capability subclass and land capability class.

### (a) Land Capability Unit

It is a grouping of one or more individual soil mapping units (a portion of the landscape that has similar characteristics and qualities and whose limits are fixed by precise definitions) having similar potentials and continuing limitations or hazards.

The soils in a capability unit are sufficiently uniform to:

(i) Produce similar kinds of cultivated crops and pasture with similar management practices;

(ii) Require similar conservation treatment and management under the same kind and condition of vegetative cover; and

(iii) Have comparable potential productivity.

The land capability unit is expressed by putting an Arabic number (1, 2, 3, etc.) at the end of land capability class or subclass. For example: IIIe1, IIIe2.

## (b) Land Capability Subclass

Subclasses are the groups of capability units which have the same major conservation problem or same kind of limitation, such as Erosion and runoff (e), Excess water or wetness (w), Root zone or soil limitations (s) and Climatic limitations (c). The suffixes e, w, s, or c are added to the capability class symbol to indicate capability subclass. For example: IIIe, IIw, IIs, etc. No subclasses are made in class I land.

## (c) Land capability class (LCC)

Capability classes are the groups of capability subclasses or capability units that have the same relative degree of hazards or limitations.

Land capability classes are usually divided into two major groups:

(i) Land suitable for cultivation; and

(ii) Land unfit for cultivation but suitable for permanent vegetation like pastures, orchards and forests.

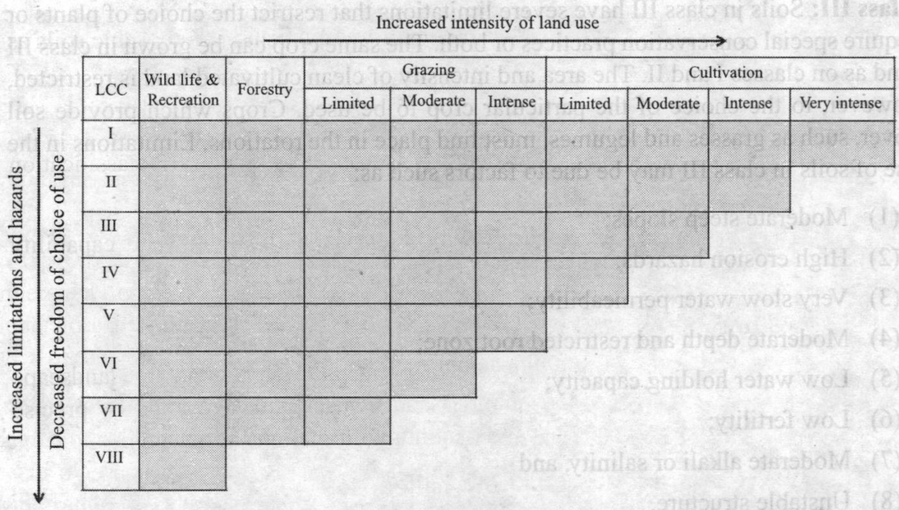


**Fig 7.** Land capability classes and the intensity of their use (Shaded portions show the use of suitability)

## A brief description of characteristics and safe use of land and soils in each LCC:

**Class I:** Soils in this class have few or no limitations that restrict their use. They can be cropped very intensively, used for pastures, grazing lands, woodlands, on even for wildlife preserves. The soils are deep, having good water holding capacity, well drained and the land is nearly level. These soils are either naturally fertile or possess such characteristics which encourage good response of crops to fertilizer application. The soils in class I need only ordinary crop management practices to maintain their productivity such as fertilization, liming, etc. Crop rotations may also be followed.

**Class II:** Soils in this class have some limitations that reduce the choice of plants or require moderate conservation practices. These soils may be used for the same crops as class I. However, they are capable of intensive cropping systems, or with the same cropping systems they require some conservation practices. The use of soils in class II may be limited by one-or more factors such as:

(1) Gentle slope;

(2) Moderate erosion hazards;

(3) Adequate soil depth;

(4) Less than ideal soil structure and workability;

(5) Slight to moderate alkali or saline conditions; and

(6) Somewhat restricted drainage.

The management practices required for class II land include terracing, strip cropping, contour tillage, rotations involving legumes and grasses. The management practices employed in class I land are also generally required for soils in class II.

**Class III:** Soils in class III have severe limitations that restrict the choice of plants or require special conservation practices or both. The same crop can be grown in class III land as on classes I and II. The area and intensity of clean cultivated land is restricted, however, to the choice of the particular crop to be used. Crops which provide soil cover, such as grasses and legumes, must find place in the rotations. Limitations in the use of soils in class III may be due to factors such as:

(1) Moderate steep slopes;

(2) High erosion hazards;

(3) Very slow water permeability;

(4) Moderate depth and restricted root zone;

(5) Low water holding capacity;

(6) Low fertility;

(7) Moderate alkali or salinity, and

(8) Unstable structure.

Soils in class III often require special conservation practice. Those mentioned for class II land must be employed frequently in combination with restrictions in kinds of crops. Improvement in drainage may also be needed.

**Class IV:** Soils in this class can be used for cultivation but there are very severe limitations on the choice of crops. Also, very careful management may be required. The alternative uses of these soils are more limited than for class III. Close growing crops must be used extensively and row crops cannot be grown safely in most cases. The choice of crops may be limited by excess moisture as well as by erosion hazards. The most limiting factors on these soils may be one or more of the following:

(1) Steep slopes;

(2) Severe erosion susceptibility;

(3) Severe past erosion;
(4) Shallow soils;
(5) Low water holding capacity;
(6) Poor drainage; and
(7) Severe alkali or salinity.

Soil conservation practices must be applied more frequently than on soils in class III.

**Class V:** Soils in this class are limited in their safe use by factors other than erosion hazards. These limitations are:

(1) Subject to frequent overflow;
(2) Growing season too short for crop plants;
(3) Stony or rocky soils; and
(4) Ponded areas where drainage is not feasible.

These soils not fit for cultivation, but only for grazing and wild life & recreation.

**Class VI:** The soils have severe limitations that restrict their use largely to pasture or range, woodland or wildlife. The physical limitations are the same as those for class IV land but severity of limitation is more. These lands fit to use moderate grazing, forestry and wildlife.

**Class VII:** Soils in class VII land have very severe limitations which restrict their use to grazing, woodland or wildlife. The physical limitations are the same as in class VI except that they are too severe to pasture use making limited use for grazing. These lands can be use for forestry and wild life.

**Class VIII:** The soils in this class of land are such that they should not be used for any kind of commercial plant production being severely eroded. The lands included in this class are tidal lands, swamps, river wash, sand dunes and barren mountain tops. These lands should be used only for recreation, wildlife, water supply or aesthetic purposes. A land capability class can be upgraded if the limitations are overcome.

**Representation of Land capability classes on the soil map:** On a soil map, the land capability classes are usually shown by standard colours.

**Table 6:** Colour representation of Land capability classes on soil map.

| Land capability class | Colour represented |
|---|---|
| I | Green |
| II | Yellow |
| III | Pink |
| IV | Blue |
| V | Dark Green |
| VI | Orange |
| VII | Red |
| VIII | Purple |

## Guidelines for categorising Land capability classes

In deciding about the land capability class of a given soil and landscape unit the weightage to various properties/features are given as per the following guidelines. Similarly, other soil, landscape, environmental or land use factors that are of importance be given due weightage in arriving at land capability classification. On the basis of consideration for each individual property of a soil mapping unit, several capability classes of the same unit will be arrived at. But the actual capability class for the unit will be decided on the basis of the class of the most limiting property. For example, a moderately deep, well drained, loam soil occurs in an arid climate, then its capability class will be decided as follows: Depth of soil- class III, drainage - class I, texture - class I, climate - class IV; so capability class of this soil will be class IVc , i.e class IV lands with climate as limitation.

**Table 7:** Soil characters with limits to qualify for Land capability classes.

| | | | |
|---|---|---|---|
| (a) | Soil Characters | | |
| (i) | Depth of solum | | |
| | D5. Very deep | >100 cm | LCC-I |
| | D4. Deep | 50- 100 cm | LCC-II |
| | D3. Moderately deep | 25- 50 cm | LCC-III |
| | D2. Shallow | 10- 25 cm | LCC-IV |
| | D1. Very shallow | <10 cm | LCC-V, VI |
| | D0. No soil cover | | LCC- VII, VIII |
| (ii) | Soil texture | | |
| | Medium | loam, sandy loam, silt loam | LCC-I |
| | Light | loamy sand or sand | LCC-II |
| | Heavy | Clay loam, silty clayloam, sandy clay loam | LCC-III |
| | Very light/ very heavy | Sand or Clay | |
| | When the above textural classes are qualified by adjectives like stony, gravely, etc., the LCC is brought down to next step. For example stony loam texture belong to LCC II; gravelly sand belong to LCC IV, so on | | |
| (iii) | Soil drainage | | |
| | Moderately well (D3) or well (D4) | | LCC I |
| | Imperfect (D2) or excessively well (D5) | | LCC II |
| | Poor (D1) | | LCC III |
| | Very poor (D0) | | LCC IV, V |
| (b) | External land features | | |
| (i) | Slope | | |
| | A.Nealry Level | 0- 1% | LCC I |
| | B.Gently slopping | 1- 3% | LCC II |
| | C.Moderately sloping | 3- 5% | |
| | D.Sloping | 5- 10% | LCC III |

| | | | |
|---|---|---|---|
| | E.Extremely sloping | 10- 15% | LCC IV |
| | F.Moderately steep | 15- 25% | |
| | G.Steep slope | 25- 33% | LCC V or VI |
| | H.Very steep | 33- 50% | |
| | I.Extremely steep | >50% | LCC VII, VIII |
| (ii) | **Erosion** | | |
| | None to slight erosion ($e_1$) | | LCC I |
| | Moderate erosion ($e_2$) | | LCC II & III |
| | Severe erosion ($e_3$) | | LCC III & IV |
| | Very severe erosion ($e_4$) | | LCC VI & VII |
| (c) | Environmental factors | | |
| (i) | Climate | | |
| | Well distributed rain, humid climate | | LCC I |
| | Occasional dry spell, humid climate | | LCC II |
| | Subhumid climate | | LCC II & III |
| | Semiarid climate | | LCC III & IV |
| | Arid climate | | LCC IV |
| (ii) | Flood | | |
| | No flood | | LCC I |
| | Occasional floods during rainy season | | LCC II |
| | For considerable period under flood | | LCC III & IV |
| | Almost all the year under flood | | LCC V |
| | When provision for drainage introduced the capability class will change | | |
| (d) | Present land use | | |
| (i) | Cultivated land | | |
| | Triple cropped | | LCC I |
| | Double cropped | | LCC I |
| | Single cropped | | LCCII |
| (ii) | Terraced land | | |
| | Bench terrace well maintained | | LCC I |
| | Moderately maintained terraces | | LCC II |
| | Poorly bunded and poorly maintained terraces | | LCC III, IV |

## Soil conservation measures require in relation of Land capability classification

The system of land classification based upon the soil and site characteristics enables us to plan and match soil conservation measures according to the intensity and nature of the problem. As we moves from LCC I to LCC IV, the intensity of the problem increases and greater intensive land treatment would be required to keep it under permanently productive condition. The land classes V to VIII have severe hazards in respect of the problems given above and are, therefore, unsuitable for agriculture. As we moves from land classes V to VIII increasing hazards are noticed in respect of use for pasture and afforestation.

The permissible intensity of grazing will also have to be cut down for land classes VII and VIII. Classification of land into units enables us to divide the land into specific land management groups and spell out the quantum of treatment needed to obtain maximum productivity from each piece of land.

**Table 8:** Land capability classes, subclasses and units.

| Major land use suitability | Land capability class | Land capability sub class (kind of limitation) | Land capability unit (distinctive physical character) |
|---|---|---|---|
| Suitable for cultivation | LCC I- few limitations, wider land use. Very good land for every sand point | - | - |
| | LCC II- moderate limitations. Good land for all round stand point | - | - |
| | LCC III- severe limitations for regular cultivation. | IIIe- limited by erosion hazard, moderately slopping land | 13-C-2 moderately sloping, slightly acidic soils on lime stone |
| | | | 9-C-2 moderately sloping, highly acid soils on sand stone or shale |
| | | IIIw- limited by excessive water, needs drainage for cultivation | |
| | | IIIs- limited by low soil moisture retention capacity, loamy sand/ sandy soils | |
| | | IIIes- very slowly permeable clay pan subsoils | |
| | LCC IV-Very severe limitations, suited for occasional cultivation or limited cultivation | | |
| Not suitable for cultivation | LCC V- Not suited for cultivation because of wetness, stoniness, over flow etc. Few limitations for grazing or forestry use | Groupings of rangelands, pastures or forest sites. | Land management groups based on permanent physical factors such as range sites or forest sites. |
| | LCC VI- Too steep slopes, stony, arid climate, wetness limitations for cultivation. Moderate limitations for Grazing and forestry | | |
| | LCC VII- Very steep slopes, rough, arid climate, wetness limitation for cultivation. Severe limitations for grazing and forestry | | |
| | LCC VIII- Extremely rough, arid, swampy nature, not suitable for cultivation, grazing, forestry. Suitable for wild life, watersheds and recreation. | | |

## 2.5.2.4. Land suitability classification:

The objective of various soil-site evaluation studies has been to predict and classify land for plant growth. Observations on growth inhibiting factors for certain species and tolerance of others to extremely adverse conditions have been evaluated by many scientists.

### Suitability criteria

Most of the plant species need well drained, moderately fine to medium texture soils, free of salinity and having optimum physical environment. Soil resource maps based on several parameters, can aid in predicting the behaviour and suitability of soils for growing field crops, horticultural crops, forest species and other plantation crops once the suitability criteria is established. Within limits, it may also find application in transfer of technology to other areas with comparable soil-site characteristics. The FAO land suitability classification system has four different categories: Orders, Classes, Subclasses and Units. There are two orders(S and N) which reflect the kind of suitability (S for suitable and N for unsuitable land).

### Structure of the suitability classification

The framework has the same structure, i.e. recognizes the same categories, in all of the kinds of interpretative classification (see below). Each category retains its basic meaning within the context of the different classifications and as applied to different kinds of land use. Pour categories of decreasing generalization are recognized:

**Table 9:** Structure of Land suitability classification

| Land Suitability Orders | Reflecting kinds of suitability. |
|---|---|
| Land Suitability Classes | Reflecting degrees of suitability within Orders. |
| Land Suitability Subclasses | Reflecting kinds of limitation or main kinds of improvement measures required, within Classes. |
| Land Suitability Units | Reflecting minor differences in required management within Subclasses. |

**Land Suitability Orders**

Land suitability Orders indicate whether land is assessed as suitable or not suitable for the use under consideration. There are two orders represented in maps, tables, etc. by the symbols S and N respectively.

**Table 10:** Land suitability orders

| | |
|---|---|
| Order S Suitable: | Land on which sustained use of the kind under consideration is expected to yield benefits which justify the inputs, without unacceptable risk of damage to land resources. |
| Order N Not Suitable: | Land which has qualities that appear to preclude sustained use of the kind under consideration. |

Land may be classed as Not Suitable for a given use for a number of reasons. It may be that the proposed use is technically impracticable, such as the irrigation of rocky steep land, or that it would cause severe environmental degradation, such as the cultivation of steep slopes. Frequently, however, the reason is economic: that the value of the expected benefits does not justify the expected costs of the inputs that would be required.

**Land Suitability Classes**

Land suitability Classes reflect degrees of suitability. The classes are numbered consecutively, by arabic numbers, in sequence of decreasing degrees of suitability within the Order. Within the Order Suitable the number of classes is not specified. There might, for example, be only three, Si, $S_2$ and $S_3$. The number of classes recognized should be kept to the minimum necessary to meet interpretative aims; five should probably be the most ever used.

If three Classes are recognized within the Order Suitable, as can often be recommended, the following names and definitions may be appropriate in a qualitative classification:

**Table 11:** Land suitability classes in Land suitability order 'Suitable'

| | |
|---|---|
| Class S1 Highly Suitable: | Land having no significant limitations to sustained application of a given use, or only minor limitations that will not significantly reduce productivity or benefits and will not raise inputs above an acceptable level. |
| Class S2 Moderately Suitable: | Land having limitations which in aggregate are moderately severe for sustained application of a given use; the limitations will reduce productivity or benefits and increase required inputs to the extent that the overall advantage to be gained from the use, although still attractive, will be appreciably inferior to that expected on Class S1 land. |
| Class S3 Marginally Suitable: | Land having limitations which in aggregate are severe for sustained application of a given use and will so reduce productivity or benefits, or increase required inputs, that this expenditure will be only marginally justified. |

In a quantitative classification, both inputs and benefits must be expressed in common measurable terms, normally economic. In different circumstances different variables may express most clearly the degree of suitability, e.g. the range of expected net income per unit area or per standard management unit, or the net return per unit of irrigation water applied to different types of land for a given use.

Where additional refinement is necessary it is recommended that this should be achieved by adding classes, e.g. S4, and not by subdividing classes, since the latter procedure would contradict the principle that degrees of suitability are represented by only one level of the classification structure, that of the suitability class. This necessarily change e the meanings of class numbers, e.g. if four classes were employed for classifying land with respect to arable use and only three with respect to forestry, Marginally Suitable could refer to S4 in the former case but S3 in the latter.

An alternative practice has been adopted in some countries. In order to give a constant numbering to the lowest Suitable class, classes have been subdivided as, e.g. S2.1,

S2.2. This practice is permitted within the Framework, although for the reason given in the preceding paragraph it is not recommended.

Suitability Class S1, Highly Suitable, may sometimes not appear on a map of a limited area, but could still be included in the classification if such land is known or believed to occur in other areas relevant to the study.

Differences in degrees of suitability are determined mainly by the relationship between benefits and inputs. The benefits may consist of goods, e.g. crops, livestock products or timber, or services, e.g. recreational facilities. The inputs needed to obtain such benefits comprise such things as capital investment, labour, fertilizers and power. Thus an area of land might be classed as Highly Suitable for rainfed agriculture, because the value of crops produced substantially exceeds the costs of farming, but only Marginally Suitable for forestry, on grounds that the value of timber only slightly exceeds the costs of obtaining it.

It should be expected that boundaries between suitability classes will need review and revision with time in the light of technical developments and economic and social changes. Within the Order 'Not Suitable', there are two classes are present.

**Table 12:** Land suitability classes in Land suitability order 'Not suitable'

| | |
|---|---|
| Class N1 Currently Not Suitable: | Land having limitations which may be surmountable in time but which cannot be corrected with existing knowledge at currently acceptable cost; the limitations are so severe as to preclude successful sustained use of the land in the given manner. |
| Class N2 Permanently Not Suitable: | Land having limitations which appear so severe as to preclude any possibilities Of successful sustained use of the land in the given manner. |

Quantitative definition of these classes is normally unnecessary, since by definition both are uneconomic for the given use. The upper limit of Class N1 is already defined by the lower limit of the roast suitable class in Order S.

The boundary of Class N2, Permanently Not Suitable, is normally physical and permanent. In contrast, the boundary between the two orders, Suitable and Not Suitable is likely to be variable over time through changes in the economic and social context.

**Land Suitability Subclasses:** The sub classes reflect kinds of limitations or main kinds of improvement measures required within classes. They are indicated in the symbol using lower case letters viz., c : Climatic conditions, t : Topographic limitations, w : Wetness limitations, n : Salinity(and/or alkalinity)limitations, f : Soil fertility limitations not readily to be corrected, s : Physical soil limitations(influencing soil/ water relationship and management).

**Land suitability units:** This grouping is used to identify land development units having minor differences in management requirements. This can indicate the relative importance of land development works. It is indicated by Arabic numerals, enclosed in parenthesis, following the subclass symbol. Example of total unit : The whole unit is indicated by a symbol; for example : S2w(2). Here "S" represents Order (Suitable); the number 2 after the letter S represents Class 2(moderately suitable); "w" represents Subclass w(wetness limitation); and (2)represents Unit 2.

Sys and Verheye(1975) proposed the following capability index(Ci ) based on nine parameters for crop production in the arid and semi-arid regions.

**Ci =A.B.C.D.E.F.G.H.I.**

Where,

A = rating for soil texture (taken as 100 for best texture, say loam)

B = rating for calcium carbonate (taken as fraction of 1(one)

C = rating for gypsum (taken as fraction of 1)

D = rating for salinity (taken as fraction of 1)

E = rating for sodium saturation (taken as fraction of 1)

F = rating for drainage (taken as fraction of 1)

G = rating for soil depth (taken as fraction of 1)

H = rating for slope (taken as fraction of 1)

I = rating for profile development (taken as fraction of 1) .

**Table 13:** Rating based on kind and degree of limitation (adopted from Sys, 1976)

| Land characteristics | 100 | 90 | 80 | 60 | 40 | 20 |
|---|---|---|---|---|---|---|
| A: Texture | loam | Loam, silty clay loam, siltloam | Fine-loamy, fine, clayey, sandy clay loam | Fine loamy sand | Fine sand | Coarse sand |
| | 1 | 0.9 | 0.8 | 0.6 | 0.4 | 0.2 |
| B: % CaCO3 | <10 | 10-20 | 20-30 | 30-50 | >50 | - |
| C: % Gypsum | <0.5 | 0.5-2.0 | 2.0-5.0 | 5.0-10.0 | 10.0-25.0 | >25 |
| D: Salinity EC (dSm-1) | <4.0 | 4.0-8.0 | 8.0-16.0 | 16.0-30.0 | >30.0 | |
| E: Sodium saturationESP % | <5.0 | 5.0-15.0 | 15.0-25.0 | | 25.0-40.0 | >40.0 |
| F: Drainage | moderate | slow | Slow or rapid | | v.slow or v.rapid | - |
| Internal drainage (cm/hr) | 3.5-7.0 | 0.8- 3.5 | 0.2- 0.5; 7.0-10.0 | | 0.1-0.2; 10-15 | <0.1; >25 |
| Natural drainage | well | moderate | Imperfect | | poor | v.poor |
| G: Soil depth | Very deep | deep | moderate | shallow | veryshallow | |
| | 100+ | 50-100 | 25-50 | 10-25 | <10 | - |
| H: % slope | 0-2 | 2-5 | 5-8 | 8-15 | 15-25 | >25 |
| I: Profile development | Mature profile | Moderate maturity | poor | Very poor | immeture | |

For example a soil has loam texture, has 23 percent calcium carbonate, 4.0 percent of gypsum, 4.8(dS/m) of salinity content, low in sodium saturation, is well drained, very deep. has well defined epipedon and matured soil horizons with gentle slope. The Capability index(Ci) of the soil as per the scheme will work out to be

**Ci** =100 x 0.8 x 0.8 x 0.9 x 1.0 x 1.0 x 1.0 x 1.0 x 1.0 = 57.6

The Capability index of the above soil is 57.6. It comes under the Ci range of 45 to 60; so the soil has moderate limitation for economic production of crops.

**Table 14:** Scheme for evaluating the degree of limitation (Sys,1976)

| Extent of limitation | Ci range | Remarks |
|---|---|---|
| No limitation (0) | Ci 80 or more | The characteristics(quality) are optimal for plant growth |
| Slight Limitation(1) | Ci 60 to 80 | The characteristics are nearly optimal for the land utilization type and Limitation affect productivity for not more than 20 per cent with regard to optimal Yield |
| Moderate (3) | Ci 45 to 60 | The characteristics have moderate influence on crop yield decline; Limitation however, benefits can still be made and the yield remain economical. |
| Severe (4) | Ci less than 30 | Such limitations will not only decrease the yields below the Limitation profitable level, but may inhibit the use of the soil for the considered land utilization |

The limitation approach has been successfully used to provide a qualitative land evaluation based on general characteristics which are made available after a quality soil survey and general study of other soil resources in the area.

## Review questions

1. Differentiate between Rocks and minerals.
2. Differentiate between volcanic rocks and plutonic rocks
3. Give detailed classification of Rocks
4. Define Mineral and classify the minerals.
5. Write about characterization of minerals
6. Write important characters of Quartz, pyrophyllite, ampbibolite, serpentine, Muscovite, Gypsum, Gibbsite, Calcite, Dolomite, Augite and Apatite.
7. Explain various types of chemical weathering.
8. Define soil genesis and explain soil forming factors.
9. Differentiate between podzolisation and laterisation.
10. Discuss fundamental pedogenic processes
11. Explain soil profile wit help of neat sketch.
12. Differentiate between surface soil and subsoil
13. What are the limitations of genetic system of soil classification?
14. Discuss the evolution of USDA Soil Taxanomy.

15. Differentiate between soil moisture regimes and soil temperature reimes.
16. Define Epipedon and differentiate between Mollic and Ochric epipedon.
17. What is endopedon? Describe any 6 endopedons.
18. Explain the nomenclature of Soil Taxonomy.
19. Describe the salient features of Vertisols.
20. Write simplified Key to soil orders.
21. Give brief account of Land capability classification.

## Objective question

1. The smallest volume that can be called as soil is
   a) Ped b) Clod
   c) Pedon d) Fragment
2. A branch of soil science which deals with the soil from the crop production point of view
   a) Pedology b) Soil geology
   c) Ecology d) Edaphology
3. The unconsolidated material overlying the bed rock is called as:
   a) Solum b) Regolith
   c) Soil Profile d) All of the above
4. The density of earth as a whole in g $cm^{-3}$ is:
   a) 2.6 b) 4
   c) 5.5 d) 9
5. The process of desilication is known as:
   a) Laterization b) Podzolization
   c) Salinization d) Humification
6. The metamorphic rock formed due to action of high pressure is known as
   a) Dynamo metamorphic b) Thermo metamorphic
   c) Hydro metamorphic d) Viscous metamorphism
7. The soil having more than 30% organic matter is placed in
   a) Oxisols b) Arid sols
   c) Histosols d) Alfisols
8. The main center of NBSS & LUP is located at
   a) Palampur b) Nagpur
   c) Kanpur d) Jodhpur
9. The progressive accumulation of $Na^+$ ion on exchange complex is called as
   a) Salinization b) Calcification
   c) Alkalization d) Gleization

10. Hematite is converted to limonite by the process of

a) Hydration b) Hydrolysis

c) Oxidation d) Carbonation

11. The horizon A + B are collectively called as

a) Solum b) Regolith

c) Organic d) Illuvial

12. The element N is not found in

a) Atmosphere b) Hydrosphere

c) Lithosphere d) Planets

13. A surface horizon that is cemented mostly by silica is known as

a) Duripan b) Fragipan

c) Placic d) Glossic

14. Ultisols is comparable with Alfisols except for having

a) High base saturation b) Low base saturation

c) Uniform soil profile d) High water table

15. Who is the founder of modern Pedology?

a) Dokuchiev b) C.F. Marbut

c) Baldwin d) Coffey

16. Assertion a): The Thermal metamorphism is also known as additive metamorphism Reason (R): There is an addition of magmatic material to the metamorphosed rock.

a) Both a) and (R) are correct and (R) explains a)

b) Both a) and (R) are correct but (R) doesn't explains a)

c) a) is correct and (R) is incorrect

d) a) is incorrect and (R) is correct

17. Advanced stage of weathering represents in

a) Oxisols b) Ultisols

c) Alfisols d) Podzols

18. Which is NOT a Fundamental soil forming process?

a) Humification b) Salinization

c) Elluviation d) Illuviation

19. What are the common favorable conditions for Podzolization?

a) Cold & Humid climate b) Siliceous (sandy) material

c) Acid producing vegetation d) All the three

20. Lowest category of soil taxonomy classification is
    a) Series b) Family
    c) Suborder d) Sub group
21. Division of the factors of soil formation into active and passive groups was proposed by
    a) H. Jenny b) C.F. Marbut
    c) J.S. Joffe d) V.V. Dokuchaiev
22. A thick (>50 cm), manmade surface horizon produced by long and continued manuring
    a) Anthropic b) Agric
    c) Plaggen d) Ochric
23. Silica: sesquioxide ration for black cotton soil is:
    a) 2.0 to 2.5 b) 1.0 to 1.5
    c) 3.0 to 3.5 d) 4.0 to 4.5
24. Which of the following Weathering processes justifies the statement "clay begets clay"
    a) Hydration b) Carbonation
    c) Hydrolysis d) Oxidation
25. The most abundant metal found on Earth Crust is:
    a) Oxygen b) Silicon
    c) Aluminum d) Iron
26. Banded or laminated Character is the peculiar feature of:
    a) Igneous rocks b) Sedimentary Rocks
    c) Metamorphic Rocks d) None
27. Exfoliation or Onion type weathering is due to the changes in:
    a) Temperature b) Pressure
    c) Wind d) water
28. Study of Origin of rocks is known as:
    1) Pedology b) Petrography
    3) Petrology d) Petrogenesis
29. Rocks having glassy structure are called as
    a) Intrusive b) Extrusive
    c) Volcanic d) Both 2 & 3
30. An example for basic rock is
    a) Granite b) Pumice
    3) Basalt d) Gneiss

31. The fine grained rock with similar composition of Granite is
    a) Basalt b) Gneiss
    c) Rhyolite d) Gabbro
32. The rocks in which stratification is clearly seen are:
    a) Igneous rocks b) Sedimentary Rocks
    c) Aqueous rocks d) Both B & C
33. An example of in-situ sedimentary rock is:
    a) Bauxite b) Sandstone
    c) Gypsum d) Shale
34. Rocks formed under anaerobic conditions are:
    a) Peat b) Lignite
    c) Anthracite d) All three
35. Serpentine is formed from:
    a) Sandstone b) Marble
    c) Dolomite d) Shale
36. Very fine grained metamorphic rock is
    a) Marble b) Quartzite
    c) Gneiss d) Slate
37. Magnesium containing Olivine is
    a) Forsterite b) Fayalite
    c) Dolomite d) Haematite
38. Which of the following is specific Pedogenic process?
    a) Humifiation b) Alkalization
    c) Elluviation d) Illuviation
39. Zone of Illuviation is:
    a) O- horizon b) A- horizon
    c) B- horizon d) E- horizon
40. List (I) List (II)

| List (I) (weathering stage) | | List (II) (soil material) |
|---|---|---|
| i. Initial | ( ) | 1. Unweathered parent material |
| ii. Juvinile | ( ) | 2. Weathering has just started |
| iii. Virile | ( ) | 3. Easily weatherable minerals are not seen |
| i. Senile | ( ) | 4. Only most resistant minerals are seen |

Codes

| | i | ii | iii | iv |
|---|---|---|---|---|
| A | 1 | 3 | 4 | 2 |
| B | 4 | 3 | 1 | 2 |
| C | 3 | 4 | 1 | 2 |
| D | 1 | 2 | 3 | 4 |

**Answer key**

| | | | | | | | | | | | | | | | |
|---|---|---|---|---|---|---|---|---|---|---|---|---|---|---|---|
| **1** | c | **6** | a | **11** | b | **16** | a | **21** | a | **26** | c | **31** | c | **36** | c |
| **2** | d | **7** | c | **12** | d | **17** | a | **22** | c | **27** | c | **32** | c | **37** | d |
| **3** | b | **8** | b | **13** | a | **18** | a | **23** | c | **28** | a | **33** | d | **38** | a |
| **4** | c | **9** | c | **14** | c | **19** | b | **24** | c | **29** | d | **34** | a | **39** | b |
| **5** | a | **10** | a | **15** | b | **20** | d | **25** | b | **30** | d | **35** | d | **40** | d |

# 3

# Soil Physics

Soil physics is the study of soil's physical properties and processes. Soil physics deals with the dynamics of physical soil components and their phases as solid, liquids, and gases. Physical properties of soil include texture, structure, density, porosity, consistency, temperature, colour and water content. The physical properties depend on the amount, size, shape and arrangement and mineral composition of its particles , kind and amount of organic matter and the volume and form of its pores and the way they are occupied by water, air at a particular time. Various soil physical properties are soil texture, soil structure, soil consistence, soil colour, soil water, soil air and soil temeperature.

## 3.1. Soil texture

The relative proportion of the various size groups or fractions of primary particles particles in the soil is called soil texture. Using the names of the predominant size fractions makes up texture designations and the word "loam" is used whenever all the three major size fractions occur in sizable proportions. Thus the term "silty clay" describes a soil in which the clay characteristics are outstanding and which also contains much silt. A "silty clay loam" is similar to the silty clay except that it contains relatively more sand and therefore is somewhat mellower soft.

The U.S. Department of Agriculture has developed a system of texture designation that is based on that department's soil particle classificatiun. This system is generally accepted in the United States and several other countries. Conventionally the particles smaller than 2.0 mm diameter are considered as soil and above 2.0 mm size are non-soil.

**Tabel 1:** Classification of soil particles based on their size.

| Soil seperates | USDA system (mm) | ISSS system (mm) |
|---|---|---|
| Very coarse sand | 2.00 – 1.00 | - |
| Coarse sand | 1.00 – 0.50 | 2.00 – 0.20 |
| Medium sand | 0.50 – 0.25 | - |
| Fine sand | 0.25 – 0.10 | 0.20 – 0.02 |
| Very fine sand | 0.10 – 0.05 | - |
| Silt | 0.05 – 0.002 | 0.02 – 0.002 |
| Clay | < 0.002 | < 0.002 |
| Gravel / Pebbles | 2 – 75 | |
| Cobbles(Round) / Flags (Flat) | 75 – 250 | |
| Stones | 250 - 600 | |
| Boulders | more than 600 | |

USDA : United States Department of Agriculture
ISSS: International Society of Soil Science

The twelve main textural classes and their compositions are shown on the U.S. Department of Agriculture soil – texture triangle (Fig.). The various textural class names are (1) sand (2) loamy sand (3) sandy loam (4) loam (5) silt loam (6) sandy clay loam (7) silty clay loam (8) clay loam (9) sandy clay (10) silty clay (11) silt and (12) clay.

If a soil contains more than 85% of sand, then it is called sandy soil, 80% silt it is called soil silt and only 40% of clay is required to calls soil clay. Generally the best agricultural soils are those that contain 10 to 20 percent clay, 5 to 10 percent organic matter and the rest divided about equally between sand and silt. It has been proposed to group the various soil textures in the manner shown in the following table.

**Table 2:** Soil Texture classification:

| | |
|---|---|
| Coarse-Textured soils (Light soils/Sandy soils) | Sand, Loamy sand, Sandy loam |
| Medium-Textured soils | Loam, Silt loam, Silt |
| Fine-Textured Soils (Heavy soils/ clay soils) | Sandy clay loam, Clay loam, silty clay loam, Sandy clay, Silty clay, Clay |

Formerly coarse-textured soils have been called "light" soils and fine textured soils as "heavy" soils. These terms have been abandoned because they tend to be confusing.

**Table 3:** Range of soil separates in different textural classes:

| Textural class | Ranges of soil separates (%) | | |
|---|---|---|---|
| | Sand | Silt | Clay |
| Sand | 85 - 100 | 0 - 15 | 0 - 10 |
| Loamy sand | 70 - 90 | 0 - 30 | 0 - 15 |
| Sandy loam | 40 - 80 | 0 - 50 | 0 - 20 |
| Loam | 23 - 52 | 28-50 | 7-27 |
| Silt loam | 0 -50 | 50-88 | 0-27 |
| Silt | 0-20 | 80-100 | 0-12 |
| Sandy clay loam | 45-80 | 0-28 | 20-35 |
| Clay loam | 20-45 | 15-53 | 27-40 |
| Silty clay loam | 0-20 | 40-73 | 27-40 |
| Sandy clay | 45-65 | 0-20 | 35-45 |
| Silty clay | 0-20 | 40-60 | 40-60 |
| Clay | 0-45 | 0-40 | 40-100 |

The following definitions of the basic soil-textured classes are (essentially those given by the USDA (1960).

**Sand:** Sand is loose and single-grained. The individual grains can readily be seen or felt squeezed in the hand when dry it will fall apart when the pressure is released squeezed when moist, it will form a cast, but will crumble when touched, the water holding capacity of sand is low and because of the large spaces between the separate particles, water and air passes through rapidly.

**Sandy loam:** A sandy loam is a soil containing much sand but which has enough silt and clay to make it somewhat coherent. The individual sand grains can readily be seen and felt squeezed when dry, it will form a cost which will readily fall apart, but if squeezed when moist a cast can be formed that will bear careful handling without breaking.

**Loam:** A loam is a soil having a relatively even mixture of different grades of sand, silt and clay. It is mellow with a somewhat girths feel, yet fairly smooth and slightly plastic. Squeezed when dry, it will form a cast that will bear careful handling, while the cast formed by squeezing the moist soil can be handled quite freely without breaking.

**Silt loam:** A silt loam is a soil having a moderate amount of the fine grades of sand and only a small amount of clay, over half of the particles being of the silt when dry it may appear cloddy but the lumps can be readily broken and when pulverized it feels soft and floury (when wet the soil readily runs together and puddles either dry.

**Clay loam:** Clay loam is a fine textured soil that usually breaks into clods or lumps that are hard when dry. When the moist soil is pinched between the thumb and finger it will form a thin "Ribbon" which will break readily, barely sustaining its own weight. The moist soil is plastic and will form a cast that will bear much handling.

**Clay:** A clay is a fine textured soil that usually forms very hard lumps or clods when dry and is quite plastic and usually sticky when wet when the moist soil is pinched out between the thumb and fingers it will form a long flexible "Ribbon" clay expands and contracts greatly on wetting and drying and water holding capacity as high.

### 3.1.1. Determinations of Soil Texture by Feel method

The three main soil components can be recognized in the field by these simple tests. Sand feels gritty; Silt, when pressed with a finger, shows finger prints; and Clay when streaked out moist between two thumbnails shows a shiny surface.

The common field method of determining the textural class of a soil is by its feel. This is ascertained by rubbing a sample of the soil, usually in a moist to wet condition, between the thumb and fingers. The way a wet soil "sticks out", that is, develops a continuous ribbon when pressed between the thumb and fingers – indicates the amount of clay present. The longer and smoother the ribbon formed, the higher the clay content. Sand particles are gritty; silt feels like flour or taken powder when dry and is only slightly plastic and sticky when wet. Persistent cloddiness of dry soils generally is characteristic of silt and clay.

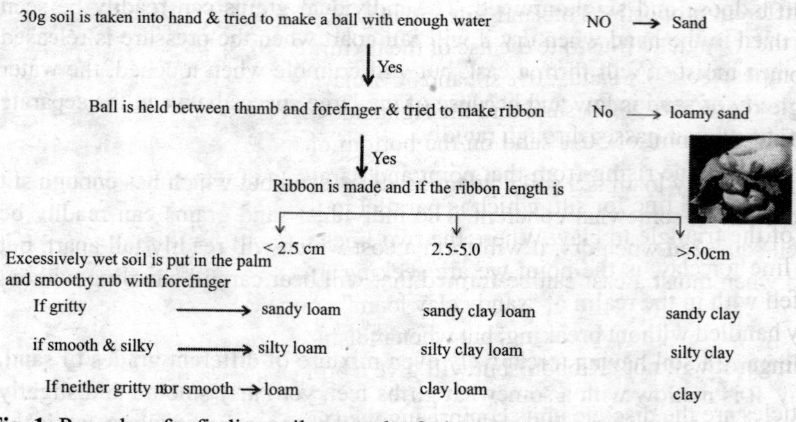


**Fig. 1:** Procedure for finding soil texture by feel method

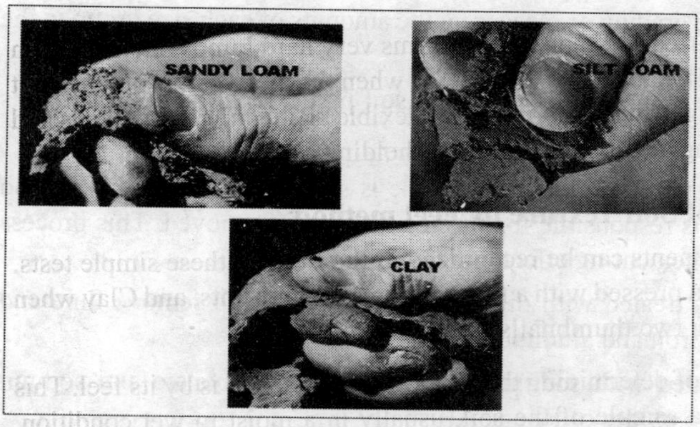


**Fig. 2:** Ribbon characters of different textural classes

### 3.1.2. Determinations of soil texture by laboratory method

The U.S. D.A. has developed a method for naming the soils based on laboratory-determined particle size analysis. The 12 main texturecharres and their compositions about in the soil texture triangle. The overall textural designation of a soil called the texture class, which is determined on the basis of the mass ratio of three fractions). Soils with different proportions of sand, silt and clay are assigned to different classes as shown in the triangle to illustrate the use of the textural triangle, let us assume that the soil is composed of 50% sand, 20% silt and 30% clay. Note that the lower left apex of the triangle represents 100% sand and the right side of the triangle represents 0% sand. Now find the point of 50% sand on the bottom of the triangle and follow the diagonally leftward line rising from that point and parallel to the zero line for sand. Next identify the 20% line for silt, which is parallel to the zero line for silt, namely, the left eye of the triangle to clay. Where the two lines intersect each other, as well as the 30% line for clay, is the point we are seeking. In this particular example, it happens to fell with in the realm of "sandy clay loam" or moist it will form casts that can be freely handled without breaking, but when moisture and squeezed between the thumb and finger it will not "Ribbon" but will give a broken appearance.

Primary particles are the discrete units comprising the soil solids. These are generally clustered together as secondary particles or aggregates. For quantitative determination of particle sizes, they have to be separated from one another by physical and chemical processes. Once the particles are separated (or the soil sample is dispersed into its ultimate particles). Several methods can be used to measure the proportions of each particle size fraction generally the fractions are separated by serving method and the finer ones by settling in a visions medium by sedimentation.

The determination of the relative distribution the size groups of ultimate soil particle are called mechanical Analysis." In mechanical analysis (particle size analysis) two steps are necessary.

1. Separation of all particles from each other incomplete dispersion of soil into ultimate particles and stabilizing the suspension and

2. Fractionation or the grading or measuring the amounts of each size group in the soil.

**1. Dispersion:** In achieve complete dispersion of soil particles all binding materials are removed from the soil

i. To remove surface tension effect, the sample is stirred in water and boiled so that the air, which is responsible for surface tension, is removed. This process also eliminates the cementing effect of dehydration.
ii. Then the sample is treated with dilute hydrochloric acid to remove lime and colloidal oxides of iron and aluminum.
iii. The sample is treated with sodium citrate, sodium bicarbonate and sodium dithionite to remove oxides of Fe and Al.
iv. Then it is treated with hydrogen peroxide to remove organic matter. $H_2O_2$ oxides the organic matter present as cementing agent.
v. Then the soil solution is filtered under section to reduce the concentration of electrolytes or flocculating agents.
vi. Finally the soils treated with sodium hexa meta phosphate or sodium metaphosphate or sodium carbonate (to increase the zeta potential of the particles` to remove complex polyvalent cations and replace with monovalent cations especially Na.

**2. Fractionation:** After the complete dispersing of soil sample next step in the particle size distribution analysis the grading of the particles in to a number of size groups to know the percentage distribution of these groups of various sizes in the soil mass. Any process that is used to separate the primary particles in to various size groups is referred to as the fractionation. The coarser particles are fractionated into desired size groups through direct sieving and the finer particles are quantitatively determined by sedimentation methods including pipette, hydrometer and centrifugation approximate size range determined by different methods.

**Table 4:** Methods of separation of soil particles

| Methods of Analysis | Approximate size range in mm. |
|---|---|
| Direct Sieving | 100 – 0.05 |
| Electron microscope | 0.005 – 0.0000 |
| Sedimentation | 1.1– 0.0005 |
| Centrifugal seperation | 0.01 – 0.00005 |

a) **Direct Sieving:** In direct sieving the dispersed soil suspension is passed through a next of sieves of different sizes. The particles of different sizes are separated and retained on different sieves. Suitability of this method is restricted to coarse fractions only. The diameter of the sieve opening gives the diameter of a spherical particle, which would just pass through the sieve aperture.

Sieving is the oldest and most widely used method for particles larger than 0.05mm size (size of sand and it is the smallest size of the particle that can be separated with this method). (The standard test sieve consists of a woven wire screen with square apertures mounted in a shallow frame. Sieve tests can be carried out by hand or in a sieve-shaking machine designed to impart necessary shaking, vibrating or jolting motion to the sample in the screen. The nest of the sieve is assembled with the coarsest mesh at the top and the finest at the bottom, with the bottom sieve mounted on a receiver pan. A weighed sample is transferred on the top sieve and the sieve nest is shaken for a prescribed time. Each sieve is then subjected to an end point test, which involves sieving for 2-minute period until the weight of material passing through sieve in 2 minutes is less than 0.2% of the original sample weight.

b) **Sedimentation**: For fractionating the finical particles, a well-dispersed soil suspension forming a liquid is sampled at different time intervals and the concentration of soil particles at a given depth is known as a function time. It is known that the rate of fall of particles through a liquid medium is a function of (1) size, density and shape of the particles. The larger the particle, the faster is the rate of settling. The forces that act on a particle as it settles through a liquid medium are the gravity or centrifugal forces, the drag or fractional and/or the buoyant forces.

**Stoke's Law**

In 1851, G.G. Stokes formulated the law governing the rate of settling of spherical particles in a viscous medium. The resistance offered by a liquid to the fall of rigid spherical particles verils with the medium of the particle and not with its surface. When a particle is dropped undersigned, it begins to move down in the liquid. The rate of fall of a particle through a liquid medium is a function of size, density and shape of the particle and the viscosity and the density of the medium.

**Derivation of Stoke's Law**

For Laminar flow, there are three forces that can be calculated to act on a solid sphere moving vertically through a fluid .

Gravity force = $\frac{4}{3}\ \pi r^3\ (\Delta\rho)\ g$

where, 4/3 $\pi r^3$ is volume of the particle in $cm^3$

$\Delta\rho$ = is density difference of particle and fluid = $\rho p - \rho f$

g is acceleration due to the gravity

Falling sphere (spherical soil particle in liquid medium) with velocity $V_t$ was influenced by Buoyance force or drag force

Drag force = form drag + friction drag

$2\pi r \eta v_t + 4\pi r \eta v_t = 6\pi r \eta v_t$

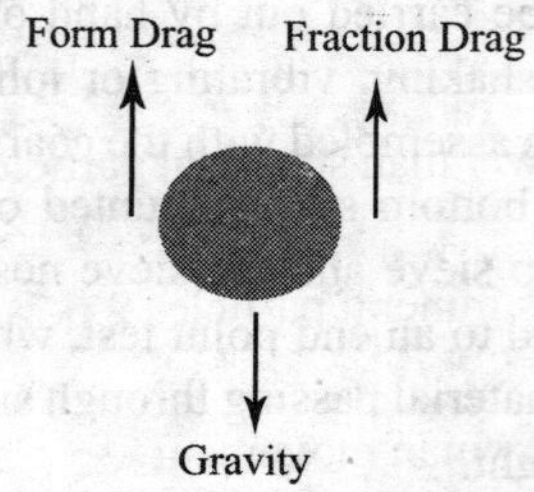


**Fig. 3:** Forces acting on a falling soil particle in water

The net resultant force on the falling spherical solid particle in liquid medium is obtained by equating the two forces

$6 \pi r \eta vt = 4/3 \pi r^3 (\rho p - \rho f) g$

$$Vt = \frac{2(\rho\rho - \rho f)gr^2}{9\eta}$$

For separation of remaining particle sedimentation technique is used Sedimentation: in the process of sedimentation the relation between particle density and rate of fall is expressed by Stokes' law.

According to Stokes' Law the terminal velocity of a spherical particle settling under the influence of gravity in a fluid of a given density and viscosity is proportional to the square of the particle's radius.

$$Vt = \frac{2(\rho s - \rho f)gr^2}{9\eta}$$

v = Settling velocity (cm/sec.) ;

R = Radius of the particle

$\rho s$ = Density of the solid particle (Mg/m3) ;

$\rho$**f** = Density of the fluid (Mg/m3)

$\eta$ = *Viscosity of the fluid*

## Assumptions of Stoke's Law

i. Particles must be spherical, smooth and rigid.

ii. Particles must be of uniform density.

iii. Particles must be sufficiently large (>0.001mm) as compared to molecules of fluid so as to avoid thermal (Brownian) motion of the fluid molecules.

iv. The particles should not interfere with one another and should settle independently (unhindered fall).

v. The suspension must be still with out any turbulence.

## Limitations of Stoke's Law

a) Some colloidal particles are plate shaped and fall slower than spherical particles of the same mass.

b) Soil particles are not all of the same density. silicates mineral particles 2.6- 2.7 $gm/cm^3$; iron oxides 5.0 $gm/cm^3$ or even more.

c) So soil particles smaller than 0.0002 mm exhibit Brownian movement

d) Many fast falling sand and silt particles may drag finer particles down along with them.

e) Particles >0.08 mm diameter cause turbulence in the settling fluid.

f) Stokes law does not account for differences in type of clay.

## Specific methods of mechanical analysis (by sedimentation)

The most commonly used methods based on Stoke's law are

1. Pipette method (Robinson, 1922) and
2. Hydrometer method (Bonyoucos 1927)

The pipette method is preferred for accurate work, while the hydrometer method is faster and still accurate enough for most purposes.

**1. Pipette method:** In the pipette method (fig. 4) a sample of the soil suspension is taken at a given depth below the surface at a predetermined time. In this way the sample contains all the fractions still in suspension at that depth. By sampling at different intervals after the stirred suspension has come to rest the mechanical composition of the soil can be calculated. For water at 25ºC and soil particles of 2.65g/cc density in the gravitational field of 980 $cm/Sec^2$. Stoke's law becomes $v=40285\ r^2$. This means that a particle of 0.05mm in diameter (sand) falls at a velocity of 0.25 cm/Sec. That is the same as 5cm in 20 sec. or 10cms in 40 sec. For particles of 0.01mm in diameter, the rate of sedimentation is 0.01 cm/sec or 5cms in 8 min and 20sec. This period makes it possible to sample fine silt plus clay in a satisfactory manner by pipette technique.

**Fig. 4:** Robinson (International) Pipette method

**2. Hydrometer method:** This method was introduced by G.Bonyoucos in 1927 and is applicable for determining the size distribution of particles having diameters less than 0.074mm. The coarser particles bigger than this size are determined by direct sieving. The method of analysis is based upon the extended principle of hydrometer (Figure 5) used in measuring the specific gravity of fluids. It is based on the principle of continuous reduction of the density of the soil suspension with time at the rate the particles drop below the level of the hydrometer at any moment the density of the suspension is lowest near the top and increases toward the bottom. By knowing the density of the suspension with a calibrated hydrometer for known times different fractions are calculated. This method is very popular for rapid determination of soil texture.

**Fig. 5:** Hydrometer method of soil texture determination

Bouyoucos hydrometer is made up of glass or length 28.5cm., having a stem length is 15 cms. A dead Mercury is provided the bottom of the bulb to make it direct in suspension. It is calibrated at 19.4°C. Temperature correction of 0.2 (0.3) to the observed reading is to be affected for ever 1°C rise or fall over this value.

**Calculating Settling velocities of soil particles**

Let us consider a uniform suspension of soil particles in water in a cylinder in which particles are settling under the action of gravity. Assuming that a single particle with radius, r, has settled down from the position at the surface to that B at a depth of 'h'cms. In to seconds, then $v = h/t = kr^2$. If this kept constant, as is usually done, at 10cm. then

$$t = 10/ kr^2$$

The time t, required by particles larger than a certain radius r, can be calculated. Thus for h = 10cm, particles larger than 20 microns in diameter (r = 10 microns) will settle at 20°C just below the plane B in 4 minutes and 48 seconds. The suspension above B will then comprise particles of size less than 20 microns, i.e. Clay plus silt. If the time allowed at 20°C is 8 hours, the soil suspension above plane B will consist of particles less than 2 microns, ie. Clay. At 25°C, the corresponding sedimentation time will be 4 min. 15 seconds and 7 hours 5 min. for silt plus clay and clay, respectively.

If the concentration or density of suspension in a thin section at plane B is determined, this provides a method for determining clay and silt. This can be done by either pipetting out a small aliquot from this plane or by determining the density of suspension with the help of a Bonyoucos hydrometer. In the pipette method, the percentage is calculated by oven-drying an aliquot and in the hydrometer method; this is read directly from the stem of the hydrometer, which is calibrated in concentration. Then, the entire clay and silt fractions are separated by repeated decantation, sedimentation and suspension. The residue which settles down consists of the fine sand and coarse sand which are separated by passing through a sieve.

**Soil textural diagram or Textural triangle (ISSS):** Considering the amounts of sand, silt and clay soil textural class can be determined with the help of **textural triangle** (figure 6). There are 12 textural classes in the textural triangle.

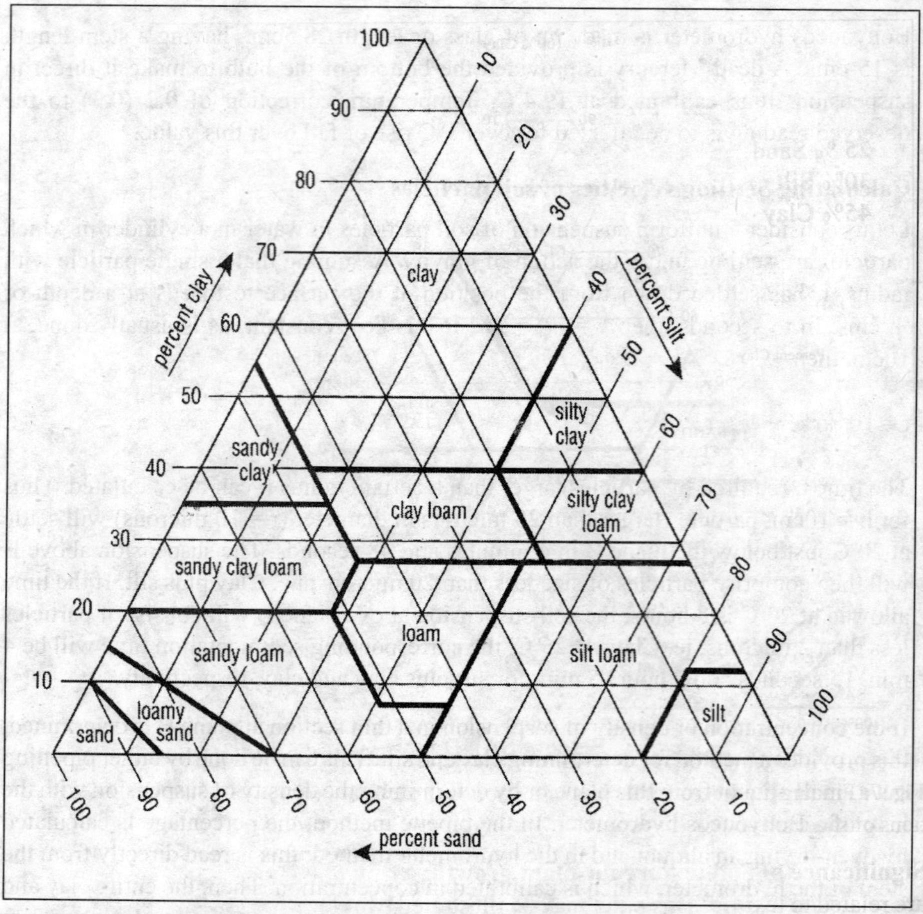


**Fig. 6:** Textural diagram or Textural triangle

Finding soil texture using textural triangle: After determining the percent sand, silt and clay in soil, the proportions are interpolated in the textural triangle by following guidelines shown in figure7. The base of the triangle represents percent sand, left side of triangle represents percent clay and right side of triangle represent percent silt.

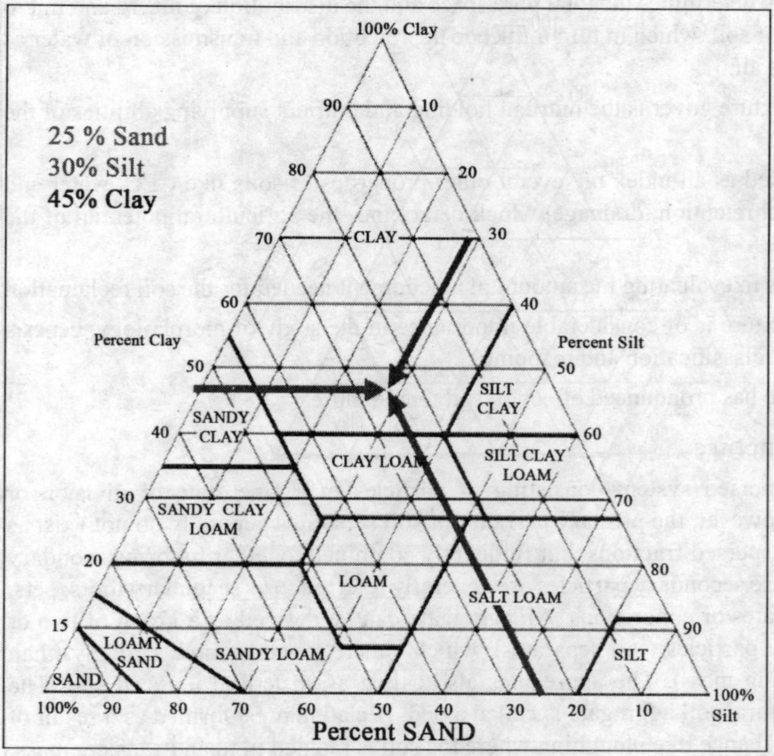


**Fig. 7:** Finding the soil textural class of the soil with the help of textural triangle using proportions of sand, silt and clay

**Significance of Soil texture for plant growth:** Many of the important soil properties are related to texture. This point may be illustrated by referring to the clayey and sandy soils. Clayey soils show high water holding capacity, high plasticity, stickiness and swelling where as sandy soils are conspicuous by the absence of these properties. The most important was in which soil texture affects plant growth is water and with it the nutrient supply. The available water capacity of soils is related to soil texture, water infiltration is more rapid in coarse textured soils as compared with fine textured soil, hence plants which require good drainage will thrive well on a coarse rather than on fine textured soil water is more available to plants in coarse textured soils than in fine textured soils. The importance of soil texture is emphasized in the following points.

- Knowledge of texture of the soil is important to assess the value of the land
- Land capability and methods of soil management are determined by soil texture.
- Texture determines the suitability of land to raise different crops.
- Texture determines its suitability for foundation of buildings and construction of roads.

- Texture determines the total pore space and the distribution of macro and micro pores of soil, which in turn influence the retention and transmission of water as well as air.
- Soil texture governs the nutrient holding and nutrient supplying abilities of the soils.
- It is used as an index of several other properties of soils like CEC, water and nutrient retention, drainage, which determines the agricultural potential of the soils.
- It helps in evaluating the amount of amendment needed for the soil reclamation
- Soil texture is of considerable importance in the study of morphology, genesis of soil, classification and mapping.
- Texture has pronounced effect on soil temperature.

## 3.2. Soil structure

Soil is a dispersed system consisting of particles in a fine state of division or dispersion. However, the primary particles or soil separates generally do not exist as completely dispersed fractions, but frequently combined together to form secondary particles. These secondary particles are generally referred to as secondary units, pets, clods, aggregates or concretions. An aggregate can be defined as a group of two or more primary particles (soil separates), which cohere together more strongly (than the surrounding mass). The aggregates also called as structural units of soil. The individual natural soil aggregate is called a ped. A clad may be formed as a result of physical disturbance like ploughing where the soil is molded in to an (transient mass. (A concretion is the result of local concentration of compounds, which irreversibly cement the soil particles together).

The arrangement of the individual soil particles in to a pattern is called soil structure. Soil structure as such is not a plant-growth factor, but it influences practically all plant-growth factors viz., water supply, aeration, availability of plant nutrients, microbial activity.

### 3.2.1. Genesis of soil structure

Soil structure formation is two step process viz., flocculation and cementation. The bonding of soil particles in to structural units is the genesis of soil structure.

**Floculation:** Flocculation or coagulation of called is an irreversible clustering together of colloidal particles in the form of loose and irregular shapes or floccules. This behaviour is due to the presence of electrical double layer when the electro-kinetic potential is lowered sufficiently, a collision between particles results in a mutual attraction and the formation of floccules.

**Cementation**:- Flocculation does not guarantee stable aggregate formation. Flocculation is an electro-kinetic phenomenon. While stable aggregate formation requires a cementation or binding together have flocculated particles. Organic matter, inorganic colloids including silicate clays, hydrous oxide clays and sesquoxides, cations and calcium carbonate are generally the binding or cementing agents for the formation of aggregates. These can be broadly grouped together as biological, chemical and physical in nature. Non-single mechanism, however, is responsible for formation of aggregates but several of these mechanisms may work together.

There is a vast difference between flocculation and aggregation. Though both lead to the formation of compound particles, they are two separate processes. While flocculation is brought about by the coalescence of colloidal particles, aggregations require the cementation of flocculated particles. Flocculation is only the first step in the process of aggregate formation. Aggregation is, therefore, something more than flocculation involving a combination of different factors such as hydration, pressure, dehydration, etc., Flocculation not only precedes aggregation but also helps the formation of stable aggregates.

The mechanisms of soil structure formation are broadly grouped to three 1) Biological mechanisms, 2)Chemical mechanisms and 3. Physical mechanisms

**1) Biological mechanisms of structure formation**: When the organic matter of plant or animals origin is incorporated in soil and the conditions of temperature and water are favourable there is a rapid rise in the population of bacteria, fungi and actinomycites. The microorganisms use plant and animal residues as source of energy. A large increase in the population of organisms produces an intense microbial activity. The degradation products accumulate in the soil and large amount of organic materials of microbe organic origins are synthesized Fungi and actinomycetes (ramify and) produce considerable amount of mycelia soils rich in organic matter are recognized to be well granulated and aggregated. Fungi and actinomycites mechanically bind the aggregates with the help of their mycelia, while bacteria may cement the particles with gum they produce Algae, fungi, actinomyces and bacteria form living matter that will keep the soil particles together more effectively than any exchangeable ion could.

Earth worm casts are well known for their stability. It also prepares channels that are useful for infiltration, drainage and aeration. The casts excreted by earthworms consist of practically decomposed organic matter intermixed with soil. The annual earthworm cast production may range from 7.5 – 18.12 tons per acre. The presence of calcium hamates and gum producing bacteria in the earthworm's intestine also helps in the cementation of soil particles.

Vegetation affects aggregate formation and stability in three ways. First the (1) plant residues incorporated in the soil are the sources of energy for micro-organic activity and formation of organic complexes, which promote soil aggregation (2) Second, the root systems of the vegetation influences soil aggregation through mechanical and biological effects and (3) the third the vegetation provides a protective canopy against the beating action of raindrops and dispersion of aggregates.

**2) Chemical mechanisms of structure formation:** The chemical mechanisms responsible for aggregation including a)clay and exchangeable cation interactions and b)cementing action of inorganic colloidal substances.

**a) Clay and exchangeable ions:** Generally clay assists in aggregating soil by acting as a cement and also by its ability of swelling and shrinking with moisture changes. Thin films of clay cover small aggregates of particles of certain soils and serve to hold these aggregates together. Such films are called clay skins.

Exchangeable ions have also a pronounced effect on the flocculation of clays. Generally calcium, magnesium and Potassium have a flocculating effect on clay, while the hydrogen and especially sodium have a peptizing effect.

$$\underset{\text{Ca, Mg, K}}{\xleftarrow{\text{Flocculating}}} \qquad \underset{\text{H, Na}}{\xrightarrow{\text{peptizing}}}$$

Sodium tends to peptize (deflocculates) clay and to create a reaction in which several of the plant nutrients become unavailable, consequently depressing the growth of roots and b)microbes. Thefore, a poor structure results.

**Inorganic cementing substancess**: Sesquioxies (Fe and Al) form irreversible and slowly reversible colloids. These help to form water-stable aggregates. The trivalent iron serves as flocculating agent while gelatinous iron hydroxide acts as a cementing agent. The iron and aluminum hydrous oxide coatings on quartz particles are the cause of high strength of bonding between clay and quartz calcium carbonate precipitating out around soil particles also acts as a cement.

**3) Physical mechanisms of structure formation**:- The major physical mechanisms responsible for aggregate formation are shrinkage and swelling, freezing and thawing and colloidal clay effects.

i) **Shrinking and swelling**: Shrinkage and swelling of certain soils caused by alternate welting and drying. Therefore, variation in soil water content significantly affects the changes in aggregation wetting and drying of soil profile is also responsible for different types of structural pads. When a soil is dried, initially the decreasing in volume equals the volume of water lost till air enters the soil after which the shrinkage is less than the volume of water withdrawn. However, the volume of rewetted soil is greater than the original volume due to swelling and occlusion of air in the pore space. The decrease in soil volume on drying results in to the shrinking of soil mass and cementation of soil particles by surface tension effect. The cleavage planes due to drying and predominantly in vertical direction, columnar structure is the result of drying the soil profile.

ii) **Surface tension effect**: Surface tension resulting from the presence of water and air surfaces keeps the soil particles together and orients them. On the other hand, surface tension forces water in to spaces previously filled by air and thus causes minor explosions when the air escapes. Thus the break down of soil colloids by wetting and drying is due to dispersion of the cement substances, reduction in cohesion between the particles, explosions due to entrapped air and stress and strains set up due to differential swelling.

iii) **Freezing and thawing**: Freezing and thawing have a similar effect on aggregate formation and disruption as the alternate wetting and drying. (Freezing may cause aggregation or disruption depending upon the nature of crystallization of ice, which depends on the time rate of cooling. Slow cooling and freezing of moist soil is beneficial for soil structure; water begins to freeze in the larger cavities. On slow cooling large size crystals are formed which partially melt during thawing and serve nuclei for further freezing. Due to expansion of water during freezing the soil mass around the ice crystal is compressed. As in this process, water is drawn from the surrounding soil mass to the ice crystal it also causes dehydration. These combined effects produce aggregation. Slow freezing holds the small aggregates together, but breaks up the bigger soil masses by cracking them at the natural "surfaces of weakness". If, however, cooling is rapid resulting into the formation of larger number of crystals, soil aggregates break due to integrated effect of expansion. Fast cooling and freezing of a wet soil causes ice-crystal formation through out the soil mass. Frequent freezing and thawing of the soil surface result in an accumulation of water in the surface to such an extent that the soil may become thoroughly dispersed.

iv) **Action of clay:** By virtue of high surface charge and high specific surface area, clay particles play a key role in the formation of soil aggregates in which the constituents aggregated are clay, silt and sand. Obviously, the degree of aggregation depends on the clay content. The bonding may be clay – clay particles or clay- silt plus sand through some forces or agents called cementing agents.

v) **Action of roots**: The action of roots is twofold. When the rootless penetrate in to soil, tremendous pressure as exerted on the soil particles to make run for them and thus causes soil particles to become compact and aggregated mechanically., root penetration, all these factors are affected by soil structure. Consequently, poor soil structure may be an indirect factor limiting plant growth.

### 3.2.2. Classification of soil structure

Many soil scientists have attempted to classify soil structure, but to date no universally accepted classification exists, probably because no satisfactory method of measuring soil structure has been devised. The following classification is adopted from various sources. It is greatly simplified and omits many sub-types.

1. **Simple structure**: Natural cleavage planes absent or indistinct.

a) **Single grain structure:** This occurs normally in sands and silts of low organic matter content. In sands single-grain structure allows for aeration and maximum of capillary water movement.

b) **Massive Structure:** Massive structure is similar to single-grain structure, except that it is coherent. Examples of massive structure are (dense soil crusts), plow pans and fragipans.

2. **Compound Structure:** Natural cleavage planes are distinct. The shape of the individual peds of soil of compound structure can be described according to the

relative lengths of vertical and horizontal axes and by the contours of their edges. The forms of peds are:

a) Cube like structure: Vertical and horizontal axes of similar length.

b) Prism like structure: Columnar structure, prismatic- vertical axis longer than horizontal axes. Vertical cleavage planes predominant.

c) Platy structure: Horizontal axes longer than vertical axes. Horizontal cleavage planes predominant.

The contours of the edges of the peds are: a)Angular, corners and edges are sharp and distinc; b)Sub-angular, corners are rounded, edges are sharp; c)Granular, both corners and edges are rounded.

**Compound Structure**: in compound structure i) The natural cleavage plains are distinct and ii) Described with relative length of horizontal and vertical axes and shape of peds.

**Various structural types based on Shapes of Aggregates:** There are four important shapes of structural aggregates (peds) associated with compound structure viz., spheroidal, plate like, prism like and block like (Figure 8).

**Spheroidal structure:** Small rounded peds with irregular faces and are usually separated from each other in a loosely packed arrangement. When spherical peds are more porous, the structure is called as Crumb, while less porous peds are called granular. Usually granular peds are <1cm, while crumb peds are<0.5 cm in diameter. The diameter typically ranged from 1.0 to 10.0 mm. Granular and crumb structures are characteristic of many surface soils (A horizon) particularly those high in organic matter. They are prominent in grassland soils and soils that have been worked by earth worms. ***Crumby structure***: If the spheroidal peds are more porous, the structure is called as Crumb, ***Granular structure***: If the steroidal peds are less porous, the peds are called granular. Usually granular peds are <1cm, while crumb peds are<0.5 cm in diameter. The diameter typically ranged from 1.0 to 10.0 mm. Granular and crumb structures are characteristic of many surface soils (A horizon) particularly those high in organic matter. They are prominent in grassland soils and soils that have been worked by earth worms.

**Platy like structure:** Relatively thin horizontal peds or plate line units. Horizontal dimension is more than vertical dimension. There are two types in this group, i)Platy: The thicker units are called Platy ; ii)Laminar: The thinner ones are called Laminar. The platy types are often inherited from parent materials, especially those laid down by water or ice. Sometimes, compaction of clayey soils with heavy machinery can create platy structure. It is found in surface layers of some virgin soils.

**Prism like structure:** The vertically oriented pillar like peds are with varying heights. Vertical dimension is more than horizontal dimension. Columnar structure: pillars have distinct rounded tops, and in invariably found in subsoils high in sodium (nitric horizon). Prismatic structure: If the tops of the pillars are relatively angular and flat horizontally, the structure is designated as Prismatic. These structures are associated with swelling types of clay and commonly occur in sub_surface horizons in arid and semi-arid regions

**Block Like:** Blocky peds are irregular, roughly cube -like structures. Both vertical and horizontal dimensions are equal. *Angular Blocky:* When the edges of the blocks are sharp and rectangular faces distinct, they are called angular blocky, and *Subangular blocky:* when faces and edges are some what rounded they are referred as subangular blocky. These types are usually found in B horizons, where they promote good drainage, aeration and root penetration.

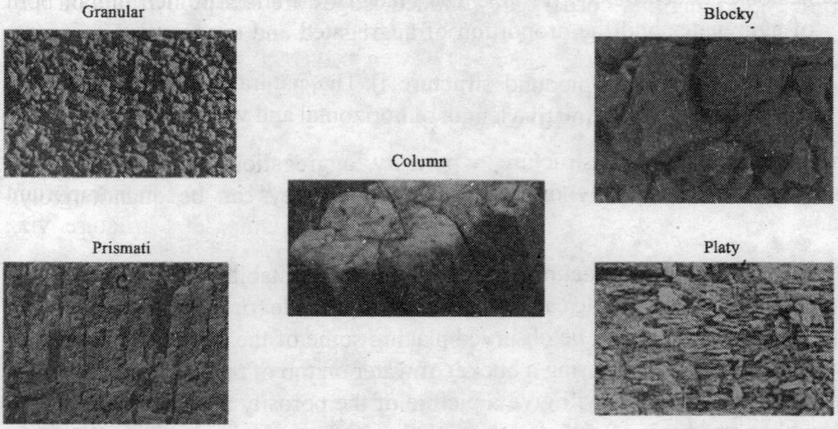


**Fig. 8:** Different shapes of Soil structure

Structural CLASS: Based on the size of the individual peds, the primary structural types are differentiated in to five sub-classes.

**Table 5:** Size ranges of peds according to shapes in soil structural classes

| Size or class | Granular (mm) | Platy (mm) | Blocky (mm) | 0Prismatic (mm) |
|---|---|---|---|---|
| Very fine /very thin | <1 | <1 | <5 | <10 |
| Fine or thin | 1-2 | 1-2 | 5-10 | 10-20 |
| Medium | 2-5 | 2-5 | 1-20 | 20-50 |
| Coarse or thick | 5-10 | 5-10 | 20-50 | 50-100 |
| Very coarse/very thick. | >10 | >10 | >50 | >100 |

**Structual Grades:** Grade determines the degree of distinctness and durability of individual peds. **Structure less or simple structure (0):** When there is no observable aggregation Eg. Single grain, soil particles in sand dune areas and massive structure (no cleavage lines) in puddle soil from a rice field. **Weak strecture (1):** Poorly formed, weak, non-durable, indistinct peds and break in to a mixture and many broken peds and much unaggregated material. **Moderate (2)***:* Moderately well developed peds which are fairly durable and distinct. **Strong (3):** Very well formed peds, which are quite durable and distinct.

The overall structure of a soil is a resultant of arrangement and bonding of primary soil particles in structural units and subsequent arrangement and bonding of these units. The shape and arrangement, the size and distinctness and durability of these visible pads or aggregates are used for describing the field soil structure. A combination of these three qualities is used for naming the structure. In the field terminology for structure, shape and arrangement of pads is designated as type of structure, size of pads as class; and degree of distinctness as grade. The table shows the type, class and grades of structure encountered in the field practice. The grade of structure is the degree of aggregation expressing the difference between the cohesion within the aggregates and the adhesion between the aggregates. The grade of the structure depends on the durability of aggregates and the proportion of aggregated and exaggerated material types, classes and grades.

### 3.2.3. Evaluation of soil structure

The four phenomena of soil structure – porosity, aggregation, cohesiveness and permeability can be studies by direct observation or they can be quantitatively measured.

**Qualitative field tests:** Determination of soil structure and its stability can be performed with simple tools. The soil profile can be exposed with a spade (or pick) and aggregate size, shape and distribution can be observed placing some of the aggregates in water will show their stability and pouring a bucket of water on top of the ground (or on top of an exposed lower horizon) will give a picture of the porosity and permeability of the soil. Another rapid test of soil structure can be performed by use of a soil prove (Hoofer tube). After preparing a straight cut of the profile in the probe with a knife, the entrance of a chalk suspension in to the soil is used as an indication of the porosity. An eight to ten power-magnifying lens serves to detect the shapes of small aggregates and the presence of clay skins. A soil penetrometer registers the pressure required to push a metal rod or disk in to the ground and thus measures an aspect of cohesiveness.

**Quantitative structure analysis:** The pore space distribution of a soil can readily be determined by saturating a sample with water and removing it by use of successively increased tension. In this way pores from the largest sizes down to those of a diameter of 0.0002mm can be determined. The volume weight of a soil whose true density is known permits the calculating of the total pore space. Dry sieving; their stability and size distribution of stable aggregates can find the distribution of aggregates by wet sieving or by determining the settling rate of the aggregates in water. (Aggregate stability can be measured also by the effect of water impact on individual aggregates or on the entire soil surface). The shape of aggregates can be determined approximately by measurement of the vertical and horizontal axes.

To measure the various phases of permeability, a number of methods exist to determine the infiltration capacity, percolation capacity, the resistance to mechanical penetration and to airflow and the consistence of the soil. The degree of shrinking of a soil paste is determined by placing it in to a flat stainless steel cylinder and allowing it to dry out slowly. The decrease in diameter can readily be measured with a caliber.

A very valuable approach to soil – structure research is soil micro morphology, the study of soils with the aid of various types of microscopes. In this way it is possible to determine sizes, arrangement and percentage distribution of paras and of solids and to obtain an insight in to the types of cementing materials. In addition to the optical microscopes and the transmission electron microscopes, more recently the scanning electron microscope has been used for soil structure studies. It provides reflected micrographs of considerable clarity up to 50,000 magnifications

### 3.2.4. Management of soil structure

The aim of soil structure management is to get porosity, aggregation and permeability to air, water and roots into as nearly optimal condition as possible and to do this to a sufficient depth that the majority of the crop roots find such favourable conditions that the maximum yields possible under the given conditions of climate and plant nutrition result.

**Soil conservation:** The sand aim of soil-structure management is to decrease its detachability and transport ability by water or wind and to increase its infiltration and percolation capacities so that run off and erosion are kept at a minimum.

**Principal approaches**: In order to reach these aims several important approaches have to be followed Regulation of moisture of aeration and of consistence of soil, deepening of the productive soil; erosion control; and destruction or horizontal structure units.

Since most soils are flocculated except some of the sodium soils it is our task to separate masses of soil, not to aggregate them. Once aggregates of desirable size have been obtained they have to be protected from further breakdown, which would eventually result in a massive soil.

The actual methods that follow these avenues to bring about desirable soil structure are manifold and varied. Most of them serve several purposes at once. The methods will only be enumerated here. Proper land use, increasing plant growth addition of organic matter, fertilization tillage, (sub soil improvement, mulching, drainage, irrigation (specific soil conservation methods, protection from compaction when wet from rainfall impact and from rapid frost and possibly the use of soil conditioners.

### Structure management of different soils

**Coarse textured soils:** There is no practical way to greatly improve the structure of coarse-textured soils, although organic matter additions can be helpful. Organic matter will not only act as a binding agent for the particles, but also will increase the water holding capacity farm manures and crop residues are good sources for the organic matter.

**Fine textured Soils**: The structural management of a fine-textured soil such as clay or sandy clay is not as simple as that of a sandy soil. Because of the high plasticity and if a clay soil is tilled when wet, the aggregate are broken down, the individual particles tend to act independently the macro pores disappear, and the soil is said to be puddle. In this condition it is nearly impervious to air and water when the soil dries, it usually becomes hard and dense. On the other hand, if a clay or sandy clay soil is plowed too dry, large hard clods, which are difficult to work into a good seedbed, are turned up. Obviously, the tillage of clay soils must be timed carefully with respect to their moisture content.

Some clay soils of tropical regions are much more easily managed than these just described. The clay fraction of these soils is dominated by hydrons oxides of Fe and Al, which are not as sticky, plastic and difficult to work when properly fertilized and managed, these oxide type clays can be productive. For example they are among the worlds best souls for pine-apple production.

As was the case for sandy soils, organic matter maintenance plays a key role in the management not only of clay soils, but of much larger areas of medium-textured soils such as silt loams and loams. The return of crop residues and inclusion of sod crops helps to maintain good soil structure.

### 3.2.5. Significance of soil structure on plant growth

The soil as a medium for plant growth provides the environment for the germination and emergence of seedlings and for the development and functioning of the plants root system. Not only must the soil furnish the nutrients required fr the metabolic processes of plant growth but it must also provide favourable air, water for proper functioning of the plant. In too many instances, the nutrient uptake from the soil is limited by a lack or excess of water, a deficiency of oxygen, or inadequate root proliferation resulting from high bulk densities that impede the growth of roots. The combination of increased density and decreased aeration not only restricts the proliferation of roots and their normal absorption of nutrients and water but also impedes microbiological activity. Unfavourable soil structure therefore becomes the time being factor in crop production and should be considered a parameter of soil fertility.

## 3.3. Density of soil

soil density is defined as weight per unit volume of soil and expressed in units of g/c $m^3$ or Mg/ $m^3$. There are two types of soil densities viz., i)particle density and ii)bulk density.

### i) Particle density of soil (ρp)

One means of expressing soil mass (weight) is in terms of the density of the solid particles making up the soil. Density is usually defined as the mass of a unit volume of soil solids and is called particle density ($\rho_p$). In the metric system particle density can be expressed in terms of mega grams per cubic meter (Mg/$cm^3$). Thus 1 $m^3$ of soil solids weighs 2.6 Mg. the particle density is 2.6 Mg/ $m^3$. (Since 1 mg = 1 million grams and 1 $m^3$ =1 million cubic centimeters, the particle density can also be expressed as 2.6 g/c $m^3$).

**Table 6:** Particle densities of soil constituents (in g/cm$^3$)

| | | | |
|---|---|---|---|
| Humus | :1.3-1.5 | Clay : | 2.2-2.6 |
| Orthoclase : | 2.5-2.6 | Quartz: | 2.5-2.8 |
| Calcite : | 2.6-2.8 | Muscovite: | 2.7-3.0 |
| Biotite : | 2.8-3.1 | Apatite : | 3.2-3.3 |
| Pyrite : | 4.9-5.2 | Hematite: | 4.9-5.3 |

## Factors affecting Particle density

The sizes of the particles of a given mineral and the arrangement of the soil solids have nothing to do with the particle density. Particle density depends on the chemical composition and crystal structure of the mineral particles and is not affected by pore space.

Although there is considerable range in the density of the individual soil minerals for most mineral soils it usually vary between 2.60 and 2.75 Mg/ m$^3$. This narrow range occurs because quartz, feldspar, mica and the colloidal silicates with densities within this range usually make up the major portion of mineral soils. When unusual amounts of minerals with high particle density, such as magnetite, garnet, epidotic, zircon, tourmaline, or horn bunds are present, the particle density may exceed 2.75 Mg/ m$^3$.

Organic matter weighs much less than an equal volume of mineral solids, having a particle density of 1.1 to 1.4 Mg/ m$^3$. Consequently, mineral surface soils, which almost always have higher organic matter content than the sub soils. Usually have lower particle densities than do sub soils. Some mineral top soils high in organic matter (15-20%) may have particle densities as low as 2.4 Mg/ m$^3$ or even below clearly the amount of organic matter in a soil markedly affects the particle density (nevertheless) for general calculations, the average arable mineral surface soil (3.5% organic matter) may be considered to have a particle density of about 2.65 Mg/ m$^3$.

## ii) Bulk density (ρb)

It is the mass per unit volume of dry soil (volume of solid and pore spaces). The bulk density of a soil is always smaller than its particle density. Loose and porous soils have low bulk densities as compared to compacted soils. Bulk density is of importance than particle density in understanding physical behaviour of soils. Generally in normal soils bulk density ranges from 1.0 to 1.60 Mgm$^{-3}$; Finer the texture of the soil, lesser is the bulk density. Sand dominated soils records 1.7 Mgm$^{-3}$; Organic peat soils records 0.5 Mgm$^{-3}$; Compacted sub soils records 2.0 Mgm$^{-3}$

## Factors affecting bulk density of soil

1) More is the pore space, per unit volume of soil, less is the bulk density.
2) Higher is the compactness; more will be the bulk density.
3) Higher is the depth of soil, more will be the bulk density.
4) Finer is the texture of the soil, lesser is the bulk density.

5) High organic matter contents, lead to reduced bulk density.
6) Crumb soil structure shows low bulk density than that of platy structure.
7) Tillage temporarily reduces the bulk density.

## Importance of bulk density

1. Bulk density of the soil provide information not only on total pore space but the macro and micro pore space also, which in turn governs the soil – water – air relationship, thereby facilitate better crop growth.
2. Infiltration, permeability, percolation of water and water retention in soil system, have direct relation with bulk density of soil.

## Methods of determination of bulk density

1) Core sampler method.
2) Clod saturation method
3) Clod coating method
4) Weighing bottle method

Among the four methods, Core sampler method is widely used.

## Pore space of soil (f)

The pore space of a soil is that portion of the soil volume occupied by air and water. The amount of pore space is determined largely by the arrangement of the solid particles. If the particles lie close together, as in sands or compact sub soils, the total porosity is low. If they are arranged in porous aggregates, as is often the case in medium-textured soils high in organic matter, the pore space per unit volume will be high.

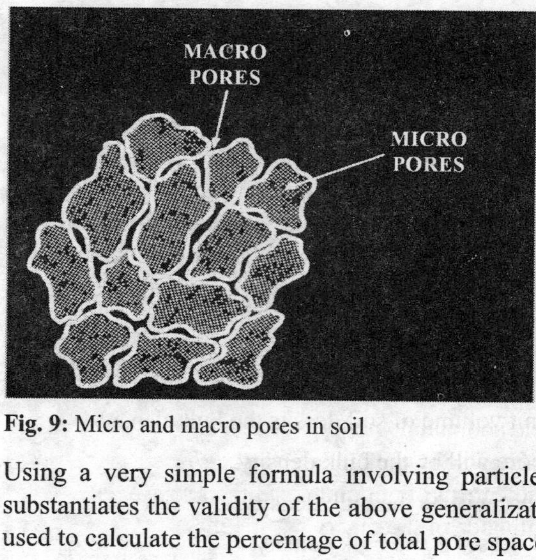


**Fig. 9:** Micro and macro pores in soil

Using a very simple formula involving particle density and bulk density figures substantiates the validity of the above generalizations. The derivation of the formula used to calculate the percentage of total pore space in soil follows:

Let ρb = bulk density  Vs = Volume of solids

ρp = particle density  Vp = Volume of pores

Ws = Weight of soil (solids)  Vs + Vp = total soil volume

By definition

$$\rho p = \frac{W_s}{V_S} \text{ and } \rho b = \frac{W_s}{V_s + V_p}$$

Solving for Ws gives = Ws = ρp x Vs and Ws = ρb (Vs + Vp)

Therefore ρp x Vs = ρb (Vs + Vp)

$$\text{And } \frac{\rho b}{\rho p} = \frac{Vs}{Vs+Vp}$$

$$\text{Since } \frac{Vs}{Vs+Vp} \text{ X } 100 = \%\text{ solid space}$$

$$\text{Then \% solid space} = \frac{\rho b}{\rho p} \text{ x } 100$$

Since % pore space + % solid space = 100,

And % pore space = 100-% solid space, then

$$\%\text{ pore space} = 100 - \left[\frac{\rho b}{\rho p}\times 100\right]$$

Using this formula, a sandy soil having a bulk density of 1.50 and a particle density of 2.65 will be found to have 43.4% pore space. A silt loam for which the corresponding volves are 1.30 and 2.65 possesses 50.9% air and water space.

## Factors influencing total pore space

The total pore space is quite variable among soils. Sandy surface soils show a range of 30 - 50% whereas medium to fine textured soils vary from 40 to 60% and even more in cases of high organic matter and marked granulation. Pore space also varies with depth; some compact sub soils have as little as 25 – 30%. This accounts in part for the inadequate aeration and resistance to root penetration of such horizons.

## Size of pores

Two types of individual pore space "macro" and "micro" – occur in soils although there is no clear cut demarcation, pores less than about 0.06mm in diameter are considered micro pores and those larger, macro pores. The macro pores characteristically allow the ready movement of air and percolating water. In contract, the micro pores are mostly filled with water in a moist field soil and do not permit much air movement in to or out of the soil. The water movement also is slow. Thus, even though a sandy soil has relatively low total porosity, the movement of our and water through such a soil is surprisingly rapid because of the dominance of the macro pores.

**Calculation of porosity**

% solid space = Bulk density / particle density x 100

% pore space + % solid space = 100

% pore space = 100 – % solid space

= 100 – (B.D / P.D) x 100

= 100 [1 - BD/PD]

***Model calculation-1***

A soil core was taken for the determination of porosity. The measurements were:

Cylinder volume 73.5 cc

Dry soil weight 87.8g;

Particle density 2.65 g $cc^{-1}$

Soil weight 87.8g

Bulk density = Weight of soil / Volume of soil

= 87.8/73.5 = 1.19 g/cc

% pore space = 100 [1- (1.19 / 2.65)] = 55.10

***Model calculation-2*** Calculate porosity of a soil having bulk density of 1.5 Mg $m^3$ and particle density of 2.65 Mg $m^3$.

$$\text{Percent porosity} = 1 - \frac{\text{bulk density}}{\text{Particle density}} \text{ X } 100$$

$$= 1 - \frac{1.5}{2.62} \times 100 = 43.4\%$$

***Model calculation-3*** The porosity of a soil sample is 40 percent and its bulk density is 1.5 Mg $m^3$. Calculate its particle density and void ratio.

$$\% \text{ Porosity} = 1 - \frac{\text{bulk density}}{\text{particle density}} \times 100 \text{ or}$$

$$40 = 1 - \frac{1.5}{x} \text{ x } 100$$

$$\text{or } 0.40 = 1 - \frac{1.5}{x}$$

$$\text{or } \frac{1.5}{x} = 1 - 0.40 = 0.60$$

$$\text{or } x = \frac{1.5}{0.6} = 2.5 \text{ Mg.m}^3$$

$$\text{The void ratio} = \frac{\text{Volume of soil pores}}{\text{Volume of soil solids}} = \frac{40}{100-40}$$

$$= \frac{40}{60} = 0.667$$

## 3.4. Soil crusting

The surface soil aggregates are most vulnerable to destruction, especially by heavy rains. The dispersed surface tends to seal over and prevent water infiltration. Then when the soil dries, a hard crust is formed that impedes the emergency of seedlings. Keeping some cover on the land to minimize the impact of raindrops can minimize crust formation. Also a light tillage while the soil is still moist will break up the crust before it hardens.

Soil crusting is phenomena associated with deterioration of soil structure, where the natural soil aggregates break and disperse. If rapid drying follows dispersion the soil solids rearrange in to crust. When the raindrops strike the exposed soil surface the kinetic energy thus dissipated disintegrates the aggregates and the depressive action of water leaves the surface few millimeters of the soil in to a mono grain state. The dispersed particles are carried in to the soil with the infiltrating water clogging the soil pores. Later when drying starts surface tension forces pull the soil particles together, tending to form a dense and strong soil layer known as soil crust. The thickness of the rain formed soil crust depends upon the size of the raindrops. Larger drops having more energy destroy the original structure to a great depth. These crusts are about 5mm thick.

The soil crust formed by raindrop impact consists of two distinct parts. First, there is a thin skin approximately 0.1mm thick formed on the surface due to the compaction that results from the impact. Second, the dispersed particles that arise from impact are washed in to the soil with the infiltrating water and clag the pores immediately beneath the surface to give a layer of decreased porosity. The water permeability of the "washed-in" zone is reduced about 200 times below that of the undisturbed soil beneath; that of the surface skin is reduced 2000 times (aggregates from virgin soils with rather stable structure slake but do not disperse under raindrop impact. Consequently there is no washed/in zone and the water permeability of the crust is a function of the permeability of the thin surface skin. It seems, therefore, that dispersion of aggregates rather than slaking is responsible for soil crusting).

A soil crust has several adverse effects on soil properties, in particular on infiltration rate, bulk density and thermal capacity. The severity of these adverse effects is manifested mainly in the germination and emergence of plant seedlings. However, this depends on the crust strength and its thickness aside from the emergence force of the seedlings and the management practices followed to overcome such adverse effects.

### Evaluating crust strength

Evaluation of the strength of soil crusts can be made in three ways. The most popular appears to be the modulus of rupture test to simulate crusting. The resistance to penetration of a soil probe can also be used. The determination of crust strength can be approached from the point of view the emerging seedling. Morton and Bushel developed a penetro-meter to simulate a mechanical seedling and measured the force required to push it upward through a 3 inch layer of soil compacted to different bulk densities using various sizes of probes to represent different seedling diameters, they found that the emergence-energy increased directly with seedling diameter, degree of compaction, initial soil moisture content and depth of planting.

Arndt designed an instrument that records the force required to bring about emergence of a mechanical probe that is buried prior to the formation of a surface crust. This type of probe does not penetrate the crust as it is forced upward. In the case of a wet crust, there is a conical rupture as a cone-shaped mass of soil is forced out of the crust ahead of the probe. In a dry soil, first there is a cracking of the crust, followed by the formation of a demo-shaped structure of tilted, broken pieces of crust. (The modulus of rupture values, which reflect tensile strengths, were only about 20 percent of the force required for emergence of the in sintu probe through the crust. This probe produces both compression and shear stress).

**Seedling emergence:** Restriction of seedling emergence can take place in two ways. First and foremost in the effect of mechanical impedance due to crust strength resistance at the surface may be so high that the seedlings grow in to horizontal direction and fail to emerge.

It was repeated that the emergence of grain sorghum seedlings decreased if the soil strength, as measured with a penetration, exceeded 3 bars and ceased when the strength reached 13 to 18 bars; this depends on the type of soil. A crust strength of 1273m bars prevented the emergence of bean seedlings in Pachappa fine sandy loam on the other hand, critical modulus of rupture values ranged from 1200 to 2500 milli bars for the emergence of sweet corn on Pachappa loam that contained exchangeable sodium. It is necessary to specify the moisture content in establishing critical crust strength for restricting seedling emergence. For example, modulus of rupture values for a silt loam soil vary for both crop and moisture content of the crust.

## 3.5. Soil compaction

In soil mechanics literature, several terms like compaction, consolidation, compression etc. are used to describe the change in the soil volume and increase in density. Compaction refers to the increase in density of a soil by a dynamic load consolidation is the gradual decrease in volume of voids with increase indemnity under the action of continuously acting static load over a period of time. The compressibility of soil means reduction in the volume of soil (dry, wet, or saturated) by application of an external load or force on the soil. Compression is thus, defined as change in the volume of soil under applied stress soil compaction is of great agricultural significance, while consolidation and compression are of interest in civil engineering.

When a soil is subjected to certain pressure, it tends to compact, i.e, its bulk density increases, as the soil particles, bind and soil porosity reduces, with the major reduction occurring in non-capillary porosity. Under very high pressure the soil particles may themselves be compressed but only slightly. However, this is not common under normal agricultural or engineering operations. In agriculture compaction is due to the machines used for tillage of land and inter-culturing, harvesting and threshing of crops. While the former compaction is useful, the latter has adverse effects on the normal growth of plants due to reduced aeration and increased bulk density of the soil compaction of the soil may also be due to the grazing of animals, human activities in the field and intense rainfall as well as irrigation.

The compact soil layer at the bottom of the zone of ploughing is termed as the plough sole. Since all agricultural implements may create some kind of sole under favourable soil conditions, these have been named on the implement used, the plough soles, sub-soiler soles, disc harrow soles, traffic soles etc. The last one is caused by the tractor tyers and vehicular traffic.

The artificial compaction is a achieved by rolling, tamping, vibrating or compressing the soil by any means. The degree of compaction depends upon soil water content, the compactive energy, nature of soil and the amount of manipulation. The state of compaction affects the soil-air-water-temperature relationship (profound and determines the physical, chemical and biological properties of soil). The objective of common tillage practices is to achieve an optimum state of compaction favourable for crop production.

**Compaction Process:** As seen above, soil changes from one state of compaction to another under different kinds of forces. In the compaction process, soil particles are packed together in a close state of contact indicated by a change in bulk density, porosity etc. under a dynamic load sufficient to increase the density of soil, the following changes may occur.

1. A compression of soil solids.
2. A compression of liquid and gas within the pore space.
3. A change is liquid and gas contents in the pore spaces.
4. A re-arrangement of soil solids.

Thus, on dynamic loading the soil solids are re-arranged and a compression of liquid and gas within the soil pores takes place (figure 10). The pore water is displaced and the contact area between soil solids increases depending upon the deformation of particles.

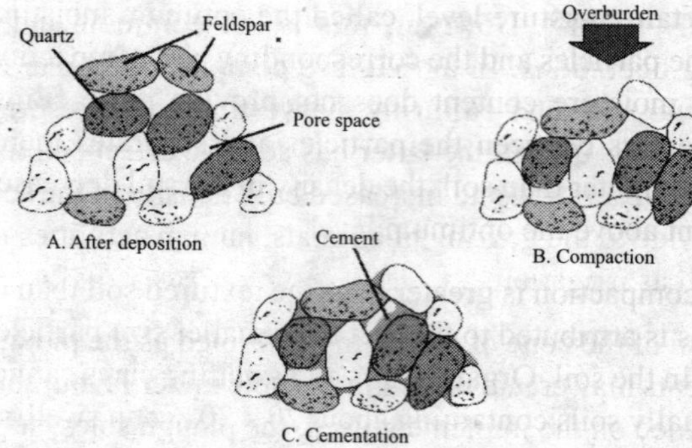


**Fig. 10:** Soil compaction

The deformation is elastic in nature and particles rebound on release of the applied load. Due to the increased contact area, the menisci formed by water in soil pores assume different shapes. The liquid may also be forced from the pore space between the particles when soil is compacted. The amount of liquid forced out in to the voids depends upon the magnitude of the load applied, water content, types of soil particles and the binding forces between liquid and solid particles. If the amount of water displaced does not exceed the capacity of the pore space, the forces of surface tension draw the water back when the load is released.

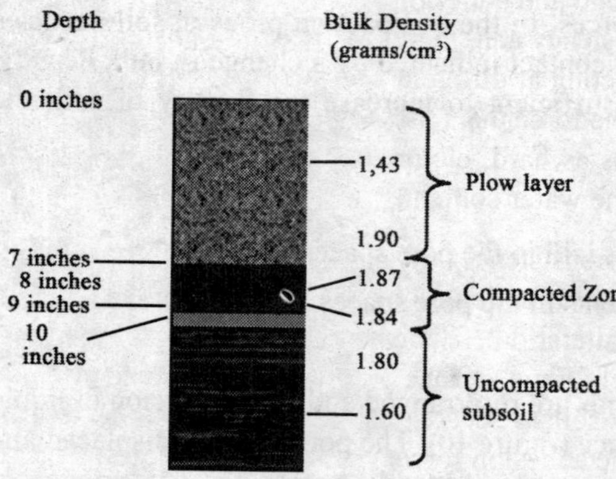


**Fig. 11:** Increased bulk density values of compacted layers

## Factors affecting soil compaction

**Soil wetness**: For a given pressure, compaction of a soil depends on its wetness compaction increases with an increase in soil moisture content and attains a maximum value at a certain level of moisture called the optimum moisture content of the soil; thereafter it decreases with further increase in moisture content. A dry soil cannot be compacted much as water enters the soil it forms films around soil particles. The water in the films serves as a lubricant whereby particles slip together under pressure. With an increase in moisture content, water films surround more and more particles and compaction increases. At a certain moisture level, called the optimum moisture content, water films surround all the particles and the corresponding soil compaction is maximum. Further increase in moisture content does not provide more films, however, rather the excess water enters between the particles and separates them. Since water is non-compressible and lighter than soil, the density of the soil decreases with an increase of moisture content above the optimum level.

**Soil texture:** For a given load soil compaction is greater in a fine textured soil than in a coarse-textured mineral soil. This is attributed to the fact that smaller size particles can fit in easily in their pore space in the soil. Organic soils and swelling clays cannot be artificially compacted much usually soils containing about 20 – 30% non swelling clay particles and less than 5% organic matter attains maximum compaction and is therefore considered suitable for making earthern dams.

## 3.6. Soil consistence

Soil consistence is a term used to describe the resistance of a soil at various moisture contents to mechanical stress or manipulations. It is usually defined as a term to designate the manifestations of the physical forces of cohesion and adhesion acting within the soil at various moisture contents.

Consistency is the physical property of soil that is manifested by its resistance to flow. When the water content in the soil is such that, the soil is not freely flowing, then the forces of cohesion (molecular attraction0 and adhesion (surface tension) between soil particulars come in to play. Soil consistency can, therefore, be regarded as a result of the forces of cohesion and adhesion acting at different water contents. It refers to the resistance offered by the soil against the force that tends to deform it. Soil consistency is commonly described by such terms as hard, cloudy, friable, soft, mellow, sticky, plastic, puddle etc., depending upon the water content.

The two forces responsible for soil consistence are cohesion and adhesion which act within the soil mass. The molecular attraction (cohesion force) is due to the Zeta Potential or Vander Wabl's forces of alteration. This is large only if the soil particles are close together and have a relatively large surface area in common. The effect of molecular attraction is greatest in dry soils. It decreases as water enters between the soils particles, keeping them apart. The adhesion force of particles that binds is due to surface tension and hence depends on the presence of both water and air.

### 3.6.1. Forms of soil consistence

The following forms of soil consistence are formed, depending on the moisture content.

1. Dry soil, hard or harsh consistence; soil particles when once apart do not stick together again. Soil consistence in dry soils depends on the amount of surface contacts per unit volume.
2. Moist soil, friable consistence; soil sticks together only lightly and it is soft.
3. Wet soil, plastic consistence, soil can be molded without loosing its coherence.

4. Very wet or saturated soil, viscous consistence, soil may either flow under pressure or merely by gravity. Soil adheres to objects and it is sticky.

The consistence of soil is generally described at three soil moisture levels; wet, moist and dry. Terms used to describe soil consistence at these three moisture levels are shown in the following table.

**Terms used to describe the consistence of soils:** Conditions of least coherence are represented by terms at the top of each column greater coherence characterizes terms as you move down the column.

**Table 7:** Forms of soil consistence in wet, moist and dry conditions of soil

| Wet soil | | Moist soil | Dry soil |
|---|---|---|---|
| Stickiness | Plasticity | Loose | Loose |
| Non-sticky | Non-plastic | Very friable | Soft |
| Slightly sticky | slightly plastic | Friable | Slightly hard |
| Sticky | Plastic | Firm | hard |
| Very sticky | Very Plastic | Very firm | Very hard |
| | | Extremely firm | Extremely hard |

## Field examples

**Wet Soils:** Wet clay soils are commonly termed slightly sticky, sticky, or very sticky and their plasticity would be indicated as slightly plastic, plastic or very plastic.Wet sand, on the other hand, would be non-sticky and non-plastic. Gradiations in stickiness and plasticity would be found with wet medium-textured soils such as clay loams, silt loams and some loams.

**Moist Soil:** The consistences of moist soil various from loose to extremely firm. Sandy soils (and some high in silt) are mostly loose, whereas soils high in sticky claysare usually extremely firm. Moist soils that are easily crushed and yet show some coherence when pressed together are termed friable or very friable, conditions that are prized for ease of manipulation and management.

**Dry Soils**: Dry soils with little coherence also will be loose. Those with high coherence are extremely hard and are so resistant to pressure that clouds cannot be broken by hand. The degree of cementation of soil materials, independent of soil moisture, also is considered in identifying soil consistence. Terms such as weakly cemented, strongly cemented and indurate are used to define categories of cementation.

The state of consistencies between solid and liquid range has been divided in to a number of distinct stages limits have been imposed to indicate (conventionally) the soil water content limits for various states of consistency. These limits are also referred to as "Alter berg" limits and are defined as follows.

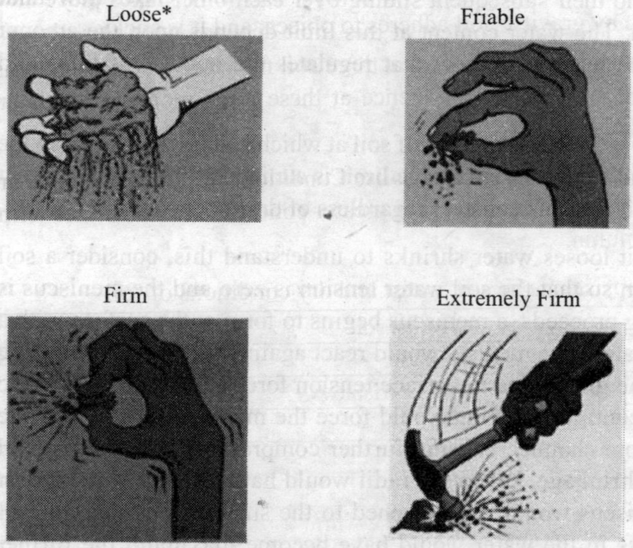


*Soil with "single grained" structure always have loose consistence

**Fig. 12:** Moist consistency of soil

### 3.6.2. Atterberg limits of soil consistence

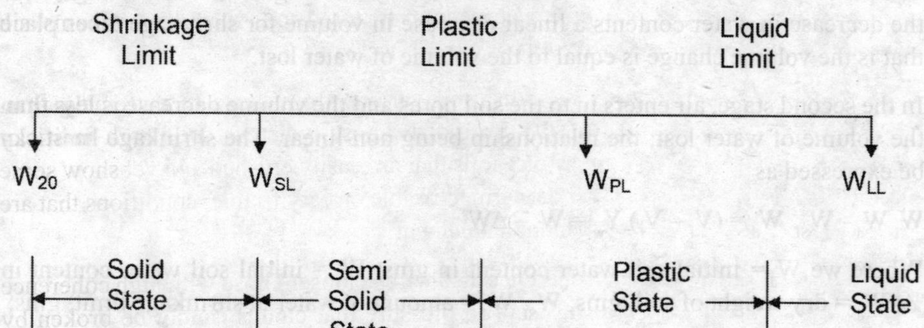


**Fig. 13:** States and limits of soil consistency

**Liquid limit ($W_{LL}$):** It is the water content at which soil is practically liquid but possesses a small sharing strength. This is also termed as upper plastic limit. At this point the water film becomes so thick that cohesion is decreased and soil mass flows under an applied force. In fact water films coalesce to fill up the majority of the soil pore space. The soil transforms at this limit from liquid in to plastic state, thereby looking its ability to flow as a liquid.

**Plastic Limited ($W_{PL}$):** It is the minimum water content at which a soil is plastic (and the soil threads of 1/8" diameter start to crumble). This is also known as lower plastic limit. At this point sufficient water is available to provide a film around each particle.

Orientation of particles and their subsequent sliding over each other takes place and the cohesion is maximum. The water content at this limit depends upon the amount and the nature of colloidal material present that regulates the number of films and quantity of water adsorbed.

**Shrinkage limit ($W_{SL}$):** It is the water content of soil at which soil transforms from the semi-solid state to the solid state. The shrinkage limit is attained at that water content at which the volume of soil remains constant regardless of drying.

When a moist soil dries it looses water shrinks to understand this, consider a soil mass saturated with water, so that the soil water tension is zero and the meniscus is a plane surface. As drying proceeds, a meniscus begins to form at the surface with a resulting tension in the water. The menisci would react against the grains producing a compression between the grains due to surface tension forces and reduction in the volume of the soil mass continued drying would force the meniscus down in to the smaller portions of the pore channel, causing further compression between the soil grains and the resultant shrinkage. When the radii would have become the smallest possible, that is, the meniscus would have reached to the smallest possible pore of all passages and the stress in the water would have become maximum the further recession of the menisci would cause no increase in stress and shrinkage. At this point, the water menisci at the air-water interfaces may actually break. The water content at this stage is termed as the shrinkage limit.

The process of drying is therefore characterized by two stages. In the first stage, with the decrease in water contents a linear decrease in volume for shrinkage takes place that is the volume change is equal to the volume of water lost.

In the second stage, air enters in to the soil pores and the volume decrease is less than the volume of water lost; the relationship being non-linear. The shrinkage limit can be expressed as

$$W_i\ W_d - W_{SL}\ W_d = (V_i - V_f)\ Y_w = W_i - \Delta W$$

Where we $W_d$ = initial soil water content in gms; $W_i$ = initial soil water content in %; $W_d$ = dry weight of soil, gms; $W_{SL}\ W_d$ = amount of water at shrinkage limit, gms;

$Y_w$ = Unit weight of water, 1 gmcm$^{-3}$; $V_i$ = initial soil volume before drying, Cm$^3$;

$V_f$ = final soil volume before drying Cm$^3$.

The above equation gives the loss of water during shrinkage. The water content percentage at shrinkage limit can therefore be expressed as:

$$W_{SL} = W_i - \left[\frac{V_i - V_f}{W_d}\right] - Y_w = W_i - \Delta W$$

Where $\Delta W$ = percent water loss (dry wt.) on drying from initial water content $W_i$ to the water content which corresponds to the shrinkage limit of the soil.

The above discussion reveals that shrinkage caused by drying results in compaction.

**Plasticity**:- It is defined as the property, which enables clay to change its shape without cracking when it is subjected to a deforming stress. Soil colloids possess the property of (like gels) plasticity, i.e. they can be molded in to any form or shape. The shape can be changed without producing cracks and it remains unchanged when dried. Sands and silts can be molded when they are wet, but the molded form falls in to pieces when they are dried, therefore they are not plastic.

**Plasticity Index:** A useful index derived from the consistency is the plastic index, defined as the difference between the liquid and plastic limits. It is generally taken to be an indication of a soils clayeyness or potential plasticity and is used as such in engineering classification systems for soils. The plasticity index depends not only on the clay content of a soil but also on the nature of the clay present, whether of the swelling or non-swelling type, as well as on the adsorbed cations, the organic matter content, and on the pre-treatment of the sample.

Soil stickiness and plasticity influenced by soil texture (Figures 14 & 15), with increasing fineness of soil texture stickiness and plasticity increases.

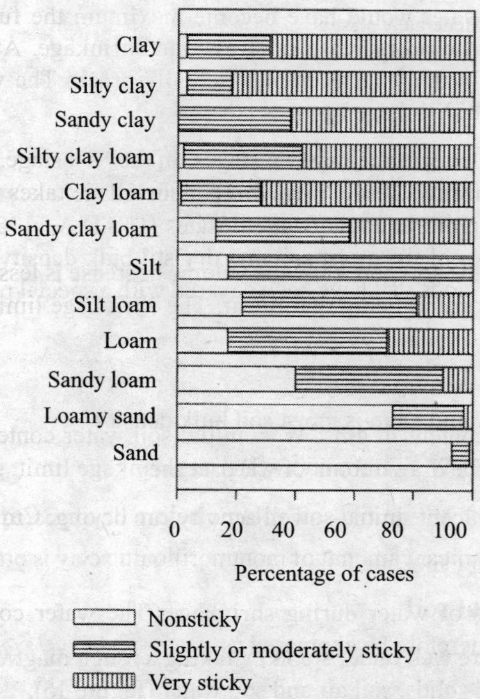


**Fig. 14:** Influence of soil texture on soil stickiness

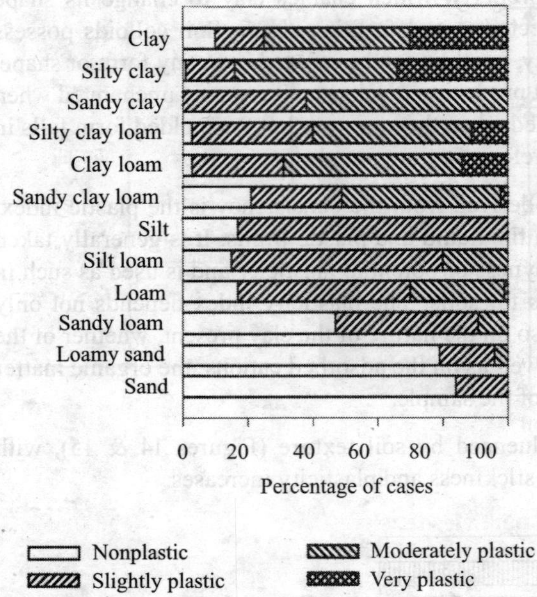


**Fig. 15:** Influence of soil texture on plasticity

## COLE

The co-efficient of linear extensibility, COLE, has been adopted as a standard measurement in soil survey investigations in the United States (U.S.D.A., 1967). It is calculated from the bulk densities of the moist soil and dry stil bulk density. The measurements are made on natural clods that have been covered with a special plastic coating.

$$COLE = \frac{Dbd}{Dbm} - 1$$

Where, Dbd is drysoil bulk density and Dbm is moist soil bulk density

Some uses made and inferences to be drawn from COLE data are

1. If COLE exceeds 0.09, significant shrink – swell activity can be expected.
2. If COLE exceeds 0.03, a significant amount of montmorillonitic clay is present.

## 3.7. Weight- Volume relations of soil

Weight – volume relations of soil are well under stood by taking a rough diagram of all the three phases of soil viz., soil solids, soil air and soil water (Figure 16).

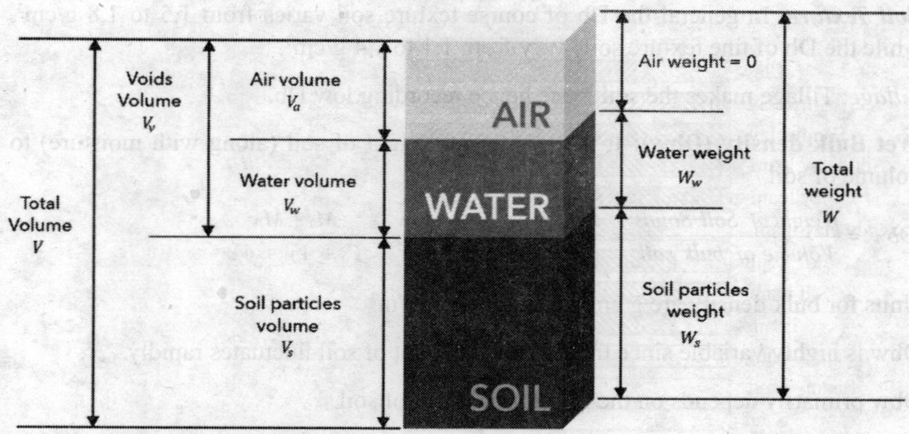


**Fig. 16:** Weight- volume relations of soil

## Weight relations

**a) Density of soil (True density/ particle density (Dp)):** It is the weight of unit volume soil (volume does not include pore space) or it is the ratio of Mass of soil solids to volume of soil solids.

$$Dp = \frac{Weight\ of\ Soil\ Solids}{Volume\ of\ Soil\ solids} = \frac{Ms}{Vs}$$

*Units for true density are g/cm³ or Kg/m³ or Mg/m³*

The particle density of soil depends on the physical and chemical composition of soil. In most of agricultural & horticultural soils dominant minerals are quartz and feldspars, whose density (Dp) is 2.65 $g/cm^3$. For all practical purposes the Dp of soil is taken as 2.65 $g/cm^3$.

## Bulk Density (Db) of Soil (dry bulk density)

It is the weight of unit volume bulk soil (volume include pore space) or it is the ratio of weight of soil solids to volume of soil bulk soil.

$$Dp = \frac{Weight\ of\ Soil\ Solids}{Volume\ of\ bulk\ soil} = \frac{Ms}{Vt} = \frac{Ms}{Vs + Vf} = \frac{Ms}{Vs + Vw + Va}$$

**Units** for bulk density are $g/cm^3$ or $Kg/m^3$ or $Mg/m^3$. Factors influencing dry bulk density of soil:

***Soil Structure*:** Loose (aggregated) soils record less Db than compact soils.

***Soil Organic matter*:** High organic matter in the soil reduces Db by two ways. 1. Organic matter encourages soil aggregation thus making the soil loose. 2. The particle density of organic matter is much less (1.3 to 1.5 $g/cm^3$) than inorganic matter (2.65 $g/cm^3$) hence low Db in with increase in soil organic matter content.

***Soil Texture***: In general the Db of course texture soil varies from 1.5 to 1.8 g/cm$^3$, while the Db of fine texture soils vary from 1.1 to 1.4 g/cm$^3$.

*Tillage*: Tillage makes the soil loose hence recording low Db.

**Wet Bulk density (Dbw):** It is the ratio of weight of soil (along with moisture) to volume of soil.

$$Dbw = \frac{Weight\ of\ Soil\ Solids}{Volume\ of\ bulk\ soil} = \frac{Ms + Mw}{Vt} = \frac{Ms + Mw}{Vs + Vf} = \frac{Ms + Mw}{Vs + Vw + Va}$$

Units for bulk density are g/cm$^3$ or Kg/m$^3$ or Mg/m$^3$

Dbw is highly variable since the moisture content of soil fluctuates rapidly.

Dbw primarily depends on the moisture content of soil.

## b) Mass wetness of soil (w)

It is also known as gravimetric water content of soil. It is expressed as dry weight basis of soil. '*w*' is the ratio of mass of water to mass of oven dry soil.

$$w = \frac{Weight\ of\ water}{Weight\ of\ Soil\ solids} = \frac{Mw}{Ms}$$

It is unit less, hence expressed as percentage by multiplying the value with 100. '*w*' of the soil depends on 1. Soil texture, 2. Soil Organic matter and 3. Soil structure.

a) Soil texture: Fine textured soils retain more water than course textured soils.
b) Soil structure: well aggregated soil accommodate more water than compact soil, however in fine textured soils the gravitational water is more under good aggregation than that in dispersed condition.
c) Organic matter: well-decomposed organic matter increases water retention by soil. '*w*' is directly proportional to soil organic matter content.

## Volume relations of soil

**a. Specific volume of soil (Vb)**: It is the ratio of total volume of soil to dry weight of soil.

$$Vb = \frac{Total\ volume\ of\ soil}{Dry\ weight\ of\ soil} = \frac{Vt}{Ms}$$

*Units for Vb are cm$^3$/g or m$^3$/Mg.*

*Vb* is an index for degree of compaction or looseness of soil. The value of Vb for course textured soils may vary from 0.56 to 0.67 cm$^3$/g and for fine textured soils 0.71 to 0.91 cm$^3$/g.

**b. Pore space or porosity (*f*):** It is the ratio of total volume of pores to the total volume of soil.

$$f = \frac{Vf}{Vt} = \frac{Va + Vw}{Vs+Va+Vw}$$

Porosity is generally expressed in fraction or percentage. Porespace is an index for relative volume of the pore space in the soil and it varies from 0.3 to 0.6 or 30 to 60%. In course-textured soils the porosity varies from 32.1- 43.4% and in fine textures soils the corresponding range is 47.2- 58.5%. Although the size of individual pores may be greater in course-textured soils, total porosity of these soils is less when compared to fine textured soils.

Determination of total pore space is a difficult task hence deriving the total porosity of soil from Db and Dp is possible.

$$Db = \frac{Ms}{Vf + Vs} \quad Dp = \frac{Ms}{Vs}$$

$$Ms = (Vf+Vs)Db \quad Ms = Vs \times Dp$$

$$(Vf+Vs)Db = Vs \times Dp$$

$$\frac{Db}{Dp} = \frac{Vs}{Vf + Vs} \text{ it is Fraction of solid space}$$

Fraction of porosity = 1 – fraction of solid space

$$= 1 - \frac{Vs}{Vf+Vs}$$

$$= 1 - \frac{Db}{Dp}$$

$$\text{Pore space or porosity \%} = \left[1 - \frac{Db}{Dp}\right] 100$$

**c. Size of the pores**

Based on the size the soil pores are divided to Macro and Micro pores for easy understanding of pore functions. Macro pores are those pores having size (diameter) of >0.06 mm and Micro pores are those having size of <0.06 mm.

Solid Space: It is the ratio of soil solid volume to total soil volume.

$$\text{Solid space} = \frac{Vs}{Vt} = \frac{Vs}{Vs + Vf} = \frac{Vs}{Vs + Vw + Va}$$

## Capillary and Non- Capillary pores

The pore space in soil contains both capillary and non-capillary pores. Under saturated condition both capillary and non-capillary pores of soil are filled with water. As the soil dries the non-capillary pores will first become empty giving their spaces to air. The non-capillary porosity of soil is assessed by measuring the volume of water retained in the soil at 50 cm tension (or 50 cm Pressure). Capillary porosity is calculated by subtracting non-capillary porosity from total porosity.

Capillary pores hold water against gravitational force and supply the same to plant roots, while non-capillary pores loose water due to gravitational force, simultaneously gains air and thus supply oxygen to plant roots.

**d. Void Ratio**

It is the ratio of total volume of pores to total volume of soil solids. It is an index of relative pore volume of the soil.

$$Void\ ratio = \frac{Vf}{Vs}$$

**e Volume wetness of soil:** It id the ratio of water volume to total soil volume.

$$Ø = \frac{Vw}{Vt} = \frac{Vw}{Vf+Vs} = \frac{Vw}{Vs+Va+Vw}$$

The *q* is expressed as fraction or percentage.

The mass wetness (*w*) multiplies with Db gives to volumetric wetness (*q*).

**f. Air filled porosity or Aerating porosity (*n or fa*):** It is the volume of air space to the volume of soil.

$$n\ or\ fa = f - q\ or = \frac{Va}{Vs+Vw+Va}$$

## 3.8. Soil colour

The colour of an object depends upon the kind of light which it is reflecting to the eye. Soils reflect light of a great variety of wave-lengths. It may be possible to distinguish the great many colours, but, describing them accurately is a bit difficult, because of confused colour vocabulary.

### Importance of soil colour

Soil color does not have any direct influence on plant growth, but indirectly influences the plant growth through its influence on soil temperature and soil moisture.Soil colour is an indicator of the climatic condition under which a soil was developed (pedochromic) or of its parent material (litho chromic). Soil color is also taken as criteria for assessing soil productivity. Soil color is the important parameter, used to classify the soils. Practically all colours occur in soils, except pure blue and pure green. Generally, soil colors are not pure but mixtures, such as grey, brown and rust. Frequently, two or three colours occur in patches, which is called as "mottling".

The colour of the soil is a composite of the colours of its components viz., mineral matter, organic matter, water, air.

***Soil Colour*** *= f (colour of mineral matter + colour of organic matter + colour of moisture + colour of air)*

The effect of these components on the color of the composite soil is roughly proportional to their total surface, which is equal to their specific surface times their volume percentage in the soil.

Colour of soil is proportional to (Specific surface of mineral matter x % volume of mineral matter + Specific surface of organic matter x % volume of organic matter + Specific surface of soil moisture x % volume of soil moisture+ Specific surface of soil air x % volume of mineral air). Colloidal material i.e., iron oxides and humus will have greatest impact on soil color. Humus – brown or dark brown; Iron oxides – red, rust – brown, or yellow depending upon degree of hydration; Reduced iron – blue green; Quartz - white

Lime stones – white, gray or sometimes olive green.

## Measuring soil color

**Colour components:** As the, a standard system for accurate colour description has been developed using Munsell color chart. In this system, a small piece of soil is compared to standard colour chips in a soil colour book (figure 17). Each colour chip is described by the three components of colour i.e., Hue, Value and Chroma

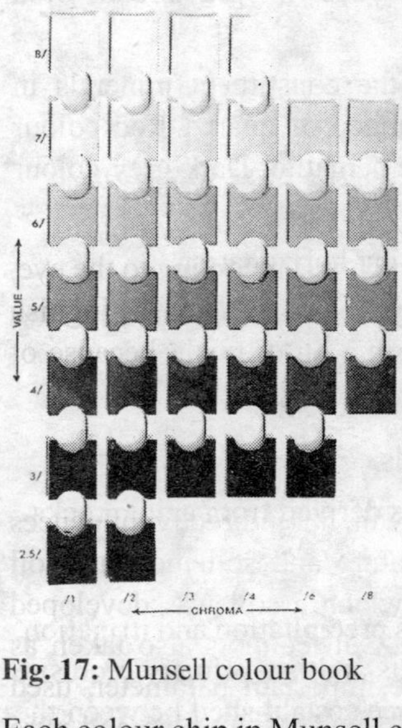


**Fig. 17:** Munsell colour book

Each colour chip in Munsell color chart is described by the three components of colour i.e., hue, value and chroma

1) **Hue** refers to the **dominant** spectral colour or quality which distinguishes red from yellow etc.
2) **Value or brilliance** expresses apparent lightness as compared to absolute white. It refers to relative brightness or darkness of colour with in a scale of (0 –10) as compared to absolute white. It refers to gradations white to black (lightness or darkness).
3) **Chroma** defines the gradations of purity of colour , or the apparent degree of departure from neutral grays to white (intensity or brightness) (strength of colour), with in a scale ranging from (0 – 20).

The numerical notation $2.5\ YR\ \frac{5}{6}$

Where in, hue of 2.5 YR, value of 5 and chroma of 6.

## Significance of Soil Colour

- Colour is taken as a diagnostic criterion for classifying soils e.g. the comprehensive system uses color as formative element in its nomenclature as alb (white), ochr (light coloured), umbr (dark), sombr (dark).
- Soil color can be guide to the climatic soil group, to the parent material, or to the physiographic location. Humid temperate regions – Grayish; Tropics and sub tropics – red and yellow
- The productivity of the soils is assessed based on soil color e.g. dark color is an indication of high productivity. Black > brown > rust brown > gray brown > red > gray >yellow > white.
- Colour indicates the presence / domination of the constituent minerals in the soil e.g. Light colour results from preponderance of quartz, Red colour indicates domination of iron rich compounds like hematite, Dark grey colour for Smectite, Dark colour for humous.
- Colour is used to describe soil profile.- distinguishing horizonation
- Colour indicates the process involved in soil formation i.e. leaching and erosion etc.
- Soil color is indicative of presence of excessive soluble salts.
- Soils in depression are darker than adjacent uplands.
- Soils derived from basic rocks are darker than soils derived from acidic rocks

## 3.9. Soil Water

Soil serves as a regulated reservoir for water, as it receives precipitation and irrigation. Water is an excellent solvent of most of the plant nutrients and is a primary requisite for plant growth. The water that fills part of the pore spaces established between the solid particles of soil is known as soil water. It is the liquid phase of the soil and is important as the other two phases of the soil system. It has profound influence on the life of the plant, at all the stages.

All the water required by the plants come from the soil. Water supplied to the soil is first adsorbed and held by soil colloids until the adsorption capacity of the soil is satisfied. If more moisture is available then it moves downwardl by gravity and drains off. It is mainly clay or other colloidal particles that are responsible for adsorbing water. Silt and sand particles do not adsorb water or very little, if at all. The water so held on the surface of the clay particle is believed to be different from ordinary liquid water. It is non-liquid, more dense and viscous compared to ordinary water.

### 3.9.1. Water retention in soil

The electrostatic forces exist on colloidal particles would attract water molecules. Water molecule by virtue of dipolar nature, gets attached to colloidal particle with it positively charged end, leaving its negative charge end outward (figure 18). In this fashion water molecules will be oriented or arranged on the surface of clay particles. Now, another layer of water molecules will be attached to the negative end of water molecules, in a successive manner. This orientation results in thickness of water layers around colloidal particles. As the water molecular layer gets thicker, the orientation becomes weaker, and at a certain distance from the particle surface, the water molecules cease to orient and liquid water or capillary water begins to appear.

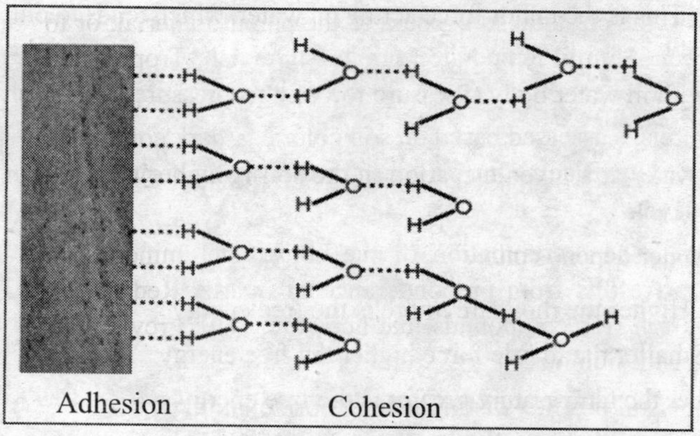


**Fig. 18:** Water retention in soil by the forces of cohesion and adhesion

### Forces of soil water retension

- Matric force/ adhesive force- adhesion and, surface tension
- Cohesive forces: Attraction of water molecules for each other
- Osmatic forces: attraction of water molecules to ionic forms of salts.
- Gravitational forces: pulls the water from the soil pores depending on the slope.
- Pressure forces : pushed the water up through soil pores due to impeding pressure.

### 3.9.2. Soil water energy concept

The retention of water in soil, its movement, uptake and translocation in plants, its loss to the atmosphere are all related to the energy. Two kinds of soil water energy are potential, kinetic energy.

**Kinetic Energy:** This energy is by virtue of its motion and quantitatively it is ½ $mv^2$ where 'm' is the mass and 'v' is the velocity. Since the movement of water in soil is very slow its K.E. is also very less.

**Potential Energy:** This is the energy, which a body has by virtue of its position in a force field. Thus the potential energy is measured by the force required to move a body against the force field and is equal to the product of force and distance moved (i.e., F x d) .

### Factors effecting free energy of water

- Attraction of soil colloids to water by adhesive and cohesive force (matric forces)
- Attraction of ions and other solutes to water (Osmotic force),
- These two forces reduce the free energy of water.
- Gravitational Force: This is the major force acting on water, which tends to pull the water downwardly.
- Pressure force on /with in water body (Pressure force): this pressure force push the water.
- The free energy of water at a given elevation in the soil is higher than that of pure water at lower level.

So the free energy of soil water depends on

a) **Moisture content** : Higher the moisture ,more is the free energy.

b) **Matric potential**: Smaller the matric force higher the free energy.

c) **Temperature**: Higher the temperature greater is the free energy.

d) **Dissolved salts**: Lower the content dissolved more is the free energy.

e) **Reference point:** higher is the elevation from reference point higher is the free energy.

**Soil water potential:** The term 'Potential' may be defined as the amount of work done (or potential energy stored) per unit mass in bringing 'm' mass of water from any distance to the reference point (figure 19). The work need to be done on soil water to change it to pure (free) water at reference state against the forces acting on it is called soil water potential.

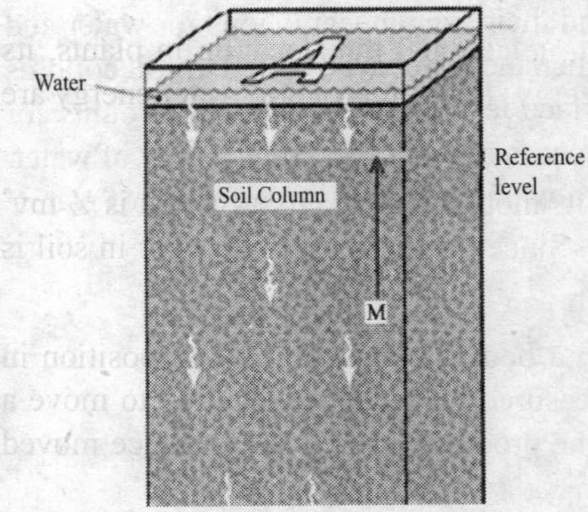


**Fig. 19: Concept of soil water potential**

Total Water Potential : The water in a system at rest , the sum of all potentials or the total potential at a point will be either constant or zero ( This depends on the reference taken).

Total Potential ($\Psi_T$) = Ψm + Ψg + Ψo + Ψp .....

$\Psi_T$ = Total Potential;

Ψm = Matric Potential;

Ψg = Gravitational potential;

Ψo = Osmotic potential;

Ψp = Pressure potential *(Ψ is psi)*

If the total potential of water is constant through out the given soil, then the water is said to be in 'static equilibrium' i.e., the net force acting on water to move is zero. If this is not zero, water movement occurs in the direction of decreasing free energy..

**Gravitational Potential (Ψg)** : The force of gravity acts on soil water, the attraction being towards the earth's center but up to water table. The gravitational potential at any point in a gravitational field is defined as the 'work done per unit mass in carrying any mass 'm' infinitesimal quantity to that point in question'.

$$(\Psi g) = \frac{work}{mass} = \frac{mgh}{m} = gh$$

The point of reference is usually chosen with in the soil at a lower point to soil surface so that gravitational potential is always positive. That is why the gravitational water always moves downwardly up to water table. If the reference point selected in the soil at higher point, the gravitational potential is negative. This gravitational force is important as it removes excess water from the upper rooting zone following heavy irrigation or precipitation.

**Matric Potential (Ψm):** Matric potential is the result of two phenomena i.e., adhesion and capillarity. The attraction of soil solids and their exchangeable ions for water and the loss of energy of water (heat of wetting) when the water is adsorbed to soil colloids (figure 20). In addition to this attraction, surface tension of water also accounts for capillary force. The net effect of these two forces is to reduce the free energy of water. Matric potential is always negative, because of which it reduces free energy of water.

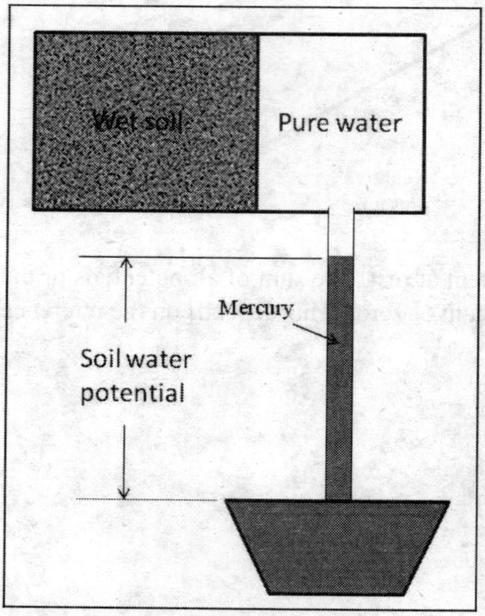


**Fig. 20:** Matric potential of soil water

**Osmotic Potential (Ψo) :** The osmotic potential is attributed to the presence of solutes in the soil. The solutes may be inorganic salts or organic compounds. They reduce the free energy of water. Osmotic potential is the result of hydration of ions_in the soil solution. Due to dipolar nature of water molecules, the ions or compounds are attached on both sides, reduces the free energy of water. "Osmotic potential or solute potential is the work to be done to detach these ions from water molecules". More the attachment, more the osmotic potential and less is the free energy. Unlike the matric potential, osmotic potential has little effect on the mass movement of water. But its effect is prominent in the uptake of water by plant roots. In soils of high soluble salts osmotic potential may be greater than in plant roots. Osmatic potential is also negative potential.

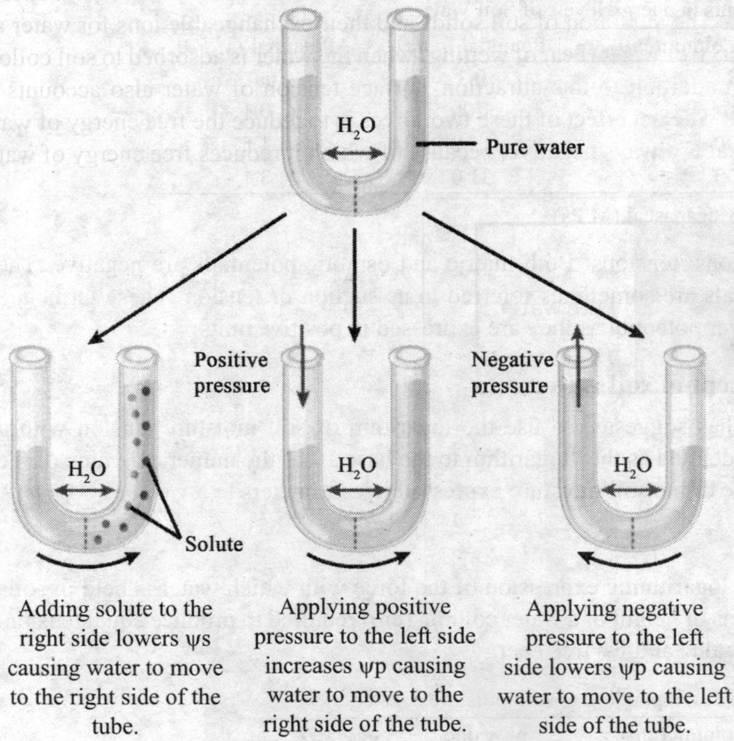


**Fig. 21:** Depiction of osmatic and pressure potentials

**Pressure Potential (Ψp) :** It is due to the weight of water at a point under consideration or vapour pressure (barometric pressure) exerted on water. With the change in elevation, the pressure potential in a soil with reference to water table may be either positive (below water table) or negative (above water table). The positive pressure potential is due to the weight of water column resting on any point below the water table and negative pressure potential is caused due to capillarity and adsorptive forces. If the point is beneath the water table, the potential is equal and opposite to the gravitational potential. This pressure potential is otherwise known as '*Submergence potential' or 'Peizometric potential or 'Pneumatic Potential'*. Pressure potential due to water weight may be measured with 'manometer' or 'Peizometer'.

### 3.9.3. Units of soil water potential

A common means of expressing potential is in terms of height of water column in entimeters. This can also be expressed in terms of standard *atmospheric pressure (atmosphere)* at sea level. This is equal to 14.7 lb/in. or ; 760 mm of Hg; or 1020 cm of water. ie, 1 atmosphere = 14.7 lb/sq.in. or ; = 760 mm of Hg column; or = 1020 cm of water colum or = 1 bar or =0.1 MPa or = 100 KPa . The unit 'bar' which is approximately equal to one atmosphere is also used.

**Table 8:** Equivalents in energy levels of Soil Water

| Height of Water Column*(cm)* | Potential *(bars)* | Potential *(M Pa)* |
|---|---|---|
| 306 | 0.3 | 0.03 |
| 1020 | 1.00 | 0.1 |
| 15,300 | 15.0 | 1.5 |
| 31,700 | 31.0 | 3.1 |

Note:10 bars = 1 Megapascal (M Pa)

Soil water Suctions /tensions: Both matric and osmotic potentials are negative. The negative potentials are sometimes referred to as suction or tension. These terms are advantageous over potential as they are expressed in positive units.

### 3.9.4. pF Concept of soil water

Schofield 1935 has suggested to use the logarithm of soil moisture tension with a symbol pF. It is defined as the "logarithm to the base 10 of the numerical value of the negative pressure of the soil moisture expressed in centimeters".

pF= log(- h)

where, pF is the logarithmic expression of the force with which water is held in soils expressed in terms of height of a water column (cm) required to produce equal tension. p = logarithmic value and F= free energy.

**Table 9:** pF values corresponding to soil moisture constants

| Height of water column (cm) | pF value |
|---|---|
| 1 | 0 |
| 346 | 2.53 (field capacity) |
| 15849 | 4.18 (wilting point) |
| 31623 | 4.5 (Hygroscopic coefficient) |

**Classification of soil water on the basis of pF:** Free water : Tension free (pF =0); Gravitational water : pF 0.0 to 2.53; Capillary water : pF 2.53 to 4.5; Permanent wilting : 4.18; Hygroscopic water : pF 4.5 to 7.0; inter layer water: pF more than7.0

### 3.9.5. Soil water Retention curves or Soil moisture characteristic curves (SMCC)

The curves obtained from the graph, which are plotted using the water retained by soil at different matric suction values (Figure 22), are called 'soil moisture retention curves' or 'soil moisture characteristic curves'. The retention term is considered to cover both the process i.e., uptake and release. The desorption curve starts at saturation and shows retentivity at progressively increasing suction value. SMCC are nonlinear indicating that for a given change in water content the change in matric potential will be different.

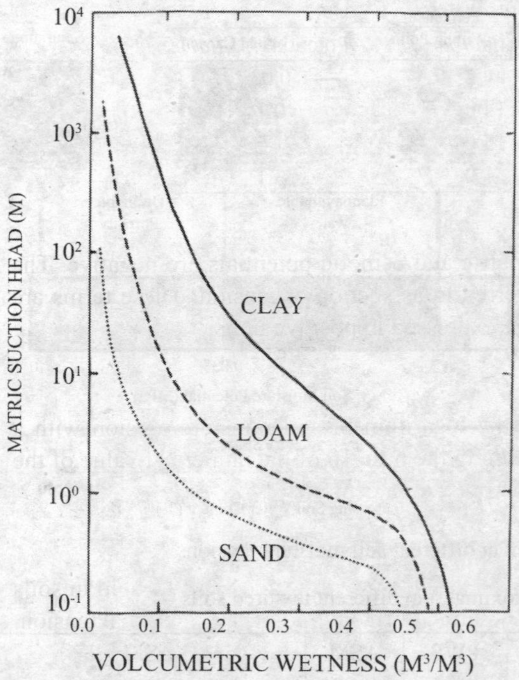


**Fig. 22:** Soil moisture characteristic curves

Soil moisture characteristic curves reveals the facts viz., Gradual decrease in soil water content with increased tension and vice versa; Clay soils hold more water at a given tension than loam or sandy soils; At a given moisture content water is held more tenaciously in clay soils compared to loams or sands; Much of water in clay soils is held more tightly (can not be removed by plants); At 15 bars matric tension sandy loam soil contains 5 %, silty clay loam soil contains 12 % moisture; Well granulated soil has more total pore space than compact soil. Compact soils hold water with greater suction than loose soils; The curves are important in understanding the movement of water in soil and uptake by plants.

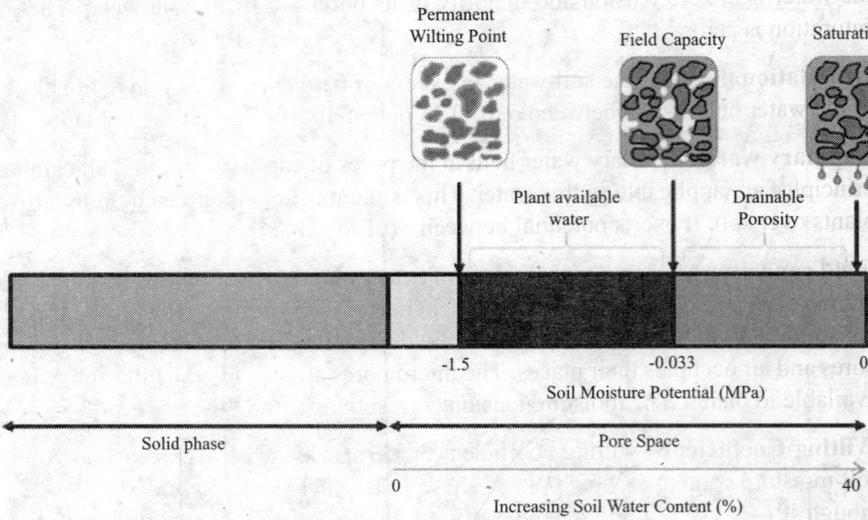


**Fig. 23:** Changes in soil moisture content at different soil moisture tensions

**Tabele 10:** Soil moisture retention (approximate) in different textured soils

| Soil texture | Saturation capacity (%) | FC (%) | PWP (%) | ASM | ASM | |
|---|---|---|---|---|---|---|
| | | | | | % | mm/m |
| Clayey | 60 | 40 | 20 | 20 | 200 | Clayey |
| Clay loam | 50 | 30 | 15 | 15 | 150 | Clay loam |
| Silt loam | 45 | 22 | 12 | 10 | 100 | Silt loam |
| Sandy loam | 42 | 14 | 6 | 8 | 80 | Sandy loam |
| Loamy sand | 40 | 10 | 4 | 6 | 60 | Loamy sand |
| Sandy | 38 | 6 | 2 | 4 | 40 | Sandy |

**Importance of soil water:** Soil water dissolves and carries nutrients to the plant. It helps in soil formation as it influences biosphere and transports water with in soil profile and soil erosion, forms alluvium/colluviums. It makes soil biological active by supporting flora and fauna. It carries dissolved oxygen in to the soil. It buffers soil from getting too cold or too hot.

### 3.9.3. Classification of soil water

Soil water classification is conveniently as three types viz., 1)Physical classification, 2)Biological classification and 3)pF based classification.

**i) Physical classification of soil water**

**Saturated water(free water)** : A soil whose pore spaces are completely filled with water and the water is at zero tension. Hence it is define a soil as saturated when the water is at zero tension and majority of its pores are filled with water. The pF at saturation is zero.

**Gravitational water**: the soil water in excess of field capacity is gravitational water. or the water held in soil between o to -0.33 potential. It is limited use to plants

**Capillary water**: capillary water held in the pores of capillary size and all capillarity principles are applicable to this water. This is the most of water taken up by growing plants over soil. It exerts potential between - 0.1 to -31

**Field capacity:** After two or three days of heavy rain or irrigation to the soil the water drains off rapidly, later this rapid movement becomes slow and negligible. At this stage the soil is said to be at field capacity. At field capacity the water drained out of macro pores and air occupies their places. The micro pores are still filled with water, which is available to plants. The moisture tension corresponds to 0.33 bar or pF of 2.53.

**Wilting Coefficient**: (wilting coefficient or Permanent wilting percentage): It is the soil moisture content at which plants show wilting symptoms and can't recover even though it is kept in humid chamber. It occurs at a moisture tension of 15 bar or pF value 4.18. At this point a little percentage water remains in small capillary pores and around the soil particles but is no use for plants.

**Hygroscopic coefficient**: Hygroscopicity is the ability of a body to adsorb moisture from the atmosphere. Soils high in expanding type of clay minerals and organic matter have a high hygroscopic coefficient. This hygroscopic water is not available to plants, but available to certain microbes. It occurs at a soil moisture tension of 31 bar or pF value 4.5

**Aeration porosity limit (Non-capillary porosity)**: Aeration porosity of a soil is defined as that part of the pore space volume that is free of water which can be established by creating a tension of water column of 50cms. This corresponds to a pF of 1.7 or 1/20 atmospheres tension. Aeration porosity is the volume of the pores whose diameter is more than 0.06 mm.

**Best tillage range** : To attain proper tilth, the soil should contain enough moisture during tillage operations to get fine tilth. The pF range of 2.8 to 4.4 is considered optimal for tillage operations in medium textured soils.

**Biological classification of soil water:** Available water : Water held by soil at potential ranging between -15 bars (PWP) to -1/3 bars (FC), is considered as plant available water. It occurs at pF value 2.53 to 4.18.Unavialable water : Water held by soil at potential lower than -15 bars (PWP). It occurs at pF value above 4.18

**pF value based classification**: It is based on the soil water tension expressed as pF scale, which in turn is logarithmic value of soil water tension expression in centimeter water column (table.11).

**Table.11:** Classification of Soil water based of pF

| pF value | Soil water class |
|---|---|
| 0 | Free water (Tension free ) |
| 2.53 – 0.0 | Gravitational water |
| 4.5 – 2.53 | Capillary water |
| 4.18 | Permanent wilting |
| 7.0 | Hygroscopic water |
| >7.0 | Water of inter layer water |

### 3.9.7. Soil Moisture Constants

Knowledge about amount of water held by the soil at various tensions is required to calculate i) the amount of water available to plants, ii) the amount of water retained by the soil before percolation starts and iii) the amount of water that is favourable for irrigation.

Soil moisture constants are expressed in terms of the energy with which the water is held at that moisture content. For practical use, it is necessary to compare moisture tension with some reference points. These reference points are known as 'Moisture Constants'. Soil moisture constant at which the soil water potential is given in the table.

**Table 12:** Soil moisture contents and respective soil water potential.

| Soil moisture constant | Soil water potential |
|---|---|
| Saturation | Zero "0" |
| Field capacity | -1/3 bars |
| Wilting coefficient | -15 bars |
| Hygroscopic coefficient | -31 bars |
| Air dry soil | -1,000 bars |
| Oven dry soil | -10,000 bars |

### 3.9.8. Measurement of soil moisture

Two basic principles by which moisture in the soil is determined viz.,

1. The amount of water in the given soil and
2. The tension under which the water is held by the soil.

Soil moisture is measured by direct and indirect methods. Direct methods include thermo gravimetric method, lysimeter method, spirit burning method and infrared balance method. Indirect methods are resistance blocks method, neutron probe method, tensiometer method and pressure plate apparatus method.

**Thermo-Gravimetric method** : This is a standard method in which the soil sample is collected from the field (moist condition) and weighed. The weighed sample is oven dried at about 105-110°C for overnight and weighed again. The weight difference is considered as water content.

$$\% \text{ moisture} = \frac{\text{Weight of fresh soil} - \text{weight of oven dry soil}}{\text{Weight of oven dry soil}} \times 100$$

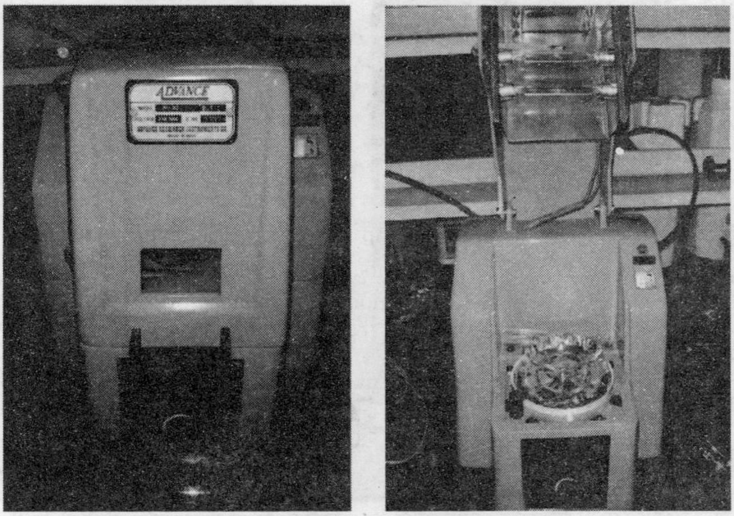

cost; Minimum
ation with which
curacy is subject
nd it is laborious

rated by adding
il sample at field
porating the soil
c matter content.

250 watt infrared
ation emitted by
irectly calibrated
ut 5 minutes.

**Fig. 24:** Infrared moisture meter

## Lysimeter method

By isolating the crop root zone from its environment and controlling the processes that are difficult to measure, the different terms in the soil water balance equation can be determined with greater accuracy. This is done in lysimeters where the crop grows in isolated tanks filled with either disturbed or undisturbed soil.

A requirement of lysimeters is that the vegetation both inside and immediately outside of the lysimeter be perfectly matched such that same height and leaf area index (figure 25). Lysimeters provide the direct estimates of water balance components viz., E, T, ET, drainage, runoff, effective rainfall etc in a cropped field against which other methods can be tested and calibrated. As lysimeters are difficult and expensive to construct and as their operation and maintenance require special care, their use is limited to specific research purposes

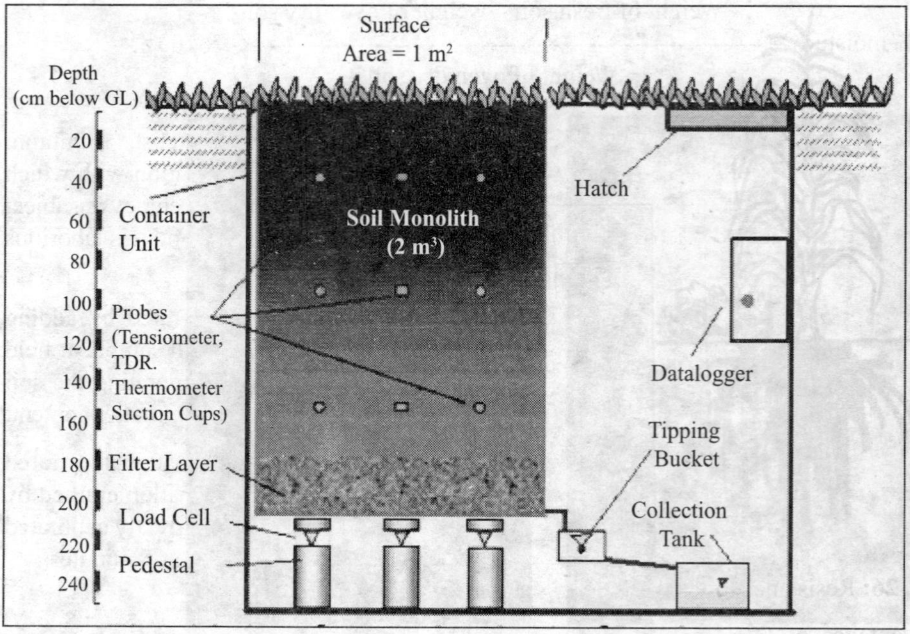


**Fig. 25:** Lysimeter

## Electric resistance method or electrical conductance method

In this method two platinum electrodes are embedded in gypsum block at a definite distance apart (figure 26). These blocks are buried in the soil at a required depth and conductivity is measured with a modified Wheatstone bridge. The electrical resistance of the soil decreases with increase in water content. The soil water content is obtained with the help of a calibration curve, for the same block of electrical resistance against known soil water content.

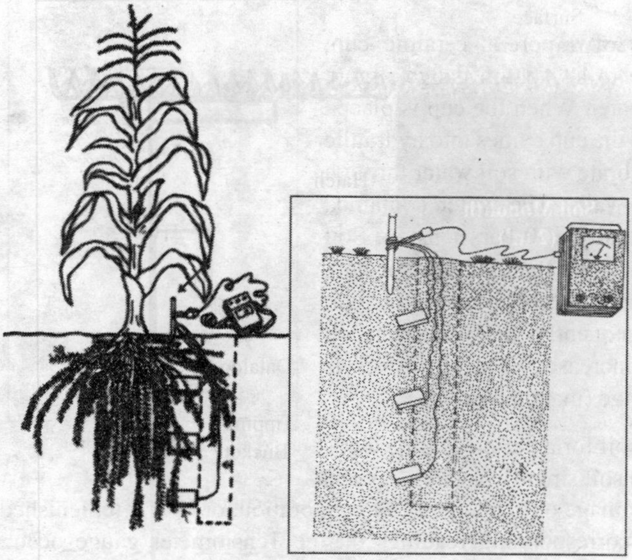

**Fig. 26:** Resistance blocks

Advantages of resistance blocks method: This method is useful to measure the percentage of soil moisture from field capacity to wilting percentage. Constraints of resistance blocks method: This method is not an appropriate method in soils containing high salt concentration.

**Neutron scattering method** : The neutron moisture meter consists of Americium and Beryllium or Radium and Beryllium as the neutron source and Boron tri fluoride (BF3) gas as a detector (Figure 27). The fast moving neutrons emitted in to the soil collide with hydrogen atoms of water molecules, lose their energy and slowed down. The slow neutrons are caught by $BF_3$ gas and counted with the help of electronic microprocessor. The more slowed down neutrons return to the probe the greater is the water content of soil.

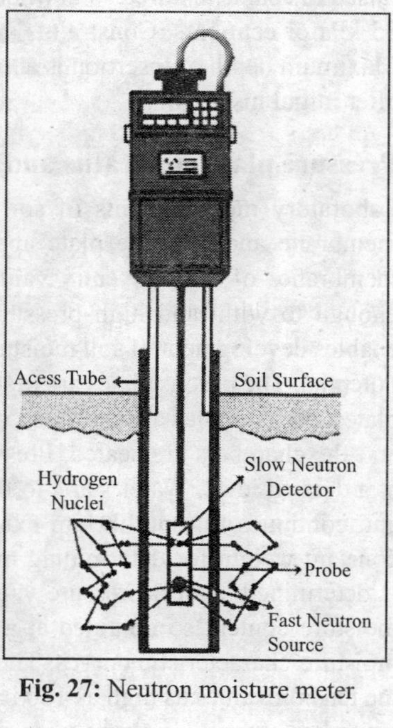


**Fig. 27:** Neutron moisture meter

Advantages of Neutron moisture meter method: This method is rapid, less laborious, non-destructive and repeated measurements can be made at the same depth. Constraints: This instrument has to be calibrated for each soil. This method is not suitable for moisture determination of surface soil

## Tensiometer method

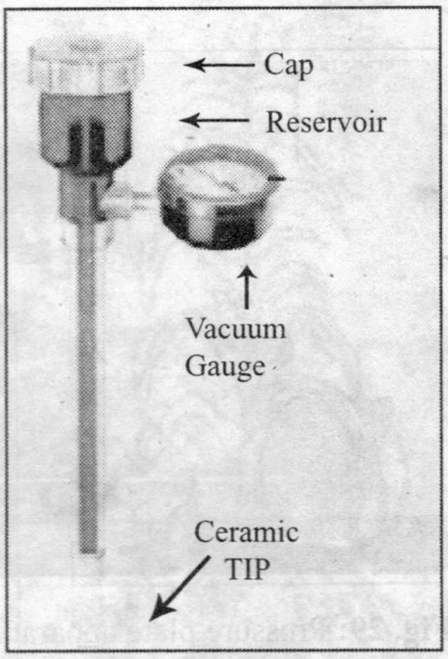


**Fig. 28:** Tensiometer

The tensiometer consists of a porous ceramic cup, connected through a tube to a vacuum gauge (figure 28), all parts filled with water. When the cup is placed in the soil, the water inside the cup comes into hydraulic contact and tends to equilibrate with soil water through the pores in the ceramic walls. The water contained in the tensiometer is generally at 0 bars tension. Soil water, being generally at more than '0' bar tension, hence certain amount of water move from the air tight tensiometer to soil. Consequently, the water tension inside the tensiometer increases. The thension is indicated by a vacuum gauge (manometer).

A Tensiometer left in the soil for a period of time tends to track the changes in the soil's matric suction. As soil moisture is depleted by drainage or plant uptake or evaporation, or as it is replenished by rainfall or irrigation, corresponding readings on the Tensiometer gauge occur. Suction measurements by tensiometry are generally limited to matric suction values below 0.85 bars. Hence under field conditions the sensitivity of tensiometer is about 0.85 bars or 85 kPa.

Advantages of Tensio meter method are repeated measurements at the same location can be taken, non-destructive method and Suitable for scheduling irrigations to crops raised in coarse textured soils where majority of ASM is between 0 – 0.85 bars or 0 to 85 kPa or centibars. Constraints are measurements limited to 0.85 bars suction only Maximum depth of insertion is about 50 cm only, requires few hours for equilibration after initial installation

## Pressure plate apparatus and Pressure membrane apparatus method

Laboratory measurements of soil water potential are usually made with pressure membrane and pressure plate apparatus. It consists of ceramic pressure plates or membranes of high air entry values contained in airtight metallic chambers strong enough to withstand high pressure of 15 bars or more (figure 29). The apparatus enables development of soil moisture characteristic curves over a wide range of matric potential. The porous plates are first saturated and then soil samples are placed on these plates. Soil samples are saturated with water and transferred to the metallic chambers and the chambers are sealed. Pressure is applied from a compressor and maintained at a desired level. Water starts to flow out from saturated soil samples through outlet and continues till equilibrium exists. Soil samples are taken out and oven dried to constant weight for determining moisture content on weight basis. Moisture content is determined against pressure values varying from −0.1 to −15 bars. The values of moisture content so obtained at a given applied pressure are used to construct soil moisture characteristic curves. The Pressure membrane apparatus is used to measure the tension values as high as 100 bars. Helpful to measure energy moisture relation of a number of samples over a wide energy ranges in a short time.

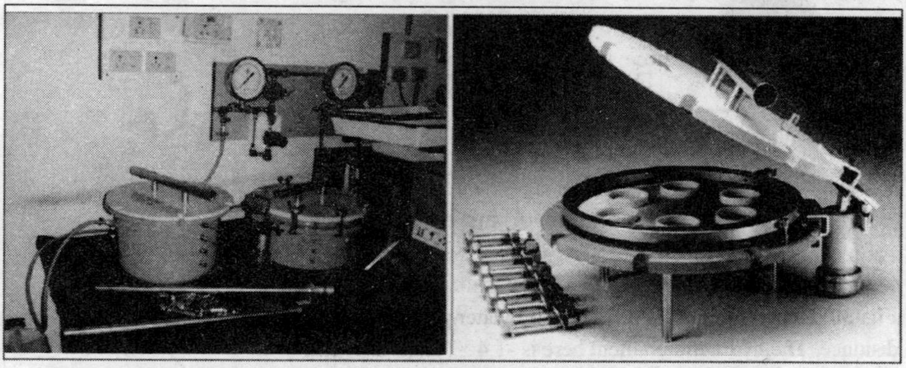

**Fig. 29:** Pressure plate apparatus

**Psychrometer method :** The total soil water potential is determined with the help of soil psychrometer. As the liquid water in soil is always in equilibrium with its vapour in the ambient air, the energy status of water vapour (vapor potential) must be equivalent to the energy status of soil water. At constant temperature and gravitational potential, the vapour potential is largely contributed by matric and osmotic potentials. The vapour potential of soil can be precisely estimated with the help of a thermocouple psychrometer.

## 3.9.9. Movement of water in soil

Water is a highly dynamic component in soil system. It moves in all three phases: solid, liquid and vapour. In saturated or flooded situation, water moves in liquid phase, while in a partially dry or unsaturated soil, it moves in both liquid and vapour phases. The flow of liquid water is due to a gradient in matric potential from one soil zone to another. The direction of flow is from a zone of higher matric potential to one of lower matric moisture potential. Saturated flow takes place when the soil pores are completely filled with water. Unsaturated flow occurs when the pores are only partially filled with water. In each case, moisture flow is due to energy – relationships with in soil.

## Concept of water movement in soil

Water movement through soil is proportional to the product of the driving force and the conductivity of soil for water.

Q = c DK

Q = Flow velocity ; c = Proportional factor; D = Driving force; K = Conductivity of the medium.

The driving force (D) in case of water is a pressure gradient. Water moves from a position of high pressure to a position of low pressure in both saturated and unsaturated conditions, including in vapour state. The permeability of soil for water vapour is proportional to the volume of water free pore space, regardless of size of pores. Free path plays significant role rather than size of pores.

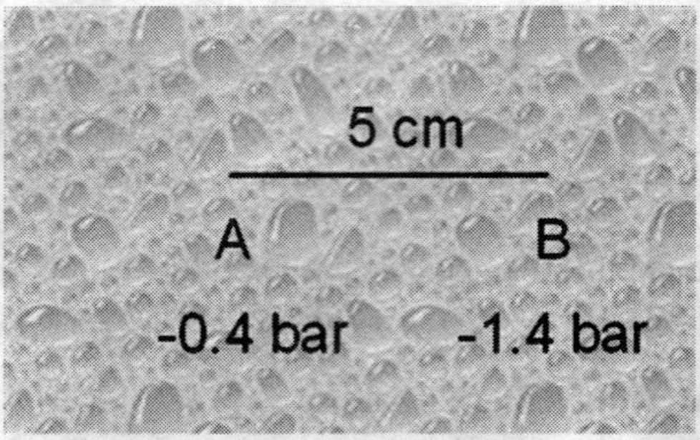


**Fig. 30:** Representation of different soil water potential at two adjacent places in soil at certain distance. The hydraulic gradient here is -1.4 – (-0.4)/ 5 = -0.2.

**Soil water movement under saturated conditions**: After heavy rain or irrigation application, pores of soil are often filled entirely with water. This is said to be the saturated condition of the soil. The flow of water under saturated condition is determined by hydraulic force (hydraulic gradient) driving the water through the soil (commonly gravity) and the hydraulic conductivity, or the ease with which the soil pores permit water movement.

V = Kf

V = Volume of water move per unit time; f = water moving force (driving force : ΔH/L); K = Hydraulic conductivity.

The driving force, known as hydraulic gradient, is the difference in height of water above and below the soil column. Besides, downward flow of water in saturated condition, hydraulic force also will cause horizontal and upward flow. This flow is usually not quite rapid, because the force of gravity does not assist horizontal flow and hinders upward flow. Downward flow is more in coarse textured soils, whereas horizontal move is equally significant in fine textured soils.

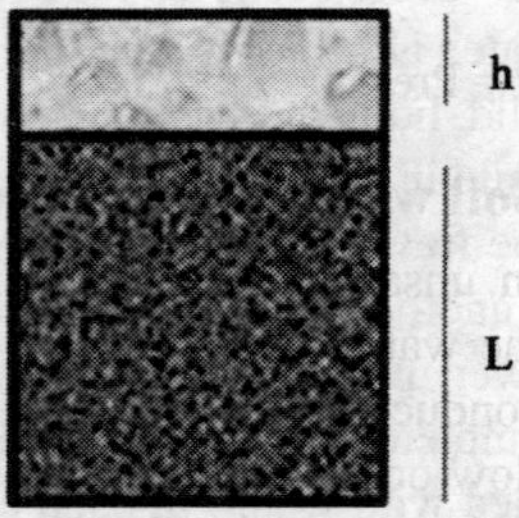


**Fig. 31:** Representation of hydraulic head and pressure head in vertical movement

## Poiseuille expression of saturated water flow

According to Poiseuille's law, the rate of flow of a liquid through a narrow tube is proportional to the fourth power of the radius of the tube and to the pressure; and inversely proportional to the viscosity of the liquid and the length of the tube.

$$Q = \frac{PR^4}{8\eta L}$$

Where

Q = Volume of flow cc/sec;

P = Pressure difference dynes/cm$^2$

R = Radius of the tube, cm;

L = Length of tube, cm

$\eta$ = Viscosity of liquid dynes – sec/cm$^2$ (poises)

The rate of saturated flow in soils of various texture is in the sequence : Sand > Loam > Clay. Viscosity of water increases more than one percent with each $^0$C drop in the temperature.

## Darcy's theorem of saturated water movement in soil

The law of "Darcy" states that the velocity of flow of a liquid through a porous medium is proportional to the force causing the flow and to the hydraulic conductivity of the medium

$$Q = c\frac{KAP}{L}$$

Where

Q = Flow velocity (cm$^3$/sec)

c = Dimensionless proportionality constant;

K = Hydraulic conductivity (cm$^3$/hr/cm$^2$

A =Cross sectional area (cm$^2$) of the porous medium;

P = Pressure gradient = $\Delta$H/L.

## Soil water movement under unsaturated conditions

In unsaturated condition, only micro pores would accommodate water and allow the water to move slowly. The relationship between matric potential and hydraulic conductivity governs the water movement under unsaturated condition. Saturation flow occurs at or near zero potential, while most of the unsaturated flow occurs at a potential of less than -0.1 bar

Under saturated condition (high potential levels) hydraulic conductivity is higher in sand than in clay. Under unsaturated condition (less potential values) hydraulic conductivity is higher in clay than in sand. In both saturated and unsaturated conditions, flow is majorily governed by hydraulic conductivity of the soil and the diving force. The diving force in unsaturated condition is primarily the matric potential gradient or the difference in matric potential of the moist soil. Movement will be from a zone of thick moisture films (higher matric potential e.g. -0.001 MPa) to one of the thin films (lower matric potential e.g. -0.1MPa)

**Vapour movement in soils :** Two types of water vapour movement occur in soils, internal and external. Internal movement takes place within the soil i.e., in soil pores. External movement occurs at the land surface and the water vapour is lost by evaporation. Water vapour moves from one point to another within the soil in response to difference in vapour pressure. Thus, water vapor moves from high vapour pressure (under saturated condition where in air is nearly 100% saturated with water vapour) zone to lower vapor pressure zone (dry soil).

As the quantity of water vapor is small, its movement in soil is not of much practical significance. However, in drought situation, its movement may be of considerable significance especially in supplying moisture to drought resistant desert plants.

Electrolytes in soil as well as the applied fertilizers will reduce the vapour pressure, locally, thereby creates the pressure gradient, which makes water vapour to move towards the pockets of high salt concentration and fertilizers. This is useful for making nutrients available to plants.

## 3.9.10. Infiltration, percolation and permeability

**Infiltration** : Infiltration is the entry of water at the soil air interface due to sorption and vertical flow of water though the soil profile. This process is of great practical significance, as it determines how much of rain storm or irrigation water enters the soil and how much over flows the land surface as run-off. Irrespective of soil texture, the infiltration rate in a dry soil would be high, initially and the infiltration rate reduces exponentially with time and attains a steady rate after a long lapse of time. It is expressed as millimeters per hour. Infiltrate rate is high and constant in nonswelling clay soil i.e., laterite soils than swelling clay soils i.e. black cotton soils.

**Percolation**: The down ward movement of water through soil is percolation. Percolation occurs when the water is under pressure or when the tension is smaller than about ½ atmosphere or when the hydraulic gradient is 1.0 or less. Importance of Percolation in soil development and land management: i) Percolation removes some valuable nutrients away from soil i.e. nitrates and calcium. ii) In coarser textured soils, which are porous in nature (macropores) exhibits greater percolation capacity and iii) As evaporation and transpiration use much of water that enters the soil, the amount of water that percolates through it decreases with depth. The greatest amount of percolation goes on in the top few centimeters.

**Permeability** : The ease with which water pass through a bulk mass of soil or a layer of soil is the permeability of the soil. The permeability of different soils are in the order of *Course soils > medium soils > Fine soils*

**3.9.11. Soil Drainage:** Drainage means removal of excess water from surface or sub-surface of soil body by means of some water conveying devices. Excess water may be due to i) over irrigation, ii) accumulation of monsoon runoff in low lying areas for want of an outlet and iii) seepage from reservoirs, canals and ditches.

The drainage problem are two types viz., firstly surface drainage problem : Surface drainage problems arise in the flat areas of land that are subjected to ponded water. Uneven land, low capacity of disposal channels, above ground level water bodies are the chief causes for ponding of water in an area. Secondly, sub-surface drainage problems: The sub-surface drainage problems may arise due to a)low soil permeability, b)humid climate and c) rise of ground water causing development of a shallow water table.

## Importance of Drainage

1) Drainage is essential, as the crop growth will be drastically affected by continuous soil saturation with water.
2) Soil saturation may encourage certain diseases and parasites to crop plants.
3) Poor drainage leads to the development of salt affected soils.
4) High water table may limit root penetration.
5) Soil saturation stops the gaseous exchange causing oxygen deficiency and accumulation of $CO_2$ to toxic levels in root zone.
6) Gowing season for winter crops is shortened in poorly drained soils, since wet soil requires more heat to warm up than dry soil, due to high specific heat of water.
7) Under poor drainage conditions many organic and inorganic compounds will be reduced to toxic levels and inhibit crop growth.
8) De-nitrification loss of N occurs in anaerobic conditions.
9) Some micronutrients will become unavailable.

During rain or irrigation, the fields become wet. The water infiltrates into the soil and is stored in its pores. When all the pores are filled with water, the soil is said to be saturated and no more water can be absorbed; when rain or irrigation continues, pools may form on the soil surface (Fig.32 ).

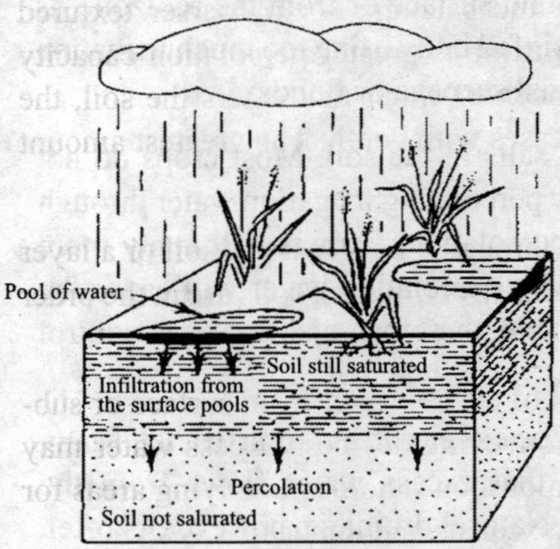


**Fig. 32:** During heavy rainfall the upper soil layers become saturated and pools may form. Water percolates to deeper layers and infiltrates from the pools.

Part of the water present in the saturated upper soil layers flows downward into deeper layers and is replaced by water infiltrating from the surface pools. When there is no more water left on the soil surface, the downward flow continues for a while and air re-enters in the pores of the soil. This soil is not saturated anymore. However, saturation may have lasted too long for the plants' health. Plant roots require air as well as water and most plants cannot withstand saturated soil for long periods (rice is an exception). Besides damage to the crop, a very wet soil makes the use of machinery difficult, if not impossible.

The water flowing from the saturated soil downward to deeper layers, feeds the groundwater reservoir. As a result, the groundwater level (often called groundwater table or simply water table) rises. Following heavy rainfall or continuous over-irrigation, the groundwater table may even reach and saturate part of the rootzone (see Fig. 33). Again, if this situation lasts too long, the plants may suffer. Measures to control the rise of the water table are thus necessary.

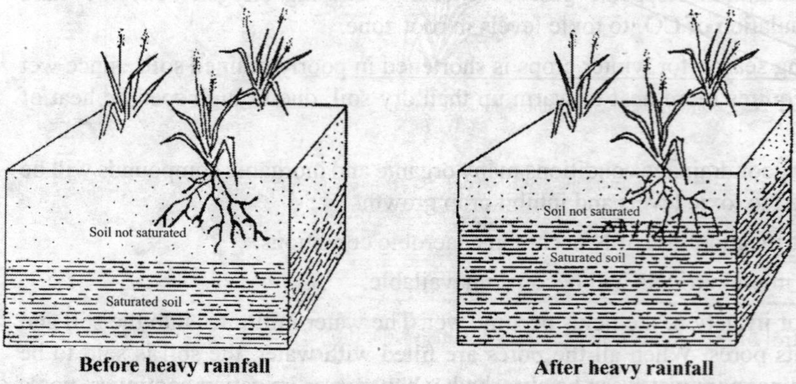


**Fig. 33:** After heavy rainfall the groundwater table may rise and reach the rootzone

The removal of excess water either from the ground surface or from the rootzone, is called drainage. Excess water may be caused by rainfall or by using too much irrigation water, but may also have other origins such as canal seepage or floods.

In very dry areas there is often accumulation of salts in the soil. Most crops do not grow well on salty soil. Salts can be washed out by percolating irrigation water through the rootzone of the crops. To achieve sufficient percolation, farmers will apply more water to the field than the crops need. But the salty percolation water will cause the water table to rise. Drainage to control the water table, therefore, also serves to control the salinity of the soil.

**Different types of soil drainage :** Drainage can be either natural or artificial. Many areas have some natural drainage; this means that excess water flows from the farmers' fields to swamps or to lakes and rivers. Natural drainage, however, is often inadequate and artificial or man-made drainage is required. There are two types of artificial drainage: surface drainage and subsurface drainage.

**Surface drainage:** Surface drainage is the removal of excess water from the surface of the land. This is normally accomplished by shallow ditches, also called open drains. The shallow ditches discharge into larger and deeper collector drains. In order to facilitate the flow of excess water toward the drains, the field is given an artificial slope by means of land grading.

**Subsurface drainage:** Subsurface drainage is the removal of water from the rootzone. It is accomplished by deep open drains or buried pipe drains.

**i. Deep open drains:** The excess water from the rootzone flows into the open drains (see Fig. 34). The disadvantage of this type of subsurface drainage is that it makes the use of machinery difficult.

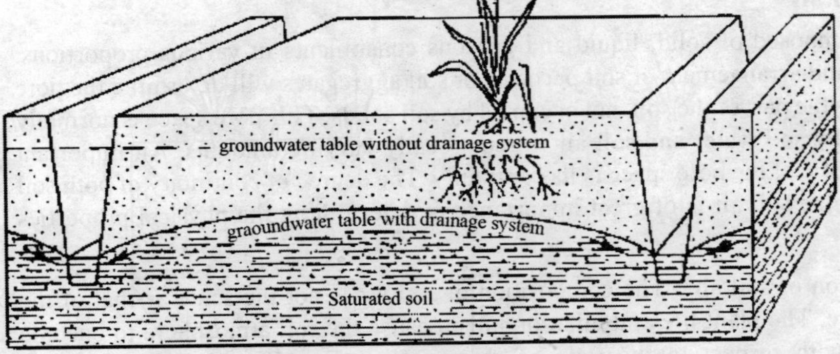


**Fig. 34:** Control of the groundwater table by means of deep open drains

**ii. Pipe drains:** Pipe drains are buried pipes with openings through which the soil water can enter. The pipes convey the water to a collector drain (see Fig. 35).

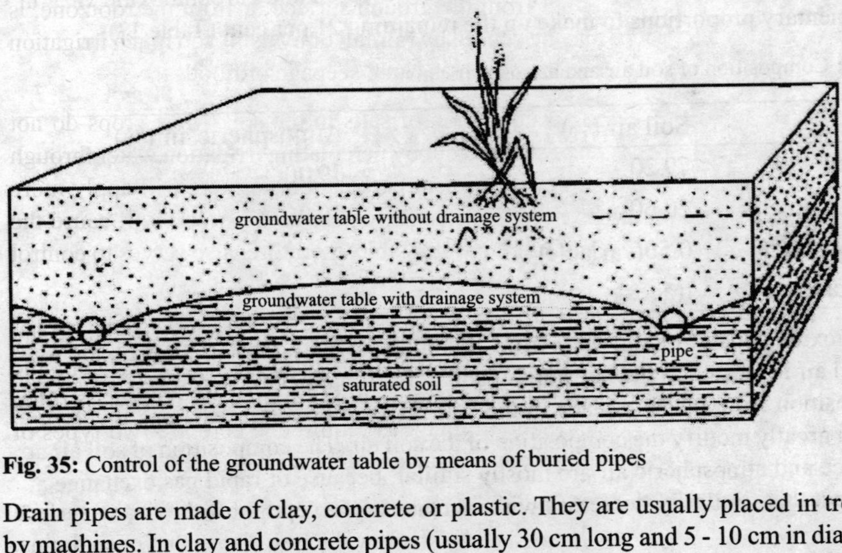


**Fig. 35:** Control of the groundwater table by means of buried pipes

Drain pipes are made of clay, concrete or plastic. They are usually placed in trenches by machines. In clay and concrete pipes (usually 30 cm long and 5 - 10 cm in diameter) drainage water enters the pipes through the joint. Flexible plastic drains are much longer (up to 200 m) and the water enters through perforations distributed over the entire length of the pipe.

Open drains use land that otherwise could be used for crops. They restrict the use of machines. They also require a large number of bridges and culverts for road crossings and access to the fields. Open drains require frequent maintenance (weed control, repairs, etc.). In contrast to open drains, buried pipes cause no loss of cultivable land and maintenance requirements are very limited. The installation costs, however, of pipe drains may be higher due to the materials, the equipment and the skilled manpower involved.

## 3.10. Soil air

Soil is composed of solid, liquid and gaseous constituents in various proportions. The size and arrangement of soil particles or soil aggregates will determine the pore space. It is the part of the soil, not occupied by soil solids. This pore space is normally occupied by soil water and soil air in reciprocally varying amounts. An important characteristic of the pore space is its continuity. The degree of continuity of both soil water and the soil air is of great importance in determining the physical properties of soil.

Composition of soil air : The composition of soil air is not the same as that of the atmosphere. The plant life and microorganism cause the soil atmosphere to become dynamic with respect to the ratio of oxygen to carbon-di-oxide. The principal components of soil air are nitrogen, oxygen, inert gases, carbondioxide, water vapour and hydrogen. Methane, hydrogen sulfide etc., are present in negligible quantities. The proportion of nitrogen plus argon is usually found to be equal to that in the atmosphere and is 79 per cent by volume. Oxygen and carbondioxide vary in complementary proportions to make up the remaining 21 per cent (Table 13).

**Table 13:** Composition of soil air and atmospheric air

| Gases | Soil air (%) | Atmospheric air (%) |
|---|---|---|
| Nitrogen | 79.20 | 79.00 |
| Oxygen | 20.60 | 20.97 |
| Carbondioxide | 0.50(variable) | 0.033 |
| Other gases | traces | traces |

Carbondioxide content of soil vary from 10-10000 times to that of atmospheric air. Since soil air is located in the net work of soil pores separated by soil solids, it varies in composition from place to place in the soil. In local pockets, reactions involving gases can greatly modify the composition of the soil air. The composition of soil air at the surface and atmospheric air are mostly similar, because of rapid gas exchange at surface, but this similarity decreases with increase in the depth of soil.

The composition of air is the resultant of two sets of processes that function simultaneously viz : chemical and biological reactions that result in the production of $CO_2$ at the use of $O_2$; a nd physical processes that affect the interchange of gaseous constituents between the soil voids and the atmosphere. First process accentuates the differences, whereas the second one minimizes the differences.

### 3.10.1. Gaseous exchange in soil

The exchange of gases in between soil air and atmospheric air is referred to as soil aeration. This is very important for growth of plants and soil microorganisms and thin biological activities. Two mechanisms have been identified in the interchange of gases between soil voids and the atmosphere i.e. mass flow and diffusive flow.

**Mass flow** : Mass flow of gases in to and out of the soil occurs whenever there is a difference in total pressure between the soil air and atmosphere. This difference is mainly due to expansion and contraction of soil gases due to changes in temperature and barometric pressure, air replacement through rainfall, irrigation and drainage.

**Diffusion** : Diffusion is the random movement of molecules (molecular motion) of a gas or of a liquid. Net movement of gases by diffusion occurs when the partial pressures of individual gases in two neighbouring systems are different, but the total pressure is the same in both. Diffusion accounts for >90% of gaseous exchange.

In soils, due to biological activity, there is a continuous production of $CO_2$ at the exhaustion of oxygen. This results in the increased partial pressure of $CO_2$ and reduced partial pressure of oxygen. It leads to continuous interchange of $CO_2$ and $O_2$ in between soil and atmosphere.

According to Fick's law, diffusion is a function of the concentration gradient, the diffusion co-efficient of the medium, and the cross sectional area.

$$dQ = DA (dc/dx)dt$$

where dQ is the mass flow (moles) during the time at across area A (sq.cm), dc/dx the concentration gradient [moles/cc (cm) ], and D the proportionally constant or diffusion coefficient (sq.cm/sec).

### 3.10.2. Effect of soil air on plant growth, soil properties and nutrient availability

Soil air modify soil properties and thereby governs the nutrient and water availability and ultimately the plant growth.

1) Under poor aerated conditions

   a) Development of plant roots will be restricted or inhibited.

   b) Absorption of water and nutrients will be decreased.

   c) Toxic substances will be formed.

   d) Many nutrients like Fe and Mn will be at their toxic levels.

e) Many toxic organic acids like lactic, butyric and citric acid etc., will be found under anaerobic conditions and cause injury to plant roots thereby minimize the ability of roots to absorb nutrients and water

f) Zn and Mn availability will be reduced

2) Soil aeration influences the activity of soil micro organism.

3) Soil aeration affects the rate of decomposition of organic matter.

4) It influences BNF and many nutrient transformations.

5) under poor aeration conditions.

6) Incidence and virulence of diseases controlled by soil aeration.

### 3.10.3. Characterization of soil aeration

There are various parameters that can be used for characterizing soil aeration.

1) The volume percentage of soil air or air capacity: Indicates the part of pore space filled with air. Minimum of 10-12 % is required.

2) Gaseous composition: The principal components of soil air are nitrogen, oxygen, inert gases, carbondioxide, water vapour and hydrogen. Methane, hydrogen sulfide etc., are present in negligible quantities. The proportion of nitrogen plus argon is usually found to be equal to that in the atmosphere and is 79 per cent by volume. Oxygen and carbon dioxide vary in complementary proportions to make up the remaining 21 per cent.

3) Oxidation reduction potential: Indicates whether the soil is in oxidized or reduced state. It is a measure of tendency of a system to reduce or oxidize expressed as negative and positive potentials, respectively.

4) Composition of soil for its reduced components.

5) The oxygen diffusion rate: Determines the rate at which oxygen can be replenished if it is used by respiring plant roots or microbes.

**Measurement of ODR by platinum electrode method:** The ODR meter consists of 1. platinum microelectrode, 2. reference electrode, 3. galvanometer/voltmeter, 4.the ammeter, 5. variable resistor and 6. battery (Fig. 36).

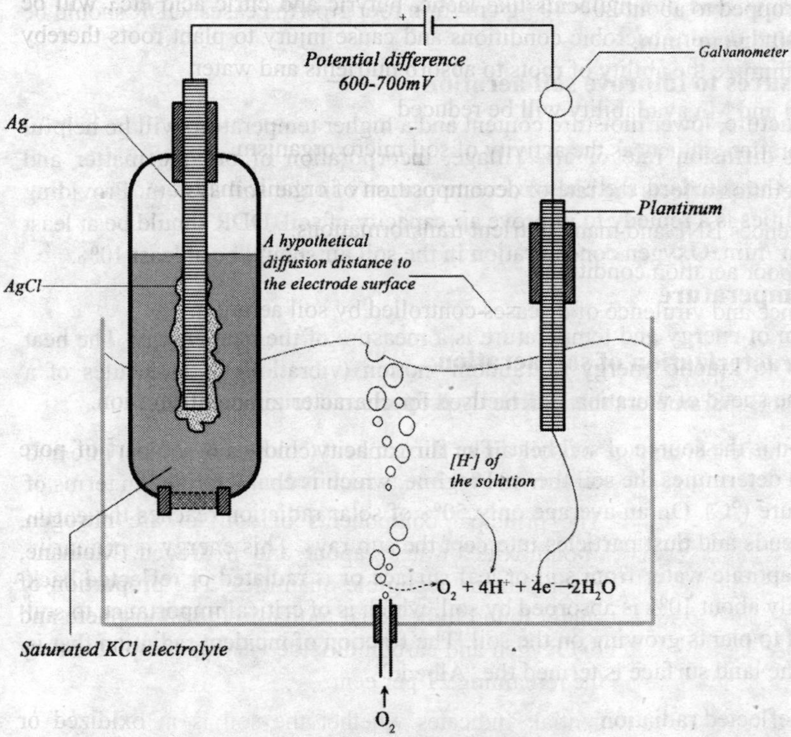


**Fig. 36:** ODR meter

When a certain electrical potential is applied between a reference electrode and a platinum electrode in the soil, oxygen gets reduced at a platinum surface resulting in a current flowing in between the electrodes. This current flow is proportional to the rate of oxygen reduction, that is, rate of oxygen diffusion in to soil. Each molecule of oxygen which diffuses to the surface of the electrode takes up four electrons and reacts with hydrogen ions to form water in an acid solution, or, it reacts with water to form a hydroxyl ion in an alkaline solution.

In acid soils:

$O_2 + 2H^+ + 2e^- = H_2O_2$

$H_2O_2 + 2H^+ + 2e^- = 2H_2O$

$O_2 + 4H + 4e^- = 2H_2O$

In neutral or alkaline soils:

$O_2 + 2H_2O + 2e^- = H_2O_2 + 2OH$

$H_2O_2 + 2H^+ + 2e^- = 2H_2O$

$O_2 + 2H_2O + 4e^- = 4OH^-$

If the ODR dropped to about $20 \times 10^{-8}$ g/cm²/min root growth ceases. ODR should be at least $30 \times 10^{-8}$ g/cm²/min.

### 3.10.4. Measures to improve soil aeration

Open soil structure, lower moisture content and a higher temperature will be helpful in increasing diffusion rate of air. Tillage, incorporation of organic matter and mulching are the standard methods to create porosity in the soil system. Providing drainage facilities is a remedy to improve air capacity of soil. ODR should be at least $30 \times 10^{-8}$ g/cm²/min. Oxygen concentration in the soil air should be at least 10%.

## 3.11. Soil temperature

Heat is a form of energy and temperature is a measure of the heat energy. The heat energy refers to kinetic energy or random motion (vibration) of molecules of a substance. The speed of vibration is directly proportional to temperature.

Solar radiation is the source of soil heat. The flux of heat (colories or joules) into and out of the soil determines the soil thermal regime, which is characterized in terms of soil temperature (°C). On an average only 50% of solar radiation reaches the earth, because of clouds and dust particles intercept the sun rays. This energy is primarily utilized to evaporate water from soil or leaf surface or is radiated or reflected back to the sky. Only about 10% is absorbed by soil, which is of critical importance to soil processes and to plants growing on the soil. The fraction of incident radiation that is reflected by the land surface is termed the "Albedo".

$$\text{Albedo} = \frac{\text{Reflected radiation}}{\text{Incidental radiation}}$$

Albedo ranges from 0.1 to 0.2 for dark coloured, rough soil surfaces to as high as 0.5 or more for smooth, light coloured surfaces.

### 3.11.1. Thermal properties of soil

**Specific heat**: It the amount of heat required to rai se the temperature of one gram substance to one degree high. Units of specific heat are Cal per gram per degree centigrade.

$$\text{Specific heat } (c) = \frac{\Delta h}{\Delta T}$$

ΔQh is quantity of heat supplied to the body, ΔT is increase in temperature of the body and **c** is specific heat

**Heat capacity or Thermal capacity** (C'): The "heat capacity" of a soil is defined as the ratio of heat supplied to a body to the corresponding rise in its temperature. In other words amount of heat required to raise the temperature of one cubic centimeter substance to one degree high. Units for heat capacity are Cal per cubic centimeter per degree centigrade.

Heat capacity (c') = specific heat × density

$C' = C \times \rho$

The heat capacity is expressed as quantity of heat required to raise the temperature of unit volume of soil by 1°C and is known as ***"Volumetric heat capacity" or simply the "heat capacity"*** (gram calorie)

Units for Cv : $J/m^3/°C$ or $Cal/cm^3/°C$

Volumetric heat capacity = Specific heat × Density of soil

**Thermal conductivity**

Thermal conductivity (λ) is defined as the quantity of heat passing in a unit time through a unit area of soil under a unit temperature difference between the faces. λ is expressed as Cal $cm^{-1}sec^{-1}°C^{-1}$ or Joules / meter / second / °C

Heat flu $qh = -\lambda \frac{\Delta T}{\Delta x}$

$\lambda = -qh \frac{\Delta x}{\Delta T}$ Therefore, thermal conductivity (λ)

qh = heat flux or quantity of heat flowing through a unit area in a unit time Cal/cm2/ sec

λ = thermal conductivity; Cal $cm^{-1}sec^{-1}°C^{-1}$

ΔT/ Δx = temperature gradient °C/ cm

Actual heat conductance depends on 1. thermal conductivity and 2. on the heat gradient. Thermal conductivity of a soil depends on its 1.water content, 2.texture, 3.structure, 4.mineralogical composition, 5.organic matter content and 6.compaction. Thermal conductivity of water = $1.42 \times 10^{-3}$ Cal $cm^{-1}sec^{-1}°C^{-1}$, Air = $0.062 \times 10^{-3}$, Quartz = $20.0 \times 10^{-3}$, Quartz: water : air = 333 : 23 :1

Thermal conductivity increases with increased water content, till about 50% of the soil saturation. TC is observed to decrease with a reduction in particle size. Thermal conductivity of soil varies in the order : Sand > loam > clay > peat.

**Fouriers's Law:** Fouriers's law states that the negative gradient of temperature and the time rate of heat transfer is proportional to the area at right angles of that gradient through which the heat flows. Fourier's law is the other name of the law of heat conduction. Newton's law of cooling and Ohm's law are a discrete and electrical analog of Fourier's law.

$$\frac{Q}{\Delta t} = -kA\frac{\Delta T}{\Delta x}$$

Where,

- A is the area of the cross-sectional surface
- $\Delta T$ is the temperature difference between the endpoints
- $\Delta x$ is the distance between two ends
- $Q/\Delta t$ is heat flux

**Thermal Diffusivity ($m^2/s$): It** is defined as the ratio of thermal conductivity to heat capacity .

Thermal diffusivity =

$$Thermal\ Diffusivity\ (\alpha)\frac{\lambda}{c'} = \frac{Thermal\ conductivity}{Volumetric\ heat\ capacity}$$

$$Thermal\ Diffusivity\ (\alpha) = \frac{Thermal\ conductivity}{specificheat \times density\ of\ soil}$$

It is the measure of the rate at which changes in temperature occur in the body. Units for thermal diffusivity are $m^2/s$

**Thermal retentivity:** The ability of the body to retain heat is thermal retentivity. It is equal to the reciprocal of thermal diffusivity.

$$\text{Thermal Retentivity} = \frac{1}{\alpha} = \frac{1}{\text{Thermal Diffusivity}}$$

$$\text{Thermal Retantivity} = \frac{\text{specificheat} \times \text{densityofsoil}}{\text{Thermal conducitivity}}$$

### 3.11.2. Heat Transfer in Soils

The major physical processes responsible for the transport of heat are conduction, convection and radiation.

a. **Conduction**: It may be defined as " the flow of heat through an unequally heated body from the places of high temperature to lower temperature". It is through the momentum of molecules. The molecules of the body nearer to the source absorb heat and increase their kinetic energy and begin to vibrate about their positions faster than before. They collide with other molecules and in this process they share their energies with the adjacent molecules so that other molecules increase their K.E and behave similarly to transfer heat energy from one end to another.

b. **Convection** : May be defined as the motion of the hot body itself carrying its heat with it. In this process absorption of heat causes increase in volume there by decrease in density of molecules. Such a decrease in density make the hotter molecules to go up, giving their places to the cooler molecules which also behave like that after absorbing heat energy. This process occurs only in fluids.

c. **Radiation**: Radiation may be defined as the process by which heat is transmitted from one place to other with out aid of intervening medium. In this process heat is transferred through the space in the form of electromagnetic waves.

### 3.11.3. Management of soil temperature

The soil temperature depends upon the heat flux in to the soil and the heat transfer processes occurring in the soil and in between the soil and atmosphere, which in turn, depend upon the thermal characteristics of the soil solids, gases and water. The soil thermal regimes greatly modify the microclimate of the area and exercise a major influence on growth and development of plants, particularly during germination and early seedling development stages. Soil temperature can be modified by a) soil temperature regulating energy balance on the soil surface b) Modification of soil thermal properties and c) heating the soil through artificial means.

#### a) Soil temperature regulation by energy balance on the soil surface

The source of heat for soil is the solar radiation. The rate at which the radiant energy reaches the earth's atmosphere from the sun is called the solar constant, which has the value of about 2 cal cm -2 min-1.

The solar radiation, which reaches the soil surface, is partly reflected, partly used for heating soil, partly used as latent heat for evaporation and partly reradiated as long wave radiation. About 90% of long wave radiation is absorbed by the water vapours present in the earth's atmosphere, which partly reradiate to the earth's surface. It implies that the energy available for heating the soil is function of the incoming and outgoing radiation. If a greater fraction of incoming radiation is absorbed by the soil than the fraction reflected or reradiated the soil will be heated. The energy balance on the soil surface and therefore be modified through tillage and shaping of fields, mulching and vegetation; and shading and row spacing.

Tillage and shaping of fields : Since both slope and direction affect soil temperature and heat flux, soil thermal regimes can be modified by specific tillage practices such as ridges and shaping. Comparatively high temperature is recorded on ridges as compared to furrows. Besides, the slope facing west would be comparatively warmer. Soil temperature is highest in east-west oriented ridges and decreases for other ridge directions in the order. North – south ridges > East – west ridges > N-S conventional > Listing

Mulching and vegetation: From the energy point of view, mulching is the application or creation of a soil cover that constitutes a barrier to the transfer of heat or water vapour, where as the vegetation intercepts a considerable part of the incoming radiation. Vegetation and mulching affect the soil thermal regimes by 1) interception of the incoming radiation 2) changes in albedo conditions and 3) reduction in latent heat transfer by evaporation. The mulches commonly used include : soil, stubble, straw, weed or trash, gravel, plastic (black, transparent, opaque etc.) paper (different textures and colors) asphalt and aluminum foils. In general, light coloured mulches will increase albedo of the surface, thereby reduces soil temperature. Black plastic mulch – reduces out going temperature (heating); Paper and straw mulches – increases outing temperature (cooling); Aluminum foil mulch – sharply increases out going temperature (cooling); Transparent plastic mulch – green house effect; Opaque mulches – thermal insulators (decreases maximum temperature and increases minimum temperature).

Shading and row spacing : Artificial shading is often practiced either to reduce the intensity of the direct radiation falling on the soil surface or to conserve the soil crop canopy heat. Increased width in between rows, increases the temperature in rows and also shows lot of variation in temperature in between rows. In regard to row direction, east – west direction planted rows showed more variation in temperature between the rows.

**b) Modification of soil thermal properties** : Soil thermal properties can be modified by changing soil physical conditions through tillage practices, soil compaction, irrigation and drainage.

Soil tillage : Tillage will loosen soil, increase the soil porosity and decrease the soil thermal conductivity and heat capacity.

Soil compaction: Soil compaction increases contact between the soil particles thus help in heat conductance.

Irrigation and drainage: Since the specific heat is of water is 3 to 4 times higher than soil, irrigation to warm soil gives cooling effect. Likewise draining the soil in temperate areas will warm-up the soil more quickly.

**c) Heating the soil through artificial means:** In some intensive cropping establishments like green houses, glass houses etc to control soil temperature, the soil warming are done by electric heating coils.

**Importance of soil temperature for plant growth:** Too low or too high temperatures affect the germination of seeds. Different crops have different optimum temperature for germination. Absorption of water and nutrients is impaired under low temperatures. Temperature influences nutrient availability by affecting the weathering of minerals and decomposition of organic matter. Low soil temperature results in white succulent roots with less branching. Low temperature enhances disease incidence by parasitic fungi. Soil microbial activity and decomposition of organic matter is restricted below 10 o C and ceases below $5^0$C. Biological nutrient transformations like nitrification, ammonification etc are affected by very high or low temperatures.

## Review questions

1. Define soil texture and give classification of soil primary particles as per USDA and ISSS systems.
2. How you determine soil texture by feel method.
3. Define Stokes law and derive Stokes equation and write assumptions and limitations of stokes law.

4. How you achieve complete dispersion of soil during soil mechanical analysis.
5. Write significance of soil texture
6. Define soil structure and discuss physical, chemical and biological mechanisms of soil structure formation.
7. Give detailed account of structure classification based on structural types.
8. Differentiate between bulk density and particle density of soil.
9. Give the importance of soil porosity.
10. Discuss weight volume relations of soil.
11. What is soil consistence, explain the forces influencing soil consistence.
12. Discuss the forms of soil consistence and Atterberg's limits of of soil consistence
13. How you manage soil structure? Discuss use of different mulches in soil structure management.
14. Differentiate between soil crusting and soil compaction.
15. How you measure soil colour? write significance of soil color.
16. Explain how water retained in soil.
17. Explain potential concept of soil water and discuss the components of total soil water potential.
18. Describe soil moisture components.
19. List out different method followed for measurement of soil water and explain thermo- gravimetry and tensiometer methods.
20. Differentiate between saturate flow and unsaturated flow of water in soil.
21. Explain Darcy and Poisuille theorems of soil water flow.
22. Define drainage and explain the importance soil drainage.
23. What is soil aeration? Discuss methods to characterize soil aeration.
24. Define ODR and explain platinum electrode method of ODR measurement.
25. Discuss management of soil aeration.
26. Describe thermal properties of soil and give detailed account of soil temperature management.

## Objective questions

1. The type of soil structure common in sodic soils is

   a) Platy b) Columnar

   c) Granular d) All of these

2. Thermal conductivity is highest in

   a) Sandy soils b) Clayey soils

   c) Loam soils d) Peat soils

3. The cation that brings about deflocculating of soil colloids is
   a) Calcium  b) Sodium
   c) Magnesium  d) Aluminium
4. Free energy of saturated soil water is
   a) Positive  b) Negative
   c) Zero  d) None of these
5. The amount of heat existing in a body is called
   a) Specific heat  b) Heat capacity
   c) Thermal conductivity  d) Thermal diffusivity
6. The number of textural classes present in ISSS model of soil textural triangle are
   a) 25  b) 10
   c) 5  d) 12
7. The gram per litre type of hydrometer used in mechanical analysis is
   a) ASTM 152H  b) ASTM 152B
   c) ASTM 152M  d) ASTM156H
8. The reciprocal of thermal diffusivity is
   a) Thermal retentivity  b) Thermal capacity
   c) Thermal diffusivity  d) Thermal conductivity
9. The concentration of carbon dioxide in soil air is
   a) 0.30 %  b) 0.03%
   c) 3.0%  d) 3.5%
10. With an increase in organic matter content of the soil there will be decrease in
    a) Friability  b) Bulk density
    c) Porosity  d) Water holding capacity
11. The bluish green colour of soil particles is due to presence of
    a) Humus  b) Oxidized iron
    c) Reduced iron  d) Silica
12. According to stokes law, the sedimentation velocity is inversely proportional to
    a) Radius of the particle  b) Density of the particle
    c) Density of the liquid medium  d) Viscosity of the liquid medium
13. The common mechanism by which gaseous exchange takes place in soils is
    a) Diffusion  b) Mass flow
    c) Radiation  d) Conduction

14. In Munsell colour chart, the relative purity of colour is

a) Hue b) Value

c) Chroma d) None of these

15. The type of soil structure considered most favourable for plant growth is

a) Angular Blocky b) Columnar

c) Platy d) Crumby

16. Author of reference book ***The Nature and Properties Soil*** is

a) J. S. Kanwar b) B. P. Gildhyal

c) Nyle. C. Brady d) J. L. Sehgal

17. Author of reference book ***Soil Physics*** is

a) J. S. Kanwar b) B. P. Gildhyal

c) Nyle. C. Brady d) J. L. Sehgal

18. Plasticity index is

a) difference in water content between saturation limit and plastic limit

b) difference in water content between shrinkage limit and plastic limit

c) difference in water content between sticky limit and plastic limit

d) difference in water content between liquid limit and plastic limit

19. Water retained between field capacity and permanent wilting point is known as

a) Unavailable water b) Hygroscopic water

c) Free water d) Available water

20. The constraint associated with black clay soil is

a) High permeability b) good drainage

c) Narrow workable moisture d) Low draft power

21. The most important cementing agent in red and laterite soils is

a) Organic matter b) Calcium carbonate

c) Hydrous oxides of Fe and Al d) Clay particles

22. In pF scale the Field Capacity is designated as

a) 2.53 b) 3.23

c) 1.53 d) 2.93

23. Matric potential of soil water is due to

a) Dissolved salts b) Solute forces and over burden forces

c) Adsorptive and capillary forces d) Gravitational forces

24. The clay colloid which increases plasticity index is

a) Kaolinite b) Montmorillonite

c) Illite d) Beidellite

25. For normal plant activities Oxygen Diffusion Rate (ODR) should be at least
   a) $3 \times 10^{-8}$ g $cm^{-2}$ $min^{-1}$
   b) $0.3 \times 10^{-8}$ g $cm^{-2}$ $min^{-1}$
   c) $30 \times 10^{-8}$ g $cm^{-2}$ $min^{-1}$
   d) $300 \times 10^{-8}$ g $cm^{-2}$ $min^{-1}$
26. In neutron moisture meter the chemical used for detecting slowed down neutrons is
   a) Americium-Beryllium
   b) Radium- Beryllium
   c) Boron trifluoride
   d) None of these three
27. The rate of water flow in soils in saturated conditions is in the following order
   a) Sand > Loam > Clay
   b) Clay > Loam > Sand
   c) Loam > Clay > Sand
   d) Loam > Sand > Clay
28. For normal plant activities Oxygen concentration in the soil air should be at least
   a) 1 %
   b) 10%
   c) 15%
   d) 20%
29. The soil consistence suitable for tillage operations is
   a) Sticky
   b) Plastic
   c) Friable
   d) Hard
30. The moisture tension of adjacent points in soils measured at point A is 2.0 bar and at point B is 8.5 bar, the water moves
   a) From A to B
   b) From B to A
   c) Both directions
   d) None of these three
31. Pressure potential (Ψp) of soil water is otherwise known as
   a) Submergence Potential
   b) Peizometric potential
   c) Pneumatic potential
   d) All the three
32. One atmosphere pressure is equal to
   a) 14.7 lb/in
   b) 760 mm Hg
   c) 1020 cm water column
   d) All the three
33. Solar Constant is
   a) 2 Cal $cm^{-2}$ $min^{-1}$
   b) 12 Cal $cm^{-2}$ $min^{-1}$
   c) 20 Cal $cm^{-2}$ $min^{-1}$
   d) 0.2 Cal $cm^{-2}$ $min^{-1}$
34. Transparent plastic mulch is associated with
   a) Increase outgoing temperature
   b) Green house effect
   c) Increase outgoing moisture
   d) Radiation effect
35. The soil tillage that records high Soil temperature is
   a) N-S ridges
   b) E-W ridges
   c) N-S conventional
   d) Listing

36. Specific heat of soil minerals is

a) 1.0 cal /g
b) 0.46 cal /g
c) 0.2 cal /g
d) 0.53 cal /g

37. **Sedimentation velocity of soil particles in a liquid medium is calculated by**

a) Darcy equation
b) Stokes equation
c) Ficks equation
d) Fourier equation

38. Any soil can exhibit plasticity, if it contains clay % more than

a) 10
b) 15
c) 20
d) 25

39. The mineral magnetite is an example of the following metal element

a) Mg
b) S
c) Fe
d) Mn

40. The soil physical property which does not change with normal cultural practices is

a) Texture
b) Structure
c) Aeration
d) Temperature

41. Heavy minerals are those minerals having specific gravity more than

a) 1.53 $g/cm^3$
b) 2.83 $g/cm^3$
c) 1.26 $g/cm^3$
d) 2.96 $g/cm^3$

42. If the soil bulk density is 1.5 $g/cm^3$, then the % total porosity of the soil is

a) 50.3
b) 45.5
c) 43.4
d) 35.3

43. The concentration of $CO_2$ in soil air is

a) 3.0%
b) **0.30%**
c) 0.03%
d) 0.003%

44. About 90% of gaseous exchange between soil and atmosphere occur by the process of

a) Mass flow
b) Diffusion
c) Conduction
d) Radiation

45. The tendency of a system to reduce or oxidize expressed as negative and positive potentials respectively is measured a

a) Redox potential
b) Gravimetric potential
c) Matric potential
d) Gravitational potential

46. The percent of soil volume which is not occupied by solids is referred to as

a) Pore space
b) Void space
c) Water space
d) A or B

47. The root growth ceases when the ODR dropped to

a) 40 g $cm^{-2}min^{-1}$.
b) 30 g $cm^{-2}min^{-1}$
c) 20 g $cm^{-2}min^{-1}$
d) 10 g $cm^{-2}min^{-1}$

48. The soil physical property which does not change with normal cultural practices is

a) soil texture
b) particle density
c) Soil structure
d) A & B

49. The current flow between platinum and reference electrodes is

a) Directly proportional to the rate of oxygen reduction.
b) Inversely proportional to the rate of oxygen reduction.
c) Directly proportional to the rate of $CO_2$ reduction.
d) Inversely proportional to the rate of $CO_2$ reduction

50. In ODR meter, each molecule of oxygen which diffuses to the surface of the electrode takes up

a)1 electron and reacts with four hydrogen ions to form water in acid soils

b) 2 electrons and reacts with four hydrogen ions to form water in acid soils

c) 3 electrons and reacts with four hydrogen ions to form water in acid soils

d) 4 electrons and reacts with four hydrogen ions to form water in acid soils

51. Oxygen concentration in the soil air should be at least

a) 10%
b) 15%
c) 20%
d) 25%

52. For satisfactory plant growth the ODR should be at least

a) 10 x 10-8 g/cm²/min.
b) 20 x 10-8 g/cm²/min.
c) 30 x 10-8 g/cm²/min.
d) 40 x 10-8 g/cm²/min.

53. To improve air capacity of wet soil, the remedy is

a) Intercultivation
b) Fertilisation
c) Irrigation
d) Drainage

54. On an average, the percent of solar radiation reaches the earth, because of clouds and dust particles intercept the sun rays.

a) 20
b) 30
c) 40
d) 50

55. The fraction of incident radiation that is reflected by the land surface is termed the
   a) Reradiation  b) Difraction
   c) Absorbtion  d) Albedo
56. Soil microbial activity and decomposition of organic matter is restricted when soil temperature drops below
   a) 5⁰C  b) 10⁰C
   c) 15⁰C  d) 20⁰C
55. The most favourable structural type for plant growth is
   a) Columnar  b) Platy
   c) Crumby  d) Prismatic
56. The mineral that is most resistant to weathering is
   a) Quartz  b) Feldspars
   c) Olivines  d) Micas
57. The strength of soil crust is measured with the help of
   a) pH meter  b) Radiometer
   c) Voltmeter  d) Penetrometer
58. In mechanical analysis of soil the chemical used to destroy the calcium carbonate is
   a) Dil HCl  b) $H_2O_2$
   c) NaOH  d) Sodium citrate
59. Textural Triangle was developed by
   a) IISS  b) CRIDA
   c) ICRISAT  d) ISSS
60. The most important mechanism for renewal of soil air is
   a) Diffusion  b) Mass flow
   c) Contact exchange  d) Root interception
61. With an increase in organic matter content of the soil there will be increase in
   a) Particle density  b) Water retention
   c) Compaction  d) Bulk density
62. Worlds Soils Day is celebrated on
   a) 5th March  b) 5th June
   c) 5th October  d) 5th December
63. According to stokes law, the sedimentation velocity is directly proportional to
   a) Square of the radius of the particle
   b) Difference in the density of the particle to density of medium

c) Acceleration due to gravity
d) all the threee

64. In Munsell colour chart, the dominat spectral colour is
a) Hue b) Value
c) Chroma d) None of these

65. The potential associated with black clay soil is
a) High permeability b) Good drainage
c) High water retention d) Low draft power

66. The soil consistence suitable for tillage operations is
a) Sticky b) Plastic
c) Friable d) Hard

67. The average value of particle density in cultivated mineral soil is
a) 1.65 g/cm3 b) 2.65 g/cm3
c) 3.65 g/cm3 d) 4.65 g/cm3

68. The chemical dispersing agent used for mechanical analysis
a) Sodium hexa meta phosphate b) Sodium Bicarbonate
c) Sodium polyphosphate d) Any one of the three

69. The '*grade*' of soil structure refers to
a) Size of aggregates b) Stability of aggregates
c) Shape of aggregates d) Pattern of aggregates

70. Providing drainage facilities in wet soil is a remedy to improve of soil
a) Soil moisture b) Air capacity
c) Thermal retentivity d) Hydraulic conductivity

71. Standard methods to create porosity in the soil system
a) Tillage b) Incorporation of organic matter
c) Mulching d) All three

72. Diffusion rate of soil air improved by
a) Open soil structure b) Lower moisture content
c) Higher temperature d) All three

73. The albedo for smooth, light coloured soil surfaces is
a) >0.5 b) 0.25
c) <0.25 d) 0.3

74. Soil temperature affects

a) Germination of seeds
b) Absorption of water and nutrients
c) Decomposition of organic matter
d) All the three

75. Black plastic mulch

a) Reduces out going temperature
b) Increases outing temperature
c) Green house effect
d) Thermal insulators

76. Soil thermal characteristics can be modified by

a) Tillage practices
b) Soil compaction
c) Irrigation
d) All the three

Find the right option from assertion and reasons given below:

**77.*Assertion (A)*** : Soil structure helps in evaluating the amount of amendment needed for the soil reclamation of acid soil

***Reason (R)*** : The amount of amendment needed for reclamation of acidic soils depends on buffering capacity of soils which in turn depends on the clay content.

**Options**

a) Both A and R are true and R judiciously explains A
b) Both A and R are true and R does not judiciously explains A
c) Only A is true
d) Only R is true

**78.*Assertion (A)*** : Dry soils heats up quickly as compared to that of moist soils

***Reason (R)*** : Specific heat of soil forming minerals is less than that of water

**Options**

a) Both A and R are true and R judiciously explains A
b) Both A and R are true and R does not judiciously explains A
c) Only A is true
d) Only R is true

**79.*Assertion (A)*** : Red soils have more CEC than black soils

***Reason (R)*** : Montmorillonite clay minerals have more CEC than Kaolinite clay minerals

**Options**

a) Both A and R are true and R judiciously explains A
b) Both A and R are true and R does not judiciously explains A
c) Only A is true
d) Only R is true

**80. *Assertion(A)*:** Soil moisture content between field capacity and wilting coefficient is available to plants

***Reason (R):*** Soil moisture below Field capacity is lost by gravity and beyond wilting coefficient is not available to plants.

**Options**

a) Both A and R are true and R judiciously explains A
b) Both A and R are true and R does not judiciously explains A
c) Only A is true
d) Only R is true

**81.*Assertion(A)*** : Particle density is not influenced by soil compaction

***Reason (R)*** : Compaction is the process of increase in dry bulkdensity of soil

**Options**

a) Both A and R are true and R judiciously explains A
b) Both A and R are true and R does not judiciously explains A
c) Only A is true
d) Only R is true

**82.*Assertion (A)*** : Clay soils hold more water at given soil moisture tension than loamy and sandy soils

***Reason (R)*** : Clay soil has less surface area and surface tension and more micro porosity than loamy and sandy soils

**Options**

a) Both A and R are true and R judiciously explains A
b) Both A and R are true and R does not judiciously explains A
c) Only A is true
d) Only R is true

**83.*Assertion (A)*** : The energy of free water at reference point is zero

***Reason (R)*** : At reference state no force is acting on soil water, hence it has maximum possible energy of zero bar.

**Options**

a) Both A and R are true and R judiciously explains A
b) Both A and R are true and R does not judiciously explains A
c) Only A is true
d) Only R is true

**84. *Assertion (A)*** : The electrostatic forces exist on colloidal particles of soil would attract water molecules.

***Reason (R)*** : Water molecules by virtue of dipolar nature, gets attracted to colloidal particles.

**Options**

a) Both A and R are true and R judiciously explains A

b) Both A and R are true and R does not judiciously explains A

c) Only A is true

d) Only R is true

**85.*Assertion (A)*** : Soil serve as regulated reservoir for water.

***Reason (R)*** : Soils holds a fraction of water that received from rain or irrigation. The plants that grow on soil use water that held between FC and PWP. The water held in soil beyond FC is lost by gravitational force.

**Options**

a) Both A and R are true and R judiciously explains A

b) Both A and R are true and R does not judiciously explains A

c) Only A is true

d) Only R is true

**86.*Assertion (A)*** : In general light colour soil mulches reduces soil temperature.

***Reason (R)*** : Light colour mulches on soil increases albedo.

**Options**

a) Both A and R are true and R judiciously explains A

b) Both A and R are true and R does not judiciously explains A

c) Only A is true

d) Only R is true

87.***Assertion (A)*** : Provision of drainage facilities to soil is remedy to improve air capacity of soil.

***Reason (R)***: Soil drainage removes excess soil water from soil, thus makes air entry in to soil

**Options**

a) Both A and R are true and R judiciously explains A

b) Both A and R are true and R does not judiciously explains A

c) Only A is true

d) Only R is true

**Match the Following**

**88.**

| **List (I)** **(Soil moisture content)** | | **List (II)** **(Soil moisture potential)** |
|---|---|---|
| i. Field capacity | ( ) | 1. Zero bars |
| ii. Wilting coefficient | ( ) | 2. - 31 bars |
| iii. Saturation | ( ) | 3. – 1/3 bars |
| iv. Hydroscopic Coefficient | ( ) | 4. -15 bars |

**Codes**

| | i | ii | iii | iv |
|---|---|---|---|---|
| a) | 1 | 3 | 4 | 2 |
| b) | 4 | 3 | 1 | 2 |
| c) | 3 | 4 | 1 | 2 |
| d) | 3 | 2 | 4 | 1 |

**89. List –I** | **List-II**

| (Properties) | | (Instruments used) |
|---|---|---|
| i. Soil Texture | ( ) | 1. Pycnometer |
| ii. Soil Strength | ( ) | 2. Hydrometer |
| iii. Particle Density | ( ) | 3. Thermometer |
| iv. Soil Temperature | ( ) | 4. Pentrometer |

**Codes**

| | i | ii | iii | iv |
|---|---|---|---|---|
| a) | 2 | 1 | 4 | 3 |
| b) | 3 | 1 | 4 | 2 |
| c) | 2 | 4 | 1 | 3 |
| d) | 3 | 4 | 2 | 1 |

**90. List (I)** | **List (II)**

| **(Soil property)** | | **(Units of measurement)** |
|---|---|---|
| i. Bulk density | ( ) | 1. cm / hr |
| ii. Infiltration rate | ( ) | 2. g /cm$^3$ |
| iii. Heat capacity | ( ) | 3. cal / cm$^3$/ $^0$C |
| iv. ODR | ( ) | 4. g /cm$^2$/ min$^1$ |

**Codes**

| | i | ii | iii | iv |
|---|---|---|---|---|
| a) | 1 | 3 | 4 | 2 |
| b) | 2 | 1 | 3 | 4 |
| c) | 2 | 4 | 1 | 3 |
| d) | 2 | 3 | 4 | 1 |

**91. List (I)** | **List (II)**

| **(Soil particle)** | | **(diameter as per USDA)** |
|---|---|---|
| i. Clay | () | 1. 2.0 – 75 mm |
| ii. Sand | () | 2. 0.05 – 2.0 mm |
| iii. Silt | () | 3. < 0.002 mm |
| iv. Gravel | () | 4. 0.002 – 0.05 mm |

**Codes**

| | i | ii | iii | iv |
|---|---|---|---|---|
| a) | 1 | 3 | 4 | 2 |
| c) | 4 | 3 | 1 | 2 |
| c) | 3 | 2 | 4 | 1 |
| d) | 3 | 1 | 4 | 2 |

**92. Expand the following abbreviations**

1. ASTM : American Standards Testing Material
2. COLE : Coefficient of Linear Extensibility
3. USDA: United States Department of Agriculture
4. ISSS: International Society of Soil Science
5. ODR: Oxygen diffusion rate
6. MWD: Mean weight diameter
7. GMD: Geometric mean diameter
8. PI : Plasticity index

**93. Write mathematical formule for the following**

1. Stoke's Equation
2. Poisuellie's Equation
3. Darcy's Equation
4. Fick's Equation
5. Thermal conductivity
6. Thermal diffusivity
7. Thermal retentively
8. Heat capacity
9. Specific heat
10. Plasticity index
11. Mean weight diameter (MWD)
12. Geometric mean diameter (GMD)
13. Poisuelle equation
14. Darcy equation
15. Total soil water potential

**1) Stoke's Equation**

$$vt\frac{2(\rho p - pf)gr^2}{9\eta}$$

**2) Poisuellie's Equation**

$$Q = \frac{PR}{8\eta L}$$

**3) Darcy's Equation**

$$Q = c\frac{KAP}{L}$$

**4) Fick's Equation**

$$dQ = DA(dc / dx)dt$$

**5) Thermal conductivity**

$$\lambda = -qh\frac{\Delta x}{\Delta T}$$

**6) Thermal diffusivity**

$$\text{Thermal diffusivity } (\alpha) = \frac{\lambda}{c'} = \frac{\text{Thermal conductivity}}{\text{Volumetric heat capacity}}$$

$$\text{Thermal diffusivity } (\alpha) = \frac{\text{Thermal conductivity}}{\text{Specific heat} \times \text{density}}$$

**7) Thermal retentively**

$$\text{Thermal Retentivity} = \frac{1}{\alpha} = \frac{1}{\text{Thermal Diffusivity}}$$

$$\text{Thermal Retentivity} = \frac{\text{Specific heat x density}}{\text{Thermal conductivity}}$$

**8) Heat capacity**

Volumetric heat capacity = Specific heat x Density of soil

**9) Specific heat**

$$\text{Specific heat} = \frac{\Delta Qh}{\Delta T}$$

**10) Plasticity index (PI)**

PI = upper plastic limit – Lower plastic limit

**11) Mean weight diameter(MWD)**

= Σ(wixi)/ Σwi

**12) Geometric mean diameter (GMD)**

$$\text{GMD} = \exp\left[\sum_i^n -1W_i\ \bar{X}i / \sum_i^n -1\ W_i\right]$$

**13) Poisuellie's Equation**

$$Q = \frac{PR^4}{8\eta L}$$

## 14) Darcy's Equation

$$Q = c\frac{KAP}{L}$$

Total Potential $(\psi_T) = \psi_m + \psi_g + \psi_o + \psi_p$

**Answer key**

| | | | | | | | | | | | | | | | | | | | |
|---|---|---|---|---|---|---|---|---|---|---|---|---|---|---|---|---|---|---|---|
| 1 | b | 11 | c | 21 | c | 31 | d | 41 | b | 51 | a | 61 | b | 71 | d | 81 | b | 91 | c |
| 2 | a | 12 | d | 22 | a | 32 | d | 42 | c | 52 | c | 62 | d | 72 | d | 82 | c | | |
| 3 | b | 13 | a | 23 | c | 33 | a | 43 | b | 53 | d | 63 | d | 73 | a | 83 | a | | |
| 4 | c | 14 | c | 24 | b | 34 | b | 44 | b | 54 | d | 64 | a | 74 | d | 84 | a | | |
| 5 | b | 15 | d | 25 | c | 35 | b | 45 | A | 55 | d | 65 | c | 75 | a | 85 | a | | |
| 6 | d | 16 | c | 26 | c | 36 | c | 46 | d | 56 | b | 66 | c | 76 | d | 86 | b | | |
| 7 | a | 17 | b | 27 | a | 37 | b | 47 | c | 57 | d | 67 | b | 77 | d | 87 | a | | |
| 8 | a | 18 | d | 28 | b | 38 | b | 48 | d | 58 | a | 68 | d | 78 | a | 88 | c | | |
| 9 | a | 19 | d | 29 | c | 39 | c | 49 | a | 59 | d | 69 | b | 79 | d | 89 | c | | |
| 10 | b | 20 | c | 30 | a | 40 | a | 50 | d | 60 | a | 70 | b | 80 | a | 90 | b | | |

# 4

# Soil Chemistry

Soil Chemistry is the study of chemical composition of soil in relation to crop needs. It deals with the chemical constitution of the soil, the chemical properties and the chemical reactions in soils. Traditional soil chemistry focuses on chemical and biochemical reactions in soils that influence nutrient availability for plant growth, and potential environmental consequences associated with inorganic and organic fertilization. Soil chemistry has increasingly focused on soil genesis and the environment over the past few decades. Understanding the reactions and biogeochemical processes of potential pollutants and contaminants in soils will enable a more accurate prediction of fate and toxicity of contaminants, and development of remediation strategies. The overall goal of soil chemistry research is a more fundamental understanding of chemical and biochemical reactions in soils related to plant growth, sustainability while maintaining soil and environmental quality. Soils are the medium in which crops grow to provide food and cloth to the world. The chemistry of soil plays the primary role in all concepts of soil ecology.

## 4.1. Chemical composition of earth crust

Most of the hard, naturally formed substance of the earth is referred to as 'rock' out of 106 elements known eight are sufficiently abundant as to constitute 98.6 % (by weight) of the earth crust (up to 16m) Out of eight elements, two are in great abundance and comprise ¾ the total composition of the crust. The other six elements are metals.

**Table 1:** Average chemical composition of the earth's crust

| Element | Percentage | Oxide | Percentage |
|---|---|---|---|
| O | 46.5 | $SiO_2$ | 59.07 |
| Si | 27.6 | $Al_2O_3$ | 15.22 |
| Al | 8.1 | $Fe_2O_3$ | 3.10 |
| Fe | 5.1 | FeO | 3.71 |
| Ca | 3.6 | CaO | 5.10 |
| Mg | 2.1 | MgO | 3.45 |
| Na | 2.8 | $Na_2O$ | 3.71 |
| K | 2.6 | $K_2O$ | 3.11 |
| Ti | 0.6 | $TiO_2$ | 1.03 |
| P | 0.12 | $P_2O_5$ | 0.30 |

| Element | Percentage | Oxide | Percentage |
|---|---|---|---|
| S | 0.06 | $MnO_2$ | 0.11 |
| Cl | 0.05 | $H_2O$ | 1.30 |
| C | 0.04 | | |

## Elemental composition of soils

Similar to the chemical composition of earth crust, a large proportion of the inorganic material of soils is composed of four elements; oxygen, silicon, aluminium and iron. Nearly 90% of the mineral mater of most soils consists of the combined oxides of silicon, aluminium and iron. During chemical weathering and leaching, the proportions of these chemical constituents and others get changed largely. The chemical composition of some wider soil types are presented in table. It can be observed higher silica contents in alluvial soils and red soils while very low silica in laterite soils and medium silica in black soils. The sesquioxide content particularly iron oxides low in alluvial and red soils, while high in laterite soils (table 2). It is also noted that base cations like Ca, Na, K are absent in laterite soils.

**Table 2:** Chemical composition of some major soils of India (modified from Fundamentals of Soil Science of ISSS)

| Horizon | Depth (cm) | Chemical composition % | | | | | | | | |
|---|---|---|---|---|---|---|---|---|---|---|
| | | $SiO_2$ | $Al_2O_3$ | $Fe_2O_3$ | $TiO_2$ | CaO | MgO | $MnO_2$ | $Na_2O$ | $K_2O$ |
| Alluvial soils of Indo-Gangetic plains (Alfisol) | | | | | | | | | | |
| Ap | 0-18 | 76.8 | 14.2 | 3.1 | 0.6 | 0.4 | 0.5 | 0.1 | 1.2 | 3.0 |
| A/B | 18-36 | 73.8 | 16.2 | 3.6 | 1.2 | 0.3 | 0.3 | 0.1 | 0.8 | 3.1 |
| Bt1 | 36-88 | 70.1 | 19.7 | 3.9 | 1.5 | 0.3 | 0.3 | 0.1 | 0.7 | 3.1 |
| Bt2 | 88-133 | 69.1 | 20.4 | 4.2 | 1.6 | 0.3 | 0.3 | 0.1 | 0.7 | 3.1 |
| Bt3 | 133-175 | 67.4 | 20.6 | 4.9 | 1.4 | 0.3 | 0.7 | 0.1 | 0.8 | 3.2 |
| B/C | 175-200 | 68.9 | 19.3 | 4.3 | 1.3 | 0.5 | 0.8 | 0.1 | 1.0 | 3.6 |
| Red soils on Granite-Gneiss parent material of West Bengal (Alfisol) | | | | | | | | | | |
| Ap | 0-13 | 72.0 | 13.0 | 5.1 | 1.6 | 2.3 | 0.7 | - | - | 7.8 |
| Bt1 | 13-32 | 63.0 | 27.2 | 6.9 | 1.7 | 2.0 | 0.7 | - | - | 7.1 |
| Bt2 | 32-49 | 65.3 | 19.0 | 8.8 | 0.4 | 1.0 | 0.7 | - | - | 7.8 |
| Bt3 | 49+ | 57.0 | 23.1 | 12.9 | 0.3 | 1.23 | 0.7 | - | - | 7.7 |
| Laterite soils on serpentine rock in Indonasia (Oxisol) | | | | | | | | | | |
| A | 0-100 | 7.6 | 10.9 | 65.1 | 0.3 | - | 0.3 | 0.3 | - | - |
| B1 | 100-200 | 3.7 | 13.8 | 65.4 | 0.3 | - | 0.4 | 0.8 | - | - |
| B2 | 200-500 | 2.5 | 12.1 | 68.7 | 0.2 | - | 0.9 | 0.8 | - | - |
| B3 | 500-650 | 7.3 | 9.8 | 60.4 | 0.1 | - | 6.4 | 0.9 | - | - |
| C1 | 650-700 | 19.0 | 8.4 | 50.0 | 0.1 | - | 7.6 | 0.7 | - | - |
| C2 | 700-750 | 34.0 | 5.6 | 34.9 | 0.1 | - | 11.4 | 0.5 | - | - |
| R | 750+ | 39.4 | 2.5 | 8.4 | Tr | - | 33.2 | 0.1 | - | - |
| Balck soils of Deccan trap in Bijapur dist of Karnataka | | | | | | | | | | |

| Horizon | Depth (cm) | Chemical composition % | | | | | | | | |
|---|---|---|---|---|---|---|---|---|---|---|
| | | $SiO_2$ | $Al_2O_3$ | $Fe_2O_3$ | $TiO_2$ | CaO | MgO | $MnO_2$ | $Na_2O$ | $K_2O$ |
| AP | 0-20 | 44.2 | 14.8 | 12.9 | 1.2 | 7.6 | 4.6 | - | 0.8 | 1.1 |
| A12 | 20-45 | 42.6 | 13.9 | 13.1 | 1.2 | 9.0 | 4.8 | - | 0.9 | 1.0 |
| A12 | 45-62 | 40.2 | 13.4 | 10.1 | 1.0 | 14.0 | 4.4 | - | 0.8 | 0.5 |
| AC | 62-95 | 42.1 | 13.3 | 11.6 | 1.2 | 11.8 | 5.3 | - | 0.7 | 0.7 |

Soil has three categories of particles based on their size, sand, silt and clay. Since sand and silt are dominantly quartz ($SiO_2$) and other primary minerals known for their resistance to weathering, these two fractions have relatively low chemical activity. The primary minerals that may contain nutrient elements in their chemical makeup are so insoluble as to make their nutrient supplying capacity insignificant.

Silicate clays, chemically vary widely. Some are relatively simple aluminosilicates, others contain in their crystal structure varying quantities of iron, magnesium, potassium, and other elements. The surface of all silicate clays hold small but significant quantities of cations such as $Ca^{++}$, $Mg^{++}$, $K^+$, $Na^+$ and $Al^{3+}$. These cations are exchangeable and can be releases for absorption by plants. In highly weathers soils, such as those found in hot- humid tropics, oxides of Fe and Al are dominant even in clay size fraction.

## 4.2. Soil chemical properties

Some soil chemical properties are discussed in the following heads.

1. Ion exchange (Anion and cation exchange capacity)
2. Soil reaction
3. Mineral formation and transformation processes and pedogenesis
4. Sorption and precipitation reactions in soil
5. Chemistry of puddle rice soils
6. Thermodynamic theories

### *4.2.1. Ion exchange*

Ion adsorption and subsequent exchange are important processes that take place between soil colloidal particles (clays, organic matter, sesquioxides, and amorphous minerals) and various ions. Soil colloids serve very much as a modern bank. They are the sites within the soil where ions of essential plant nutrients are held and protected from excessive loss by leaching. Subsequently, the nutrients can be "withdrawn" from the colloidal "bank" sites and taken up by plant roots. In turn, these elements can be "deposited" or returned to the colloids through the addition of commercial fertilizers, lime, manures, and plant residues.

The charges associated with soil particles attract ions (simple and complex) of opposite charge. In temperate region soils, negative charges generally predominate on the soil particles (colloids), hence adsorbed cations are present in larger quantities than anions. In more highly weathered soils (e.g. in tropics) where 1:1 type clays and Fe/Al oxides are the most dominant type of colloids, anion adsorption and exchange is relatively more prominent.

## Ion exchange reactions

The ion exchange phenomenon was first identified by Harry Stephen Thompson in England during 1850. When soil was leached (washed) with ammonium sulphate, calcium sulphate was detected in the leachate. The ammonium ion in the solution replaced calcium in the soil.

$(NH_4)_2SO_4$ + Soil Ca → Soil $(NH_4)_2$ + $CaSO_4$

The process by which ions are exchanged between solid and liquid phases and /or between solid phases if in close contact with each other is called ion exchange. The common exchangeable cations are $Ca^{2+}$ $mg^{2+}$ $H^+$ ,$K^+$, $NH_4^{\ +}$ and Na. The common anions are $So_4^{\ 2-}$, $Cl^-$ $Po_4^{-3}$ and $NO_3^-$.

The ion exchange in soil is due to the presence of residual positive and negative charges on the soil colloids. These residual charges result due to the process of isomorphous substitution and ionisation of hydrogen and hydroxide functional groups. The negative charges attract positively charged ions and the positive charges attract negatively charged ions from soil solutions coming in contact with colloids. The ions thus attracted are reversible and are on equivalent proportions.

Exchange of cation is called cation exchange and exchange of anion is called anion exchange. The cation exchange phenomenon was first discovered by Thomasway (1850) Ion exchange is the second most important reaction is nature. The first one is photosynthesis by green plants.

The capacity of the soil to hold cation is called cation exchange capacity. The unit is C mol $(P^+)$/kg. The capacity to hold anion is called Anion exchange capacity (AEC). The unit of expression is C mol $(e^-)$ / kg

## Mechanism of cation exchange

Clay colloids have negative charges. Cations are attracted to the clay particles. These cations are held on the held on the clay surfaces electro statistically. They are held by small particles of clay and organic matter. These small particles are called Micelle (Micro cell). The cations that can be replaced on exchange site by other cations are called exchangeable cations. They are weekly held and they are in direct contact with the soil solution. They can be exchanged fairly easily. Ions that are held very lightly with the colloid may be traped between layers of clay micelle. They do not pass to the soil solution very easily. They are called non exchangeable cations. When any cation is added to the soil such as $Ca^{++}$, $K^+$ or $NH_4$ through fertilizers and soil amendments they are exchanged with those ions held on the colloid. When calcium is added to an acid soil the following reaction takes place.

Clay micelle – $2H^+ + Ca^{2+} \rightarrow$ Clay micelle –$Ca^{2+} + 2H^+$

Similarly when $H^+$ is added to the soil solution through the decomposition of organic matter or through acidic materials $Ca^{2+}$ is replaced from the exchange complex by $H^+$ ions.

Clay micelle –$Ca^{2+} + 2H^+ \rightarrow$ Clay micelle – $2H^+ + Ca^{2+}$

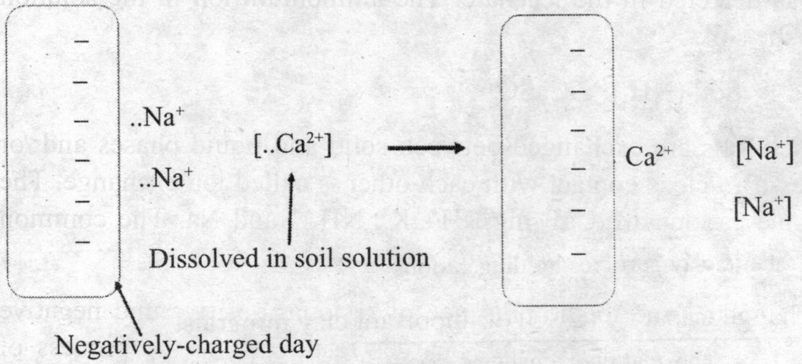


**Fig. 1:** $Ca^{2+}$ exchange displaces exchangeable $Na^+$

Clay complex- $2Na^+ + Ca^{2+} \longrightarrow$ Clay complex- $Ca^{2+} + 2Na^+$

**Cation exchange capacity:** Cation exchange capacity (CEC) is a measure of the total negative charges within the soil that adsorb plant nutrient cations such as calcium ($Ca^{2+}$), magnesium ($Mg^{2+}$) and potassium ($K^+$). As such, the CEC is a property of a soil that describes its capacity to supply nutrient cations to the soil solution for plant uptake. The nutrient cations plants use in the largest amounts are potassium ($K^+$), calcium ($Ca^{2+}$) and magnesium ($Mg^{2+}$). Other cations adsorbed on exchange sites are ammonium ($NH_4^+$), sodium ($Na^+$), hydrogen ($H^+$), aluminum ($Al^{3+}$), iron ($Fe^{2+}$ or $Fe^{3+}$), manganese ($Mn^{2+}$), copper ($Cu^{2+}$) and zinc ($Zn^{2+}$). Micronutrient cations such as zinc, copper, iron and manganese are typically present at very low concentrations in soils.

In most soil reports, CEC is expressed as milliequivalents (meq) of charge (number of charges) per 100 grams of soil (meq/100 g or as cmol/kg when using International Scientific Units). The number of milliequivalents is used rather than a weight (grams) of adsorbed cations because CEC represents the total number of charges, which is a better standard of comparison of different soils because each cation species has a different weight and soils differ in the proportions of the different cation species. Table 3 and 4 shows the typical cation exchange capacities of soil clay minerals and soils of various textures.

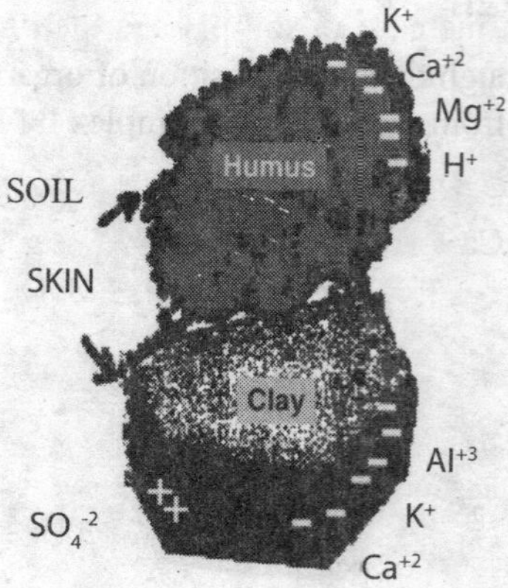


**Fig. 2:** Inorganic and organic clay particles holding cations

**Table 3:** Cation Exchange Capacity (CEC) of important clay minerals

| Soil clay mineral | Cation exchange capacity |
|---|---|
| Kaolinite | 7-10 C mol ($P^+$) / kg |
| Montmorillonite | 80-100 C mol ($P^+$) / kg |
| Vermiculite | 100-150 C mol ($P^+$) / kg |
| Illite | 25-30 C mol ($P^+$) / kg |
| Chlorite | 25-30 C mol ($P^+$) / kg |
| Fe & Al oxides | 5.0 C mol ($P^+$) / kg |
| Humus | 200-400 C mol ($P^+$) / kg |

**Table 4:** Cation Exchange Capacity (CEC) of different textural classes.

| Soil texture | Cation exchange capacity |
|---|---|
| Sand | 0-5 C mol ($P^+$) / kg |
| Sandy loam | 5-10 C mol ($P^+$) / kg |
| Loam | 10-15 C mol ($P^+$) / kg |
| Clay loam | 15-30 C mol ($P^+$) / kg |
| Clay | 30- 40 C mol ($P^+$) / kg |

## Factors influencing CEC

i. **Soil Texture**: CEC increases with fineness of the soil particles. This means increasing clay content will increase the CEC. As the particle size decreases the surface area increases per unit volume. This naturally increases the net charge and the CEC.

ii. **Organic matter**: In general CEC increases with increase in organic matter content. The pH dependent charges in the organic matter cause variation in CEC.

iii. **Nature of clay:** CEC of clay minerals vary soil dominated with montmorillonite and vermiculite have higher CEC than those dominated with kaolinite, chlorite or illite.

iv. **Soil reaction**: In general CEC increases with increase in soil pH. As the pH increases the pH dependent charge increases. In Humus most CEC is pH dependent.

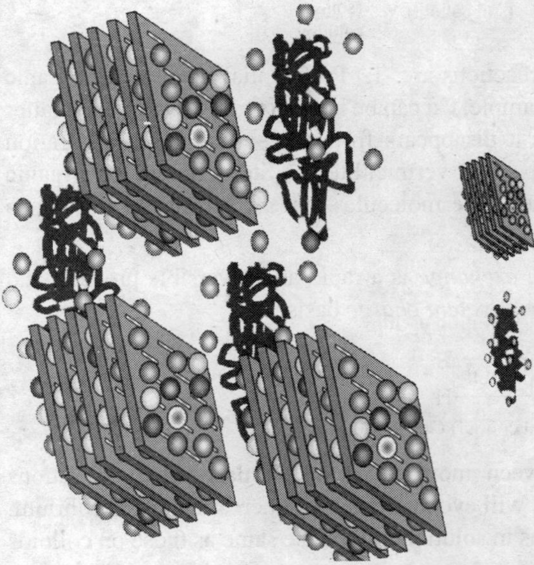


**Fig. 3:** Depiction of inorganic and organic collidal particles holding cations. The higher the soil CEC the greater ability it has to store plant nutrients. CEC increases with increase of clay particles, organic matter and soil pH.

## Replacing power of ions

Replacing power of ions increases with atomic weight. Divalent cations have more replacing power than monovalent ions. Hydrogen is an exemption. Hydrogen ions are adsorbed more strongly than other monovalent or divalent ions. The replacing power of cations varies with the type of ion, size, degree of hydration, valence, concentration and the kind of clay mineral involved. As it is controlled by number of factors, no single order of replacement can be given. All other factors being equal the replacing power of monovalent cations increases in the following order: Li < Na < K < Rb < Cs < H and for divalent cations: Mg < Ca < Sr < Ba. In case of mixture of monovalent and divalent cations as they exist in normal soils the replacing power increases in the following order: Na < K < $NH_4$ < Mg < Ca < H. This means Na is more easily replaced than K and K more easily than $NH_4$. In general the power of replacement is H > Ca > Mg > $NH_4$ > K > Na

## Principles governing cation exchange reactions

1. Reversibility: Cation exchange reactions, the swapping of two cations from soil solution to being bound to a colloid is a reversible reaction. This means the ratio of cations in solution will be the same as those on colloids In general, cation

exchange reactions are reversible unless some special reactions also occur.

$$\boxed{\text{Micelle}}\ Na^{+} + H^{+} \rightleftharpoons \boxed{\text{Micelle}}\ H^{+} + Na^{+}$$

(soil colloid) (soil solution) (soil colloid) (soil solution)

Exceptions for reversibility reactions are: 1.) Preferential bonding by organic matter to trace metals. For example, Cu can be held so tightly by organic matter through covalent bonds that it disappears from the soil solution. 2.) Cation fixation e.g. $K^{+}$ or $NH_4^{+}$ is fixed by vermiculite. 3.) Steric hindrance Organic ions/molecules, e.g., some pesticide molecules that are physically too big to approach exchange sites.

2. **Charge equivalence:** *Cation exchange* is a rapid and reversible process, and cations are exchanged on an *equivalent charge* basis.

$$\boxed{\text{Micelle}}\ Ca^{2+} + 2H^{+} \rightleftharpoons \boxed{\text{Micelle}}\begin{matrix} H^{+} \\ H^{+} \end{matrix} + Ca^{2+}$$

(soil colloid) (soil solution) (soil colloid) (soil solution)

3. **Ratio law:** Exchanges between more than one similarly charged cations between solution and colloid, will eventually even out creating an equilibrium. This means the ratio of cations in solution will be the same as those on colloids

$$\boxed{\text{Micelle}}\ 20\ Ca^{2+} + 5\ Mg^{2+} \rightleftharpoons \boxed{\text{Micelle}}\begin{matrix} 16\ Ca^{2+} \\ 4\ mg^{2+} \end{matrix} + 1\ Mg^{2+} + 4\ Ca^{2+} \quad \text{Ratio: 4Ca: IMg}$$

(colloid) (soil solution) (colloid) (soil solution)

4. **Anion effects on mass action:** An exchange reaction will be more likely to proceed to the right if released ion is prevented from reacting in the reverse direction - processes which accomplish this are precipitation, volatilization and strong association with an anion

$$\boxed{\text{Micelle}}\begin{matrix} H^{+} \\ H^{+} \end{matrix} + CaCO_3 \rightleftharpoons \boxed{\text{Micelle}}\ Ca^{2+} + H_2O + CO_2\uparrow$$

(added) Water (gas)

$$Al^{3+} > Ca^{2+} > Mg^{2+} > K^{+} = NH_4^{+} > Na^{+}$$

5. **Cation selectivity:** The orders of cation selectivity depend on the valance of the ion.
6. **Complementary ion effect:** In complex mixture of ions, likelyhood of displacement of an ion depends on strength of sorption of ion and neighboring ions. The complementary ion effect is shown in Figure 3.complementary ion effect.

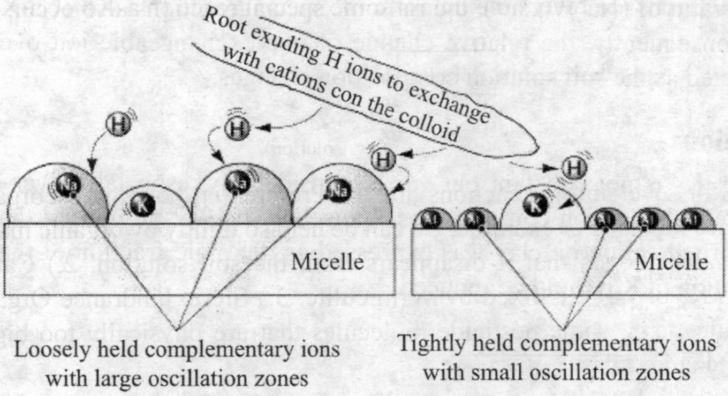


**Fig. 4:** Complementary ion effect

## Cation Exchange Equations

Several exchange equations have been proposed, each has its own advantage and disadvantage, and the choice of a particular equation is often based on familiarity of individual reader with that equation.

### 1) Freundlich equation

**$x/m = KC^{1/n}$**

It relates the concentration of a given cation in equilibrium solution (C) to the amount adsorbed (x/m) leading to the coefficients K and n derived from experimental data.

### 2) Kerr equation or mass-action equation.

$$\text{Ca-X} + 2\text{Na}^+ \rightleftharpoons 2\text{Na-X} + \text{Ca}^{2+}$$

The reaction coefficient is

$$K_k = \frac{(\text{Na}-\text{X})^2 (\text{Ca}^{2+})}{(\text{Ca}-\text{X})\ (\text{Na}^+)^2} \quad (1)$$

Here Na-X and Ca-X are exchangeable cations. Ca and Na are ions is solution.

All the terms here are in activity units, even though Kerr originally used concentrations in place of activities. The coefficient $K_k$ remains fairly constant only over a narrow range of concentration. Since activities of exchangeable cations cannot be measured or calculated precisely, the above equation is often written as:

$$\text{(in mmole/g)}\ \frac{(\text{Na}-\text{X})^2}{(\text{Ca}-\text{X})} = K_k \times \frac{(\text{Na}^+)^2}{(\text{Ca}^{2+})} \quad (2)$$

The activities of ions in solution can be estimated from Debye-Huckel theory or by specific-ion-electrode measurement. $K_k$ is taken by averaging several values.

Knowing $K_k$ and the ratio of $(Na^+)^2/(Ca^{2+})$, the ratio of exchangeable $(Na\text{-}X)^2/(Ca\text{-}X)$ can be estimated. Consequently, the relative change of one exchangeable ion over another can be predicted as the soil solution composition changes.

### 3) Vanselow equation

As mentioned above, $K_k$ is not constant but varies considerably, especially over a wide range of exchangeable cation compositions. Vanselow in 1932 proppposed that if activity of an adsorbed cation on clay was expressed as the mole fraction of that cation, then the variation of $K_k$ would be reduced.

Ex: Activity of [Na-X] $= \dfrac{[Na-X]}{[Na-X]+[Ca-X]}$

Here, all the units are in mmole/g Act.

$$[Ca\text{-}X] = \frac{[Ca-X]}{[Na-X]+[Ca-X]}$$

substituting these values into eq. 2 gives:

$$\frac{[Na-X]^2}{[Ca-X][Na-X+Ca-X]} = \frac{K_v\left(Na^+\right)^2}{\left(Ca^{2+}\right)} \qquad (3)$$

For homovalent exchange, the 2 equations are the same; for heterovalent exchange, the Vanselow equation differs from the Kerr equation by the factor of [Na-X + Ca- X].

### 4) Gapon equation

The most commonly used cation exchange equation in soils is t Gapon equation which has a square root term:

$$(Ca)_{1/2} - X + Na^+ \qquad Na\text{-}X + \tfrac{1}{2}\, Ca^{2+}$$

and $$\frac{[Na-X]}{[Ca_{1/2}X]} \rightleftarrows = K_G \frac{[Na^+]}{[Ca^{2+}]^{1/2}} \qquad (4)$$

Where [NaX] and [$Ca_{1/2}$X] are exchangeable cation concentrations and in meq/g, (or mmol of charge/g) and [$Na^+$], [$Ca^{2+}$] are solution cation concentrations in mmole/liter.

## Importance of cation exchange capacity (CEC) of soil

Cation exchange capacity is the most important reaction in nature after photosynthesis. It is an important reaction in correcting soil acidity and basicity. Soil physical properties altered by CEC. Through cation exchange mechanism (large cation exchanger), the percolating waters will be purified (Ground water pollution is checked). Plants absorb exchangeable cations by interchange or contact between the root hairs and colloidal complex. The amounts of cations in the soil solution are intimately related to the exchangeable cations. CEC offers buffering capacity to soils.

**Base saturation:** The percentage of Cation exchange capacity (CEC) that is satisfied by the base forming cations (Ca, Mg, Na and K) is called base saturation percentage.

$$\% \text{ Base saturation} = \frac{\text{Exchangeable base forming cations (C mol / kg) x 100}}{\text{CEC (C mol / kg)}}$$

Aluminium and hydrogen are considered as acid forming ions. Calcium, Magnesium, Potassium and Sodium are considered as base forming ions. Base forming substances are called as besoids and acid forming substances are called as acidoids. The percentage of sodium in the total CEC is called Exchangeable sodium percentage (ESP) = [(Na/CEC) x 100]. These parameters are considered while deriving fertilizer prescription and amendments for problem soils. Knowledge of base saturation percentage is useful in many ways.

i. Base saturation helps in determining the quantity of lime required to raise the pH of acid soils.

ii. It indicates the proportion of plant nutrients in CEC. It is an index of soil fertility.

iii. Degree of saturation of a particular cation in CEC indicates the ease with which the cation can be released for plant nutrition. For example if calcium saturation is more Ca can be very easily replaced from the exchange complex.

iv. Base saturation is an indication of soil fertility. The ease with which adsorbed cations are released to plants depends on the degree of base saturation. If bases saturation of soils is >80%, it is considered as highly fertile soils, if it is 50-80% the soil is medium fertile, while the base saturation is <50% , the soil is considered as low fertile in nature.

## Factors affecting base saturation of soil

- Liming of acid soils increases percent base saturation.
- The degree of base saturation is higher for arid soils than humid region soils.
- Soils formed from lime stones or basic igneous rocks has higher BS than soils formed from sand stones or acid igneous rocks.
- Soils with dominance of organic colloids or 1:1 type clay colloids can supply much lower nutrient cations to plants than soils with dominance of 2:1 colloids.
- The percent base saturation does provide numerical values of the amount of exchangeable hydrogen and aluminum ions present in soils and this helps in predicting the amount of lime needed to neutralize soil activity.
- Kind of adsorbed cations also affects the soil pH. With same base saturation percentage, the soil dominated by $Na^+$ cation presents higher pH values than the soil dominated by $Ca^{2+}$ and $Mg^{2+}$ cations.

## Model calculation of base saturation and exchangeable acidity

**Example 1:** Laboratory analysis of 100 g soil sample recorded exchangeable $Ca^{++}$ content as 10.0 m.eq., exchangeable $Mg^{++}$ as 3.0 m.eq., exchangeable $Na^+$ is 2.0 m.eq., and exchangeable $K^+$ is 3.0 m.eq.,, exchangeable $Al^{+++}$ as 3 m.eq.,, exchangeable $H^+$ as 4 m.eq. Calculate the percent base saturation and base un-saturation or exchangeable acidity.

Exchangeable Calcium = 10.0 m.eq. / 100g

Exchangeable Magnesium = 3.0 m.eq. / 100g

Exchangeable Sodium = 2.0 m.eq. / 100g

Exchangeable Potassium = 3.0 m.eq. / 100g

Exchangeable Potassium = 3.0 m.eq. / 100g

Exchangeable Potassium = 4.0 m.eq. / 100g

Total CEC = 10.0 m.eq. / 100g + 3.0 m.eq. / 100g + 2.0 m.eq. / 100g + 3.0 m.eq. / 100g + 3.0 m.eq. / 100g + 4.0 m.eq. / 100g = 25.0 m.eq. / 100g

Total Exchangeable bases = 18 m.eq. / 100g

$$\text{Percent Base Saturation} = \frac{18}{25} \times 100 = 72$$

Exchangeable acidity or Base un-saturation = 100 – 72 = 28 %

**Example 2:** The CEC of a soil is 20m.eq./100g soil. The estimated amount of $Ca^{++}$ is 0.160g/100g; $Mg^{++}$ is 0.048g/100g; $Na^{+}$ is 0.046g/100g and $K^{+}$ is 0.117g/100g soil. Calculate the percent base saturation and base un-saturation or exchangeable acidity.

$$\text{Exchangeable Calcium} = \frac{0.160 \times 1000}{20} = 8.0 \text{ m.eq. / 100g}$$

$$\text{Exchangeable Magnesium} = \frac{0.048 \times 1000}{12} = 4.0 \text{ m.eq. / 100g}$$

$$\text{Exchangeable Sodium} = \frac{0.046 \times 1000}{23} = 2.0 \text{ m.eq. / 100g}$$

$$\text{Exchangeable Potassium} = \frac{0.117 \times 1000}{39} = 3.0 \text{ m.eq. / 100g}$$

Total Exchangeable bases = 17 m.eq. /100g

$$\text{Percent Base Saturation} = \frac{17}{20} \times 100 = 85$$

Exchangeable acidity or Base- unsaturation = 100 – 85 = 15 %

## Anion exchange

Replacement of one anion by another anion on the positively charged colloids is called anion exchange, positive charges are due to OH of iron and aluminium, 1:1 clays and allophone (a morphous clays). Anion exchange is $p^H$ dependent. Lower the $p^H$ greater is the anion exchange. Soils with Kaolinite dominant clay have higher anion exchange capacity than montmorillonite or illite. The relative order of anion exchange is $OH^- > H_2PO^-_4 > SO^-_4 > NO^-_3 > Cl$.

### Importance of anion exchange

a. The phenomenon of anion exchange is important for the release of fixed phosphorous in the soil.

   In acid soils the phosphorus is fixed as insoluble Al-Phosphate. Liming the acid soils release fixed phosphorous. Here the OH ion replaces $H_2PO_4^-$ from $Al(OH)_2 H_2PO_4$.

   $Al(OH)_2 H_2 PO_4 + OH^- \longrightarrow Al(OH)_3 + H_2PO_4^-$.

b. Similarly the availability of other nutrients like $NO^-_3$, $SO^-_4$ and $Cl^-$ are influenced by anion exchange.

   Soil $-NO_3 + Cl^- \longrightarrow$ Soil $-Cl + NO_3^-$

   Soil solution Soil solution

### Significance of ion exchange

Next to photosynthesis ion exchange is the most important reaction in the world. Plants take up their food material from the soil through ion exchange only. Plant roots which are in contact with the soil solution exchange the nutrients from soil solution for $H^+$ ions present on the surfaces of root hairs. A soil is considered to be fertile when the base saturation percentage is more than 80. Each percent of humus contributes about 2 C mol /kg of CEC. Montmorillonite contributes about I C mole and Kaolinite contributes about 0.08 C mol/kg for every one percent.

### Availability of applied nutrients

When fertilizers are applied to supply plant nutrients elements like K, Ca, Mg and $NH_4$ dissolve in soil solution. These nutrients in soil solution are exchanged for other cation like $H^+$ present in the exchange complex. If there is no cation exchange the applied nutrients would be lost in drainage water. Similar is the case with anion radicals like $PO_4$, $NO_3$, $SO_4$ etc. Soils with high CEC can absorb higher amounts of nutrients. Hence, in clay soils we can apply larger quantities of fertilizers in a single dose. Sandy soils have very low CEC and in such soils fertilizers should be applied in splits.

**Effect of adsorbed cations:** When the soil exchange complex has calcium the soil will have desire able physical properties. The activity of soil micro organisms, ammonification and nitrification processes also are determined by the cations of exchange complex.

**Toxic ions:** When the exchange complex had adsorbed metals like Cadmium, Nickel and lead they are toxic to the crop plants.

**Effect on soil pH:** Clays with $H^+$ are acidic and with $Na^+$ are alkaline. The acidic and alkaline nature of soil has its own effect on soil properties.

### *4.2.2. Soil reaction*

Soil reaction is property of soil functionally related to the concentration of $H^+$ and $OH^-$ ions in the solid and liquid parts of the soil. If $H^+$ ions are predominant, the soil reaction is acid; if $OH^-$ ions are predominant, it is alkaline. If the concentrations are equal, the soil reaction is neutral. In other words soil reaction is a measure of the acidity or basicity (alkalinity) of a soil. The terms Soil reaction and soil pH are used synonimical for the negative logarithm (base 10) of the activity of hydronium ions ($H_3O^+$)in a solution.

| | | |
|---|---|---|
| Acidic: | $[H_3O^+] > [OH^-]$ | $pH < 7$ |
| Neutral: | $[H_3O^+] = [OH^-]$ | $pH = 7$ |
| Basic: | $[H_3O^+] < [OH^-]$ | $pH > 7$ |

In soils, it is measured in a slurry of soil mixed with water (or a salt solution, such as 0.01 M $CaCl_2$), and normally falls between 3 and 10, with 7 being neutral. Acid soils have a pH below 7 and alkaline_soils have a pH above 7. Ultra-acidic soils (pH < 3.5) and very strongly alkaline soils (pH > 9) are rare

Soil pH is considered a master variable in soils as it affects many chemical processes. It specifically affects plant nutrient availability by controlling the chemical forms of the different nutrients and influencing the chemical reactions they undergo. The optimum pH range for most plants is between 5.5 and 7.5; however, many plants have adapted to thrive at pH values outside this range.

Model calculation 1

What is the pH of stomach acid, a solution of HCl with a hydronium ion concentration of $1.2 \times 10^{-3} M$?

$$pH = -\log[H_3O^+]$$
$$= -\log(1.2\times10^{-3})$$
$$= -(-2.92)$$
$$= 2.92$$

Model calculation 2

Calculate the hydronium ion concentration of blood, the pH of which is 7.3 (slightly alkaline).

$$pH = -\log[H_3O^+]=7.3$$

$$\log[H_3O^+] = -7.3$$

$$[H_3O^+] = 10^{-7.3} \text{ or}$$

$$[H_3O^+] = \text{antilog of } -7.3$$

$$[H3O^+] = 5\times10^{-8}M$$

## Key equations in pH calculations

$$pH = -\log[H_3O^+]$$

$$pOH = -\log[OH^-]$$

$[H_3O^+] = 10^{-pH}$

$[OH^-] = 10^{-pOH}$

$pH + pOH = pK_w$ (ionization constant) = 14.00 at 25 °C

**Table 5:** Classification of soils based on pH

| pH range | Category (rating) |
|---|---|
| Less than 4.5 | Extremely acidic |
| 4.5 – 5.0 | Very strongly acidic |
| 5.1 – 5.5 | Strongly acidic |
| 5.6 – 6.0 | Moderately acidic |
| 6.1 – 6.5 | Slightly acidic |
| 6.6 – 7.5 | Neutral |
| 7.5 – 7.8 | Mildly/ slightly alkaline |
| 7.9 – 8.4 | Moderately alkaline |
| 8.5 – 9.0 | Strongly alkaline |
| More than 9.1 | Very strongly alkaline |

**Classification of soil acidity:** Soil acidity is classified as four types viz., a) Active acidity, b) Exchangeable acidity, c) Reserve acidity and d)Potential acidity

a) **Active acidity:** Active acidity may be defined as the acidity developed due to concentration of hydrogen and aluminium ions present in the soil solution. The hydrogen ions in the soil solution contribute to active acidity which is measured in terms of pH of the soil. This acidity is very small, implying that only a low amount of lime would be required to neutralise the active acidity. Inspite of smaller concentration, active acidity is important since the plant root and microbes around the rhizosphere are influenced by it.

b) **Exchangeable acidity:** Exchangeable acidity may be defined as the acidity developed due to adsorbed hydrogen ($H^+$) and aluminium ($Al_3^+$) ions on soil colloids. In strongly acidic soils, the concentration of exchangeable aluminium and hydrogen ions contribute to exchangeable acidity. However, this exchangeable hydrogen ($H^+$) and aluminium ($Al_3^+$) ions concentration is meagre in moderately acidic soils.

Total acidity =Active acidity + Exchangeable acidity

c) **Reserve acidity:** It is defined as the acidity developed due to non-exchangeable $H^+$ and $Al^{3+}$ which were associated with oxides and hydroxides of Al & Fe, with organic matter and clays account for the reserve acidity. Reserve acidity is measured by titrating a soil suspension with NaOH up to a certain pH, normally up to 8.0

d) **Potential acidity:** It refers to the acidity caused by $H^+$ held in different chemical combinations, $H^+$ adsorbed on the surfaces of soil clay particles and $H^+$ adsorbed on the surfaces of organic colloids. The potential acidity is much higher as compared to active or exchangeable acidity.

## Influence of soil pH on nutrient availability

- In strongly acidic soils aluminium toxicity has direct effects on plant growth; however, by limiting root growth, it also reduces the availability of plant nutrients. Because roots are damaged, nutrient uptake is reduced, and deficiencies of the macronutrients (nitrogen, phosphorus, potassium, calcium and magnesium) are frequently encountered in very strongly acidic to ultra-acidic soils (pH<5.0).
- The effect of pH on phosphorus availability varies considerably, P availability is maximum near neutrality (soil pH 6.5–7.5), and decreased at higher and lower pH.
- Potassium and Sulphur show decreases availability in moderate to severe acidity.
- Under acidic soil pH Calcium and Magnesium become deficient and Fe, Mn become toxic, while in alkaline soil pH, zinc, iron, copper and manganese become deficient.
- Molybdenum availability is increased at higher pH; this is because the molybdate ion is more strongly sorbed by clay particles at lower pH.
- At acidic soil pH the bacteria activity is very low, hence the BNF, phosphorous solubility are low, however the VAM (VA mycorrhiza) activity is high hence it improve the availability of soil less mobile nutrient elements like P,Zn,Cu etc.
- Some plants, such as azaleas, rhododendrons, blueberries, white potatoes and conifer trees, tolerate strong acid soils and grow well.
- Soil pH & water availability: Strongly alkaline soils are sodic and dispersive, with slow infiltration, low hydraulic conductivity and poor available water capacity.
- In high pH alkali soils, because poor aeration plant growth is severely restricted, soil also become hard on drying with high soil strength

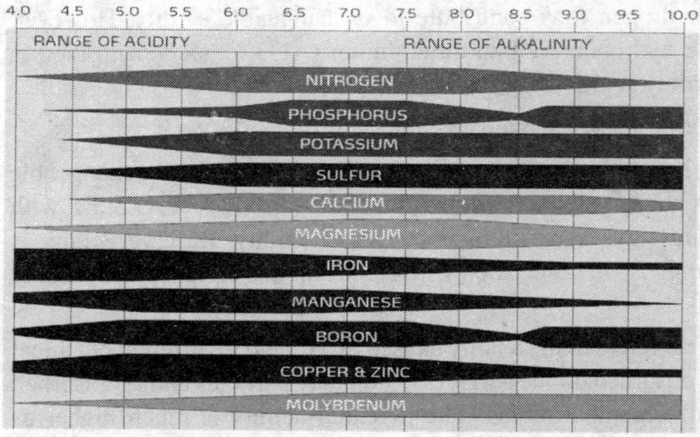


**Fig. 5:** Influence of soil pH on nutrient availability to plants

## Factors affecting soil pH

a) **Nature of soil colloids**: When the soil colloids (1: 1 type and sesquioxides) have $H^+$ as adsorbed ion on the exchange complex, the soil reactions becomes acidic.

b) **Percentage base saturation (PBS) and kind of adsorbed base:** Soils having low PBS show acidic reaction.

c) **Soil solution:** More dilute soil solution of a base unsaturated soil, more dissociation of hydrogen ions into the soil solution.

d) **$CO_2$ concentration**: With increase in concentration of $CO_2$ of soil solution the pH lowers due to the formation of carbonic acid.

e) **Climate**: In general, soils formed in the areas of high rainfall are acidic, while those formed in areas of low rainfall are alkaline.

f) **Soil management:** As a result of intense cultivation basic cations are lost from the soil through crop removal and leaching making the soil acidic. Use of acid forming fertilizers like ammonium sulphate, Urea make the soil acidic.

g) **Oxidation – Reduction** state of soil under anaerobic as in water logged soils the ferric and manganic ions are reduced to ferrous ($Fe^{2+}$) and manganous ($Mn^{2+}$) states, thus increasing pH, while in oxidised conditions sulphides are converted to $SO_4^{=}$ thus lowering the pH upon formation of acid.

## Buffering capacity of soils

Buffering refers to the resistance to a change in soil pH. Buffering in soils is their resistance against any change in the concentration of hydrogen or other cations in the soil solution. In soils, the clay and humus act as buffer systems. The buffering capacity is greater in clay soils than in sandy soils.

In soils, the buffering action may be due to the following buffering agents:

- Carbonates, bicarbonates and phosphates present in the soil.
- Colloidal complex with associated cations.
- Organic acids being released continuously in the soil act as weak acids.
- $H^+$ released from the soil exchange complex provides the reserve acidity.

## Factors affecting buffering capacity of soils

- **Amount of soil colloids**: Soils containing large amounts of clay and organic matter are said to be highly buffered.
- **Cation exchange capacity**: Keeping all the factors constant, buffering capacity of the soil increased with increase in cation exchange capacity.
- Soils having higher **amounts of calcium carbonate**, organic acids and phosphates show higher buffering capacity.
- **Nature of Clay**: Soils containing large amounts of 1:1 type clays are generally less strongly buffered than soils in which the predominant clay minerals are 2:1 type.

- The **degree of buffering** is highest between the soil pH of 4.5 and 6.0 and drops off below and above this range.

## Importance of soil buffering capacity in Agriculture

i) Buffering stabilizes of soil pH, there by alleviate the problems of physical, chemical and biological which are sensitive to change in soil pH.

ii) Helpful in calculations of amount of amendment necessary to affect a certain change in soil reaction.

iii) Deficiencies of certain plant nutrients and excess of others in toxic amounts would seriously upset the nutritional balance in the soil.

iv) The greater the buffering capacity of the soil, the larger must be the amounts of amendments to affect a given change in pH.

## 4.2.3. Mineral formation, transformation and pedogenesis

Pedogenesis (from the Greek *pedo*-, or *pedon*, meaning soil, earth, and *genesis*, meaning origin, birth) (also termed soil development, soil evolution, soil formation, and soil genesis) is the process of soil formation as regulated by the effects of place, environment, and history. Biogeochemical processes act to both create and destroy order (anisotropy) within soils. These alterations lead to the development of layers, termed soil horizons, distinguished by differences in color, structure, texture, and chemistry. These features occur in patterns of soil type distribution, forming in response to differences in soil forming factors.

The study of pedogenesis is important to understanding soil distribution patterns in current (soil geography) and past (paleopedology) geologic periods. Soil develops through a series of changes. The starting point is weathering of freshly accumulated parent material. Primitive microbes feed on simple compounds (nutrients) released by weathering, and produce acids which contribute to weathering. They also leave behind organic residues.

New soils increase in depth by a combination of weathering, and further deposition. An estimated 1/10 mm per year rate of soil production from weathering fits observations rates. New soils can also deepen from dust deposition. Gradually soil is able to support higher forms of plants and animals, starting with premitive species, and proceeding to more complex plant and animal communities. Soils deepen with accumulation of humus primarily due to the activities of higher plants. Top soils deepen through soil mixing. As soils mature, they develop layers as organic matter accumulates and leaching takes place. This development of layers is the beginning of the soil profile. Soil formation is influenced by five classic factors that are intertwined in the evolution of a soil. They are: parent material, climate, topography (relief), organisms, and time. When reordered to climate, relief, organisms, parent material, and time, they form the acronym CROPT (Figure 6).

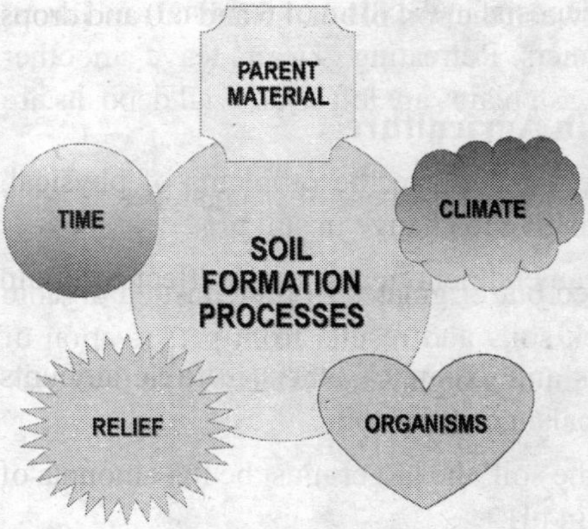


**Fig. 6:** Factors influencing pedogenesis

The mineral material from which a soil forms is called parent material. Rock, whether its origin is igneous, sedimentary, or metamorphic, is the source of all soil mineral materials and the origin of all plant nutrients with the exceptions of nitrogen, hydrogen and carbon. As the parent material is chemically and physically weathered, transported, deposited and precipitated, it is transformed into a soil. Typical soil parent mineral materials are: Quartz: $SiO_2$; Calcite: $CaCO_3$; Feldspar: $KAlSi_3O_8$; Mica (biotite): $K(Mg,Fe)_3AlSi_3O_{10}(OH)_2$.

Parent materials are classified according to how they came to be deposited. Residual materials are mineral materials that have weathered in place from primary bedrock. Transported materials are those that have been deposited by water, wind, ice or gravity. Cumulose material is organic matter that has grown and accumulates in place. Residual soils are soils that develop from their underlying parent rocks and have the same general chemistry as those rocks. The soils found on mesas, plateau, and plains are residual soils. In the United States as little as three percent of the soils are residual.

Most soils derive from transported materials that have been moved many miles by wind, water, ice and gravity.

- Aeolian processes (movement by wind) are capable of moving silt and fine sand many hundreds of miles, forming loess soils (60–90 percent silt), common in the Midwest of North America, north-western Europe, Argentina and Central Asia. Clay is seldom moved by wind as it forms stable aggregates.
- Water-transported materials are classed as either alluvial, lacustrine, or marine. Alluvial materials are those moved and deposited by flowing water. Sedimentary deposits settled in lakes are called lacustrine. Lake Bonneville and many soils around the Great Lakes of the United States are examples. Marine deposits, such as soils along the Atlantic and Gulf Coasts and in the Imperial Valley of California of the United States, are the beds of ancient seas that have been revealed as the land uplifted.

- Ice moves parent material and makes deposits in the form of terminal and lateral moraines in the case of stationary glaciers. Retreating glaciers leave smoother ground moraines and in all cases, outwash plains are left as alluvial deposits are moved downstream from the glacier.
- Parent material moved by gravity is obvious at the base of steep slopes as talus cones and is called colluvial material.
- Cumulose parent material is not moved but originates from deposited organic material. This includes peat and muck soils and results from preservation of plant residues by the low oxygen content of a high water table. While peat may form sterile soils, muck soils may be very fertile.

## Weathering

The weathering of parent material takes the form of physical weathering (disintegration), chemical weathering (decomposition) and chemical transformation. Generally, minerals that are formed under high temperatures and pressures at great depths within the Earth's mantle are less resistant to weathering, while minerals formed at low temperature and pressure environment of the surface are more resistant to weathering. Weathering is usually confined to the top few meters of geologic material, because physical, chemical, and biological stresses and fluctuations generally decrease with depth. Physical disintegration begins as rocks that have solidified deep in the Earth are exposed to lower pressure near the surface and swell and become mechanically unstable. Chemical decomposition is a function of mineral solubility, the rate of which doubles with each 10°C rise in temperature, but is strongly dependent on water to effect chemical changes. Rocks that will decompose in a few years in tropical climates will remain unaltered for millennia in deserts. Structural changes are the result of hydration, oxidation, and reduction. Chemical weathering mainly results from the excretion of organic acids and chelating compounds by bacteria and fungi, thought to increase under present-day greenhouse effect.

**Physical disintegration** is the first stage in the transformation of parent material into soil. Temperature fluctuations cause expansion and contraction of the rock, splitting it along lines of weakness. Water may then enter the cracks and freeze and cause the physical splitting of material along a path toward the center of the rock, while temperature gradients within the rock can cause exfoliation of "shells". Cycles of wetting and drying cause soil particles to be abraded to a finer size, as does the physical rubbing of material as it is moved by wind, water, and gravity. Water can deposit within rocks minerals that expand upon drying, thereby stressing the rock. Finally, organisms reduce parent material in size and create crevices and pores through the mechanical action of plant roots and the digging activity of animals. Grinding of parent material by rock-eating animals also contributes to incipient soil formation.

Chemical decomposition and structural changes result when minerals are made soluble by water or are changed in structure. The first three of the following list are solubility changes and the last three are structural changes.

1. Solution: The solution of salts in water results from the action of bipolar water molecules on ionic salt compounds producing a solution of ions and water, removing those minerals and reducing the rock's integrity, at a rate depending on water flow and pore channels.
2. Hydrolysis: Hydrolysis is the transformation of minerals into polar molecules by the splitting of intervening water. This results in soluble acid-base pairs. For example, the hydrolysis of orthoclase-feldspar transforms it to acid silicate clay and basic potassium hydroxide, both of which are more soluble.
3. Carbonation: In carbonation, the solution of carbon dioxide in water forms carbonic acid. Carbonic acid will transform calcite into more soluble calcium bicarbonate.
4. Hydration: It is the process of inclusion of water in a mineral structure, causing it to swell and leaving it stressed and easily decomposed.
5. Oxidation: Oxidation of a mineral compound is the inclusion of oxygen in a mineral, causing it to increase its oxidation number and swell due to the relatively large size of oxygen, leaving it stressed and more easily attacked by water (hydrolysis) or carbonic acid (carbonation).
6. Reduction: the opposite of oxidation, means the removal of oxygen, hence the oxidation number of some part of the mineral is reduced, which occurs when oxygen is scarce. The reduction of minerals leaves them electrically unstable, more soluble and internally stressed and easily decomposed. It mainly occurs in waterlogged conditions.

Of the above, hydrolysis and carbonation are the most effective, particularly in regions of high rainfall, temperature and physical erosion. Chemical weathering becomes more effective as the surface area of the rock increases, thus is favoured by physical disintegration. Saprolite is an example of a residual soil formed from the transformation of granite, metamorphic and other types of bedrock into clay minerals. Often called weathered granite, saprolite is the result of weathering processes that include: hydrolysis, chelation from organic compounds, hydration (the solution of minerals in water with resulting cation and anion pairs) and physical processes that include freezing and thawing. The mineralogical and chemical composition of the primary bedrock material, its physical features, including grain size and degree of consolidation, and the rate and type of weathering transforms the parent material into a different mineral. The texture, pH and mineral constituents of saprolite are inherited from its parent material. This process is also called arenization, resulting in the formation of sandy soils (granitic arenas), thanks to the much higher resistance of quartz compared to other mineral components of granite (micas, amphiboles, feldspars).

### 4.2.4. Sorption and precipitation reactions in soil

Generally ions and solution phases depending on the soil and material properties. The concept is illustrated in Figure7. The solution phase is the reactive soil and provides ions and compounds for plant and microbial uptake, leaching and other process illustrated in Figure8.

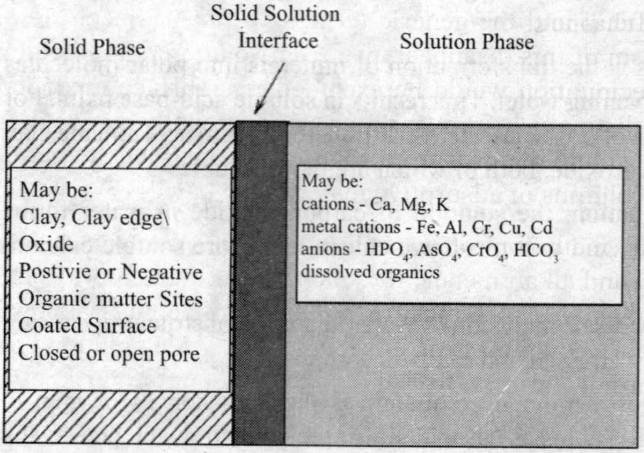


**Fig. 7:** Schematic depiction of the solid- solution interface

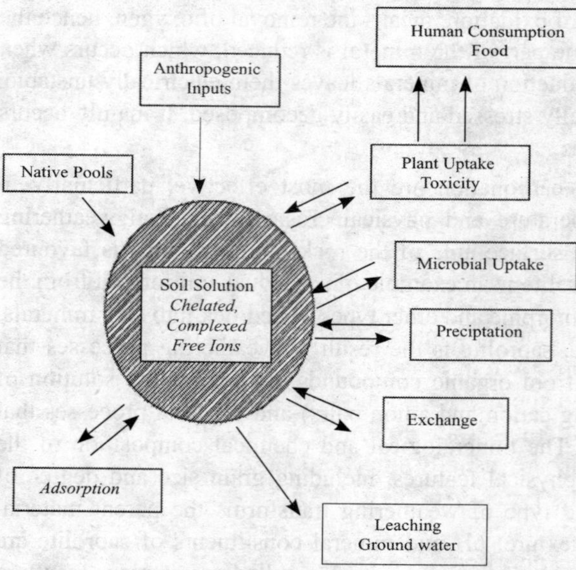


**Fig. 8:** The relation of solution phase to sources and sinks for ions and compounds

Adsorption is a general term that refers to the disappearance of solutes from solution with the presumption of adsorption on a solid phase. Adsorption is the accumulation at the solid- solution interface, and may result from either physical or chemical interaction with the colloidal surface. Physical adsorption is a relatively weak bonding to the surface while chemical (chemisorption) is a stronger interaction which involves ionic or covalent bonding in addition to van der Waal's and dispersion forces operative in physical adsorption. Adsorption refers to attraction and bonding onto a surface, while absorption is a process in which the solute is taken up into a structure or across a membrane. In some cases the distinction is difficult and the generic term **sorption** has been used. Nothing about the mechanism of this disappearance from solution is implied by the term sorption. Bulk precipitation would be excluded, but surface mediated precipitation is difficult to distinguish from sorption. Operationally, sorption is determined by the extent of solute removal from solution in either batch studies or in leaching studies with columns of adsorptive materials.

Adsorption reactions are the main control on solution composition for cations other than the alkaline earth and alkali metals, for anions other than chloride and nitrate, and for organic compounds. A typical technique is to supply a known concentration of **sorbate** to a known mass of **adsorbent**. After the solution and solid have come to equilibrium, (at a known constant temperature and known solid to solution ratio) solution concentration is then measured and the difference between the initial concentration and final equilibrium concentration adjusted for the solution volume is assumed to be the amount of sorption per unit mass of sorbent. Algebraically this is:

$$\text{Amount adsorbed} = q = \frac{\left(c_f - c_i\right) v_{solution}}{M_{sorbant}} \tag{1}$$

Units for the sorption onto the solid phase depend on the units of concentration and mass, however mg $kg^{-1}$ (ppm), mmol $kg^{-1}$, and : mol $g^{-1}$ are commonly used units for q. Some times the adsorption is expressed in terms of surface area and the units are moles/meter squared. In order to determine the solution concentration, the solid and solute must be separated. Centrifugation, and filtration are the commonly used techniques.

Knowing the amount of sorbate per unit mass of sorbate, **(q)** a plot of equilibrium concentration (**$C_{eq}$**) vs **q** is constructed. This plot is called an adsorption sorption isotherm. The name refers to the constant temperature maintained during the sorption process. Figure 9 depicts four common isotherms types. The data contained in these isotherms are usually analyzed to ascertain whether they conform to specific isotherms types. Another way to consider these isotherms is to calculate a distribution coefficient, or the partitioning between the solution and solid phase. Distribution coefficients are simply the ratio of solid phase concentration to solution phase concentration. These coefficients increase as adsorption increases.

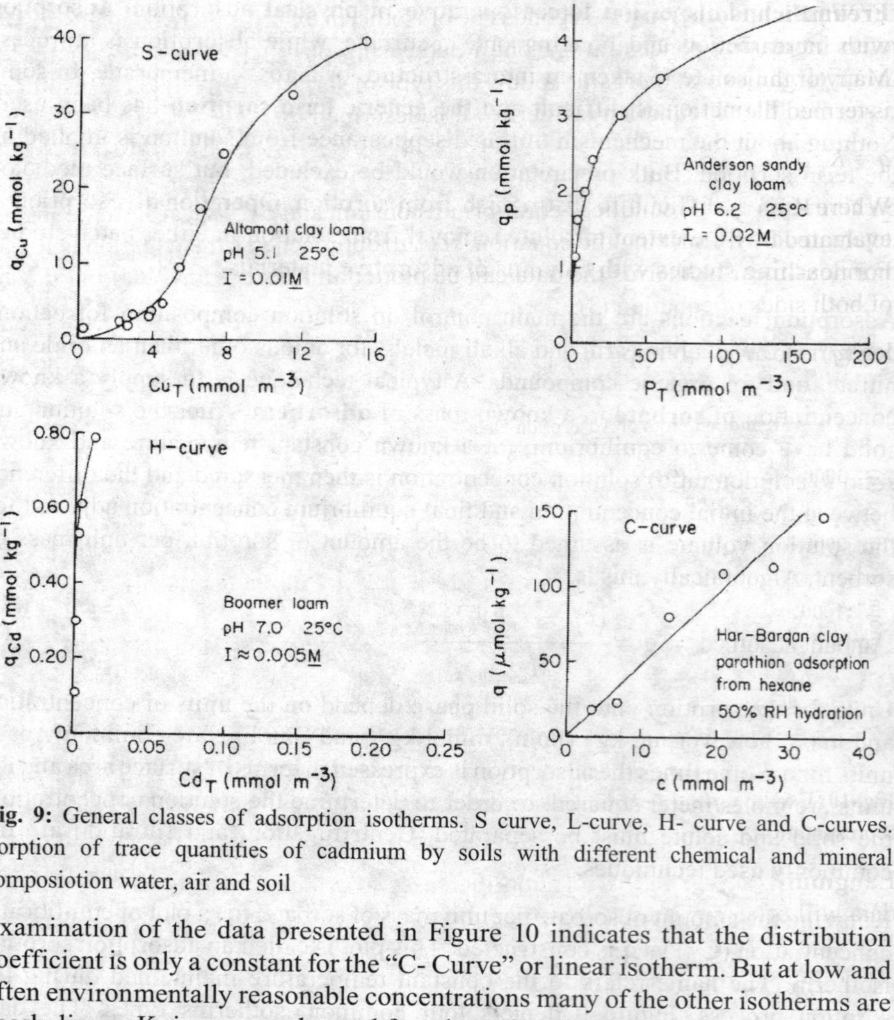


**Fig. 9:** General classes of adsorption isotherms. S curve, L-curve, H- curve and C-curves. Sorption of trace quantities of cadmium by soils with different chemical and mineral composiotion water, air and soil

Examination of the data presented in Figure 10 indicates that the distribution coefficient is only a constant for the "C- Curve" or linear isotherm. But at low and often environmentally reasonable concentrations many of the other isotherms are nearly linear. $K_d$ is commonly used for adsorption studies of organic compounds $K_d$ is defined by equation

$$K_d = \frac{\text{Concentration in the solid phase}}{\text{Concentration in the solution phase}} \quad (2)$$

## Common sorption isotherms types (descriptors of partitioning)

**Linear Isotherms** - The simplest form is the linear relationship between

q and $c_{eq}$. which is $q = a + b\,(c_{eq})$ (3)

In this instance linear regression may be used to find the slope (b) and the intercept (a). Many sorbates exhibit linear isotherms at low concentrations.

**Freundlich Isotherms:** A second type of isotherm exhibits increasing adsorption with increasing concentration, but a decreasing positive slope as $c_{eq}$ increases. Many organics and inorganic follow this type of sorption behavior. This isotherm is termed the "Freundlich Isotherm". It is described by:

$$q = K_F \, c^{1/n} \quad (4)$$

Where $K_F$ is the "Freundlich" equilibrium constant and 1/n is an arbitrary constant evaluated by linearizing the equation. If (1/n) approaches 1 the equation is linear. For non-linear isotherms the data can be plotted in linear form by taking the log of both sides of equation (3):

$$\log(q) = \log K_F + 1/n \log c_{eq} \quad (5)$$

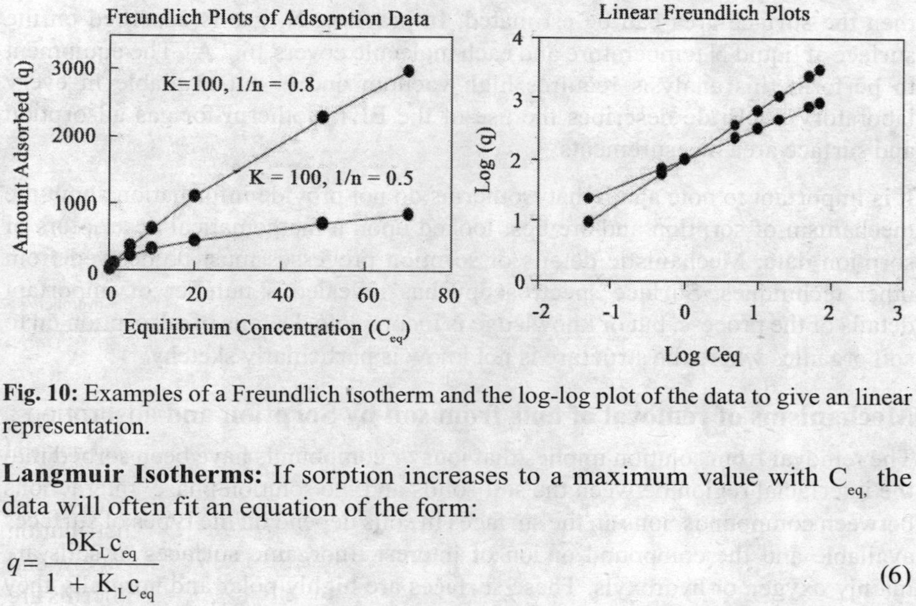


**Fig. 10:** Examples of a Freundlich isotherm and the log-log plot of the data to give an linear representation.

**Langmuir Isotherms:** If sorption increases to a maximum value with $C_{eq}$, the data will often fit an equation of the form:

$$q = \frac{bK_L c_{eq}}{1 + K_L c_{eq}} \quad (6)$$

This formulation is called a Langmuir Equation. Note that if $K_L \, c_{eq} << 1$, the equation is linear. As mentioned, at low concentration a Langmuir isotherm may appear to be nearly linear. The non-linear form can be evaluated by transforming to the linear equation:

$$\frac{c_{eq}}{q} = \frac{1}{K_L} = \frac{n}{bK_L} \quad (7)$$

## Brunauer, Emmett and Teller (BET) Isotherms

Another important isotherm type describes multi-layer sorption and was developed by Brunauer, Emmett and Teller (1938). After its authors, the isotherm is called a BET equation and has the form:

$$q = \frac{Bc_{eq}\, b}{\left(c_s - c_{eq}\right)\left[1 + (B-1)\left[\frac{c_{eq}}{c_s}\right]\right]} \qquad (8)$$

Like the other isotherms, the BET has a linear form:

$$\frac{c_{eq}}{\left(c_s - c_{eq}\right)} = \frac{1}{Bb} + \left[\frac{(B-1)}{Bb}\right]\left[\frac{c_{eq}}{c_s}\right] \qquad (9)$$

In the BET equations: B = a term for the energy of interaction with the surface; b = the monolayer capacity and; $C_s$ = the concentration of the solute at saturation.

The BET equation is important because it is used to measure surface area. If the monolayer capacity can be determined, and the area of a monolayer can be found, then the surface area can be estimated. In practice, $N_2$ gas is adsorbed on the surface at liquid N temperature and each molecule covers 16.2 $A^0$. The equipment to perform this analysis requires high vacuum and is not available in every laboratory. McBride describes the use of the BET isotherm for gas adsorption and surface area measurements.

It is important to note again that isotherms do not provide information about the mechanism of sorption and are best looked upon a mathematical descriptors of sorption data. Mechanistic details of sorption processes must be derived from other techniques. Surface spectroscopy has revealed a number of important details of the process, but or knowledge is incomplete. Details of adsorption on to soil organics where the structure is not know is particularly sketchy.

## Mechanisms of removal of ions from soil by Sorption and adsorption

The removal from solution implies that ions or compounds have been sorbed into the interfacial region between the soil solids and the solution phase. Interactions between compounds, ions an the surfaces in soils depend on the types of surfaces available and the compound or ion of interest. Inorganic surfaces in soils are mainly oxygen or hydroxyls. These surfaces are highly polar and normally they carry either a positive of negative charge. Organic surfaces can also be charged and have surfaces that range from strongly polar to nonpolar.

**Sorption of inorganic compounds on soil surfaces**. Inorganic compounds are adsorbed mostly by chemical interaction with soil surfaces. These interactions range from purely electrostatic (*e.g.* exchange) to strong covalent bonding. Exchangeable ions are considered to be fully hydrated and completely dissociated from the surface. The real picture of exchange is more complex and even the alkali metals may be adsorbed into a **Stern Layer** in which the ions are partially dehydrationed. Figure11 depicts the interaction of Li, Na and K with a soil surface. Sodium and Li are strongly hydrated and too large to form specific bonds with the mineral surface. However, $K^+$ can be strongly bound by many soil minerals.

This specific binding is not considered in Figure

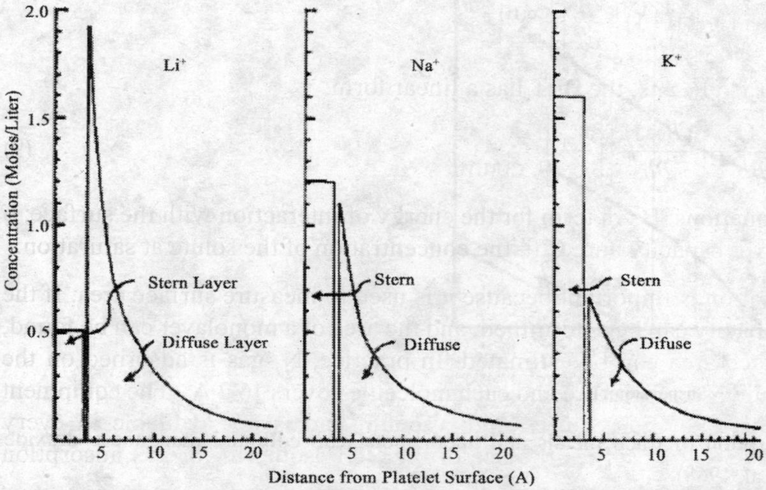


**Fig. 11:** A calculated distribution of cations near a clay mineral surface (after Shainberg and Kemper, 1966)

In contrast to the alkaline earth and alkali metals, that are easily replaced by other exchangeable cations, phosphorus is strongly adsorbed on soil surfaces. This sorption is believed to occur via displacement of surface hydroxyl groups and the formation of mono and bidentate surface complexes with covalent bonding character. Electrostatic exchange of alkali metals and covalent bonding of P are the end members of a continuous array of progressively stronger interactions with the colloidal surfaces. In this array of possibilities, the nature of the binding site on a particular sorbant may be different, the sorbant may change and the different sorbates can display a range of interactions.

A term has been coined which separates strong adsorption from weaker interactions. **Specific adsorption or sorption** is an operational term related to the electrophoretic mobility of the sorbant. Sorbates which can cause a charge reversal in the sorbant are said to be specifically adsorbed. Implicit in this definition is that the sorbates are entering a "Stern or Stern-like layer" at the interface. Also implicit in this is an interaction stronger than the electrostatic interaction ascribed to exchange. Figure 12 depicts the difference among cations in their ability to be adsorbed on colloidal $MnO_2$ in relation to pH. Those ions adsorbed below the point of zero charge are concentrated at the interface in opposition to electrostatic forces and are therefore considered to be specifically adsorbed.

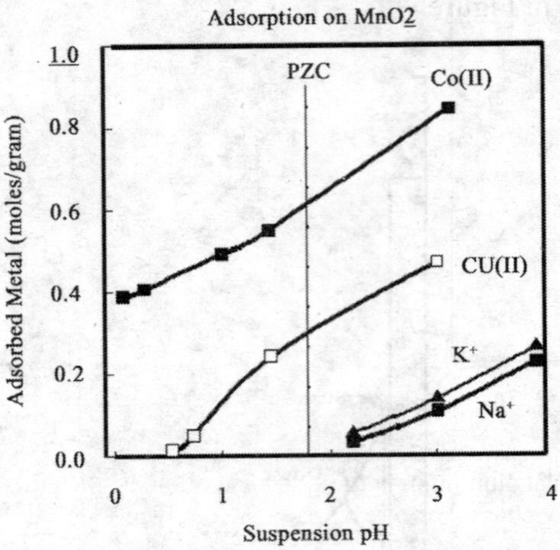


**Fig. 12:** Adsorption of trace metals and alkali metals on colloidal manganese dioxide (*After* Murry et al. 1968)

Ions adsorbed into the interfacial region by specific adsorption have the capability to change the potential at the interface between the solid and the solution and thereby change the electrophoretic mobility of the particle. If the particle undergoes a charge reversal, by definition, the ions causing this charge reversal are considered to be specifically adsorbed. Figure 13 and 14 depict the effects of Al on the electrophoretic mobility of $SiO_2$. At high pH where the surface is covered with Al, the Si surface appears to behave like aluminum oxide. If ions are specifically adsorbed in the surface they may be bonded by covalent linkages and will be dehydrated. Ions with strong ionic interactions with the surface may also be partially dehydrated.

Another way of distinguishing the bonding between a surface site and an ion is to determine whether they form inner sphere or outer sphere complexes. Inner sphere complexes do not have waters of hydration between the ion and the surface, while the outer sphere complexes are hydrated. This is very similar to the distinction between an ion pair and a complex.

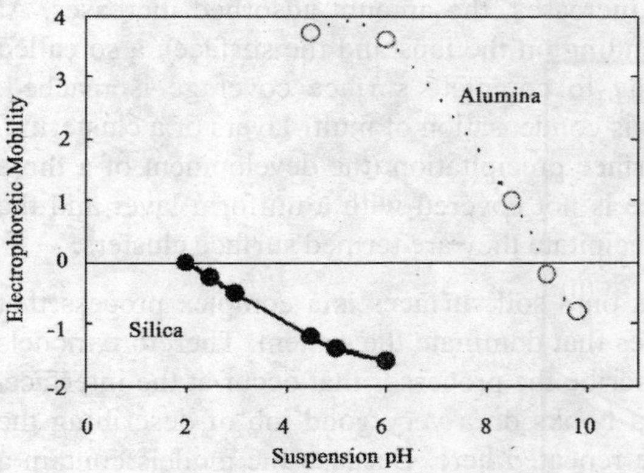


**Fig.13:** The electrophoretic mobility of colloidal aluminum hydroxide and colloidal silica in relation to pH. After Stiger, 1975

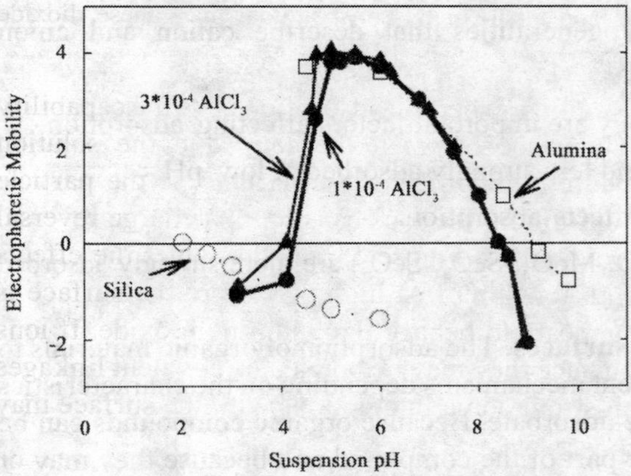


**Fig. 14:** The electrophoretic mobility of colloidal aluminum hydroxide and colloidal silica in relation to pH and added $AlCl_3$. After Stiger, 1975

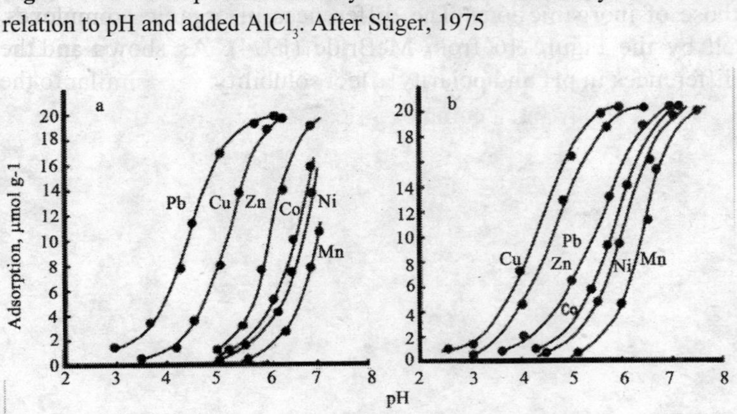


**Fig. 15:** The effects of pH on metal adsorption on (a) hematite and (b) goethite. Data from McKenzie (1980) taken from Sparks (1995).

When the level of solution ions increases, the amount adsorbed increases. At some point in the isotherm (depending on the ions and the surface), a so called monolayer capacity corresponding to complete surface coverage is reached. Any adsorption beyond this point is condensation of multi-layers or a clustering. Multi layers are equivalent to surface precipitation (the development of a three dimensional phase). If the surface is not covered with a uniform layer and the precipitate occurs as islands of precipitate they are termed surface clusters.

Obviously, the adsorption of ions onto soil surfaces is a complex process that varies among the ions and surfaces that dominate the system. Therefore models have been used extensively to describe the processes that occur at the interface. Several recent soil chemistry text books do a very good job of describing the various models and it will not be repeated here. Because the models contain a number of adjustable parameters they can all normally be "manipulated" to fit the experimental data. Our modeling efforts have outstripped our ability to understand the surfaces that adsorb ions.

None-the-less there are a few generalities that describe cation and anion adsorption in soils.

1. Surface area and mineralogy are important factors affecting adsorption.
2. Metals are more soluble and less strongly adsorbed at low pH.
3. The identity of the metal affects absorption..
4. Anions (eg $PO_4$, $AsO_4$, $AsO_3$ $MoO_4$, $SeO_4$, $SeO_3$) are more strongly adsorbed at low pH.

**Adsorption of organics on soil surfaces.** The adsorption of organic materials to soils can occur via several different mechanisms depending on the characteristics of the adsorbing surface and the adsorbate. Because organic compounds can be either polar or nonpolar in all or part of the compound and because they may or may not be charged, the interactions of organics with soil surfaces can be quite different from those of inorganic ions. The differences in organic compounds is illustrated well by the Figure 16 from McBride (1994). As shown in this representation, differences in pH and polarity affect solubility.

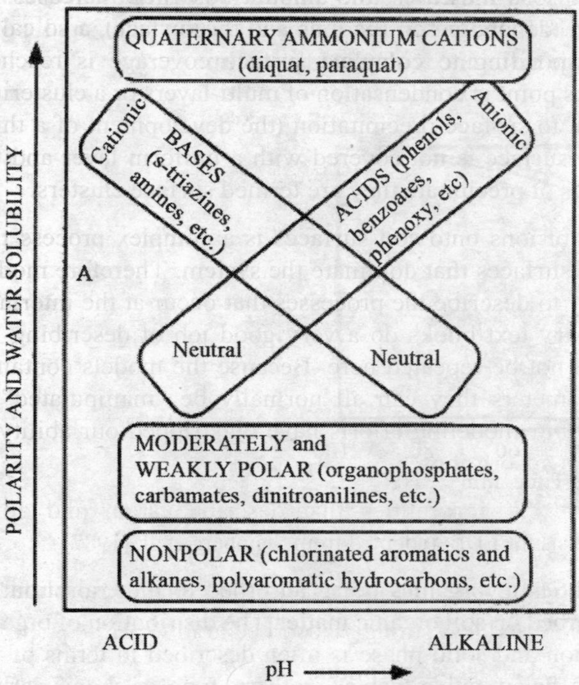


**Fig. 16:** Schematic representation of ranges in polarity and solubilities of organic compounds in relation to pH. (From McBride 1994).

Several mechanisms of adsorption for organic compounds on mineral and organic surfaces is listed in table 6. The variations in adsorption for organic compounds by soils is large.

The reversibility of adsorption is an important consideration. For many ions and compounds, the desorption does not follow the same isotherm as adsorption.

**Table. 6:** The mechanisms of organic matter interaction with solution ions.

| Mechanism | Principal organic function groups involved. |
|---|---|
| Cation exchange | Amines, ring NH, heterocyclic N |
| Protonation | Amines, heterocyclic N, carbonyl, carboxylate |
| Anion exchange | Carboxylate |
| Water bridging | Amino, carboxylate, carbonyl, alcoholic OH |
| Cation bridging | Carbonylate, amines, carbonyl, alcoholic OH |
| Ligand exchange | Carboxylate |
| Hydrogen bonding | Amines, carbonyl, carboxyl, phenylhydroxyl |
| Van der Waals interactions | Uncharged, nonploar organic functional groups |

This difference occurs with inorganic ions, but the desorption of organic compounds is especially subject to hysteresis. Hysteresis is defined as a difference in partitioning during adsorption and desorption. Figure 17 illustrates an example of hysteresis for desorption of EDB. In this case, slow diffusion from small pores is a likely explanation for the differences observed over time.

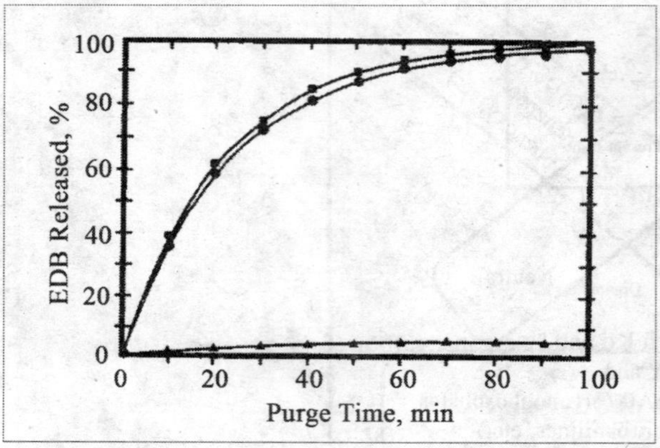


**Fig.17:** An example of hysteresis for EDB. (adopted from Steinberg et al. (1987)

In addition to the possibilities of organics being adsorbed to mineral surfaces, these compounds can be sorbed on soil organic matter. The distribution of organic compounds between solution and solid phase is often described in terms of the partition or distribution coefficient rather than an isotherm type, as shown earlier. Some organic surfaces are not as polar as mineral surfaces and provide sorption possibilities for less polar organics. For the nonpolar organics, water is a strong competitor for sites on minerals surfaces and the adsorption of organics decreases with increasing water content. For volatile organic compounds (VOC's) in dry soils, the distribution coefficient is can be much greater than in moist soils.

Figure 18 shows that the log of $K_d$ increases dramatically as soil moisture content decreases. This particular $K_d$ is calculated for the partitioning between the vapor phase and the solid phase. Since organic compounds (particularly the less polar ones) are associated with organic materials, the distribution coefficient can be normalized for organic matter content. The resulting distribution coefficient often shows much less variability among soils.

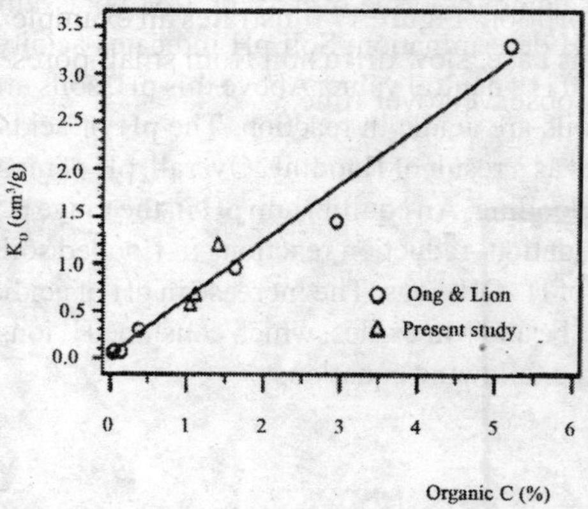


**Fig. 18:** The relationship between Kd and % organic carbon (from Petersen et al. 1995)

## 4.2.5. Chemistry of puddle paddy soils

Lowland rice, also known as irrigated rice or flooded rice, is grown on leveled lands with bunds and with irrigation facilities. Under normal conditions, lowland rice fields are flooded with water about 3 to 4 weeks after sowing. The water level of about 10 to 15 cm is maintained during the crop growth cycle and is drained before harvest. Flooding or water logging eliminates oxygen from the rhizosphere and causes changes in the soil chemical properties.

The most important change in the soil as a result of flooding is the conversion of the root zone of the soil from an aerobic environment to an anaerobic or near-anaerobic environment where oxygen is absent or limiting. Oxygen deficiency or exclusion in submerged soils can occur within a day after flooding. The oxygen movement through the flooding water is usually much slower than the rate at which oxygen can be reduced in the soil. This situation may result in the formation of two distinctly different layers being formed in a waterlogged soil. On the top is an oxidized or aerobic surface layer where oxygen is present, with a reduced or anaerobic layer underneath in which no free oxygen is present. A thin oxidized layer (usually 1 to 20 mm in thickness) normally found at the interface between water and soil. In addition, flooding also has major effects on the availability of macro and micronutrients. Some nutrients are increased in availability to the crop, whereas others are subject to greater fixation or loss from the soil as a result of flooding.

### Chemical changes occur in puddled rice soils

The most important chemical changes that occur in flooded or submerged rice soils are to pH, redox potential, and ionic strength or electrical conductivity. These changes occur as a result of oxygen depletion. Soil pH is an important chemical property because of its influence on soil microorganisms and availability of nutrients to plants. It is determined by a pH meter using a glass electrode and in a specific soil: solution ratio. Usually distilled water or 0.01 M calcium chloride ($CaCl_2$) or 1 M potassium chloride (KCl) solution is used for soil pH determination. Soil pH indicates acidity, alkalinity, or neutrality of a soil. Soil pH 7.0 is a neutral value. Above this pH, soils are designated as alkaline, and below this, soils are acidic in reaction. The pH of acidic soils increases and alkaline soils decreases as a result of flooding. Overall, pH of most soils tends to change toward neutral after flooding. An equilibrium pH in the range 6.5 to 7.5 is usually attained. A majority of oxidation–reduction reactions in flooded soils involve either consumption or production of $H^+/OH^-$ ions. The increase in pH of acidic soils is mainly determined by reduction of Fe and Mn oxides, which consume $H^+$ ions. These reduction processes are shown in the following equations:

$$Fe_2O_3 + 6H^+ + 2e^- \leftrightarrow 2Fe^{2+} + 3H_2O$$

$$MnO_2 + 4H^+ + 2e^- \leftrightarrow Mn^{2+} + 2H_2O$$

The decrease in the pH of alkaline soils is associated with the microbial decomposition of organic matter, which produces $CO_2$, and the produced $CO_2$ reacts with $H_2O$ to form carbonic acid, which dissociates into $H^+$ and bicarbonate ($HCO_3^-$) ions. Patrick and Reddy (1978) reported that the decrease in pH of alkaline and calcareous submerged soils is associated with sodium carbonate ($Na_2CO_3$)–$H_2O$–$CO_2$ and calcium carbonate ($CaCO_3$)– $H_2O$–$CO_2$ systems, respectively. Figure 19 shows the change in soil pH of lowland rice collected from four locations. It can be seen from Figure 19 that soil pH increases with flooding and stabilized around 56 days after flooding in all the soils. However, the magnitude of pH change differs from soil to soil. The pH of most agricultural soils is in the range of 4 to 9 (Fageria 2009). The most suitable pH for growth of annual crops such as soybean, corn, dry bean, and wheat in lowland soils is around 6.5 (Fageria and Baligar 1999).

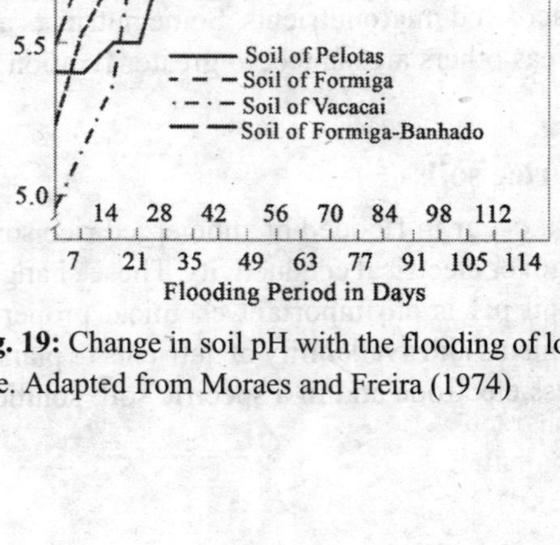


**Fig. 19:** Change in soil pH with the flooding of lowland rice. Adapted from Moraes and Freira (1974)

## Oxidation–Reduction potential

Oxidation–reduction or redox potential has significant influence on chemistry of iron and other nutrients in the submerged soils. It is the best single indicator of the degree of anaer- obiosis in the flooded soil and allows reasonable predictions to be made concerning the behavior of several essential plant nutrients. Oxidation is the donation and reduction is the acceptance of electrons from other substances. Oxidizing agents accept electrons from other substances and thereby reduce themselves. Reducing agents donate electrons to other substances. For example, $Fe^{++}$ is an electron donor or a reducing agent when it oxidized to $Fe^{+++}$. Hydrogen peroxide ($H_2O_2$) is an oxidizing agent when it accepts electrons from organic matter and oxidizes it to $CO_2$ (Bohn, McNeal, and O'Connor 1979).

Oxidation–reduction potential is measured in milli volts, and symbol used for this chemical change in flooded soil is Eh. Oxidized soils have redox potentials in the range of + 400 to + 700 millivolts, whereas waterlogged soils' redox potential is generally in the range of -100 to -300 millivolts (Patrick and Mahapatra 1968). Important oxidation– reduction processes that occur in the waterlogged soils are presented in Table 7. Some of the oxidized soil components that undergo reduction after oxygen is depleted are reduced sequentially; that is, all of the oxidized components of one system will be reduced before any of the oxidized components of another system begin to be reduced. As the $O_2$ depletes from the waterlogged soils, reduction processes occur in sequence. Nitrate and manganese compounds are reduced first, then ferric compounds are reduced to the ferrous form, and at last sulfate is reduced to sulfide. Redox potential decreased with flooding of rice soils (Figure 20).

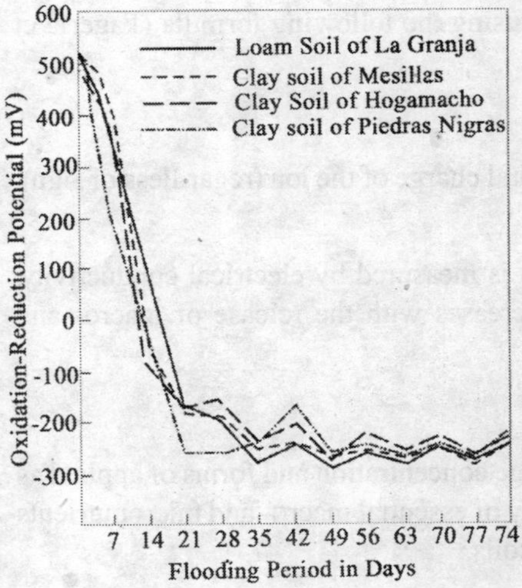


**Fig. 20:** Influence of flooding on redox potential of some Mexican soils. Adapted from Moraes and Freira (1974).

A rapid decline in redox potential is characteristic of soils with low contents of reducing Fe and Mn and high organic matter content. Iron and Mn compounds serve as buffers against the development of reducing conditions in the soil (Patrick and Mahapatra 1968). The critical redox potentials for Fe reduction and consequent dissolution are between +300 mV and +100 mV at pH 6 and 7, and -100 mV at pH 8, while at pH 5 appreciable reductions occur at +300 mV (Gotoh and Patrick 1976). Oxidation–reduction or potential reduction values for oxidized and submerged soils and reduction processes are given in Table 7.

**Table 7:** Range of oxidation–reduction potential values in oxidized and submerged soils and at which reduction processes occur. Adapted from Patrick (1966), Patrick and Reddy (1978), Marschner (1995), and Fageria et al. (2008).

| Soil moisture/reduction processes | Redox potential (mV) |
|---|---|
| Well-oxidized soils | +700 to +400 |
| Moderately reduced soils | +400 to +200 |
| Reduced soils | +100 to −100 |
| Highly reduced soils | −100 to −300 |
| $NO_3^-$ to $N_2$ | +280 to +220 |
| $Mn^{4+}$ to $Mn^{2+}$ | +280 to +220 |
| $Fe^{3+}$ to $Fe^{2+}$ | +180 to +150 |
| $SO_4^{2-}$ to $S^{2-}$ | −120 to −180 |
| $CO_2$ to $CH_4$ | −200 to −280 |
| $O_2$ to $H_2O$ | +380 to +320 |
| Absence of free $O_2$ | +350 |

## Ionic Strength

Ionic strength is defined as the measure of the electrical environment of ions in a solution. Ionic strength can be calculated by using the following formula (Fageria et al. 2008):

$$\text{Ionic strength} = 1/_2\ \Sigma M_i Zi^2$$

where Mi is the molarity of the ion, $Z_i$ is the total charge of the ion (regardless of sign), and $\Sigma$ is a symbol meaning the "sum of."

The concentration of ions in the soil solution is measured by electrical conductivity. The ionic strength of the submerged soil increases with the release of macro- and micronutrients in the soil solution.

## Nutrient Availability

Reducing conditions in flooded rice soils change concentration and forms of applied as well as native soil nutrients. Hence, availability of essential macro- and micronutrients is significantly influenced in the flooded rice soils.

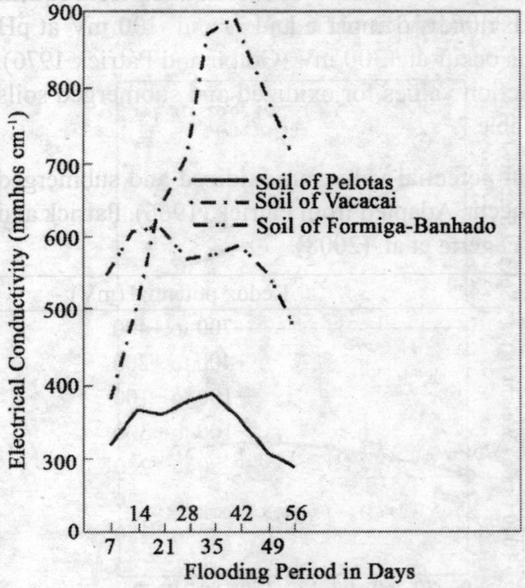


**Fig. 21:** Influence of flooding on electrical conductivity of some Mexican soils. Adapted from Moraes and Freira (1974)

## a) Nitrogen availability

Nitrogen improves grain weight and reduces grain sterility. Figure 22 shows influence of N on yield components of lowland rice. Grain yield in rice is a function of panicles per unit area, number of spikelets per panicle, 1000-grain weight, and spikelet sterility or filled spikelets. Therefore, it is very important to understand the management practices that influence yield components and consequently grain yield. Nitrogen application up to 210 kg $ha^{-1}$ influenced panicle length significantly ($P < 0.01$) and the relationship between N applied and panicle length was linear (Figure 22). The number of panicles $m^{-2}$ and 1000 grain weight also increased significantly and quadratically with the application of N fertilizer. Spikelet sterility, however, decreased significantly and linearly with increasing N rates. Nitrogen treatment account for about 96% variation in panicle length, about 91% variation in panicles $m^{-2}$, about 75% variation in spikelet sterility, and about 73% of variation in 1000 grain weight.

A major part of N in the flooded rice soils is lost through leaching and denitrification. The major biological reaction involving nitrate in flooded soil is denitrification. Denitrification is the biological process in which nitrate reduces to N gas or nitrous oxide, or both. It was reported that denitrification losses of 50% or more of applied N are common in flooded rice soils. Frequent fluctuations in moisture content of a field as a result of flooding and drainage create favourable conditions for denitrification. Nitrogen converted to the nitrate form during the period when the soil is drained is lost through denitrification when soil is flooded. Deep placement of N in the flooded rice reduces N lost through denitrification. Nitrate produced in the surface oxidized layer of a waterlogged soil can easily move downward by diffusion and percolate into the underlying reduced layer, where it is rapidly denitrified.

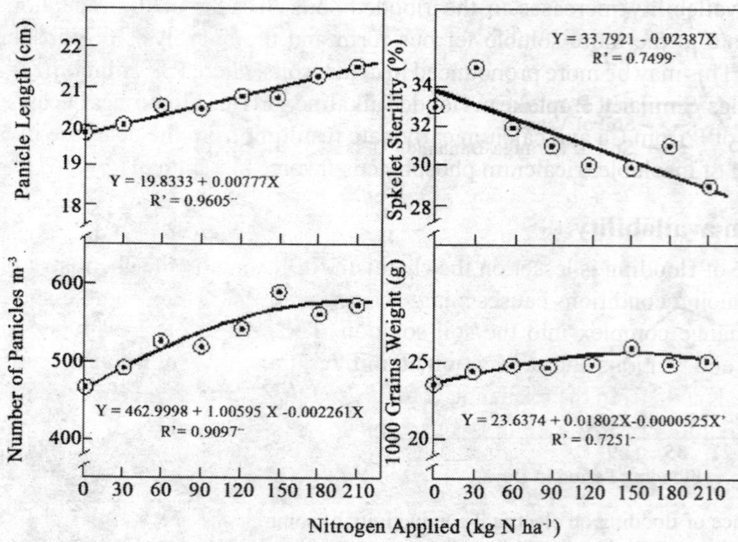


**Fig. 22:** Influence of nitrogen on number of panicles, panicle length, thousand-grain weight, and spikelet sterility of lowland rice. (adopted from Fageria and Baligar,2001).

Even with best management practices such as adequate rate, forms, methods, and timing of application, the utilization of added N is generally poorer in flooded rice soils. Patrick and Mahapatra (1968) reported that 30 to 40% applied N is recovered by lowland rice as compared to an availability of 50 to 60% when applied to upland crops.

In aerated soils, most of the N is in the form of $NO_3^-$ because of the nitrification process. In waterlogged soils, absence of $O_2$ inhibits the activity of the *Nitrosomonas* microorganisms that oxidize $NH_4^+$, and therefore N mineralization stops at the ammonium ($NH_4^+$) form. Accumulation of $NH_4^+$ in the waterlogged soils would mean that the N is not lost from the soil–plant system, as is the case in denitrification. This may only happen if rice fields are constantly flooded during the crop growth cycle. If availability of water is not under farmers' control because of lack of rainfall or storage facility, the situation may change in the transformation and availability of N to plants. Hence, if N is applied in the reduced soil layer and the water level is maintained in the rice field constantly, N uptake may improve in flooded rice.

## b) Phosphorus availability

Phosphorus deficiency is one of the most important yield-limiting factors in annual crops grown on highly weathered acidic soils of the tropics. The P deficiency is associated with low natural P as well as with high P-fixation capacity of these soils. Added soluble P is usually rapidly adsorbed on the surfaces of iron (Fe) and aluminum

(Al) oxides, which are followed by immobilization in other forms and within soil particles.

Phosphorus availability increases in the flooded soils because of the reduction of ferric phosphate to the more soluble ferrous form and the hydrolysis of phosphate com- pounds. This may be more pronounced in acidic soils where P is immobilized by Fe and Al oxides. Similarly, P uptake in flooded alkaline soils also improves because of the liberation of P from Ca and calcium carbonate resulting from the decrease in pH. The formation of insoluble tricalcium phosphate is favored at a high pH.

**d) Potassium availability**

The influence of flooding is lesser on the chemistry of K than on the chemistry of N and P. The reducing conditions causes a larger fraction of the K ions being displaced from the exchange complex into the soil solution. The release of a relatively large amount of Fe and Mn ions and production of ammonium ions result in displacement of some of the K ions from the exchange complex to the soil solution. This may leads to greater availability of K to rice in flooded soils).

**e) Sulfur availability**

In flooded soils, $SO_4^{2-}$ ion is reduced to hydrogen sulfide ($H_2S$) by anaerobic microbial activities. Furthermore, in flooded soils, $Fe^{3+}$ reduction to $Fe^{2+}$ precedes $SO_4^{2-}$ reduction; $Fe^{2+}$ will always be present in the soil solution by the time $H_2S$ is produced, so that $H_2S$ will be converted to insoluble iron sulfide (FeS). This reaction protects microorganisms and higher plants from the toxic effects of $H_2S$ . Overall, availability of S is reduced in flooded soils due to formation of insoluble FeS.

**f) Calcium and Magnesium availability**

Calcium (Ca) and magnesium (Mg) deficiencies are rare in lowland rice. Rice is highly tolerant to soil acidity. In highly acidic soils, dolomitic lime can be added to supply Ca and Mg. Only a small amount of these elements are removed in the grain, and unless the straw is removed from the field, the total removal is small. Changes in Ca and Mg concentrations are minimum in flooded soils.

**g) Micronutrients availability**

The $Fe^{3+}$ reduces to $Fe^{2+}$ and $Mn^{4+}$ reduces to $Mn^{2+}$; hence uptake of these elements increased in the flooded rice soils. The reduction processes of Fe and Mn are shown under the section on pH changes. The greater concentration of $Fe^{2+}$ (>300 mg $kg^{-1}$) may be toxic to rice plants under certain conditions. For most crops the critical deficiency soil Fe concentration range was 2.5–5.0 mg $kg^{-1}$ of di-ethylene triamine pentaacetic acid (DTPA)–extractable Fe but is also influenced by soil pH. In field and pot experiments, the degree of bronzing in a given variety showed a high significant correlation with yield. Iron toxicity in rice plants as indicated by bronzing of leaves, was reported when soluble Fe in the soil solution was more than 300–500 mg $kg^{-1}$ by DTPA extracting solution. Fe toxicity level in the soil is also dependent on the extracting solution used to extract the Fe from the soil. Values of Fe toxivity for the Mehlich- 1 extracting solution are greater than for the DTPA solution.

Iron toxicity in lowland rice has been reported in South America, Asia, and Africa. Metal toxicity in crop plants can be expressed in two ways. One is when metal is absorbed in greater amounts and becomes lethal to the plant cells. This is known as direct toxicity of metals. Another metal toxicity is associated with inhibition of uptake and utilization of essential nutrients by plants. This is known as indirect metal toxicity. Indirect toxicity creates nutrient imbalance in plants. This type of Fe toxicity is more common in lowland rice than direct toxicity. The most important nutrient deficiencies observed in irrigated or flooded rice in Brazil are P, K, and Zn. The yield reduction of rice cultivars due to Fe toxicity depends on tolerance or susceptibility of cultivars to toxicity.

Zinc and copper (Cu) concentrations generally decreased after flooding rice soils. The decrease in concentration with the flooding may be associated with increase in soil pH after flooding. Little is known about the behavior of B and Mo in the submerged soils. Boron concentration seems to remain more or less constant after submergence of rice soils. Molybdenum concentration in rice soils was found to increase after submergence, possibly because of the increased pH. In flooded soils, Si generally tends to increase after submergence. This increase is probably due to the release of adsorbed and occluded Si from oxyhydroxides of Fe and Al as well as to the effect of the increased pH resulting from submergence. Decompositing rice straw with its high silica content may also contribute to the increased Si content of the soil solution of flooded soils.

### 4.2.6. Thermodynamic theories

Thermodynamics is the study of relation between work, heat and chemical reactions or with the physical changes of the state which are confined to the laws of thermodynamics. Some general terms like heat, energy, and work were done are often used in thermodynamics. Let us learn a bit about basic thermodynamics and understand these terms.

#### Internal Energy

It referred to the energy content within the system. The energy represents the overall energy of the system and may include many forms of energy such as potential energy, kinetic energy etc. In a chemical reaction, we know about the energy transformations and basic thermodynamics provides us with the information regarding energy change associated with the particles of the system.

#### Factors Affecting the Internal Energy

- Heat passes into or out of the system,
- Work is done on or by the system or matter enters or leaves the system.

**Work:** Work done by a system is defined as the quantity of energy exchanged between a system and its surroundings. Work is completely governed by external factors such as an external force, pressure or volume or change in temperature etc.

**Heat:** Heat in thermodynamics is defined as the kinetic energy of the molecules of the substance. Heat and the thermodynamics together form the basics which helped process designers and engineers to optimize their processes and harness the energy associated with chemical reactions economically. Heat energy flows from higher temperature to lower temperature.

## Thermodynamic System and the Surrounding

Thermodynamics defines a system as the part of the universe under study that is the part where observations are made. The surrounding and universe interact with each other and depending on the type of the system, exchange of matter and energies occur. The system's classification entirely depends on the movement of energy and matter in or out of the system. There are two types of system:

1. **Open System:** Those systems where the exchange of energy, as well as matter takes place, are treated as an open system. For example, water is boiled on a stove without it being covered, the container behaves as an open system because it receives heat energy from an external source and the matter being released are water vapours.
2. **Closed System:** A closed system is referred to the system where only energy can be exchanged with the surrounding and not the matter. It can also be said that a closed system has a constant amount of matter and only the energy of the system can be changed. For instance, when we keep a shield bottle of water in the fridge, the loss of energy to the surrounding will result in the temperature of the water inside it going down even though the quantity of water in the bottle remains the same.
3. **Isolated System:** In case of isolated systems, neither energy nor matter can be exchanged between the system and the surrounding. An example of an isolated system is a Thermos flask.

The major aspects on which environment can affect the thermodynamics of a system are, internal energy, enthalpy, entropy, gibbs free energy,

The laws of thermodynamics define a group of physical quantities, such as temperature, energy, and entropy, that characterize thermodynamic systems in thermodynamic equilibrium. The laws also use various parameters for thermodynamic processes, such as thermodynamic work and heat, and establish relationships between them. They state empirical facts that form a basis of precluding the possibility of certain phenomena, such as perpetual motion. In addition to their use in thermodynamics, they are important fundamental laws of physics in general, and are applicable in other natural sciences.

Traditionally, thermodynamics has recognized three fundamental laws, simply named by an ordinal identification, the first law, the second law, and the third law. A more fundamental statement was later labelled as the zeroth law, after the first three laws had been established.

The zeroth law of thermodynamics defines thermal equilibrium and forms a basis for the definition of temperature: If two systems are each in thermal equilibrium with a third system, then they are in thermal equilibrium with each other. First law of thermodynamics: When energy moves into or out of a system, the system's internal energy changes in accordance with the law of conservation of mass. Second law of thermodynamics: The state of the entropy of the entire universe, as an isolated system, will always increase over time. Third law of thermodynamics: Entropy of a perfect crystal at absolute zero is zero. The laws of thermodynamics were the most important lesson for people understanding the mechanism behind phase change of matter.

The first law of thermodynamics states that, when energy passes into or out of a system (as work, heat, or matter), the system's internal energy changes in accord with the law of conservation of energy.

The second law of thermodynamics states that in a natural thermodynamic process, the sum of the entropies of the interacting thermodynamic systems never decreases. Another form of the statement is that heat does not spontaneously pass from a colder body to a warmer body.

The third law of thermodynamics states that a system's entropy approaches a constant value as the temperature approaches absolute zero. With the exception of non-crystalline solids (glasses) the entropy of a system at absolute zero is typically close to zero.

The first and second law prohibit two kinds of perpetual motion machines, respectively: the perpetual motion machine of the first kind which produces work with no energy input, and the perpetual motion machine of the second kind which spontaneously converts thermal energy into mechanical work.

**Gibb's free energy:** Gibbs free energy is equal to the enthalpy of the system minus the product of the temperature and entropy. The equation is given as;

$G = H - TS$ ;

Where, G = Gibbs free energy; H = enthalpy; T = temperature; S = entropy

Gibbs free energy is a state function hence it doesn't depend on the path. So change in Gibbs free energy is equal to the change in enthalpy minus the product of temperature and entropy change of the system.

$\Delta G=\Delta H-\Delta(TS)$

If the reaction is carried out under constant temperature $\{\Delta T=O\}$

$\Delta G=\Delta H-T\Delta S$

This equation is called the Gibbs Helmholtz equation

$\Delta G>0$; the reaction is non-spontaneous and endergonic

$\Delta G<0$; the reaction is spontaneous and exergonic

$\Delta G=0$; reaction is at equilibrium

1. According to the second law of thermodynamics entropy of the universe always increases for a spontaneous process.
2. $\Delta G$ determines the direction and extent of chemical change.
3. $\Delta G$ is meaningful only for reactions in which the temperature and pressure remain constant. The system is usually open to the atmosphere (constant pressure) and we begin and end the process at room temperature (after any heat we have added or which is liberated by the reaction has dissipated).
4. $\Delta G$ serves as the single master variable that determines whether a given chemical change is thermodynamically possible. Thus if the free energy of the reactants is greater than that of the products, the entropy of the world will increase when the reaction takes place as written, and so the reaction will tend to take place spontaneously. $\Delta S$ universe = $\Delta S$ system + $\Delta S$ surroundings
5. If $\Delta G$ is negative, the process will occur spontaneously and is referred to as exergonic.
6. Therefore spontaneity is dependent on the temperature of the system.

Free energy change criteria for predicting spontaneity is better than entropy change criteria because the former requires free energy change of system only whereas the latter requires entropy change of system and surroundings.

The second law of thermodynamics is defined on the following basis; All spontaneous processes are thermodynamically irreversible. It is impossible to convert heat completely into work without wastage. The entropy of the universe is continuously increasing. The total entropy change i.e, entropy change of the system + entropy change of surroundings is positive.

**Table 8:** Entropy and Enthalpy on Gibb's free energy,

| | $\Delta H<0$ | $\Delta H>0$ |
|---|---|---|
| $\Delta S>0$ | Spontaneous at all temperatures ($\Delta G<0$) | Spontaneous only at high temperature (when $T\Delta S$ is high) |
| $\Delta S<0$ | Spontaneous at low temperature (when $T\Delta S$ is small) | Not spontaneous at any temperature ($\Delta G>0$) |

- Spontaneous – is a reaction that is considered to be natural because it is a reaction that occurs by itself without any external action towards it.
- Non-spontaneous – needs constant external energy applied to it in order for the process to continue and once you stop the external action the process will cease.

**Standard-state free energy of reaction ($\Delta G°$):** The free energy of reaction at standard state conditions.

$\Delta G° = \Delta H° - T\Delta S°$

**Standard-state conditions**

1. The partial pressure of any gases involved in the reaction is 0.1 MPa.
2. The concentrations of all aqueous solutions are 1 M.
3. Measurements are also generally taken at a temperature of 25C (298 K).
4. Pure liquids: the liquid under a total (hydrostatic) pressure of 1 atm.
5. Solids: the pure solid under 1 atm pressure.

The change in the free energy of the system that occurs during a reaction measures the balance between the two driving forces that determine whether a reaction is spontaneous. As we have seen, the enthalpy and entropy terms have different sign conventions. In favorable spontaneous reactions $\Delta H°<0$ and $\Delta S°>0$; while in unfavourable reaction $\Delta H°>0$ and $\Delta S°<0$

## Calculating the change in Gibbs free energy

Even though $\Delta G$ is temperature-dependent, we assume to take $\Delta H$ and $\Delta S$ are independent of temperature when there is no phase change in the reaction. So if we know $\Delta H$ and $\Delta S$, we can find out the $\Delta G$ at any temperature. Methods to follow for calculation of Gibb;s free energy are Estimating $\Delta H$ reaction using bond enthalpies; Calculating $\Delta H$ using standard heats of formation $\Delta fH°$; Calculating $\Delta H$ and $\Delta S$ using standard values. Unit of $\Delta G$ is equal to $\Delta H$; Unit of $\Delta H$ is (KJ) /(mole reaction); Unit of $\Delta S$ is (KJ) /(mole reaction)K

Examples of spontaneous reactions are, rusting of iron is an example of a spontaneous reaction that occurs slowly, little by little, over time. Melting of ice also another example of a spontaneous reaction that occurs slowly, little by little, over time.

**Relationship between Free Energy and Equilibrium Constant**

The free energy change of the reaction in any state, $\Delta G$ (when equilibrium has not been attained) is related to the standard free energy change of the reaction, $\Delta G°$ (which is equal to the difference in the free energies of formation of the products and reactants both in their standard states) according to the equation.

$\Delta G = \Delta G° + RT\ InQ$

Where Q is the reaction quotient.

At equilibrium,

$\Delta G=0$ and Q become equal to the equilibrium constant. Hence the equation becomes,

$\Delta G° = -RT\ In\ K(eq)$

$\Delta G° = -2.303\ RT\ \log K(eq)$

- $R = 8.314\ J\ mol^{-1}\ K^{-1}$ or $0.008314\ kJ\ mol^{-1}\ K^{-1}$.
- T is the temperature on the Kelvin scale

In a reversible reaction, the free energy of the reaction mixture is lower than the free energy of reactants as well as products. Hence, free energy decreases whether we start from reactants or products i.e,$\Delta G$ is -ve in backward as well as forward reactions.

## Relationship between Gibbs free energy and EMF of a Cell

In the case of galvanic cells, Gibbs energy change $\Delta G$ is related to the electrical work done by the cell.

$\Delta G = -nFE(cell)$

Where

n = no. of moles of electrons involved

F = the Faraday constant

E = emf of the cell

F=1 Faraday =96500 coulombs

If reactants and products are in their standard states,

$\Delta G° = -nFE°cell$

**Table 9:** Non-spontaneous, spontaneous and equilibrium reactions

| $\Delta G°$ | Reaction | Keq |
|---|---|---|
| + | Non-spontaneous | <<1 |
| – | Spontaneous | >>1 |
| 0 | At equilibrium | 0 |

**Review question**

1. What is soil chemistry? List out some chemical properties of soil.
2. What is ion exchange? Explain cation exchange in soils.
3. Discuss the significance of CEC.
4. What is Base saturation (BS)?, calculate the base saturation of soil when the estimated amount of $Ca^{++}$ is 0.160g/100g; $Mg^{++}$ is 0.048g/100g; $Na^{+}$ is 0.046g/100g and $K^{+}$ is 0.117g/100g soil. CEC of a soil is 20m.eq./100g soil.
5. What is soil pH?, calculate the pH of soil when its hydronium ion concentration of $1.2 \times 10^{-3}M$.
6. Discuss factors affecting soil pH.
7. What is buffering capacity of soil? Discuss factor affecting soil buffering capacity.
8. Give brief account on sorption and precipitation reactions in soil
9. Briefly discuss nutrient availability in puddle soils

## Objective questions

1. The units of expression of AEC are
   a) C mol ($e^-$) / kg
   b) C mol ($P^+$)/kg
   c) Milli mhos/ cm
   d) Desi Simens / m
2. The sum total of the exchangeable cations that a soil can adsorb is called
   a) Cation adsorption capacity
   b) Cation equivalent capacituy
   c) Cation saturation capacity
   d) Cation exchange capacity
3. The orders of cation selectivity depend on
   a) valance of the ion
   b) atomic weight on an ion
   c) solubility of an ion
   d) atomic number of an ion
4. Loosely held ion with large oscillation zone on the soil colloids with exchange first is referred as
   a) Ratio law
   b) Additive ion effect
   c) Complementary ion effect
   d) Supplementary ion effect
5. CEC of sandy loam textured soil approximately is
   a) 20-25 C mol ($P^+$) / Kg
   b) 5-10 C mol ($P^+$) / Kg
   c) 25-35C mol ($P^+$) / Kg
   d) 35-40 C mol ($P^+$) / Kg
6. The CEC of illite clay mineral is approximately
   a) 80- 120 cmol ($P^+$) / Kg
   b) 100-150 cmol ($P^+$) / Kg
   c) 25-30 cmol ($P^+$) / Kg
   d) 200-400 C mol ($P^+$) / Kg
7. The CEC of soil is influenced by
   a) Soil texture
   b) Soil pH
   c) Nature of clay mineral
   d) All three
8. Increase in soil pH result in
   a) Increase in CEC
   b) Decrease in CEC
   c) Do not affect CEC
   d) Increases up to pH 7 thereafter decreases
9. The CEC of soil increases with increase in organic matter content due to
   a) Isomorphous substitution increases with increase in organic matter.
   b) pH dependent charges in the organic matter cause variation in CEC
   c) Organic matter content of soil directly related to microbial populationinsoil
   d) A&B
10. The replacing power of cations varies with
    a) The type of ion
    b) The size of ion
    c) The valence of ion
    d) All the three

11. Replacing power of monovalent cations increases in the following order
    a) Li < Na < K < Rb < Cs < H
    b) < K < Rb < Cs < H<Li < Na
    c) H< Na < K < Rb < Cs < Li
    d) Cs < H < Na < K < Rb< Li
12. Replacing power of divalent cations increases in the following order
    a) Mg < Ca < Sr < Ba
    b) Sr < Ba< Mg < Ca
    c) Ca < Sr < Ba< Mg
    d) Ba< Mg< Ca < Sr
13. The percentage of Cation exchange capacity (CEC) that is satisfied by the base forming cations (Ca, Mg, Na and K) is
    a) Base: acid ratio
    b) Base unsaturaton
    c) Base replacement capacity
    d) Base saturation
14. The acid forming ions in soil are
    a) Fe &Mn
    b) Al &H
    c) Ca &Mg
    d) Na &K
15. The percentage of sodium in the total CEC is known as
    a) RSC
    b) SAR
    c) ESP
    d) adjusted SAR
16. If bases saturation of soils is >80%, it is considered as
    a) Low fertile
    b) Moderately fertile
    c) Highly fertile soils
    d) Unfertile
17. With same base saturation percentage of soil, high sodium saturation causes
    a) Low soil pH
    b) Higher soil pH
    c) Higher EC
    d) Low EC
18. The relative order of anion exchange is
    a) OH > $H_2PO_4$ > $SO_4$ > $NO_3$ > Cl
    b) $SO_4$ > $NO_3$ > Cl> OH > $H_2PO_4$
    c) OH> $NO_3$ > $SO_4$ > Cl > $H_2PO_4$
    d) Cl > $H_2PO_4$ > $SO_4$ > OH > $NO_3$
19. In general each percent increase in soil humus contributes about
    a) 0.2 C mol /kg of CEC
    b) 4 C mol /kg of CEC
    d) 2 C mol /kg of CEC
    d) 6 C mol /kg of CEC
20. It is recommended to apply larger quantities of fertilizer nutrient in a single dose in
    a) Fine textured soils
    b) Coarse textured soil
    c) Medium textured soils
    d) Sandy soils

21. The soil property considered as master variable in soils
    a) Soil colour
    b) Soil EC
    d) Soil consistence
    d) soil pH
22. The pH of stomach acid, a solution of HCl with a hydronium ion concentration of $1.2 \times 10^{-3} M$ is
    a) 2.92
    b) 0.292
    c) 3.92
    d) 0.392
23. The acidity of soil developed due to non-exchangeable $H^+$ and $Al^{3+}$ which were associated with oxides and hydroxides of Al & Fe is
    a) Replaceable cidity
    b) Active acidity
    c) Exchangeable acidity
    d) Reserve acidity
24. The acidity developed due to adsorbed hydrogen ($H^+$) and aluminium ($Al_3^+$) ions on soil colloids is
    a) Exchangeable acidity
    b) Reserve acidity
    c) Active acidity
    d) Potential acidity
25. Soils with low buffering capacity expresses
    a) Low soil pH
    b) High soil pH
    c) Low EC
    c) High EC
26. Water logging of acid soil results in
    a) Decrease in soil pH
    b) Increasing in soil pH
    c) Increase in EC
    d) Increase in CEC
27. In soils, the buffering action may is due to
    a) Carbonates, bicarbonates and phosphates present in the soil.
    b) Colloidal complex with associated cations.
    c) Organic acids being released continuously in the soil.
    d) All the three
28. The degree of buffering is highest at the soil pH range of
    a) 3.0 to 5.0
    b) 4.5 to 6.0
    c) 6.5 to 8.0
    d) 7.0to 8.5
29. The soils that develop from their underlying parent rocks and have the same general chemistry as those rocks are
    a) Residual soils
    b) Transported soils
    c) Marine soils
    d) Alkaline soils
30. Plant availability of the following nutrient elements increased in the flooded rice soils
    A) Ni & Cl
    b) Zn &Cu
    c) B&Mo
    d) Fe &Zn

31. The following nutrient concentrations in soil generally decreased after flooding rice soils.

a) Zn &Cu
b) Fe &Zn
c) B&Mo
d) Ni & Cl

32. In annual crops grown on highly weathered acidic soils of the tropics the most important yield-limiting factors is

a) Iron deficiency
b) Phosphorus deficiency
c) Nitrogen deficiency
d) Mn deficiency

33. The reason for improvement of N availability in flooded rice

a) Absence of $O_2$ inhibits the activity of the *Nitrosomonas* microorganisms that oxidize $NH_4^+$, and therefore N mineralization stops at the ammonium ($NH_4^+$) form.

b) Absence of $O_2$ in flooded soils retards denitrification and hence accumulationof nitrtate form of N .

c) Organic matter decomposition in flooded soils preserve the soil nitrogen from losses.

d) Leaching losses of N kept minimum in flooded soils, thus the available H content is high

34. A major part of N in the flooded rice soils is lost through

a) Leaching
b) Denitrification
c) Volatilization
d) A&B

35. Denitrification is

a) Hydration process
b) Oxidation process
c) Reduction process
d) Hydrolysis process

36. In soil $NO_3^-$ is converts to $N_2$ at a redox potential of

a) +180 to +150 mV
b) +280 to +220 mV
c) +380 to +320 mV
d) −120 to −180 mV

37. The rdox potential of well oxidized soils is

a) +700 to +500
b) +180 to +150 mV
c) +380 to +320 mV
d) −120 to −180 mV

38. waterlogged soils' redox potential is generally in the range of

a) +700 to +500 mV
b) -250 to -300 mV
c) +180 to +150 mV
d) +380 to +320 mV

39. The most important chemical changes that occur in flooded or submerged rice soils are

a) pH
b) Redox potential
c) Ionic strength
d) All the three

40. Soil constituents responsible for buffering capacity is
    a) Human
    b) Clay
    c) Salts
    d) All

## Answer key

| | | | | | | | | | | | | | | | |
|---|---|---|---|---|---|---|---|---|---|---|---|---|---|---|---|
| **1** | a | **6** | c | **11** | a | **16** | c | **21** | d | **26** | b | **31** | a | **36** | b |
| **2** | d | **7** | d | **12** | a | **17** | b | **22** | a | **27** | d | **32** | b | **37** | a |
| **3** | a | **8** | a | **13** | d | **18** | a | **23** | d | **28** | b | **33** | a | **38** | b |
| **4** | c | **9** | b | **14** | b | **19** | d | **24** | a | **29** | a | **34** | d | **39** | d |
| **5** | b | **10** | d | **15** | c | **20** | a | **25** | a | **30** | d | **35** | d | **40** | d |

# 5

# Soil Mineralogy

Soil minerals are naturally formed inorganic compounds with definite chemical composition and characteristic geometric (crystalline) form. They are classified as primary and secondary minerals, silicates and non-silicates minerals, and crystalline and non-crystalline minerals. Primary minerals have not been significantly altered chemically since deposition or crystallization from molten magma and are usually found in the sand and coarse silt fractions. Secondary minerals form as a result of the alteration of primary minerals by weathering and are found in the clay and fine silt fractions. Most soil minerals contain silica as their major structural constituent (silicate minerals), while non-silicate minerals are composed mainly of oxides, carbonates, and sulfates. Crystalline minerals are composed of atoms arranged in a three dimensional periodic pattern, while the structural periodicity is absent in non-crystalline. Most mineral structures are bonded with ionic and covalent bonds to form coordinated polyhedrons with a unique composition and inter-atomic distance characteristics, which affect their overall physicochemical properties. Ionic isomorphous substitutions occurring within the mineral crystal lattice and broken bonds exposed on structural edges result in net positive or negative surface charge that plays an important role in adsorption/desorption, ion exchange processes which are important for plant nutrient availability and reclamation of soil problems. The nature of counter ions which neutralize the surface charge has significance on dispersion/flocculation phenomena which ultimately impacts soil erosion, land stability, and groundwater quality.

## 5.1. Classification and distribution of primary minerals

Primary minerals are divided into following classes depending upon their chemical composition, and dominant ionic group.

- Native elements
- Sulfides
- Oxides and hydroxides
- Carbonates, nitrates, and borates
- Sulfates, chromates, and molybdates
- Phosphates, arsenates, and vanadates and
- Silicates.

The above classes of minerals are subdivided further according to chemical and structural similarities. Silicate minerals are most commonly found and are dominate in most soils. Some of minerals, such as the native elements, molybdates, nitrates, borates, chromates, arsenates and vanadates are sparsely/ rarely found in soil environments.

Primary minerals have no significant alteration in chemical or structural properties due to their crystallization within igneous or metamorphic rocks or their deposition in sedimentary rocks. They are inherited from the parent material, and are usually found in the sand and coarse silt fraction of soils. Common primary minerals in soil environments include: silicates, oxides of iron (Fe), zircon (Zr) and titanium (Ti), and phosphates (P).

Secondary minerals are weathered, re-crystallized or transformed products of primary minerals under favourable conditions. Secondary minerals are mainly found in the clay and fine-silt fractions because the particle size of primary minerals usually decreases during weathering. Typical secondary minerals found in soils include alumino-silicates, oxides and hydroxides, carbonates, sulfates, and amorphous minerals. The separation of soil minerals in to primary and secondary categories is not necessarily and mutually exclusive since some of the minerals can occur in both classes (quartz, mica, carbonates). Non-silicate minerals commonly found in soils are listed in Table 1.

**Table 1:** Commonly found non- silicate minerals in soils

| Mineral class | Mineral | Chemical formula |
|---|---|---|
| Oxides and hydroxides of Aluminum | Gibbsite | $Al(OH)_3$ |
| | Bauxite | $Al_2O_3$ |
| Oxides and hydroxides of iron | Hematite | $Fe_2O_3$ |
| | Goethite | FeOOH |
| | Lepidocrocite | FeOOH |
| | Maghemite | $Fe_2O_3$ |
| | Hematite | $Fe_2O_3$ |
| | Ferrihydrite | $Fe_5O_7(OH)_24H O$ |
| | Magnetite | $Fe_3O_4$ |
| Halides | Halite | NaCl |
| Sulphates | Gypsum | $CaSO_4.2H_2O$ |
| | Jarosite | $KFe_3 (SO4)_2 (OH)_6$ |
| Sulphides | Pyrite | $FeS_2$ |
| Carbonates | Calcite | $CaCO_3$ |
| | Dolomite | $CaMg(CO_3)_2$ |
| Phosphates | Apatite | $Ca_{10} (PO_4)_6 (OH, F, Cl)$ |
| Native elements | | |
| Manganese | Birnessite | $(Na^+,Ca^{2+}, Mn^{2+}) Mn_7 O_4$ 2.8H O |
| | Lithiophorite | $LiAl_2 Mn_2^{4+} Mn^{3+}O_6 (OH)_6$ |
| Titanium | Rutile | $TiO_2$ |
| | Anatase | $TiO_2$ |
| | Ilmenite | $Fe^{2+} TiO_3$ |

Mineral occurrences in soil environments are the result of inheritance from parent materials, precipitation from soil solution (neoformation), or alteration of existing minerals into new phases (diagenesis). Climatic plays paramount effect in these processes and in the overall distribution of minerals in soils of different climatic regions and different soil horizons. Therefore, the mineralogical composition of an individual soil is determined mostly by the make-up of the parent material(s), and the intensity and duration of the weathering. Sand and coarse silt fractions tend to be composed mainly of inherited primary minerals, while clay fractions are dominated by secondary minerals, either formed in-situ or transported from other elsewhere. The diversity of climate variations generates some distinct soil mineral distribution trends.

## Oxides and Hydroxides

The most common soil occurrences of this group minerals are oxides and hydroxides of Fe and Al, often referred to as sesquioxides (aftere Schwertmann and Fitzpatrick). Typical minerals of this group are goethite, hematite, and gibbsite, which are abundant in intensely weathered tropical soils. They exhibit amphoteric surface charge properties depending on pH conditions. They also act as chemical binding agents playing an important role in soil aggregation and structural formation. The presence and abundance of Fe-minerals exert a significant influence on the color of the soil due to their pigmentation properties. Under oxidizing conditions, they exhibit brownish-yellow to bright-red hues, while under reducing conditions they turn gray due to $Fe^{3+}$to $Fe^{2+}$transitions. These color changes are used as soil development and wetness indicators in pedogenic (degree of weathering) and land use interpretations (soil drainage classes, septic systems, wetland delineations).

Manganese oxides and hydroxides are also commonly found in soils in the form of concentric brown/black nodules or black coatings of soil structural units, reflecting current or past periodic water saturation conditions. Birnessite and lithiophorite are the most prevalent Mn soil minerals. They are often associated with Feoxide minerals. Another group of minerals found in small quantities in soils includes titanium oxides, such as rutile, anatase, and ilmenite. They are usually found in sand and silt fractions as common accessory minerals of igneous and metamorphic rocks. Because of their increased weathering stability they are frequently used as pedogenic indicators of soil maturity and uniformity.

## Halides, Sulphates and Carbonates

These groups of minerals occur mainly in arid/semiarid regions or in immature soils where weathering is minimal. They contain soluble soil minerals and are easily broken down by physical and chemical weathering. Halite, gypsum, calcite, and dolomite constitute the most common occurrences in this group. Their solubility decreases in the anionic sequence $Cl^- > SO_4^{2-} > CO_3^{2-}$. Halite is found only in salt affected soils. Gypsum is also common in saline soils either under arid or hydromorphic conditions, but it could also form in weathering interfaces where sulfide and carbonate minerals dissolve. Calcite and dolomite minerals are also prevalent in arid and semiarid climatic zones and these minerals can also be found in more diverse soil environments where the dissolution of calcareous parent materials provides opportunities for secondary carbonate formation in lower soil depths. These minerals are also introduced in soil environments through land reclamation of acid soils like liming.

### Sulphides

Pyrite ($FeS_2$) is the most common mineral of this sulphide group. Generally sulphides forms under prolonged anaerobic conditions typical in wetlands, which explain its common association with coal deposits. Pyrite is unstable and weathers readily into other sulfate minerals and sulfuric acid under oxidizing conditions. The released acidity constitutes a serious environmental problem in reclaimed coal-mined areas and in drained pyritic soils as incase of acid sulphate soils in marine coasts.

### Phosphates

Phosphate minerals are found in only a few soils, with the orthophosphate ion ($PO_4^{2-}$) forming a tetrahedron as the basic structural unit. Apatite is the most frequently reported phosphate mineral of igneous or sedimentary origin. Along with their weathering products, usually Al and Fe-phosphates, apatites are considered a natural source of P for plants.

### Silicates

Silicates constitute the largest and most important group of soil minerals. They occur as primary minerals resulting from the breakdown of parent materials or secondary minerals formed from weathering of primary minerals. The fundamental structural building unit of silicates is the silica tetrahedron, in which one $Si^{4+}$is coordinated with four apical $O^{2-}$. Different arrangements of individual Si-tetrahedra linked by shared $O^{2-}$ form more complex structures, consisting of single tetrahedral (neso-silicates), double tetrahedral (soro-silicates), rings (cyclo-silicates), single or double chains (ino-silicates), sheets (phyllo-silicates), or three-dimensional frameworks (tecto-silicates).

The mineral part of soil is made up of many types of minerals. A Primary mineral comes directly from rock like pyrite (fools gold). Some common minerals founding the soils are presented in Figure.1. These minerals break down into different types of clay minerals. All of these different types of materials have different abilities to hold onto different nutrients and chemicals.

## 5.2. Secondary minerals (clay minerals)

Clay minerals occur under a fairly limited range of geologic conditions. Most clay minerals form where rocks are in contact with water, air, or steam. Examples of these situations include weathering boulders on a hillside, sediments on sea or lake bottoms, deeply buried sediments containing pore water, and rocks in contact with water heated by magma (molten rock). All of these environments may cause the formation of clay minerals from pre-existing minerals. Extensive alteration of rocks to clay minerals can produce relatively pure clay deposits that are of economic interest (for example, bentonites and primarily montmorillonite used for drilling muds and clays used in ceramics).

The transport and deposition of clays and clay minerals produced by eroding older continental and marine rocks and soils are important parts of the cycle that forms sedimentary rocks. The ancient sedimentary rock record is composed of about 70 percent mudstones (which contain about 50 percent clay-sized fragments) and shales (which are coarser than mudstones but which may contain clay-sized particles). Clearly, clays and clay minerals are critical components of both ancient and modern sedimentary environments.

Diagenesis is the in-place alteration of a mineral to more stable forms, excluding surficial alteration (which is weathering); diagenesis occurs, for example, when minerals stable in one depositional environment are exposed to another by burial and compaction. Common silicate materials such as quartz, feldspars, and volcanic glasses, as well as carbonates, noncrystalline iron oxides, and primary clay minerals, are transformed during diagenesis into more stable clay minerals mainly by dissolution and recrystallization.

The formation of bentonite (beds containing smectite-group clay minerals including montmorillonite) and fuller's earth (a type of clay mineral deposit that has high capacity to absorb water) may occur primarily by diagenesis, although some deposits may also form by hydrothermal processes. Bentonite beds usually form from altered volcanic ash, but other types of rock may also serve as sources.

## Weathering of rocks and primary minerals

Weathering of rocks and soil is the primary way that clays and clay minerals form at the Earth's surface today. The weathering process involves physical disaggregation and chemical decomposition that change original minerals to clay minerals; weathering is uneven, and many stages of breakdown may be found in the same clay sample. Factors governing rock weathering and soil formation include the initial type of rock, the ratio of water to rock, the temperature, the presence of organisms and organic material, and the amount of time. The types of clay minerals found in weathering rocks strongly control how the weathered rock behaves under various climatic conditions (such as humid-tropical, dry-tropical, and temperate conditions).

Kaolinite is found in most weathering zones and soil profiles. Montmorillonites, which are chemically more complex than kaolinites, are common in the lower parts of weathering profiles, nearer the rock, where chemistry exerts a strong control on mineralogy. Complex mixed-layer clay minerals (such as illite-smectites) are abundant in clay assemblages that develop from mica-bearing precursor rocks, such as the granite plutons that occur in temperate regions of the Northeastern United States. For example, a large component of soils formed by weathering of granites may consist of metastable muscovite, biotite, and chlorite. These minerals will alter progressively to clay minerals.

## 5.3. Properties of Secondary minerals (Clay minerals)

Clay minerals refers to a group of hydrous aluminosilicates that predominate the clay-sized (<2 mm) fraction of soils. These minerals are similar in chemical and structural composition to the primary minerals that originate from the Earth's crust; however, transformations in the geometric arrangement of atoms and ions within their structures occur due to weathering. Primary minerals form at elevated temperatures and pressures, and are usually derived from igneous or metamorphic rocks. Inside the Earth these minerals are relatively stable, but transformations may occur once exposed to the ambient conditions of the Earth's surface. Although some of the most resistant primary minerals (quartz, micas, and some feldspars) may persist in soils, other less resistant minerals (pyroxenes, amphiboles, and a host of accessory minerals) are prone to breakdown and weathering, thus forming secondary minerals. The resultant secondary minerals are the culmination of either alteration of the primary mineral structure (incongruent reaction) or neoformation through precipitation or recrystallization of dissolved constituents into a more stable structure (congruent reaction). These secondary minerals are often referred to as phyllosilicates because, as the name implies (Greek: *phyllon,* leaf), they exhibit a platy or flaky habit, while one of their fundamental structural units is an extended sheet of $SiO_4$ tetrahedra.

There are several different types of clay minerals that come from primary minerals. Clay minerals that are older and more weathered hold on to less nutrients. Some young soils are very active, and they materials that cause them to shrink and swell. These materials are high activity clays, and these soils have the ability to hold a lot of nutrients. In younger soils, the clay minerals are new. In well weathered soils, certain minerals, like Gibsite, Hematite, and Anatase form oxides. These oxides are the leftover metals from intense weathering, have a very low Cation Exchange Capacity and cannot hold many soil nutrients.

Soil Clay minerals exhibit colloidal properties such as adsorption, electrical charge, veried shapes, swelling and shrinkage, optical properties, brownian movement, plasticity, etc.

i) **Size:** Colloids are extremely small in size. The upper limit in diameter for the colloidal state is generally considered to be about one micron. Some with upper limit of 2μ exhibit colloidal properties but they are not technically colloids.

ii) **shape:** Soil colloidal size clay particles exhibit varied shapes such as, Kaolinite–Hexagonal crystals, Halloysite–Rod shaped crystals, Montmorillonite – flakes like crystals.

iii) **Surface area:** The colloids expose a large surface area per unit mass. The external surface area of one gram of clay colloid is at least 1000 times that of one gram of sand. Certain silicate clays with 95 expanding plate like makeup offer internal surfaces besides external surface area. The total surface area of soil colloids ranges from 10m$^2$ /g for clays with only external surfaces to more than 800 m$^2$ /g for clays with extensive internal surfaces. The colloidal surface area in the upper 15 cm of a hectare of a clay soil could be as high as 7,00,000 sq.kms.

iv) **Electrical properties:** colloidal clay minerals has both external and internal characteristically carry negative and / or positive charges. For most colloids electronegative charges predominate, although some mineral colloids in very acid soils have a net positive charge.

v) **Sorption properties:** Adsorption of cations / anions and water (Figure.1) is the important consequence of the charges on colloids. The negatively charged colloids attract hundreds of thousands of positively charged ions (cations). The adsorption of ions is governed by the type and nature of ion, ion concentration and the type of colloidal particle. Higher is valency, higher the adsorption (excepting $H^+$). With same valency, the ion with more atomic weight is preferred. Concentration of the ion in the solution also determines the adsorption rate.

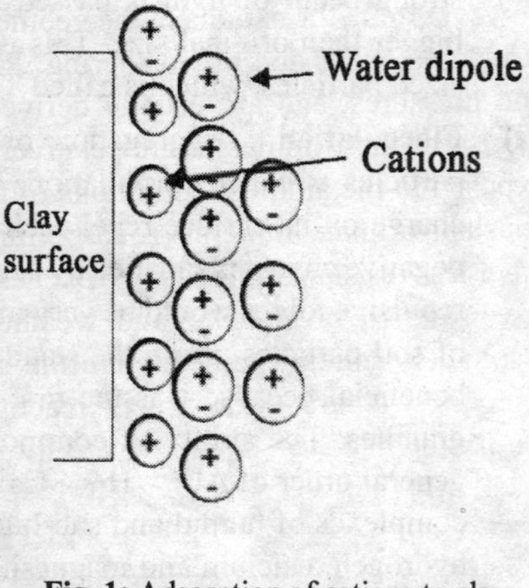


**Fig. 1:** Adsorption of cations to clay surface

vi) **Plastic properties:** Plasticity is the capability of being molded. Soil containing more than about 15% clay exhibits plasticity that is pliability and the capability of being molded. This property is due to the plate like nature of the clay particles and the combined lubricating and binding influence of the adsorbed water. Thus the particles of plastic soils easily slide over each other, much like panes of glass with films of water between them. Plasticity is of practical importance because of its influence on tillage operations. The clayey soils with smectite type clay minerals present a significant problem, by not allowing to obtain a stable granular structure.The plastic properties of clay minerals are in the order of Montmorillonite > Illite > Kaolinite.

vii) **Swelling and shrinkage:** Some clays swell when wet and shrink when dry. After prolonged dry period, soils rich in smectite minerals develop wide deep surface cracks that, at first, allow rain water to infiltrate rapidly, later after swelling very poor infiltration. Adsorbed ions for water and air entrapped are primarily responsible for swelling.

viii) **Cohesion and adhesion**: Cohesion indicates the tendency of clay particles to stick together. This tendency is due primarily to the attraction of the clay particles for the water molecules held between them. Hydrogen bonding between the clay surfaces and water and also among water molecules is the attractive force responsible for cohesion.

ix) **Brownian movement**: The continuous rapid zigzag movement executed by a colloidal particle in the dispersion medium is called Brownian movement (Robert Brown, 1927). In a colloid suspension, the colloidal particles are under constant rapid zigzag motion. The Brownian movement is mainly responsible for the coagulation or flocculation of colloidal particles.

x) **Tyndal Effect** : Dust particles float in air form a colloidal suspension. If a strong beam of light is passed through such suspension the particles appear bigger than original size. This is due to the diffusion of light by the colloidal sized particles, which is called "Tyndal Effect".

xi) **Flocculation** : Aggregation or clumping together of individual, tiny soil particles is called flocculation. Clay particles by virtue of carrying negative charge on its surface repel each other and disperse in the medium. When the negative charge is satisfied by cations, which are tightly held on clay surface, the repulsive forces would be very minimum. It leads to coagulation or flocculation of soil particles. From the stand point of agriculture, flocculation is generally beneficial because it is the first step in the formation of stable aggregates or granules. The ability of common cations to flocculate soil colloids is in the general order of $Al^{3+} > H+ > Ca^{2+}, Mg^{2+} > K^{+} > Na^{+}$. Incidentally the colloidal complexes of humid and sub-humid region soils are dominated by aluminum, hydrogen, calcium and magnesium and those of semi-arid regions are high in Calcium and magnesium ions.

xii) **Deflocculating** : It is the dispersion of colloidal particles, due to the repulsion of negatively charged particles. Under deflocculated condition particles move away from each other, and act independently. Dispersion is encouraged by the large number of water molecules associated with each micelle and with the adsorbed cations. Highly hydrated monovalent cations like Na+ do not effectively reduce the electro negativity of the micelle and are loosely held. It makes the individual micelles to repel each other continuously and stay in dispersion. Dispersion can be reduced by (i) decreasing the pH of the medium (ii) replacing the sodium with $H^{+}$ , $Ca^{2+}$, $Mg^{2+}$ and $Al^{3+}$ like cations and (iii) increasing the salt concentration in the soil solution.

## 5.4. Classification of Clay minerals

Clay minerals are broadly classified to silicate clays and non silicate clay minerals (table 2 and fig 2). Silicate clays contain Si as integral part.

**Table 2:** Brief classification of clay minerals

| Clay minerals | |
|---|---|
| Silicate clay minerals | Non- silicate clay minerals |
| a)Layer silicate clays/Phyllo silicate clays | a) Fe abd Al oxide clays |
| b)Allophone and associated amorphous clays | |

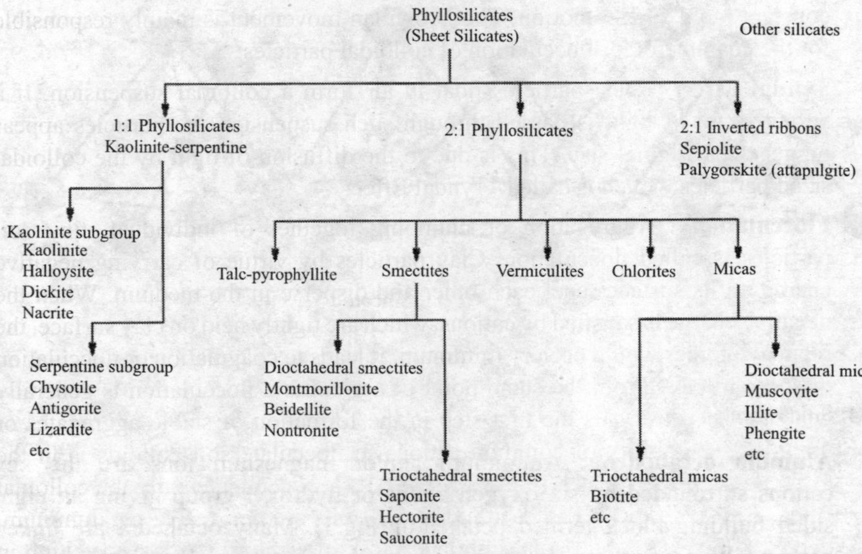


**Fig. 2:** Classification clay phyllosilicate

## 5.5. Structures of Aluminosilicates clay minerals

The most important silicate clays are known as phyllosilicate (Gr. Phyllon, leaf) because of their leaf or plate like structure. Two types of structural units are basic in the layer lattice structure of most clay minerals that are tetrahedral and octahedral unit.

a) **Silica tetrahedron:** The basic building block for the silica dominated sheet is a unit composed of one silicon atom surrounded by four oxygen atoms. It is four sided configuration called "Silica Tetrahedron". An interlocking array of a series of these silica tetrahedra tied together horizontally gives a "Tetrahedral sheet" (figure 3 &4).

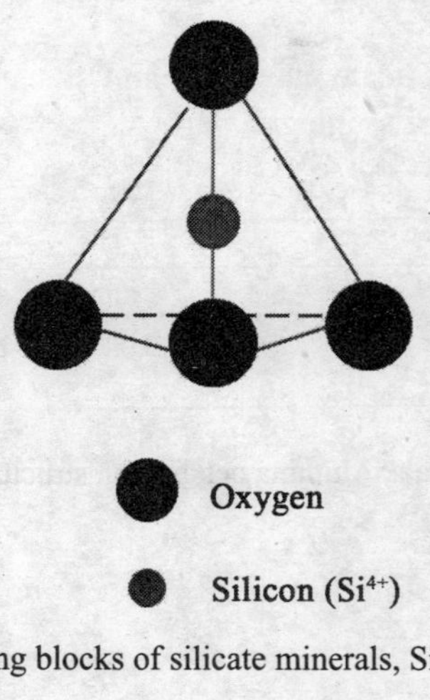


**Fig. 3:** Basic building blocks of silicate minerals, Silica tetrahedron structure

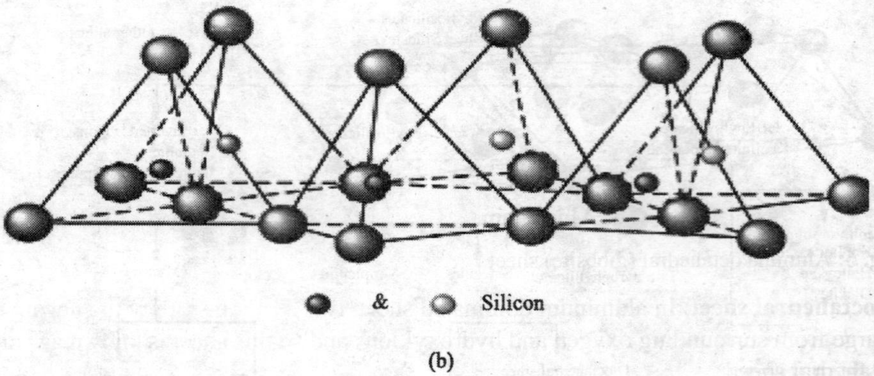


Fig. 4: Silica sheet

b) **Alumina octahedron:** Aluminium and/or magnesium ions are the key cations surrounded by six oxygen atoms or hydroxyl group giving an eight sided building block termed octahedron(Fig.5). Many octahedra are linked horizontally to form an octahedral sheet. An aluminum-dominated sheet is known as a di-octahedral sheet, whereas one dominated by magnesium is called a tri-octahedral sheet. The distinction is due to the fact that two aluminum ions in a di-octahedral sheet satisfy the same negative charge from surrounding oxygen and hydroxyls as three magnesium ions in a tri-octahedral sheet. The tetrahedral and octahedral sheets are the fundamental structural units of silicate clays. These different sheets are bonded together to form crystalline units composed of alternating sheets of Si tetrahedra and Al (or Mg) octahedra. The Si tetrahedral sheet is chemically bonded to the one or two adjacent Al (or Mg) octahedral sheet(s) via shared oxygen atoms. The specific nature and combination of sheets, called layers, vary from one type of clay to another and control the physical and chemical properties of each clay.

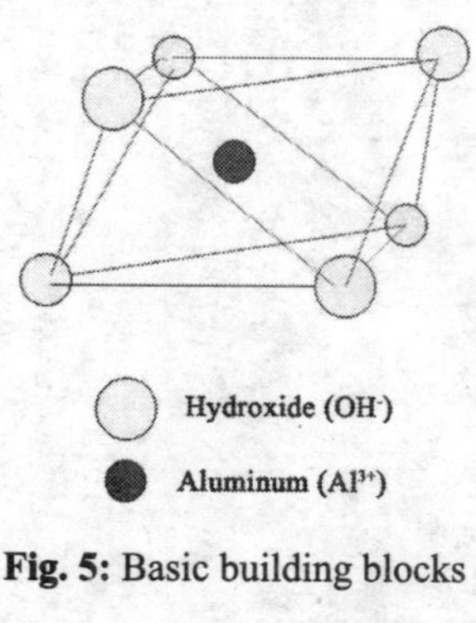


**Fig. 5:** Basic building blocks of silicate minerals, Alumina octahedron structure

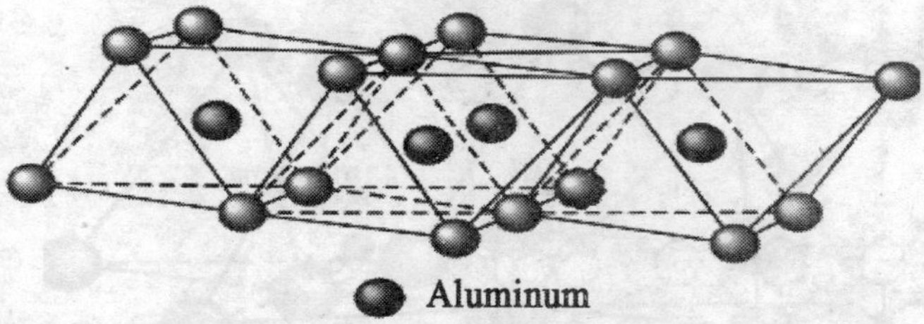


**Fig. 6:** Alumina octahedral (gibbsite) sheet

**Dioctahedral sheet**: In aluminum dominated sheet, two $Al^{3+}$ ions satisfy the negative charge from surrounding oxygen and hydroxyl ions and so this sheet is known as "di-octahedral sheet".

**Triocta hedral sheet:** In Magnesium dominated sheet, three $Mg^{2+}$ ions satisfy the negative charge and so this sheet is known as tri-octahedral sheet. The tetrahedral and octahedral sheets are bound together with in crystals by shared oxygen atoms in to different layers.

## 5.6. Classification of silicates minerals

On the basis of the number and arrangement of tetrahedral (silica) and octahedral (alumina-magnesia) sheets contained in the crystal units or layers, silicate clays are classified into three different groups:

(a) 1 :1 Type layer silicate clay minerals Eg. Kaiolinite

(b) 2:1 Type layer silicate clay minerals Eg. Momtmorrillonite illite

(c) 2: 1: 1 (or) 2:2 Type layer silicate clay minerals Eg. Chlorite

(d) Hydroxyl Aluminum inter layered vermiculites and semities clay minerals

(e) Mixed and interstratified layers type

## 5.7. Structure and properties of silicate clay minerals

### a) 1:1 layer silicate clay minerals

In soils, kaolinite group is the most prominent 1:1 clay mineral, which includes kaolinite, hallosite, nacrite and dickite. These have one Si tetrahedral and one Al octahedral sheet per crystalline unit The tetrahedral and octahedral sheets in a layer of a kaolinite crystal are held together tightly by oxygen anions, which are mutually shared by the silicon and aluminum cations in their respective sheets (figure 7). These layers, in turn, are held together by hydrogen bonding. Consequently, the structure is fixed and no expansion ordinarily occurs between layers when the clay is wetted. Cations and water do not enter between the structural layers of a 1:1 type mineral particle. The effective surface of kaolinite is restricted to its outer faces or to its external surface area. Kaolinite crystals usually are hexagonal in shape. Because of the strong binding forces between their structural layers, kaolinite particles are not readily broken down into extremely thin plates. Kaolinite exhibits very little plasticity, cohesion, shrinkage, and swelling.

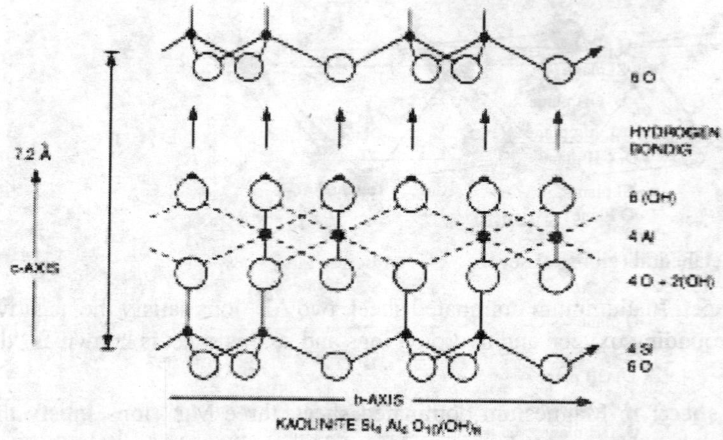


**Fig. 7:** Structure of Kaolinite (1:1 layer silicate mineral)

## b) 2:1 layer silicates clay minerals

The crystal units (layers) of these minerals are characterized by an octahedral sheet sandwiched between two tetrahedral sheets. Three general groups have this basic crystal structure. Two of them, *smectite* and *vermiculite* are expanding type minerals, while the third *mica group* (*illite*), is non-expanding

## 2:1 Expanding clay minerals

the smectite group of minerals like montmorillonite, beidellite, nontronite , saponite etc. This group of minerals are noted for their interlayer expansion and swelling when wetted. The water enters the interlayer space and forces the layers apart. Montmorillonite is the most prominent member of this group in soils. The flake-like crystals of smectite are composed of an expanding lattice 2:1 type clay mineral. Each layer is made up of an octahedral sheet sandwiched between two tetrahedral (silica) sheets (Figure 8). The layers are loosely held together by very weak oxygen – oxygen and cation-to-oxygen linkages. Exchangeable cations and associated water molecules are attracted between layers causing expansion of the crystal lattice.

The internal surface exceeds the external surface of clay crystal. In montmorillonite, magnesium replaces replaced aluminum in some sites of octahedral sheet. Likewise, some silicon atoms in the tetrahedral sheet may be replaced by aluminum. These substitutions give rise to a negative charge.

These minerals show high cation exchange capacity, swelling and shrinkage properties. Wide surface cracks commonly form in smectite dominated soils (e.g., Vertisols) when dried. The dry aggregates or clods are very hard, making such soils difficult to till.

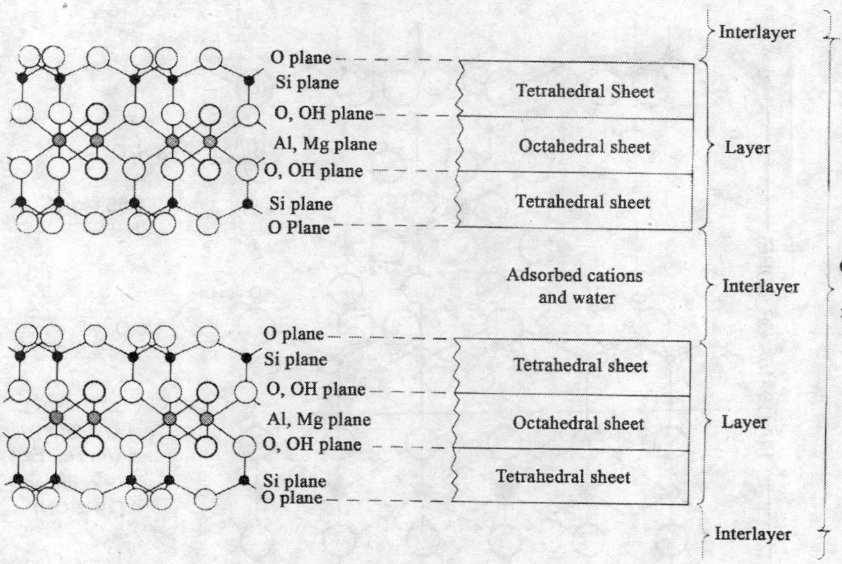


**Fig. 8:** Stucture of Montmorillonite (expanding 2:1 type) clay mineral

**Vermiculites:** Vermiculites have structural characteristics similar to those of montmorillonite in that an octahedral sheet is found between two tetrahedral sheets (figure 9). In the tetrahedral sheet of most vermiculite, aluminum is substituted by silicon in most of the sites. This accounts for most of the very high net negative charge associated with these minerals. Water molecules, along with magnesium and other ions, are strongly adsorbed in the interlayer space of vermiculites. They act primarily as bridges holding the units together rather than as wedges driving the units apart.

The degree of swelling is less for vermiculites than for smectite. Therefore, vermiculites are considered limited expansion clay minerals, expanding more than kaolinite but much less than the smectite. The cation exchange capacity (CEC) of vermiculite is higher than all other silicate clays because of very high negative charge in the tetrahedral sheet. Vermiculite crystals are larger than those of the smectite but much smaller than those of kaolinite.

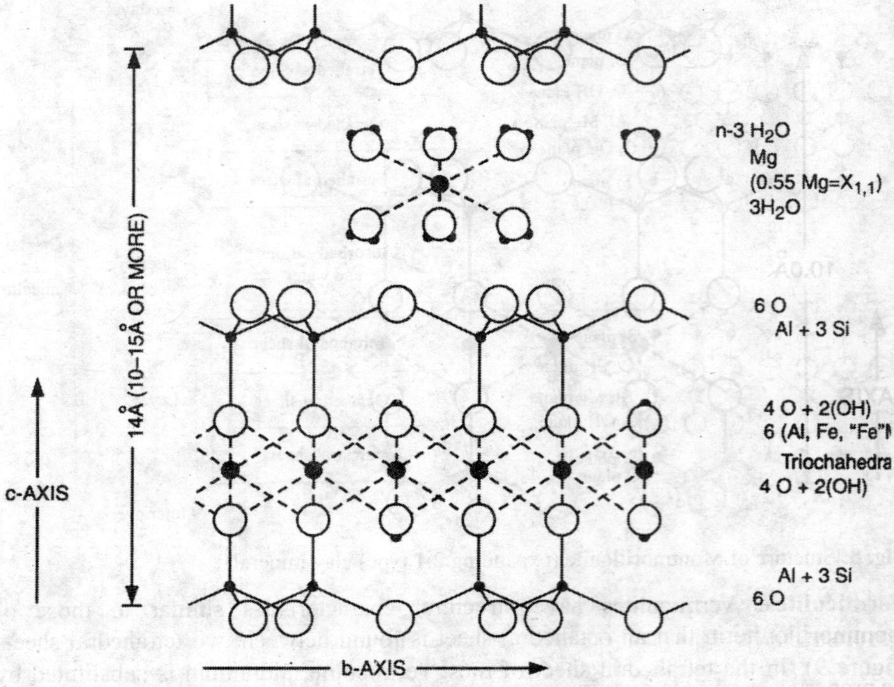


**Fig. 9:** Structure of Vermiculite clay mineral

## 2:1 Non-expanding minerals

Micas are the type of minerals in this group-*muscovite* and *biotite.* Weathered minerals similar in structure to these micas are found in the clay fraction of soils. They are called *fine-grained micas* or *illite*. The basic structure of illite or micaceous mineral is similar to that of montmorillonite. However, the particles are much larger than those of the smectite. Some of the silicon ions are replaced by aluminium ions in the tetrahedral sheet (20% ). This results in a net negative charge in the tetrahedral sheet which is compensated by potassium ions (figure10). The potassium as a binding agent, preventing expansion of the crystal, hence, fine-grained micas are quite non-expanding.

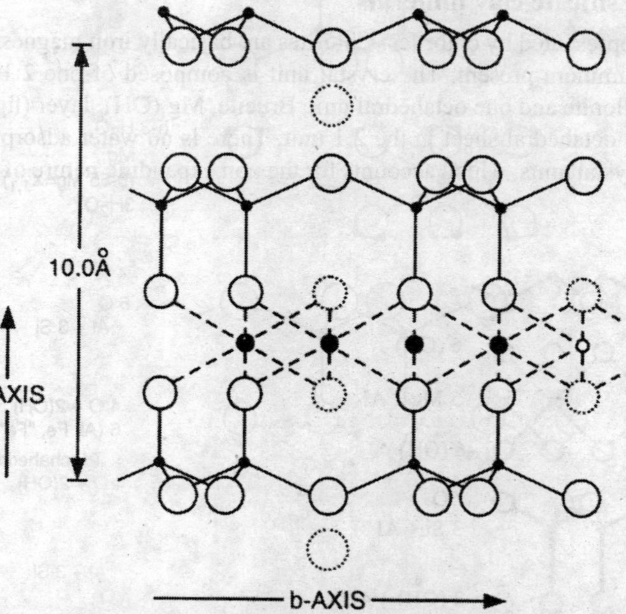


**Fig. 10:** Structure of illite (Non-expanding 2:1 type) clay mineral

The properties such as hydration, cation adsorption, swelling, shrinkage and plasticity are less intense in fine grained micas. The specific surface area varies from 70 to 100 $m^2 g^{-1}$, about one eighth that for the smectite. Table 3 summarizes the properties of kaolinite, montmorrillinite, illite and vermiculite minerals.

**Table 3:** Differential properties of some silicate clay minerals

| Property | Kaolinite | Monotmorillonite | Illite | Vermiculite |
|---|---|---|---|---|
| Shape | Hexagonal crystals | Irregular flakes | Flakes | Plates, flakes |
| Structure size (microns) | 1:1 lattice | 2:1 | 2:1 | 2:1 |
| | 0.5 to 5.0 | 0.01 to 1.0 | 0.1 to 2.0 | 0.1to5.0 |
| Specific surface ($m^2$/g) | 5-20 | 700-800 | 100-200 | 500-600 |
| External and internal surface | Low | Very High | Medium | High |
| Cohesion and plasticity swelling capacity | Low | High | Medium | High |
| Cation exchange capacity (cmol($P^+$) $kg^{-1}$) | 3-15 | 80-120 | 15-40 | 100-180 |
| Anion exchange capacity | High | Low | Medium | Low |
| Inter layer bonding | H- bonding | o-o bonding | o-o bonding | o-o bonding |
| Cohesion | low | high | medium | medium |
| Plasticity | low | high | medium | medium |
| Swelling & Shrinkage | low | high | medium | medium |

### c) 2: 1: 1 Type layer silicate clay minerals

This silicate group is represented by chlorites. Chlorites are basically iron magnesium silicates with some aluminum present. The crystal unit is composed of one 2:1unit like mica or montmorillonite and one octahedral unit, Brucite, $Mg_3(OH)_6$ layer (figure 11). Mg dominates the octahedral sheet in the 2:1 unit. There is no water adsorption between the chlorite crystal units, which accounts for the non expanding nature of this mineral.

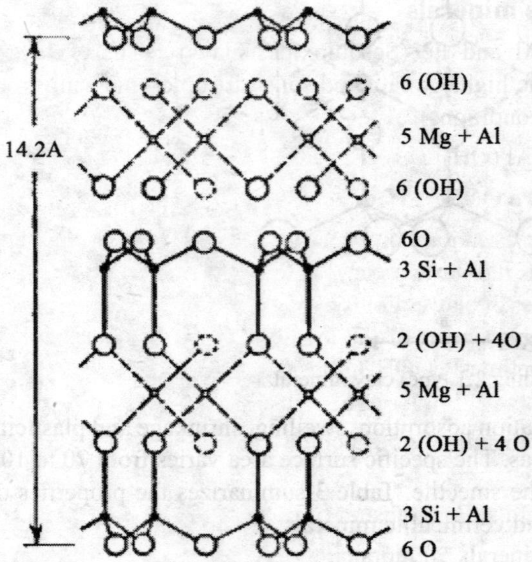


**Fig. 11:** Structure of Chlorite clay mineral

### d) Hydroxy-Aluminium interlayered Vermiculites and Smectites

Chlorites exist in soils as primary minerals that weather to form vermiculite and smectite. In contrast, hydroxy-Al interlayered vermiculites or smectites are considered to be exclusively secondary minerals forming as intermediate mineral weathering products, or from deposition of hydroxy-Al polymeric components within the interlayer space of expanding minerals. These hydroxy-Al polycations with the general formula $Al_6(OH)16^{2-}{}_{16}$ balance a portion of the charge but they are not exchangeable. Because the level of hydroxy-Al occupancy within the interlayer space is variable, the CEC of the expandable 2:1 clays is reduced as a function of the quantity and valence of the hydroxy-Al polymer residing within the interlayer.

### e) Mixed and interstratified layers type clay minerals

Specific groups of clay minerals do not occur independently of one another. In a given soil, it is common to find several clay minerals in an intimate mixture. Furthermore, some mineral colloids have properties and composition intermediate between those of any two minerals. Such minerals are termed mixed layer or interstratified because the individual layers within a given crystal may be of more than one type. Terms such as "chlorite-vermiculite" and mica - smectite are used to describe mixed layer minerals.

## 5.8. Non-silicate minerals

The soil minerals that donot contain silicon are non silicate minerals. Oxides and hydroxides of iron and aluminium, carbonates, sulphates, apatities are some of the non silicate minerals present in soil.

## Classification of non-silicate minerals

i) Oxides, Hydroxides of Al and Fe (Sesquioxides clays) : These clays are commonly dominant in the highly weathered soils of tropics and semi tropics (humid-hot-well drained conditions)

Gibbsite : $Al_2O_3 . XH_2O$ / Al $(OH)_3$

Goethite : $Fe_2O_3$. $XH_2O$ / Fe OOH

These clays tend to occur as amorphous coatings on soil particles imparting characteristic colors (Pink, red, brown etc.). Sesquioxide clays are non-sticky and non-plastic. These types of clay colloids carry positive and negative charges (variable charges) depending on pH of the medium. Soils rich in sesquioxide clays (Hematite-$Fe_2O_3$, Goethite-FeOOH, Gibbsite-$Al_2O_3$ $3H_2O$) pose a problem of phosphorus availability.

ii) Carbonates containing minerals: Eg. Calcite (CaCO3), Dolomite (CaCO3 MgCO3)

iii) Sulphates containing minerals: Eg. Gypsum (CaSO4)

iv) Phosphorus containing minerals : Fluorapatite $Ca_{10}$ $(PO_4)_6$ $F_2$ ; Apatite: $Ca_{10}$ $(PO_4)_6^{2+}$ ; Variscite $AlPO_4.2H_2O$; Vivianite $Fe_3(PO_4)_2$; Strengite $FePO_4.2H_2O$

v) Boron containing minerals: Tourmaline -Na $(Mg,Fe)_3$ $Al_6(BO3)_3Si_6O_{18}$

vi) Iron containing minerals: Limonite – $Fe_2O_3$. $3H_2O$ ; Viviamite $Fe_3$ $(PO4)_2$ ; Ilmenite - $FeTiO_3$ ; Siderite - $FeCO_3$; Pyrite - $FeS_2$ ; Hematite: $Fe_2O_3$; Chalcopyrite - $CuFeS_2$ ; Magnetite - $Fe_3O_4$

vii) Zinc containing minerals : Sphalerite - ZnS; Smithsonite - :$ZnCO_3$;

viii) Mg containing minerals : Epsom - $MgSO_4.7H_2O$ ; Magnesite - $MgCO_3$; Dolomite - $CaCO_3$. Mg $CO_3$; Epsomite - $MgSO_4.7H_20$

ix) Cu containing minerals: Chalcorite - $Cu_2S$; Covellite - CuS ;Chalcopyrite - $CuFeS_2$

x) Mn containing minerals: Pyrolusite - $MnO_2$ ; Manganite - MnOOH ; Rhodocrosite - MnCO3

xi) Mo containing minerals : Molybdenite : $MoS_2$; Wulfenite: $PbMoO_4$; Powellite $CaMoO_4$

## Amorphous minerals

### a) Sesquioxide minerals / Oxides and hydroxides of Iron and Aluminum:

The most common examples of iron and aluminum oxides are gibbsite ($Al_2O_3.3H_2O$) and geothite ($Fe_2O_3.H_2O$). Less is known about these clays than about the layer silicates. Under conditions of extensive leaching by rainfall and long time intensive weathering of minerals in humid warm climates, most of the silica and alumina in primary minerals are dissolved and slowly leached away. The remnant materials, which have lower solubility are called sesquioxides. Sesquioxides (metal oxides) are mixtures of aluminum hydroxide, $Al(OH)_3$, and iron oxide, $Fe_2O_3$, or iron hydroxide, $Fe(OH)_3$. Some of these clays have crystalline structure.

These minerals carry a small negative charge at high pH values which attract cations. However, their capacity to adsorb cations is less than kaolinite. These clays do not swell, not sticky and have high phosphorus adsorption capacity.

### b) Allophane and imogolite minerals

These colloidal matter without an orderly three dimensional crystalline structure are referred to as amorphous minerals or short range minerals. Allophone is a general name for amorphous alumino silicate gels with a general composition approximating $Al_2O_3$ $2SiO_2$ $H_2O$. Allophone is most prevalent in soil developed from volcanic ash. These clays have a variable charge, which is governed by soil pH. Allophone has high CEC (150 cmol kg-1) in neutral to mildly alkaline conditions and even exhibits high AEC in acidic environment. Zeolite and $TiO_2$ (Rutile) are the other important amorphous minerals. Imogolite $SiO_2.Al_2O_3$. $2.5H_2O$ (135 C mol $kg^{-1}$) found in weathered volcanic ash on pumice beds.

These are amorphous, hydrated alumino-silicate. Typically, these clays occur where large amount of weathered products existed. These clays are common in soils forming from volcanic ash (e.g. Allophane). Its approximate general composition is $Al_2O_3.2SiO_2.H_2O$. These clays have high anion exchange capacity or even high cation exchange capacity.

## 5.9. Genesis of clay minerals

The silicate clays are developed from the weathering of a wide variety of minerals by the two distinct process.

i) **Alteration-:** A slight physical and chemical alteration of certain primary minerals. Changes in particle size. Alteration of muscovite mica to fine grained mica is the good example. As weathering occurs muscovite mineral is broken down in size to the colloidal range, part of the potassium is lost and some silicon is added from weathering solutions. Net result is the less rigid crystal structure and an electronegative charge. The fine mica colloid has a 2:1 structure, only have been altered in this process.

ii) **Recrystallization:** Decomposition of primary minerals with subsequent recrystallization of certain of their products in to the silicate clays

Complete breakdown of clay structures and re-crystallization of clay minerals from product of this breakdown. It is the result of much more intense weathering than that required for alteration. Formation of Kaolinite (1;1) from solutions containing soluble aluminum and silicon that came from the breakdown of the primary minerals having 2;1 type structure

## Relative stages of weathering

Weathering: According to Velde (1995) the contact of rocks and water produces clays, either at or near the surface of the earth.

Rock +Water → Clay

The carbon dioxide gas can dissolve in water and form carbonic acid, which will become hydrogen ions H+ and bicarbonate ions, and make water slightly acidic.

$$CO_2 + H_2O \rightarrow H_2CO_3 \rightarrow H^+ + HCO_3^-$$

The acidic water will react with the rock surfaces and tend to dissolve the K ion and silica from the feldspar. Finally, the feldspar is transformed into kaolinite.

Feldspar + hydrogen ions + water → clay (kaolinite) + cations, dissolved silica

$$2KAlSi_3O_8 + 2H^+ + H_2O \rightarrow Al_2Si_2O_5(OH)_4 + 2K^+ + 4SiO_2$$

Fine grained micas and magnesium rich chlorites represent earlier weathering stages of the silicates and kaolinite and ultimately iron and aluminum oxides the most advanced stages

**iii) Genesis of individual clays:** General conditions for the formation of the various layer silicate clays and oxides of iron and aluminum. Fine-grained micas, chlorite, and vermiculite are formed through rather mild weathering of primary aluminosilicate minerals, whereas kaolinite and oxides of iron and aluminum are products of much more intense weathering (figure 12) Conditions of intermediate weathering intensity encourage the formation of smectite. In each case silicate clay genesis is accompanied by the removal in solution of such elements as K, Na, Ca, and Mg.

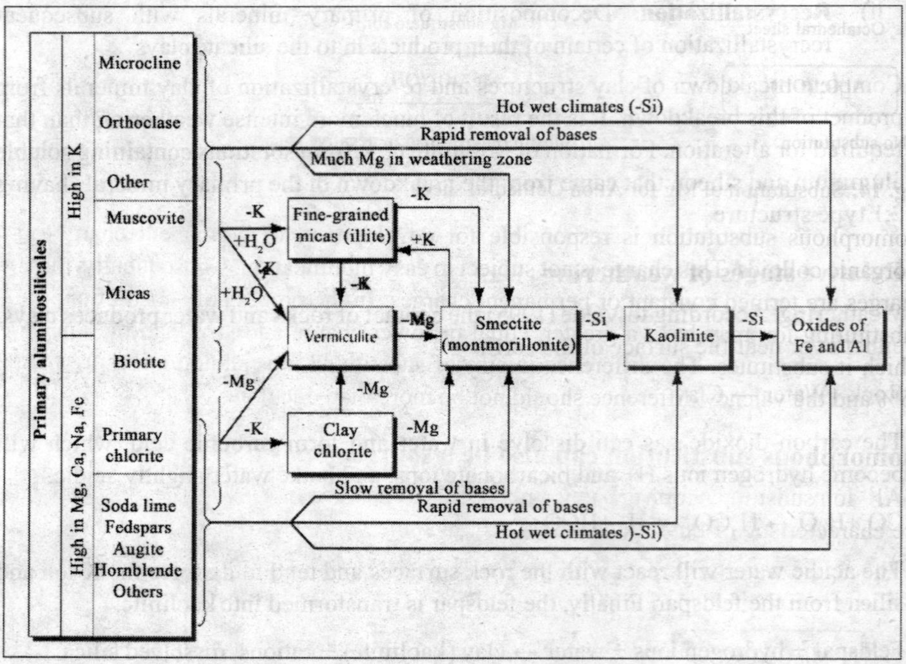


**Fig. 12:** Genesis of clay minerals (Adapted from Brady and Weil)

## 5.10. Charge development on soil clay mineral particles

Two types of charge arise on inorganic colloids i.e., 1. Permanent charge or Constant charge; 2. Temporary charge or variable charge or pH dependent charge.

**a) Permanent charge:** permanent charge development is by the process of isomorphous substitution: Substitution of one ion with another ion of comparable size (within 15%) and charge variation of one unit without disturbing the crystal lattice.

**On the tetrahedral sheet:** The weathering of wide variety of rocks and minerals permits cations of comparable size to substitute for silicon. Eg. $Al^{3+}$ion replace $Si^{4+}$ in the tetrahedral without changing the basic structure of the crystal (Figure 13).

**Fig.13:** Substitution of Al for Si on tetrahedral sheet.

**On the octahedral sheet:** Isomorphous substitution also occurs in the octahedral sheet. Iron ($Fe^{3+}$ 0.064 nm) and zinc (0.074 nm) or $Fe^{2+}$ (0.070 nm) can fit in to the position of either $Al^{3+}$ or $Mg^{2+}$ (0.06), as a central ion in the octahedral sheet (figure 14).

Octahedral sheet

$Al_2O_2(OH)_2$

No substitution

Mg substituted for Al

$MgAlO_2(OH)_2^-$

Substitution resulted in negative charge

**Fig. 14:** Substitution of Mg for Al on Octhedral sheet

Isomorphous substitution is responsible for development of permanent charge on inorganic colloids. This charge is not subject to easy modification. Consequently, these charges are termed constant or permanent charges. In the process of substitution, the substituting ion may have a greater, equal or lower charge and size than the ion for which it substitutes. The difference in size of substituted ions should not more than 15% and the valency difference should not be more than one unit .

## Isomorphous substitution can also be a source of positive charge

If $Al^{3+}$ ion substitutes for $Mg^{2+}$ ion, a net positive charge results. Such positive charges are characteristic of the trioctahedral sheet in some vermiculites and chlorites (figure 15).

$Mg_3O_2(OH)_2$

Trioctahedral sheet

$Mg_2AlO_2(OH)_2^+$

Net positive charge because

$Al^{3+}$ substituted for $Mg^{2+}$

**Fig. 15:** Substitution of Al for Mg on Octhedral sheet

Though there is a scope for development of both positive and negative charges on silicate clays, in general the net charge is always negative.

## b) Temporary charge or variable charge or pH dependant charge on clay particles

It is the second source of charge on some silicate clays (kaolinite) and on humus, allophone and Fe, Al hydroxides. Since these charges are "pH dependant", they are termed "variable, temporary and pH- dependant" charges. In this process +ve charges developed at low pH and -ve charge developed at high pH.

The zero point charge (ZPC) has been used as on index of the positive and negative charges on soil colloids. The ZPC is the pH at which negative and positive charges of a colloid are equal. The contribution of edge OH groups on crystal edges gives rise to charge on certain colloids. At high pH, $H^+$ of $OH^-$ on crystal lattice dissociates and combines with hydroxyl ions to form water molecules. This leaves negative charge on colloidal surface.

$>Al-OH + OH^- \rightarrow >Al-O^- + H_2O$

At low pH, protonation of exposed hydroxyl groups results in positive charge development.

$>Al-OH + H^+ \rightarrow >Al^+ + H_2O$

Kaolinites tend to stack in C-dimension, increasing the relative area of edges in relation to basal plane area. For these reasons, pH dependency of charge is more important for kaolinite than for smectites or vermiculite. In 2:1 type minerals there will be 5-10% pH dependant charge, while in 1:1 type mineral 50% pH dependant charg. Table 4 summarizes the charge particulars in different clay minerals.

## c) Development of positive and negative charges on crystal edges

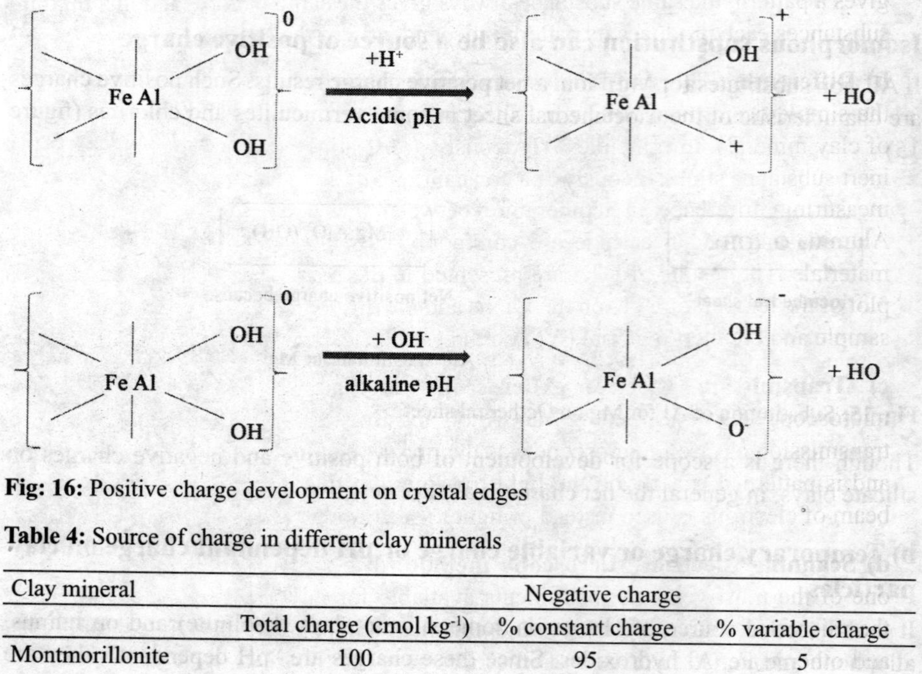


**Fig: 16:** Positive charge development on crystal edges

**Table 4:** Source of charge in different clay minerals

| Clay mineral | | Negative charge | |
|---|---|---|---|
| | Total charge (cmol $kg^{-1}$) | % constant charge | % variable charge |
| Montmorillonite | 100 | 95 | 5 |
| Illite | 30 | 80 | 20 |
| Kaolinite | 8 | 5 | 95 |
| Gibbsite | 4 | 0 | 100 |
| Goethite | 4 | 0 | 100 |
| Organic colloid | 200 | 10 | 90 |

## 5.11. Identification of clay minerals

Soils contain various minerals in clay size particles. Different method employed in identification of clay minerals.

**a) X-Ray Diffraction method:** X-ray diffraction (XRD) technique is one of the most commonly used methods to identify the minerals present in a soil. Minerals may be considered to be made up of regularly repeating unit cells, each unit cell having dimensions of a, b, and c along the three coordinate directions x, y, and z. In clay minerals, the c spacing, which is the dimension of the unit cell across the thickness of the layer, which usually varies from mineral to mineral, while the other two dimensions remain more or less constant among different clay minerals.

The first X-ray diffraction experiment was performed by Friedrich and Knipping using a crystal of copper sulfate in 1912. Diffraction was first proposed by English physicists W.H. Bragg and his son W.L. Bragg as a means of analyzing the structure of crystals. Both father and son were awarded the Nobel Prize in physics in 1915 for their work in determining crystal structures. It was found later that every crystalline substance gives a pattern, the same substance always gives the same pattern, and in a mixture of substances each produces its pattern independently of the others.

**b) Differential Thermal Analysis method:** Differential thermal analysis, DTA, is the simplest and most widely used thermal analysis technique for the identification of clay minerals. In principle, DTA consists of heating a test sample and a thermally inert substance simultaneously at a constant rate of about 10°C/min and continuously measuring difference in temperature between the sample and the inert material. Alumina and silicon carbide are commonly used as the thermally inert reference materials. The results of DTA are presented in the form of a thermogram, which is a plot of the temperature (T) on the x-axis and the difference in temperature between the sample and the inert material ($\Delta T$) on the y-axis.

**c) Transmission Electron Microscopy method:** The transmission electron microscope was developed by Max Knoll and Ernst Ruska in Germany in 1931. The transmission electron microscope was the first type of electron microscope developed and is patterned exactly on the light transmission microscope except that a focused beam of electrons is used instead of light to see through the specimen.

**d) Scanning Electron Microscopy method:** The scanning electron microscope is one of the most versatile instruments available for the examination and analysis of the microstructure, morphology, and chemical composition characterization of soils and other materials. Electron microscopes are scientific instruments that use a beam of energetic electrons to examine objects on a very fine scale. Zworykin et al. first described a true SEM in 1942 with a resolving power of 50 nm. Modern SEMs can have resolving power better than 1 nm.

**e) Infrared Spectroscopy method:** Infrared (IR) spectroscopy is one of the most common spectroscopic techniques used by organic and inorganic chemists. It involves the absorption measurement of different IR frequencies by a sample positioned in the path of an IR beam. Infrared (IR) spectroscopy deals with the infrared region of the electromagnetic spectrum with a longer wavelength and lower frequency than visible light. Infrared radiation spans a section of the electromagnetic spectrum having wavenumbers from roughly 13000 to 10 $cm^{-1}$, or wavelengths from 0.78 to 1000 μm. It is bound by the red end of the visible region at high frequencies and the microwave region at low frequencies.

The IR region is commonly divided into three smaller areas: near IR, mid IR, and far IR. For IR spectroscopy, the most frequently used is the mid IR region having wave number between 4000 and 400 $cm^{-1}$ (or wave length between 2.5 and 25 μm). IR absorption positions are generally presented as either wave numbers or wavelengths. Wave number defines the number of waves per unit length usually expressed as number per cm.

## 5.12. Influence of minerals on soil properties

### a) Influence on soil physical soil properties

Most physical soil properties are determined largely by the mineralogical composition of the soil. Soil texture, structure, moisture characteristics, pore size distribution, plasticity, shrink-swell potential, soil strength', and erodibilty can be directly or indirectly related to· soil mineral properties. Particle size distribution results from a physical degradation of soil minerals due to imperfect lattice energy distributions in the silicate layers causing strains in the mineral structure. Soil strength, aggregation and plasticity are affected by the various attractive and repulsive forces exerted by various minerals. The overall structure of the soil is the· product of the arrangement and bonding of individual soil particles into aggregates caused by ionic, organic, water and clay mineral surface interactions. The individual shape and stability of soil aggregates, however, varies with soil mineralogical composition. Smectitic minerals induce formation of spherical aggregates, whereas kaolinite promotes a platy structure. Structural stability of soil aggregates is a function of surface charge (CEC) of the minerals contained. It decreases in the order montmorillonite > illite > kaolinite soils. Soil water retention characteristics are a!so affected by the clay mineralogical make-up of the soils. This is due to variations in specific surface area and solute-water-solid interactions of specific olay minerals. Smectite soils retain much more water than illitic or kaolinitic soils, as is illustrated in Table 4. Because of this, water adsorption characteristics can be used for help in identifying soil mineralogical composition. Shrink-swell characteristic of a soil is another important physical property directly related to soil mineral content. It is due to charge imbalances in the crystal lattice and surface charge density of clay minerals. Volume expansion and shrink-swell potential are much higher in smectitic than in illitic or kaolinitic soils because of their higher surface change density. Soil erodibility is also affected by mineralogical composition, because the mineralogy of the soil influences such properties as aggregate stability, permeability and soil shear strength. Smectite soils, despite their higher aggregate stability, are more susceptible to surface losses because of their slow permeability. Total erosional losses however, are generally greater from illitic or kaolinitic soils because they require less water to reach saturation. At that point their shear strength is reduced to near zero, and they erode with surface water flow (Table 5).

**Table 5:** Effect of clay mineral composition on some physical soil properties. (Karathanasis, 1985)

| Dominant mineralogy | Volume expansion (%) | Water content (%) | | | |
|---|---|---|---|---|---|
| | | At different tensions (in bars) | | | At zero shear strength |
| | | 0.1 | 0.33 | 15 | |
| Smectitic (montmorillinitic) | >150 | >50 | >35 | >18 | >20 |
| Illitic or mixed | 70-150 | 35-50 | 22-35 | 9-18 | 5-20 |
| Kaolinitic | <50 | <22 | <15 | <9 | <5 |

### b) Influence on nutrient availability

Two aspects which relate soil mineralogical composition to nutrient availability are noteworthy. First, soil minerals are the source from which most plant nutrients are released in available forms into soil solution. Such release comes about through mineral transformation. weathering, or dissolution - reprecipitation reactions, i.e. K-release from weathering of micas; P and Ca release from phosphate and carbonate minerals. Secondly, due to their ion exchange properties and surface reactivity, soil minerals control the ionic equilibrium of the soil solution. This is the equilibrium on which mineral uptake necessary for plant growth depends. It is regulated by ion exchange and fixation processes. The available nutrient supply in soils is considered to be the sum of both soluble and exchangeable forms. Roots of living plants absorb nutrients from the soil's solution phase which is in equilibrium with the exchange phase, with the plant root itself being a cation exchanger. The capacity of the soil to replenish the solution nutrient concentration after it has been lowered by plant nutrient uptake depends the mineralogioal composition of the soil. This is because a great portion of the exchange capacity of the soil which determines the ability of the soil to supply growing plants with nutrients is related to crystalline and sub-crystalline mineral components. The magnitude of the ion exchange reaction is determined by the type of ions involved, nature of the soil solution and the strength of attractio~ between a particular mineral and a specific cation. Generally minerals with high cation exchange capacity (CEC) such as smectite, have a greater attraction for polyvalent (Ca ,Mg ) than monovalent ($K^+$, $Na^+$) ions. The reverse is true for minerals with low CEC such as kaolinite. Equally important is the role of soil minerals in controlling anion exchange and fixation phenomena. Minerals with low CEC and low crytallinity often have higher affinity for anions (e.g., $H_2P0_4^-$), especially under aoid pH conditions. The greater the negative charge per unit of surface area of the mineral (e.g., smectite) the more anions are excluded from the particle surface and lost by leaching from the rooting zone.

### c) Impact of chemical reactions involving soil minerals on controlling the availability of essential plant nutrients

Plants require 17 elements to complete their life cycle, and an additional four elements have been identified as essential for some plants. With the exception of C, H, and O, which plants obtain from air and water, plants derive the remaining 14 elements from the soil or through fertilizers, manures, and amendments. The bulk of the soil solid fraction is constituted by soil minerals, which exert significant direct and indirect influences on the supply and availability of most nutrient elements. The main processes involved in the release and fixation of nutrient elements in soils include dissolution-precipitation and adsorption-desorption.

Soil minerals serve as both sources and sinks of essential plant nutrients. As primary minerals that originally formed at high temperatures and pressures in igneous and metamorphic rocks are weathered in soils, they release plant nutrients into the soil solution. New minerals form in the aqueous phase of soil environments. These secondary minerals serve as sources of nutrients themselves, or they precipitate or adsorb essential elements, keeping them from being taken up readily by plants. In many cases, secondary minerals serve as important reservoirs where nutrients are held strongly enough to prevent leaching, yet weakly enough to allow plants to draw on them to meet their nutritional needs. In some soils and in certain topsoil layers, the soil organic matter contains and releases plant nutrient elements.

## d) Primary minerals and soil fertility

Sedimentary rock covers 75-80% of the Earth's crust, and it forms parent materials for a large majority of soils. Soil parent material has a significant direct influence on the nutrient element contents of the soil; this influence is more pronounced in young soils and diminishes somewhat with increasing soil age and soil weathering. In order to better understand the effect of soil parent materials on the soil elemental composition, it is useful to review the mineralogical composition of common rocks that make up the soil parent material (Table 6). Primary minerals form at elevated temperatures from cooling magma during the original solidification of rock or during metamorphism, and they are usually derived from igneous and metamorphic rocks in soil. In most soils, feldspars, micas, and quartz are the main primary mineral constituents, and pyroxenes and hornblendes are present in smaller amounts.

**Table 6:** Average mineralogical and nutrient element composition of common rocks on the Earth's land surface (adopted from Klein & Hurlbut 1999).

| Mineral constituent | Igneous rock | Shale | Sandstone | Nutrient element constitutents | |
|---|---|---|---|---|---|
| | | % | | Major | Minor |
| Feldspars | 59.5 | 30.0 | 11.5 | K, Ca, Na | Cu, Mn |
| Amphiboles & pyroxenes | 16.8 | – | - | Mg, Fe, Ca | Ni, Co, Cu, Mn, Zn, Mo |
| Micas | 3.8 | – | - | K, Ca, Na, Mg, Fe | Ni, Mn, Co, Zn, Cu |
| Titanium minerals | 1.5 | – | - | Ti, Fe, Ca | Co, Ni |
| Apatite | 0.6 | – | - | Ca, P | |
| Clay | – | 25.0 | 6.6 | K, Mg, Fe, Ca, Na | |
| Iron oxides | – | 5.6 | 1.8 | Fe | Mn, Zn, Ni, Co |
| Carbonates | – | 5.7 | 11.1 | Ca, Mg, Fe | |
| Other minerals | – | 11.4 | 2.2 | – | – |

Primary minerals including K-feldspars (orthoclase, sanidine, and microcline), micas (muscovite, biotite, and phlogopite), and clay-size micas (illite) are widely distributed in most soil types, except in highly weathered and sandy soils. These primary minerals act as an important reservoir for K, with over 90% of K in soils existing in the structure of these minerals. Significant amounts of Ca, Na, and Si and smaller amounts of Cu and Mn are also present in the feldspars. Micas and illite are the most important source of K in many soils, and they also contain Mg, Fe, Ca, Na, Si, and a number of micronutrients. Amphiboles and pyroxenes are vital reservoirs of Mg, Fe, Ca, Si, and most of the micronutrients. Carbonate minerals, including those derived from soil parent material and those formed in soil through pedogenic processes, serve as both a source and a sink for Ca and Mg in soils.

The physical, chemical, and biological weathering of primary minerals releases a number of nutrient elements into the soil solution. Weathering rates and pathways of primary minerals are highly variable and depend on several factors, including mineral properties and climatic conditions. Although the weathering rates of primary minerals for certain elements may not be fast enough to meet plant nutrient requirements on a short-term basis, particularly in managed cropping systems, mineral weathering is an important and long-term source of several geochemically derived nutrients. The nutrient supply capacity of a soil through weathering of primary minerals diminishes as the extent of soil weathering increases.

## e) Secondary minerals and soil fertility

In contrast to the primary minerals, secondary minerals in soils are usually formed by low-temperature reactions during the weathering of primary minerals in the aqueous environment at the Earth's surface. Secondary minerals primarily control nutrients through adsorption-desorption, dissolution-precipitation, and oxidation-reduction reactions.

Adsorption reactions involving minerals are often more important in controlling plant nutrient element availability than the release of nutrient elements by mineral weathering. Phyllosilicates with a permanent charge (e.g., vermiculite and smectite) offer exchange sites that hold a number of essential nutrients in their cationic form (cation exchange capacity), such as $Ca^{2+}$, $Mg^{2+}$, $K^{+}$, and $Na^{+}$; the nutrients are retained by outer-sphere complex formation (Figure 15) and may be easily taken up by plant roots. On the other hand, variable charge minerals (e.g., Fe oxides) retain some nutrients (P, Zn) by forming inner-sphere complexes, and such nutrients are not readily available to plants. Important reactions relevant to specific nutrient elements are discussed below.

**i) Nitrogen:** Plants usually take up the nitrate ($NO_3^-$) and ammonium ($NH_4^+$) forms of soil nitrogen. In soils, N applied through fertilizers and mineralized N from organic matter mostly ends up in the $NO_3^-$ form. Due to the limited anion exchange capacity of most soils, leaching of applied N in the form of $NO_3^-$ ions is a common water quality problem, particularly in agricultural regions. It also represents an important economic inefficiency, because producers apply excessive amounts of fertilizer to compensate for the leaching. Highly weathered soils, such as oxisols and ultisols, are the exception. The mineralogy of oxisols and ultisols is dominated by minerals with variable surface charge, mainly kaolinite and Fe and Al oxides, which provide these soils with the capacity to retain large amounts of $NO_3$-N, particularly in the subsoil horizons. It was reported that 150-300 kg $NO_3$-N $ha^{-1}$ (up to a depth of 2 m) in a Brazilian oxisol in a maize-soybean cropping system and 17,000-32,000 kg $NO_3$-N $ha^{-1}$ to a depth of 10 m under different land uses in oxisols from northern Queensland in Australia. The anion exchange capacity of the Australian oxisols was large, with values as high as 41 mmol $kg^{-1}$. The adsorbed nitrate is too deep and is likely inaccessible to most field crops, nevertheless, it does not leach into groundwater.

In contrast to above oxisols and ultisols, soils in temperate regions generally have permanent charge minerals (e.g., smectite and vermiculite) with high cation exchange capacity and the ability to retain ammonium ($NH_4^+$) ions. Indeed, a large proportion of the $NH_4$-N is retained in the interlayers of 2:1 phyllosilicates and is not readily exchangeable, causing it to be referred to as fixed $NH_4$. The process of $NH_4$-fixation is similar to that of K-fixation, which is demonstrated in Figure 17. Vermiculite, illite, and interstratified minerals with 2:1 layers are involved in the fixation of $NH_4^+$ ions in soils. With the exception of sandy soils, the amount of fixed $NH_4^+$ in the soil ranges from 350-3,800 kg $NH_4$-N $ha^{-1}$ in the top 30 cm of soil; vermiculite and partially weathered illite generally have a greater capacity to fix $NH_4^+$ in soils than the smectite group of minerals. The different behavior and capacity of 2:1 phyllosilicates in fixing $NH_4^+$ ions is related to the magnitude and origin of negative charge in these minerals. $NH_4$-fixation generally increases with the increasing amount of layer charge in the 2:1 phyllosilicates, and the fixation is greater in minerals with charge originating in the tetrahedral sheet than in minerals with charge originating in the octahedral sheet.

**ii) Phosphorus:** Phosphorous is primarily taken up by plants in the form of phosphate ions ($HPO_4^{2-}$ and $HPO_4^-$) from the soil solution. The concentration of P in soil water is generally very low (< 0.01% of the total soil P), with the bulk of the soil P existing as organic P, insoluble compounds of P with Al, Fe, and Ca, and phosphate adsorbed to Fe and Al oxides and phyllosilicates. Phosphate ions from dissolved chemical fertilizers react rapidly in most soils, resulting in P fixation in the soil. These soil reactions involve both adsorption and precipitation processes.

**(i) Adsorption Reactions:** Adsorption reactions of phosphate ions on mineral surfaces predominantly involve the formation of inner-sphere complexes on the variable charge surfaces of Fe and Al oxides and kaolinite. Phosphate ions adsorbed by such processes are only slowly available to plants. Phosphate is also known to be sorbed by calcite in calcareous soils, with the sorption occurring via the replacement of $CO_3^{2-}$ on the calcite surfaces.

**(ii) Precipitation Reactions:** In strongly acidic soils, the precipitation reactions involving soluble phosphate from fertilizer results in the formation of insoluble Al, Fe, or Mn phosphates. In contrast, in calcareous soils, insoluble Ca phosphates are formed, which are gradually converted to insoluble carbonated hydroxyapatite. General chemical reactions of phosphate in acidic and calcareous soils are shown below.

In acid soils: $Al^{3+} + H_2PO_4^- + 2H_2O \rightarrow Al(OH)_2H_2PO_4 + 2H^+ \rightarrow AlPO_4.2H_2O$

Soluble slighty soluble insoluble

In calcareous soils: $Ca(H_2PO_4)_2 \rightarrow Ca_2(HPO_4)_2 \rightarrow Ca_3(PO_4)_2 \rightarrow [3Ca_3(H_2PO_4)_2]CaCO_3$

Decreasing solubility $\rightarrow\rightarrow\rightarrow$

**iii) Potassium:** Among the essential elements, K is usually the most abundant in soils. Total K in soils varies from 0.5-2.5% of the soil mass, and most of the K exists in mineral form (K-feldspars and micas). Potassium is released following the weathering or dissolution of K minerals in soils, as shown in the following examples:

## Potassium release by weathering of minerals

a) Weathering of mica

$K(Mg_2Fe^{2+}(Si_3Al)O_{10}(OH)_2 + 0.2H_4SiO_4 + 1.25Mg^{2+} + 1.1\ H_2O \rightarrow$

Biotite

$Mg_{0.35}(Mg_{2.9}Fe_{0.1}{}^{3+})(Si_{3.2}Al^{0.8})\ O_{10}(OH)_2 + 0.9FeOOH + 0.2Al(OH)_3 + 1.5\ H^+ + K^+$

Vermiculite Goethite Gibbsite

b) Weathering of K- feldspar

$2KAlSi_3O_8 + 9H_2O + 2H^+ \rightarrow Al_2Si_2O_5(OH)_4 + 4H_4SiO_4 + K^+$

Microcline Kaolinite

Among these two reactions, K release by the weathering of mica is generally more important in supplying K to plants in unfertilized soils.

Phyllosilicates retain and release K for plants from non-exchangeable or fixed (i.e., exchanged very slowly and only when the K concentration in soil water drops below a threshold value) and exchangeable forms. Potassium ions present on the exchange sites are adsorbed by outer-sphere complexation and are readily available for plant uptake. On the other hand, illite, vermiculite, and interstratified 2:1 clay minerals release fixed or non-exchangeable K from interlayer sites through cation exchange and diffusion processes at slower rates than the exchangeable K. Similar to $NH_4^+$ ions, K supplied through fertilizers or other amendments can be fixed in the interlayers of the 2:1 minerals. The non-exchangeable or fixed K can be potentially released back into soil solution if the solution K concentration falls below a certain threshold value.

## iv) Secondary Nutrients

Among the secondary nutrients, Ca and Mg are taken up by plants in their cationic forms, $Ca^{2+}$ and $Mg^{2+}$. These cations are retained at negatively charged sites of phyllosilicates via electrostatic attraction (outer-sphere complexation). The precipitation of secondary carbonates, such as calcite ($CaCO_3$), magnesium calcite ($Ca_{1-x}Mg_xCO_3$), and gypsum

($CaSO_4 \cdot 2H_2O$), is common in soils of arid and semi-arid environments. Secondary carbonates are considered to be important scavengers of some nutrients through incorporation in the mineral structure (e.g., Mn) or inner-sphere complexation (e.g., P and Zn) at the mineral surface.

Sulphur is taken up by plants as sulfate ($SO_4^{2-}$), and this is the most common inorganic S form in soils. Fe and Al oxides and kaolinite provide adsorption sites for $SO_4^{2-}$ in most soils, even if these minerals are present in small amounts. Sulfate ions are believed to be adsorbed by these minerals by forming both inner- and outer-sphere complexes. In calcareous soils, $SO_4^{2-}$ may be sorbed on $CaCO_3$ by forming a $CaCO_3$-$CaSO_4$ co-precipitate, which renders $SO_4^{2-}$ unavailable to plants. Sulfide ($S^-$, $S^{2-}$) minerals form under reducing environments (e.g., freshwater and tidal marshes) where $SO_4^{2-}$ ions are reduced to form minerals such as pyrite ($FeS_2$). Such reduced forms of S are oxidized following exposure to air, releasing $SO_4^{2-}$, $H^+$, and $Fe^{3+}$ ions to the soil solution.

## v) Micronutrients

Among the micronutrients, Fe, Mn, Cu, Zn, and Ni are taken up by plants in their cationic forms, and B, Mo, and Cl are taken up by plants in their anionic forms. Fe and Mn are often present in large quantities in most soils, and adsorption reactions play little role in controlling their plant availability in soils. Oxidation and precipitation reactions predominantly control the soil solution concentration of Fe and Mn. Goethite, hematite, and ferrihydrite are the most commonly occurring secondary Fe oxides in soils. Due to the microcrystalline size of Fe oxides, these minerals possess high specific surface areas and provide numerous adsorption sites for both cationic and anionic elements in all varieties of soils. The two most stable Fe oxides, goethite and hematite, are known to have substantial structural substitution of trace elements, including Mn, Ni, Zn, and Cu. Manganese minerals are not as abundant and common as Fe oxides. Often, they exist in soils as mineral coatings, as nodules, or as finely dispersed particles in the soil matrix. Both Fe and Mn oxides are common mineral constituents in many soils and are important substrates for the retention of many macronutrients and micronutrients. Plant availability of both Fe and Mn is greatly reduced in calcareous soils due to the extremely low solubility of Fe and Mn oxides and of Mn carbonates. In such situations, plants induce biochemical responses, such as release of reducing and chelating compounds and acidification of rhizosphere, which can increase the availability of Fe, Mn, and other micronutrients.

Copper, Zinc, and Nickel are adsorbed by Fe and Al oxides by forming inner-sphere complexes at low solution concentrations. However, at higher solution concentrations, precipitation of metal hydroxides occurs (Ginder-Vogel & Sparks 2010). Adsorption of $Cu^{2+}$, $Zn^{2+}$, and $Ni^{2+}$ occurs by outer-sphere complex formation on negatively charged surfaces of 2:1 phyllosilicates and perhaps by inner-surface complex formation on kaolinite surfaces. In alkaline soils, adsorption of Zn on calcite and co-precipitation of Cu in calcite may also occur.

Boron and Molybdenum are taken up by plants as $H_3BO_3$ and $MoO_4^{2-}$. Limited evidence suggests that B species (i.e., $B(OH)_3$ and $B(OH)_4^-$) in soils are adsorbed by forming inner-sphere complexes on the surfaces of Fe and Al oxides (Su & Suarez 1995). Similarly, $MoO_4^{2-}$ is strongly adsorbed by metal oxides. Chlorine is taken up by plants in the chloride ($Cl^-$) form, and adsorption reactions involving $Cl^-$ ions are similar to those involving $NO_3^-$ ions. Therefore, a relatively high potential for exchange-based adsorption of $Cl^-$ ions occurs in highly weathered soils because their mineralogy is dominated by variable-charge minerals, such as kaolinite and Fe and Al oxides. In certain soil environments, such as those with restricted leaching or those with low-lying areas in arid climates, $Cl^-$ may exist in precipitated mineral forms, such as NaCl, $CaCl_2$, and $MgCl_2$.

## Review questions

1) Briefly discuss on classification and distribution of primary minerals with special emphasis on halides, sulphates, sulphides, carbonates and phosphate minerals.
2) Explain the genesis of clay minerals
3) List out the properties of clay minerals and explain swelling and shrinkage, surface are, electrical properties.
4) Give brief classification of clay minerals
5) Explain the basic structure of alumino silicate clays with neat sketch
6) Describe the structure ands properties of Kaolinite mineral
7) Differentiate between kaolinite, montmorrilinite and illite minerals.
8) Differentiate between organic and inorganic soil colloids
9) Explain the charge development on silicate clay minerals

## Objective questions

1. The sandstone on metamorphosis is converted to
   a) Quartzite  b) Slate
   c) Schist  d) Gneiss
2. The mineral that is most resistant to weathering is
   a) Micas  b) Feldspars
   c) Olivines  d) Quartz
3. Anion exchange capacity is more in
   a) Illite  b) Kaolinite
   c) Smectite  d) Vermiculite
4. The minerals which form the characteristic component of rock is called
   a) Accessory minerals  b) Essential minerals
   c) Primary minerals  d) Secondary minerals

5. An example for 1:1 type of clay mineral is
   a) chlorite b) Illite
   c) Montmorillonite d) Halloysite
6. Iron rich humus poor 'Plinthite' is present in
   a) Alluvial soils b) Laterite soils
   c) Red soils d) Black Soils minerology
7. Author of reference book Chemistry of the soil
   a) J. S. Kanwar b) Fireman E Bear
   c) Nyle. C. Brady d) J. L. Sehgal
8. Publisher of reference book Soil Minerology with environmental applications
   a) Indian Society of soil Science b) Indian Society of Agril Chemistry
   c) Soil Science Society of America d) International society of soil Science
9. The peeling of the outer layer of rocks due to differential expansion and contraction between the outer and inner parts of the rocks is referred to as
   a) Digenesis b) Defoliation
   c) Exfoliation d) Stratification
10. The clay colloid which increases plasticity index is
    a) Kaolinite b) Montmorillonite
    c) Illite d) Beidellite
11. Specific heat of soil minerals is
    a) 1.0 cal /g b) 0.46 cal /g
    c) 0.2 cal /g d) 0.53 cal /g
12. The mineral magnetite is an example of the following metal element
    a) Mg b) S
    c) Fe d) Mn
13. A non- ferromagnesian mica mineral which is resistant for weathering is
    a) Biotite b) Pyroxene
    c) Amphibole d) Muscovite
14. On metamorphosis, the mineral 'Shale' changes to
    a) Marble b) Quartzite
    c) Gneiss d) Slate
15. Heavy minerals are those minerals having specific gravity more than
    a) 1.53 g/cm³ b) 2.83 g/cm³
    c) 1.26 g/cm³ d) 2.96 g/cm³

16. On hydrolysis Muscovite converts to
   a) Vermiculite   b) Montmorrilonite
   c) illite   d) Kaolinite
17. Charge density is high in ------- positions of a clay mineral
   a) Planar positions   b) edge positions
   c) Interlayer positions   d) B &C
18. CEC on Vermiculite is
   a) 300 cmol $kg^{-1}$   b) 3-15 cmol $kg^{-1}$
   c) 70-120 cmol $kg^{-1}$   d) 100-150 cmol $kg^{-1}$
19. Sequioxide minerals are
   a) Non-sticy and non-plastic   b) high P fixation capacity
   c) Expanding type   d) A &B
20. Hydrogen bonding is present in the interlayer spaces of
   a) Kaolinite   b) Illite
   c) Montmorrillonite   d) Imogolite
21. Illite mineral is
   a)1:1 type crystalline mineral   b) Non crystalline amorphous mineral
   c)2:1 expanding type mineral   d) 2:1 non-expanding type mineral
22. The shape of kaolinite mineral is
   a) Rod shate   b) Fluffy
   c) Hexagonal   d) Needle shape
23. The specific surface of organic colloid is
   a) 700 $m^2/g$   b) 900 $m^2/g$
   c) 20 $m^2/g$   d) 100 $m^2/g$
24. An example for variable charge mineral is
   a) Goethite   b) Montmorrilonite
   c) Illite   d) Vermiculite
24. Weak o-o binding is found in interlayer of
   a) Kaolonite   b) Montmorrilonite
   c) Illite   d) A & B
25. In illite mineral the lattice structure is non expanding due to
   a) Presence of hydrogen bonding   b) Presence of potassium ions
   c) presence of brusite layer   d) A & B

26. In montmorillonite mineral isomorphous substitution occur in
a) Tetrahedral sheet b) Octahedral sheet
c) Both A & B d) Neither A nor B

27. Cloud is colloidal system of
a) Liquid in liquid b) Liquid in gas
c) Liquid in solid d) Solid in liquid

28. The soil particles of the following size possesses colloidal properties
a) less than 1.0 μm b) 1-10 μm
c) 10-100 μm d) 100-1000 μm

29. The clay mineral has high internal surface area is
a)Kaolinite b) Montmorillonite
c) Illite d) Chlorite

30. The source of negative charge on clay colloids comes from
a) ionisable hydrogen ions b) isomorphous substitution
c) A & B d) hydrolysis

31) Swelling and shrinkage property is a character of
a) Kaolinite b) Montmorillonite
c) Hallosite d) Illite

32. The floculation ability is high for the following cation
a) $Al^{3+}$ b) $Ca^{2+}$
c) $Mg^{2+}$ d) $K^{+}$

33. An example for tecto silicate mineral is
a) Kaolinite b) Illite
c) Muscovite d) Quartz

34. An example for tri-octahedral mica is
a) Biotite b) Muscovite
c) Illite d) Phengite

35. The constant zig-zag movement of soil colloidal particles when suspended in water is reffered as
a) Molecular movement b) Brownian movement
c) Celestial movement d) Centripetal movement

36. Tyndals effect of soil colloidal suspension is due to
a) Difraction of light b) Reflection of light
c) Radiation of light d) Diffusion of light

37. In acidic conditions the organic colloids of soil exhibit

a) Positive charge
b) Negative charge
c) No charge
d) Isoelectric charge

38. ***Assertion (A)*** : Dry soils heats up quickly as compared to that of moist soils

***Reason (R)*** : Specific heat of soil forming minerals is less than that of water

**Options**

a) Both A and R are true and R judiciously explains A
b) Both A and R are true and R does not judiciously explains A
c) Only A is true
d) Only R is true

39. ***Assertion (A)*** : Red soils have more CEC than black soils

***Reason (R)*** : Montmorillonite clay minerals have more CEC than Kaolinite clay minerals

**Options**

a) Both A and R are true and R judiciously explains A
b) Both A and R are true and R does not judiciously explains A
c) Only A is true
d) Only R is true

40. ***Assertion (A)***: Laterisation occurs in tropical climates

***Reason (R):*** Tropical climate favours accumulation of sesquioxides and forms Plinthite which is hard on drying

**Options**

a) Both A and R are true and R judiciously explains A
b) Both A and R are true and R does not judiciously explains A
c) Only A is true
d) Only R is true

41. ***Assertion(A)*** : Quartz is resistant to physical and chemical weathering

***Reason (R)*** : The frame work of quartz is very densely packed electrically neutral and Prevents any form of substitution.

**Options**

a) Both A and R are true and R judiciously explains A
b) Both A and R are true and R does not judiciously explains A
c) Only A is true
d) Only R is true

42. ***Assertion (A)*** : The electrostatic forces exist on colloidal particles of soil would attract water molecules.

***Reason (R)*** : Water molecules by virtue of dipolar nature, gets attracted to colloidal particles.

**Options**

a) Both A and R are true and R judiciously explains A
b) Both A and R are true and R does not judiciously explains A
c) Only A is true
d) Only R is true

43. ***Assertion (A)*** : Feldspars are most abundant among rock forming minerals in the earth crust

***Reason (R)*** : Pyroxenes and Amphiboles are excellent host minerals for micronutrients in soils

**Options**

a) Both A and R are true and R judiciously explains A
b) Both A and R are true and R does not judiciously explains A
c) Only A is true
d) Only R is true

44. List (I) List (II)

| Mineral group | | Example |
|---|---|---|
| i. Ortho silicate mineral | ( ) | 1. Quartz |
| ii. Inosilicate | ( ) | 2. Muscovite |
| iii. Tecto silicate mineral | ( ) | 3. Olivine |
| iv.Phyllosilicate mineral | ( ) | 4. Amphibolite |

**Codes**

| | i | ii | iii | iv |
|---|---|---|---|---|
| a | 3 | 2 | 1 | 4 |
| b | 3 | 4 | 1 | 2 |
| c | 4 | 2 | 1 | 3 |
| d | 4 | 3 | 2 | 1 |

45. List –I List-II

| (Type of rocks) | | (Examples) |
|---|---|---|
| i. Volcanic rock | ( ) | 1. Andesite |
| ii. Subacid Rock | ( ) | 2. Trachyte |
| iii. Sedimentary Rock | ( ) | 3. Limestone |
| iv. Metamorphic Rock | ( ) | 4. Serpentine |

**Codes**

| | i | ii | iii | iv |
|---|---|---|---|---|
| a | 2 | 3 | 4 | 1 |
| b | 3 | 1 | 4 | 2 |
| c | 2 | 4 | 1 | 3 |
| d | 1 | 2 | 3 | 4 |

46. Weak 0-0 bounding is found in inter layer of
    a) Kadimite
    b) Montinfoilonik
    c) Illite
    d) A and B

## Answer key

| | | | | | | | | | | | | | | | | | | | |
|---|---|---|---|---|---|---|---|---|---|---|---|---|---|---|---|---|---|---|---|
| **1** | s | **6** | b | **11** | c | **16** | d | **21** | d | **26** | c | **31** | b | **36** | d | **41** | a | **46** | b |
| **2** | f | **7** | b | **12** | c | **17** | d | **22** | c | **27** | b | **32** | a | **37** | a | **42** | A | | |
| **3** | b | **8** | a | **13** | d | **18** | d | **23** | b | **28** | a | **33** | d | **38** | a | **43** | b | | |
| **4** | b | **9** | c | **14** | d | **19** | d | **24** | a | **29** | b | **34** | a | **39** | d | **44** | b | | |
| **5** | f | **10** | b | **15** | b | **20** | a | **25** | b | **30** | c | **35** | b | **40** | a | **45** | c | | |

# 6

# Soil Organic Matter

Soil organic matter is any material produced originally by living organisms (plant or animal) that is returned to the soil and goes through the decomposition process. At any given time, it consists of a range of materials from the intact original tissues of plants and animals to the substantially decomposed mixture of materials known as humus. Soil organic matter (SOM) is the organic component of soil, consisting of three primary parts including plant residues and living soil organisms, decomposing (active) organic matter, and stable organic matter (humus). Most soil organic matter originates from plant tissue. Plant residues contain 60–90 percent moisture, remaining dry matter consists of carbon (C), oxygen, hydrogen (H) and small amounts of sulphur (S), nitrogen (N), phosphorus (P), potassium (K), calcium (Ca) and magnesium (Mg). Although present in small amounts, these nutrients are very important from the viewpoint of soil fertility management. Soil organic matter exists as four distinct fractions, with each varying in properties and decomposition rates (table 1).

**Table 1:** Size, composition, turnover rate and decomposition stage of the four soil organic matter fractions.

| Fraction | Size | Turnover time | Composition |
|---|---|---|---|
| Dissolved organic matter | < 45 μm (in solution) | Dissolved organic matter generally turns over very rapidly (minutes to days) | Made up of soluble root exudates, simple sugars and decomposition by-products. It generally constitutes less than one per cent of total soil organic matter. |
| Particulate organic matter | 53 μm – 2 mm | From months to decades | Composed of fresh and decomposing plant and animal matter with an identifiable cell structure. Makes up between 2-25 per cent of total soil organic matter. |
| Humus | < 53 μm | Decadal (from tens of years up to hundreds of years) | Made up of older, decayed organic compounds that have resisted decomposition. Includes both structural (e.g. proteins, cellulose) and non-structural (e.g. humin, fulvic acid) organic molecules. Often makes up more than 50 per cent of total soil organic matter. |
| Resistant organic matter | <53 μm and in some soils <2 mm | Ranges from hundreds to thousands of years | Resistant organic matter is relatively inert material made up primarily of chemically resistant materials or remnant organic materials such as charcoal (burnt organic material). This pool can constitute up to 30 per cent of soil organic matter. |

Organic matter is different to organic carbon in that it includes all the elements that are components of organic compounds. Soil organic matter is difficult to measure directly, so laboratories tendto measure soil organic carbon and use a conversion factor to estimate how much organic matter is held within a soil. Soil organic carbon is a component of soil organic matter, with about 58 per cent of the mass of organic matter existing as carbon. Therefore, if we determine the amount of soil organic carbon in a sample and multiply it by 100/58 (or 1.72) we can estimate the proportion of organic matter in the soil sample:

Organic matter (%) = total organic carbon (%) × 1.72

## 6.1. Sources and composition of soil organic matter

The original source of soil organic matter is plant tissue (leaves, roots and left outs of harvested crops). Animals are the secondary sources of organic matter (their bodies when their life cycles are consummated). On an average fresh organic matter contain 75% water and 25% dry matter (Figure 1). Carbon and oxygen are the most dominant element in organic matter constituting 42% each. Hydrogen and ash constitute 8% each. Among the chemical compounds, cellulose constitutes 45%, followed by lignin 20%, hemicelluloses 18%, protein 8%, fats and waxes 2% and polyphenols 2%.

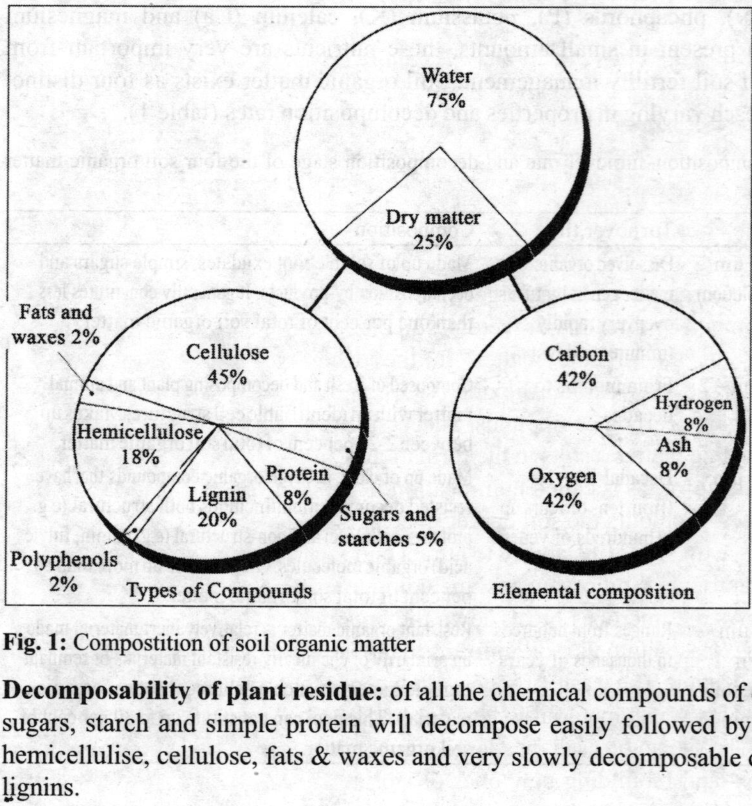


**Fig. 1:** Compostition of soil organic matter

**Decomposability of plant residue:** of all the chemical compounds of organic matter, sugars, starch and simple protein will decompose easily followed by crude protein, hemicellulise, cellulose, fats & waxes and very slowly decomposable compounds are lignins.

## 6.2. Organic matter decomposition

When plant residues are returned to the soil, they undergo decomposition. Decomposition is a biological process that includes the physical breakdown and biochemical transformation of complex organic molecules of dead material into simpler organic and inorganic molecules. The continual addition of decaying plant residues to the soil surface contributes to the biological activity and the carbon cycling process in the soil. Breakdown of soil organic matter and root growth and decay also contribute to these processes. Carbon cycling is the continuous transformation of organic and inorganic carbon compounds by plants and micro- and macro-organisms between the soil, plants and the atmosphere (Fgure2).

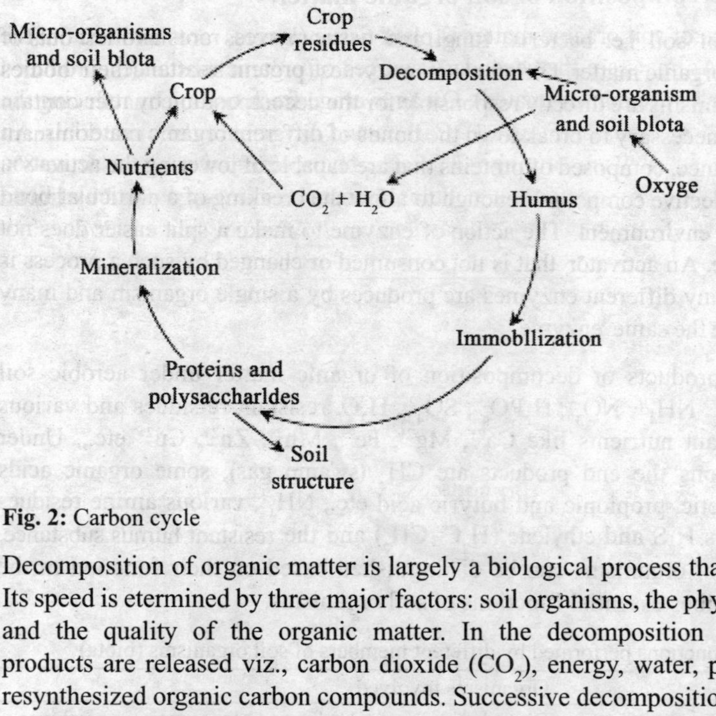


**Fig. 2:** Carbon cycle

Decomposition of organic matter is largely a biological process that occurs naturally. Its speed is etermined by three major factors: soil organisms, the physical environment and the quality of the organic matter. In the decomposition process, different products are released viz., carbon dioxide ($CO_2$), energy, water, plant nutrients and resynthesized organic carbon compounds. Successive decomposition of dead material and modified organic matter results in the formation of a more complex organic matter called humus. This process is called humification. Humus affects soil properties. As it slowly ecomposes, it colours the soil darker; increases soil aggregation and aggregate stability; increases the CEC (the ability to attract and retain nutrients); and contributes N, P and other nutrients. Soil organisms, including micro-organisms, use soil organic matter as food. As the microbes break down the organic matter, any excess nutrients (N, P and S) are released into the soil in forms that plants can use. This release process is called mineralization. The waste products produced by micro-organisms are also soil organic matter. This waste material is less decomposable than the original plant and animal material, but it can be used by a large number of organisms. By breaking down carbon structures and rebuilding new ones or storing the C into their own biomass, soil biota plays the most important role in nutrient cycling processes and, thus, in the ability of a soil to provide the crop with sufficient nutrients to harvest a healthy product. The organic matter content, especially the more stable humus, increases the capacity to store water and store (sequester) C from the atmosphere.

When organic tissue is added to soil three general reactions take place. First the bulk of the material undergoes enzymatic oxidation with $CO_2$, $H_2O$ and energy and heat as the major products. Secondly, the essential elements such as nitrogen, phosphorous and sulphur are released and / or immobilized by a series of specific reactions relatively unique for each element. Finally, compounds very resistant to microbial actions are formed either through modification of compounds in the original plant tissue or by microbial synthesis.

The native flora of soil i.e. bacteria, fungi and actinomycetes are involved in the decomposition of organic matter. Different soil enzymes (protein substances) produced by these microorganisms are directly responsible for the decomposition by reducing the activation energy, necessary to breakdown the bonds of different organic materials. An enzyme is a substance, composed of proteins that are capable of lowering the activation energy of other selective compounds enough to allow the breaking of a particular bond under a particular environment. The action of enzyme to make a split easier does not use up the enzyme. An activator that is not consumed or changed by such a process is called catalyst. Many different enzymes are produces by a single organism and many organisms produce the same enzyme.

The simple end products of decomposition of organic matter under aerobic soil situations are $CO_2$, $NH_4$+, $NO_3^-$, $H_2PO_4^-$, $SO_4^{2-}$, $H_2O$, resistant re sidues and various other essential plant nutrients like $Ca^{2+}$, $Mg^{2+}$, $Fe^{2+}$, $Mn^{2+}$, $Zn^{2+}$, $Cu^{2+}$ etc.,. Under anaerobic conditions the end products are $CH_4$ (swamp gas), some organic acids (R-COOH) eg. lactic, propionic and butyric acid etc., $NH_4^+$, various amine residues ($R\text{-}NH_2$), toxic gas $H_2S$ and ethylene ($H_2C{=}CH_2$) and the resistant humus substance. Different groups of organisms can be distinguished based on their function in the soil. Table 2 describes essential functions of soil organisms.

**Tabel 2:** Essential functions performed by different members of soil organisms (biota)

| Functions | Organisms involved |
|---|---|
| Maintenance of soil structure | Bioturbating invertebrates and plant roots, mycorrhizae and some other micro-organisms |
| Regulation of soil hydrological processes | Most bioturbating invertebrates and plant roots |
| Gaseous exchange and carbon sequestration (carbon accumulation in soil) | Mostly microorganisms and plant roots, some carbon protected in large compact biogenic invertebrate aggregates |
| Soil detoxification | Mostly microorganisms |
| Nutrient cycling | Mostly microorganisms and plant roots, some soil and litter feeding invertebrates |

| | |
|---|---|
| Decomposition of Organic matter | Various saprophytic and litter feeding invertebrates (detritivores), fungi, bacteria, actinomycetes and other microorganisms |
| Suppression of pests, parasites and diseases | Plants, mycorrhizae and other fungi, nematodes, bacteria and various predators. |
| Source of food and medicine | Plant roots, various insects (crickets, beetle larvae, ants, termites), earthworms, vertebrates, microorganisms and their by-products |
| Symbiotic and asymbiotic relations with plants and their roots | Rhizobia, mycorrhizae, actinomycetes, diazotrophic bacteria and various other rhizosphere micro organisms and ants |
| Plant growth regulating (positive and negative) | Direct effects: plant rros, rhizobia, mycorrhizae, actinomycetes, pathogens, phytoparasitic nematodes, rhizophagous insects, plant growth promoting rhizosphere microorganisms, biocontrol agents<br>Indirect effects: most soil biota |

Fresh residues consist of recently deceased micro-organisms, insects and earthworms, old plant roots, crop residues, and recently added manures. Crop residues contain mainly complex carbon compounds originating from cell walls (cellulose, hemicellulose, etc.). Chains of carbon, with each carbon atom linked to other carbons, form the "backbone" of organic molecules. These carbon chains, with varying amounts of attached oxygen, H, N, P and S, are the basis for both simple sugars and amino acids and more complicated molecules of long carbon chains or rings. Depending on their chemical structure, decomposition is rapid (sugars, starches and proteins), slow (cellulose, fats, waxes and resins) or very slow (lignin). During the decomposition process, microorganisms convert the carbon structures of fresh residues into transformed carbon products in the soil. There are many different types of organic molecules in soil. Some are simple molecules that have been synthesized directly from plants or other living organisms. These relatively simple chemicals, such as sugars, amino acids, and cellulose are readily consumed by many organisms. For this reason, they do not remain in the soil for a long time. Other chemicals such as resins and waxes also come directly from plants, but are more difficult for soil organisms to break down. Humus is the result of successive steps in the decomposition of organic matter. Because of the complex structure of humic substances, humus cannot be used by many micro-organisms as an energy source and remains in the soil for a relatively long time.

**Mineralisation**: Mineralization is the conversion of an element from an organic form to an inorganic as a result of microbial decomposition. The heterogeneous group of heterotrophic soil micro-organisms takes part in mineralization of organic nitrogen and converts it to inorganic nitrogen. The reactions go uninterrupted, as long as carbon source is available for microbes.

**Aminisation**: Process of conversion of proteins (enzymatic break down) to Aminoacids

*Enzymatic reaction* *Enzymatic hydrolysis*

Protein → Polypeptides → Aminoacids, Amines + $CO_2$ + Energy

**Ammonification :** Process of conversion of amino acids to ammonia

$$RNH_2 + HOH \xrightarrow{\text{Enzymatic hydrolysis}} NH_3 + R-OH + \text{Energy}$$

$$NH_3 + H_2CO_3 \xrightarrow{\text{Enzymatic hydrolysis}} NH_4^+ + CO_2 + H_2O$$

**Nitrification:** Microbial oxidation of ammonia to nitrate

$$NH_4^+ + O_2 \xrightarrow{\text{Nitrosomonas, Notrosococas}} NO_2 + 4H^+$$

$$NO_2 + O_2 \xrightarrow{\text{Notrobacter}} NO_3^-$$

**Immobilisation:** The conversion of the element from the inorganic to the organic form in microbial tissues or in plant tissues, thus rendering the element not readily available to other organisms or to plants, is called immobilization.

## 6.3. Soil Humus

Humus is a complex and rather resistant mixture of brown or dark brown amorphous and colloidal organic substances that results from microbial decomposition and synthesis and has chemical and physical properties of great significance to soils and plants

**Synthesis of Humus :** The formation of humus is strictly a bio-chemical process and rather complicated one. The compounds that make up humus have been synthesized by microorganisms in the soil from the products of the break down or alteration of the original tissue. Biochemical reactions of humus formation are two steps viz., Decomposition and Synthesis.

The simpler compounds resulted from break down of organic matter may take part immediately in human forming processes, a bio-chemical synthesis. These simpler compounds are metabolized in to new compounds in the body tissue of the soil organisms. The new compounds are subjected to further modification and synthesis as the microbial tissue is subsequently attacked by other soil micr oorganisms.

Additional synthesis reactions involve break down products of lignin as the phenols and quinones. These decomposition products, present initially as separate molecules called 'monomers' are enzymatically stimulated to join together in to polymers. By this process of polymerization (linkage) polyphenols and polyquinones are formed. These high molecular weight compounds interact with nitrogen containing amino compounds and give rise to a significant component of resistant 'Humus'.

## 6.4. Fractionation of soil organic matter

Soil humus, the complex array of substances left after extensive chemical and biological break down of fresh plant and animal residues, makes up 60-70% of the total organic carbon in soil. Because of the complexity, humus is often divided through solubility separations (figure 3).

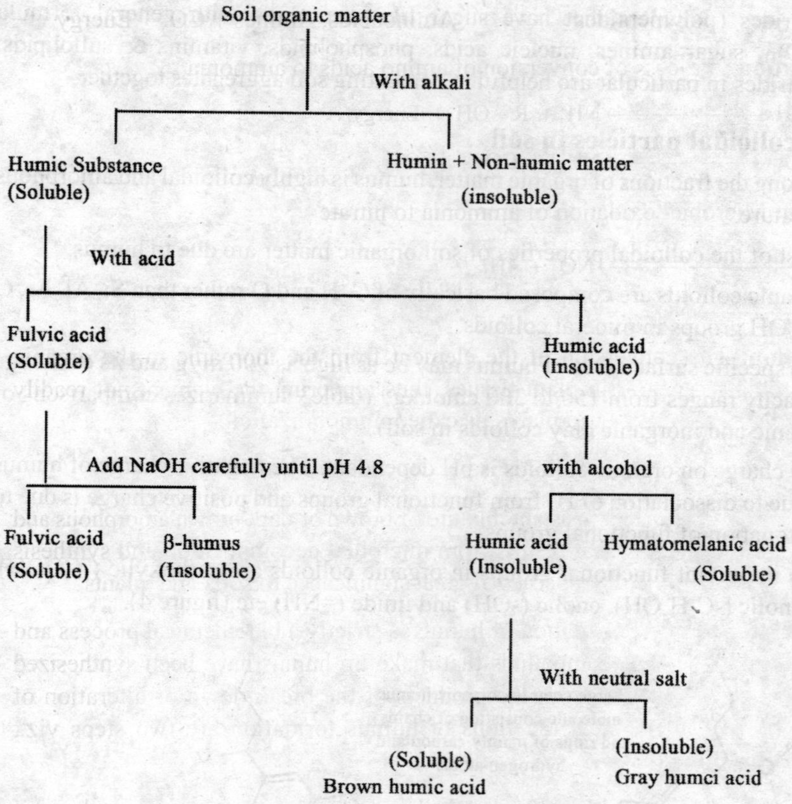


**Fig. 3:** Extraction and fractionation of organic matter or humus

**Humic group :** The humic substances makes up about 60 - 80% of soil organic matter and comprised of the most complex materials (Aromatic, ring type structures that include polyphenols and polyquinones) with no sharply defined physical and chemical properties. They are resistant to microbial attack. They are amorphous, dark in color and have high to very high molecular weights from a few hundred to several thousand.

On the basis of resistance to degradation and of solubility in aids and alkalis, humic substances are classified in to three chemical groupings.

1 Fulvic acid: - lowest molecular weight, lightest in color, soluble in both acid and alkali and most susceptible to microbial attack.

2 Humic acid:- Medium in molecular weight and color, soluble in alkali but insoluble in acid and intermediate in resistance to degradation.

3 Humin:- Highest in molecular weight, darkest in color, insoluble in both acid and alkali and most resistant to microbial attack.

**Non-Humic Group :** Comprises 20-30% of the organic matter in soils. These are less complex and less resistant to microbial attack. They are comprised of specific organic compounds with definite physical and chemical properties. These include polysaccharides (polymers that have sugar like structures with general formula of $Cn(H_2O)n$, sugar amines, nucleic acids, phospholipids, vitamins & sulfolipids. Polysaccharides in particular are helpful in cementing soil aggregates together.

## Organic colloidal particles in soil

- Among the fractions of organic matter, humus is highly colloidal and amorphous in nature.
- Most of the colloidal properties of soil organic matter are due to humus.
- Organic colloids are composed basically of C, H and O rather than Si, Al, Fe, O and OH groups in mineral colloids.
- The specific surface area of humus may be as high as 900 $m^2/g$ and its exchange capacity ranges from 150 to 300 $cmol.kg^{-1}$ (table3 summarizes comparisons of organic and inorganic clay colloids in soil).
- The charge on organic colloids is pH dependant. The negative charge of humus is due to dissociation of $H^+$ from functional groups and positive charge is due to protonation of functional groups.
- The important functional groups in organic colloids are carboxylic (-COOH), phenolic ($-C_6H_4OH$), enolic (-OH) and imide (=NH) etc (figure 4).

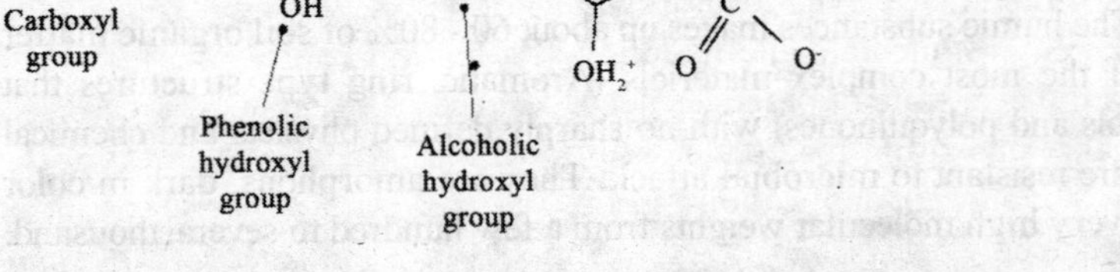


**Fig. 4:** Hypothetical diagram showing the principal chemical grops responsible for the high amount of negative charge of humus particle. The three groups include –OH that can lose its hydrogen ion by dissociation and thus become negatively charges. Note that the alcoholic, phenilic and carboxylic groups on the right side of the diagram are shown in their dissociated state, while those on the seft side still have their associated hydrogen ions.

**Table 3:** Comparisons of organic and inorganic colloids

| Property | Organic Colloids | Inorganic Colloids |
|---|---|---|
| Origin | Decomposition of organic matter | Weathering of rocks and minerals |
| Composition | C, H, O (N, P, S) | Si, Al, Fe, Mg, O, OH |
| Shape | Amorphous | Crystalline except allophone and sesquioxide clays |
| Stability | Unstable in nature | Stable |
| Origin of charge | pH dependant | Mostly due to isomorphous substitution and partly by soil pH |
| CEC | 150-300 c mol $kg^{-1}$ | 3-150 c mol $kg^{-1}$ |
| Water and nutrient holding capacity | High | Low |
| Plasticity | Low | High |
| Surface area $m^2/g$ | Around 900 | Variable (10 to >700) |

## 6.5. Charge development on organic colloids

The primary source of charge on organic colloids is considered to be the gain or loss of $H^+$ ion from functional groups on the surface of colloids. The functional groups include hydroxyl (-OH), carboxyl (-COOH), phenolic ($-C_6H_4OH$) and amine ($-NH_2$). The charge that develops from such groups depends largely on the pH of the ambient solution, which regulates the degree of protonation or de-protonation of the group.

$$-CO-OH + OH^- \quad -CO-O^- + H_2O$$

No charge -ve charge

The soil colloids that are capable of developing pH dependant charge include 1. layer silicates, 2. oxides and hydroxides of Fe and Al and 3. organic matter.

## 6.6. Importance of Organic matter

Once a land manager begins working towards enhancing soil organic matter, a series of soil changes and environmental benefits follow (figure 5). The rate and degree of these changes and the best suite of practices needed to achieve results vary with soil and climate. Initially, managing for greater soil organic matter may require higher pesticide, herbicide, or nutrient applications. In time, productivity and environmental quality will be enhanced.

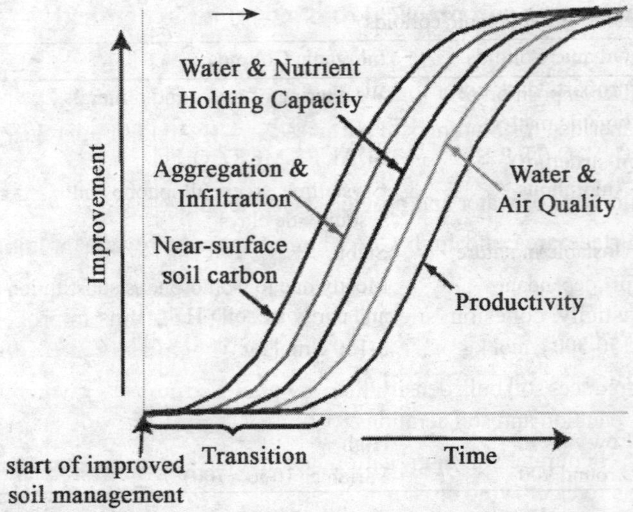


**Fig. 5:** Improvement of soil properties start with soil organic matter

## Soil organic matter improve soil properties

- Surface structure becomes more stable and less prone to crusting and erosion.
- Water infiltration increases and runoff decreases when soil structure improves.
- Soil organic matter holds 10 to 1,000 times more water and nutrients than the same amount of soil minerals.
- Beneficial soil organisms become more numerous and active with diverse crop rotations and higher organic matter levels.

## Soil organic matter improve air quality, water quality, and agricultural productivity

- Sediment and nutrient loads decline in surface water as soon as soil aggregation increases and runoff decreases.
- Dust, allergens, and pathogens in the air immediately decline.
- Ground and surface water quality improve because better structure, infiltration, and biological activity make soil a more effective filter.
- Crops are better able to withstand drought when infiltration and water holding capacity increase.
- Organic matter may bind pesticides, making them less active. Soils managed for organic matter may suppress disease organisms, which could reduce pesticide needs.
- Crop health and vigor increase when soil biological activity and diversity increase.
- Wildlife habitat improves when residue management improves.

**Soil organic matter influences soil physical, biological and chemical properties**

1. Soil organic matter imparts dark color to soils
2. It supplies polysaccharides for binding soil particles for formation of aggregates (genesis of good soil structure)
3. It increases infiltration rate of water and provides better drainage
4. Due to the high surface area and high CEC, humus increases water holding capacity of soil.
5. Humus reduces plasticity, cohesion, stickiness etc in clay soils, thus improves workability of soil.
6. Soil Organic matter reduces soil bulk density, there by influence porosity favorably for optimum water retention and soil aerations.
7. Soil organic matter encourage soil aggregation and stable aggregate formation and granulation thus reduces wind erosion losses
8. It provides mulching (raw organic matter) and lowers soil temperature during summer. Acts as an insulator and retards heat movement between atmosphere and soil
9. Soil organic matter reduces alkalinity in soils by releasing organic acids and $CO_2$
10. With high adsorption capacity, the soil humus accounts for 30 -90% of the adsorbing power of mineral soils (Carboxylic group – 54% ; phenolic & enolic groups – 36%; imide group – 10%)
11. It acts as a buffering agent and reduces the likelihood damage from acids and alkalis.
12. With its solubilising effect, decomposition of organic matter releases carbondioxide which on solubility in water forms carbonic acid. Carbonic acid solubilises the mineral nutrients thus increases the availability of nutrients to plants:
13. Acts as a store house for nutrients. Organic matter is the source of 90-95% of nitrogen in unfertilized soils. Also supplies available 'P', 'S' and micro nutrients like Fe, Mn, Cu and Zn etc.,
14. Adsorbs temporarily the heavy metal pollutants and cleans the contaminated waters.
15. Serves as a source of energy for macro and micro organisms in soils and helps in performing various beneficial functions in soils (N - fixation, mineralization etc.)
16. Soil humus acts as a chelate and increases the availability of micro nutrients
17. Various organic substances like vitamins, antibiotics and growth promoting substances namely auxins are produced by different micro organisms during decomposition of organic matter. Also some fungi-toxins are produced to control diseases.

**Table 4:** Influence of soil organic matter on soil properties.

| Soil attribute or function | Mode | Impact |
|---|---|---|
| Soil architecture | Microbial by-products such as bacterial glues and fungal hyphae can increase soil aggregation and stability of soil structure. | Can improve water infiltration, soil porosity and the exchange of water and oxygen. |
| Soil fertility | Decomposition of organic matter determines nutrient supply. | Increased plant productivity. Decreased input costs. |
| Water infiltration | Microbial by-products such as bacterial glues and fungal hyphae increase potential for soil aggregation. Increasing organic matter can be associated with development of water repellence on soil with less than 15 per cent clay. | Decreases moisture loss from run-off, evaporation. More water enters the soil profile. Increased soil porosity helps the exchange of water and oxygen. |
| Water holding capacity (or 'bucket size') | Soil organic matter can hold several times its own weight in water. | Increased plant available water and decreased deep drainage below the root zone. Supports higher productivity. |
| Soil biological processes | Organic matter is the primary source of energy required by soil microorganisms for growth and reproduction. | Influences nutrient cycling and availability, soil diversity,resilience and impact of stress events (i.e. pathogen impact, recovery). |
| Buffers soil pH (helps maintain acidity at a constant level) | Organic matter is alkaline in its nature. | Large amounts of residues or a concentration of burnt residues can increase soil pH. |
| Soil resilience — defined as the ability of a soil to recover to its initial state after a stress event | Increasing organic matter supports an increasing diversity of microorganisms, many of which can undertake similar roles allowing for some redundancy. | Increased ability of soil to recover soil function after a disturbance. |

## 6.6. Soil organic matter and Soil health

Soil organic matter plays an important role in integrating many aspects of soil health. It serves as food and energy source for soil microorganisms. Organic matter increases water and nutrient retention in soil, it imparts buffering capacity to soil, it increases cation exchange capacity of soil, adsorb pollutants. Soil organic matter can be divided into labile and stable pools, each of which has different characteristics and functions in the soil. In agricultural soils, organic matter can range from 1 to 8 percent depending on climate, soil type, and soil management practices.

The labile pool of organic matter, which accounts for 5–20 percent of the total pool of soil organic matter, includes the living biomass of soil organisms and plant roots, fine particles of organic detritus, and relatively simple organic compounds such as polysaccharides, organic acids, and other compounds that are synthesized by microbial activity or are by-products of decomposition processes. Labile organic matter is readily decomposed by microbes and is the principal energy source that fuels the soil food web. It is the principal reservoir of organic nitrogen that can be readily mineralized and made available for plant use.

Polysaccharides in labile organic matter also enhance aggregate stability. When microbial consumption of labile organic matter is greater than the input of fresh organic matter into the soil, labile organic matter levels will decline. Excessive tillage of the soil can speed the decline of labile organic matter by oxygenating the soil, which increases microbial activity, and by exposing organic matter that had been protected within soil aggregates.

The stable pool of organic matter, which accounts for 60–95 percent of the total pool of soil organic matter, consists of organic compounds that are relatively resistant to decomposition because of either their chemical structure, their adsorption to clay particles, or their protection within microaggregates. Stable organic matter contributes cation exchange capacity and water-holding capacity to soil. The pool of stable organic matter is increased or depleted slowly as only a small portion of the labile organic matter those cycles through the food web is stabilized into forms that are resistant to decomposition.

The quantity of organic matter in a given soil is the result of a balance between organic matter inputs, such as crop residues, manure, and compost, and the rate of organic matter decomposition. Organic matter inputs can be influenced by crop management, such as the use of cover crops, crop rotations, and residue management, as well as soil management, such as using organic forms of nutrients like compost and manure. The quantity of labile organic matter generally responds to changes in management practices more quickly than the quantity of stable soil organic matter, so changes in labile organic matter levels can serve as a leading indicator of long-term trends in total organic matter levels.

## 6.7. Carbon: Nitrogen ratio (C:N ratio)

Both mineralization and immobilization are accomplished by microbes under the influence of temperature, moisture and pH etc. These processes are more influenced by C/N, C/P and C/S ratios of decomposing plant residues (table 5). An average proportion of C/N/P/S in soil humus is appropriately 140:10:1.3:1.3. The ratios recorded for C and P are some what more variable than for C and N, or C and S.

**Table 5:** C:N ratio, C:P ratio and C: S ratios of soil and net mineralsiation (NM) and net immobilistion (NI)

| C:N Ratio | C:P Ratio | C:S Ratio | N.M | N.I |
|---|---|---|---|---|
| <20:1 | <200:1 | <200:1 | Yes | No |
| >30:1 | >300:1 | >400:1 | No | Yes |
| 20-30:1 | 200-300:1 | 200-400:1 | Neither gain nor loss | |

The C:N ratio of organic inputs influence the amount of soil nitrogen made available to plants. Organic residues with a C:N ratio of between 25:1 and 30:1 have sufficient nitrogen available for microbes to decompose them without needing to use soil nitrogen stores. Residues with a lower C:N ratio (< 25:1) such as pulses and legume pastures will generally result in more rapid decomposition of organic residues and tend to release plant-available nitrogen. Residues with a higher ratio (> 30:1) such as cereal crops will decompose more slowly and result in less plant-available nitrogen being released.

Plant material contains about 45 per cent carbon and depending on residue type between 0.5-10 per cent nitrogen. The ratio of organic carbon to total nitrogen is referred to as the carbon to nitrogen (C:N) ratio. This ratio indicates the proportion of nitrogen and other nutrients relative to carbon in that material. Organic matter varies widely in its C:N ratio (see Table.6) and reflects how readily organic matter decomposes, providing an indication of both the amount and rate of nitrogen release that might be expected to result from decomposition. To grow and reproduce soil microbes requires a balanced amount of carbon and nitrogen that reflects a relatively low C:N ratio (generally less than 15:1). In plant residues such as wheat stubble, which have a high C:N ratio (120:1) and contain relatively more carbon than nitrogen, soil microbes must find another source of nitrogen to fully digest wheat stubble and this often results in soil nutrient reserves being immobilised and soil becoming nitrogen deficient. Such nitrogen deficiency is often seen in the field where stubble has been incorporated during sowing operations and nitrogen availability decreases in the soil because it becomes tied-up, or immobilised in the microbial biomass. Therefore, residues with a high C:N ratio are considered nutrient poor and can take years to decompose. In contrast, fresh legume residues with a low C:N ratio (less than 25:1) have proportionally less carbon per unit of nitrogen enabling them to decompose faster and release surplus nitrogen for plant use.

**Table 6:** C:N ratios of some organic materials

| Material | C:N ratio or C/N |
|---|---|
| Microbial tissues | 6-12 |
| Sewage, Sludge | 5-14 |
| Soil Humus | 10-12 |
| Annual Manures | 13-25 |
| Legume residues and green manures | 13-25 |
| Cereal residues and straws | 60-80 |
| Forest Wastes | 150-500 |
| Poultry manure | 5:1 |
| Humus | 10:1 |
| Cow manure | 17:1 |
| Legume hay | 17:1 |
| Green compost | 17:1 |
| Lucerne | 18:1 |
| Field pea | 19:1 |
| Lupins | 22:1 |
| Grass clippings | 15 – 25:1 |
| Medic | 30:1 |
| Oat hay | 30:1 |
| Faba bean | 40:1 |
| Canola | 51:1 |
| Wheat stubble | 80 – 120:1 |
| Newspaper | 170 – 800:1 |
| Sawdust | 200 – 700:1 |

The quantity of carbon declines with the advancement of decomposition process. Similarly, the inorganic element released during mineralization process is also lost, by several means i.e. plant utilization, leaching, volatilization, conversion to insoluble compounds etc. It results in a stabilized C:N ratio to the soil i.e. 10:1 or 12:1

## 6.8. Significance of C : N ratio on soil fertility

C : N ratio determines the microbial activity and their proliferation. C:N ratio of organic residues determines the rate of decomposition of organic residues. C : N ratio determines the net availability of nutrients to plants. The supply of nitrogen, phosphorus and sulphur are largely determined by C:N ratio of organic residues.

## 6.9. Indicators of soil organic matter

Sensory indicators (sight, smell and touch) and some basic soil properties can provide a valuable and cheap way of identifying areas in which changes in soil organic matter may have occurred.

**Soil colour**: Soil colour is influenced by the mineral soil derived from the parent material, the amount and condition of organic matter, the presence of iron oxide and soil aeration. Soils with high organic matter content are often darker in colour and when dry leave hands dirty and dusty.

**Ground cover**: Soils with a high level of ground cover generally have a greater potential to generate soil organic matter because the cover promotes biological activity and helps protect the soil from wind, rain and temperature extremes. Soil organic matter increases with the length of time that a soil has actively growing plants in it and the risk of losses due to erosion decrease. Stubble levels greater than 80 per cent are generally required, with bare soils less able to buffer changes in temperature and more prone to erosion.

**Cover crops, green manures and pastures**: Cover crops provide soil cover and prevent soil erosion, while promoting the production of soil organic matter. When a cover crop is grown to decrease the risk of nutrient leaching or to retain nutrients that would otherwise move deeper into the soil profile, it is referred to as a 'catch-crop'. A green manure is grown primarily to manage weed populations and improve soil fertility, but may also be an option for a crop that has failed. Plant growth is stopped shortly after flowering and the residues either incorporated, surface mulched, or desiccated (brown manure). Pastures and perennial crop phases can potentially generate more soil organic matter than annual cropping sequences (where not limited by subsoil constraints and where summer rainfall supports continued growth). This is often associated with more below-ground biomass production deeper in the soil profile. Crop rotations that include cover crops, perennial grasses and legumes are an important factor in soil organic matter management and can be adapted to any cropping system. Increasing the diversity of residues and quantity of soil organic matter using legume and cereal cover crops in low disturbance systems has the potential to increase the biomass and diversity of soil biota. Crop rotations that maximise soil carbon inputs and maintain a high proportion of labile carbon are important in maintaining a sustainable cropping system.

**Root architecture and root exudates**: Roots are influenced by the physical structure of a soil and commonly follow cracks or bio-pores resulting in concentrated areas of belowground carbon. Root architecture and biomass varies between plant species, genotype and even cultivars within the same species. These differences can alter the amount and spatial distribution of below-ground organic matter vertically and horizontally. Increasing root biomass influences soil organic matter: i) directly by increasing organic inputs to soil and ii) indirectly by influencing the location of roots and production of root exudates that may stimulate mineralisation. Root exudates and other by-products are also more readily absorbed and protected by soil aggregates and where concentrated are more likely to persist in the particulate organic matter and humus fractions than shoot-derived soil organic carbon.

Plant Residues: Plant type, species and rotational sequence influence the population size and diversity of decomposer organisms due to differences in their chemical composition and lignin content, which in turn affects the rate at which soil organic matter is decomposed.

**Presence of Earthworms and other Soil Organisms in soil**: Earthworms modify the physical, chemical and biological properties of soil and contribute to nutrient cycling, soil aeration and water infiltration. The presence of earthworms is an indicator of soil health. In high rainfall zones of eastern Australia, more than 10 earthworms per spadeful (20 cm x 20 cm x 10 cm deep) is indicative of an active biological system. However, earthworms are rarely found in sandy soils, which are low in calcium and often have low pH (less than 4.5) not suited to their survival. Termites, ants, beetles and collembolan (commonly called 'springtails') help aerate and mix soils and are considered important indicators of soil organic matter. However, as soil biota is seasonally active its absence does not always indicate an unhealthy soil.

## 6.10. Organic matter loss from the soil

While soil forms and regenerates very slowly, it can degrade rapidly and could in essence be considered a non-renewable resource. Soil organic matter is in a constant state of turnover whereby it is decomposed and then replaced with new organic material. The balance between these additions and losses determines the relative flux and amount of soil organic matter present at any point in time.

**Direct Losses:** Soil erosion In Australia, annual soil losses from erosion are negligible under a good pasture, but can be up to eight tonnes per hectare under planted crop. Erosion risk is strongly influenced by the amount of ground cover and the highest risk scenarios are most often associated with bare fallow, under which typical soil losses in a single year can reach between 60-80 tonnes per hectare. While less common, wind and water erosion resulting from single, high-intensity storms can erode up to 300 tonnes per hectare. Since a 1 mm depth of soil weighs between 10-15 tonnes per hectare (assuming a bulk density of 1.0 to 1.5g/cm3 ), erosion events in cropped soils represent a significant loss of topsoil along with its associated carbon and nutrient-rich fractions (Hoyle et al. 2011). Soil physical attributes associated with high organic matter content such as more stable soil aggregates, greater porosity, improved water infiltration and improved workability at high moisture content (plastic limit) all contribute to a lower risk of soil loss from erosion.

**Indirect Losses**: Losses of soil organic carbon occur primarily when organic matter is decomposed and mineralized to carbon dioxide. The rate at which organic matter is decomposed is driven by factors that regulate microbial activity, including climate (soil moisture and temperature), soil disturbance and the management of organic inputs. Climate In moist soils, organic matter breaks down more rapidly as average temperatures increase. As a general rule for every 10°C rise in average temperature between 5°C and 40°C the rate of mineralization will nearly double where carbon substrates are not limited (see Figure 6,1; Hoyle et al. 2006). Therefore, it is more difficult to store large amounts of organic carbon in soils subject to high temperatures and even more difficult in soils exposed to high temperatures and extended periods of adequate soil water. In cooler environments, decomposition does not occur year-round and is constrained at low temperature.

Drying soils increasingly inhibit microbial activity and therefore decomposition of organic matter because there are fewer substrates and nutrients for microbial growth and reproduction. Soil moisture between 20-60 per cent of water holding capacity is considered optimal for microbial activity, with wetter soils inhibiting biological activity due to low oxygen availability. As soils warm up in spring, the microbial biomass increases in size as well as activity. In general, population size and activity of soil microorganisms are highest during spring and lowest during winter. This means warm, moist environments can support high levels of microbial activity and soil organic matter can be lost quickly in these systems if organic inputs stop. Conversely, in soils with very low levels of soil microbial activity organic carbon can slowly accumulate and build to relatively high levels, despite being in an environment of poor productivity. For example, in highly acidic, waterlogged or clay soils, organic matter can accumulate but does not break down. Highly alkaline and in particular sodic soils do not support high organic carbon stocks.

**Factors that control organic matter decomposition**: (1) Physical protection. Organic matter can be protected inside soil aggregates limiting access to it by microorganisms and their enzymes. Micro-aggregates (53–250 mm) slow the turnover of soil organic matter, withstand physical disturbance and protect carbon more effectively than larger macro-aggregates. (2) Chemical protection; Organic matter can become adsorbed on to mineral surfaces protecting it from decomposition. (3) Drought; Low soil moisture results in thinning or absent water films in soil, slowing the flow of extracellular enzymes and soluble carbon substrates. Organic compounds in dry or hydrophobic soils are isolated from degradation by water-soluble enzymes. (4) Flooding; Flooding slows the diffusion of oxygen and constrains aerobic decomposition of organic matter. (5) Freezing; The diffusion of substrates and extracellular enzymes within the soil below 0°C is extremely slow and this, in turn, slows the decomposition of organic matter. Soil disturbance and cultivation can accelerate the decomposition of organic matter, increasing its rate of mineralisation. Cultivation and soil disturbance exposes previously protected organic matter to soil biota increasing its decomposition. Minimum tillage has the greatest potential to maintain, or perhaps increase levels of organic matter in Australian cropping soils over the long-term, especially in surface soils. The increasing use of soil management practices such as mouldboard ploughing is likely to have a profound effect on the amount and distribution of soil organic matter.

## 6.11. Management of soil organic matter

To increase soil organic matter, the rate at which organic matter is added to the soil must exceed the rate at which it is lost via microbial decomposition, erosion or leaching. Cropping and pasture management practices that generate adequate amounts of high-quality residues are critical to rebuilding and sustaining soil organic matter.

While soil organic matter can be increased it is important to consider the economic cost of doing so. For example, soil organic matter can be increased substantially by adding high rates of organic amendments such as manure and compost, but this is likely to involve significant transport costs. Increasing soil organic matter will be more economic in farming systems and climates that support high production and generate on-farm supplies of organic soil amendments.

Any practice that increases the production of biomass carbon (via photosynthesis) and slows the return of this carbon back into the atmosphere as carbon dioxide (via respiration, fire or erosion) will increase soil carbon reserves. In this way the soil becomes a carbon 'sink' - removing atmospheric carbon dioxide from circulation and locking it away in organic carbon structures within the soil. Each soil has a finite capacity to store carbon, with heavier soils generally able to store more than lighter sandy soils.

a) **Increased crop biomass:** Optimizing agronomic management practices and water use efficiency to increase grain yield should result in higher plant biomass (as long as harvest index is maintained). The adoption of management practices such as low soil disturbance seeding (e.g. conservation tillage), stubble retention and increasing the frequency of crops and pastures (e.g. use of cover crops) can also contribute to higher soil organic matter through greater capture of carbon from the atmosphere to the soil and mitigation of losses from the soil.

b) **Crop and pasture stubble retention**: Leaving stubble standing will slow the decomposition rate, promote water conservation and decrease soil erosion (about 60 per cent of organic matter in the top 30 cm of cropped soils is stored in the top 10 cm and this can take years to replace). Frequent burning of crop residues decreases the amount of organic matter entering different soil fractions in Australian agricultural soils. The most significant effect of burning is a change in the composition of the residue and a resulting decrease inparticulate organic matter in the topsoil (0-10 cm) and a lower potential for soil nitrogen supply. This loss of particulate organic matter is reflected in a decrease in biological activity, microbial biomass and the biodiversity of soil organism Burning stubble results in the rapid loss of carbon, nitrogen (up to 80 per cent), phosphorous (about 25 per cent), sulphur (about 50 per cent) and potassium (20 per cent) from crop residues and contributes to the loss of surface soil via erosion. Nutrients that remain in the ash are also at risk of loss from wind or water erosion events.

c) **Pasture management**: Both short-term (up to 5 years) and long-term perennial pasture systems can have a positive effect on the quantity and quality of soil organic matter in Australian agricultural systems. This is because:

   i They have a higher root to shoot biomass compared to annual crops (meaning they produce more biomass underground where it is less susceptible to loss).

   ii Perennials grow for a longer proportion of the year.
   They are subject to less soil disturbance than cropping.

   iii They have a slower rate of decomposition associated with less available soil moisture.

d) **Controlled Grazing:** Overgrazing reduces ground cover and pasture species composition, increases erosion and nutrient loss, causing a subsequent loss in animal performance and a longer recovery period for pastures. Cell or rotational grazing allows pasture soil to be rested. This can help regenerate soil structure and improve permeability on self-mulching soils through clay swelling and shrinkage (limited to specific clay soils) and soil biological activity. Most importantly it prevents pastures from being overgrazed and soils from becoming compacted. Removing or reducing stocking rates (particularly cattle) when the soil is wet will help decrease the risk of compaction as will maintaining high stocking rates for shorter duration.

e) **Growing catch-crops:** Catch-crops and pastures are used strategically and at times opportunistically to take up nutrients in the soil that would otherwise leach, provide ground cover, use excess soil moisture, control weeds and supply nitrogen to subsequent crops.

e) **Growing cover crops**: Cover crops established through summer help anchor soil and lessen the impact of raindrops, minimizing wind and water erosion. They are also grown to take-up excess nutrients in the soil and prevent leaching between catch-crops, or fix nitrogen in the soil for future use (e.g. leguminous species). In addition, they can be used in a similar way to a break crop to control pests such as plant parasitic nematodes. The use of cover crops in rotation is an effective way to increase soil organic matter and, depending on the plant species, deliver other beneficial functions to farming systems. Combined with low soil disturbance systems, increasing the diversity of residues and quantity of soil organic matter using legume and cereal cover crops has the potential to increase the population and diversity of soil biota. However, in very dry climates growing a cover crop is not always beneficial because there is not enough soil moisture to support the following catch crop.

f) **Practicing green manuring:** A green manure is a crop or pasture that is grown and returned to the soil in situ to increase organic matter inputs, nitrogen supply or control weeds and manage erosion. The term green manure reflects the stage of plant growth (flowering) at which the crop is manured. Different types of green manuring include, 1) Green manuring incorporates a green crop or pasture into the soil using discs, plough or other mechanical means. It is generally sprayed with a non-selective herbicide before incorporation. 2) Desiccation with a non-selective herbicide and allowing the residues to be incorporated naturally over time. Also termed brown manuring. 3)Using a slasher or mower to cut residues which are left in a layer on the soil surface. Also termed mulching. Spraying with a non-selective herbicide before slashing will decrease the risk of regrowthSoil type influences the choice of green manure crop, while the farming system may constrain how and when the green manure phase is implemented. Careful cultivation is recommended to minimise damage to soil structure. Lighter soils at risk of erosion should be treated with care, with mulching or brown manuring the safer options on these soils.

g) **Use of organic amendments:** Many organic soil amendments, which reportedly offer agronomic benefits, are now available to Australian farmers. Until recently, organic amendments have largely been used in intensive horticultural industries and organic farming systems. Increasingly, however, these products are being targeted at broad acre farming systems to supply plant nutrients, control pests and diseases and improve soil health. However, many products come with little independent or local evidence of their actual benefits.

h) **Conservation Agriculture:** Conservation tillage is often promoted as a way to increase or maintain soil organic matter with one study showing no-tillage increased soil carbon sequestration rates by about 0.4 tonnes per hectare annually compared to conventional tillage. However, recent evidence suggests that in Australia conservation tillage results in, at best, only a slow increase in soil organic matter (Chan et al. 2011), and that perceived differences in soil organic matter stocks often did not exist when considered over soil depths greater than 30 cm.

i) **Erosion control**: Soil forms at a very slow rate, typically about one millimetre (or 14 t/ha) every 100 years. Only a small amount of erosion is required to exceed this rate of soil formation and typical soil losses can be up to 50 t/ha of soil per year from bare fallow, 8 t/ha per year under a crop and 0.24 t/ha under pasture. Organic matter that is concentrated towards the soil surface would also be lost. Organic residues that form mulch on the soil surface protect the soil from raindrop impact, minimize the risk of wind and water erosion, reduce evaporation and buffer the soil from extreme temperatures. Mulch also promotes root growth in the topsoil where nutrients tend to be concentrated and protect seedlings from wind damage. Under grazing, an initial ground cover of 70 per cent is desirable. A decreased risk of erosion in cropping systems requires a minimum 50 per cent of ground covered by standing stubble (minimum 10 cm height). Most significant erosion events occur on land before the opening rains where soils are bare, dry and loose due to cultivation or grazing Incorporate pasture phases, increasing cropping intensity to grow more plant biomass, increasing plant production and retaining stubble will decrease the risk of crusting and hard-setting soils, which increase the risk of erosion. On gypsumresponsive soils (i.e. sodic soils) gypsum may also be required in combination with reduced tillage to manage the effects of high levels of sodium on soil structure and stability.

## Review question

1. What is soil organic matter? Write source and composition of organic matter
2. Discuss briefly on decomposition of organic matter
3. Differentiate between mineralization and immobilization .
4. How you fractionate organic matter and describe various fractions of organic matter.
5. Differentiate between organic and inorganic colloids.
6. What is C:N ratio and write its importance..
7. Write benefits of soil organic matter.
8. Discuss general properties of organic colloids.

## Objective questions

1. Most dominant element in plant tissue is
   a) Carbon b) Oxygen
   c) Hydrogen d) Nitrogen
2. The dominant compound in plant residue on dry weight basis is
   a) Fats b) Lignins
   c) Proteins d) Carbohydrates
3) The primary source of essential elements in organic matter are
   a) Lignins b) Proteins
   c) Carbohydrates d) Fats
4. When plant residue added to soil, the compound subject for very slow decomposition is
   a) Lignins b) Cellulose
   c) Protein d) Carbohydrates
5. Decomposition of plant residue in soil is
   s) Reduction b) Hydrolosis
   c) Hydration d) Oxidation
6. Under optimum conditions about ____ kg/ha of $CO_2$ can be evolved per day
   a) 100 b) 200
   c) 300 d) 400
7. The breakdown of complex organic compounds to release inorganic elements is termed as
   a) Humification b) Immobiloisation
   c) Mineralisation d) Vernalisation
8. Humus is a heterrogenous mixture of
   a) simple organic compounds b) complex organic compounds
   c) complex inorganic compounds d) simple inorganic compounds
9. The humic substance with lowest molecular weight and susceptible for microbial attach is
   a) Fulvic acid b) Humic acid
   c) Humin d) Hymatomelanic acid
10. The humic substance with highest molecular weight and resistant for microbial attach is
    a) Fulvic acid b) Humic acid
    c) Humin d) Hymatomelanic acid

11. The humic substance that si soluble in alkali and insoluble in acid is
    a) Humin
    b) Fulvic acid
    c) Humic acid
    d) Hymatomelanic acid
12. The non-humic group comprises of about ___ % of organic residue
    a) 10- 20
    b) 20- 30
    c) 30- 40
    d) 40-50
13. The non-humic substances which is not found in plants but enhance soil aggregate stability is
    a) Humic acid
    b) Poly sacharides
    c) Fulvic acid
    d) Polyuronides
14 The water holding capacity of humus on mass basis is
    a) 4 – 5 times high compared silicate clays
    b) 6 – 7 times high compared to silicate clays
    c) 9 – 10 times high compared to silicate clays
    d) 12 - 13 times high compared to silicate clays
15. Stable organo- metal complexes with metal cations are formed by ------ to help in respective nutrient availability to soils
    a) Poly saccharides
    b) Fulvic acid
    c) Humic acid
    d) A & B
16. The CEC of humus is ---- time s higher than mineral colloids
    a) 50-60
    b) 30- 50
    c) 40-60
    d) 20- 30
17. Addition of organic matter to soil results in
    a) Increase in granulation
    b) Reduction in plasticity
    c) Increase in WHC
    d) All three
18. The C;N ratio of fulvic acid is
    a) 20:1
    b) 200 :1
    c) 50: 1
    d) 100 : 1
19. The C:N ratio of microbial tissue is
    a) 6 :1
    b) 20 : 1
    c) 30 : 1
    d) 40 : 1
20. Among the microbes, the lowest C :N ratio is associated with
    a) Protozoa tissue
    b) Fungal tissue
    c) Bacterial tissue
    c) Actinomycetes tissue

21. During decomposition process, the C:N ratio when the % of total nitrogen removed equals to % total carbon lost is
    a) 30- 35 :1 b) 20- 25 :1
    c) 10- 12 : 1 d) 40- 45 :1
22. The decay rate of organic matter is high in
    a) Cool climates b) Warm climates
    c) Temperate climates d) Tundra climates
23. The total organic matter of soil is high in
    a) Deciduous forests b) Grassland vegetation
    d) Coniferous forests d) Wetlands
24. Application of fertilsiers to soil results in
    a) No change in SOM b) Decrease in SOM
    c) Increase in SOM d) Decrease in soil nutrients

**Answer key**

| | | | | | | | | | |
|---|---|---|---|---|---|---|---|---|---|
| **1** | a | **6** | a | **11** | c | **16** | d | **21** | c |
| **2** | d | **7** | c | **12** | b | **17** | d | **22** | b |
| **3** | b | **8** | b | **13** | d | **18** | d | **23** | b |
| **4** | a | **9** | a | **14** | a | **19** | a | **24** | c |
| **5** | d | **10** | c | **15** | d | **20** | c | | |

# 7

# Soil Biology

Study of living organisms in soil is referred to as soil biology (Greek word bios= life, logos = study). The soil is teeming with millions of living organisms which make it a living and a dynamic system. The soil is a complex arrangement of soil particles and pore spaces filled with air and water. In these pore spaces, plant roots and millions of organisms, ranging from microscopic to macroscopic in size live and proliferate. The organisms in the soil, not only help in development of soils but carryout a number of transformations facilitating the availability of nutrients to the plants. In the absence of the activities of these organisms, in soil, life on earth would have come to a halt, as all available nutrients would have ended up locked in the organic form, disrupting the nutrient cycles.

Understanding soil biology is important since, organisms, both animals (fauna/micro-fauna) and plants (flora/micro-flora) are important in the overall quality, fertility and stability of soil. These soil organisms are responsible for the formation of humus, a product of organic matter degradation and synthesis. Soil organisms aid in the physical manipulation, mixing, and formation of soil & its structural stabilisation. Soils contain a vast number and wide range of organisms, a greater proportion of these belong to the plant kingdom and are important in the myriad of biochemical reactions and intricate biological processes that take place within the soil.

Weight of soil organisms per unit volume of soil is called microbial biomass. Soil organisms' number is influenced by many factors like climate, vegetation, and the physical and chemical characteristics of soils. The species composition in an arid desert is different from that in a humid forest area, which in turn differs from that in a cultivated soil. Acid soils are populated by quite different species of soil organisms from those in alkaline soils. Similar differences are expected from tropical forest areas to cool temperate regions also. Forested areas support more diverse fauna than do grasslands, but the activity of flora and fauna is higher in grass lands. Cultivated fields are generally lower than undisturbed native lands in numbers and biomass of soil organisms especially the fauna. Some autotrophic bacteria are, nitrifying bacteria, iron bacteria, hydrogen bacteria, sulphur bacteria, manganese bacteria, methane bacteria and so on.

### 7.1. Role of soil organisms

Soil organisms (biological component) of the soil play major roles in nutrient cycling & release (breakdown of organic compounds), biochemical weathering of minerals &

soil development, ameliorating soil physical & chemical properties and without this living component, the mere accumulation of the mineral fraction would not be "soil". Majority of soil organisms are plants (micro-flora), but animals are equally important (have more physical role). Most are microscopic, i.e., microflora and microfauna.

**Table 1:** Major groups of organisms of common occurrence in soils are.

| | | | |
|---|---|---|---|
| Animals (fauna) | Macro | Live mainly on plant materials | Small mammals, insects, millipedes, mites, slugs, snails, earthworms |
| | | Mostly predatory | Moles, Insects, mites, centipedes, spiders |
| | Micro | Predatory, parasitic or live on plant tissue | Nematodes, Protozoa, Rotifers |
| Plants (flora) | Macro | Roots of higher plants | |
| | Micro | Algae Green algae, Blue-green algae (BGA); Diatoms | |
| | | Fungi Mushrooms, Yeasts, Molds | |
| | | Actinomycetes of many kinds | |
| | | Bacteria Aerobic | Autotrophic |
| | | Anaerobic | Heterotrophic |

**Table 2:** Biomass of major groups of soil organisms in different vegetation.

| | Biomass (g m$^{-2}$) | | |
|---|---|---|---|
| | Grassland | Oak Forest | Spruce Forest |
| Herbivores | 17.4 | 11.2 | 11.3 |
| Detritivores: Large | 137.5 | 66.0 | 1.0 |
| Small | 25.0 | 1.8 | 1.6 |
| Predators | 9.6 | 0.9 | 1.2 |
| Total | 189.5 | 79.9 | 15.1 |

**Table 3:** Numbers and biomass of Soil microflora and microfauna

| Organisms | Values commonly found in surface soil | | |
|---|---|---|---|
| | No./sq. meter | No. per gram | Biomass (kg/HFS) |
| **Microflora** | | | |
| Bacteria | $10^{13}$-$10^{14}$ | $10^{8}$ - $10^{9}$ | 440 - 4400 |
| Actinomycetes | $10^{12}$ - $10^{13}$ | $10^{7}$– $10^{8}$ | 440 - 4400 |
| Fungi | $10^{10}$ - $10^{11}$ | $10^{5}$ – $10^{6}$ | 550 - 5500 |
| Alage | $10^{9}$ - $10^{10}$ | $10^{4}$ – $10^{5}$ | 55 - 550 |
| **Microfauna** | | | |
| Protozoa | $10^{9}$- $10^{10}$ | $10^{4}$- $10^{5}$ | 16 -160 |
| Nematoda | $10^{6}$- $10^{7}$ | 10 - $10^{2}$ | 11 -110 |
| Earthworms | 30 -300 | - | 16 -160 |
| Others | $10^{3}$- $10^{5}$ | - | 110 -1100 |

## 7.2. Classification of soil organisms

Soil organisms are broadly classified two groups viz., soil flora and soil fauna. These groups further divided to macro flora, micro flora, macro fauna and micro fauna.

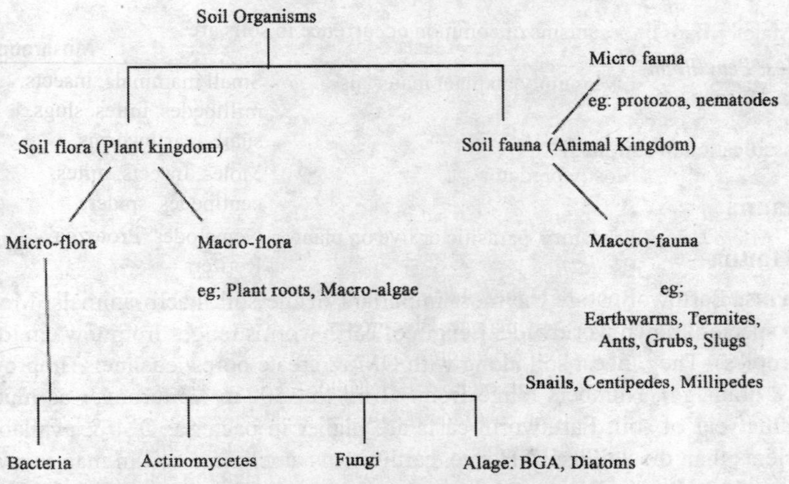


**Fig. 1:** Classification of soil organisms

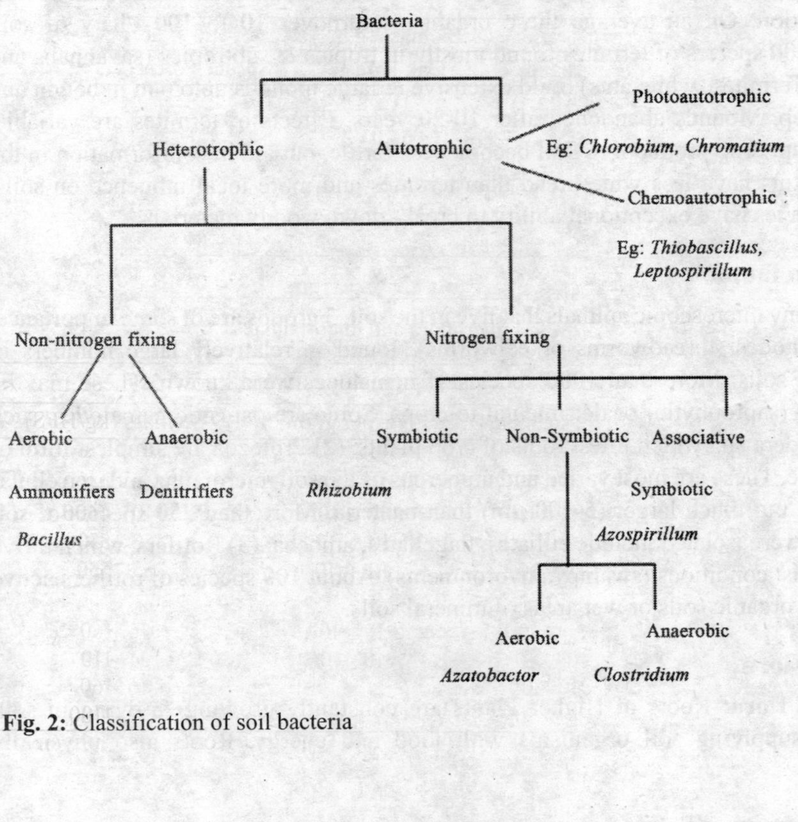


**Fig. 2:** Classification of soil bacteria

Fungi

- Moulds
  *Aspergillus, Pencillium*
  *Fusarium, Mucor*
- Yeasts
- Mushrooms

**Fig. 3:** Classification of soil fungi

## A) Soil fauna

### i)Macro fauna

**Earthworms:** Earthworms are the most important of the soil macro-animals. More than 1800 species known worldwide. Length of earth worms ranges from few cm to 3 m long (tropics). They ingest soil along with OM & create pores/ channels (improve aeration & drainage). Numbers range from <1/m$^2$ to >500/m$^2$ & turnover as much as 250 mt/ha/year of soil. Earthworm casts are higher in bacteria, OM & available plant nutrients than the soil itself. Hence, earthworms ameliorate soil in many ways, physically and biochemically

**Termites and Ants:** These play important in organic matter breakdown and soil manipulation. On an average these organisms turnover 10 to 100 t/ha/y of soil. About 2000 species of termites found mostly in tropics & subtropics (savannahs and forests). Termites (white ants) build extensive & large mounds upto 6 m in height and >6 m deep. Mounds abandoned after 10-20 years. Effects of termites are variable, may disrupt crop production. Soil become less fertile, but aid in soil formation in the tropics. Ants have less widespread than termites and more local influence on soils. Some species have exceptional ability to break- down woody materials.

### ii) Micro fauna

Of the many microscopic animals that live in the soil, 3 groups are of some importance. (1) Nematodes (threadworms or eelworms), found in relatively large numbers in nearly all soils. More than 1000 species of nematodes were known. These may be beneficial (saprophytic) or detrimental to crops. Some are parasitic nematodes, such as Heterodera sp., which infest roots of crop plants. (2) Protozoa are simplest form of animal life. These are most varied and numerous of the soil microfauna and unicellular in nature, but much larger (5-100 μm) than bacteria. More than 250 species of soil protozoa were isolated include cilliates, flagellates, amoeba. (3) Rotifers, which thrive under moist conditions (swampy envoronments). About 100 species of rotifiers active mainly in organic soils or wet areas of mineral soils.

## B) Soil Flora

**i) Macro flora:** Roots of Higher Plants are constantly growing & dying in soil, thereby, supplying soil organisms with food and energy. Roots also physically modify soils by creating stresses & strains enhancing aggregation; create channels in soil which aid in air and water movement in soil. They constantly release exudates, mucilages, enzymes, as well as, dead cells – materials leading to the formation of humus to significant depths in the soil. Upon harvesting of crops, 15-40% of above ground weight of plants is left in the ground as root mass. Rhizosphere is the zone immediately surrounding active young roots. Here microorganisms are ~100 time greater in number. Organic acids solubilize plant nutrients making them readily available for plant uptake.

## ii) Micro flora

**Soil Alage:** Chlorophyll containing organism. Aerobic. Photoautotrophic Ex:- Blue green algae (Anabaena, Nostoc, Oscillatoria, Tolypothrix etc). Soil algae are found near the surface of soil (need light for photosynthesis). They are active mostly under wet conditions. Several hundred species identified; four major groups in soils:

- Blue-green algae (BGA): most common; grow in grasslands and wet soils such as paddy fields
- Green algae: survive under non-flooded conditions and under low pH (acidic) soils
- Yellow-green algae: less common in soils.
- Diatoms: tend to occur in old gardens under drier conditions.

Blue-green alage (BGA) growing within leaves of *Azolla* (aquatic fern) in paddy fields fix nitrogen.

**Soil Fungi:** Like bacteria and actinomycetes, fungi contain no chlorophyll. They are longer than bacteria and heterotrophic in nature (Saprophytic in nature). Fungi are broadly grouped in to yeasts, molds and mushrooms, out of which molds and mushrooms are important in soils. Important molds in soils are Penicillium, Mucor, Fusa rium and Aspergillus. In prefer acid medium (4.5-6.5). Some fungi can tolerate even pH 9.0. Mushroom fungi are found in forests and grasslands. Mushroom fungi are not widely distributed like molds, but are of much significance especially in breaking down of woody tissue. A mutually beneficial (symbiotic) association between numerous fungi and the roots of higher plants is called Mycorrhizae (fungus root), is useful in adsorption of nutrients. There are over 690 species of soil fungi identified. Important in transforming soil constituents (all types of organic compounds such as cellulose, starches, gums and lignin). Fungi include yeasts, molds & mushrooms. **Yeasts** are uncommon in soil habitats. **Molds** are filamentous and microscopic to semi-microscopic and develop in all ranges of soil pH (acidic to alkaline). Molds especially thrive in acidic surface soils (e.g., forests). Common genera of moulds are *Penicillium, Mucor, Fusarium, Aspergillium*. **Mushroom fungi** are found in moist areas with high organic matter. These are common in forests and grassland soils. Mush rooms decompose woody tissue (lignins). These fungi have mass of hyphae below the ground surface

Mycorrhizae fungi: Mycorrhizae is symbiotic fungus-root association markedly increases availability of certain essential nutrients such as P, Zn, Cu, Ca, Mg, Mn & Fe to plants, especially in infertile soils. Economic importance in production of wheat, potato, beans, sugarcane, cassava & dryland rice. Ectomycorrhizal association is more with tree roots (pine, birch, oak, spruce, fir, etc.). Endomycorrhizal association is more with annual & perennial crops + trees (apple, maple, poplar, coffee, rubber, etc). Example for endomycorrhiza is VAM, vesicular arbuscular mycorrhizae.

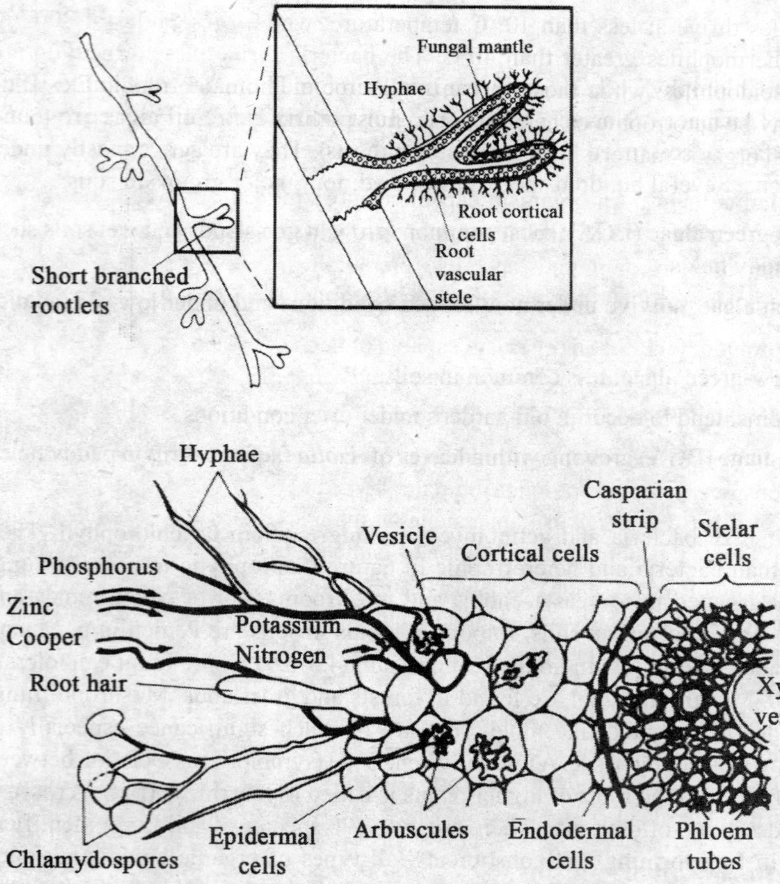


**Fig. 4:** Mycorrhizal association

**Actinomycetes:** These are thread / filamentous bacteria and next to bacteria in abundance. A common genus of actinomycetes is streptomyces. These are filamentous and similar to molds, but mycelial threads are smaller than fungal hyphae. Unicellular and similar in size to larger bacteria. Actinomycetes grow best in moist but well aerated soil conditions. Most suitable pH for actinomycetes activity is 6.0 to7.5 (near neutral). Actinomycetes remain active even under low moisture conditions.

**Bacteria:** Bacteria are single cell organisms. They are known for rapid proliferation. They adjust quickly to changes in environment. Size will seldom exceed 4 -5 μm (length). Shape may be round, rod like or spiral. In the soil, the rod shaped ones seem to predominate. Bacterial population varies from few billion to three trillion in each kilogram of soil. Bacteria undego rapid reproduction and show response to environment changes. Bacteria participate vigorously in all organic transactions in the soil; they are most numerous and often exceed both fungi and actinomycetes. They have nearly exclusive action in nitrification, sulfur oxidation & N-fixation in soil.

Psychrophiles thrive at less than $10^0$ C temperature, while, Mesophiles at $20^0$C to $40^0$C and thermophiles greater than $40^0$C. The bacteria thrive in acid medium are known as acidophiles, while those thrive in alkaline medium are alkaliphiles. Soil bacteria may be autotrophic or heterotrophic, most are heterotrophs (get energy and carbon from organic matter).

**Autotrophic bacteria :** An organism capable of utilizing $CO_2$ or carbonates as the sole so urce of carbon and obtaining energy for life processes from the oxidation of inorganic elements or compounds such as iron, sulphur, hydrogen, ammonium and nitrites (chemoautotrophic) or from radiant energy (photoautotrophic)

**Heterotrophic bacterial :** An organism capable of deriving energy for life processes only from the decomposition of organic compounds and incapable of using inorganic compounds as sole sources of energy or for organic synthesis.

**Role of soil organisms on organic matter decomposition in soil:** Detritivores (primary consumers-microflora/fauna) initiate the decomposition of detritus (organic debris). They utilise the energy in plant residues as main food source. Microflora mainly act through chemical decomposition. Action of fauna in decomposition process is both physical and chemical. Secondary consumers such as mites, snails, beetles, springtails, earthworms, etc join in the process. These include carnivores that feed on animals and microphytic feeders that consume bacteria, fungi, algae. Tertiary consumer further prey upon the primary or secondary consumers. Even in the latter cases, however, microflora still play a major role as they are active within the digestive tract of many animals. Moreover, microflora are mainly responsible for further decomposing organic material in animal feces and dead bodies of animals. Hence, microflora are known as the ultimate decomposers.

## 7.3. Benefits of soil organisms

1 Soil organisms decompose organic matter.

2 Inorganic transformations (Ammonium ion to Nitrate ion, Sulphur to sulphate ion) are supported by soil organisms.

3 Nitrogen fixation by variety of soil organisms such as Rhizobium, Azatobacter, Azospirillum etc.

4 Solubilisation of insoluble phosphorous compounds through production of organic acids in soil.

Microbial respiration → $CO_2$

$CO_2 + H_2O \rightarrow H_2CO_3$

5 Solubilisation of insoluble Sulphur compounds (S- oxidizing and reducing bacteria)

6 Formation and development of soil, humification is one of the fundamental pedogenic processes.

7 Production of soil enzymes, growth promoting substances and antibiotics in soils.

8 Some of the soil organisms Detoxification of soil pollutants.

9. Protect plant roots from invasion by soil parasites and pathogens. Soils that contain larger microbial biomass tend to have higher rates of microbial activity and as such, some organisms may have the ability to out compete other organisms including plant pathogens. This type of suppression of plant pathogens is known as general suppression. Soils that support specific antagonistic microorganisms suppresses plant pathogens by production of antibiotics.

## Involvement of soil organisms in different processes and nutrient cycles

**Mineralisation**: Mineralization is the conversion of an element from an organic form to an inorganic as a result of microbial decomposition.

The heterogeneous group of heterotrophic soil micro-organisms takes part in mineralization of organic nitrogen and converts it to inorganic nitrogen. The reactions go uninterrupted, as long as carbon source is available for microbes.

**Aminisation**: Process of conversion of proteins (enzymatic break down) to Aminoacids

$$\text{Protein} \xrightarrow{\textit{Enzymatic reaction}} \text{Polypeptides} \xrightarrow{\textit{Enzymatic hydrolysis}} \text{Aminoacids, Amines} + CO_2 + \text{Energy}$$

**Ammonification :** Process of conversion of amino acids to ammonia

$$RNH_2 + HOH \xrightarrow{\textit{Enzymatic hydrolysis}} NH_3 + R-OH + \text{Energy}$$

$$NH_3 + H_2CO_3 \xrightarrow{\textit{Enzymatic hydrolysis}} NH_4^+ + CO_2 + H_2O$$

**Nitrification:** Microbial oxidation of ammonia to nitrate

$$NH_4^+ + O_2 \xrightarrow{\textit{Nitrosomonas, Notrosococas}} NO_2 + 4H^+$$

$$NO_2 + O_2 \xrightarrow{\textit{Notrobacter}} NO_3^-$$

**Immobilisation:** The conversion of the element from the inorganic to the organic form in microbial tissues or in plant tissues, thus rendering the element not readily available to other organisms or to plants, is called immobilization.

## Biological nitrogen fixation

Biological nitrogen fixation is the bio -chemical process by which elemental nitrogen is combined in to organic forms. It is carried out by a number of organisms including several species of bacteria, a few actinomycetes and blue green algae. The quantity of nitrogen fixed globally each year is enormous (175 million Mg), which is more than the nitrogen applied through chemical fertilizers. Although nitrogen is fixed by a number of different organisms a common mechanism appears to be involved in all the cases.

$$N2 + 6H + 6e^- \xrightarrow{\text{Nitroginase}} 2NH3$$

$$NH3 + \text{organic acids} \longrightarrow \text{Amino acids} \longrightarrow \text{Protein}$$

The site of nitrogen reduction is the enzyme nitrogenase, a two-protein complex consisting of a large ion, a molybdenum containing member, and a smaller companion containing iron.

**Different groups of $N_2$ fixers** : There are three main groups of $N_2$ fixers as given below

## I. Symbiotic nitrogen fixers

Symbiotic nitrogen fixers reduce nitrogen in association with plants by forming some specialized structures in plants.

### A. Legume plants – rhizobium symbiosis

Some plants of leguminosae family form a symbiotic association with bacteria with the genus *Rhizobium* which fix atmospheric nitrogen. *Rhizobia* infect the root moving to the root cortex through an infection thread which results in the formation of a tiny outgrowth called root nodule. *Sesbania rostrata* forms nodules on the stem as well. This association is host specific. The *Rhizobia* turn to bacteroids which have nitrogenase enzyme embedded into leghaemoglobin. The symbiotic association of the host and the bacterium is mutually beneficial to both organisms. Specific species of bacterium infects the roots of a particular group of legumes. This host specificity of different *Rhizobium* speces or the group of legume crops nodulated by single *Rhizobium* sp. is referred to as 'cross inoculation group'.

**Table 4:** Host specificity in Rhizobacteeria

| *Rhizobium* sp. | Plants nodulated |
|---|---|
| *R. meliloti* | *Medicago sativa* (alfalfa), *Melilotus* sp. (sweet clover), *Medicago* sp., *Trigonella foenum grakum* (fenugreek) |
| *R. trifolii* | *Trifolium* sp. (red clover, white clover) |
| *R. leguminosarum* | *Pisum sativum* (garden and field peas), *Vicia faba* (broad bean), *Lens esculenta* (lentils), *Lathyrus* sp. |
| *R. phaseoli* | *Phaseolus vulgaris*, *P. coccineus* |
| *R. lupini* | *Lupinus* sp., *Ornithopus sativus* |
| *R. japonicum* | *Glycene max* (soybean) |
| *Rhizobium* sp. | *Vigna ungiculata* (cow pea), *V. radiata* (mung bean), *Arachis hypogeae* (groundnut), *Cajanus cajan* (redgram) *Cicer arietinum* (chickpea), *Sesbania sp.* |

## B. Symbiosis with non legumes

i) **Non legume – *Frankia* symbiosis:** The member actinomycetes, the *Frankia* forms symbiosis with forest trees belonging to the families *Casuarinaceae* and *Myricaceae.*

ii) **Azolla** – Anabaena symbiosis : Azolla is a fresh water fern found floating on the surface of water. All the species of *Azolla* have an algal symbiont called *Anabaena azollae* in a specialized cavity in the upper leaf surface. In rice fields the symbiosis can fix $N_2$ up to 30-40 kg N $ha^{-1}$.

II) **Non specific associative nitrogen fixers :** Rhizobacteria like *Azospirillum*, *Acetobacter*, *Azotobacter*, *Flavobacterium*, *Pseudomonas* etc. These bacteria are capable of using exudates of roots as the source of energy.

III) **Free living nitrogen fixers :** Many soil bacteria are capable of reducing molecular $N_2$ to $NH^+$ without forming any association with plants. They supply about 10-15 kg N $ha^{-1}$. They also function as plant growth promoting rhizobacteria (PGPR).

$$N_2 + 6\,H^+ + 6\,e^- \xrightarrow{\text{Nitrogenase}} 2\,NH_3$$

Ammonia is the end product of the nitrogenase catalysed reaction. The intermediate products are as follows

$$\underset{\text{Nitrogen}}{N_2} \xrightarrow{+2H+} \underset{\text{Diimide}}{[NH{=}NH]} \xrightarrow{+2H+} \underset{\text{Hydrazine}}{[NH_2{=}NH_2]} \xrightarrow{+2H+} \underset{\text{Ammonia}}{2\,NH_3}$$

However biological nitrogen fixation requires, an active nitrogenase enzyme system, $Mg^{2+}$ and a continuous supply of ATP, a strong reducing agent i.e., suitable $e^-$ donor and low oxygen tension.

**Denitrification :** In flooded soils the oxygen supply to soil is virtually cut - off because very less diffusion rate. Under such conditions, the facultative anaerobes Viz: *Pseudomonas*, *Bacillus denitrificans*, *Micrococcus*, *Achromobacter* and *Paracoccus* are responsible for this. Autotrophs like *Thiobacillus denitrificans*, Thiobacillus thioparus, during respiration, utilize the $NO_3^-$ as terminal acceptor and reduce it to NO, $N_2O$ and finally to $N_2$. In this fashion the nitrate leached from oxidized layer and the nitrate nitrified at the root rhizosphere will be denitrified.

The losses due to denitrification may range from 10 to 40 percent of the applied nitrogen. Of the denitrified gas, about 90 % is in the form of $N_2$ . $N_2O$ is lost at pH below 7.0 and causes atmospheric pollution. Denitrification is influenced by high organic matter content, moisture regime, soil type, Eh of soil, high temperature, supply of nitrate and cropping.

The most probable biochemical pathway leading to these losses is indicated as

$$2NO_3^- \xrightarrow{-2[O]} 2NO_2^- \xrightarrow{-2[O]} 2NO \xrightarrow{-[O]} N_2O \xrightarrow{-[O]} N_2$$

Facultative anaerobic bacteria belonging to the genera *Pseudomonas*, *Bacillus denitrificans* and *Paracoccus* are responsible for this. Autotrophs like *Thiobacillus denitrificans* also brings about denitrification.

**Phosphorous solubilisation :** The phosphorous element which is released during mineralization is likely to be converted to an insoluble inorganic compound. Some of the bacteria and other micro organisms are known to solubilise these compounds and make the phosphorous available to plants.

Phosphorous solubilising bacteria such as *Bacillus megatherium*, Pseudomonas and fungi such as *Aspergillus awamori*, Penicillium bilajii. Phosphorous absorbers are VAM (Versicular arbiscular mycorrhizae), fungi *Glomus fasciculatum* and *Gigaspora margarita*. Among different phosphorus solubilisers fungi is more efficient, but not exploited commercially, as they do cause some diseases to crop plants also. VAM is useful to explore more area and utilize phosphorus like immobile elements.

## 7.4. Harmful activities of soil organisms

Rodents, snails, slugs, termites, insect larvae a nd nematodes cause great damage to plants. Mainly fungi, but bacteria and actinomycetes ,also, are responsible for many soil borne diseases of crop plants. Soil organisms offer lot of competition to higher plants in utilizing the soil available nutrients, particularly in the presence of carbon sources in the soil. Under conditions of poor drainage, depletes the available oxygen and hampers the normal growth of plants. Under anaerobic conditions, bacteria transform some of the nutrients to unavailable form (potassium and zinc etc.) and some nutrients to be available in toxic levels ( iron and manganese etc.)

## 7.5. Soil biota and pedosphere interaction with the atmosphere, biosphere, hydrosphere, and lithosphere

Soil biota consist of the micro-organisms (bacteria, fungi, archaea and algae), soil animals (protozoa, nematodes, mites, springtails, spiders, insects, and earthworms) and plants living all or part of their lives in or on the soil or *pedosphere*. Millions of species of soil organisms exist but only a fraction of them have been cultured and identified. Microorganisms (fungi, archaea, bacteria, algae and cyanobacteria) are members of the soil biota but are not members of the soil fauna. The soil fauna is the collection of all the microscopic and macroscopic animals in a given soil. Soil animals can be conventionally grouped by size classes, (1) macrofauna (cm; enchytraeids, earthworms, macroarthropods), (2) mesofauna (mm; microarthropods, mites and collembolan), and (3) microfauna (μm; protozoa, nematodes). The size of a soil organism can restrict its location in the soil habitat. Smaller members of the microfauna like nematodes are basically aquatic organisms that live in the thin water films or capillary pores of aggregates preying or grazing on other aquatic microfauna such as amoebas. Soil protozoa are also land-adapted members of aquatic microfauna that can dwell in water films in field moist soils. Water films are created by the adsorption of water to soil particles. Soil has a direct effect on the environmental conditions, habitat and nutrient sources available to the soil biota. The term pedosphere is often used interchangeably with soil and captures the concept that the soil is a habitat where the integration of spheres occurs. These spheres include the lithosphere, atmosphere, hydrosphere, and the biosphere (Figure 5). Numerous biogeochemical processes regulated by soil biota occur in the pedosphere. Studies of the pedosphere range in scale from the field (km) to a soil aggregate (μm to nm).

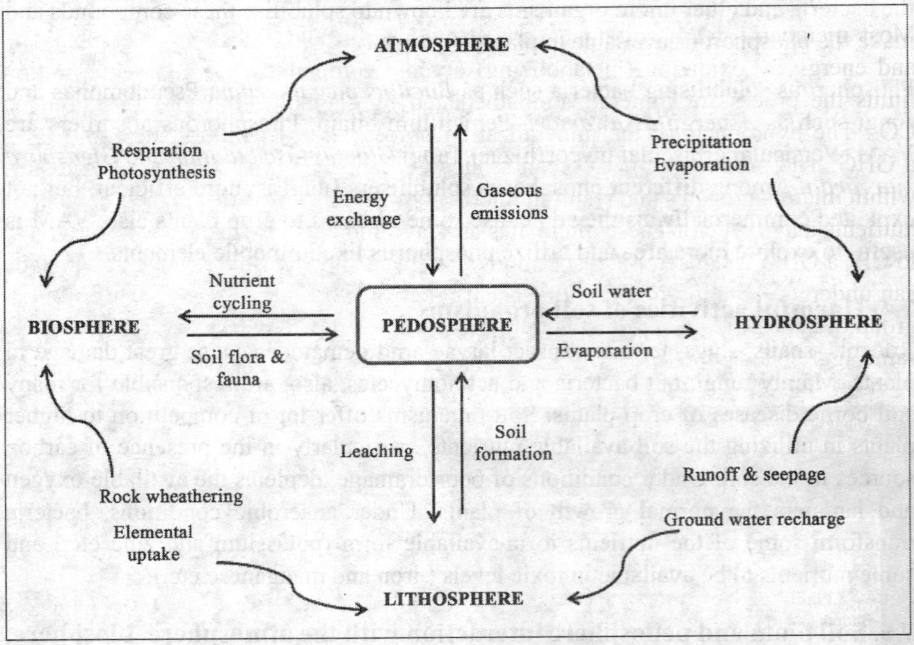


**Fig. 5:** Interactive processes linking the pedosphere with the atmosphere, biosphere, hydrosphere, and lithosphere

## 7.6. Soil food web

The first trophic level of soil food web includes photosynthesizers that get energy from the sun. Organisms that eat photosynthesizers make up the second trophic level. Third trophic level organisms eat those in the second level, and so on. It is a simplified way of thinking of the food web. In fact, some organisms eat members of several trophic levels (figure6).

The soil food web consists of the community of organisms that live all or part of their lives in the pedosphere and mediate the transfer of nutrients among the living (biotic) and non-living (abiotic) components of the pedosphere through a series of conversions of energy and nutrients as one organism and or substance is consumed by other organisms. The mesofauna (collembolan, mites) play a role in nutrient turnover by shredding materials into smaller pieces with higher surface area providing greater access for microfauna (bacteria, fungi, mycorrhizae) that recycle the majority of carbon. All food webs contain several trophic levels or feeding positions in a food chain (Figure6). The term grazing is used when organic C is obtained from living things. Soil organisms are part of the detrital food chain if their organic C is derived from dead materials. The detrital food chain creates new soil organic matter and cycles nutrients from existing OM. Biological systems and organisms contain fairly constant elemental ratios of carbon:nitrogen:phosphorus:sulfur (C:N:P:S). These ratios and mass balances (net change = input + output + internal change) allow scientists to determine biochemical shifts between organisms or ecosystems.

Most members of the soil fauna are chemoheterotrophs, meaning they obtain C and energy by oxidizing (metabolizing) organic compounds. Carbon sequestration limits the process of mineralization mediated by chemoheterotrophs that produce $CO_2$. The byproducts of the mineralization process are metabolites, heat and $CO_2$, a GHG. The production of $CO_2$ can reduce $O_2$ concentrations creating anoxic sites within microaggregates that result in microenvironments that differ in their content of nutrients and oxygen). These microsites are environments in which $CO_2$ is converted to $CH_4$ a GHG via anaerobic respiration by archaea known as methanogens. Methane can undergo oxidation to $CO_2$ in adjacent microsites. This process is regulated by a group of bacteria known as methanotrophs that transform $CH_4^+$ back to $CO_2$ all in the same aggregate.

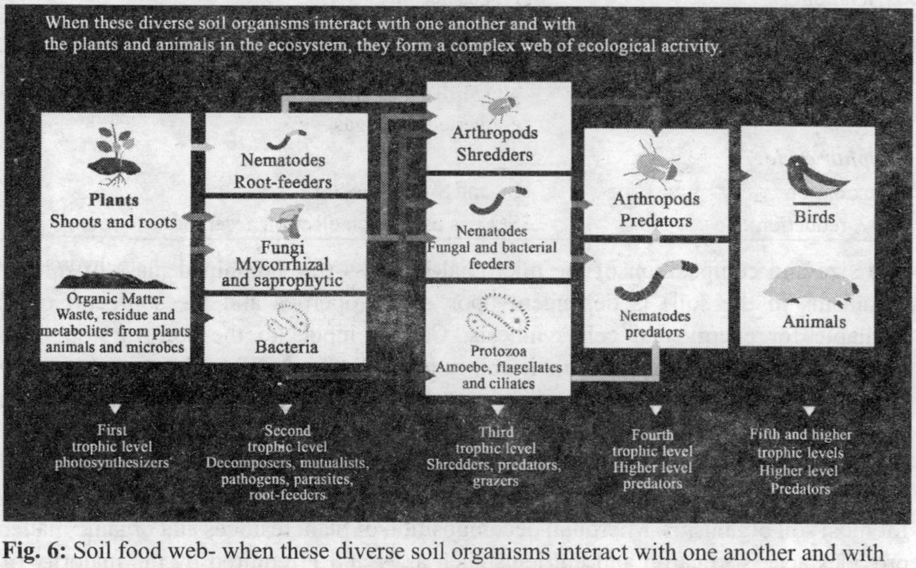


**Fig. 6:** Soil food web- when these diverse soil organisms interact with one another and with plants and animals in the ecosystem, they form a complex web of ecological activity

## 7.7. Role of the soil biota in biogeochemical cycles and nutrient transformations

Soil organisms serve numerous roles in the pedosphere. Their most critical function is the regulation of biogeochemical transformations (Table 4). Five functions mediated by the soil biota are (1) the formation and turnover of soil organic matter (SOM) that includes mineralization and sequestration of carbon, (2) nutrient cycling, (3) disease transmission and prevention, (4) pollutant degradation, and (5) improvement of soil structure. The byproducts of metabolic oxidation or reduction of carbon and nitrogen compounds in soils include green house gases (GHG). The dominant soil GHGs consist of carbon dioxide ($CO_2$), methane ($CH_4$), and nitrous oxide ($N_2O$). Soil management practices such as N fertilization and tillage can stimulate specific microbial activities such as autotrophic nitrification, denitrification and mineralization that regulate green house gas (GHG) emissions by the oxidation and reduction of carbon and nitrogen.

**Table 5:** Examples of physiological processes catalyzed by microorganisms in biosphere habitats, (adopted from Madsen, E. L. (2005).

| Carbon cycle | Nature of process |
|---|---|
| Photosynthesis | Light-driven $CO_2$ fixation into biomass |
| Respiration | Oxidation of organic C to $CO_2$ |
| Cellulose decomposition | Depolymerization, respiration |
| Methanogenesis | $CH_4$ production |
| Aerobic $CH_4$ oxidation | $CH_4$ becomes $CO_2$ |
| Anaerobic $CH_4$ oxidation | $CH_4$ becomes $CO_2$ |
| ***Nitrogen cycle*** | |
| $N_2$ fixation | $N_2$ gas becomes $NH_3$ |
| $NH_4^+$ oxidation | $NH_3$ becomes $NO_2^-$, $NO_3^-$ |
| Anaerobic $NH_4^+$ oxidation | $NO_2^-$ and $NH_3$ becomes $N_2$ gas |
| Denitrification | $NO_3^-$ is used as an electron acceptor and converted to $N_2$ gas |
| ***Sulphur cycle*** | |
| $S_2$ oxidation | $S^{2-}$ and $S^0$ become $SO_4^{2-}$ |
| $SO_4^{2-}$ reduction | $SO_4^{2-}$ is used as an electron acceptor |

The size and composition of the microbial biomass (the combined mass of micro-organisms in the soil) is dependent upon soil properties and the source(s) of C available for energy and cell synthesis. Carbon inputs to the soil vary in their biochemical composition (e.g., their ability to be decomposed) and nutrient content. Carbon turnover, decomposition and microbial activity often lead to increases in organic matter and soil aggregation. Different ecosystems vary in their potential to support soil organisms (Table 5) and sequester carbon in organic matter. Organic carbon constitutes the chemical backbone of organic matter and is the energy source for most soil organisms. Microbial decomposition of plant residues and organic matter provides access to caron and nutrients such as N and P required by the majority of living organisms. Mineralization of organic N to ammonium ($NH_4^+$) and additions of N fertilizers that contain $NH_4^+$ stimulate nitrification a process driven by nitrifying bacteria and archaea that transform $NH_4^+$ to nitrate ($NO_3^-$). Nitrate can then undergo an additional microbially mediated step, denitrification. Denitrifiers include bacteria and archaea. Both nitrification and denitrification are pathways that produce nitrous oxide ($N_2O$).

**Table 5:** Typical numbers of soil organisms in health soil ecosystems

| | Agricultural land | Prairie | Forest |
|---|---|---|---|
| **Organisms** | **Organisms per gram (tea spoon) of soil** | | |
| Bacteria | 100 millions to 1 billion | 100 millions to 1 billion | 100 millions to 1 billion |
| Fungi | Several meters | 10's to 100's of meters | 1.60 km (conifers) |
| Protozoa | 1000's | 1000's | 100,000's |
| Nematodes | 10- 20 | 10's to 100's | 100's |
| | Organisms per sqare meter | | |
| Arthopods | <10 | 45- 200 | 900- 2,300 |
| Earthworms | 4- 25 | 8- 42 | 8- 42 (0 in conifers) |

## 7.8. Role of microorganisms on soil aggregate formation

Soil texture (fineness or coarseness) affects plant rooting, soil structure and organic matter content. Soil texture and structure determine the pore-size distribution, soil water holding capacity and the amount of water to air-filled pore space in soil aggregates that provide habitat for soil organisms. Aggregates can be broadly classified into macro-aggregates (>250 μm) and micro-aggregates (20-250 μm). An aggregate is a naturally formed assemblage of sand, silt, clay, organic matter, root hairs, microorganisms and their glue like secretions mucilages, extracellular polysaccharides, and hyphae (filaments) of fungi as well as the resulting pores. Soil aggregates often contain fine roots that grow into soil pores (Figure7) associating aggregates with the rhizosphere, the zone of soil under the influence of plant roots. Persistent binding agents like organic matter and metals stabilize micro-aggregates. The temporary binding agents (polysaccharides and hyphae) produced by soil organisms aid in the formation of macro-aggregates contained within the more stable micro-aggregates. These macro-aggregates function as ecosystems or arenas of activity. Thus, an aggregate is a unit of soil structure that could be considered as a very small-scale model of a pedosphere. One can visualize all the interactions of soil gases, water, organisms and organic and inorganic constituents at the microscale (figure 7).

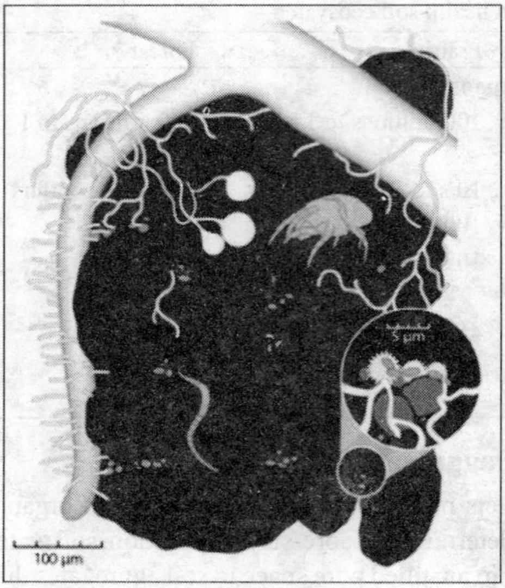


**Fig. 7:** A soil aggregate or ped is a naturally formed assemblage of sand, silt, clay, organic matter, root hairs, microorganisms and their secretions, and resulting pores

## 7.9. Arenas of activity

The ability of microorganisms to recycle carbon can provide indirect health benefits to plant communities. Soils that contain larger amounts of organic matter and microbial biomass tend to have higher rates of microbial activity and as such, some organisms may have the ability to out compete other organisms including plant pathogens. This type of suppression of plant pathogens is known as general suppression. Soils that contain high levels of organic matter may also support specific antagonistic microorganisms that have an explicit means of suppressing pathogens by production of antibiotics. This is an example of specific suppression. Soils that exhibit such properties are termed suppressive soils.

Microorganisms also interact directly with plants through symbiotic relationships that provide nutrients to plants while supplying carbon to the microorganisms. An example of a symbiotic relationship between a soil microbe and a higher plant is the interaction of the bacterium known as a ***Rhizobium*** that induces the formation of nodules on roots of soybean plants in which it fixes atmospheric nitrogen for the plant by using carbohydrate supplied by the plant.

## 7.10 Common approaches of measuring of soil biological activity

Organic management focuses on soil management with the intent to maintain and enhance soil biological activity. There are a variety of ways that biological activity is measured in soils. Few, if any, of these measures are used routinely to test soils in the way that standard chemical assays are used. The various methods of assessing soil biology and biological activity differ in their accuracy, difficulty, and suitability for use.

The following table is a summary of measures prepared by the Soil Organic Matter, Formation, Function and Management Committee describing the characteristics of widely used measures.

**Table 7:** Common measures of soil microbial activity and microbial biomass.

| | Bacteria/ Fungi ratio (B/F) | Mycorrhizae | Microbial abundance | Nematodes |
|---|---|---|---|---|
| What's measured | Direct counts of active or total biomass using staining and microscopy | Plant associated fungi that assist with nutrient and water uptake | Direct counts of bacteria and fungi grown on media | abundance, diversity, and/ or community structure |
| Sensitivity | Low | High | Low | High, functional level types |
| Cost | Medium to high; labor-intensive | Medium | Labor costs high | Labor costs medium |
| Requires special equipment | Microscope | Microscope | Colony counter | Microscope |
| Accepted interpretation for soil test application | No; there is considerable interest in developing F/B ratios as an index but weaknesses in the method make many sceptical about its utility | Some | High | High; food web integrator; reflects nutrient enrichment |

| | Earthworms | Cellulose decomposition, paper disk | Cotton strip or buried bag | Smell |
|---|---|---|---|---|
| What's measured | Abundance and diversity | Mass loss | fiber strength | subjective |
| Sensitivity | High | Very high | Very high | High |
| Cost | Labor costs medium | Low | Low | Low |
| Requires special equipment | Microscope | Microscope | Colony counter | Microscope |
| Difficulty | Low | Low | Low | Low |
| Requires special equipment | No | No | No | No |
| Accepted interpretation for soil test application | Low | | | Low |

| | ß-Glucosidase | Phenol Oxidase | FDA | Dehydro-genase | Phosphatase | Arylsulfatase | Urease | Amidase |
|---|---|---|---|---|---|---|---|---|
| Sensitivity | High | Low | High | Moderate | High | High | High | High |
| Cost | Medium | Medium | Medium | Medium | Medium | Medium | Medium | Medium |
| Requires special equipment | Spectrometer | Spectrometer | Spectrometer | Spectrometer | Spectrometer | Spectrometer | Spectrometer | Spectrometer |
| Accepted interpretation for soil test application | Not yet; needs correlating with other properties | | | Biological activity index | Not yet; needs correlating with other properties | | | |
| Difficulty | Medium | Medium | Medium | Medium | Medium | Medium | Medium | Medium |
| Scope and/or relevance | General | General | General | General; must use fresh soil samples | General; P cycle | General; S cycle | General; urea fertilizer use negates results | General |

## Objective questions

1. Weight of soil organisms per unit volume of soil is
   a) Specific weight b) Biomass
   c) Microbial count d) A&C
2. Biological diversity is more in
   a) Forest soils b) Grassland soils
   c) Arid soils d) barren soils
3) Microbial activity is more in
   a) Forest soils b) Grassland soils
   c) Arid soils d) Barren soils
4. Autotrophic Bacteria is
   a) Nitrifying bacteria, b) Iron bacteria,
   c) Sulphur bacteria d) A, B & C
5. Most soil microflora are
   a) Visible b) Microscopic
   c) Submicroscopic d) Parasitic
6. More number of micro flora commonly found in surface soil are
   a) Bacteria b) Fungi
   c) Actinomycetes d) Algae
7. More biomass of micro flora commonly found in surface soil are
   a) Bacteria b) Fungi
   c) Actinomycetes d) Algae
8. More number of micro fauna commonly found in surface soil are
   a) Protozoa b) Nematoda
   c) Earthworms d) Nematodes
9. Most important of the soil macro-animals
   a) Earth worms b) Ants
   c) Termites d) Nematodes.
10. Earthworms casts are higher in
    a) Bacteria b) Fungi
    c) Actinomycetes d) Algae
11. macro fauna important for Important in organic matter breakdown & soil manipulation
    a) Earth worms b) Ants
    c) Termites d) Nematodes.

12. Importance of plant roots in soil

a) Roots also physically modify soils by creating stresses & strains enhancing aggregation; create channels in soil which aid in air and water movement insoil.

b) Roots constantly release exudates, mucilages, enzymes, as well as, dead cells – materials leading to the formation of humus to significant depths in the soil

c) Roots secret organic acids solubilize plant nutrients making them readily available for plant uptake.

d) A, B and C

13. Soil algae are found near the surface of soil due to

a) Algae need light for photosynthesis.

b) Algae make symbiotic association with plant roots

c) Algae parasitize on surface plant roots

d) Algae uses oxygen from atmosphere

14. The soil algae found in grasslands and wet soils such as paddy fields is

a) Blue green algae b) Yellow green algae

c) Green algae d) Red algae

15. The following are not fungi

a) Yeasts b) Molds

c) Mushrooms d) Rhizobium

16. Mushroom fungi are

a) Found in forests and grasslands.

b) Not widely distributed like molds,

c) Much significance especially in breaking down of woody tissue

d) A,B &C

17. The fungus uncommon in soil habitats

a) Yeasts, b) Molds

c) mushrooms d) A & B

18. Filamentous bacteria

a) Rhizobium b) Azatobacter

c) Actinomycetes d) yeasts

19. The soil microorganism remain active even under low moisture conditions

a) Bacteria b) Fungi

c) Actinomycetes d) Protozoa

20. Mesophyle bacteria thrive at soil temperatue range of

a) Less than 10$^0$C b) 10-20$^0$C

c) 20-40$^0$C d) Above 40$^0$C

21. An example for chemoautotrophic bacteria is
    a) Thiobacillus b) Chlorobiaum
    c) Rhizobium d) Azotobacter
22. An example for heterotrophic, non-nitrogen fixing aerobic bacteria is
    a) Bacillus b) Pseudomonas
    c) Thiobacillus d) Chlorobium
23. An example for non-symbiotic nitrogen fixing heterotrophic bacteria is
    a) Azatobactor b) Rhizobium
    c) Azospirilllum d) Pseudomonas
24. conversion of inorganic form of nitrogen to organic form is known as
    a) Immobilization b) Mineralisation
    c) Mobilisation d) Fixation

**25. Match the following**

| List (I) (terms of humification) | | List (II) (description) |
|---|---|---|
| i. Mor | ( ) | 1. Humus rich B horizon in podzoles |
| ii. Mull | ( ) | 2. Surface horizon of grass lands |
| iii. Sward | ( ) | 3. Forest soil surface horizon with mixed mineral and humus |
| iv. Ortede | ( ) | 4. Surface horizon of acid litter and humus |

**Codes**

| | i | ii | iii | iv |
|---|---|---|---|---|
| a | 1 | 3 | 4 | 2 |
| b | 4 | 3 | 2 | 1 |
| c | 3 | 4 | 1 | 2 |
| d | 3 | 2 | 4 | 1 |

**Answer key**

| | | | | | | | | | |
|---|---|---|---|---|---|---|---|---|---|
| **1** | b | **6** | a | **11** | a | **16** | d | **21** | a |
| **2** | a | **7** | d | **12** | d | **17** | d | **22** | a |
| **3** | b | **8** | a | **13** | a | **18** | c | **23** | a |
| **4** | d | **9** | a | **14** | a | **19** | c | **24** | a |
| **5** | b | **10** | a | **15** | d | **20** | c | **25** | a |

## Review questions

1. Give brief classification of soil organisms
2. Briefly describe microrrhizal fungi with heat sketch
3. Discuss the benefits and harms of soil organisms
4. Briefly describe soil food web
5. Explain the role of soil organisms on soil aggregation

# 8

# Soil Degradation

Soil degradation is defined as a decrement in the soil health status resulting in a diminished capacity of the ecosystem to provide goods and services for its beneficiaries. Degraded soils have a deteriorated health status such, that they do not provide the normal goods and services of the particular soil in its ecosystem. Soil degradation is the physical, chemical and biological decline in soil quality. It can be the loss of organic matter, decline in soil fertility, and structural condition, erosion, adverse changes in salinity, acidity or alkalinity, and the effects of toxic chemicals, pollutants or excessive flooding. The terms soil degradation and land degradation are many times used synonymously. Various agencies in India (Table.1) estimated the extent of soil degradation as 107 M ha (MOA, 1994) to 187.7 Mha (NBSS & LUP, 1994).

**Table 1:** Soil degradation assessment in India by different organizations

| Agency | Estimated extent (Mha) | Criteria for delineation |
|---|---|---|
| National Commission on Agriculture (NCA, 1976) | 148.09 | Based on the secondary data |
| Ministry of Agriculture (1978) (Soil and Water Conservation Division) | 175.00 | Based on the NCA's estimates. No systematic survey was undertaken |
| Society for Promotion of Wastelands Development (SPWD) (Bhumbla and Khare, 1984) | 129.58 | Based on the secondary estimates |
| Ministry of Agriculture (MOA, 1985) | 173.64 | Land degradation statistics for states |
| Ministry of Agriculture (MOA, 1994) | 107.43 | Elimination of duplication of area. Area reclaimed counted |
| NBSS&LUP (1994) | 187.70 | Mapping on 1 : 4 million scale based on the Global Assessment of Soil Degradation(GLASOD) guidelines |
| NBSS&LUP (2004) (revised) | 146.82 | 1 : 1 million scale soil map |
| National Wasteland Development Board (1985) | 123.00 | |

Indian council of Agricultural research (ICAR) has estimated the extent of degraded lands in Agro-eco-regions of India (Table 2) and reported highest extent in Northern plain and central highlands including Aravallis, hot semi-arid ecoregion, with alluvium derived soils (14.960 Mha) followedby Western plain, Kachch and part of Kathiawar peninsula, hot arid eco-region, with desert and saline soils (13.913 M ha)

**Table 2:** Extent of Wastelands and Degraded lands in India in respect of Agro-ecoregion wise (ICAR)

| S.No | Agro-ecoregion | Extent of degraded lands (million ha) |
|---|---|---|
| 1 | Western Himalayas cold arid ecoregion, with shallow skeletal soils and length of growing period (GP) <90 days | 0.019 |
| 2 | Western plain, Kachchh and part of Kathiawar peninsula, hot arid ecoregion, with desert and saline soils | 13.913 |
| 3 | Deccan plateau, hot arid ecoregion, with red and black soils | 2.560 |
| 4 | Northern plain and central highlands including Aravallis, hot semi-arid ecoregion, with alluvium derived soils and GP 90–150 days | 14.960 |
| 5 | Central (Malwa) highlands, Gujarat plains and Kathiawar peninsula, hot semi-arid ecoregion, with medium and deep black soils and GP 90–150 days; | 7.700 |
| 6 | Deccan plateau, hot semi-arid ecoregion, with shallow and medium (with inclusion of deep) black soils and GP 90–150 days | 11.270 |
| 7 | Deccan (Telangana) plateau and Eastern Ghats, hot semi-arid ecoregion, with red and black soils and GP 90–150 days | 4.986 |
| 8 | Eastern Ghats, Tamil Nadu uplands and Deccan (Karnataka) plateau, hot semi-arid ecoregion, with red loamy soils and GP 90–150 days | 5.685 |
| 9 | Northern plain, hot sub-humid (dry) ecoregion, with alluvium derived soils and GP 90–150 days; | 4.272 |
| 10 | Central Highlands (Malwa, Bundelkhand and Eastern Satpura), hot sub-humid ecoregion, with black and red soils and GP 150–180 (to 210) days | 8.288 |
| 11 | Eastern plateau (Chhattisgarh), hot subhumid ecoregion, with red and yellow soils and GP 150–180 days; | 5.925 |
| 12 | Eastern (Chhotanagpur) plateau and Eastern Ghats, hot subhumid ecoregion, with red and lateritic soils, and GP 150–180 (to 210) days | 8.193 |
| 13 | Eastern plain, hot subhumid (moist) ecoregion, with alluvium derived soils and GP 180–210 days | 4.177 |
| 14 | Western Himalayas, warm subhumid (to humid with inclusion per-humid) ecoregion, with brown forest and podzolic soils and GP 180 to210+ days; | 5.695 |
| 15 | Bengal and Asom plain, hot subhumid (moist) to humid (inclusion of per-humid) ecoregion, with alluvium derived soils and GP 210+ days | 4.735 |

| S.No | Agro-ecoregion | Extent of degraded lands (million ha) |
|---|---|---|
| 16 | Eastern Himalayas, warm per-humid ecoregion, with brown and red hill soils and GP 210+ days; | 2.523 |
| 17 | North-Eastern Hills (Purvanchal), warm perhumid ecoregion, with red and lateritic soils and GP 210+ days; | 7.519 |
| 18 | Eastern Coastal plain, hot subhumid to semi-arid ecoregion, with coastal alluvium derived soils and GP 90–210+ days | 1.851 |
| 19 | Western Ghats and Coastal plain, hot humid–per-humid ecoregion, with red, lateritic and alluvium derived soils and GP 210+ days; | 6.062 |
| 20 | Islands of Andaman-Nicobar and Lakshadweep, hot humid and perhumid island ecoregion, with red loamy and sandy soils and GP 210+ days. | 0.077 |
| | Total Degraded lands in India | 120.410 |

Ministry of Agriculture, Government of India 1985 estimated soil degradation classes(Table 3), which revealed that loss of top soil by erosion is major extent of 128.9 M ha, followed by degraded forest lands (19.49 M ha), water logging (8.52 M ha), salt affected soils (7.61 M ha), shifting cultivation (4.91 M ha).

## Causes of soil degradation

**a) Deforestation** : It causes soil degradation on the account of exposing soil minerals by removing trees and crop cover, and which support the availability of humus and litter layers on the soil surface. Vegetation cover mainly promotes the binding of the soil together and soil formation, hence when it is removed it considerably affects the capabilities of the soil such as aeration, water holding capacity, and biological activity.

**Table 3:** Causes of soil degradation

| Causes of Degradation | Area (million hectares) | Percentage of total area |
|---|---|---|
| Water erosion | 107.12 | 61.7 |
| Wind erosion | 17.79 | 10.24 |
| Ravines | 3.97 | 2.28 |
| Salt accumulation (saline , alkali) | 7.61 | 4.38 |
| Water logging | 8.52 | 4.90 |
| Mines & quarry wastes | - | - |
| Shifting cultivation | 4.91 | 2.82 |
| Degraded forest lands | 19.49 | 11.22 |
| Special problems | 2.73 | 1.57 |
| Coastal sandy areas | 1.46 | 0.84 |
| Total | 173.64 | 100.0 |

*Source*: Ministry of Agriculture, Government of India (1985)

When trees are removed by logging, infiltration rates become elevated and the soil remains bare and exposed to erosion and buildup of toxicities. Some of the contributing activities contain logging and slash and burn techniques used by individuals who invade forest areas for farming, rendering the soils unproductive and less fertile in the end.

**b)Misuse or excess use of fertilizers**: The excessive use and misuse of fertilizers kill organisms that assist in binding the soil together. Most agricultural practices involving the use of fertilizers and pesticides entail misuse or excessive application, thereby contributing to the killing of soil's beneficial bacteria and other micro-organisms that help in soil formation. The complex forms of the fertilizer's chemicals are responsible for denaturing essential soil minerals, giving rise to nutrient losses from the soil. Thus, the misuse or excessive use of fertilizers increases the rate of Soil degradation by destroying the soil's biological activity and builds up of toxicities through incorrect fertilizer use.

**c)Water logging**: Soils become water-logged when the water balance of an area is disturbed as of excess recharge. The important sources of water are heavy rains, overland water flow towards the basin, seepage from canals and distribution system, and tidal flooding. Conditions cause waterlogging is natural basins without an outlet for water, low permeability of subsurface horizons, low intake rate of surface soils and obstructions to the natural flow of rainwater, etc.

**d)Industrial and mining activities**: Soil is mainly polluted by industrial and mining activities. As an example, mining destroys crop cover and releases a myriad of toxic chemicals that are mercury into the soil thereby poisoning it and rendering it unproductive for any other purpose. Industrial activities, release toxic effluents and material wastes into the atmosphere, land, rivers, and groundwater that eventually pollute the soil and as such, it impacts on soil quality. Altogether, industrial and mining activities degrade the soil's physical, chemical properties, and biological properties.

**d) Unscientific crop rotation**: Due to a shortage of land, the farmers have adopted intensive cropping patterns of commercial crops in place of balanced cereal-legume rotations. During the last two decades, the area under food crops decreased, and under non-food crops increased. Intensive cultivation leads to the removal of large quantities of nutrients from the soil which effects to loss of soil fertility.

**e) Unscientific cultivation practices**: Some agricultural practices are environmentally unsustainable. The tillage on agricultural lands is one of the major factors since it breaks up the soil into finer particles, expose organic particles (causing rapid oxidation), disturbing fungal filaments, exposing bacterial colonies to hot sun, ultimately detaching soil particles which increase erosion rates. The soil quality decline is exuberated more and more as a result of the mechanization of agriculture that gives room for deep plowing, plant cover reduction, and the formation of the hardpan. Other improper cultivation activities, for example, farming on steep slope and mono-cropping, row-cropping, and surface irrigation wear away the natural composition of the soil and fertility and prevent soil from regenerating.

**Examples of Soil degradation**: Soil degradation means the physical, chemical, and biological decline in soil quality. It can be the loss of organic matter, soil erosion, acidity or alkalinity, and the effects of toxic chemicals, excessive flooding. Soil degradation can involve;

- Water erosion (includes sheet, rill and gully erosion)
- Wind erosion
- Salinity (includes dry land, irrigation, and urban salinity)
- Loss of organic matter
- Fertility decline
- Soil acidity or alkalinity
- Structure decline (includes soil compaction and surface sealing)
- Soil contamination (including the effects of toxic chemicals and pollutants).

## Effects of Soil degradation

a) **Decline of soil quality:** Soil quality decline is one of the main causes of land degradation and also considered to be responsible for 84% of the ever-diminishing acreage. Year after year, huge acres of land lost due to soil erosion and contamination. The world's agricultural land (About 40%) is severely diminished in quality because of soil erosion and the use of chemical fertilizers, which prevent the land from regenerating. The decline in soil quality as a result of agricultural chemical fertilizers further leads to water pollution and land pollution thereby lowering the land's worth on earth.

b) **Drought and aridity:** Drought and aridity both are problems highly influenced and amplified by soil degradation. The contributing factors to soil quality decline such as overgrazing, poor tillage methods, and deforestation are the leading causes of desertification characterized by droughts and arid conditions. In the same context, Soil degradation may bring about loss of biodiversity.

c) **Loss of arable land:** Because Soil degradation contributes to land degradation, it means that it creates a significant loss of arable land. As stated earlier, about 40% of the world's agricultural land is lost on the account of soil quality depreciation mainly caused by agrochemicals and soil erosion. Most of the crop production practices affect the topsoil loss and the damage of soil's natural composition that makes agriculture possible.

d) **Increased flooding:** The land is normally altered from its natural landscape when it rids its physical composition from Soil degradation. For this reason, the transformed land is unable to soak up water, making flooding frequent. Soil degradation takes away the soil's natural capability of holding water thus contributing to more and more cases of flooding.

e) **Pollution and clogging of waterways:** Most of the soil eroded from the land together with the chemical fertilizers utilized in agricultural fields is discharged into waterways. With time, the sedimentation procedure can clog waterways, resulting in water scarcity. The agricultural fertilizers and pesticides damage marine and freshwater ecosystems and limit the domestic uses of the water for the populations that depend on them for survival.

Soil degradation is broadly divided to three type's viz., physical, chemical and biological degradation. *Soil physical (except soil erosion) and chemical degradation (soil acidity, soil salinity, soil sodicity, calcareousness etc.) will be discussed in later chapters. In this chapter Soil erosion and soil biological degradation are discussed.*

## 8.1. Soil Erosion

In general, soil erosion implies the physical removal of topsoil by various agents, including rain, water flowing over and through the soil profile, wind, glaciers or gravitational pull. Soil and water are the most precious natural resources that support and sustain the anthropogenic activities. Soil erosion in India is amongst the leading areas of concern as it affects cultivation and farming in the country in adverse and unfavorable ways. Soil erosion leads to deprivation of physical characteristics of soils and damages plant and crops. Soil loss due to erosion has great consequences because it leads to loss of its productivity. Soil erosion occurs though out the world but it is a very common feature and more serious problem in dry areas. Soil erosion disturbs agricultural, environmental and ecological functions performed by the soil. Soil erosion results in depletion of soil fertility, decreased moisture storage capacity and consequently in decreased crop productivity. In addition to loss of soil fertility and crop yields, soil erosion also increases environmental pollution, increasing the sediment load in streams and rivers, thereby disturbing the aquatic life, particularly fish. In the long run, soil erosion affects socio-economic conditions of the society by causing floods, silting up of water reservoirs and disruption of communication systems. The soil covering the Earth surface has taken millions of years to develop. The rate of soil formation is very slow (during every 100 to 400 years, only 1 cm soil is formed) and the enough soil depth is formed in 3000 to 12000 years to have a productive land. Thus, when soil, a non-renewable natural resource is ruined then it will be lost entirely (Lal 2001). Globally, out of 22% of the land suitable for sustaining agricultural productivity, around 5 to 7 Mha are being lost annually due to soil degradation, consequently, threatening food security of the world. Soil and water resources conservation and management is important for the welfare of the people.

In India almost 130 million hectares of land, i.e., 45 % of total geographical surface area, is affected by serious soil erosion through gorge and gully, shifting cultivation, cultivated wastelands, sandy areas, deserts and water logging. The average annual rate of soil erosion in the country is 16.35 tonnes per hectare i.e. 5334 million tonnes per year. Of this, about 29 percent is lost to the sea, ten percent is deposited in reservoirs (leading to a reduction in the storage capacity of reservoirs by 1 to 2 percent annually) and the remaining 61 percent continues to get shifted from one place to another. This different kinds and increasing extent of erosion contribute to rapid conversion of cultivable lands into wastelands. For example, recent evidence shows that about 63.85 million hectares (20.17 percent) of the total geographical area (328.72 million hectares) of the country has been converted into wastelands. The National Bureau of Soil Survey & Land Use Planning (NBSS & LUP) estimates that soil erosion extends to about t 119.2 million hectare area in India. The loss of top layer of the soil can lead to severe depletion and degradation of the soil leading to loss of soil quantity and fertility resulting in fall in agricultural production and productivity. The annual loss in output of main crops in India because of soil erosion has been estimated to be 7.2 million tonnes which is about 4 to 6.3 percent of annual agricultural production of the country and ranges to a loss in terms of replacement cost from 1 to 1.7 % of the GDP. As high as 74 million tonnes of major nutrients are lost to soil erosion in India per year.

The state wise extent of soil erosion from wind and water is presented in Table 4. The figures are adopted from National Bureau of Soil Survey & Land Use Planning(NBSS&LUP 2005). The NBSS&LUP estimates are higher in case of water erosion and lower in case of wind erosion when compared to GLASOD.

**Table 4:** Extent of erosion in India unit- (1000 ha) (Source: NBSS&LUP, 2005)

| S.No. | Name of the States | Water Erosion | Wind Erosion |
|---|---|---|---|
| 1 | Andhra Pradesh & Telangana | 11518 | 0 |
| 2 | Arunachal Pradesh | 2372 | 0 |
| 3 | Assam | 688 | 0 |
| 4 | Bihar+ Jharkhand | 3024 | 0 |
| 5 | Goa | 60 | 0 |
| 6 | Gujarat | 5207 | 443 |
| 7 | Haryana | 315 | 536 |
| 8 | Himachal Pradesh | 2718 | 0 |
| 9 | Jammu & Kashmir | 5460 | 1360 |
| 10 | Karnataka | 5810 | 0 |
| 11 | Kerala | 76 | 0 |
| 12 | Madhya Pradesh + Chhattisgarh | 17883 | 0 |
| 13 | Maharastra | 11179 | 0 |
| 14 | Manipur | 133 | 0 |
| 15 | Mizoram | 137 | 0 |
| 16 | Meghalaya | 137 | 0 |
| 17 | Nagaland | 390 | 0 |
| 18 | Orissa | 5028 | 0 |
| 19 | Punjab | 372 | 282 |
| 20 | Rajasthan | 3137 | 6650 |
| 21 | Sikkim | 158 | 0 |
| 22 | Tamil Nadu | 4926 | 0 |
| 23 | Tripura | 121 | 0 |
| 24 | Uttar Pradesh + Uttarakhand | 11392 | 212 |
| 25 | West Bengal | 1197 | 0 |
| 26 | Delhi | 55 | 0 |
| 27 | Union Territories | 187 | 0 |
| | Grand Total | 93680 | 9483 |
| | Grand Total(Million ha) | 93.68 | 9.48 |

In Indian condition, the control of soil erosion is a challenging task in the sense that the onset of monsoon often coincides with the kharif sowing and transplanting. In this stage of kharif crop when canopy cover is minimal, major part of the land is exposed to the rainfall let the land prone to soil erosion. In agriculture, soil erosion refers to the removal of topsoil by the natural physical forces of water and wind at a greater rate than it is formed or through forces associated with farming activities such as tillage. Erosion removes the topsoil first and once this nutrient rich layer is lost, the potential of soil to support plant growth is reduced. Without soil and plants the land becomes desert like and unable to support life.

## Causes of Soil Erosion

Soil erosion is influenced by political, economic, social conditions, climate, land use and management and topography. Poverty level directly relates with soil erosion in developing countries. There is no way to measure conservation practices for poor farmers that have limited or lacked resources. The risk of soil erosion is decreased by the elimination of implementing conservation practices and for year after year food production on small agriculture farms (0.5 - 2 ha) compels farmers to use over exploiting practices by subsistence farming. The causes of soil erosion are deforestation, intensive cultivation, over grazing, cultivation on steep slopes, mismanagement of soil and rapid urbanization.

### i) Deforestation

Energy fluxes, erosion control, moderation of climate and ecosystem stabilization are the essential ecosystem services provided by forest. The major causes of denudation are urbanization, unnecessary logging and clear-cutting, construction of roads and highways, frequent fires and expansion of farming to marginal lands. As the human population continues to increase, there is a clear need for more food. In addition, the increases demand of agricultural products has created incentives to convert forests to farmland and pastures. The land use for agriculture or other purposes is done by deforestation that causes the permanent destruction of forests. The land is swept into river by erosion without vegetative cover. So, the cycle of soil loss continues by the movement of farmers in the forest, clearing more forest for cultivation.

### ii) Intensive cultivation

Industrial agriculture that is also termed by the intensive farming or cultivation is attributed by maximum use of inputs such as low fallow ratio, labor and capital per unit land. Higher yields are produced with the use of less land and less labor that capable the farmer by more intensive agriculture. Human health and farm productivity is affected by increased environmental impacts by the potential degradation of water and soil resources. Even when there is not excessive soil erosion, soil quality is reduced by depletion in organic matter and natural supplies of trace elements as the result of intensive cropping. Wide range of plant and animal species in soil maintain the fertility of soil with the diverse contributions and recycling of nutrients in natural ecosystem. When fertilizers are not used, than some trace elements are depleted as a result of single mono crop species grown and no diversity and rotation is replaced year after year. The organic content of the soil also decreased if there is no replacement of mined nutrient over time when no crop residues or organic matter is added.

### iii) Overgrazing

In many livestock farms, the same piece of land for a long time is mostly concentrated by the herds of cattle and sheep. Soil displacement during cattle traffic, repeated crushing or trampling and overgrazing is resulted by this confinement. Soil erosion on steep slope or hillsides is increased when the protective cover is reduced by removing or thinning of grasses. Acceleration of water and wind erosion, degradation of soil structure and reduction in organic matter content of soil is resulted by overgrazing. Reduction in root proliferation and growth, soil compaction, decrease of drainage and water infiltration rate by cattle trampling. Soil erosion in heavily grazing areas increase runoff by increasing stocking rate. Soil erosion is increased in wet and clayey soils by surface runoff and compaction on overgrazing lands. Siltation and sediment-related pollution of downstream water bodies also increase the soil erosion of pasture lands. The soil susceptible to wind erosion in surface soils due to disintegration of particles by animal traffic in dry regions. Loss of topsoil and nutrients by the conversion of natural ecosystem that caused higher rates of erosion by increasing continuous grazing that initially damage the land.

### iv) Cultivation on steep slopes

Cultivation in slopes by land tillage makes the soil vulnerable for erosion. The moving water will scour soil particles and starts the process of erosion. As the intensity of rain increases, the runoff increases and the force exerted on soil particles also increases. As the slope steepness increases, the velocity of runoff and force on soil particles also increases. Amount of rainfall, slope steepness, vegetation and soil type are the major factors causing slope erosion. Terracing on the slopes decreases the erosion by decreasing speed of runoff and crops which require heavy irrigation i.e. rice can be grown on these terraces.

### v) Mismanagement of soil

The common cause of soil erosion is the expansion of agriculture on poor quality soils and irrigation water, indiscriminate chemical inputs use, and clean cultivation degrade soils. Crop residues are removed for fodder and biofuel and industrial uses, this practice leaves the soil bared from protective cover below a critical level and soil becomes vulnerable to erosion. Runoff is increased by mismanagement of soil causing soil erosion, and ultimately transporting nutrients and pesticides off-site in to water bodies causing water pollution.

### vi) Urbanization

There is significant effect of urbanization because most of the productive agricultural land near cities has been converted into residential and commercial area. As a result, agricultural area is decreasing which ultimately putting pressure on available land area to produce more and more from shrinking land resources which again resorting to intensive landuse Despite decrease in agricultural land, the limited land is used intensively for cultivation which results in decreased soil quality over the time and become susceptible for erosion.

## Factors affecting Soil Erosion

**i)Slope:** Slope is the major factor to control soil erosion. Length and steepness of slope are the main factors that affect soil erosion. As the steepness increases the erosion increases similarly, as the length of slope increases the eroded effect of running water increases. The water conservation practices such as terraces and buffer strips reduce the intensity of flowing water by reducing the slope. Runoff velocity of water and discharge is more from channels that have relatively more smooth surfaces. On the other hand, construction of water catchments and minimizing the soil slope reduces water runoff and thus decrease the erosion.

**ii) Soil structure:** The arrangement or aggregation of soil particles is termed as soil structure. Intensive cultivation and large compaction results in deterioration of soil structure and particles binding and thus make them susceptible to erosion. Soil structure results from of symmetrical arrangements of soil particles, which keep pore spaces, micro and macro-organisms, and different sized aggregates, shapes and stability within a limit. The resilience of soil to erosion is largely depends upon its structure. The soils with poor structure are weekly aggregated, easily compacted and have high runoff with low infiltration. The quantitative measurement of soil structure is difficult therefore water infiltration, air permeability, and soil organic matter dynamics are usually related to soil structure development. Measurement of properties of aggregate is also a helpful way if soil structural stability at the aggregate level determines the macroscale structural attributes of the whole soil to withstand erosion. There are numerous techniques for characterization and modeling of soil structure. Advanced techniques for soil structure modeling aim to capture the heterogeneity of soil structure and correlate these quantifications with various processes such as erosion. The focus on soil-based techniques, coupled with the characterization of aggregates, can provide additional insight into soil structure dynamics. Current technologies include *tomography*, *neural networks* and *fractal theory*. Tomography useful for investigation of soil interior architectural design and allows for three-dimensional visualization of soil structures. By using this method, the geometry and distribution of macro pores and microporous networks in the soil can be examined, which facilitates the flow of air and water. The use of neural networks is another way to observe the structural properties of the soil to conserve water, store organic matter and resist erosion. Soil debris and its sensitivity to soil erosion are controlled by fractal theory in the process of cultivation. This theory involves the study of the complexity of soil particle arrangement, tortuosity and soil pore abundance, which is the key to explain the process of water flow through the soil. These relatively new technologies can help quantify the structural properties of the soil.

**iii) Soil Organic matter:** The cementing agent that binds the soil particles together in the aggregate formation is organic matter. Organic matter plays important role in prevention or soil erosion. The fundamental source of energy for soil organisms is organic matter. It is both of animal and plant origin. Soil protection from compaction and erosion, improvement in soil structure, water and nutrient holding capacity increases and healthy communities of soil organisms are supported with the frequent addition of organic matter. Crop rotations that contains high plants residues, leaving crop residues in the field growing cover crops, using minimum or no tillage systems, mulching, growing perennial forage crops, using optimum nutrient and water management strategies for healthy plants production with large number of residues and roots, growing cover crops and applying compost or manure are the practices that increase organic matter addition in soils.

**iv) Vegetation cover:** Loss of protective vegetation through fire, ploughing and overgrazing makes soil susceptible to being wash away by water and wind. The vegetative cover provides natural measure in reducing soil erosion losses. The water is slowed down by the plants as it flows over the land and ground is soaked by allowing much of the rain to do this. Plant roots that hold the soil in place prevent the soil being swept or blown away by wind. Soil's ability to erode is reduced by plants that protect the soil from the abrasive effect of raindrops. The flow of water is slowed down by the plants in wetland and on the banks of river and the roots prevents erosion by binding the soil.

**v) Land use:** Scientific land use based on land capability classification. Based on the limitations of the land when put to use, the land use plan should be prepared such that in lands that suits for crops will pun in use for cultivation of crops duly following required soil conservation measures. The best soil protector against soil erosion is grass due to its highly dense cover. The considerable obstruction to surface wash is small grains such as wheat. During the early stages of growth, little cover is provided by the row crops such as potatoes, maize that also encourages erosion. The areas that are most subjected to erosion are fallowed areas where entire residue has been incorporated into the soil and no crop is grown. The land should be protected with vegetative cover for which proper cropping pattern and crop rotation should be followed with an objective of arresting soil erosion.

## Types of Soil Erosion

Soil erosion can be classified into two major types, i.e., accelerated and geological erosion. The normal process of weathering is geological erosion that usually happens as a part of natural soil-forming mechanisms at low rates in all soils. It is not affected by human activities as well as it happens at the period of long geological time. The processes is slow but constant and soil loss is compensated by the soil formation through pedogenic processes. On the contrary, in accelerated erosion, soil erosion becomes a main anxiety and a specific threshold level is exceeded by the erosion rate and soil loss through erosion exceeds the soil formation through pedogenic processes. Anthropogenic activities such as slash-and burn agriculture, intensive and uncontrolled grazing, deforestation and burning of biomass and intensive plowing are main factors which trigger accelerated soil erosion. The soil becomes less productive after the loss of fertile topsoil even by applying the same farm inputs. Although soil erosion cannot be eliminated but there are ways to minimize excessive erosion and its adverse effects on agricultural production and environment protection.

**i) Natural or geologic erosion:** Geologic erosion sometimes referred to as natural or normal erosion, represent erosion under the cover of vegetation. It includes soil as well as soil eroding processes that maintain the soil in favourable balance, suitable for the growth of most plants. The rate of erosion is so slow that the loss of soil is compensated by the formation of new soil under natural weathering processes. It ranges from very little in undisturbed lands to extensive in steep arid lands. Geological erosion takes place, as a result of the action of water, wind, gravity and glaciers and it takes place, at such slow rates that the loss of soil is compensated for the formation of new soil under natural weathering processes. It is sometimes referred to as normal erosion.

**ii) Accelerated erosion(**exogeneic erosion) caused by the disturbances of people (cutting forests, cultivating lands, constructing roads and buildings etc.) and is increasing as the population increases. In this erosion, the removal of soil takes place at a much faster rate than that of soil formation. It is also referred to as abnormal erosion. Accelerated soil erosion can be divided into two categories depending on the eroding agents namely water erosion and, wind erosion. Soil erosion due to the agents like water and wind is mostly prevalent and tangible. The erosion caused through chemical and geological agents is a slow process and continues to years and often it is non- tangible.

## 8.1.1. Water Erosion

The soil erosion caused by water as an agent is called water erosion. In water erosion, the water acts as an agent to dislodge and transport the eroded soil particle from one location to another (Fig. 1).

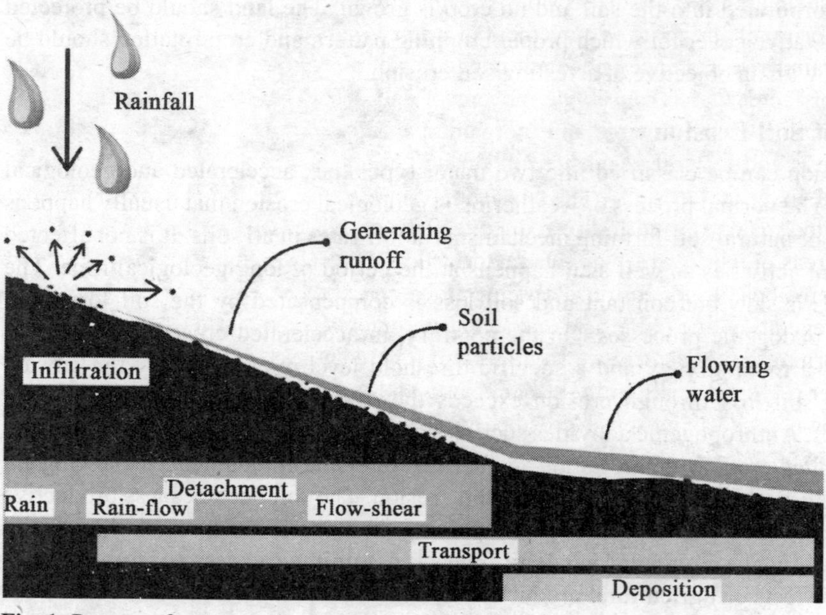


**Fig. 1:** Process of water erosion

### Extent of water erosion in India and South Asia

The extent of water erosion in Indian subcontinent are presented in Table 4. In India, 32.8 Mha area in is affected by water erosion which accounts for 18% of the land. However, the water erosion extent estimated by different sources varies from 87 to 111 Mha in India.

**Table 4:** GLASOD assessment: areas affected by water erosion (Unit: 1000 ha) in South Asian region

| Region | Light | Moderate | Strong | Total | Total as % of land |
|---|---|---|---|---|---|
| Afghanistan | 8560 | 2597 | 0 | 11156 | 29% |
| Bangladesh | 0 | 1504 | 0 | 1504 | 15% |
| Bhutan | 36 | 0 | 4 | 40 | 10% |
| India | 2936 | 17217 | 12620 | 32773 | 18% |
| Iran | 14 504 | 11896 | 0 | 26400 | 45% |
| Nepal | 520 | 1072 | 0 | 1592 | 34% |
| Pakistan | 6080 | 1124 | 0 | 7204 | 28% |
| Sri Lanka | 72 | 157 | 845 | 1074 | 46% |
| India, dry region | 1177 | 0 | 1676 | 2853 | - |
| India, humid region | 1759 | 17217 | 10944 | 29920 | - |
| Dry zone | 30320 | 15617 | 1676 | 47613 | 32% |
| Humid zone | 2387 | 19951 | 11791 | 34130 | 20% |

### Types of water erosion

The most severe type of soil erosion is water erosion. Detachment of soil particles from its original place due to movement of water is called water erosion. Water from runoff, rain, irrigation and snowmelt may contribute to soil erosion but rainwater is the major factor which causes the movement and detachment of soil particles. The transportation of soil organic and inorganic particles with the water flowing along the slope is subsequently deposited in surface water bodies and at lower landscape positions in water erosion. The new soil reservoirs, streams are filled by these transported materials and inturn reduce their reservoir capacity. In humid and sub-humid areas of world which are characterized by repeated rainstorms, the dominant form of erosion is wind erosion. The same problem is noticed in the soil that is bare and has no vegetation like in the arid and semiarid regions that have limited precipitation in the form of intense storms (torrential rain). There are many types of water erosion viz., inter rill, splash, rill, gully, stream bank, and tunnel erosion. Inter rill erosion is also known as sheet and splash erosion, but these two differ in the underlying fluvial processes.

#### a) Splash erosion

This type of soil erosion is because of the action of raindrop. The kinetic energy of falling raindrop dislodges the soil particle and the resultant runoff transports soil particles. Splash erosion (Fig. 2) is the first stage of soil erosion by water. It occurs when raindrops hit bare soil. The explosive impact breaks up soil aggregates so that individual soil particles are 'splashed' onto the soil surface. The splashed particles can rise as high 0.60 meter above the ground and move up to 1.5 meter from the point of impact. The particles block the spaces between soil aggregates, so that the soil forms a crust that reduces infiltration and increases runoff.

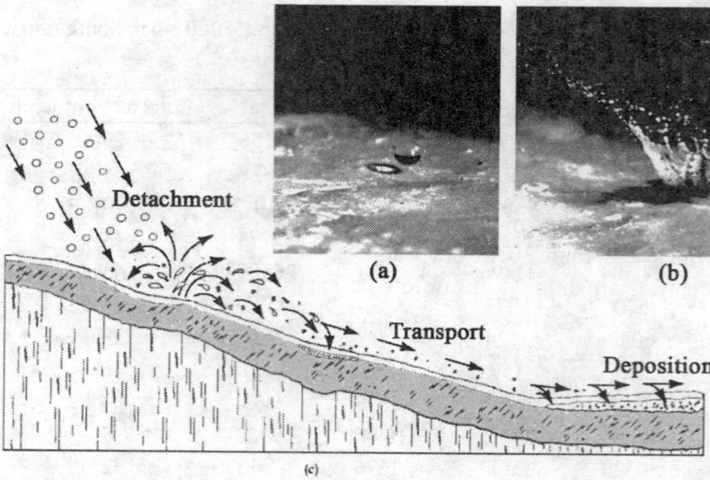


**Fig. 2:** Picture describing Splash erosion of the soil
(*Source*: http://restoringutopia.blogspot.in/2010/ 07/like-hollow-point-bullets-from-sky.html)

**b) Sheet erosion:** Sheet erosion (Fig. 3) is the removal of soil in thin layers by raindrop impact and shallow surface flow. This action called skimming and is prevalent in the agricultural land. It results in loss of the finest surface soil particles that contain most of the available nutrients and organic matter in the soil. Soil loss is so gradual that the erosion usually goes unnoticed, but the cumulative impact accounts for large soil losses. This type of soil erosion is mainly responsible for loss of soil productivities. Soils most vulnerable to sheet erosion are overgrazed and cultivated soils where there is little vegetation to protect and hold the soil. Early signs ofsheet erosion include bare areas, water puddling as soon as rain falls, visible grass roots, exposed tree roots, and exposed subsoil or stony soils.

Vegetation cover is vital to prevent sheet erosion because it protects the soil, impedes water flow and encourages water to infiltrate into the soil. The surface water flows that cause sheet erosion rarely flow for more than a few meters before concentrating into rills.

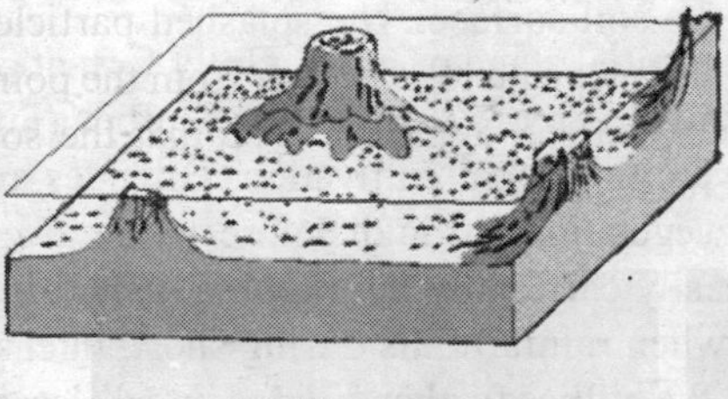

(a) Sheet erosion

**Fig. 3** Sheet erosion (modified from Kilders 2015)

**c)Rill erosion:** Rills formation is the intermittent process of transforming to gully erosion. The advance form of the rill is initial stage of gully formation. The rills are shallow drainage lines less than 30cm deep and 50 cm wide. They develop when surface water concentrates in depressions or low points through paddocks and erodes the soil. Rill erosion is common in bare agricultural land, particularly overgrazed land, and in freshly tilled soil where the soil structure has been loosened. The rills can usually be removed with farm machinery. Rill erosion is mostlyoccurs in alluvial soil and is quite frequent in Chambal river valley in India. The typical rill formation is shown in Fig. 4.

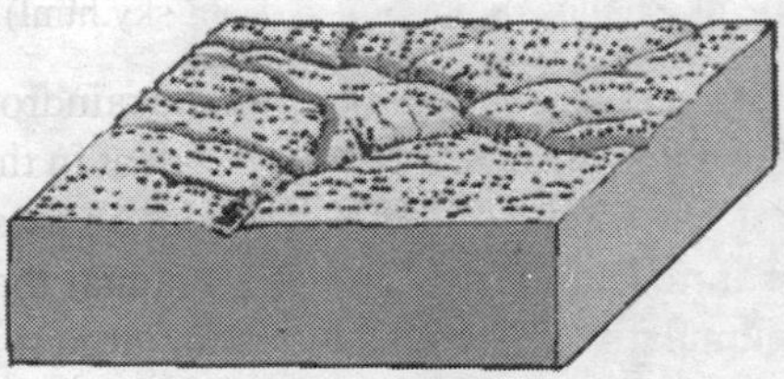

(b) Rill erosion

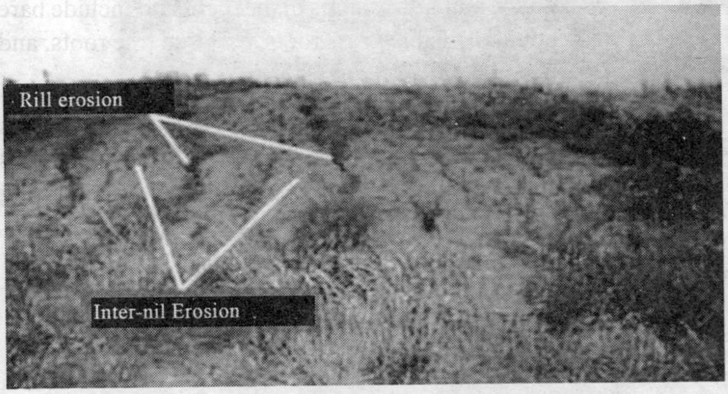


**Fig. 4:** Rill Erosion (*Source*: http://passel.unl.edu/UserFiles/Image/siteImages/Rillerosion-LGjpg.jpg

### d) Gully erosion

The advance stage of rills is transformed into initial stage of gully. Gully formation are initiated when the depth and width of the rill is more than 50 cm. Gullies (Fig. 5) are deeper channels that cannot be removed by normal cultivation. Hillsides are more prone to gullying when they cleared of vegetation, through deforestation, overgrazing or other means. The eroded soil is easily carried by the flowing water after being dislodged from the ground, normally when rainfall falls during short, intense storms. Depending upon the depth and width, the gullies further divided into 4 classes namely Gl, G2, G3 and G4 (Table 5). Gullies reduce the productivity of farmland where they incise into the land, and produce sediment that may clog downstream water bodies. Because of this, much effort are required to invested into the study of gullies within the scope of geomorphology, in the prevention of gully erosion, and in restoration of gullied landscapes. The total soil loss from gully formation and subsequent downstream river sedimentation can be sizable.

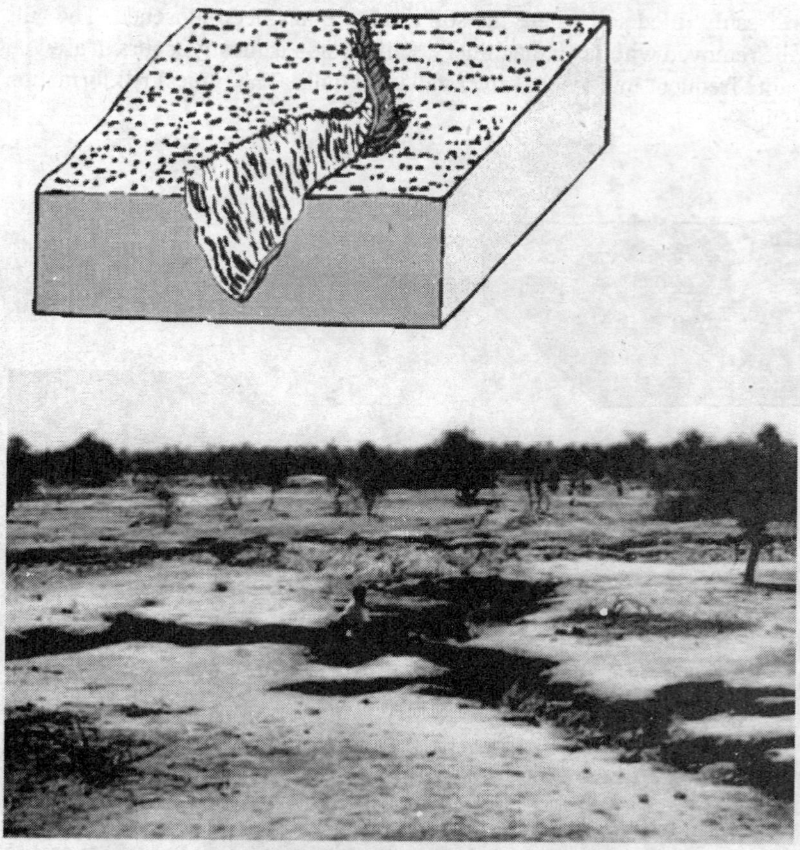

**Fig. 5:** Gully Erosion (*Source*: CSWCRTI, Dehradun)

**Table 5:** Classification of gully

| Particulars | Description of symbols of gully | | | |
|---|---|---|---|---|
| | G1 | G2 | G3 | G4 |
| Depth (m) | Up to 1.0 | 1.0-3.0 | 3.0-9.0 | > 9.0 |
| Width (m) | <18.0 | <18.0 | 18 | >18.0 |
| Side slope (%) | <6.0 | <6.0 | 6.0-12.0 | > 12.0 |

## e) Tunnel erosion

Tunnel erosion (Fig.6) occurs when surface water moves into and through dispersive sub soils. Dispersive soils are poorly structured so they erode easily when wet. The tunnel starts when surface water moves into the soil along cracks or channels or through rabbit burrows and old tree root cavities. Dispersive clays are the first to be removed by the water flow. As the space enlarges, more water can pour in and further erode the soil. As the tunnel expands, parts of the tunnel roof collapse leading to potholes and gullies. Indications of tunnel erosion include water seepage at the foot of a slope and fine sediment fans downhill of a tunnel outlet. This type of erosion is more frequent in foothills where elevation is between 500-750 meter.

**Fig. 6:** Tunnel Erosion (*Source*: http://www.ccma.vic.gov.au/soilhealth/photos.htm)

## f) Stream bank erosion

Stream bank erosion (Fig.7) occurs where streams begin cutting deeper and wider channels as a consequence of increased peak flows or the removal of local protective vegetation. Stream bank erosion is common along rivers, streams and drains where banks have been eroded, sloughed. This is quite prevalent in alluvial river and streams. Generally, stream bank erosion becomes a problem where development has limited the meandering nature of streams, where streams have been channelized, or where stream bank structures (like bridges, culverts, etc.) are located in places where they can actually cause damage to downstream areas. Stabilizing these areas can help protect watercourses from continued sedimentation, damage to adjacent land uses, control unwanted meander, and improvement of habitat for fish and wildlife.

(a) (b)

**Fig. 7:** Stream Bank Erosion (Source: http://www.flickr.com/photos/sarfrazh/3849502770/)

## g) Coastal erosion

The waves energy, geology and geomorphology are the three major factors that affect the coastal erosion. Waves are the cause of coastal erosion. Wave energy is the result the speed of the wind blowing over the surface of the sea, the length of fetch and the wind blowing time. The geology of the coastline also affects the rate of erosion. If the coast is made of a more resistant type of rock (granite), the erosion rate will be lower than if the coast is made of a less resistant type of rock (boulder clay). The geomorphology (or shape) of the coastline further affects the rate of erosion. Headlands cause wave refraction, making waves converge and combining their energy. Wider, shallower bays, allow waves to diverge, losing energy due to friction with the sea bed. A wider beach cause more wave energy to be lost due to friction before the waves can break. A narrower beach will mean that the breaking point of the waves is closer to the coastline, and less energy will have been lost due to friction before they break. Similarly, pebble beaches will allow more water to infiltrate and cause more wave energy to be lost due to friction, while sandy beaches allow less infiltration and cause less friction and so allow for a higher rate of erosion. If the beach gradient is steep, this will encourage steeper, higher-energy waves. The coastal erosion is a major concern for India as about 40% of the Indian coasts are subjected to severe erosion that has the potential to change the coast line. The typical coastal erosion is presented in Fig. 8.

**Fig. 8:** A view of eroded Goa beach (*Source*:http://www.concierge.com/ideas/beachisland/tours/)

## Prediction of Water Erosion

Soil losses from cultivated fields by sheet and rill erosion are predicting by Universal soil loss equation (USLE) developed by Wischmeir and Smith (1978). USLE considers all the variables as the soil erosion is influenced by several factors. Soil management by erosion losses reduce to permissible limits is done by information of USLE equation variables. In Europe and USA, this equation was successfully applied and validated in various fields. The equation is as follows

$A = R \times K \times LS \times C \times P$

Where

A = soil loss in metric tonnes per hectare (t $ha^{-1}$)

R = rainfall and runoff factor or rainfall erosivity factor

K= soil erodibility factor

LS = slope length and steepness factor (compared to reference values of 22.6 m and 9%), dimensionless.

C = crop management factor – a ratio which compares soil loss from an experimental field with that from a field with standard treatment, dimensionless.

**Rainfall erosivity factor (R)**: It measures the erosive force of rainfall and runoff. A heavy annual precipitation received in a number of gentle rains may cause little erosion, while a lower yearly rainfall descending in a few torrential downpours may result in severe damage. This account for the marked erosion recorded in semiarid regions. The R factor, sometimes called the rainfall erosion index, takes into account the erosive effects of storms. The total kinetic energy of each storm (related to intensity and total rainfall) plus the average rainfall during the 30-min period of greatest intensity is considered. The sum of the indexes for all storms occurring during a year provides an annual index. An average of indexes for several years is used in USLE. The following equation is used for calculation of R-factor

$$R = [1/100 \times (210.3 + 89 \log Ii)hi]\ I_{max}\ 30$$

Where, Ii = Intensity of rain in a given period (cm/hr); hi = Amount of precipitation in that period (cm); Imax30 = Maximum intensity during a 30 minutes period (cm/hr)

**Soil erodibility factor (K)**: Soil erodibility factor (K), is the rate of susceptibility of soil particles to erosion per unit of rain erosivity factor (R). This factor represents both susceptibility of soil to erosion and the rate of runoff. Although soil resistance to erosion depends in part on topographic position, slope steepness, cover and the amount of disturbances created by man (e.g. during the tillage), physical and chemical properties of soil are also the most important determinates. Erosivity varies with soil texture, aggregate stability, shear strength infiltration capacity and organic matter content. Values of K are given in Table 6. for different soil groups. The value of K (Foster et al., 1981) can be determined from nomogram (fig. 9) or it can be calculated by the following regression equation:

$$K = 2.8\times10^{-7}\ M^{1.14}\ (12\text{-}a) + 4.3\times10^{-3}\ (b\text{-}2) + 3.3\times10^{-3}\ (c\text{-}3)$$

Where, M = Particle size parameter (% silt + % very fine sand) (100-% clay) a = % organic matter, b = Soil structure code (very fine granular,1; medium or coarse granular,3; blocky, platy or massive, 4), c = Profile permeability class (rapid, 1; moderate rapid, 2; moderate, 3; slow to moderate 4; slow, 5; very slow, 6).

**Table. 6:** Soil erodibility factor (K) values for different soil groups and physiographic units

| SI No.. | Physiographic-soil unit | Texture (surface) | O.C. (%) | Soil erodibility factor (K) |
|---|---|---|---|---|
| 1 | Steep to very steep sloping Shivalik hills | sil/sl | 0.44-0.96 | 0.30-0.38 |
| 2 | Gentle to undulating piedmont plain | si | 0.52-0.62 | 0.32 0.35 |
| 3 | Very gentle to nearly level alluvial plain | sil/sl | 0.44-0.66 | 0.38-0.42 |
| 4 | Nearly level alluvial plain | l/sil | 0.22-0.59 | 0.34 0.44 |
| 5 | Flood plain | ls/sl | 0.24-0.37 | 0.44-0.48 |

si: Sandy loam, sil: silty loam, Is: loamy sand.

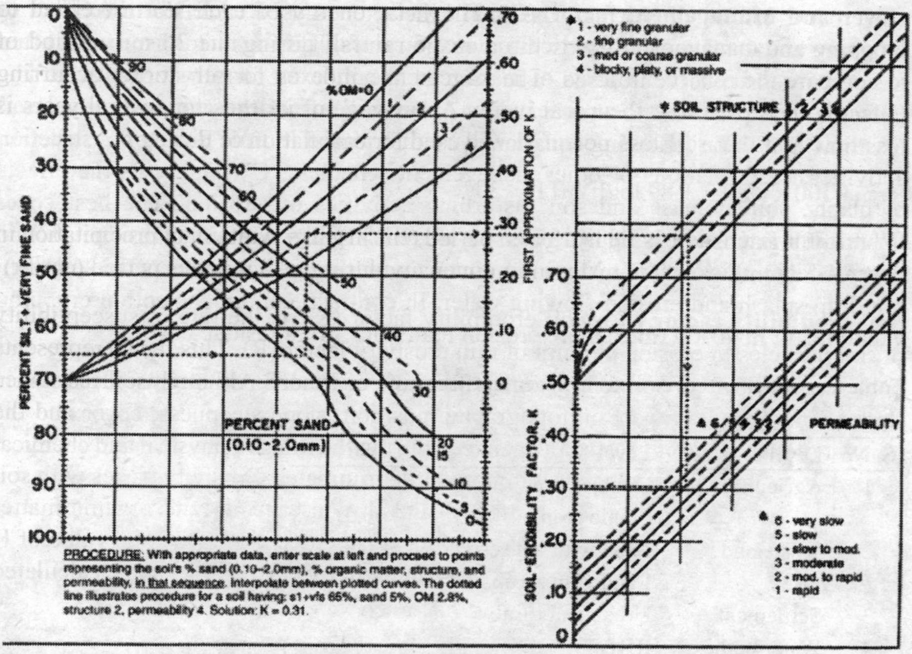


**Fig. 9:** Finding soil erodibility factor in Nomograph interpolating soil properties viz., percent si+vfs, % soil organic matter, soil structure and permeability.

**Slope length factor (L):** The length of the slope, on which the overland flow occurs, affects the rate of soil erosion. On larger slope lengths, there is a higher concentration of overland flow, and also a higher velocity of flow which triggers a higher rate of soil erosion. Zingg (1940) found that soil loss has a non-linear relationship with the land slope length, that is, soil loss α (Lp) m, where Lp is the actual slope length, and m is the ratio of the soil loss from the field plot length to the soil loss from the unit with a slope length of 22.13 meters. The slope length factor is determined by using the following formula.

L= (Lp)m / 22.13

Where, Lp = The actual unbroken length of the slope (meters) measured up to the point where the overland flow terminates, and 'm' is an exponent which is equal to 0.5 for slopes ≥ 5%, 0.4 for 4%, 0.3 for 3%, and 0.2 for 1%. The values for m are 0.5 for 5 to 10% slope and 0.6 for >10% slope.

**Slope gradient factor (S)**: On steep slope the flow velocity is high, which causes showering and cutting of soil. Also soil erosion due to splash is high, because splashed particles on steep slopes are thrown to larger distances down the slope on an inclined plane and the damage due to raindrop impact is greater on soil crust. The slope gradient factor(S) expresses the ratio of soil loss from a plot of known slope to soil loss from a unit plot under identical conditions.

**Cover and Management factor (C)**: The C-factor is used to reflect the effect of cropping and management practices on erosion rates. It is the factor used most often to compare the relative impacts of management option on conservation plants. The C-factor indicates how the conservation, plan will affect the average annual soil loss how and that soil-loss potential will be distributed in time during construction activities, crop relations or other management schemes. "C" represents the effects of plants, soil biomass and soil disturbing activities on erosion. The best cover for minimum erosion is dense forest or tall, thick grass with dead residue ground cover 4-5 cm. thick. Plant material in contact with the surface protects the soil from raindrop splash and erosion flowing water. In contrast, continuous cotton cropping would result in 40-60 times more erosion than thick grass residue cover.

**Table 7:** Values of cover and management factor(C) under different land use/ land covers classes

| S. No | Level -1 | Level- 2 | C- factor |
|---|---|---|---|
| 1 | Agriculture | Double crop | 0.4 |
| | | Single crop | 0.5 |
| 2 | Wasteland | Land with out scrub | 0.8 |
| | | Land with scrub | 0.95 |
| 3 | Settlement | Urban and rural | 0.2 |
| 4 | Water bodies | River | 0 |
| | | Reservoir/stream | 0 |
| 5 | Plantation | Plantation | 0.1 |

**Conservation practice factor (P)**: The P-factor reflects the impact of support practices on the average annual erosion rate. It is the ratio of soil loss with contouring and / or strip cropping to that with straight row forming up-and-down slope. As with the other factors, the P-factor differentiates between cropland and rangeland or permanent pasture. Both option allow for terracing or contouring, but the cropland options contains a strip cropping routine whereas the rangeland/permanent pasture option contains another mechanical disturbance routine.

**Table 8:** Conservation Practice Factor (P) on Different Slope Gradients

| S. No | Slope Percentage | P factor |
|---|---|---|
| 1 | 1.0 – 2.0 | 0.6 |
| 2 | 3.0 – 5.0 | 0.5 |
| 3 | 6.0 – 8.0 | 0.5 |
| 4 | 8.0 – 12.0 | 0.6 |
| 5 | 13.0 – 16.0 | 0.7 |
| 6 | 17.0 – 20.0 | 0.8 |
| 7 | 21.0 – 25.0 | 0.9 |

## 8.1.2. Wind Erosion

Wind erosion is the detachment, transportation and redeposition of soil particles by wind. A sparse or absent vegetative cover, a loose, dry and smooth soil surface, large fields and strong winds all increase the risk of wind erosion. Air movement must attain a certain velocity (with enough speed to generate visible movement of particles at the soil level) before it can generate deflation and transport of particles. Winds with velocities of less than 12-19 km/hr seldom impart sufficient energy at the soil surface to dislodge and put into motion sand-sized particles. Drifting of highly erosive soil usually starts when the wind attains a forward velocity of 25-30 km/hr. Wind erosion tends to occur mostly in low rainfall areas when soil moisture content is at wilting point or below, but all drought-stricken soils are at risk. Often the only evidence of wind erosion is an atmospheric haze of dust comprising fine mineral and organic soil particles that contain most of the soil nutrients. Actions to minimize wind erosion include improving soil structure so wind cannot lift the heavier soil aggregates; retaining vegetative cover to reduce wind speed at the ground surface; and planting windbreaks to reduce wind speed. Also, be ready for severe wind erosion seasons which tend to be the summers following dry autumns and winters. The most familiar result of wind erosion is the loss of topsoil and nutrients, which reduces the soil's ability to produce crops. Topsoil loss can be seen as rocky or gravelly knolls,thin soils mixed with lighter coloured subsoil, or the presence of calcium carbonate in surface soils.

Soil productivity is affected by wind erosion in various ways. Areas of erosion and deposition within a field increase the variation in soil characteristics, requiring more costly and less efficient soil management practices. Wind removes the smaller clay particles and organic matter from the soil while coarser materials are left behind. The continued loss of fine particles reduces soil quality. In shallow soils and soils with a hardpan layer, wind erosion also results in decreased root zone depth and water-holding capacity. Such changes may occur slowly and go unnoticed for many years especially if mixing by tillage masks the effects of wind erosion.

## Process of wind erosion

The process of wind erosion comprises of three basic stages namely saltation, suspension and surface creep. Fig. 10 describes the process.

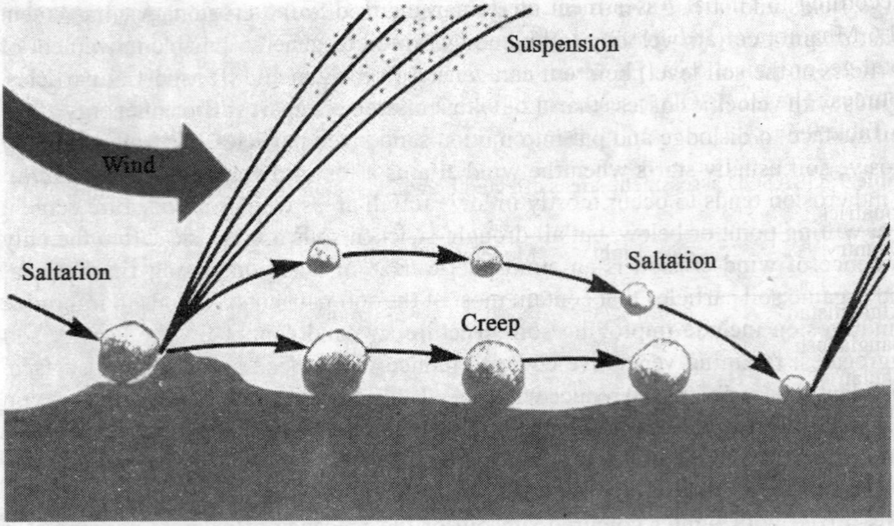


**Fig. 10:** Process of wind erosion (Source:http://www.weru.ksu.edu/new_weru/images/CreepSaltSusp.jpg)

(i) **Saltation**: Saltation occurs when the wind lifts larger particles off the ground for short distances, leading to sand drifts. Fine and medium sand-sized particles are lifted a short distance into the air, dislodging more soil as they fall back to the ground.

(ii) **Suspension**: Suspension occurs when the wind lifts finer particles into the air leading to dust storms. Very fine soil particles are lifted from the surface by the impact of saltation and carried high into the air, remaining suspended in air for long distances.

(iii) **Surface Creep**: The movement of large soil particles along the surface of the soil after being loosened by the impact of saltating particles.

## Extent of wind erosion

Several factors, other than the wind velocity itself, contribute to wind erosion. These fall into two main groups of closely interrelated elements: those inherent in the properties of the soil per se and those associated with soil cover. A rough soil structure, especially at the surface, effectively reduces the movement of soil particles. Arid regions, however, are dominated by smooth, pulverized and structure less top soils. Soil texture also influences soil erodibility; soils of fine texture are, for example, particularly susceptible to wind erosion.

Measurements of dust in the air up to three metres above the soil surface at Jodhpur, India, showed that on a stormy day the amount of dust blowing varied between 50 and 420 kg/ha. In the Jaisalmer region of India, where wind speeds generally are higher, average soil loss of 511 kg/ha was recorded.

According to Global Assessment of Human-induced Soil Degradation (GLASOD), 21.6 Mha area of Indian soil is affected by wind erosion, which account for 6% of total geographical area. However, area varied from 12.9 to 38.7 Mha from various sources. The GLASOD assessment of extent of area under wind erosion is presented in Table 9.

**Table 9:** GLASOD assessment: areas affected by wind erosion (Unit: 1000 ha) – South Asian Countries

| Country | Light | Moderate | Strong | Total | Total as percent of land |
|---|---|---|---|---|---|
| Afghanistan | 1 873 | 0 | 209 | 2 082 | 5% |
| Bangladesh | 0 | 0 | 0 | 0 | 0% |
| Bhutan | 0 | 0 | 0 | 0 | 0% |
| India | 0 | 1 754 | 9 042 | 10 796 | 6% |
| Iran | 6 559 | 25 730 | 3 085 | 35 374 | 60% |
| Nepal | 0 | 0 | 0 | 0 | 0% |
| Pakistan | 3 998 | 6 742 | 0 | 10 740 | 42% |
| Sri Lanka | 0 | | 0 | 0 | 0% |
| India, dry region | 0 | 1 754 | 9 042 | 10 796 | - |
| India humid region | 0 | 0 | 0 | 0 | - |
| Dry zone | 12 430 | 34 225 | 12 337 | 58 992 | 39% |
| Humid zone | 0 | 0 | 0 | 0 | 0% |
| Region | 12 430 | 34 225 | 12 337 | 58 992 | 18% |

## Prediction of wind erosion

Wind erosion equation (WEQ): E= f(IKCLV)

where E is the estimated average annual soil loss expressed in tons per acre per year;

I is the soil erodibility;

K is the soil ridge roughness factor;

C is the climatic factor;

L is the equivalent unsheltered distance across the field along the prevailing wind erosion direction; and

V is the equivalent vegetative cover.

## Agricultural and environmental consequences of soil erosion

The valuable top soil is the most productive part of soil profile for agricultural purposes is removed by soil erosion. Production cost will be high and yield will decrease resulted by the loss of this top soil. Gullies and rills make the cultivation of paddocks impossible and these are produced by the erosion when the top soil is removed.

- Top soil that is rich in organic matter and nutrients is removed
- The depth of soil that is available for water storage for crop growth and for rooting is reduced.
- Increase runoff by reducing the infiltration of water into soil.
- Need to repeat field operations and loss of seeds.
- Wind erosion wraps the young plants with sand
- Extra cultivations are needed to level out the eroded surfaces
- Sediment deposition onto roads, in roadside drains and on neighbouring properties.
- Excess inputs of phosphorus, pesticides and nitrogen damage the quality of lakes, coastal water and watercourses.
- Spawning grounds of fish damage by sediments in rivers.
- Greater flood hazard downstream is caused by the deposition of sediment and increased runoff.

## 8.2. Solutions to soil degradation

### 8.2.1. Reduced forestation

Avoiding or reducing deforestation completely is an uphill task. However, deforestation can be cut down and this can create an impressive method of reshaping and restoring forests and vegetation cover. As populations grow, individuals can be sensitized and educated regarding sustainable forest management and reforestation efforts. Also, preserving the integrity of guarded areas can reduce demonstration. Thus, there is a necessity for individuals all over the world to respect forest cover and reduce some of the human-driven actions that encourage logging. With the avoidance of deforestation, soil's ability to naturally regenerate can be restored. Governments, international organizations, and other environmental stakeholders want to ensure there are appropriate measures for making zero net deforestation a reality to inhibit Soil degradation.

### 8.2.2. Curb intensive farming

Tilling, multiple harvests, and agrochemicals have boosted crop yields at the expense of sustainability. Therefore, sustainable farming practicesshould be adopted

### 8.2.3. Land reclamation

The outcomes of erosion and quality decline are widely irreversible. Still, soil organic matter and plant nutrients could be replenished. To restore the lost soil mineral matter and organic content, it would need what is known as land reclamation. Though, land reclamation encompasses activities centered towards restoring the previous organic matter and soil's vital minerals. This includes activities such as the addition of plant residues to degraded soils and improving range management. Salinized soils could be restored by salt level correction reclamation projects and salinity control. One of the simplest but forgotten methods of land reclamation is the planting of vegetation such as trees, crops, and flowers over the affected soils. Plants act as protective covers as they are useful at making the soil stronger by stabilizing the land surface.

### 8.2.4. Preventing salinization

The costs of preventing salinization are incredibly cheaper than the reclamation projects in some salinized areas. Thus, actions such as reducing irrigation, planting salt-tolerant crops, and improving irrigation efficiency will have high payoffs because the inputs and the labor-demanding aspects associated with reclamation projects are zero. Preventing salinization in the first place is an environmentally friendly means of offering a solution to Soil degradation.

### 8.2.5. Establishing windbreaks

Artificial and natural windbreaks, for example, shrubs, reduce the erosion effects of wind. Plants have the added benefit of "anchoring" the soil, reducing the effects of erosion from water.

### 8.2.6. Soil Conservation

Soil is to be conserved and protected from erosion to maintain /enhance and sustain the agricultural productivity in tandem with the prevailing agro-socio-economic practices in the respective region.

**Objectives of soil conservation**

1. To sustain the production from natural resources to meet the basic requirements of food, shelter and clothing of growing population To preserve topsoil to reduce deterioration in soil fertility and the water holding capacity, thus sustaining productivity.
2. To check the formation of rills and gullies due to soil erosion in the field, which adversely affect the productivity.
3. To increase the groundwater recharge, by sustaining the soil moisture retention capacity of the soil.
4. To maintain the land productivity and prevent shrinkage of arable area.
5. To reduce the dredging work due to sedimentation in creeks, rivers, lakes and reservoirs.
6. To protect water bodies from non-point source pollution.
7. To minimize the flooding risk that affects the sustainability and livelihoods of humans, animals and plants.
8. To control the deterioration of ecosystem due to soil loss, which leads to interruption of nutrient cycle, loss of soil fertility, extinction of flora and fauna and soil erosion etc. ultimately resulting in biological impoverishment and human sufferings.
9. To facilitate environmental system that affects the plant growth and rejuvenation of forests.

Soil erosion is prevented by several agronomic and biological properties. Crop rotations, agro-forestry and soil synthetic conditioners, reduced tillage, riparian buffers, cover crops, vegetative filter strips, residue, canopy cover management and no-till are important among these. This Chapter discusses the importance of (1) soil amendments (e.g. manures) (2) soil conditioners (e.g., polymers) (3) crop residues (4) cover crops for soil erosion reduction. There are differences among these biological practices in relation to their mechanisms of erosion control. Biological measures such as buffers or thin films (e.g., conditioners), conditioner application in direct contact with the soil surface, crop residues using manure protect the soil from erosion. The raindrop is intercepted above the soil surface by the protective effect of canopy cover and standing vegetation reduces soil erosion. Growing vegetation produce the mulching effect.

## Soil Conservation measures

Contour cultivation, crop choice, manuring, mulching, crop rotation, intercropping, strip cropping and mixed cropping are included in soil conservation measures.

### 1. Crop management

Soil fertility is improved and wind and water reduce soil erosion by good crop management practices. Keep soil covered is fundamental principle of conservation agriculture. Soil protection from erosion by leaving crop residues on soil surface after harvesting is also helpful approach.

a) **Crop selection:** If the gap is too long between harvesting one crop and sowing of the next crop than the additional cover crops may be required. The stability of the conservation agriculture system is increased by cover crops and erosion impacts are reducing by the improvement of soil properties and this biodiversity in the agro-ecosystem are promoted for their capacity. The more effective crops in soil erosion are perennials than annual crops. The most effective are sugar cane, fodder grasses, sweet potatoes and tea.

b) **Early planting:** The protection of the ground against raindrop impact is ensured by the crop shoots from the ground within one or two weeks after the onset of the rains.

c) **Crop rotation:** The practice of growing a series of dissimilar types of crops in the same space in sequential seasons is crop sequencing or crop rotation for benefits such as such as avoiding pathogen and pest buildup that occurs when one species is continuously cropped. Soil nutrient depletion is avoided by the crop rotation that balance the nutrient demand of various crops. The replenishment of nitrogen with the use of green manure and legumes in sequence with cereals and other crops is a traditional component of crop rotation. Soil structure and fertility by alternating shallow-rooted and deep-rooted plants can also be improved by crop rotation. The multi-species cover crops between commercial crops is also another technique. The advantages of intensive farming with polyculture and continuous cover are combined by these techniques. So, soil fertility, reduction of diseases and pests, addition of humus and control of erosion is ensured by crop rotation.

d) **Inter-cropping:** The impact of raindrops is reduced with the soil cover by the fast-growing legumes such as cowpeas and beans early in the season before a canopy is developed by cotton or maize to shield the soil.

e) **Cover cropping:** The practice of growing crops to cover the soil surface to reduce wind and water erosion is called cover cropping. This practice creates a favorable habitat for microorganisms by regulating the soil heat and temperature. These also sources of organic matter in soil as the fallen are decomposed.

f) **Strip cropping:** This is the practice of growing different crops in alternate strips in the same field. It helps minimizing wind and water erosion. Crop rotation and minimum tillage in addition to contour strip cropping has proven to be best method to conserve soil and water.

## 2. Soil management

Soil conditions are often changed by the inappropriate land use practices which ultimately result in soil erosion. Optimum soil management aims to provide favorable conditions for plant growth through improved soil nutrient availability and aggregation. Optimum soil management practices improve infiltration of water and improve soil capacity to hold water and in result reduce runoff and erosion.

a) **Use appropriate tillage practices:** Optimum soil physical conditions for better crop production are the main objectives of tillage. It also ensures timely seedbed preparation, planting and weed control.

Tillage practices should be adopted by keeping in mind that;

- Soil is neither too fine nor powdery; and
- It breaks up the hardpan if necessary.

The main tillage methods are slash and burn, hand hoeing, ploughing and harrowing, conservation or minimum tillage, deep tillage.

b) **Applying organic manures and mineral fertilizers:** Application of manure and fertilizers provide essential plant nutrients in the soil for better crop growth. The crops with fast growth cover the soil quickly and give higher yields. Essential plant nutrients such as nitrogen, phosphorus, potassium, and sometimes Sulphur required by plants are provided by inorganic fertilizers. There is no substitute of inorganic fertilizers therefore integrated use of organic and inorganic fertilizers should be adopted. Farmyard manure, green manure and composts etc. are the main sources of organic fertilizers.

c) **Mulching and the use of crop residues:** Spreading on the bare soil surface or placement of plant materials such as dry grass, straw, dry leaves, banana leaves, sugar cane trash, and other crop residues around the stem of the plants is helpful in controlling soil erosion and moisture conservation.

d) **Agro-forestry:** Planting of trees or shrubs or protecting the naturally sustaining trees is called agroforestry. Trees decrease the magnitude of splash erosion by reducing the raindrops impacts on the soil. They regulate soil temperature by shading the soil thus reducing the water evaporation. They also minimize the wind erosion by acting as wind breaks. They also play important role in nutrient recycling in the deep soil; leguminous trees fix nitrogen that benefits food crops.

e) **Contour farming practices:** Cultivation across the slope rather than up and down is called contour farming. Soil loss as much as 50% has been reported to be reduced by contour farming on gentle. The main objective of contour ridges in semi-arid areas is water harvesting and in humid areas potato cultivation. Plant residues are placed in lines along the contour for construction of trash-lines. These trash-lines slow down the runoff and trap the eroded soil. Grass barrier strips of Napier or other fodder grasses are planted along the contour.

## 3. Physical soil conservation measures

Physical soil conservation structures are permanent features made of Earth, stones or masonry, designed to protect the soil from uncontrolled runoff and erosion and retain water where needed.

- Selection and design of structures depend on:
- Climate and the need to retain or discharge the runoff
- Farm sizes
- Soil characteristics (texture, drainage, and depth)
- Availability of an outlet or waterway
- Labour availability and cost
- Adequacy of existing agronomic or vegetative conservation measures. Below are some of the physical conservation measures:

a) **Cut-off drains:** Cut-off drains are made across a slope for intercepting the surface runoff and carrying it safely to an outlet such as a canal or stream. Their main purpose is the protection of cultivated land, compounds, and roads from uncontrolled runoff, and to divert water from gully heads.

b) **Retention ditches:** These are made along the contours to capture and retain incoming runoff water and hold it until it seeps into the ground. They are alternate to cut-off drains when there is no channel to discharge the water nearby. Sometimes these are for water harvesting in semiarid areas.

c) **Infiltration ditches:** The structure used to harvest water from roads or other sources of runoff is infiltration ditches. They comprise dug along the contour, upslope from a crop field and a ditch of 0.7-1.5m deep. Water is blocked at the other end when it is diverted from the roadside into ditch and seep into soil after it is being trapped.

d) **Water retaining pits:** Water-retaining pits allow runoff water to seep into soil after by trapping the water. The runoff normally occurs into a series of pits which are dug into ground. Banks around the pits are made by the soil from the pit. Excessive water carry from one pit to next by furrows. The amount of runoff determines the size of pit and its typical size is 2 m square and 1 m deep.

e) **Broad beds and furrows:** The runoff water is diverted into field furrows (30 cm wide and 30 cm deep) in a broad bed and furrow system. The lower end of field furrows is blocked. The water backs up into the head furrow after the filling of one furrow and flows into the next field furrow. Crops are grown on the broad bed furrows of about 170 cm wide between the fields.

## 8.2.7. Water conservation

Water loss to the atmosphere by two natural processes that are evaporation and transpiration. Water loss from soil surface by evaporation and from leaf surface by transpiration. Evaporation is the combination of evaporation and transpiration. Physical characteristics of soil and water affect the evapotranspiration as well as density, type and rate of plant growth in the field. Some other factors like wind speed, solar radiation, rooting depth, season of year, availability of soil moisture and land surface characteristics. Evapotranspiration is mainly dependent on the solar energy that is utilized to vaporize the water when the water is available.

### Water conservation

i) **Planting wind breaks:** Wind velocity is reduced by windbreaks. Reduced wind velocity reduces the evapo-transpiration rate in the area that is directly downstream of the barrier. 50 % porosity should present in wind breaks to slowed down the wind that passes through windbreaks as well as it is not deflected over the wind break. The direction of predominant wind is necessary to plant the wind breaks. A diversity of adapted species decreases the soil erosion by stabilizing the slope, decreases the risk of pest and disease problems and it provide shade for livestock. So, these provide better results. Minimized competition for water around the surrounding crops and irrigation would not be needed as much by drought tolerant species.

ii) **Keeping plant residues on the field:** Wind speed over the soil surface is reduced by crop residues that also decreases evapotranspiration; soil temperature and moisture is also reduced as it provides shades by conserving soil moisture. So, crop residues act as wind breaks over the soil surface. The residues that are not break down or decompose readily are best. Crop residues should be partially removed for fodder or fuel use and if possible, it should be left standing in the field after harvest. To retain much of the plant residues in the field, the stubbles should be cut at high point in the plant during harvesting time. The soil loss and impact of raindrop also reduces by keeping crop residues in the field.

iii) **Selecting water conserving plant species:** Too much high water utilizing crops like alfalfa cannot conserve water in the landscape. Generally, low biomass producing annual legumes like chickpeas and lentils are the crops that transpire the least amount of water.

## Water harvesting and conservation

Water harvesting techniques gather water from an area termed the 'catchment area' and channel it to the cropping area or wherever it is required. Conservation techniques conserve water within the biomass and the soil by reducing run-off and keeping the water where it falls, as much as possible.

In many farming areas, readily available water is in short supply. Although the total annual rainfall in an area may be enough to sustain farm needs, it is often distributed very unevenly so that long dry periods are interspersed with periods of intense rainfall. In many cases, a crop is unable to use a high proportion of this water, as much of it is lost through run off or leaching.This may also cause soil erosion and loss of soil nutrients.

## Principles of water harvesting and conservation

In deciding which techniques to use to make more efficient use of the available water, it is important to consider how crops receive orlosewater.Cropsreceive waterthrough rainfall,irrigationandstoredsoilwater. Theyloseitthroughrunoff, evaporation and drainage. Some key principles on effective water management are discussed hereunder.

a) **Increasing rainwater efficiency:** In many climates, rainfall is distributed unevenly in intense downpours that cannot be readily used by a crop. Storage techniques (such as external catchments or roof top collection) increase the availability of water in the drier seasons.They also harvest water from a wider area making more water available to the crop. Measures can also be taken to avoid the rainwater running off the surface during intense rainfall.

b) **Effective use of soil water reserves:** The soil stores water from rainfall providing a reserve that is available to the crop. How much water is available depends on the soil type and the rooting system of the crop. Sandy soils hold much less water than clay or silt soils, so crops will require watering more often. Deeper rooting crops, such as grasses or cereals will exploit soil water reserves more effectively than shallower rooting crops such as vegetable crops and therefore can be grown in drier periods. Good cultivation practices (e.g. not ploughing too deep or when the soil is wet) that result in a soft, friable soil will also promote deep rooting and efficient use of soil water reserves.

c) **Measures to avoid run off:** Run off is where water is not absorbed by the soil but runs across the surface away from where the crop can use it. Structures such as contour schemes, terracing, pits and bunds can reduce run-off. Run off is more likely to occur on silty or clay soils where the surface has been subjected to intense rainfall then baked in the sun to form a crust or cap. Adding mulch to break up the intensity of rainfall, or adding manure, compost or incorporating green manure residues will reduce the tendency of the soil to form a crust.

d) **Avoid wasting water through evaporation:** Water that evaporates directly from bare soil is wasteful as it is not being used for productive plant growth. It is desirable to maintain full ground cover for as much of the time as practically possible. Applying mulch to the soil will also reduce evaporation considerably. Use of drip irrigation and irrigating in the evening will also reduce the amount of water lost through evaporation.

e) **Reducing water losses through drainage:** When water drains out of the soil, not only is it wasted but essential mobile nutrients such as nitrogen are also washed out. This is more of a problem on light sandy soils. Adding organic matter in the form of compost, manures or plant residues will eventually increase the amount of water a soil can retain, but this will only have an effect if it is added over a longer period of years. Most drainage occurs during the heavy rains, especially if the soil is left bare. Growing a cash or cover crop during this period reduces these losses, as the roots lift water and nutrients back from deeper to shallower soil profiles.

f) **Planning of irrigation:** Irrigation is one way of supplementing water from rainfall and soil reserves, but can waste large amounts of water if not used carefully. A key way of making most of the water supply is to only irrigate when necessary. Many farmers irrigate the fields on regular basis whether the crop needs it or not. If water is scarce, irrigation should be restricted to the most critical periods such as germination and fruit set. Drip irrigation makes much better use of water than overhead systems as it is targeted to the roots rather than sprayed up into the air.

**Water harvesting techniques** Different water harvesting techniques and adopted depending on location specific situations.

**i) Contour farming:** Contour farming refers to field activities such as ploughing and furrowing that are carried out along contours rather than up and down the slope. They conserve water by reducing surface run off and encouraging infiltration of water into the crop area. For all contour systems the first step is to determine a contour guideline. This can be done using the "A frame method" (see appendix). From this, parallel contour guidelines can be drawn up. A number of water harvesting techniques are based along contours including: contour ploughing; contour ridges; stone lines; grass strips and terraces. The technique used depends on the steepness of the slope, soil type, conditions, crops grown etc.

a) **Contour Ploughing:** Any ploughing on a slope should be carried out along the contours rather than up and down as this reduces run off and soil erosion and increases moisture retention. Contour ploughing can be practised on any slope with a gradient less than 10%. On steeper slopes it should be combined with other measures such as terracing, strip cropping. It is not always carried out in practise because the shape and topography of the field may be considered as more important factor in determining the direction of ploughing. It is important to lay out contours properly otherwise they may channel the water and increase run off.

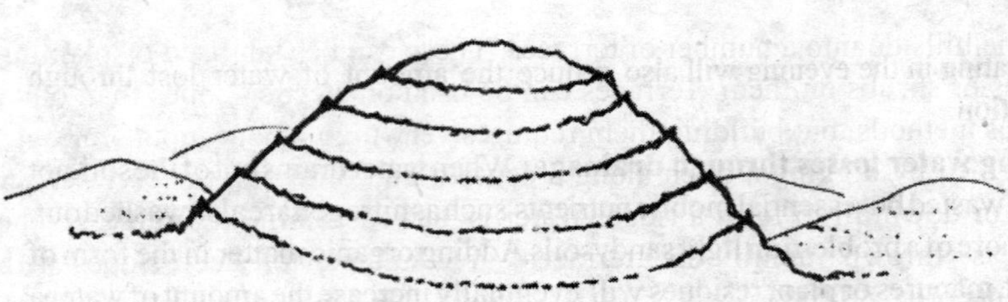

Fig. 11: Contour Ploughing

b) **Contour Furrows:** Contour furrows are small earthen banks that run along a contour. A furrow is dug next to each bank on the upper side of the slope. The distance between the ridges varies between 1 - 2 m depending on the rainfall and the slope. The aim of contour furrows is to concentrate moisture into the ridge and furrow area where the crops are planted by trapping run off water from the catchment area between them. This also decreases the risk of erosion. Plants with higher water requirements, such as peas or beans, can be planted on the higher side of the furrow whereas crops requiring less water, such as sorghum or millet, can be planted on the ridges.

Contour furrows are suitable for areas with lower rainfall (350 – 700 mm). However, the amount of water harvested is limited, so they are not suitable for very dry areas. Extreme rainfall may cause the ditches to over flow and break. This is more likely to occur on heavier soils with a lower infiltration rate, or on steeper slopes. The risk can be reduced by building higher ridges, although this increases the labour requirement.

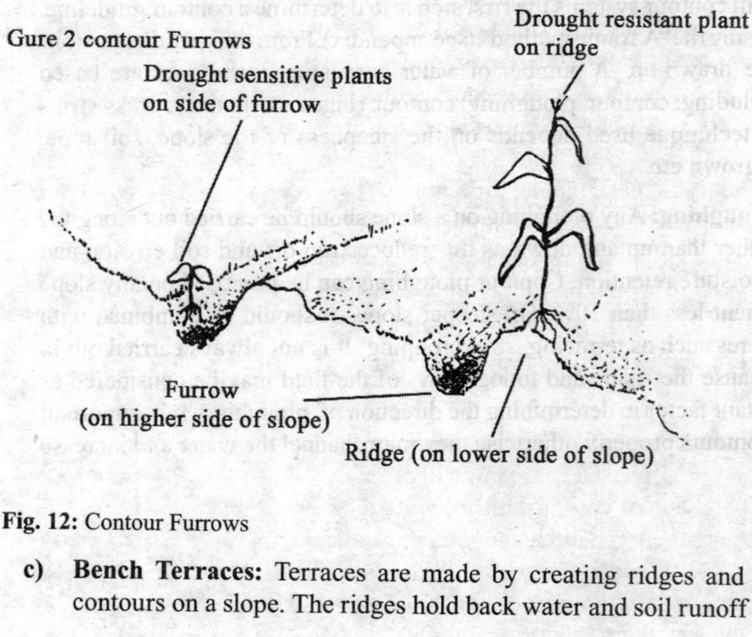


Fig. 12: Contour Furrows

c) **Bench Terraces:** Terraces are made by creating ridges and furrows along contours on a slope. The ridges hold back water and soil runoff and eventually turn the hillside into a number of terraces. These can be stabilised by planting grasses or shrubs on them. Terraces can be used on steeper slopes than other contour methods, but building them requires very high labour input. Terraces are formed by digging a ditch along a contour and throwing the earth either uphill or downhill to form a ridge. Water is stored behind the ridge. If the earth is thrown uphill the terraces are suitable for steep slopes with gradients of 30-55%.

**d) Grass Strips:** Strips of grass (up to 1m wide) planted along a contour can reduce soil erosion and runoff. Silt builds up in front of the strip and over time benches are formed. On gentle slopes the strips should be widely spaced (20-30m apart), and on steeper slopes strips are narrowly spaced (10-15m apart). The grass needs to be trimmed regularly, to prevent it competing with crops. Many grass varieties can be used, depending on what is locally available. For example, Vetiver, Napier, Guinea and Guatemale grass. Alternatively a local Veld grass can be used. The strips need to be maintained to prevent the grass from spreading and becoming a weed problem or becoming a refuge for rodents and other pests. Grass strips are most likely to be used in areas where fodder or mulch is also needed. They are not suitable for steep slopes or in very dry areas since grasses will compete with the crop. They can also be used in conjunction with other water harvesting techniques: grass strips can be planted along ditches to stabilize them, or on the rises of bench terraces to prevent erosion.

**e) Stone Lines:** Stone lines running along the contour are one of the simplest contour techniques to design and construct.The lines of stones form a semi-permeable barrier that slows the speed of run off so that spread of water over the field and infiltration is increased, and soil erosion reduced. The lines are constructed by making a shallow foundation trench along the contour. Larger stones are then put on the down slope side of the trench. Smaller stones are used to build the rest of the bund. The stone lines can be reinforced with earth, or crop residues to make them more stable. When it rains, soil builds up on the upslope side of the line, and over time a natural terrace is formed.

The stone lines are spaced 15-30m apart or a shorter distance on steeper slopes. Stone lines are suitable on gentle slopes (0.5–3%) in areas with annual rainfall of 350- 700mm. They are often used to rehabilitate eroded and abandoned land by trapping silt and are popular in dry stony areas. However, they may provide a refuge for rodents and other pests.

**ii) Retention Ditches:** Retention ditches work on a similar principle to contour furrows but on a larger scale. They are large ditches, designed to catch and retain all incoming run-off and hold it until it infiltrates into the ground, increasing the supply of water to crops planted in the ditch and reducing soil erosion. They vary from 0.3-0.6 m deep and 0.5-1.0 m wide. They are usually used on flat land where they may be spaced at 20 m or on gentle slopes the spacing can be decreased to 10-15m.When constructing the ditches, the soil is thrown to the lower side to form an embankment that prevents soil from falling back in. This structure can be stabilised further by planting grass on it. On soils with lower infiltration rate, or on slopes, the ends can be left open to allow excess water to drain out. Retention ditches are commonly used in semi-arid areas for growing crops that have high water requirements, such as bananas. They should be used on lighter, free draining soils that are deep, stable and not prone to landslides.

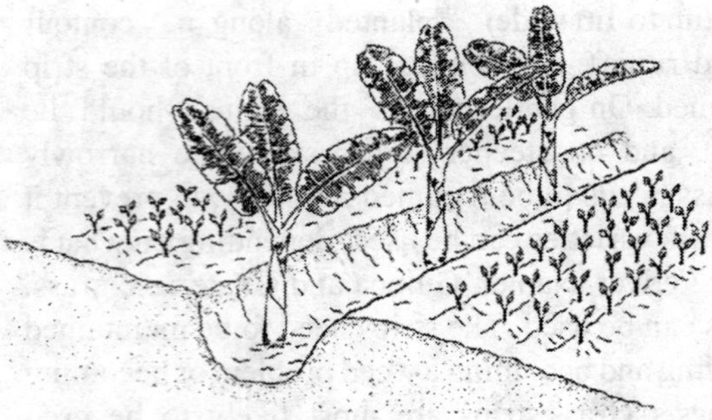

**Fig. 13 :** Retention Ditches

**iii) Planting pits**: Planting pits are a very simple form of freestanding water harvesting structure that is easy to construct. They consist of small pits in which individual or small groups of plants are sown. The pits catch run off and concentrate soil moisture around the roots. Normally the pits are 10-30 cm in diameter and 5 –15 cm deep and are spaced about 1 m apart. The earth removed from the holes is piled in a half moon shape along the lowest edge of the pit. Before planting, compost or manure is added to the pit to improve soil fertility and structure. Planting pits are particularly successful in areas of low rainfall (350–750 mm) and are suitable for crops with low water demand such as sorghum or millet. They are more suitable for heavier clay soils, which tend to form a cap and have poor infiltration. As digging the pit reduces the depth of soil, they are not suitable for shallower soils. They are only suitable for gentle slopes (less than 2% gradient).

**iv) Earth basins:** Earth basins are designed to collect and hold rainfall and are easy to construct by hand. They are square or diamond shaped basins with earth ridges on all sides. Runoff water is channelled to the lowest point and stored in an infiltration pit. The lowest point of the basin might be located in one of the corners (on sloping land) or in the middle (on flat land). Earth basins are usually used for fruit crops and the seedling is planted in or on the side of the infiltration pit.

The size of the basin depends on local rainfall and the water requirements of the trees. They are larger on flat land and smaller on sloping land. They are usually 1-2 m long, though sometimes basins of up to 30m are constructed. Grass can be planted on the bunds for reinforcement. Manure and compost can be added to the basin to improve fertility and water-holding capacity. Earth basins are suitable in arid and semi-arid areas, with annual rainfall amounts of 150mm and above. Soils should be deep, preferably at least 1.5 - 2m to ensure enough water holding capacity. The slope can be from flat up to about 5%. If earth basins are constructed on steep slopes they should be small.

**Fig. 14:** Earthen Basins

**v) Cover crops/ green manures:** Cover crops are grown to protect the soil from leaching, erosion and to improve soil fertility.They build up organic matter in the soil, improve soil structure, suppress weed growth and increase soil fertility through nitrogen fixation.They also, reduce fluctuations in temperature and improve soil moisture. Legumes, such as beans and peas, or grasses are often used.They cover the ground surface between a widely spaced perennial crop, such as young fruit trees, coffee, cacao and oil palms or between rows of grain crops such as maize. Cover crops are often combined with mulching.

Cover crops can be a source of food, fodder and mulch and may provide some cash-income. However, they may also provide a refuge for rodents and pests. The cover crop should be of a slow growing variety to minimize competition for water and nutrients with the main crop. It should be planted as soon as possible after tillage to be fully beneficial. This can be done at the same time as sowing the main crop, or after the main crop has established to avoid competition. Cover crops are not suitable for dry areas with annual rainfall of less than 500mm, as they might compete for water with the main crop. Under such conditions it might be better to keep the weeds and natural vegetation as cover.They may not do well under conditions of low phosphorous.

**vi) Mulching:** Mulching means covering the soil between crop rows or around trees with a layer of loose material such as dry grass, straw, crop residues, leaves, manure or compost. This helps to retain soil moisture by limiting evaporation, suppressing weed growth and enhancing soil structure, reducing runoff, protecting the soil from splash erosion and limiting the formation of crust. In addition, mulching reduces fluctuations in soil temperature which improves conditions for micro-organisms. It is commonly used in areas affected by drought and weed infestation.

Mulch can be spread on a seedbed or around planting holes and it can also be applied in strips. Alternative row mulching is sometimes preferred to full mulching, because it reduces the fire risk. It is most effective if applied at the start of the rains, as it intercepts and increases water take-up, but it is frequently more practical to mulch towards the end of the rains when grass is available. When crop remains are used for mulching nutrients are released more slowly, so that more manure or fertilizer has to be applied. Weeds can be a problem if some grass species are used and mulches can provide a possible habitat for pests and diseases.

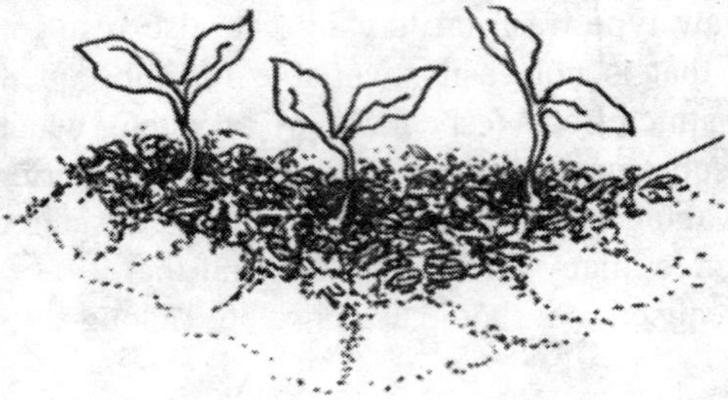


**Fig. 15:** Mulching

**vii) Drip irrigation:** Drip irrigation can conserve water especially when used in conjunction with roof top harvesting. The principle is very simple: water seeps slowly out of small holes in a pipe on the soil surface. Holes are normally located close to plants so that the water is targeted directly to the root zone. Drip irrigation comes in many forms, but at its simplest, can be constructed by puncturing a piece of garden hose at intervals and connecting this to a water supply. The end furthest from the header tank should be closed off. For smaller areas the pressure from a header tank should be more than adequate to operate the system. Larger areas that require a longer length of tubing may need to be divided into sections and irrigated at different times. Separate sets of tubes with different hole spacing may be needed to match different crop spacing. The system requires considerable work to set up, but once this is done, irrigation is relatively easy. Therefore it is more likely to be used on smaller areas of high value crops that require regular watering.

**viii) Conservation tillage**: Conservation tillage refers to a type of agriculture where soil cultivation is kept to a minimum. It aims to reduce the negative effects of conventional tillage such as soil compaction, formation of pans, disturbance of soil fauna and moisture loss.The two main features that distinguish conservation tillage systems from conventional tillage systems are minimum cultivations and permanent soil cover. Minimal cultivations vary in type. In a 'no till system' the land is prepared without the use of a conventional plough. In a 'minimum tillage' system, prong-based implements or hand hoes are used to open the soil just enough to allow a seed to be planted. Minimal cultivations reduce water losses because of a reduction in soil disturbance from tillage. In the long term the soil structure is improved. Less surface compaction and smearing at depth from the shares of the plough should increase rooting depth and therefore the drought tolerance of crops.

Permanent soil cover is commonly achieved by leaving crop residues on the ground as mulch after harvest (rather than incorporating them as in conventional tillage) or using cover crops or green manures between cash crops. This can reduce water loss and soil erosion from run off and prevent capping from heavy rainfall. Seeds or plants have to be sown or planted directly into mulch using a prong based implement or hoe.

Leaving crop residues on the soil can, however, increase the build up of pests and diseases. Leaving large amounts of straw type trash on the soil can also result in nitrogen being 'locked up' in a form that is not easily available for subsequent crops. Certain residues can have allelopathic effects (especially rye or vetch), which reduce germination of weeds but also subsequent directly sown crops for a period of up to six weeks. There are many advantages and disadvantages to be considered in conservation tillage systems. In practice, many farmers have found that it takes several years before the advantages are realised so it should be considered a long-term project.

**ix) Water harvesting from external catchment :**Water harvesting from external catchments involves diverting runoff water from an area that is not cropped to the area where crops are grown.Water is stored in a simple reservoir structure and can be applied to the crops when it is needed. The flow of water from the reservoir into the cropped area can be controlled using tied bunds that can be built up or dismantled as required.

The external catchment area should not be cultivated and may include rough grazing areas, roads or homesteads. Ideally the soil should have a low infiltration rate in order to maximise runoff and therefore vegetation should be restricted to a minimum. The stored water will be lost gradually through evaporation and seepage, this can be reduced on silt or clay soils by capping the soil using puddling boards (used to encourage flooding in rice cultivation).

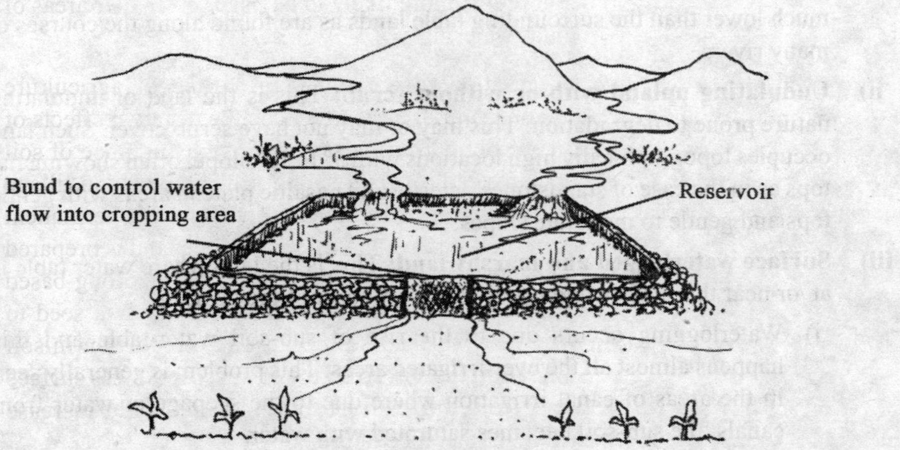


**Fig. 16:** Water Harvesting from External Catchment

## 8.3. Wastelands

Wasteland Survey and Reclamation Committee, Ministry of food and Agriculture, govt, of India (1961) has defined wastelands as those lands which are either not available for cultivation or left out without being cultivated for some reason or other. According to Bhumla and Khare (1987) the waste lands are "Those lands are wastelands (a) which are ecologicaly unstable, (b) whose top soil has nearly been completely lost, and (c) which have developed toxicity in the root zones or growth of most plants, both annual crops and trees".

Wastelands include degraded forests, overgrazed pastures, drought-struck pastures, eroded valleys, hilly slopes, waterlogged marshy lands, barren land etc. As per this definition wastelands are classified into two categories- culturable wastelands and unculturable wastelands.

1. The culturable category includes all possible wasteland areas which can be brought under vegetation (though after adequate treatment).
2. On the other hand, unculturable wastelands are the lands which cannot be developed for vegetative cover. In other words, land which is barren and cannot be put to any productive use, such as agriculture, and forest cover is unculturable wasteland.

### Types of wastelands

i) **Gullied land/ ravinous land**: the gullies are the result of the localized surface runoff affecting the friable unconsolidated material resulting in the formation of perceptible channels resulting in undulating terrain having a maximum depth of 3m. They are the first stage of excessive land dissection followed by their networking which leads to the development of ravinous land. **Ravinous land** - this category of land contains system of gullies running more or less parallel lo each other in deep alluvium and entering a nearby river flowing much lower than the surrounding table lands as are found along the courses of many rivers.

ii) **Undulating upland with or without scrub:** This is the land of undulating nature prone to degradation. This may or may not have scrub cover. Such land occupies topographically high locations with 3° to 10° slope, often showing flat tops as in the case of standstones, laterite and basaltic plateau areas with gentle tops and gentle to moderate slopes.

iii) **Surface waterlogged and marshy land:** This is the land where water table is at /or near the surface and water stands for most of the year.

   i) Waterlogging occurs due to the rise of sub-soil water-table and this happens almost all the over-irrigated areas. This problem is generally seen in the areas of canal irrigation where due to the seepage of water from canals, the sub-soil becomes saturated with water.

   ii) Marsh is the land which permanently or periodically is inundated by water and is characterized by vegetation that includes grasses and reeds. Marshes are classified into salt, brackish and fresh water categories depending on the salinity of water.

iv) **Shifting cultivation area** Such land is the result of cyclic land use consisting of felling of trees and burning of forest areas for growing crops. This results in extensive soil losses leading to land degradation.

v) **Degraded/under utilized notified forest land** Lands as notified under the forest act and those lands with various types of forest cover in which denudation of vegetation cover is less than 20% of canopy cover, are classified as degraded land.

vi) **Degraded pastures/grazing land** All those grazing lands in non-forest areas, whether or not they are permanent pastures or meadows, which have become degraded due to lack of proper soil conservation and drainage measures fall under this category.

**Table 10:** Wasteland status in India: Estimated area under the wastelands provided by different organization

| Source | Area (m.ha.) |
|---|---|
| Ministry of Agriculture and the JNU, Department of Geography (1986) | 175 |
| National Land Use and Wasteland Development Council (First Meeting 1986) | 123 |
| Society for Promotion of Wasteland Development (1982) | 145 |
| Ministry of Rural Development & NRSA (2000) | 64 |

**Table 11:** State wise wastelands of India- NRSA (Information as on year 2003)

| State | Wasteland (Area: In square km) |
|---|---|
| Andhra Pradesh | 45267.15 |
| Arunachal Pradesh | 18175.95 |
| Assam | 14034.08 |
| Bihar | 5443.68 |
| Chhattisgarh | 7584.15 |
| Goa | 531.29 |
| Gujarat | 20377.74 |
| Haryana | 3266.45 |
| Himachal Pradesh | 28336.80 |
| J & K | 70201.99 |
| Jharkhand | 11165.26 |
| Karnataka | 13536.58 |
| Kerala | 1788.80 |
| Madhya Pradesh | 57134.03 |
| Maharashtra | 49275.41 |
| Manipur | 13174.74 |

| | |
|---|---|
| Meghalaya | 3411.41 |
| Mizoram | 4469.88 |
| Nagaland | 3709.40 |
| Orissa | 18952.74 |
| Punjab | 1172.84 |
| Rajasthan | 101453.86 |
| Sikkim | 3808.21 |
| Tripura | 1322.97 |
| Tamil Nadu | 17303.29 |
| Uttarakhand | 16097.46 |
| Uttar Pradesh | 16984.16 |
| West Bengal | 4397.56 |
| Union Territory | 314.38 |
| **Total** | **552692.26** |

**Causes of formation of wastelands are**

(a) **Indiscriminate and over utilisation of forest produces:** Forests trees has been cut indiscriminately for forest products like timber, fuel, essential oils etc exposing the soil for severe erosion since most forests are occupied in sloppy lands.

(b) **Indiscriminate mining activity:** Mining for iron, bauxite, coal, gold etc makes the soil loose and exposure of sub and deeper layer soil which is poorly fertile and doesnot support vegetation. The loose soil without any vegetation cover aggravates the erosion.

(c) **Unscientific disposal of radioactive effluents:** Radioactive effluents cause aberrations in soil microbial activity and plant growth. Further the radioactive waste make entry in to the food chain.

(d) **Over-grazing:** over grazing cause poor aggregation of soil and susceptible for erosion of fertile surface soil and exposure of poor fertile subsoil.

(e) **Side effects of development projects:** Development projects like medium and large scale industries releases harmful effluents in to nearby lands causing the lands not fit or less fit for cultivation

(f) **Misuse and unscientific land management:** Very intensive cultivation in poor and marginal lands, excessive use of fertilizers and pesticides, improper use of irrigation water causes the land degradation and convert them to wastelands.

## 8.4. Soil Biological degradation

Biological sickness often results from management practices and anthropogenic influence. Soil organic carbon (SOC) is the main source of energy for soil microorganisms and a trigger for nutrient availability through mineralization. Humus participates in aggregate stability, and nutrient and water holding capacity. Organic acids (e.g., oxalic acid), commonly released from decomposing organic residues and manures, prevents phosphorus fixation by clay minerals and improve its plant availability, especially in subtropical and tropical soils. An increase in SOM, and therefore total soil carbon, leads to greater biological diversity in the soil. Biological degradation of soil refers to the impairment or elimination of one or more significant populations of microorganisms in soil, often with a resulting significant reduction in biogeochemical process within the associated ecosystem.

### 8.4.1. The complexity of microorganisms in soils

Except for occasional insects or earthworms, once visible traces of plant biomass are removed, soil appears as a lifeless mass, that is composed of mineral particles and organic residues. However, even desert soils are abundant source of living microorganisms. This seemingly lifeless matter contains complex microbial community, including bacteria, fungi, protozoa and viruses. The integrity of the aboveground and belowground ecosystems depends on the stability, resilience and function of the soil microbial community.

Soil is an interesting medium for growing microorganisms, as it contains various nutrients that the microbes need for their metabolism. Unfortunately, nutrients are not always readily available. However, it is one of the richest reservoirs of microorganisms, i.e. 1 gram of agricultural soil may contain even several billion colony forming units (CFUs) of microorganisms belonging to thousands of different species, and even though microorganisms constitute less than 0.5% of the soil mass, they have a major impact on soil properties and processes. Table 11 presents the average numbers of soil microorganisms in a "typical" temperate soil. Destruction of the soil microbiota through mismanagement or environmental pollution causes decline or even death of the aboveground plant and animal populations.

**Table 12:** Relative numbers and approximate biomass of the soil microbiota in a fertile soil.

| Organisms | Numbers Per $m^2$ | Numbers Per g | Biomass [wt kg/$ha^{-1}$] |
|---|---|---|---|
| Bacteria | $10^{13}$-$10^{14}$ | $10^{8}$-$10^{9}$ | 300-3000 |
| Actinomycetes | $10^{12}$-$10^{13}$ | $10^{7}$-$10^{8}$ | 300-3000 |
| Fungi | $10^{10}$-$10^{11}$ | $10^{5}$-$10^{6}$ | 500-5000 |
| Microalgae | $10^{9}$-$10^{10}$ | $10^{3}$-$10^{6}$ | 10-1500 |

The most characteristic feature of microbial habitats is the great micro-spatial variability of environmental parameters, like temperature or nutrient availability. Many basic requirements of heterogeneous microorganisms are satisfied by various soil microhabitats. This is the reason why, in ecological terms, a number of varying microbial niches can be described. Therefore, the microbial community is composed of diverse taxa with different nutritional demands within small microenvironments. Analysis of the spatial distribution of bacteria at micro habitat levels showed that in soils subjected to different fertilization treatments, the majority of bacteria were located in micropores of stable soil micro-aggregates (2 – 20 μm), as they contained over 80% of cells . Such microhabitats offer the most favorable conditions for microbial growth in terms of water and substrate availability, oxygen diffusion and protection against predation. The microhabitat-adapted groups of microorganisms form so-called consortia which are held together by mutually facilitating metabolic processes. The consortia are characterized by more or less sharp boundaries, and variable level of interaction with each other and with other parts of the soil biota. Numerous investigations emphasize the impact of soil structure and spatial isolation on microbial diversity and community structure. Some studies indicate that the soil particle size affects the diversity of microorganisms and community structure to a greater extent than other factors such as bulk pH and the type or amount of available organic compounds. Bacteria constitute the most numerous group of soil microbes – a teaspoon of productive soil contains between 100 million and 1 billion bacterial cells. As soil environment changes rather drastically, spore-forming bacteria tend to be the most common. When environmental conditions become too difficult for normal growth, the bacteria form spores and remain dormant until the environment returns to proper conditions. They facilitate various processes in soils, e.g. those related to water dynamics, nutrient cycling or disease suppression. Soil-dwelling bacteria may be divided into different groups based on:

- Shapes: rods (also called bacilli), sphere (also called cocci) and spiral (also called spirilla)
- Their reaction to oxygen: aerobic (bacteria that need oxygen for their survival) and anaerobic (the ones that do not require oxygen and in most cases cannot bear oxygen that is deadly for them)
- Result of Gram staining: Gram negative (stain pink and have thinner cell walls, they are the smallest ones and tend to be more sensitive to water stress) or Gram positive (stain violet, have thicker cell walls, are larger in size and tend to resist water stress)
- Source of carbon they use: autotrophs (obtain carbon from carbon dioxide – some autotrophic bacteria directly use sunlight in order to produce sugar from carbon dioxide, while others depend on various chemical reactions) or heterotrophs (they obtain carbohydrates from their environment)
- Classification based on phyla: based on morphology, barcode DNA sequences, physiological requirements and biochemical characteristics, bacteria have been classified into 12 phyla. Each phylum corresponds to a number of bacterial species and genera.

Bacteria facilitate a number of physical and biochemical alterations or reactions in soils and thereby directly or indirectly support the development of higher plants. Their performance is vital for a variety of processes that include: decomposition of cellulose or other carbohydrates (e.g. *Bacillus*, *Achromobacter*, *Cellulomonas*, *Clostridium*, *Methanococcus*), ammonification (*Bacillus*, *Pseudomonas*), nitrification (*Nitrosomonas*, *Nitrobacter*), denitrification (*Achromobacter*, *Pseudomonas*, *Bacillus*, *Micrococcus*) and nitrogen fixation (symbiotic *Rhizobium*, *Bradyrhizobium* etc., non-symbiotic *Azotobacter*, *Beijerinckia*).

On the other hand, soil fungi form three functional groups: decomposers, mutualists and pathogens. Fungi, along with bacteria, are important decomposers of hard to digest organic matter and they increase nutrient uptake of phosphorus. Mycorrhizal fungi support plants by promoting root branching and increasing nitrogen, phosphorus and water uptake. They improve plant resilience to pests, diseases or drought and improve soil structure, as fungal hyphae binds soil particles together to create water-stable aggregates. They in turn create the pore spaces in the soil that enhance water retention and drainage. The most common fungi found in soil belong to the *Penicillium* and *Aspergillus* genera together with the representatives of the Zygomycetes and the mycorrhizae-associated Ascomycetes and Basidiomycetes.

Actinomycetes are a large group of microorganisms, systematically identified as bacteria, that grow as hyphae. They decompose a wide range of substances, but they are particularly important in degrading recalcitrant (difficult to degrade) compounds such as chitin, lignin, keratin and cellulose. Moreover, they produce a number of secondary metabolites such as antibiotics i.e. streptomycin or geosmine which is responsible for "earthy" smell after soil plowing. Actinomycetes are important in forming stable humus, which enhances soil structure, improves soil nutrient storage and increases water retention in soils. The most commonly encountered soil actinomycetes belong to *Nocardia* and *Streptomyces* genera.

Algae are the most common among photosynthetic microorganisms found in soil. They are found only near soil surface, where light is readily available. The most common genera of green algae found in soil are: *Chlorella, Chlamydomonas, Chlorococcum, Protosiphon* etc. and that of diatoms are *Navicula, Pinnularia. Synedra, Frangilaria.* Their functions include the maintenance of soil fertility, increasing water retention capacity of soil, prevention of soil erosion due to the fact that they act as cementing agents in binding soil particles. They add organic matter to soil after the cell death and thus increase the amount of organic carbon, while their photosynthetic activities release large quantity of oxygen that facilitate the aeration in submerged soils or oxygenate the soil environment. Soil algae also take part in weathering rocks, thus building up the soil structure.

## 8.4.2. Beneficial role of microbes in soil

Microbes take part in soil formation, mineralize organic substances, and provide plants with bio-available compounds, source of insecticidal substances. One of the most important and most widely studied microbial groups in terms of beneficial effects to soil and plants is the group of Plant Growth Promoting Rhizobacteria (PGPR). This group includes bacterial species from genera such as *Azotobacter*,*Azospirillum*, *Bacillus*, *Burkholderia*, *Enterobacter*, *Erwinia*, *Flavobacterium*, *Pseudomonas* and *Rhizobium*. Activity of these bacteria significantly increases plant growth and yield due to a variety of mechanisms, such as phytohormone production, symbiotic and asymbiotic nitrogen fixation, production of siderophores, activity against phytopathogenic microorganisms, synthesis of antibiotics, enzymes and/or fungicidal compounds, as well as solubilization of mineral phosphates and other mineral forms of plant nutrients. PGPR may improve plant growth, salinity and metal toxicity stress tolerance, as well as they are able to produce phytohormones such as indole-3- acetic acid (IAA). Some PGPR produce the enzyme 1-aminocyclopropane-1-carboxylate (ACC) deaminase, which hydrolyses ACC, the immediate precursor of ethylene in plants. By decreasing its concentration in seedlings and thus its inhibitory effect, these PGPR stimulate seedlings' root length. Bacteria from the genus *Rhizobium* form symbiotic associations with roots of leguminous plants like clovers, peas or alfalfa. These Gram-negative, rod-shaped bacteria infect growing root hairs, forming visible nodules. In this form of symbiosis, plants supply simple carbohydrates to bacteria while bacteria convert nitrogen ($N_2$) from air into the forms ($NO^-$ or $NH^+$) that plant can use. When leaves or roots from the plant decompose, nitrogen content increases in soil. Some microbial species are capable of detergent decomposition, taking part in self-purification process of soils. Decomposers are particularly important in immobilizing or retaining nutrients in their cells, thus preventing the loss of nutrients, such as nitrogen, from the rooting zone.

## 8.4.3. Indicators of soil biological degradation

### *8.4.3.1. Soil micro organism population*

Soil is an interesting medium for growing microorganisms, as it contains various nutrients that the microbes need for their metabolism. Unfortunately, nutrients are not always readily available. However, it is one of the richest reservoirs of microorganisms, i.e. 1 gram of agricultural soil may contain even several billion colony forming units (CFUs) of microorganisms belonging to thousands of different species, and even though microorganisms constitute less than 0.5% of the soil mass, they have a major impact on soil properties and processes. Table 8 presents the average numbers of soil microorganisms in a "typical" temperate soil. Destruction of the soil microbiota through mismanagement or environmental pollution causes decline or even death of the aboveground plant and animal populations.

High concentrations of heavy metals may have inhibitory or even toxic effects on living organisms. Adverse effects of metals on soil microbes result in decreased decomposition of organic matter, reduced soil respiration, decreased diversity and declined activity of several soil enzymes. Some of the general changes in morphology, the disruption of the life cycle and the increase or decrease of pigmentation are easy to observe and evaluate. The bacterial community is more sensitive to increased concentrations of heavy metals in soils than the fungal community. The relative fungal/bacterial ratio increased with increasing metal levels. The effect of soil pH on the microbial reaction to soil pollution, i.e. that lower pH in contaminated soils enhanced the negative effect on bacteria, but not on fungi.

### *8.4.3.2 Soil organic carbon (SOC)*

A direct effect of poor SOC status is reduced microbial biomass activity, and nutrient mineralization due to a shortage of energy sources. In non- calcareous soils, aggregate stability, infiltration, drainage, and airflow are reduced. Low SOC results in less diversity in soil biota with a risk of the food chain equilibrium being disrupted, which can cause disturbance in the soil environment (e.g., plant pest and disease increase, accumulation of toxic substances).

To improve soil carbon levels, no till farming, application of manure and compost, and use of summer and/or winter cover crops, green manuring and suitable cropping pattern are to be followed. Burning, harvesting, or otherwise removing residues decreases soil organic carbon.

### *8.4.3.3. Earthworms population*

Earthworms play a key role in modifying the physical structure of soils by producing new aggregates and pores, which improves soil tilth, aeration, infiltration, and drainage. Earthworms produce binding agents responsible for the formation of water-stable macro-aggregates. They improve soil porosity by burrowing and mixing soil. As they feed, earthworms participate in plant residue decomposition, nutrient cycling and redistribution of nutrients in the soil profile. Their casts, as well as dead or decaying earthworms, are a source of nutrients. Roots often follow earthworm burrows and uptake available nutrients associated with casts. Low or absent earthworm populations are an indicator of little or no organic residues in the soil and/or high soil temperature and low soil moisture that are stressful not only to earthworms, but also for sustainable crop production. Lack of earthworms may reduce nutrient cycling and availability for plant uptake. Additionally, natural drainage and aggregate stability can be reduced when earthworm population in soil is low.

The practices that boost earthworm populations are Tillage management (no-till, strip till, ridge till), Crop rotation (with legumes) and Cover crops to reduce soil erosion, Manures and organic by-product application, Soil reaction (pH) management to neutral pH and proper irrigation or drainage to ensure aeration to earthworms.

### *8.4.3.4. Soil respiration*

Carbon dioxide ($CO_2$) release from the soil surface is referred to as soil respiration. This $CO_2$ results from several sources, including aerobic microbial decomposition of soil organic matter (SOM) to obtain energy for their growth and functioning (microbial respiration), plant root and faunal respiration, and eventually from the dissolution of carbonates in soil solution. Soil respiration is one measure of biological activity and decomposition and also known as carbon mineralization. Soil respiration reflects the capacity of soil to support soil life including crops, soil animals, and microorganisms. In the laboratory, soil respiration can be used to estimate soil microbial biomass and make some inference about nutrient cycling in the soil. Soil respiration also provides an indication of the soil's ability to sustain plant growth.

Reduced soil respiration rates indicate that there is little or no microbial activity in the soil. It may also signify that soil properties that contribute to soil respiration (soil temperature, moisture, aeration, available N) are limiting biological activity and poor decomposition of SOM. With reduced soil respiration, nutrients are not released from

SOM to feed plants and soil organisms. This affects plant root respiration, which can restrict the growth of plants. Incomplete mineralization of SOM often occurs in saturated or flooded soils, resulting in the formation of compounds that are harmful to plant roots,(e.g. methane and alcohol). In such anaerobic environments, denitrification and sulphur volatilization usually occur, contributing to greenhouse gas emissions and acid deposition. The rate of soil respiration under favorable temperature and moisture conditions is generally limited by the supply of SOM.

Soil respiration can be improved by proper irrigation and drainage methods, reduced/ minimul tillage practices, crop rotation, erosion control practices, addition of organic manures such as FYM, green manuring, green leaf manuring, crop residue incorporation, mulching practices etc should be followed. Agricultural practices that increase SOM usually enhance soil respiration.

### *8.4.3.5. Soil enzyme activity*

Soil enzymes increase the reaction rate at which plant residues decompose and release plant available nutrients. Enzymes are specific to a substrate and have active sites that bind with the substrate to form a temporary complex. The enzymatic reaction releases a product, which can be a nutrient contained in the substrate. Absence or suppression of soil enzymes prevents or reduces processes that can affect plant nutrition. Poor enzyme activity (e.g., pesticide degrading enzymes) can result in an accumulation of chemicals that are harmful to the environment; some of these chemicals may further inhibit soil enzyme activity. Organic amendment applications, crop rotation, and cover crops can be used to enhance enzyme activity. The positive effect of pasture is associated with the input of animal manure and less soil disturbance.

## Review questions

1. What is soil degradation? Write the causes for soil degradation.
2. Define soil erosion and write causes for soil erosion.
3. Discuss significance of soil erosion vis a vis Indian perspective.
4. Explain types of water erosion and write the Universal soil loss equation.
5. What are the agricultural consequences of soil erosion?
6. Discuss soil conservation measures for erosion control.
7. Define waste lands and discuss about various types of waste lands.
8. Briefly write on soil biological degradation and measures to control.
9. Discuss the indicators of biological degradation.
10. Differentiate between phyto remediation and microbial remediation.

## Objective questions

1. According to NBSS&LUP (2004), the estimated soil degradation in India is
   a) 146.82 Mha
   b) 107.43 Mha
   c) 148.09 Mha
   d) 187.70 Mha
2. According to ICAR, largest extent of soil degradation occur in the Agroecoregion
   a) Western Himalayas, warm subhumid (to humid with inclusion per-humid) ecoregion, with brown forest and podzolic soils and GP 180 to210+ days
   b) Islands of Andaman-Nicobar and Lakshadweep, hot humid and perhumid island ecoregion, with red loamy and sandy soils and GP 210+ days
   c) Central Highlands (Malwa, Bundelkhand and Eastern Satpura), hot sub-humid ecoregion, with black and red soils and GP 150–180 (to 210) days
   d) Northern plain and central highlands including Aravallis, hot semi-arid ecoregion, with alluvium derived soils and growing period 90–150 days
3. Most important cause of soil degradation in India is
   a) Salt accumulation
   b) Water erosion
   c) Water logging
   d) Shifting cultivation
4. Ill effects of soil degradation is
   a) Soil quality decline
   b) Loss of arable land
   c) Soil pollution
   d) All the three
5. The resilience of soil to erosion is largely depends upon
   a) Soil structure
   b) Soil texture
   c) Soil consistence
   d) Soil Colour
6. Poor soil structure result in
   a) Weekly aggregated,
   b) Easily compacted
   c) High runoff with low infiltration
   d) All three
7. The technique useful for investigation of soil interior architectural design and three-dimensional visualization of soil structures is
   a) Fractal theory
   b) Neural networks
   c) Tomography
   d) Regression models
8. Fractal theory of soil structure studies is
   a) Soil debris and its sensitivity to soil erosion
   b) Interior architectural design and three-dimensional visualization of soil structure
   c) Structural properties of the soil in relation to water conservation, organic matter store and erosion resistance
   d) None of the three

9. The cementing agent that binds the soil particles together and protect from erosion is
   a) Air
   b) Water
   c) Organic matter
   d) Silt particles
10. The best soil protector against soil erosion is
   a) Residue mulch
   b) Growing grass
   c) Growing crops
   d) Zerotillage practice
11. Accelerated soil erosion is also known as
   a) Exogeneic erosion
   b) Anthropogenic erosion
   c) Geogenic erosion
   d) A&B
12. According to GLASOD assessment, the extetn of water erosion in India is
   a) 32.8 Mha
   b) 87 Mha
   c) 111 Mha
   d) Mha11.2
13. According to NBSS&LUP (2005) largest extent of water erosion occur in
   a) Maharastra
   b) AndhraPradesh
   c) Madyapradesh
   d) Rajastan
14. According to NBSS&LUP (2005) largest extent of wind erosion occur in
   a) Jammu & Kashmir
   b) Rajasthan
   c) Haryana
   d) Gujarat
15. The major factor which causes the movement and detachment of soil particles is
   a) Seepage water
   b) Snowmelt water
   c) Irrigation water
   d) Rainwater
16. Sheet and splash erosion is also known as
   a) Revine erosion
   b) Inter rill erosion
   c) Rill erosion
   d) Gully erosion
17. The kinetic energy of falling raindrop dislodges the soil particle and the resultant runoff transports soil particles is known as
   a) Sheet erosion
   b) Splash erosion
   c) Rill erosion
   d) Gully erosion
18. The first stage of soil erosion by water is
   a) Rill erosion
   b) Sheet erosion
   c) Splash erosion
   d) Gully erosion
19. The splashed particles can
   a) Rise as high 0.60 meter above the ground and move up to 1.5 meter from the point of impact
   b) Rise as high 0.40 meter above the ground and move up to 1.0 meter from the point of impact

c) Rise as high 0.20 meter above the ground and move up to 0.5 meter from the point of impact

d) Rise as high 0.10 meter above the ground and move up to 1.0 meter from the point of impact

20. The splashed soil particles

a) Increases soil bulkdensity thrus reduces soil porespace and restricts soil aeration

b) Binds soil particles and form soil aggregates, so that improves infiltration and reduces runoff.

c) Makes subsurface hard pan increases soil strength and restrict root penetration.

d) Block the spaces between soil aggregates, so that the soil forms a crust that reduces infiltration and increases runoff.

21. The type of water erosion prevalent in the agricultural land is

a) Splash erosion　　b) Sheet erosion

c) Rill erosion　　d) Streamline erosion

22. Loss of the finest soil particles that contain most of the available nutrients and organic matter in the soil caused by

a) Rill erosion　　b) Splash erosion

d) Sheet erosion　　d) Streamline erosion

23. The shallow drainage lines less than 30cm deep and 50 cm wide are

a) Streamlines　　b) Gullys

c) Revines　　d) Rills

24. The types of water erosion forms that can usually be removed with farm machinery.

a) Revines　　b) Gullys

c) Rills　　d) Streamlines

25. The erosion mostlyoccurs in alluvial soil is

a) Rill erosion　　b) Gully erosion

c) Revin erosion　　d) Sheet erosion

26 The advance stage of rill erosion is

a) Sheet erosion　　b) Revin erosion

c) Gully erosion　　d) Splash erosion

27. The erosion that forms deeper channels that cannot be removed by normal cultivation is

a) Revin erosion　　d) Gully erosion

c) Sheet erosion　　b) splash erosion

28. Tunnel erosion is more frequent in foothills where elevation is between
    a) 100-250 meters b) 300-450 meters
    c) 500-750 meters d) 700-950 meters
29. The erosion that occurs where streams begin cutting deeper and wider channels as a consequence of increased peak flows
    a) Revin erosion a) Gully erosion
    b) Stream bank erosion c) Sheet erosion
30. The factor influencing coastal erosion is
    a) Waves energy b) Geology of the coast
    c) Geomorphology of the coast d) All three
31. Universal soil loss equation (USLE) was developed by
    a) Stoke &Darcy b) Lindsay & Norvell
    c) Thomas and Way d) Wischmeir and Smith
32. In Universal soil loss equation (USLE), the factor K represents
    a) Crop management factor b) Rainfall erosivity factor
    c) Slope length and steepness factor d) Soil erodibility factor
33. In Universal soil loss equation (USLE), the reference values for slope length and steepness factors are
    a) 10.5 m and 6% b) 22.6 m and 9%
    c) 42.6 m and 12% d) 32.5 m and 5%
34. In Universal soil loss equation (USLE), the units fro A are
    a) t $ha^{-1}$ b) Q $Ac^{-1}$
    c) Kg $Ac^{-1}$ d) Pounds $ha^{-1}$
35. Poor SOC status results in
    a) Reduced microbial biomass activity b) Nutrient mineralization
    c) Increase in earthworm population d) A&B
36. Beneficial role of earth worms is
    a) Responsible for the formation of water-stable macro-aggregates
    b) Improve soil porosity by burrowing and mixing soil.
    c) Earthworm casts, as well as dead or decaying earthworms, are a source of nutrients

    d)All the three
37. Soil respiration is reflected by
    a) Carbon dioxide ($CO_2$) release from the soil surface
    b) Nitrate leaching from soil
    c) Balanced nutrient availability in soil
    d) Accumulation of heavy metals in soil

38. Most common among photosynthetic microorganisms found in soil is
a) Bacteria
b) Fungi
c) Algae
d) Actinomycetes

39. The soil compound responsible for "earthy" smell after soil plowing is
a) Urease
b) Nitroginase
b) Dehydogenase
d) Geosmine

40. The substance Geosmine in soil is secreted by
a) Algae
b) Fungi
c) Bacteria
d) Actinomycetes

41. Algae found in soils at
a) Plant root surface
b) Subsurface
c) Soil surface
d) Macropores

42. Beneficial actions of soil algae are
a) Maintenance of soil fertility,
b) Increasing water retention capacity of soil,
c) Prevention of soil erosion
d) All the three

43. Size of soil micro-aggregates is
a) 20 - 30 μm
b) 0.1 – 0.5 μm
c) 2 – 20 μm
d) 30 - 40 μm

44. Generally microorganism in soil constitute
a) 2.5 % of the soil mass
b) 5% of the soil mass
c) 1.0% of the soil mass
d) less than 0.5% of the soil mass

45. Impairment or elimination of one or more significant populations of microorganisms in soil, often with a resulting significant reduction in biogeochemical process is referred as
a) Soil physical degradation
b) Soil biological degradation
c) Soil chemical degradation
d) Soil fertility degradation

46. According to Ministry of Rural Development & NRSA the extent of wastelands in India is
a) 175 mha
b) 123 mha
c) 64 mha
d) 145 mha

47. The average annual rate of soil erosion in the country is
a) 25.75 tonnes per hectare
b) 21.50 tonnes per hectare
c) 12.15 tonnes per hectare
d) 16.35 tonnes per hectare

48. In India the percent eroded soil deposited in reservoirs is
a) 15%
b) 10%
c) 20%
d) 25%

49. The annual reduction in the storage capacity of reservoirs in India due to soil erosion is
    a) 5 to 6 percent
    b) 3 to 5 percent
    c) 1 to 2 percent
    d) 6 to 8 percent
50. Annual loss of major nutrients due to soil erosion in India is
    a) 14 Million tonnes
    b) 54 Million tonnes
    c) 34 Million tones
    d) 74 Million tonnes
51. The most severe type of soil erosion is
    a) Wind erosion
    b) Water erosion
    d) Land slide erosion
    d) Geogenic erosion
52. The first stage of soil erosion by water is
    a) Rill erosion
    b) Sheet erosion
    c) Splash erosion
    d) Gully erosion
53. In splash erosion, the splashed soil particles can rise up to a height of
    a) 0.10 meter
    b) 0.30 meter
    c) 0.90 meter
    d) 0.60 meter
54. Soil forms surface crust due to
    a) Sheet erosion
    b) Splash erosion
    c) Rill erosion
    d) Gully erosion
55. The erosion type is prevalent in agricultural lands is
    a) Gully erosion
    b) Splash erosion
    c) Rill erosion
    d) Sheet erosion
56. Loss of the finest surface soil particles is due to
    a) Splash erosion
    b) Sheet erosion
    c) Rill erosion
    d) Gully erosion
57. The advance form of the rill erosion is
    a) Initial stage of Ravine erosion
    b) Initial stage of sheet erosion
    c) Initial stage of gully erosion
    d) Initial stage of splash eroson
58. Nomograph is used for determining
    a) Crop management factor
    b) Rainfall erosivity factor
    c) Soil erodibility factor
    d) Slope length factor
59. The type of erosion prevalent in alluvial river and streams is
    a) Sheet erosion
    b) Rill erosion
    c) Splash erosion
    d) Stream bank erosion
60. Coastline erosion is severe in
    a) Narrow, sandy and steep beaches
    b) Wide, pebble and shallow beaches
    c) Narrow, pebble, shallow beaches
    d) Wide, pebble, deep beaches.

61. The type of erosion is more frequent in foothills where elevation is between 500-750 meter is

a) Sheet erosion
b) Rill erosion
c) Tunnel erosion
d) Splash erosion

62. When the side slope of cully is 6 -12%, the gully symbol sig oven as

a) G3
b) G2
c) G1
d) G4

**Answer key**

| | | | | | | | | | | | | | | | | | | | | | | | | | |
|---|---|---|---|---|---|---|---|---|---|---|---|---|---|---|---|---|---|---|---|---|---|---|---|---|---|
| **1** | a | **6** | d | **11** | d | **16** | b | **21** | b | **26** | c | **31** | d | **36** | d | **41** | c | **46** | c | **51** | b | **56** | b | **61** | c |
| **2** | d | **7** | c | **12** | a | **17** | b | **22** | d | **27** | b | **32** | d | **37** | a | **42** | d | **47** | d | **52** | c | **57** | c | **62** | a |
| **3** | b | **8** | a | **13** | c | **18** | c | **23** | d | **28** | c | **33** | b | **38** | b | **43** | c | **48** | a | **53** | d | **58** | c | | |
| **4** | d | **9** | c | **14** | b | **19** | a | **24** | c | **29** | b | **34** | a | **39** | d | **44** | d | **49** | c | **54** | b | **59** | d | | |
| **5** | a | **10** | b | **15** | d | **20** | d | **25** | a | **30** | d | **35** | d | **40** | d | **45** | b | **50** | d | **55** | d | **60** | a | | |

# 9

# Problematic Soils

The soils which owe characteristics that they cannot be economically used for the cultivation of crops without adopting proper reclamation measures are known as problematic soils. Different kinds of problematic soils viz., Physical, chemical and biological problematic soils exists and pose problems of crop production.

## Classification of Problematic Soils

**Table 1:** Problematic soils are classified based on soil properties

| Soil physical problems | Soil chemical problems |
|---|---|
| Due to unfavorable textural and structural condition of soils | Due to presence of salts or beyond permissible limit of nutrients present in exch. complex |
| a) Soil hardening/ soil compaction | a) Acid soils |
| b) Highly permeable sandy soils | b) Acid sulphate soils |
| c) Slow permeable clay soils | c) Saline soils |
| d) Soil crusting | d) Sodic/Alkali soils |
| e) Fluffy paddy soil | e) Saline sodic soils |
| f) Shallow soils | f) Degraded alkali soils |
| g) Eroded soils | g) Calcareous soils |
| h) Ill drained soils in low lands | |

## 9.1. Soil physical problems

A successful farming system depends on the maintenance of the soil structure and its productivity. Crop production in any region is largely determined by the physical and climate features of the soil. The soil physical problem restricts the emergence of seedlings, root growth thereby reduce the crop yields. Physical problems manifest in many ways. These are characterized by their texture, aggregate stability, mechanical impedances, topography or erosion. Out of total cultivable area of 145 Mha, about 79 Mha has physical constraints in one form or other.

The soil physical constraints which are identified visually, nature and magnitude of the soil problem is confirmed by digging a soil profile at the site and taking undisturbed soil cores from each layer or horizon. Bulk density, hydraulic conductivity, air permeability, water retention characteristic are to be determined. pH, ECe, organic carbon, soluble salts, available P, K are to be determined for each layer. Infiltration rates of the soil with time and soil strength by a penetrometer in situ at the site are to be determined

### 9.1.1. Soil hardening/Soil compaction

It is largely due to physical process brought out either directly by compaction effects of through action of rainfall and later drying up of compacted oriented particles. About 22 Mha soil in India were reported to affect by hardening. The structure of this layer is sufficiently changed from that of soil mass below. The above layer has high BD, lower macro porosity and greater mechanical strength than soil layers below. The emergence of seedlings is affected. Incorporation of powdered groundnut shell, FYM, paddy husk increased the yield. Due to prevention of bloking of pore space by finer particles, increase in infiltration rate, increase in HC, better aggregate stability. Soil inversion (mixing of top and bottom layers) along with addition of organics improved the root growth. Mixing of finer and coarser fractions in different proportion could give good management of soil hardening through its effect on hardness strength, WHC and cracking properties. The ratio of finer and coarser fraction of 3:7 to 4:6 is found to best in reducing the soil hardening. (Soil breeding)

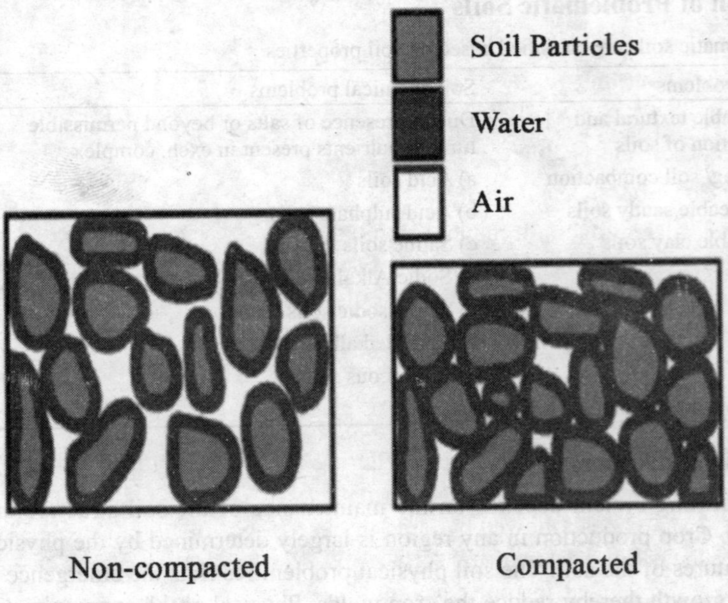


**Fig. 1:** Soil particle reorientation during Compacton

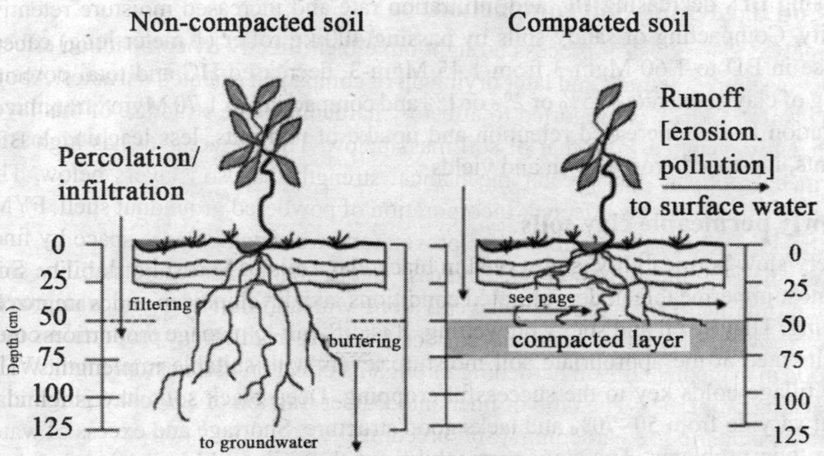


**Fig. 2:** Effect of Compaction on plant root extension (*bulk density higher than 1.9 g.cm$^{-3}$ stops the ability of plant roots to grow*).

**Causes of soil compaction**

**Natural soil compaction**: Main factors responsible are soil textural category (>35% clay) and soil morphological unit (argillic horizon, gleying, podzolisation, illimirisation).

**Human induced soil compaction:** Induced by intensive and incorrect land use (wrong agriculture/ forest management). Low amount of deep rooting structure forming plants in crop rotation eg. fodder crops. High amount of root crops (soil properties worsen plants: root system, agro techniques with high amount of crossing on the field). Low amount of organic residue incorporation in soil.

## Hard pan in the rhizosphere

Fine clay and clay is deposited in the sub soil during the process of illuviation. High clay content might have resulted in compaction and formed hard layer when combined with silica and sesquioxides followed by rapid drying, heating and high temperature. There is need to break open the hard sub surface soil by deep tillage to improve water storage besides improvement in root development. Use of chisel plough at 50 cm interval up to 30 cm depth and incorporating paddy husk at 5 tha-1 will break hard pan. 11.34 Mha in India and 2.08 lakh ha of land in Tamilnadu is affected by subsurface hard pan

## b) Highly permeable sandy soils

The light textured soil are highly susceptible to the percolation losses of water and leaching losses of nutrients. Low WHC, poor retentive capacity for nutrients, high percolation loss of water, fast evaporation rate are the major features of light textured soil. 13.75 Mha are affected by high permeability in India. To correct the textural weakness of soil, mixing 5% clay in the top 15 cm increased the crop yield by increasing BD, decreasing HC and infiltration rate and increased moisture retentive capacity. Compacting of sandy soils by passing 400 kg roller (1 meter long) caused increase in BD to 1.60 Mgm-3 from 1.45 Mgm-3, decreased HC and total porosity. Mixing of clay at the rate of 5% or 2% or 1% and compacting to 1.70 $Mgm^{-3}$ minimized percolation losses. Increased retention and uptake of nutrients, less leaching loss of nutrients, increased crop growth and yields.

### c) Slowly permeable clay soils

The very slow permeability is observed in black clay soils of Vertisol order. The soils are almost impermeable under saturated conditions and develop deep and wide cracks on drying. They swell and sticky on wetting. It is difficult to manage them unless they are cultivated at the appropriate soil moisture levels with suitable implements. The timely tillage holds key to the successful cropping. Deep black soils have high clay content varying from 50 -70% and lacks good structure. Shortage and excess of water are the twin problems. The slow permeability of the soil could be broken by deep ploughing of sub soils. This encouraged root growth and larger size of roots. Addition of FYM @ 25 t $ha^{-1}$ improved the soils.

### d) Soil crusting

It is a phenomena a associated with deterioration of soil structure where the natural soil aggregates brake and disperse. If dispersion is followed by drying the soil solids rearrange to crust. When raindrops strike the exposed soil surface, kinetic energy cause disintegration of aggregates and dispersive action of water leaves surface soil into monograin state. The dispersed particle is carried down the soil with infiltrating water clogging the soil pores. When on drying soil particles come together to form dense and strong soil layers known as soil crusts Crust can be formed in all types of soil (Red soil specific) of arid and semi-arid regions. It has high BD, low non- capillary pore space, low HC, encourages run off and soil loss. Soil crusting is observed in 10.25 Mha in India. It affects the emergence, early growth of seedlings and largely determines the crop stand. Soil factors that are associated with soil crusts formation include low OM, high salt and exch. Na, low structural stability. Since rain drop is major cause, soil crusting could be avoided by use of surface mulches which prevents dispersion of surface soil and maintains high infiltration, close growing crops and grasses, gypsum, crop rotation, residue management and chemical amendments

### e) Fluffy paddy soil

Monoculture of rice (rice – rice – rice) breaks down of aggregates and make the soil a structure less mass. Soil particles are always in a state of flux. Mechanical strength of these soils is very poor. These soils normally found in deltaic areas like Cauvery deltaic area. It is observed in the paddy soils where the working of soil by bullocks will be difficult. The fluffiness of soil will be such that knee of bullock will be completely covered. Fluffy soil is characterized with high infiltration, hydraulic conductivity but with low BD. In such soils managed by rolling of 400 kg stone for 8 passses so that soil will be compact, there is better aggregation, increase in BD and reduce in HC and infiltration

### f) Shallow soils

The shallow soils are characterised by the presence of the parent rock immediately below the soil surface at about 15 - 20 cm depth. This restricts the root elongation and spreading. Hence the crops grown in these soils necessarily are shallow rooted crops, which can exhaust the soil within 2 - 3 seasons. Therefore frequent renewal of soil fertility is a must in these soils. These soils can be managed by growing crops which can with stand the hard rocky sub soils like Mango, Ber, Fig, Country goose berry, West Indian cherry, Anona, Cashew, Tamarind etc.

### g) Eroded soils

In India almost 130 million hectares of land, i.e., 45 % of total geographical area, is affected by serious soil erosion through gorge and gully, shifting cultivation, cultivated wastelands, sandy areas, deserts and water logging. Soil erosion by rain and transportation of soil particles through rivulets that takes place in hilly areas causes severe landslides and floods. The anthropogenic activities including cutting trees for agricultural implements, firewood and timber; grazing by a large number of livestock over and above the carrying capacity of grass lands, traditional agricultural practices, construction of roads, indiscriminate quarrying and other activities, have all led to the opening of top surfaces to extreme soil erosion. In Indian condition, the control of soil erosion is a challenging task in the sense that the onset of monsoon often coincides with the kharif sowing and transplanting. In this stage of kharif crop when canopy cover is minimal, major part of the land is exposed to the rainfall let the land prone to soil erosion.

In agriculture, soil erosion refers to the removal of topsoil by the natural physical forces of water and wind at a greater rate than it is formed or through forces associated with farming activities such as tillage. Erosion removes the topsoil first and once this nutrient-rich layer is lost, the potential of soil to sustain plants is reduced. Without soil and plants the land becomes desert like and unable to support life. Soil erosion is a naturally occurring process that affects all landforms.

**Causes of Soil Erosion:** Soil erosion caused and intensified by i)Deforestation, ii)Intensive cultivation, iii)Overgrazing, iv)Cultivation of steep slopes, v) Mismanagement of soil and vi)Urbanization

### Consequences of Soil Erosion

The valuable top soil is the most productive part of soil profile for agricultural purposes is removed by soil erosion. Production cost will be high and yield will decrease resulted by the loss of this top soil. Gullies and rills make the cultivation of paddocks impossible and these are produced by the erosion when the top soil is removed.

Top soil that is rich in organic matter and nutrients is removed. The depth of soil that is available for water storage for crop growth and for rooting is reduced. Increase runoff by reducing the infiltration of water into soil. Increased cost of cultivation due to increased requirements of fertilizers and pesticides, need to repeat field operations and loss of seeds. Wind erosion blasted the young plants with sand, soures low plant stand. Extra cultivations are needed to level out the eroded surfaces damage to soil Sediment deposition onto roads, in roadside drains and on neighbouring properties. Excess inputs of phosphorus, pesticides and nitrogen damage the quality of lakes, coastal water and watercourses. Greater flood hazard downstream is caused by the deposition of sediment and increased runoff.

#### h) Ill drained soils of low lands

Ill drained simply refers to the soils ability not to allow water to pass through at a reasonable rate. Ill drained soils are associated with heavy clay content and shallow water table in land lands. They are imperfectly to poorly drained hence leaching of soluble weathering products is limited. Flooding can be a major problem in areas with higher rainfall. Proper aeration in the root zone is necessary for development of healthy growth. The crops become stunted with yellowing of leaves when the soil is saturated. In excess water, the plants usually die because of root damage caused by reduced supply of oxygen and accumulation of carbon dioxide with the related effects on the soil plant relationship.

Provision of proper drainage particularly subsurface drainage is important because not all plants prefer moist conditions. Poor drainage is easily spotted in areas where water tends to pool following a rainstorm. Where proper drainage cannot be achieved through natural means, corrugated piping called tiling may be installed below the growing surface to move moisture away from the plant's roots. Choosing the aquatic plants like rice particularly those varieties tolerant for flooding like Swarna-sub1 may be selected.

## 9.2. Soil chemical problems

Chemical problematic soil are Acid soils, Acidsulphate soil, Salt affected soils (saline, akali, saline akali and calcareous soils) .

### 9.2.1. Acid soil

Soil with low pH contain relatively high amounts of exchangeable $H^+$ and $Al^{3+}$ considered as the acid soil. The degree of acidity based on pH has been categorized to Ultra acidic soils (pH less than 3.3), Extremely acidic soils (pH 3.5 to 4.5), Very strong acidic soils (pH 4.5 to 5.0), Strong acidic soils (pH 5.1 to 5.5), Moderately acidic soils (pH 5.6 to 6.0) and Slightly acidic soils (pH 6.1 to 6.5). Soil acidity increases when $H^+$ producing processes exceed $H^+$ consuming processes.

#### 9.2.1.1. Formation of Acid soils

Hydrogen ($H^+$) and aluminium ($Al^{3+}$) producing processes of acid soil formation soils or sources of $H^+$ for acid soil formation are different in nature. They are 1) Rainfall 2) parent material 3) Fertilizer application 4) Plant root activity 5) decomposition of organic matter 6) climate 7) vegetation 8) topography 9) pedogeneic processes like podzolisation and Laterisation.

i) **Rainfall:** In high rainfall areas (humid and sub humid climates) due to leaching of bases (cation) from soil and rain water the soil become slightly acidic. This creates base unsaturation. Increases the percentages of hydrogen ($H^+$) and aluminium ($Al^{3+}$) in soil.

ii) **Parent material:** Acid soils are developed from acidic rocks like granite, genesis and quartz. When these rocks lacks bases, produces acidity in soil after decomposition by weathering.

iii) **Fertilizer Use:** Repeated application ammoniacal fertilizers like ures, ammonium sulphate, ammonium nitrate, ammonium chloride in soil leads to formation of acid soil. When Ammonium sulfate, ammonium nitrate and ammonium chloride are added to soil produces hydrogen ion during nitrification processes. Hence they are called acid forming fertilizers.

Use of mineral or organic fertilizers in agriculture increases inputs of nutrients to soils, and the form in which the nutrients are applied and their fate in the soil-plant system determine the overall effects on soil pH. Macronutrients (N, P, potassium (K), and S) have the major effects on pH as they are added in much larger quantities to soil than micronutrients.

## a) Nitrogen fertilisers

The form of N and the fate of N in the soil-plant system is probably the major driver of changes in soil pH in agricultural systems.Nitrogen can be added to soils in many forms, but the predominant forms of fertilizer N used are urea ($CO(NH_2)_2$), monoammonium phosphate ($NH_4H_2PO_4$), diammonium phosphate ($(NH_4)_2HPO_4$), ammonium nitrate ($NH_4NO_3$), calcium ammonium nitrate ($CaCO_3+NH_4(NO_3)$) ammonium sulfate ($(NH_4)_2SO_4$), urea ammonium nitrate (a mixture of urea and ammonium nitrate) and monoammonium polyphosphate ($[NH_4PO_3]_n$). The key molecules of N in terms of changes in soil pH are the uncharged urea molecule $CO(NH_2)_2$, the cation ammonium ($NH_4^+$) and the anion nitrate ($NO_3^-$). The conversion of N from one form to the other involves the generation or consumption of acidity and the uptake of ammonium or nitrate by plants will also affect acidity of soil (Figure 3).

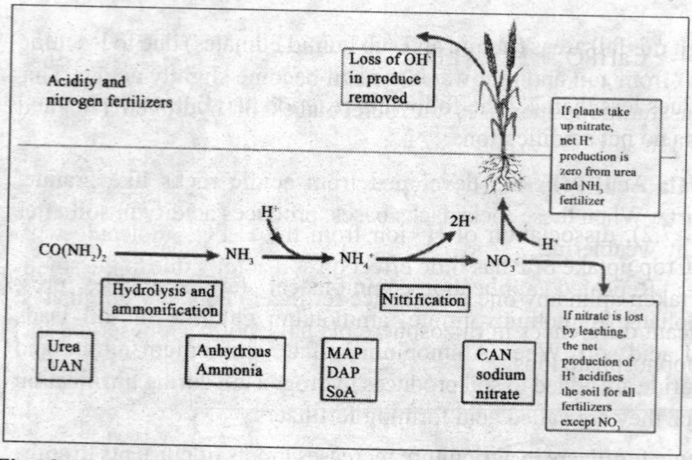


**Fig. 3:** Soil acidity and nitrogen fertilizers (modified from (Davidson 1987)). MAP = Monoammonium phosphate, DAP = Diammonium phosphate, CAN = Calcium ammonium **nitrate and sodium nitrate.**

It can be seen in Figure 3 that ammonium-based fertilizers will acidify soil as they generate two $H^+$ ions for each ammonium molecule nitrified to nitrate. The extent of acidification depends on whether the nitrate produced from ammonium is leached or is taken up by plants. If nitrate is taken up by plants the net acidification per molecule of ammonium is halved compared to the scenario when nitrate is leached. This is due to the consumption of one $H^+$ ion (or excretion of $OH^-$) for each molecule of nitrate taken up – this is often observed as pH increases in the rhizosphere. Anhydrous ammonia and urea have a lower acidification potential compared to ammonium-based products as one $H^+$ ion is consumed in the conversion to ammonium. Nitrate-based fertilizers have no acidification potential and actually can increase soil pH as one $H^+$ ion is absorbed by the plant (or $OH^-$ excreted) in the uptake of nitrate.

## b) Phosphorus fertilisers

The form of P fertilizer added to soil can affect soil acidity, principally through the release or gain of $H^+$ions by the phosphate molecule depending on soil pH (Figure 4). If phosphoric acid (PA) is added to soil, the molecule will always acidify soil as $H^+$ ions will be released - one $H^+$ ion if the soil pH is less than 6.2 and two $H^+$ ions is the soil pH is above 8.2. Monoammonium phosphate (MAP), single superphosphate (SSP) and triple superphosphate (TSP) all add P to soil in the form of the $H_2PO_4^-$ ion, which can acidify soil with a pH greater than 7.2 but has no effect on soil pH in acidic soils. The form of P in diammonium phosphate (DAP) is $HPO_4^{2-}$ which can make acidic soils (pH<7.2) more alkaline but has no effect on soil with a pH>7.2. The hydrolysis of ammonium polyphosphate (APP), where the P present as the $P_2O_7^{4-}$molecule converts to $HPO_4^{2-}$, is pH neutral and hence any acidification due to adding P can be regarded as similar to DAP. SSP or TSP are sometimes declared to cause soil acidification due to reaction products being very acidic;

$Ca(H_2PO_4)_2 + 2H_2O$ ------> $CaHPO_4 + H^+ + H_2PO_4^-$

but in soils with pH values less than 7.7 the following reaction neutralizes the acidity produced so that there is no net acidification;

$CaHPO_4 + H_2O$ ------> $Ca^{2+} + H_2PO_4^- + OH^-$

In high pH soils (pH >7.2), dissociation of $H^+$ ion from the $H_2PO_4^-$ molecule will generate some acidity. Crop uptake of P has little effect on soil acidity due to the small amounts of fertilizer P taken up in any one year, hence fertilizer chemistry dominates pH changes and significant differences in rhizosphere pH have not been observed for uptake of different orthophosphate ions.

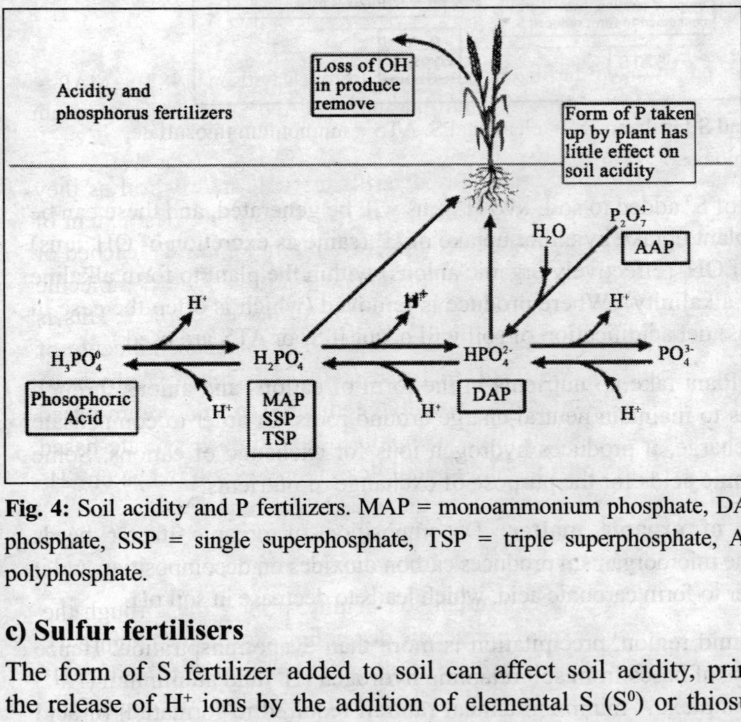


**Fig. 4:** Soil acidity and P fertilizers. MAP = monoammonium phosphate, DAP = diammonium phosphate, SSP = single superphosphate, TSP = triple superphosphate, APP = ammonium polyphosphate.

## c) Sulfur fertilisers

The form of S fertilizer added to soil can affect soil acidity, principally through the release of $H^+$ ions by the addition of elemental S ($S^0$) or thiosulfate ($S_2O_3^{2-}$, in ammonium thiosulfate - ATS) (Figure 5). However, the amounts of S added to soil and taken up by plants are generally small in comparison to N.

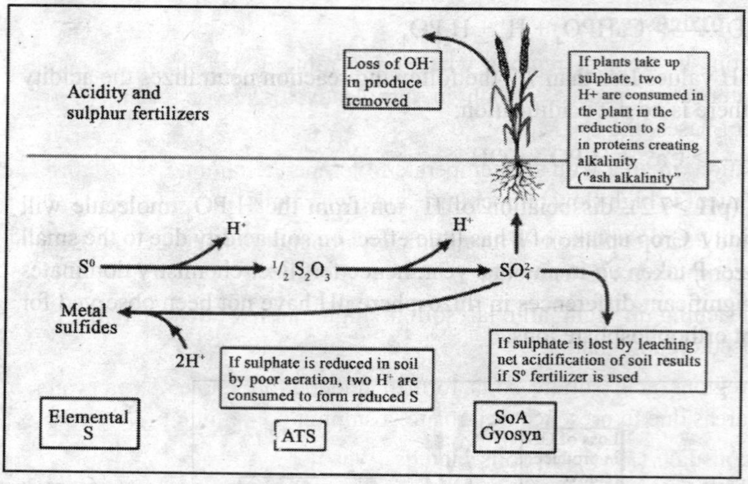


**Fig. 5:** Soil acidity and S fertilizers. $S^0$ = elemental S, ATS = ammonium thiosulfate, SoA = sulfate of ammonia

For each molecule of $S^0$ added to soil, two $H^+$ ions will be generated, and these can be balanced through plant uptake by either uptake of $H^+$ (same as excretion of $OH^-$ ions) or the generation of $OH^-$ (effectively organic anions) within the plant to form alkaline plant material (ash alkalinity). Where produce is removed (which is often the case in agricultural systems) net acidification of soil will occur if $S^0$ or ATS are used.

**iv) Root activity:** Plant take up nutrients in the form of cations and anions through root. The plants has to maintain neutral charge around roots. In order to compensate for extra positive charge, it produces hydrogen ions for exchange of cations. Some plants produce organic acids for the purpose of exchange of nutrients.

**v) Decomposition of organic matter:** Decomposition of organic matter needs microorganisms. The microorganism produces carbon dioxides on decomposition and it combines with water to form carbonic acid, which leads to decrease in soil pH.

**vi) Climate:** In humid region, precipitation is more than evapotranspiration. Hence there is leaching loss of bases from soil retaining hydrogen ($H^+$) and aluminium ($Al^{3+}$) in soil exchange complex. A minimum annual rainfall required to formation of acid soils is 750 mm annual rainfall. In temperate region, although rainfall is scanty, due to high amount of organic matter accumulation, acid soils are produced.

**vii) Vegetation cover:** In temperate regions, it is covered by acid producing conifers vegetation (heath, fir etc.) hence acid soils are produced easily. Leaf litter on ground on decomposition produces variety of organic acids which causes formation of acid soils. In coastal regions with marshy plants on death and after decay produces acid which render soil acidic

**viii) Topography:** Sloppy lands favour formation acid soils due to high amount of surface run off leading to removal of bases from soil.

### ix) Pedogeneic processes

- Laterisation: In tropical regions where high rainfall with high temperature causes laterisation process which in turn leads to intense weathering and leaching of bases like Ca, Mg
- Podsolization: in area with sub temperate to temperate climate, where organic matter content is high (low temperature and high rainfall)
- Marshy and Peaty condition: With significant amount of under decomposed and partly decomposed organic matter.
- Intense leaching in light alluvial soil in high rainfall areas having partly decomposed organic matter.
- Coastal regions acid sulfate soils formed under inundation of sea water in lowland areas due to presence of jarosite, a sulphur containing mineral.

The processes consume $H^+$ ions in soils include, Weathering of most minerals (e.g., silicates, carbonates etc.), decomposition of organic anions, reduction of oxidized forms of N, S and Fe, Roots release $OH^-$ or $HCO_3^-$ to balance internal charge when anion uptake exceeds cation uptake and Inner sphere adsorption of anions (especially sulfate) which displaces hydroxyl ($OH^-$) groups

## 9.2.1.2. Classification of acid soils

Mandal (1974) has classified acid soils of India into seven groups viz., 1. Laterite soils, 2. Lateritic and laterite red soils, 3. Mixed yellow red soils, 4. Ferruginous red soils, 5. Podzolic soils, 6. Terai soils and 7. Peaty soils. In Mandal's classification, acid sulphate soils and degraded alkali soils have not been given proper places. Nevertheless, it would be appropriate if they are classed with acid soils as the pH levels of such soils indicate that they are acidic in nature. According to the intensity of acidity, the acid soils are of the following five types viz., (1) Slight acidic (pH range 6.6 to 6.1); (2) Medium acidic (pH 6.0 to 5.6); (3) Strong acidic (pH 5.5 to 5.1); (4) Very strong acidic (pH 5.0 to 4.6) and (5) Extremely strong acidic (pH 4.5 or lower).

## Distribution of acid soils in India

About 49 million hectares are considered acidic of which 26 million hectares have pH below and 23 million hectares pH between 5.6 and 6.5. These soils include red, laterite, mixed red and yellow, brown forest, foot hill soils, peat soils and varieties of acid sulfate soils. There is considerable heterogeneity in the morphological, physical and chemical characteristics which may be attributed to various soil forming factors in the development of these soils in the various regions of the country (Fig. 6).

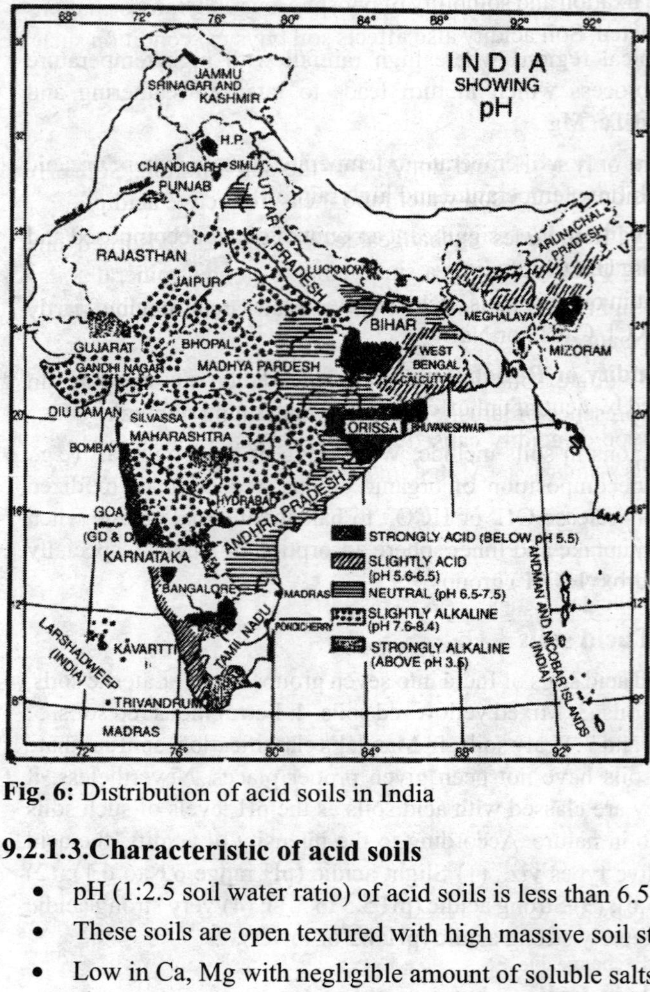


**Fig. 6:** Distribution of acid soils in India

### 9.2.1.3.Characteristic of acid soils

- pH (1:2.5 soil water ratio) of acid soils is less than 6.5
- These soils are open textured with high massive soil structure.
- Low in Ca, Mg with negligible amount of soluble salts.
- These soils appear as brown or reddish brown, sandy loams or sands
- Acid soils have kaolinitic type of clay minerals and high AEC
- Acid soils have low dispersion and flocculates in a suspension
- Acid soils have low population of bacteria and actinomycetes but very high amounts of fungi
- These soils have high base unsaturation and low base saturation.

### 9.2.1.4. Importance of soil acidity

Soil acidity determines suitability of soils for crop growth, excess $H^+$ and $Al^{3+}$ ions interfere with plant growth, nutrient balance and microbial proliferation, enzymatic activity. Soil acidity imparts toxicity of certain elements like $H^+$ and $Al^{3+}$, deficiency Ca, Mg and Mo thus affecting availability of plant nutrients in soil. Acid conditions in soil encourage fungal growth and discourage bacterial and actinomycetes growth, therefore biological nitrogen fixation and solubility of phosphorous and transformation of certain other nutrients affected. Soil acidity also affects soil physical condition such as soil structure.

### 9.2.1.5. Forms of soil acidity

1) **Active acidity**: Refers only to $H^+$ and not $Al^{3+}$ in the soil solution. Soil pH measured with suspension of 1:2.5 soil water ratio measures active acidity.
2) **Exchangeable acidity**: It includes exchangeable $Al^{3+}$ and $H^+$ adsorbed on surface of soil colloids. Usually there is a small amount in acid mineral soils but it is more abundant in organic soils. It is extracted with a neutral unbuffered salt solution, such as KCl, CaCl2 or NaCl
3) **Non-exchangeable acidity or Residual acidity:** This acidity is comprised of weak acids not replaced by neutral unbuffered salt solution and $H^+$ which bonds with $OH^-$.This is the type of acidity caused by organic matter and bound Al. Bound Al occurs in soils primarily as Al- polymers (long chain compounds) and is denoted as Al $(OH)^{2+\backslash}$

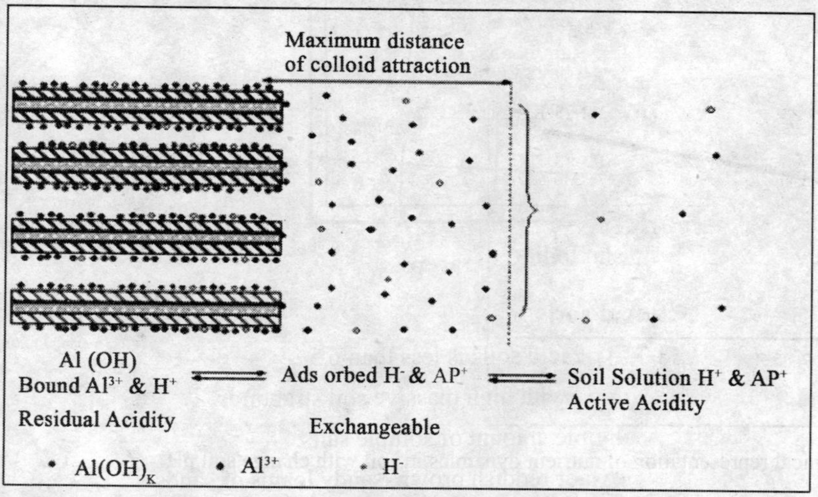


**Fig. 7:** Reserve, exchangeable and active acidities in soil

## Active acidity and reserve acidity

Active acidity refers to the activity of hydrogen ions in the aqueous phase of soil. It is the primary source of soil acidity. Small amount of lime is required to neutralize the active acidity. It is generally much less than reserve acidity (number of adsorbed H+ ions in the diffuse layer is much greater than those in true solution). Buffering capacity of soil is not dependent on the active acidity. Reserve acidity refers to the hydrogen and aluminium ions held on the soil colloids. It is the secondary source of soil acidity. Large amount of lime is required to neutralize the reserve acidity. It is commonly for greater than the active acidity. It may be 1000 times greater than the active acidity of sandy soil and 50,000 times or even 1,00,000 times greater than the active acidity of clay soil with high organic matter. Buffering capacity of soil is directly related to the reserve acidity.

### 9.2.1.6. Effect of acidity on soils and crops

**Direct effect:** Plant root system does not grow normally due to toxic Aluminum ions. Permeability of plant membranes are adversely affected due to soil acidity. Enzyme actions may be altered, since they are sensitive to pH changes

**Indirect effects:** In acid soils deficiency of Ca and Mg occur by leaching. In Watermelon and Tomato deficiency of Ca causes Blossom end rot. An important function of calcium in the plant is with the formation of cell walls. Since calcium is immobile within the plant, the fruit continues to grow but the growing tip is soft and dark due to poor cell wall formation. Al, Mn and Fe available in toxic amounts. All the micro nutrients except molybdenum are available. So 'Mo' deficiency has been identified in leguminous crops. Phosphorous gets immobilized and its availability is reduced.

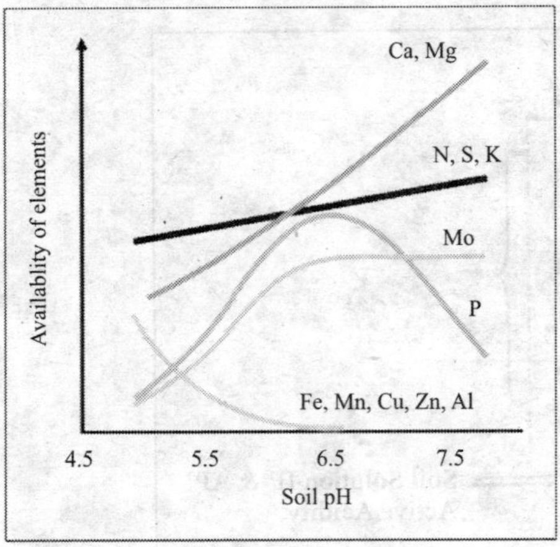


**Fig. 8:** Graphical representation of nutrient dynamics in soil with change soil pH

**Table 2:** Effects of acid soils on plant nutrient availability

| Nutrient | Action | Outcome |
|---|---|---|
| Potassium | Depleted by leaching | Health problems |
| Calcium | Depleted by leaching | Poor soil structure |
| Magnesium | Depleted by leaching | Poor soil structure |
| Phosphorous | Deficiency by fixation | Poor pasture growth |
| Molybdenum | Deficiency by fixation | Poor pasture growth |
| Aluminium | Excess | Toxic to plants |
| Iron | Excess | Ties up oether nutrients like phosphorous |

### Effect of soil acidity on activity of micro organisms

Most of the activities of beneficial organisms like *Azotobacter* and nodule forming bacteria of legumes are adversely affected as acidity increases. Nitrification process is affected in acid soils. Rhizobium activity is increased as the pH value of acid soils increases

### 9.2.1.7. Management of acid soils

1) Application of lime well ahead of planting: Reactions of Lime in acid soil when it added to acid soil

a) Lime neutralizes soil acidity.

$2H^+\text{-}X + CaCO_3 \rightarrow Ca\text{-}X + H_2CO_3 + H_2O$

b) Lime increases soil base saturation

$BS = (CEC - (Al^{3+} + H^+)/ CEC * 100$

c) Soil pH increases since lime neutralizes soluble free $H^+$ ion n

d) Solubility of Al decreases as the hydroxyl ions released by lime with form aluminium hydroxide precipitate.

$CaCO_3 + H_2O \text{ -------> } CaHCO_3 + OH^-$

$Al^{3+} + 3\ OH^- \rightarrow Al(OH)_3$

Soluble insoluble

(toxic) (not toxic)

A major reason for crop response to lime is due to the precipitation of exchangeable Al as aluminiium hydroxide.

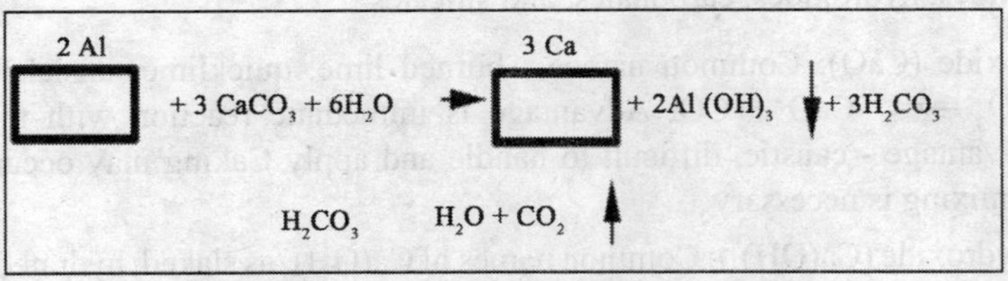


**Note:** The beneficial effects of liming on crop growth are often related to neutralization of Al and not directly to the change in pH.

**Efficiency of lime material**: The effectiveness of lime is depend on method of application, Particle size, Reactivity viz., neutralizing value and moisture content in soil

a) **Particle size of lime material:** Lime is lowly soluble in water. So it has been to powdered to neutralize soil acidity. Effectiveness depends on purity and fineness of the lime materials

b) **Neutralizing value of liming material:** The ability to neutralize acids expressed in terms of calcium carbonate equivalent. Calcium carbonate is the standard by which other materials are measured (100%) Molecular weight of $CaCO_3$ is 100; $MgCO_3$ is 84. One molecule of each will neutralize the same amount of acid but on a weight basis it only takes 84g of $MgCO_3$ to do the job of 100g of $CaCO_3$.

c) **Method of application:** Application (broadcasting) of finely ground lime (100 mesh screen) as per lime requirement and mixing it thoroughly in the soil reacts well in soil. Application of coarse lime would not react properly in soil and thus less effective.

**Table 3:** Relative neutralizing value of some common lime materials

| Liming materials | Neutralizing value |
|---|---|
| 1. Calcium carbonate | 100 |
| 2. Dolomitic lime | 95-108 |
| 3. Calcitic lime | 85- 100 |
| 4. Baked oyster shells | 80- 90 |
| 5. Marl | 50-90 |
| 6. Gypsum | none |
| 7. Basic slag | 50-70 |

### Frequency of liming

Liming is repeated after 3-4 years of first application. The frequency of liming depends on a) Texture b) N fertilizing rate c) Ca and Mg crop removal d) amount of lime added e) soil pH desired

## Liming materials

To be considered a liming material an anion must produce OH - ions to react with $H^+$ and $Al3^+$ ions. Oxides, hydroxides, carbonates, and silicates

1. Calcium oxide (CaO): Common names - burned lime, quicklime, unslaked lime $CaCO_3$ ===> CaO + $CO_2$ Advantage is immediate reaction with the soil. Disadvantage - caustic, difficult to handle and apply Caking may occur. Thorough mixing is necessary
2. Calcium hydroxide ($Ca(OH)_2$): Common names of $Ca(OH)_2$ as slaked, hydrated, builders lime CaO + $H_20$ ===> $Ca(OH)_2$ Advantage of slaked lime is quick reaction with the soil, while disadvantage is difficult and unpleasant to handle.
3. Calcitic limestone ($CaCO_3$), Dolomitic limestone ($Ca Mg (CO_3)_2$): It is Mined from deposits of arid and semiarid regions. Quality depends on amount of impurities such as clay. Good handling properties. Reaction time several months
4. Marl: It is unconsolidated deposits of $CaCO_3$. Usually contaminated with clay low in Mg
5. Slags ($CaSiO_3$): It is By-product of furnaces used for making iron, steel and elemental P.
6. Fly ash: it is By-product in thermal and coal industry rich in silica

**Lime requirement (LR)**

The amount of lime required to be added to acidic soil to raise the pH to a desired value is known as lime requirement or the amount of good quality agricultural limestone required to establish the desired soil pH range for the cropping system being used. LR are determined in the laboratory using a buffer pH in equilibrium with the soil.

**Benefits of liming of acid soils**

- Liming of acid soils prevents aluminium and manganese toxicity
- Increases phosphorus , molybdenum availability
- Improves biological nitrogen fixation by microbes
- Improves P and K efficiency
- Enhances soil exploration by roots
- Improves physical condition of soil
- Improves CEC in variable charge soil
- Improves availability of soil nutrients

**Procedure for reclamation of acid Soil**

- Bring a representative sample of soil from the field to be reclaimed.
- Determine the pH of soil in soil water extract (1:2)
- Determine the lime requirement of the soil
- Select appropriate liming material
- Calculate total quantity of the liming material after having known its neutralizing value of $CaCO_3$ equivalent.
- Spread uniformly half the quantity of lime on the field and plough/ harrow it into the soil when the soil is in friable condition.
- Spread uniformly the remaining half of the liming material and mix it thoroughly by disking and harrowing.
- While mixing, the field should be in fine tilth.

**2. Growing of Acid Tolerant Crops:** In acid soils, acid tolerant crops should be grown. Choice of crops may be done according to soil pH. Crops can be divided into following groups:

(a) Highly acid tolerant crops: Rice, potato, sweet potato, oat, castor, Echinochloa, Paspalum etc.

(b) Moderately acid tolerant crops: Barley, wheat, maize, turnip, brinjal, cow pea, mung beans, pigeon peas, pea nuts etc.

(c) Slightly acid tolerant crops: Tomato, carrot, red clover etc.

**Table 4:** Crops tolerance for different soil acidities

| Soil pH | Tolerant crops |
|---|---|
| 4.5 | Citrus, Blue berries |
| 5.0 | Tobacco, Apple, Grapes, Plum, watermelon |
| 5.5 | Cowpea, Soybean, Cotton, Wheat, Oat, Peas, Tomato, Sorghum |
| 6.0 | Peanut, Cabbage, Carrot, Onion, Radish, Spinach, Cauliflower |
| 6.5 | Alfalfa, Sugarbeat |

**3. Fertilizer management in acid soils:** Phosphate fertilization is almost essential for sustaining crop production in acid soils. As it is known that acid soils fix phosphorus in the soil. Application of rock phosphate in acid soil, soils provide an ideal situation for release of Phosphorus (P) from rock phosphate (RP). The efficiency of rock – phosphate increase with increasing fineness. A low pH (below 5.0) facilitates release of phosphate from rock phosphate and the effectiveness decreases with increasing pH above 5.0. It is advisable to apply rock-phosphate several months ahead of liming to permit dissolution of rock phosphate at low pH. Rock phosphates is preferred source of Phosphorous(P) in acid – soils as it is a cheaper source and contains 30 to 50% CaO. CaO increase available calcium (Ca) in the soil, thus raises soil pH and lowers exchangeable Al content.

The need for nitrogen fertilizer can be minimized by introducing acid tolerant N-fixing legumes in the cropping system besides improving the efficiency of fertilizer nitrogen. Higher nitrogen (N) rates are recommended in high rainfall areas to compensate leaching losses. Split application of N definitely improves recovery. The efficiency of potassium fertilizer can be increased by split application of k- fertilizer. The split application is quite essential in acid soils having sandy texture. The crop residues should not be removed from the field.

**Water management:** Acidification can be prevented by continuous flooding or saturation. Where that is not possible, the soils are drained and leached to remove the acid material. After submergence the soil pH rises due to precipitation of aluminium hydroxide and by the reduction of ferric ion.

## 9.2.2. Acid sulphate soils

Acid sulphate soils are drained coastal wetland soils that have become acid (pH<4) due to oxidation of the pyritic minerals in the soil. (or) Soils formed from weathering of sulphide bearing parent materials which results in the formation of extreme low pH (< 3.0) and precipitation of sulfate salts. Undrained soils containing pyrites need not be acid and they are called potential acid sulphate soils. Soil with sufficient sulphides ($FeS_2$ and others) to become strongly acidic when drained and aerated enough for cultivation are termed acid sulphate soils or as the Dutch refer to those soils cat clays. When allowed to develop acidity, these soils are usually more acidic than pH 4.0. Before drainage, such soils may have normal soil pH and are only potential acid sulphate soils.

### 9.2.2.1. Types of acid sulphate soils

***Potential acid sulfate soils*:** ASS which have not been oxidised by exposure to air are known as potential acid sulfate soils (PASS). They are neutral in pH (6.5–7.5), contain unoxidized iron sulphides, are usually soft, sticky and saturated with water and are usually gel-like muds but can include wet sands and gravels have the potential to produce acid if exposed to oxygen

***Actual acid sulfate soils*:** When potential acid sulfate soils are exposed to oxygen, the iron sulfides are oxidised to produce sulfuric acid and the soil becomes strongly acidic (usually below pH 4). These soils are then called actual acid sulfate soils (AASS). They have a pH of less than 4, contain oxidised iron sulfides, vary in texture and often contain jarosite (a yellow mottle produced as a by-product of the oxidation process).

### 9.2.2.2. Occurrence and distribution of acid sulfate soils in India

Geographically, the majority of acid sulfate soils occur in coastal areas, developing from recent sediments. They are usually restricted to areas relatively close to the sea, where they have formed marine and estuarine deposits. Broadly, they are distributed in:

a) Coastal areas with marine or brackish water influences.
b) Freshwater backswamps, formerly brackish.
c) Lake bottoms reclaimed by poldering and draining of water.
d) Inland continental environments with inherent pyritic parent rock or where lignite and coal have been mined for long periods of time.

Generally acid sulphate soils are found in coastal areas where the land is inundated by salt water. In India, acid sulfate soil is, mostly found in Kerala (kuttanad) and West Bengal (Sunderban). The area covered under acid sulphate soils in India 0.4 million hectares.

### 9.2.2.3. Formation of acid sulphate soils

Lands inundated with waters that contain sulphates, particularly salt waters, accumulate sulphur compounds, which in poorly aerated soils are bacterially reduced to sulphides. Such soils are not usually very acidic when first drained in water.

When the soil is drained and then aerated, the sulphide ($S^{2-}$) is oxidised to sulphate ($SO_4^{2-}$) by a combination of chemical and bacterial actions, forming sulphuric acid ($H_2SO_4$). The magnitude of acid development depends on the amount of sulphide present in the soil and the conditions and time of oxidation. If iron pyrite ($FeS_2$) is present, the oxidised iron accentuates the acidity but not as much as aluminium in normal acid soils because the iron oxides are less soluble than aluminium oxides and so hydrolyze less.

## Reactions involving the formation of acid sulphate soils

Acid sulphate soils form due to oxidation of sulphides in soils. The slow oxidation of mineral sulphides in soils is non-biological until soil pH reaches an acidity of pH 4.0, but the process is not well understood. Below pH 4.0, the bacteria, Thiobacillus ferroxidans are the most active oxidizers and the activity builds up rapidly.

**Biological**

$$2S + 3O_2 + 2H_2O \xrightarrow[\text{Oxidation}]{\text{Microbial}} H_2SO_4 \text{ (Sulphuric acid)}$$

$$H_2S{+}2O_2 \text{(Hydrogen sulphide)} \xrightarrow[\text{Oxidation}]{\text{Microbial}}$$

**Non-biological**

$$2FeS_2 + 2H_2O + 7O_2 \text{(Iron pyriyc)} \longrightarrow 2FeSO_4 \text{ (Ferrous sulphate)} + 2H_2SO_4 \text{ (Sulphuric acid)}$$

**Accelerated by bacteria (Thiobacillus ferroxidans)**

$$\underset{\text{(Ferrous sulphate)}}{4FeSO_4 + O_2 + 2H_2SO_4} \longrightarrow \underset{\text{(Ferric sulphate)}}{2\,Fe_2\,(SO_4)_3 + 2H_2O}$$

**Rapid in pH (non-biological):**

$$FeS_2 + 7\,Fe_2(SO_4)_3\ 8H_2O \longrightarrow \underset{\text{(Ferric sulphate)}}{15\,FeSO_4 + 8H_2SO_4}$$

### 9.2.2.4. Classification of acid sulphate soils

Acid sulfate soils can be classified as Sulfaquepts (Aquepts with a sulfuric horizon that has its upper boundary within 50 cm of the soil surface), Sulfie Tropaquepts (Tropaquepts with jarosite mottles and a pH 3.5 to 4 somewhere within the 50 cm depth, or with jarosite mottles and a pH<4 in some part between 50 to 150 em depth), or Sulfie Haplaquepts (comparable to Sulfie Tropaquepts but formed under a more temperate climate). The distinction between Sulfaquepts and Sulfic Tropaquepts is that the former are generally unsuitable for agriculture without costly amendment measures, whereas the latter can often be made productive easily. Potential acid sulfate soils are either Sulfaquents (Aquents with sulfidic material within 50 cm of the mineral soil surface), Sulfie Fluvaquents (Fluvaquents with sulfidic material between 50 and 100 cm depth), or Sulfihemists (Histosols with sulfidic material within 100 cm depth). Para or pseudo acid sulfate soils can be classified as· Tropaquepts or Haplaquepts. Acid sulfate soils that are dominantly organic may be Sulfohemists. Strong acidity in acid sulphate soils results toxicities of aluminium and iron, soluble salts (unless leached), manganese and hydrogen sulphide ($H_2S$) gas. Hydrogen sulphide ($H_2S$) often formed in lowland rice soils causing akiochi disease that prevents rice plant roots from absorbing nutrients.

### 9.2.2.4. Characters of acid sulphate soils

*Soil acidity:* The reported ills of acid sulfate soils may be due to the direct effect of hydrogen ions, especially below pH 3.5 to 4. However, aluminum toxicity IS probably more important in this pH range.

*Salinty*: Acid sulfate soils in tidal areas are often affected by salinity. Salinity aggravates other toxicities, both by weakening the plants and by increasing iron and perhaps aluminum in solution.

*Aluminum toxicity*: One cause of stress on the growth of certain plant species is aluminum toxicity.

A high Al level affects cell division, disrupts certain enzyme systems, and hampers uptake of phosphorus, calcium and potassium. Most plants grown on acid sulfate soils which have a pH below 4 suffer from Al toxicity.

*Iron toxicity:* Dissolved iron in excess of 300-400 ppm is toxic to rice . It was reported that about 5,000 ppm of water-soluble Fe was found two weeks after submergence in an acid sulfate soil.

*Low nutrient content:* In the absence of iron and aluminum toxicity and harmful salinity, phosphorus deficiency is the most important problem of acid sulfate soils, especially the Sulfic Tropaquepts.

*Low Base Status:* During the formation of acid sulfate soils, bases are removed as sulfate and most of the exchange complex is occupied by aluminum. Therefore, acid sulfate soils are likely to be deficient in Ca and K. Exchangeable Ca content of 3.5-5.0 *me/* 100 gm soil and exchangeable Mg of 3.0*3.2 me/ 100*gm in the top 35 cm of an acid sulfate soil in Thailand has been reported.

*Hydrogen Sulfide toxicity:* Hydrogen sulfide has been shown to be toxic to the rice plant through its suppression of the oxidizing power of the roots. It was showed that sulfate reduction occurred slowly at pH 5 and increased as the pH was increased from 5 to 7. when the pH increases above 4.5 to 5, sulfate reduction takes place, producing sulfide which is then precipitated as FeS. If the soil is low in $Fe^{2+}$ , H2S toxicity might be possible. In potential acid sulfate soil indicated that acid-soluble sulfide was detected at Eh of -55 mv or equivalent to Eh7 of -103 mv and the soil pH was 6.2. The condition of pH 6.2 could arise from liming of this acid sulfate soil. However, this soil normally contains quite high .amount of active Fe, so that free sulfide is not found.

### 9.2.2.5. Management of acid sulphate soils

Management techniques are extremely variable and depend on many specific factors viz. the extent of acid formation, the thickness of the sulphide layer, possibilities of leaching or draining the land etc.

i) **Keeping the field flooded:** Maintaining the reduced condition of flooded (anaerobic) soil inhibits acid development, which requires oxidation. This solution almost limits the use of the area to rice growing. Unfortunately, droughts occur and can in short time periods cause acidification of these soils. The water used to flood the potential acid sulphate soils often develop acidity and injure crops.

ii) **Controlling water table:** If a non-acidifying layer covers the sulphuric horizon, drainage to keep only the sulphuric layer under water (anaerobic) is possible.

iii) **Liming and leaching:** Liming is the primary way to reclaim any type of acid soil. Acid soil may require 11.45 MT per hectare of lime in a 20-year period whereas acid sulphates soils may require from several metric tons per hectare per year, upto even 224 MT per hectare (100 t/acre) within a 10-year period or less. If these soils are leached during early years of acidification, lime requirements are lowered. Leaching, however, is difficult because of the high water table commonly found in this type of soil and low permeability of the clay. Sea water is sometimes available for preliminary leaching.

iv) **$MnO_2$ Addition:** Because of the high standard oxidation- reduction potential of the $MnO_2^-$ $Mn^{2+}$ system, soil reduction is retarded by $MnO_2$ addition. Manganese dioxide improved the growth and yield of rice in an acid sulfate soil because it depresses the concentration of water-soluble $Fe^{2+}$ and $Al^{3+}$.

### 9.2.3. Salt affected soils

Salt affected soils are those soils which are not suitable for the growth of crops because of an excess of salts, exchangeable sodium, or both (or) Salt affected soils may contain, an excess of water soluble salts (saline soils), excess exchangeable sodium (sodic soils) or both an excess of salts and exchangeable sodium (saline-sodic soil).

Field diagnosis of Salt affected soils:

- Delayed/ reduced germination
- Stunted plant growth
- Foliar damage
- Presence of Salt crusts
- Water logging

Laboratory diagnosis parameters for salt affected soils:

- Soil pH
- Soil EC
- Sodium Adsorption Ratio (SAR)
- Exchangeable Sodium Percentage (ESP)
- Cation Exchange Capacity (CEC)
- Lime ($CaCO_3$) content
- Anions and Cations: eg. $Ca^{++}$, $Mg^{++}$, $Na^+$, $Cl^+$,$SO_4^{--}$, $CO3^{--}$, $HCO_3^-$
- Gypsum content
- Soil texture

Bhumbla (1977) has classified the salt lands of the country into four broad groups based on the nature of soil problem and their geographical distribution (Table 5). Sodic soils of the Indo-Gangetic plains alone account for roughly one-third of the 7 million hectares of the salt affected soils in India.

**Table 5:** Types of salt affected soils and their extent in India (adopted from Bhumbla, 1977)

| Type of soils | States in which they occur | Approx. area (million ha) |
|---|---|---|
| 1. Coastal salty soils | | |
| a) Coastal salty soils of arid regions | Gujarat | 0.714 |
| b) Deltaic coastal salty soils of the humid region | West Bengal, Orissa, Andhra Pradesh and Tamil Nadu | 1.394 |
| c) Acid salty soils | Kerala | 0.016 |
| 2. Salty soils of the medium and deep black soil regions | Karnataka, Madhya Pradesh, Andhra Pradesh and Maharashtra | 1.420 |
| 3. Salty soils of the arid and semi-arid regions | Gujarat, Rajasthan, Punjab, Haryana and Uttar Pradesh | 1.000 |
| 4. Sodic soils of the Indo-Gangetic plains | Haryana, Punjab, Uttar Pradesh, Bihar, Rajasthan and Madhya Pradesh | 2.500 |
| | Total | 7.044 |

## Distribution of salty soils in India

According to latest estimates (Mandal et al., 2010), salty soils in India are distributed almost throughout the country (Fig. 9 and Table 6) and cover an extent of about 6.7 Mha with states of Gujarat (2.23 m ha), Uttar Pradesh (1.37 m ha), Maharashtra (0.61 m ha), West Bengal (0.44 m ha) and Rajasthan (0.38 m ha) together accounting for almost 75% of saline and sodic soils in the country.

**Table 6:** Extent (ha) of salty soils in different states of India (000 ha) *Source:* Mandal et al. (2010). (Based on NRSA data of 1996 and reconciled during 2006 jointly by NRSA, CSSRI and NBSS & LUP, Nagpur)

| States | Sodic | Saline | Total |
|---|---|---|---|
| Andhra Pradesh | 196.6 | 77.6 | 274.2 |
| Andaman & Nicobar Islands | 0 | 77.0 | 77.0 |
| Bihar | 105.9 | 47.3 | 153.2 |
| Gujarat | 541.4 | 1,680.6 | 2,222.0 |
| Haryana | 183.4 | 49.2 | 232.6 |
| Jammu & Kashmir | 17.5 | 0 | 17.5 |
| Karnataka | 148.1 | 1.9 | 150.0 |
| Kerala | 0 | 20.0 | 20.0 |
| Madhya Pradesh | 139.7 | 0 | 139.7 |
| Maharashtra | 422.7 | 184.1 | 606.8 |
| Orissa | 0 | 147.1 | 147.1 |

| | | | |
|---|---|---|---|
| Punjab | 151.7 | 0 | 151.7 |
| Rajasthan | 179.4 | 195.6 | 375.0 |
| Tamil Nadu | 354.8 | 13.2 | 368.0 |
| Uttar Pradesh | 1,347.0 | 22.0 | 1,369.0 |
| West Bengal | 0 | 441.3 | 441.3 |
| Total | 3,788.2 | 2,956.9 | 6745.1 |

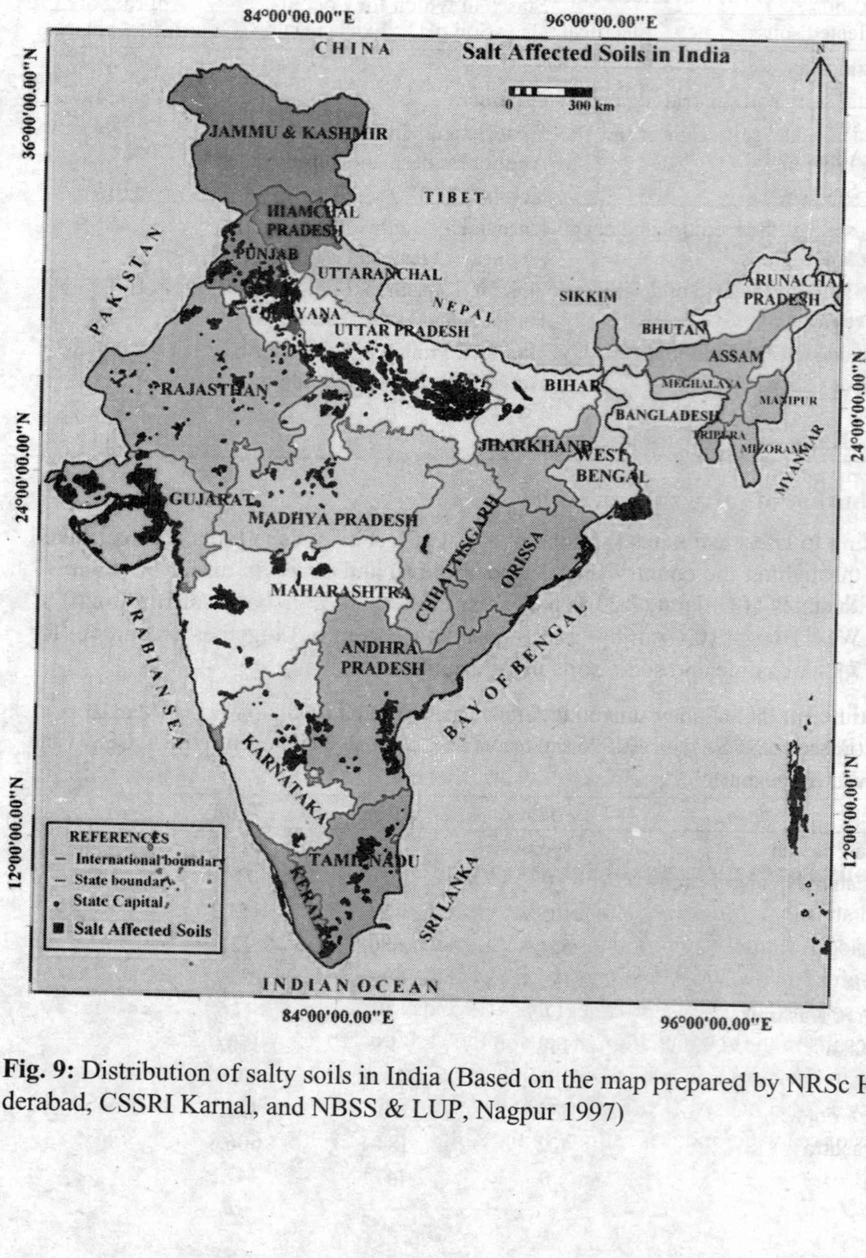


**Fig. 9:** Distribution of salty soils in India (Based on the map prepared by NRSc Hyderabad, CSSRI Karnal, and NBSS & LUP, Nagpur 1997)

## Classification of salt affected soils

Salt affected soil is classified based on three parameters viz., soil reaction (pH), Exchangeable sodium percentage (ESP) and electrical conductivity (EC). Accordingly salt affected soils is classified as Saline soils, Sodic soils (Alkali soils) and Saline -sodic soils.

**Table 7:** Characters of Salt affected soils

| Salt affected soils | Electrical Conductivity ($dSm^{-1}$) | Soil pH | Exchangeable sodium percentage (ESP) | Soil physical condition |
|---|---|---|---|---|
| Saline soil | >4 | <8.5 | <15 | Normal |
| Sodic/Alkali soils | <4 | >8.5 | >15 | Poor |
| Saline-Sodic soil | >4 | <8.5 | >15 | Normal |

### 9.2.3.1. Saline Soils

Soils, classified as saline when they contain a high enough concentration of soluble salts to interfere with normal growth and development of salt-sensitive plants. Salinity is measured in terms of electrical conductivity (EC). It is expressed as deciSiemens per meter ($dSm^{-1}$).

### Field Diagnsosis of Saline soils

- Plants appear water stresses
- Poor seed germination
- Leaf burn in crop plants
- White encrustation on surface during dry periods (hence saline soils also called white alkali soils).
- Shallow water table.

The saline soils contains toxic concentration of soluble salts in the root zone. Soluble salts consists of chlorides and sulphates of sodium, calcium, magnesium. Because of the white encrustation formed due to salts, the saline soils are also called white alkali soils.

### Reasons for occurrence of salinity in soils

In arid and semi-arid areas salts formed during weathering are not fully leached. During the periods of higher rainfall the soluble salts are leached from the more permeable high laying areas to low laying areas and where ever the drainage is restricted, salts accumulate on the soil surface, as water evaporates. The excessive irrigation of uplands containing salts results in the accumulation of salts in the valleys. In areas having salt layer at lower depths in the profile, seasonal irrigation may favour the upward movement of salts. Salinity is also caused if the soils are irrigated with saline water. In coastal areas the ingress of sea water induces salinity in the soil.

## Salinity classes and crop growth

There is no critical point of salinity where plants fail to grow. As the salinity increases growth decreases until plants become chlorotic and die. Plants differ widely in their ability to tolerate salts in the soil. Salt tolerance ratings of plants are based on yield reduction on salt-affected soils when compared with yields on similar non-saline soils. Soil salinity classes generally recognized are given in Table 6

**Table 8:** Salinity classes and crop growth

| Soil Salinity Class | EC of the Saturation extract (dS/m) | Effect on Crop Plants |
|---|---|---|
| Non saline | 0 - 2 | Salinity effects negligible |
| Slightly saline | 2 - 4 | Yields of sensitive crops may be restricted |
| Moderately saline | 4 - 8 | Yields of many crops are restricted |
| Strongly saline | 8 - 16 | Only tolerant crops yield satisfactorily |
| Very strongly saline | > 16 | Only a few very tolerant crops yield satisfactorily |

## Characteristics of Saline soils

- Saline soils are in flocculated state of physical condition (normal structure)
- Their permeability is higher than alkali soils.
- Physical condition is good and hence water can pass through them (good permeability)
- These soils have white encrustation on surface of soils.
- Soluble salts in soils create osmatic potential i.e. soil water is held tightly enough by the ions that plants cannot use it (apparent moisture stress).
- Saline soils characteristically remain moist longer than the rest of the field.
- Saline soils are usually barren but potentially productive soils (can be arable by leaching of salts)

## Major production constraints of Saline soils

Presence of salts leads to alteration of osmotic potential of the soil solution. Consequently water intake by plants restricted and thereby nutrients uptake by plants are also reduced. In this soil due to high salt levels microbial activity is reduced. Specific ion effects on plants are also seen due to toxicity of ions like chloride, sulphate, etc.

Growth reduction of cotton for salinity levels of 1, 14, 28, and 42 dS/m (left to right

Growth reduction pea caused by salinity levels of 1, 3, 7, 11, and 14 dS/m (left to right)

**Fig. 10:** Effect of soil salinity levels on crop growth

## Management of Saline soils

### 1. Removal of excess salts from the soil surface and the root zone in saline soils

**Scraping:** Removing the salts that have accumulated on the soil surface by mechanical means has had only a limited success although many farmers have resorted to this procedure. Although this method might temporarily improve crop growth, the ultimate disposal of salts still poses a major problem.

**Flushing:** Washing away the surface accumulated salts by flushing water over the surface is sometimes used to desalinize soils having surface salt crusts. Because the amount of salts that can be flushed from a soil is rather small, this method does not have much practical significance.

**Leaching:** This is by far the most effective procedure for removing salts from the root zone of soils. Leaching is most often accomplished by ponding fresh water on the soil surface and allowing it to infiltrate. Leaching is effective when the salty drainage water is discharged through subsurface drains that carry the leached salts out of the area under reclamation.

From an irrigation standpoint, the dissolved salts or salinity of irrigation water, commonly measured by electrical conductivity (EC), has received the most attention. The key to salinity control and to irrigation sustainability is leaching. Leaching is the net downward movement of water and controls the salt distribution in the profile.

When the crop root zone has become too saline for the existing plants, it requires extra water to leach out the accumulated salts by increasing LF. *Maintenance leaching* is necessary to keep the average rootzone salinity below the plant threshold EC levels. The leaching fraction (LF) is defined as

$$LF = D_{dw}/D_{iw} = C_{iw}/C_{dw} = EC_{iw}/EC_{dw}$$

and it is customary to substitute *EC* for *C* since the *EC* is much easier to measure than concentration and EC is linearly related to the total salt concentration, where

$D_{dw}$ = the equivalent depth of water passing below the rootzone

$D_{iw}$ = the equivalent depth of water applied (irrigation + rainfall)

$C_{iw}$ *(ECiw)* = Salt concentration (EC) of the applied water

$C_{dw}$ $(EC_{dw})$ = Salt concentration (EC) of the drainage water

**Leaching requirement (LR)**: LR is defined as that fraction of water that must be leached through the root zone to control salinity at specified level

### Procedure for reclamation

- Level the plots, divide into sub-plots of suitable size
- Dig a trench(45-60 cm) deep on one side of the plot to remove excess water
- Make a strong bund( 30-45 cm high) around the plot and flood it with good quality water
- Plough the soil for uniform mixing of soil and water to ensure dissolution of soluble salts
- Allow the water to stand for 2-3 days
- Remove the standing water through the channel in to trench
- Add well decomposed organics and raise salt tolerant crops

### 2. Irrigation management

Proportional mixing of good quality (if available) water with saline water and then using for irrigation reduces the effect of salinity.. Alternate furrow irrigation favours growth of plant than flooding. Drip, sprinkler and pitcher irrigation have been found to be more efficient than the conventional flood irrigation method since relatively lesser amount of water is used under these improved methods

### 3. Fertilizer management

Addition of extra dose of nitrogen to the tune of 20-25% of recommended level will compensate the low availability of N in these soils. Addition of organic manures like, FYM, compost, etc helps in reducing the ill effect of salinity due to release of organic acids produced during decomposition. Green manuring (Sunhemp, Daincha, Kolingi) and or green leaf manuring also counteracts the effects of salinity.

### 4. Crop choice

Crops are to be chosen based on the soil salinity level. The relative salt tolerance of different crops is as follows.

**Table 9:** Tolerance of crops to salinity

| *Tolerant* | *Semi- tolerant* | *Sensitive* |
|---|---|---|
| Barley | Pomegrante | Citrus |
| Sugarbeat | Wheat | Gram |
| Rape | Rice | Peas |
| Cotton | Sorghum | Groundnut |
| | Maize | Lentil |
| | Sunflower | Cowpea |
| | Potato | |

**Table 10:** Crops and their tolerance levels (threshold levels) of soil salinity

| Plant species | Threshold salinity ($dSm^{-1}$) | Plant species | Threshold salinity ($dSm^{-1}$) |
|---|---|---|---|
| Field Crops | | Vegetable crops | |
| Cotton | 7.7 | Tomato | 2.5 |
| Sugar beet | 7.0 | Cabbage | 1.8 |
| Sorghum | 6.8 | Potato | 1.7 |
| Wheat | 6.0 | Onion | 1.2 |
| Soybean | 5.0 | Carrot | 1.0 |
| Groundnut | 3.2 | Fruit crops | |
| Rice | 3.0 | Citrus | 1.7 |
| Maize | 1.7 | | |
| Sugarcaen | 1.7 | | |

## 5. Cultural practices management

Salinity reduces or slows germination and it is often difficult to obtain a satisfactory stand on saline soils or when using moderately saline water. To overcome this problem, growers plant two or three times as much seed as normal, hoping to offset the reduced germination.

With furrow irrigated crops planted on raised beds, water movement is from the furrow into the bed. Since water moves from the two furrows towards the centre of the bed, any salts present move with the water and tend to accumulate in the upper centre of the bed. Planting seeds in a single row in the centre of a raised bed places the seed exactly in the area where salts concentrate (Figure 11A), planting a double-row on a raised planting bed (Figure 11D) will place the two seed rows near each shoulder of the raised bed, away from the area of greatest salt accumulation.

There are other planting alternatives. Alternate row irrigation may help. If the beds are wetted from both sides, the salts accumulate near the top or centre of the bed (Figures 11A and 11D) but if alternate rows are irrigated, the salt can be moved beyond the single seed row (Figure 11B). The salts may still accumulate, but to a lower concentration. Off-centre, single-row planting on the shoulder of the bed closest to the watered furrow (Figure 11E) has also been used and aids germination under salty soil conditions.

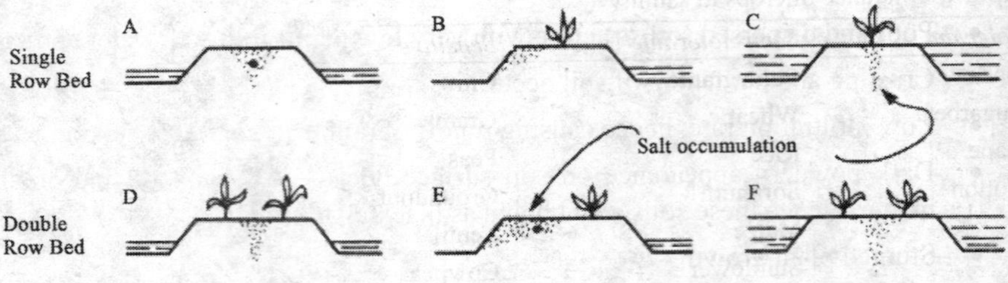


**Fig. 11:** Flat top beds and irrigation practice (Bernstein, Fireman and Reeve 1975)

Planting the seed in the centre of the raised bed / ridge may affect the germination as it is the spot of greatest salt accumulation. A better salinity control can be achieved by using sloping beds with seeds planted on the sloping side just above the water line. Alternate furrow irrigation is advantageous as the salts can be displaced beyond the single seed row. Application of straw mulch had been found to curtail the evaporation from soil surface resulting in the reduced salt concentration in the root zone profile within 30 days.

## 9.2.3.2. Alkali / Sodic soils

Alkali soils are formed due to concentration of exchangeable sodium and high pH. Because of high alkalinity resulting from sodium carbonate the surface soil is discoloured to black; hence the term black alkali is used.

## Reasons for occurrence of soil alkalinity/ sodicity

The excessive irrigation of uplands containing Na salts results in the accumulation of salts in the valleys. In arid and semi-arid areas salt formed during weathering are not fully leached. In coastal areas if the soil contains carbonates the ingression of sea water leads to the formation of alkali soils due to formation of sodium carbonates. Irrigating soils with poor drainage causes sodicity. Soils with high levels of exchangeable sodium (Na) and low levels of total salts are called sodic soils.

## Formation of Sodic soils

Soil colloids adsorb and retain cations on their surfaces. Cation adsorption occurs as a consequence of the electrical charges at the surface of the soil colloids. While adsorbed cations are combined chemically with the soil colloids, they may be replaced by other cations that occur in the soil the soil colloids. While adsorbed cations are combined chemically with the soil colloids, they may be replaced by other cations that occur in the soil solution .Calcium and magnesium are the principal cations found in the soil solution and on the exchange complex of normal soils in arid regions. When excess soluble salts accumulate in these soils, sodium frequently becomes the dominant cation in the soil solution resulting alkali or sodic soils.

## Field diagnosis of Sodic soils

- Poor and dispersed soil structure with very low permeability to water and air.
- Crusting and hardening of soil occur.
- Low infiltration rate, hence causing runoff of rain water and soil erosion.
- Dark powdery appearance on soil surface due to dispersion of fine organic matter. Hence these soils also known as black alkali soils.
- Stunted plant growth.
- Nutrient deficiencies appear on plants.

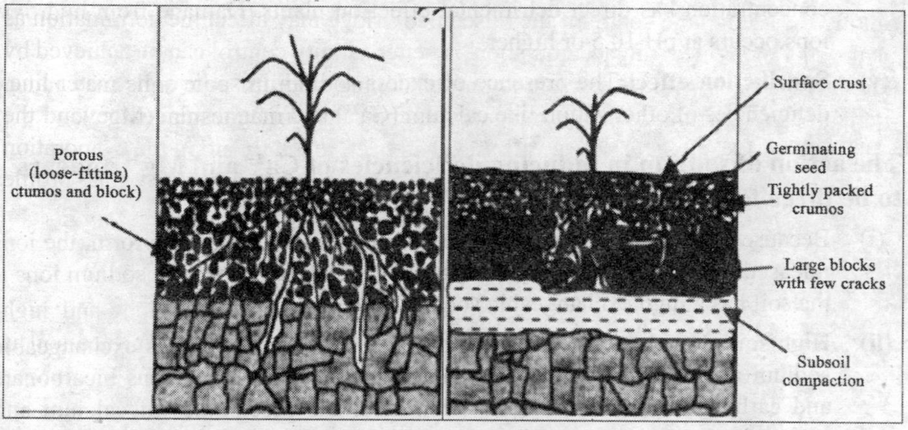


Good soil structure in normal soil Compacted soil in Sodic soil

**Fig. 12:** Development of soil crust and compacted layer in sodic soil

## Characteristics of Sodic soils

Abnormally high levels of exchangeable sodium ($Na^+$) is important characteristic feature of sodic soils. When enough $Na^+$ is adsorbed, clay particles repel each other and disperse. Sodic soils occurs when the exchangeable $Na^+$ percentage (ESP) is equal to or exceeds 15. Soil pH of sodic soils will often be above 8. Dispersed colloids become oriented as water moves into soil and eventually they plug soil pores. Poor internal drainage resulting in dry subsoil and a moist or wet surface layer. Crops fail because of excess surface water ("drown out") or for lack of water (dry sub soil) even though there may have been adequate rainfall or irrigation. If a soil is highly sodic, a brownish-black crust sometimes forms on the surface due to dispersion of soil organic matter. By the time darkened crusts are visible on the soil surface, the problem is severe, and plant growth and soil quality is significantly impacted. Plants growing on sodic soils may appear stunted and often show a burning or drying of tissue at the leaf edges, progressing inward between veins.

**Effects of Sodicity on soils and plants**: Excess exchangeable sodium in alkali soils affects both the physical and chemical properties of soils.

**(i) Dispersion of soil colloids:** The sodium ($Na^+$) ion adsorbed by clay colloids causes de-flocculation or dispersion of clay which result in a loss of desirable soil structure and helps for the development of compact soil.

**(ii) deterioration of soil physical properties:**Due to dispersion and compactness of soil, aeration, hydraulic conductivity, drainage and microbial activity are reduced.

**(iii) Caustic effect:**The caustic effect or high sodicity effect is caused by the sodium carbonate ($Na_2CO_3$) and bicarbonate ($NaHCO_3$).

**(iv) High concentration of Hydroxyl ($OH^-$) ion:**High hydroxyl ($OH^-$) ion concentration has direct detrimental effect on plants. Damage from hydroxyl ions occurs at pH 10.5 or higher.

**(v) Specific ion effect:**The presence of excess sodium in sodic soils may induce deficiencies of other cation like calcium ($Ca^{2+}$) and magnesium ($Mg^{2+}$).

## The action of sodium in inducing deficiencies of $Ca^{2+}$ and $Mg^{2+}$ appears to be three fold

(i) Because sodium is comparatively loosely held in exchangeable form, the ions released to the soil solution in a fractional exchange are mostly sodium ions if the soil has a high exchangeable sodium percentage,

**(ii)** High soil pH values (sodic soil) usually associated with excess exchangeable sodium in the absence of excess salts, the soil solution contains bicarbonate and carbonate ions that tend to form insoluble precipitates of calcium and magnesium carbonates as follows:

$$\underset{\text{Soil solution}}{Ca^{2+}} + \underset{\text{Soil solution}}{CO_3^{2-}} = \underset{\text{(Insoluble precipitation)}}{CaCO_3}$$

Exclusion of calcium and magnesium from absorption on competitive basis.

(iii) Availability of Plant nutrients. The high pH in alkali or sodic soils decreases the availability of many plant nutrients like P, Ca, N, Mg, Fe, Cu, and Zn.

**Tabel 11:** Expected loss of soil productivity due to ESP in different soils

| ESP | Loss in productivity (%) | |
|---|---|---|
| | Alluvium derived soils (Inceptisols/ Alfisols) | Black soils (Vertisols) |
| Up to 5.0 | Nil | up to 10 |
| 5.0 to 15.0 | <10 | 10- 25 |
| 15.0 to 40.0 | 10- 25 | 25- 50 |
| >40.0 | 25- 50 | >50 |

## Reclamation of alkali / sodic soils

**a. Physical amelioration:** This is not actually removes sodium from exchange complex but improve physical condition of soil through improvement in infiltration and aeration. The commonly followed physical methods include. Deep ploughing is adopted to break the hard pan developed at subsurface due to sodium and improving free-movement water. This also helps in improvement of aeration. Providing drainage is also practiced to improve aeration and to remove further accumulation of salts at root zone. Sand filling which reduces heaviness of the soil and increases capillary movements of water. Profile inversion – Inverting t he soil benefits in improvement of physical condition of soil as that of deep ploughing.

b) **Chemical amelioration:** The basic principle underlying reclamation is to adopt those ameliorative measures which replace sodium with calcium in the exchange complex followed leaching of soluble sodium from exchange complex.. This can be accomplished by the application of chemical amendments (the materials that directly or indirectly furnish or mobilize divalent cations, usually $Ca^{2+}$ for the replacement of sodium from the exchange complex of the soil) followed by leaching to remove soluble salts and other reaction products. The chemical amendments can be broadly grouped as follows:

- Direct Ca suppliers: Gypsum, calcium carbonate, phospho-gypsum, etc.
- Indirect Ca suppliers: Elemental Sulphur, sulphuric acid, pyrites, $FeSO_4$, etc.

**i) Use of Gypsum:** Because of low cost and easy availability gypsum is used widely for sodic soil reclamation. Reaction of gypsum in sodic soils in converting exchangeable sodium to soluble sodium sulphate.

$$\text{Soil-}Na_2 + CaSO4 \longrightarrow \text{Ca- Soil} + Na_2SO_4$$

Sodium sulfate being soluble it is removed by leaching

Use of amendments followed by adequate leaching are prerequisites for any reclamation measure of sodic soils. The quantity of gypsum to be applied is found out by Gypsum requirement of alkali soil

## Procedure for reclamation of sodic soils

- Level the lands and divide into sub plots
- Dig a trench across the slope of 60-75 cm
- Make a strong bund around the plot
- Apply required quantity of gypsum/ pyrite as found out by laboratory analysis and mix it with top soil by ploughing
- Flood the plot with good quality water and stand for 2-3 days
- Remove the standing water
- Raise daincha and plough at appropriate time
- Apply 25 kg $ZnSO_4$/ha
- Raise rice crop, increase plant density and add 25% extra N

- Raise salt resistance crops in the following season by adopting improved agronomic practices.

## ii) Amendments used other than gypsum for reclamation of sodic soils

**Iron pyrites** in the presence of air and water produces sulfuric acid which react with calcium carbonate to form calcium sulfate and the exchange reaction follows

$$2FeS2 + 7O2 + 2\ H2O \longrightarrow 2FeSO4 + 2H2SO4$$

**Sulfuric acid** can be used instead of gypsum on calcareous (CaCO3 containing) soil only.

$$H_2SO_4 + CaCO_3 \longrightarrow CO_2 + H_2O + CaSO_4$$

**Elemental sulfur** can also be used as an alternative to gypsum on calcareous soils Soil microbes convert sulfur into sulphuric acid

$$S + \frac{1}{2}O_2 + CO_2 + 2H_2O \longrightarrow H_2SO_4 + CH_2O$$

$H_2SO_4$ dissolves calcium carbonate and makes gypsum. Conversion to sulfuric acid takes time for several weeks in colder soils and faster in warm soils.

**Distillery spent wash:** Distillery spent wash is acidic (pH 3.8-4.2) with considerable quantity of magnesium. About 2 lakh litres of distillery spent wash can be added to an acre of sodic soil in summer months. Natural oxidation is induced for a period of six weeks with intermittent ploughing once in a month. In the second month (after 45-60 days) fresh water may be irrigated and drained. Such a treatment reduces the pH and exchangeable sodium percentage.

**Distillery effluent:** Distillery effluent contains macro and micronutrients. Because of its high salt content, it can be used for one time application to fallow. About 20 to 40 tonnes per ha of distillery effluent can be sprayed uniformly on the fallow land. It should not be allowed for complete drying over a period of 20 to 30 days. The effluent applied field has to be thoroughly ploughed two times for the oxidation and mineralization of organic matter. Then the crops can be cultivated in the effluent applied fields by conventional methods.

**Pulp and paper mill effluents:** Pulp and paper effluents contain lot of dissolved solids and stabilized organic matter and if properly treated can safely be used for irrigation with amendments viz. pressmud @ 5 tonnes $ha^{-1}$, fortified pressmud @ 2.5 tonne $ha^{-}1$ or daincha as in situ green manure

**2. Selection of tolerant crops:** Rice is preferred crop in alkali / sodic soil as it can grow under submergence, can tolerate fair extent of ESP and can influence several microbial processes in the soil. Agroforestry systems like silviculture, silvipasture etc. can improve the physical and chemical properties of the soil along with additional return on long-term basis. Some grasses like Brachariamutica (Para grass) and Cynodondactylon (Bermuda grass) etc. has been reported to produce 50% yield at ESP level above 30

**Table 12:** Relative tolerance of fruit crops to sodicity

| Tolerance to sodicity | ESP | Fruit trees |
|---|---|---|
| High | 40 – 50 | Ber, tamarind, sapota, wood apple, date palm |
| Medium | 30 – 40 | Pomegranate |
| Low | 20 – 30 | Guava, lemon, grape |
| Sensitive | <20 | Mango, Jack fruit, banana, citrus. avacode |

## 9.2.3.3. Saline- sodic soil

These soils have both soluble salts and exchangeable Na are known as saline- sodic soils.

As long as excess salts are present, soils are flocculated and pH is < 8.5. When such soils are leached, soil become strongly alkaline due to dominance of Na Such soils will have pH > 8.5, ESP > 15 and EC > 4 $dSm^{-1}$

### Formation of saline- sodic soils

These soils form as a result of the combined processes of Salinisation and alkalization. If the excess soluble salts of these soils are leached downward, the properties of these soils may change markedly and become similar to those of sodic soil.

### Management of saline- sodic soils

The reclamation / management practices recommended for the reclamation of sodic soil can be followed for the management of saline – sodic soil.

## 9.2.3.4. Degraded alkali soils

In areas of excess rainfall due to absence of Ca and Mg, part of Exch. Na (in alkali soils) is replaced by hydrogen

$Na^+$-Soil + $H_2O \longrightarrow NaOH + H^+$-Soil

$2NaOH + CO2 \longrightarrow Na_2CO_3 + H_2O$

$H^+$-Soils so formed undergo degradation to form silicic acid and forms Sesquioxide such soils will have acidic pH at the surface and subsurface will have pH > 8.5. Such soils are called degraded alkali soils.

## 9.2.3.5. Calcareous soils

Calcareous soils have often more than 15% $CaCO_3$ in the soil that may occur in various forms (powdery, nodules, crusts etc…). Soils with high $CaCO_3$ belong to the Calcisols and related calcic subgroups of other soils. They are relatively widespread in the drier areas of the earth.

**Formation of Calcareous Soils:** Calcareous soil is formed in arid as well as humid regions. It is more common in arid areas. In dry regions, calcareous soil is formed in situ whereas, it has secondary origin in water-logged areas.

(A) **Formation of Calcareous soil in situ by weathering of parent rock:** There is formation of calcium carbonate ($CaCO_3$) due to weathering of rocks. The low rainfall in arid and semi-arid regions, is not sufficient to leach out the soluble products of weathering to ground water. In dry weather, the salt moves up with the water by capillary movement to the surface where they are deposited.

(B) **Development of calcareous soil by secondary origin:** Calcareous soil of secondary origin is developed in low-lying areas or in basin shaped areas. Generally, drainage is impeded in these low-lying areas. During the rainy season, soluble calcium bicarbonate of nearby areas accumulate in the low-lying areas. During dry season, calcium bicarbonate converted into calcium carbonate and deposited on the soil surface by capillary movement.

**Distribution and Genesis of Calcareous soils :** In India, calcareous soils are distributed in the states of Rajasthan, Gujarat, Punjab, Haryana, Uttar Pradesh, Maharashtra, Karnataka, Tamil Nadu, Andhra Pradesh and parts of Madhya Pradesh and Bihar and some union territories (Fig. 1). Patches of these soils have also been occured in the state of Jammu and Kashmir, in the Nowgaon district of Assam and the Jampui hills of Tripura. Calcareous soils are developed on organogenic calcium carbonate ($CaCO_3$) in the Lakshadweep islands.

**Formation of Calcareous soils:** In an aqueous solution open to $CO_2$ gas, the dissolution and precipitation of $CaCO_3$ , the process is known as Calcification. It can be demonstrated by the following equations.

The process of precipitation and accumulation of calcium carbonate in some part of the profile is called calcification. This is a common process in arid and semi-arid regions, wherein precipitation is less than evapotranspiration (Rainfall<PET).

Whenever high carbon dioxide is produced in soils, it combines with water and forms into carbonic acid.

$$CO_2 + H_2O \longrightarrow H_2CO_3$$

This $H_2CO_3$ dissolves the calcium carbonate in soils into soluble calcium bicarbonate, which moves down along the percolating water

$$H_2CO_3 + CaCO_3 \longrightarrow Ca(HCO_3)_2$$

Again wherever a situation of high temperature and low carbon dioxide prevails, there calcium carbonate precipitates.

$$2Ca(HCO_3)_2 \longrightarrow 2CaCO_3 + H_2O + 2CO_2$$

The illuviated horizon of $CaCO_3$ is designated as ***'Calcic horizon'***

The depth of calcic horizon in hot arid zone-<50cm, in semi arid zone 100-150cm and in sub-humid zone >150cm.

In arid soils rich in CaCO3, undergoes the following reactios

$CO_2$ (gas) + $H_2O \leftrightarrow H_2CO_3$ (or $CO_2$ aq)

$H_2CO_3 \leftrightarrow H^+ + HCO_3^-$

$CaCO_3 + H^+ \leftrightarrow Ca^{++} + HCO_3^-$

The above three equations are summerised by a single equation:

$CO_2$ (gas) + $H_2O + CaCO_3 \leftrightarrow Ca^{++} + 2HCO_3^-$

The above summerised equation suggests that as the partial pressure of $CO_2$ ($pCO_2$) of a system increases, the solubility of $CaCO_3$ in that system also increases. Due to microbial respiration during the decomposition of organic materials and respiration of plant roots, the $pCO_2$ of soil air is much greater than that of atmosphere. This causes an increase in $CaCO_3$ solubility. Once the $CaCO_3$ is dissolved and in solution as $HCO_3$, carbonate precipitation is generally induced by either a lowering of $pCO_2$ or by evaporation. Water loss through evapotranspiration is considered the primary mechanism in the precipitation of pedogenic $CaCO_3$. Temperature also plays an important role in controlling water flow in the soil through its effect on evapotranspiration. This is particularly true in soils of dry (subhumid to arid) regions of India as evidenced by the presence of pedogenic $CaCO_3$.

Arid and semi-arid climate is the prime factor responsible for the depletion of $Ca^{2+}$ ions from the soil solution in the form of $CaCO_3$ with the simultaneous increase of both SAR and ESP with soil depth. The impact of arid climate in the formation of pedogenic $CaCO_3$ with concomitant development of sodicity is not realized yet in the desert soils of northwestern part of India in deteriorating physical properties of soils with ESP >15. This is due to the sandy textural class of soils with a low amount of silt and clay that ensures better leaching of bicarbonates. Therefore, $CaCO_3$ is generally observed at greater depth. However, in loamy and clayey textured soils leaching of bicarbonates has been slow and thus both $CaCO_3$ and ESP develop in upper horizons.

## Characters of Calcareous soils

Major characters of calcareous soils include. (i)Calcareous soil mainly contains calcium carbonate in free form and magnesium carbonate may remain present in small quantity. (ii) There is effervescence when calcareous soils come in contact with the acid. (iii) Soil colloids contain about 90 per cent calcium and magnesium carbonate as adsorbed bases. (iv) Such soils usually having about pH 8.0 but pH may range from 7.0 to 8.3.

Calcareous soils have different parent materials such as calcareous gneiss, basalt and alluvium. Most of the calcareous soils are Vertisols and some are Entisols and Inceptisols. Soil reaction is neutral to alkaline and usually with higher CEC. Calcium carbonate or lime content of these soils extends even upto 50%. Smectite is the dominant clay mineral in calcareous Vertisols with illite and vermiculite as associated minerals in the clay fraction. Calcareous soils of Bihar and Haryana however, have illite as a dominant clay mineral and smectite as a associated mineral.

## Potentials and limitations of calcareous soils

The potential productivity of calcareous soils is high where adequate water and nutrients can be supplied. The high calcium saturation tends to keep the calcareous soils in well aggregated form and good physical condition. However where soils contain an impermeable hard pan (petricalcic horizon) they should be deeply ploughed in order to break the pan. This should be followed by the establishment of an efficient drainage system. High calcium (and magnesium) carbonate causes high soil pH and nutrient imbalance in soils. High calcium (and magnesium) fixes soil available phosphorous. Most micronutrients are deficient in calcareous soils.

## Effects of calcareous soil

### (A) Physical properties

i) Calcareous soils remain in a flocculated condition (granulated). Soil is permeable to water and air due to large number of pore spaces.

ii) Flocculation in heavy (clay) soil increases pore spaces and consequently drainage but in light (sandy) soil it decreases water-holding capacity.

**(B) Nutrient availability**

(i) Nitrogen: alkalinity of calcareous soils causes volatilization of nitrogen in the form of Ammonia.

(ii) Phosphorus: In calcareous soil, the solubility of phosphorus is depressed which results in the decrease of P availability.

(iii) Potassium: Availability of potassium is decreased due to presence of excess calcium carbonate.

(iv) Micronutrients: Availability of zinc, Fe, Mn and Cu is reduced.

**(C) Activity of Microorganism:** Moisture is essential for microorganism activity. Microorganism activity in wet calcareous soil is normal but it is affected in dry calcareous soil.

**Management of Calcareous soils:** Management and reclamation of calcareous soil are not difficult because pH in such soil is not very high. Generally, there is no need of chemical amendments for reclamation of calcareous soil.

**(i) Tillage operation:** Where calcareous soils contain an impermeable hard pan (petricalcic horizon) they should be deeply ploughed in order to break the pan. This should be followed by the establishment of an efficient drainage system. Light (sandy) calcareous soil develops large number of pore spaces due to flocculation. These type of soils have poor water-holding capacity. Therefore, such type of soil needs compaction by plank or roller to increase the water-holding capacity.

**(ii) Application of organic manure:** When sufficient amount of farm yard manure, compost and green manure are added, the amount of carbon dioxide and acid increases and as a result pH of soil decreases.

(iii) **Irrigation management**: Furrow irrigation is better than basin irrigation on slaking calcareous soils. On undulating lands, contour and sprinkler irrigations are better options than flood irrigation. Drip irrigation may also be practiced. Calcareous soils generally have low organic matter content and lack nitrogen.

### (iv) Nutrient management

**Nitrogen management**: Nitrogen fertilizer may be applied any time from just before planting up to the time the plant is well established. Application of nitrogen through side-dressing to the growing crop is an efficient way of nitrogen application. Care should be exercised so as not to apply nitrogen close to the seed as it may prevent germination. Ammoniac sources of nitrogen and urea should not be left on the surface of calcareous soils, since considerable loss of ammonia through volatilization may occur, and they should be incorporated in the soil instead.

**Phosphorous** is often lacking in calcareous soils. Amounts to apply depend on how deficient the soil is and the crop requirements. Excess applied phosphorus may lead to deficiency of zinc or iron. To be effective on calcareous soils, applied phosphorus fertilizer should be in water soluble form. Band application of phosphate is more effective as compared to broadcast application. Application at the time of seeding has been found to be most appropriate since phosphorus is required mostly during the younger stages of plant growth. Availability of phosphorus is low in calcareous soil. To increase the availability of P, the phosphatic fertilizers should be used in the following manner:

(a) Phosphatic fertilizers should be used near the roots of plant (placement application).

(b) Use of phosphatic fertilizers in ball form also increases its availability.

(c) Split dose application of P fertilsiers also will have beneficial effect.

**Micronutrient management**: Calcareous soils usually suffer from a lack of micronutrients, especially zinc and iron. Zinc deficiency is most pronounced in maize, especially under high yield intensive cultivation systems. Zinc sulphate is an effective zinc source and is the most popular form in use. For soil application, zinc sulphate is broadcast and incorporated in soil. A single application lasts for several years. Foliar applications of zinc are used on fruit trees. Heavy applications of animal manure are helpful in preventing deficiency of iron and zinc. The presence of organic matter, either native or added improves Fe availability in soil, While that of calcium carbonate in soil affects it adversely. Groundnut grown on calcareous soil has shown tremendous response to Fe fertilization. On calcareous soils, foliar sprays of $FeSO_4$ are preferred to soil applications for a faster recovery from Fe- chlorosis. Organic amendments like FYM containing appreciable amount of Fe was found to increase the fertilizer use efficiency.

## Review questions

1. What is problematic soils? List out various problematic soils.
2. Briefly discuss about different physical problem soils and their management.
3. How acid soils are formed and write on the effects of soil acidity on plant growth.
4. Describe different types of soil acidity.
5. List out different liming materials use for reclamation of acid soils and mention their lime equivalence.
6. What is saline soil and discuss about effects of salinity on plant growth.
7. How you manage saline soils.
8. What is Leaching requirement? Give formulae for LR. Calculate the LR when the EC of irrigation water is 0.5 $dSm^{-1}$ and EC of drainage water is 2.5 $dSm^{-1}$.
9. What is sodic soil? How sodic soils are formed? How you diagnose soil sodicity.
10. Differentiate between saline, sodic and saline- sodic soils.
11. How you diagnose saline, sodic soils under field conditions.

## Objective questions

1. The structural shape of alkali soil is

   a) Plate b) Blocky
   c) Columnar d) Granular

2. Akhiochi disease is seen in the following crop

   a) Cotton b) Wheat
   c) Maize d) Rice

3. The range of soil pH for moderately acidic soil

   a) 5.5- 6.5 b) 4.5- 5.5
   c) 6.5- 7.5 d) 7.5 -8.5

4. Calcium deficiency on plant is first seen in

   a) Middle leaves b) Older leaves
   c) New leaves d) Stem

5. Mo deficiency mostly observed in

   a) Saline soils b) Alkali soils
   c) Acid soils d) Calcareous soils

6. When plant sample contains 40 ppm Zn, it is rated as

   a) Adequate b) Deficient
   c) Toxic d) None of them

7. Luxury consumption is associated with
   a) Ca b) P
   c) N d) K
8. Tip burning is a symptom of
   a) K deficiency b) P deficiency
   c) S deficiency d) Mg deficiency
9. Al-containing P-compound in soil is
   a) pyrolusite b) Strengite
   c) Vivianite d) Variscite
10. The amendment not suitable for reclamation of acid soils is
   a) Calcite b) Dolomite
   c) Basic slag d) Gypsum
11. An example of salt tolerant crop is
   a) Redgram b) Cabbage
   c) Barley d) Blackgram
12. Saline soils are reclaimed by
   a) Addition of gypsum b) Addition of Lime
   c) Addition of Pyrite d) Leaching of salts
13. Adsorbed calcium, magnesium encourage soil aggregate formation starting with the process
   a) Flocculation b) Deflocculation
   c) Exfoliation d) Desorption
14. Sodic soils are generally found in
   a) Sub-Humid climate b) Arid climate
   c) Humid climate d) Temperate climate
15. Ca deficiency is commonly seen in
   a) Calcareous soils b) Saline soils
   c) Alkali soils d) Acid soils
16. A soil which has pH less than 4.0 is referred as
   a) Acid soil b) Acid sulphate soil
   c) Saline soil d) Sodic soil
17. Molecular weight (g) of $CaCO_3$ is
   a) 80 b) 100
   c) 120 d) 140

18. The ideal green manure crop suitable for sodic soils is
    a) Sunhemp  b) Pillipesara
    c) Dhaincha  d) Cowpea
19. When base saturation of soil increases, the soil pH
    a) Decreases  b) Increases
    c) No change  d) First increases then decreases
20. One of the most useful indices for quantifying pesticide adsorption in soils is
    a) Adsorption coefficient  b) Partition coefficient
    c) Desorption coefficient  d) diffusion coefficient
21. The soils showing pink colour on land capability classification maps are belong to
    a) Class I  b) Class III
    c) ClassV  d) ClassVII
22. The process of enrichment of surface water bodies with nutrients is
    a) Elluviation  b) Podzolisation
    c) Laterisation  d) Eutrophication
23. The type of soil acidity is expressed by pH of 1: 2.5 soil : water ratio is
    a) Active Acidity  b) Exchangeable acidity
    c) Reserve Acidity  d) Potential Acidity
24. Dispersion of clay particles is due to
    a) Calcium ion  b) Magnesium ion
    c) Sodium ion  d) Potassium ion
25. The major cause of waste lands in India is due to
    a) Wind erosion  b) Water erosion
    c) Soil acidity  d) Soil salinity
26. The capacity of soil to rebound from degradation is
    a) Soil health  b) Soil strength
    c) Soil resilience  d) Soil resistance
27. Favorable climate for development of saline soils
    a) Arid  b) Humid
    c) Semiarid  d) A&C
28. High salt tolerant crop is
    a) Redgram  b) Bengalgram
    c) Wheat  d) Barley

29. Floating of small sized particles in the air stream is known as
   a) Saltation b) Surface creep
   c) Suspension d) Scourin
30. Nitrate contaminates can occur through.
   a) Surface running of water b) Drainage water
   c) Both a&b d) None of these
31. The following elements stimulate growth of algae in lakes
   a) N b) P
   b) Both A & B c) None of these
32. Waste waters of distillery and paper mills are characterised by
   a) Low pH b) High pH
   c) Neutral pH d) All of these
33. Destruction of Soil structure carried by
   a) Effluents of tanning and textile industry
   b) Effluents of distillery and paper industry
   c) Effluents of jute mills
   d) Effluents of sugar industry
34. Methemoglobinemia is caused by
   a) Amide – N b) Ammoniacl – N
   c) Nitrate – N d) Gaseous – N
35. The process of removal of pollutants from soil by using living organism is
   a) Bioleaching b) Biodegradation
   c) Biomagnification d) Bioremediation
36. The kinetic energy of Sea-waves acting on the shoreline results in loss of coastal soils is called
   a) Marine erosion b) Seashore erosion
   c) Coastal erosion d) Hidden erosion
37. The buffer method for lime requirement was proposed by
   a) Shoemaker b) Shoonover
   c) Woodruff d) A&C
38. The minimum detectable area on the ground by the detector placed on a remote sensing sensor is called
   a) Swath b) Spatial Resolution
   c) Radiometric resolution d) A or B

39. pH meter measurement in 1:2 soil water suspension is
    a) Exchangeable acidity
    b) Potential acidity
    c) Active acidity
    d) Labile acidity
40. Enrichment of water bodies with plant nutrients is
    a) Denudation
    b) Eutrophication
    c) Conglomaration
    d) Regeneration
41. The physical problem of soil due to continuous rice-rice cropping under puddle situation is
    a) Crusting
    b) Hard pans
    c) Fluffiness
    d) Excess percolation
42. One of the most useful indices for quantifying pesticide adsorption on soils is
    a) Partition coefficient
    b) Extinction coefficient
    c) Diffusion coefficient
    d) Refraction coefficient
43. White alkali soils are
    a) Acid soils
    b) Saline soils
    c) Acid sulphate soils
    d) Alkali soils
44. Phytoremediation of Cr contaminated soils are done by using
    a) Pitirogramma calmelanos
    b) Spinacia oleracea
    b) Sutera fotida
    d) Azadiracta indica
45. The land capability class does not have any subclass is
    a) Class III
    b) Class VIII
    c) Class V
    d) Class I
46. Dispersion of clay particles due to the excess presence of
    a) Na
    b) Ca
    c) K
    d) Mg
47. According to the WHO the permissible limit of nitrate N in drinking water is
    a) 4 $mgL^{-1}$
    b) 2 $mgL^{-1}$
    c) 5 $mgL^{-1}$
    d) 10 $mgL^{-1}$
48. The macronutrient generally deficient in acid soils is
    a) Mn
    b) P
    c) N
    d) Fe
49. Non- persistent pesticide is
    a) Aldrin
    b) Atrazin
    c) Aldicarb
    d) Paraquat

50. pH of potential acid sulphate soils is
a) 3.5 to 4.5 b) 4.5 to 5.5
c) 5.5 to 6.5 d) 6.5 to 7.5

51. Soil surface crusting is predominant problem in
a) Histosols b) Alfisols
c) Mollisols d) Spodisols

52. if SAR is 21, then the sodicity hazard class of irrigation water is
a) S3 b) S1
c) S4 d) S2

53. Sleeping type erosion is
a) splash erosion b) Rill erosion
c) Gully erosion d) Sheet erosion

54. One $dSm^{-1}$ is equals to
a) 1 $\mu mhos\ cm^{-1}$ b) 100 $\mu mhos\ cm^{-1}$
c) 10 $\mu mhos\ cm^{-1}$ d) 1000 $\mu mhos\ cm^{-1}$

55. The distillery and paper mill waste waters (effluents) are characterized by
a) Low pH, high BOD&COD and high NPK
b) Low pH, high BOD&COD and low NPK
c) High pH, high BOD&COD and high NPK
d) Low pH, low BOD&COD and high NPK

56. In rice paddy fields about 50% applied nitrogen can be lost through
a) Volatalization b) Denitrification
c) Leaching d) Immobilisation

57. Very strong acid tolerant plant is
a) Tobacco b) Citrus
d) Alfalfa d) Sugarbeat

58. According to FAO the subclass in the land suitability classification indicate
a) Degree of suitability b) kind of suitability
c) Kind of limitation d) Potential productivity

59. According to ICAR, number of agro ecological regions in India is
a) 17 b) 15
c) 10 d) 20

60. The capacity of the soil to rebound from changes stimulated by the disturbance or external forces is
a) Soil quality b) Soil resistance
c) Soil resilience d) Soil strength.

61. UNEP is
   a) Universal Non-Governmental External Project
   b) United Nations Extension Project
   c) Universal Non-aligned Environment Party
   d) United Nations Environment Progamme
62. Plant roots stops to grow when soil bulkdensity is
   a) less than 1.3 $g.cm^{-3}$ b) 1.3 to 1.5 $g.cm^{-3}$
   c) 1.4 to 1.6 $g.cm^{-3}$ d) more than 1.9 $g.cm^{-3}$
63. The extent of area affected by soil erosion in India is
   a) 45 % of total geographical area b) 25 % of total geographical area
   c) 15 % of total geographical area d) 65 % of total geographical area
64. In India, Acid soil occupy an extent of
   a) 18% of total geographical area b) 8% of total geographical area
   c) 50% of total geographical area d) 40% of total geographical area
65. Jarosite mineral is found in
   a) Saline Soils b) Alkalisoils
   c) Calcareous soils d) Acid sulphate soils
66. High permeable soils are
   a) Laterite soils b) Clay soils
   d) Sandy soils d) Low land soils
67. Slowly permeable water logged soils are
   a) Black clay soils b) Laterite soils
   c) Coastal sand soils d) Flood plain soils
68. The phenomena associated with deterioration of surface soil structure where the natural soil aggregates brake and disperse is
   a) Soil crusting b) Soil compaction
   c) Soil alkalization d) Soil salinization
69. Soil crust has
   a) High BD b) Low non- capillary pore space
   c) Low HC d) All the three
70. Characters of fluffy soils is
   a) High infiltration b) Hydraulic conductivity
   c) Low BD d) All the three

71. In India the area subjected to soil erosion is
   a) 130 Mha  b) 50Mha
   d) 250Mha  d) 300Mha
72. When base saturation of soil increases, the soil pH
   a) Increases  b) Decreases
   c) Non change  d) A&B
73. Calcium carbonate equivalence of hydrated lime $Ca(OH)_2$ is.
   a) 119  b) 179
   c) 136  d) 109
74. The percent of silica in acid igneous rock is.
   a) Less than45%  b) 45-50%
   c) 50-65%  d) Above 65%
75. Suitable green manure crop for alkali soils is.
   a) Sunhemp  b) Sweet clover
   c) Cow pea  d) Dhaincha
76. The amendment to acid soils, which is byproduct of steel industry is.
   a) Limestone  b) Dolomite
   b) Basic slag  d) Quicklime
77. The microorganisms surviving under acidic condition is.
   a) Bacteria  b) Fungi
   c) Actinomycetes  d) Protozoa
78. The type of acidity developed due to adsorbed hydrogen and aluminium ions in soil colloids is.
   a) Active acidity  b) Exchangeable acidity
   c) Reserve acidity  d) Potential acidity
79. The amendment suitable for reclamation of non-calcareous alkali soils is.
   a) Gypsum  b) Pyrites
   b) Lime stone  d) Sulfuric acid
80. After reclamation of sodic soils, the macro nutrient which becomes deficient is.
   a) Nitrogen  b) Phosphorous
   b) Potassium  d) Zinc
81. Buffering capacity of soil means resistance to a change in
   a) Microbial activity  b) Soil pH
   c) Soil alkalinity  d) Base saturation

82. Degree of salinity is indicated by its
    a) Total soluble salt content
    b) Organic matter content
    c) Nitrogen content
    d) Phosphorous content
83. Saline soils are reclaimed by
    a) Addition of gypsum
    b) Addition of lime
    c) Addition of pyrite
    d) leaching of soluble salts
84. The acidity of $H^+$ ions adsorbed on solid phase is known as
    a) Reserve acidity
    b) Active acidity
    c) Exchangeable acidity
    d) Total acidity
85. The process responsible for the formation of calcarious soil is
    a) Salinization
    b) Alkalinization
    c) Podzolization
    d) Calcification
86. Over liming of acid soils results in deficiency of
    a) Micronutrients
    b) Calcium
    c) Phosphorous
    d) A and C
87. An example of highly salt tolerant crop is
    a) Carrot
    b) Datepalm
    c) Cotton
    d) Cabbage
88. Due to high content of nitrates in water causes
    a) Blue baby syndrome
    b) Minimitta
    c) Itai-itai
    d) Psorosis
89. If SAR of irrigation water is 15, then irrigation water belong to the class of
    a) S1
    b) S2
    c) S3
    d) S4
90. The soil forming process in tropical regions result in acid soils is
    a) Gleisation
    b) Podzolisation
    c) Pedoturbation
    d) Biological Nitrogen fixation
91. the amendment not suitable for reclamation of acid soils is
    a) Calcite
    b) Dolomite
    c) Basicslag
    d) Gypsum
92. In Land capability classification, subclasses indicate
    a) Degree of suitability
    b) kind of suitability
    c) Degree of limitation
    d) Kind of limitation
93. The acidity Fe and Al minerals is known as
    a) Reserve acidity
    b) Active acidity
    c) Exchangeable acidity
    d) Total acidity

94. Saline soils are also called as

a) Solanchaks
b) Solanedges
c) Salt peters
d) Black alkali

95. The soils which are uneconomical for the cultivation of crops without adopting proper reclamation measures are called as

a) Waste lands
b) Polluted soils
c) Degraded soils
d) Problematic soils

96. The extent of acidic soils in India is

a) 49 ha
b) 49 mha
c) 4.9 mha
d) 4.9 ha

97. The pH of saline-alkali soil is

a) <6.5
b) < 8.5
c) > 8.5
d) > 10.0

98. The neutralizing values of the following liming materials are such that

a) $CaO>Ca(OH)_2>CaCO_3>CaMg(CO_3)_2$
b) $CaO>Ca(OH)_2> Ca Mg(CO_3)_2>CaCO_3$
c) $Ca(OH)_2>CaO>CaCO_3>CaMg(CO_3)_2$
d) $CaCO_3>CaO>Ca(OH)_2>CaMg(CO_3)_2$

99. A soil which has pH less than 6.5, ESP less than 15 and EC less than 4mmhocm$^{-1}$ at 25$^0$C is called

a) Saline soils
b) Alkali soils
c) Saline Alkali soils
d) Acid soils

100. Acid soil generally not found in

a) Sub-Humid climate
b) Arid & Semi arid climate
c) Temperate Climate
d) Humid climate

101. An example of highly salt tolerant crop is

a) Carrot
b) Barley
c) Potato
d) Cabbage

102. Acid sulphate soils are commonly called as

a) Podzols
b) Laterites
c) Cat clays
d) Wetlands

103. Lime addition to acid soils result in increase of

a) Microbial activity
b) Soil pH
c) Base saturation
d) All three

104. In alkali soils N is lost by
a) Nitrate leaching
b) Nitrification
c) Ammonia volatilization
d) Ammonium fixation

105. Acid sulphate soils are managed by
a) Addition of Pyrite
b) Maintaining ill drained condition
c) Addition of elemental S
d) Leaching of soluble salts

106. The Russian term for saline soils is
a) Chernozems
b) Solanchak
c) Savannah
d) Steppe

107. The process responsible for the formation of sodic soil is
a) Salinization
b) Alkalinization
c) Podzolization
d) Calcification

108. Over liming of acid soils results in
a) Ca deficiency
b) Fe deficiency
c) Mg deficiency
d) Mo deficiency

109. sand dunes are formed by
a) wind erosion
b) water erosion
c) sea shore erosion
d) Land slide erosion

110. Due to high content of nitrates in water causes
a) Blue baby syndrome
b) Minimitta
c) Itai-itai
d) Psorosis

111. A soil which has pH less than 4.0 is called
a) Acid soil
b) Sodic soils
c) Saline- sodic soils
d) Acid sulphate soil

112. Acid soils are generally found in
a) Humid Cool climate
b) Tropical climates
c) Arid Climate
d) A &B

113. Molecular weight of $CaCO_3$ is
a) 10g
b) 100g
c) 110g
d) 140g

114. Buffering capacity of soil means resistance to a change in
a) Microbial activity
b) Soil pH
c) Soil alkalinity
d) Base saturation

115. Salt tolerant plant is

a) Mango b) Apple

c) Citrus d) Date plam

116. If EC of irrigation water is 1.0 $dSm^{-1}$ and EC of drainage water is 5.0 $dSm^{-1}$, the leaching requirement fraction is

a) 2.0 b) 1.0

c) 0.2 d) 5.0

117. The acidity of $H^+$ ions present in soil solution is known as

a) Residual acidity b) Active acidity

c) Exchangeable acidity d) Total acidity

118. Ca deficiency is commonly seen in

a) Saline soils b) Calcareous soils

c) Alkaline soils d) Acid soil

119. The grass suitable for Acid sulphate soils is

a) Rye b) Cynodon

c) Bermuda d) A &C

120. Lime requirement of acid soil is estimated by

a) Schoonover method b) Shoe maker method

c) Asija & Subbia d) Walkley & Blak

121. Alkali soils have

a) Low pH, low base status, low Na & high Al

b) Low pH, high base status, low Na & low Al

c) High pH, high base status, high Na & low Al

d) High pH, low Base status, low Na & low Al

122. The Agro-ecosystems in Andhra Pradesh are

a) 5 b) 6

c) 9 d) 11

123. The pH of saline soil is

a) > 9.0 b) < 8.3

c) > 8.3 d) > 10.0

124. A soil which has pH more than 8.5, ESP more than 15 and EC less than 4mmhocm$^{-1}$ at 25$^0$C is called

a) Saline soils b) Alkali soils

c) Saline Alkali soils d) Acid soils

125. Alkali soils are generally found in

a) Sub-Humid climate
b) Arid & Semi arid climate
c) Temperate Climate
d) Humid climate

126. Adsorbed calcium, magnesium encourage soil aggregate formation starting with the process

a) Flocculation
b) Deflocculation
c) Exfoliation
d) Podzolization

127. Ca deficiency is commonly seen in

a) Saline soils
b) Calcareous soils
c) Alkaline soils
d) Acid soil

128. The grass suitable for Acid sulphate soils is

a) Cyprus
b) Cynodon
c) Bermuda
d) Echinoclova

129. Lime requirement of acid soil is estimated by

a) Schoonover method
b) Shoe maker method
c) Woodruff method
d) B &C

## Matching type questions

130.

| | | |
|---|---|---|
| Acid sulphate soil | ( ) | a. Gypsum |
| LCC-III | ( ) | b. Flooding |
| Saline soils | ( ) | c. Sun |
| Active sensor | ( ) | d. Yellow |
| Alkali soil | ( ) | e. elemental sulphur |
| LCC -II | ( ) | f. Purple |
| Passive sensor | ( ) | g. Leaching |
| Acid soil | ( ) | h. Pink |
| LCC-VIII | ( ) | i. Lime |
| Calcareous alkali soil | ( ) | j. RADAR |

131.

| | | |
|---|---|---|
| Saline soil | ( ) | a. ESP > 15 |
| Alkali soil | ( ) | b. pH < 4.0 |
| Salt injury | ( ) | c. EC > 4 dS $m^{-1}$ |
| Acid sulphate soil | ( ) | d. leaf burn in plants |
| Inland saline soils | ( ) | e. soil salinity and water logging |
| Costal saline soils | ( ) | f. ingress of sea water |
| Moderately saline soils | ( ) | g. EC 4-8 $dsm^{-1}$ |
| Moderate alkali | ( ) | h. 15-30 ESP |
| High alkali | ( ) | i. 30-50 ESP |
| Slight alkali | ( ) | j. <15 ESP |

132.

| | | |
|---|---|---|
| Saline Soil | ( ) | a. Gypsum |
| Alkali soil | ( ) | b. Pyrite |
| Calcarious alkali soil | ( ) | c. Leaching |
| Acid Soil | ( ) | d. Lime |

133.

| | | |
|---|---|---|
| $H_2S$ injury | ( ) | a. Blue baby disease |
| Shoonover | ( ) | b. $CaMg(CO_3)_2$ |
| Dolomite | ( ) | c. Lime requirement |
| Poor drainage | ( ) | d. Minamata disease |
| Gibbsite | ( ) | e. Gypsum requirement |
| Mercury | ( ) | f. $Al(OH)_3$ |
| Cadmium | ( ) | g. Black clay soils |
| Shoemaker | ( ) | h. Akiochi disease |
| Nitrates | ( ) | i. Itaita disease |
| Calcite | ( ) | j. $CaCO_3$ |

## Answers

130. b,h,g,j,a,d,c,i,f,e

131. c,a,d,b,e,f,g,h,I,j

132. c,a,b,d

133. hH,e,b,g,f,d,i,c,d,j

**Answer key**

| | | | | | | | | | | | | | | | | | | | | | | | | | |
|---|---|---|---|---|---|---|---|---|---|---|---|---|---|---|---|---|---|---|---|---|---|---|---|---|---|
| **1** | c | **11** | c | **21** | b | **31** | c | **41** | c | **51** | b | **61** | d | **71** | a | **81** | b | **91** | d | **101** | b | **111** | d | **121** | c |
| **2** | d | **12** | d | **22** | d | **32** | a | **42** | a | **52** | a | **62** | d | **72** | a | **82** | a | **92** | d | **102** | c | **112** | d | **122** | b |
| **3** | a | **13** | a | **23** | a | **33** | a | **43** | b | **53** | d | **63** | a | **73** | a | **83** | d | **93** | a | **103** | b | **113** | b | **123** | b |
| **4** | c | **14** | b | **24** | c | **34** | c | **44** | b | **54** | d | **64** | b | **74** | d | **84** | c | **94** | a | **104** | c | **114** | b | **124** | b |
| **5** | c | **15** | d | **25** | b | **35** | d | **45** | d | **55** | a | **65** | d | **75** | d | **85** | d | **95** | d | **105** | b | **115** | d | **125** | b |
| **6** | b | **16** | b | **26** | c | **36** | b | **46** | a | **56** | b | **66** | c | **76** | b | **86** | d | **96** | b | **106** | b | **116** | c | **126** | a |
| **7** | d | **17** | b | **27** | d | **37** | d | **47** | d | **57** | b | **67** | a | **77** | b | **87** | b | **97** | b | **107** | b | **117** | b | **127** | d |
| **8** | a | **18** | c | **28** | d | **38** | d | **48** | b | **58** | c | **68** | a | **78** | b | **88** | a | **98** | a | **108** | b | **118** | d | **128** | c |
| **9** | d | **19** | b | **29** | c | **39** | c | **49** | c | **59** | d | **69** | d | **79** | a | **89** | b | **99** | d | **109** | a | **119** | d | **129** | d |
| **10** | d | **20** | b | **30** | c | **40** | b | **50** | d | **60** | c | **70** | d | **80** | b | **90** | b | **100** | b | **110** | a | **120** | b | | |

# 10

# Soil Fertility

Soil fertility is defined as the quality that enables the soil to provide proper nutrient compounds in proper amounts and in proper balance for the growth of specified plants. Soil fertility is also defined as the ability of soil to supply adequately the nutrients normally taken from the soil by plants.

## 10.1. Concepts of soil fertility and soil productivity

- It is evident from the early writing of Theophrastus (372 – 287 bc) even before the advent of christian era, Greek and Romans realised the impact of soil on the growth of plants and made a mention about the application of organic wastes and saltpeter for the plants.
- The first experiment aimed at elucidating the increase in the weight of plant during its growth was reported by Nicholas (1401 - 1446).
- Jan baptiste van Helmont (1577-1644) attributed the increase in weight of willow shoot to water.
- but a German chemist, Glauber (1604-1668), who attributed the growth of plants to the absorption of saltpeter ($KNO_3$) from the soil.
- John Woodward (during the year about 1700) first conducted water culture experiments on spearmint and emphasised that the growth factor is some terrestrial matter but not the water.
- Jean baptiste boussingault (1802-1882) carried out field plot experiments.
- He was called as 'father of field plot technique'.
- Justus von Liebig (1803-1873) put forth the 'law of minimum' which states that the yield is governed by the limiting nutrient and is directly proportional to the factor which is minimum in the soil.

as civilization moves to beginning of the twenty first century and as World's population continues to increase, the importance of a continuing increase in food production is obvious. India alone is contributing more than a billion people. It means that our crop production must be at least doubled by the end of the century. There are two options open to us. One way is to increase the area under the plough and the other to increase production per unit area) The scope is limited for the first option. Therefore, greater attention will have to be paid to the increasing of the production per unit area, per unit time. The strategy is to maintain soil fertility and proper nutrient management without impairing soil quality (sustainable agriculture).

The overall goal of soil fertility research is a more fundamental understanding of chemical and biochemical reactions in soils related to plant growth, sustainability while maintaining soil and environmental quality. Soils are the medium in which crops grow to provide food and cloth to the world) Soil is the major factor that limits the type of vegetation and crops. Under similar climatic conditions, a loose and porous soil that retains little water will support sparse vegetation when compared to deep, fertile loam or clay. The basic need of crop production is to maintain soil fertility and soil productivity.

Soil fertility is the inherent capacity of the soil to provide essential chemical elements for plant growth while soil productivity emphasizes the capacity of soil to produce crops and is expressed in terms of yield) Soil fertility is a combination of soil properties and an aspect of soil – plant relationships, while soil productivity is an economic concept and not a property of soil. Soil fertility is vital to a productive soil. but a fertile soil is not necessarily be a productive soil. Many factors can limit production, even when fertility is adequate, for eg., soils in arid region may be fertile but not productive. Soil fertility is one factor among all the external factors that control plant growth like air heat (temp.), light, mechanical support, soil fertility and water. Plant depends on soil for all these factors except for light. Organic matter in the soil improves soil fertility by mineralization of nutrients while, organic matter also improves soil productivity by improving soil porosity, aggregation and physical condition of soil thus modifying the soil environment for crop growth.

### Soil as a source of plant nutrients

Soils are complex natural formations on the surface of the earth and consist of five main components: mineral matter, organic matter, water, air and living organisms. The rocks and minerals on weathering release nutrients into the soil. The most important part of the soil with respect to plant nutrition is the colloidal fraction which consists of inorganic colloids (clay) and organic colloids (humic substances). Most of soil colloids possess electronegative adsorption sites available for attracting cations including calcium, magnesium, potassium, ammonium etc as well as $H^+$ arising from the biological activity. Organic matter on decomposition releases nutrients. The cations adsorbed on the surface of the colloids are capable of exchanging rapidly and reversibly with those in soil solution. The principal immediate source of mineral nutrients to plant roots is ions in the soil solution. This nutrient supply is gradually depleted by absorption of nutrient ions by plant roots and continuously replenished by desorption of exchangeable ions on the clay-humus complex and break down of readily decomposable organic debris. The microbes in the soil also help in supplementing nutrients by the way of nutrient transformations. These sources represent the reserves that serve to replace but only at a relatively slow rate. For intensive cultivation of crop plants, however, application of mineral salts to soil is required)

## 10.2. Essential plant nutrients

In the nature there are nearly one hundred and three elements. Of them nearly ninety elements are taken in by the plants. In order to distinguish the elements which are essential from those which may be taken in by the plants but are not essential, arnon (1954) has laid down the following criteria)

i. The plant must be unable to grow normally or complete its life cycle in the absence of the element.
ii. The element is specific and can not be replaced by another.
iii. The element plays a direct role in plant metabolism.

It appears that an element would have to be considered essential even if it has not been possible to demonstrate that it fulfills the second criterion of essentiality. For example for many bacteria, diatoms and other algae, vitamin $b_{12}$ is known to be essential, but the essentiality of cobalt per se has not been demonstrated) according to this criterion, molybdenum and chlorine cannot be considered as essential though they are functional in plant metabolism since they can be replaced by vanadium and halides respectively. d.J. Nicholas gave more exact definition of essential elements and advanced the term "functional or metabolic nutrient" to include any mineral element that functions in plant metabolism, whether or not its action is specific) The discovery of essentiality of plant nutrients and their forms of plant absorption are presented in tables 1 and 2.

**Table 1:** Essential plant nutrient elements and their discovery of essentiality

| Element | Essentiality was established by | Year |
|---|---|---|
| H, O | --- | Since time immemorial |
| c | Priestly *et al.* | 1800 |
| N | Theodore de Saussure | 1804 |
| P, K, Mg | c Sprengel | 1839 |
| S | Sachs Knop | 1857 |
| Fe | E Gris | 1843 |
| Mn | JS Mc Hargue | 1922 |
| Zn | aL Sommer and cP Lipman | 1926 |
| cu | aL Sommer, cP Lipman and G Mc Kinney | 1931 |
| Mo | dI arnon and PR Stout | 1939 |
| Na | brownell and Wood | 1957 |
| co | ahmed and Evans | 1959 |
| ca | c Sprengel | 1839 |
| b | K Warington | 1923 |
| cl | broyer, carlton, Johnson and Stout | 1954 |
| Vanadium | dI arnon and G. Vessel | 1953 |
| Silicon | J.c. Lewin | 1962 |
| Iodine | L. Fries | 1966 |
| Selenium | SF Trelease and HM Trelease | 1938 |
| Gallicum | Ra Steinberg | 1938 |
| aluminium | K. Taubock | 1942 |

*Note: Va is essential for Scenedesmus Obliquus, Si is essential for Diatoms, I is essential for Polysiphonia group plants, Se is essential for Astragalus sp., Al is essential for Aspergillus niger.*

**Table 2:** Forms of nutrient elements absorbed by plants

| Nutrient element | Forms absorbed by plants |
|---|---|
| absorbed as single nutrient ion: | |
| Potassium | $K^+$ |
| calcium | $ca^{2+}$ |
| Magnesium | $Mg^{2+}$ |
| Iron | $Fe^{2+}$ |
| Manganese | $Mn^{2+}$ |
| copper | $cu^{2+}$ |
| Zinc | $Zn^{2+}$ |
| chlorine | $cl^-$ |
| Silicon | $Si^{4+}$ |
| cobalt | $co^{2+}$ |
| Sodium | $Na^+$ |
| **Absorbed as combined form** | |
| Nitrogen | $NH_4^+$ and $NO_3^-$ |
| Phosphorus | $H_2PO_4^-$, $HPO_4^{2-}$ |
| Sulphur | $SO_4^{2-}$ |
| boron | $H_3bO_3$, $H_2bO_3^-$, $HbO_3^{2-}$, $bO_3^{3-}$ |
| Molybdenum | $MoO_4^{2-}$ (Molybdate) |
| Hydrogen | $H_2O$ |
| carbon | $cO_3^=$ & $HcO_3^-$ |

## Key terms of soil fertility

**Plant nutrition:** Plant nutrition is defined as the supply and absorption of chemical compounds required for plant growth and metabolism. It is the process of absorption and utilization of essential elements for plant growth and reproduction.

**Nutrient:** Nutrient is chemical element/compound or ion form of element/compound required by an organism. The mechanism by which the nutrients are converted to cellular material or used for energetic purposes are known as metabolic processes.

**Essential nutrient**: It is the nutrient element that fulfills the arnon's criteria of essentiality.

**Beneficial elements**: The elements, the essentiality of which for growth and metabolism has not been unequivocally established, but which are shown to exert beneficial effects at very low concentrations are often referred to as beneficial elements, or potential micronutrients. The beneficial effect of these nutrients may be due to the ability of these elements affecting the uptake, translocation and utilization of the essential elements. They may be essential only for certain plant speces or under specific conditions. Eg : Silicon, vanadium, cobalt and aluminium.

**Functional element**: Nicholas d J (1961) advanced the term functional or metabolic nutrient to include any mineral element that functions in plant metabolism whether or not its action is specific)

To describe the level of nutrient element in plants the following terms are proposed)

i. **Deficient** : When an essential element is at low concentration that severely limits yield and produces more or less distinct deficiency symptoms.

ii. **Toxic** : when the concentration of either essential or other element is sufficiently high to inhibit the plant growth to a great extent.

## Classification of essential nutrients

Nutrients are chemical compounds needed for growth and metabolic activities of an organism. The essential plant nutrients may be divided into macronutrients (primary and secondary nutrients) and micronutrients (figure 1).

### Macronutrients

Macronutrients or major nutrients are so called because they are required by plants in larger amounts. These are found and needed in plants in relatively higher amounts than micronutrients. They include c, H, O, N, P, K, ca, Mg and S. c, H and O constitute 90 – 95 per cent of the plant drymatter weight and supplied through $cO_2$ and water. Remaining six macronutrients are further sub divided into primary and secondary nutrients.

i) **Primary nutrients:** Nitrogen, phosphorus and potassium are termed as primary nutrients because the correction of their wide spread deficiencies is often necessary through the application of commercial fertilisers of which these are the major constituents.

ii) **Secondary nutrients:** calcium, magnesium and sulphur are termed as secondary nutrients because of their moderate requirement by plants, localised deficiencies and their inadvertent addition through carriers of the primary nutrients. For example, the phosphatic fertiliser, single super phosphate (SSP) contains both ca and S. Similarly, ammonium sulphate, a nitrogenous fertiliser also supplements S.

### Micronutrients

Micronutrient is an element that is required in relatively small quantities but is as essential as macronutrients. These elements have often been called trace elements. They are again classified into micronutrient cations (eg. Fe, Mn, Zn and cu) and micronutrient anions (eg., b, Mo and cl) depending upon the form in which they are available.

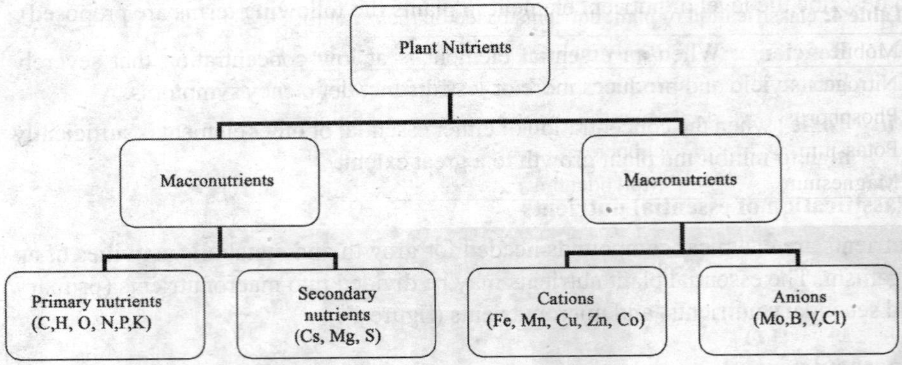


**Fig. 1:** Classification of plant nutrients

This above division of plant nutrients into macro and micro nutrients is somewhat arbitrary and in many cases, the differences between the contents of macronutrients and micronutrients are considerably less well defined) Therefore, classification of plant nutrients according to biochemical behavior and physiological functions seems more appropriate (Table 3).

**Table 3:** The classification of plant nutrients based on biochemical functions:

| Essential plant nutrient | biochemical functions |
|---|---|
| 1st group | |
| c, H, O, N, S | Major constituent of organic material, essential elements of atomic groups which are involved in enzymatic processes and assimilation by oxidation – reduction reactions. |
| 2nd group | |
| P, b, Si | Esterification with native alcohol groups in plants. Involved in energy transfer reactions |
| 3rd group | |
| K, Na, Mg, ca, Mn, cl | Non specific functions establishing osmotic potentials; enzyme activation, balance of ions, is controlling membrane permeability and electro potentials. |
| 4th group | |
| Fe, cu, Zn, Mo | Present predominantly in a chelated form incorporated in prosthetic groups |

Plant nutrients also classified based on their mobility in plant (table 4). Mobile nutrients are those when deficient in the plant, move from the matured tissue (older leaves) to the young meristem thus the deficiency symptoms are manifested on the older tissue. Immobile nutrients are those which under the situation of deficiency in the soil cannot move from older to younger tissue and hence the deficiency symptoms appear first on the younger leaves.

**Table 4:** classification of plant nutrients based on their mobility in plant

| Mobile | Partly mobile | Immobile |
|---|---|---|
| Nitrogen | Iron | calcium |
| Phosphorus | Zinc | Sulphur |
| Potassium | copper | boron |
| Magnesium | Molybdenum | |

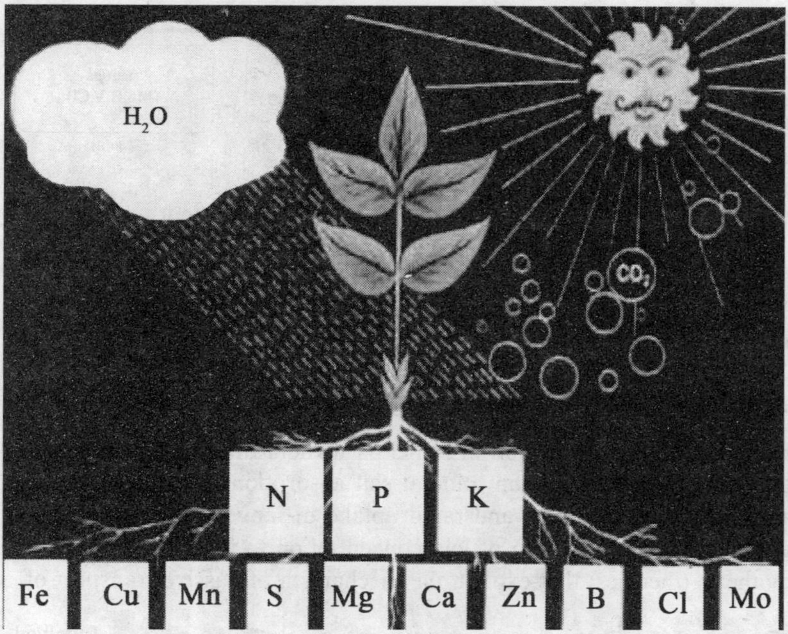


**Fig. 2:** Essential plant nutrients

## 10.3. Movement of nutrient ions from soils to roots

Mineral nutrient uptake is the process in which minerals nutrients enter the cellular material, typically following the same pathway as water. The most normal entrance portal for mineral nutrient uptake is through plant roots. during transport throughout a plant, minerals can exit xylem and enter cells that require them. Mineral ions cross plasma membranes by a chemiosmotic mechanism. Plants absorb minerals in ionic form: nitrate ($NO_3^-$), phosphate ($H_2PO4^{2-}$, $HPO4^-$) and potassium ($K^+$) ions, all have difficulty crossing a charged plasma membrane. The uptake of nutrients occurs at both the roots and the leaves. Ions may be taken up by the plant cells by two methods viz., 1. Passive absorption and 2. active absorption.

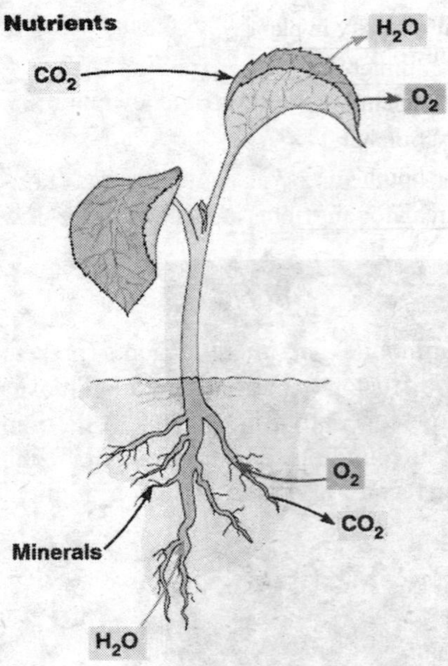


**Fig. 3:** Uptake of mineral nutrients, water and oxygen by plant roots.

**Passive absorption:** It is the absorption of minerals without direct expenditure of metabolic energy. In passive absorption mineral salt absorption is not affected by temperature and metabolic inhibitors and rapid uptake of ions occurs when plant tissues are transferred from a medium of low concentration to high concentration. The major hypothesis (theories) that explain the mechanism of passive transport of ions are:

i. Mass flow
ii. diffusion
iii. carbonic acid exchange theory
vi. Root interception and ion exchange
v. donnan Equilibrium

**1. Mass flow**: Movement of ions from the soil solution to the surface of roots is accomplished largely by mass flow and diffusion. Mass flow, a convective process occurs when plant nutrient ions and other dissolved substances are transported in the flow of water to the root in enmass that results from transpirational water uptake by the root. This depends on the rate of water flow or the water consumption of plants. Mass flow supplies an over abundance of calcium, magnesium is many soils and the most mobile nutrients such as N and S.

**Factors affecting mass flow** as mass flow involves nutrient movement with water, both the amount of water and volume of soil it comes from, affect the mass flow.

i. **Soil moisture content**: In dry soil, no mass flow of nutrients occurs because there is no water to carry them to the plant roots. .

ii. **Soil temperature**: Low temperature reduces transpiration and evaporation; resultantly reduced water flux occurs across the roots.

iii. **Size of the root system**: affects the water uptake and consequential mass flow. Root density, however is much less critical for nutrient supply by mass flow than for root interception and diffusion.

## 2. Diffusion

Most of the phosphorus, potassium (relatively immobile) and micronutrients (present in small quantities), move to root by diffusion. diffusion occurs when an ion moves from an area of high concentration to one of low concentration by random thermal motion. as plant roots absorb nutrients from the surrounding soil solution, a diffusion gradient is set up. a high root absorbing power results in a high diffusion gradient favouring ion transport.

The three principal factors influencing the movement of nutrients into the roots are the diffusion coefficient, concentration of the nutrient in soil solution and the buffering capacity of the solid phase to release nutrients into the soil solutions.

Soil moisture is a major factor that affects the relative significance of the mass flow and diffusion. diffusion becomes progressively less important as the moisture content decreases.

The amount of nutrient ion diffusing across a unit area in unit time (F) with a concentration gradient, dc/dx is given for steady state diffusion by

Fick's first law of diffusion $F = -d\ (dc/dx)$

d is the diffusion coefficient in the soil with units $cm^2/s$

The minus sign indicates movement from higher to lower concentration.

## Factors affecting diffusion

a) **Soil water:** The higher the moisture content, higher will be the diffusion coefficient, until the moisture content reaches saturation.

b) **Soil compaction:** Incremental increase in the soil compaction at the same moisture content leads to the exclusion of air; soil particles come closer, the continuity of moisture flows increases, path to be traversed by the ion by diffusion becomes less tortuous and resultantly, the diffusion coefficient exhibits an increase.

c) **Temperature:** The rate of chemical reaction doubles for every 10°c rise in temperature. The increase in temperature increases the effective diffusion coefficient of a nutrient ion.

d) **Chemical amendments**: chemical treatments modify the concentration of the nutrient ions in soil solution as given below

i. Amelioration of soil acidity through liming raises the pH of the soil. Thus, diffusion coefficient of cations decreases and anion increases as a result of liming.

ii. Application of gypsum to a sodic soil reduces the soil pH, as a result of which diffusion coefficient raises.

iii. Organic manures act as the store house of nutrients. In addition, they release low molecular weight organic substances which have the ability to form complexes with cations. This increases the diffusion coefficient of nutrients.

### 3. Carbonic acid exchange theory or carbon dioxide theory

The soil solution provides medium for exchange of ions between the root and clay micelles. carbon dioxide released in respiration of root forms carbonic acid by reacting with water of the soil solution. carbonic acid then dissociates in the soil solution to form $H^+$ and $HcO_3^-$ ions shown above expression. Hydrogen ions ($H^+$) are adsorbed to the clay micelles in exchange for cations, such as $K^+$, which are released into the soil solution. From here they may diffuse to the root, where they may be adsorbed in exchange for $H^+$ ions.

$cO_2 + H_2O \rightarrow H_2cO_3 \rightarrow H^+ + HcO_3^-$

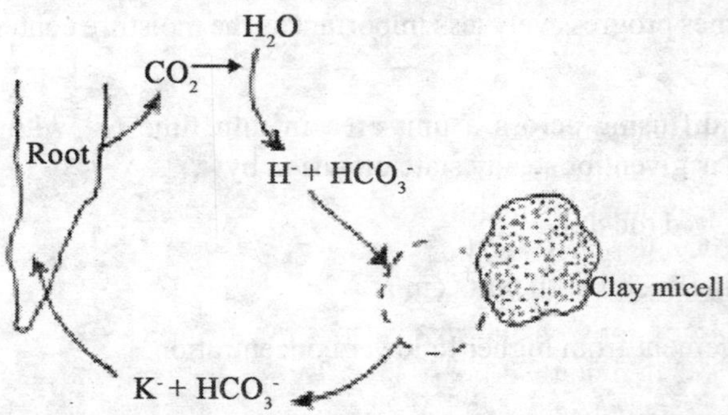


**Fig. 4:** Carbonic acid hypotheis

**4. Root interception and ion exchange**: Jenny and Overstreet (1939) propounded the 'theory of contact exchange'. Theory of contact exchange rests on the concept of overlapping oscillation spaces of adsorbed ions, or redistribution within intermingling electric double layers. contact exchange as a mechanism for nutrient movement could be pictured as

clay $M^+$ + $H^+$ Root $\rightarrow$ clay $H^+$ + $M^+$ Root and modeled as,

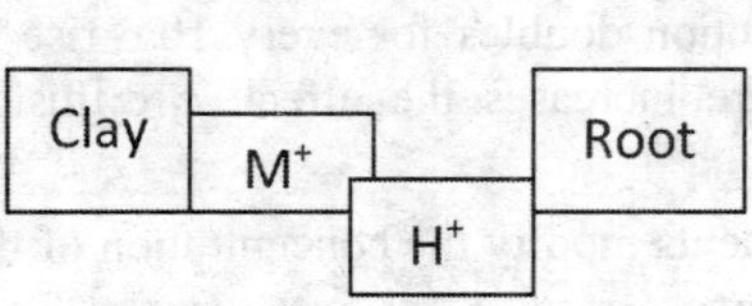


**Fig. 5:** Mechanism of root interception

Cation exchange theory on further refining gave rise to the concept of 'root interception' a term coined by Stanely a barber, which is used to describe the soil nutrients at the root surface that do not have to move to the interface to be positionally available for absorption, but are approached by the root itself in the soil.

As the root system develops and exploits the soil more completely, soil solution and soil surfaces retaining the adsorbed ions are exposed to the root mass and absorption of these ions by the contact exchange mechanism is accomplished) The quantity of nutrients that can come in direct contact with the plant roots is the amount in volume of soil equal to the volume of roots. It can be assumed that roots usually occupy 1 % or less of the soil. It is estimated that roots would contact a maximum of 3 % of the available nutrients in the soil.

It has been observed that plant roots also possess the cation exchange property, ranging from 10 to 100 cmol/ kg roots. This property could be due to the –cOOH groups of pectic substances of cell wall. Legumes have high root cEc and absorb more divalent cations, monocots have low root cEc and absorb more of monovalent cations like $K^+$. Ions attached to the surface of roots may exchange with those held on the surface of clays and organic colliods because of the contact between roots and soil particles. The mucilaginous gel around root surface could serve as contact complex. The presence of mycorrhiza, a symbiotic association between fungi and the roots of plants, enhances the uptake of nutrients particularly phosphorus.

## 5. Donnan Equilibrium

Cell membrane is composed of macromolecules of proteins and lipids that have many carboxyl groups (-cOOH) and phosphate ($HPO_3^-$) groups, from which positively charged particles like protons of hydrogen ($H^+$) can dissociate, leaving the macromolecules with negative charge. Thus the membrane is usually negatively charged) The negative charges are not diffusible because they are within the membrane structure. These negatively charged ions on the membrane called fixed ions. The negatively charged membrane is called donnan phase. Now, suppose that a solution of potassium chloride (Kcl) is present outside the donnan phase. The cations ($K^+$) will tend to diffuse through the membrane because of electronic potential difference. The cations will finally come to equilibrium with fixed negative charges of the membrane. The $cl^-$ ions with negative charge will not move into the cell because of this electronic potential. They can, however, diffuse by chemical potential difference or difference of concentration of $cl^-$ on both sides of the membrane. Some $K^+$ions will also move into the cell by chemical potential gradient. The $cl^-$ ions which remain outside the membrane will set up an electric potential difference at the membrane surface which is negative, on account of $cl^-$ ions, compared to the external solution. Such an electric potential is called donnan potential. The donnan potential will allow $K^+$ to diffuse through the membrane and repel the $cl^-$ ions (opposite charges repel each other), this equilibrium is called donnan Equilibrium. In general, donnan equilibrium may be expressed in the following equation:

Concentration of positive ions (inside) = concentration of negative ions (outside)

Concentration of positive ions (outside) = concentration of negative ions (inside)

## Active absorption

The active transport of ions from the outer space of the cell to the inner space is generally occurs against the concentration gradient and hence requires metabolic energy, this energy is obtained from metabolism of the cell either directly or in directly. The major hypotheses that explain the mechanism of active transport of ions are:

1. carrier hypothesis – transport by a carrier protein.
2. cytochrome Pump theory – transport by electrochemical gradient generated by electron transport

## 1. Carrier hypothesis

According to this hypothesis the carrier protein picks up an ion from one side of the membrane and discharges it on the other side. The picking up and discharge of the ion requires energy. Energy is obtained by hydrolysis of aTP. aTP changes into adP and energy released is used to change the conformation of the carrier which may be aTPase itself, so that the ion is picked up on one side of the membrane and released on the other. after discharge of an ion, carrier protein is reprimed to pick up an other ion. The carrier protein may carry one ion inwards and may exchange it with another ion at the inner surface of membrane, so that the other ion is carried by the same carrier outwards.

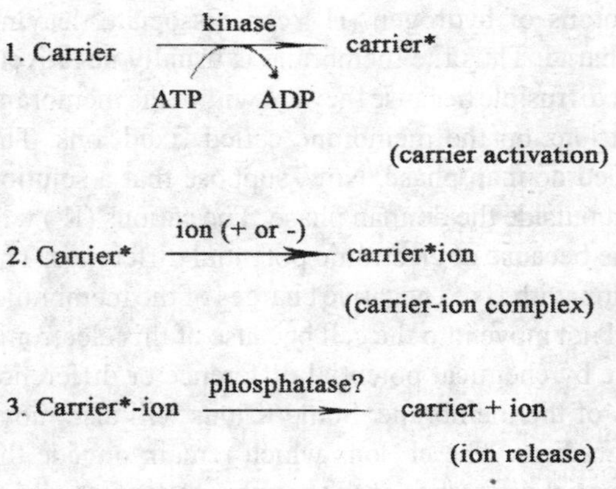


Fig. 6: Steps in carrier hypothesis

## 2. Cytochrome pump salt respiration theory

Anions could be transported across the membrane by cytochrome system. Energy is supplied by direct oxidation of respiratory intermediates. This mechanism of ion transport is based on electrochemical gradient generated by electron transport. When hydrogen is removed from a substrate in respiration and carried along an electron transport chain, it is changed into two charged species:

H (hydrogen atom) → (hydrogen ion) $H^+$ + e- (electron)

The H+ and e- are separated on opposite sides of the mitochondrial membrane by the electron carrier enzymes, which are so arranged in the inner mitochondrial membrane that they carry hydrogen ion to the outside and electron to the inside. The hydrogen atoms transported through the membrane must be derived from water by the following reaction:

$H_2O \rightarrow OH^- + H^+$

$H^+$ + e- (from electron carrier) → H

The $OH^-$ ions of water remain inside the membrane. Since $H^+$ on the outside and $OH^-$ ions on the inside of the membrane, the outside of the membrane becomes positively charged and the inside is negatively charged) This also generates a pH gradient, because outside of the membrane is more acidic because of accumulation of $H^+$ ions and inside of the membrane is basic on account of the presence of $OH^-$ ions. Proton gradients are produced both in mitochondrial membrane and the thylakoid membrane of chloroplasts. In mitochondria $H^+$ ions move outwards, while in chloroplast membranes they move inwards along electron transport chain. Proton gradients as proposed by Mitchell's chemiosmotic hypothesis must be generated in the plasmalemma of cells for active transport of solutes. Electric potential difference present in the plasmalemma suggests that charge separation occurs across this membrane also.

Electron carriers are also required for charge separation, but it is not yet known what substances in plasmalemma act as electron carriers. aTPase has been found in this membrane, which may generate proton gradient. On the establishment of proton gradient, the cations move actively in exchange for $H^+$ ions in a direction opposite to that of $H^+$ ions and $OH^-$ anions move passively to satisfy the charge balance. Similarly, when anions move actively in exchange for $OH^-$ ions, the cations move passively to maintain the charge balance. $H^+$ ions crossing the membrane react with ions to form $H_2O$.

Similarly $OH^-$ ions transported through the membrane react with $H^+$ ions to produce water. cytochrome pump hypothesis on salt absorption, anions ($a^-$) are actively absorbed via a cytochrome pump and cation ($M^+$) are passively absorbed) The rate of respiration, which is solely due to anion absorption, is called as anion respiration or salt respiration. The original rate of respiration (without anion respiration) can be observed in distilled water and is called ground respiration.

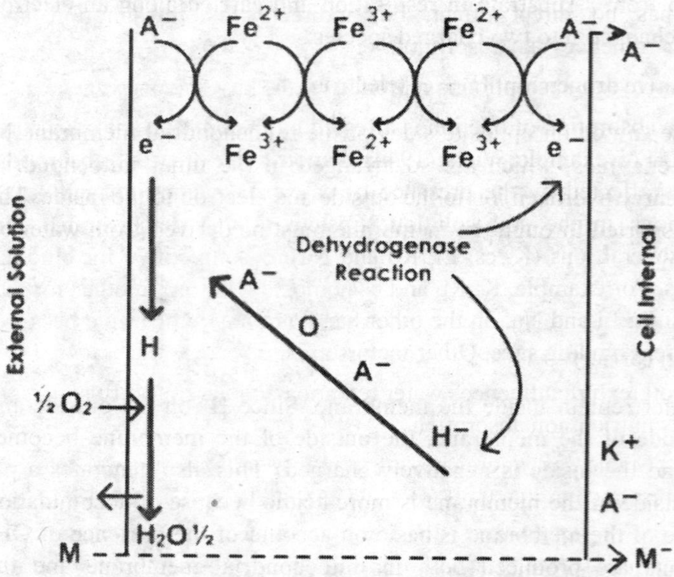


**Fig. 7:** Cytochrome pumping of anions

## Factors affecting ion absorption

Absorption of mineral salts is affected by the number of external and internal factors. External factors, Light, Hydrogen ion concentration (pH), light, Oxygen, soil texture, soil structure, climate. Internal factors include, Growth and morpho-physiological status and ageing.

## External factors

1. Temperature: absorption of mineral salt is affected by change in temperature. In general, an increase in temperature results increase in the absorption of salts up to a certain optimum level. at very high temperature the absorption is considerably inhibited) The inhibition might be due to denaturation of proteins which are directly or indirectly involved in mineral salt absorption. The change in temperature also affects the process of diffusion. The rate of diffusion depends upon the kinetic energy of diffusing molecules or ions which, in turn, dependent upon temperature.
2. Hydrogen ion concentration (pH): change in the hydrogen ion concentration (pH) of the soil solution affects the availability of ions to the plants. In general, decrease in the pH of soil solution accelerates the absorption of anions. For example, boron is taken up as the undissociated acid, $H_3bO_3$ as the $H_2bO_3^-$ ions. It is absorbed at lower pH. In contrast to the anions, increase in pH will favor the absorption of cations. However, pH values across the physiological range may damage the plant tissue and inhibit the salt absorption.

3. Light. Light has no direct effect, but indirectly by transpiration and photosynthesis, influences salt absorption.
4. Oxygen. The active salt absorption is inhibited by the absence of oxygen
5. Interaction. The absorption of one ion is affected by the presence of other ions in the medium. For example, Viets (1944) demonstrated that the absorption of potassium is affected by the presence of calcium, magnesium and other polyvalent cations in the soil solution, Epstein (1978) demonstrated the interaction of several ions (K, cs, Li, Rb and Na) as competitive for binding sites on carriers. For example, K, Rb and cs compete with one another for the same binding sites. Li and Na, on the other hand, are not competitive because they have different binding sites. Other factors are
6. Soil Texture : soil texture influences water retention and gaseous diffusion, soil heat thus affects nutrient ion absorption.
7. Soil Structure : Stable soil aggregates of granular/ crumby favours nutrient absorption by forming optimum soil- root- water contact.
8. climate : arid and semiarid climates causes water stress in the soil, thus affecting nutrient ion absorption. Humid climates on the other hand, leaches down the nutrient ions and also causes water stagnation (soil saturation) thereby affecting the nutrient ion absorption. .

**Internal factors**

1. Growth: active cell division, elongation and developmental processes promote the absorption of mineral salts.
2. Morpho-physiological status: It is the physiological and morphological trait of the plant influencing ion absorption.
3. ageing: as the root matures it increases the surface area which is fåvorable for salt absorption, but due to heavy suberization the mineral salt uptake is greatly reduced)

## 10.4. Macronutrients

### 10.4.1. Nitrogen

The ultimate source of nitrogen used by the plants is the inert gas nitrogen, which constitutes about 78 % (by volume) of the earth's atmosphere. among all the essential nutrients, nitrogen is the extensively studied one and still receiving much attention.

The ploughed layer of majority of cultivated soils contains about 0.02 – 0.04 % nitrogen. Indian soils have very low content because of low organic matter accumulation due to tropical climate. Soils developed under cool humid climate have higher N content than those formed under dry climate. On an average, black soils have more nitrogen content (0.08%) than red soils (0.03%).

### 10.4.1.1. Factors influencing the content of nitrogen in soils

a) climate

i) Temperature: Lower the temperature higher is the N content due to more organic matter addition and slow rate of decomposition.

ii) Water supply: Soil N content increases with water supply up to field capacity.

b) **Effect of cultivation:** N content decreases with cultivation due to organic matter loss through decomposition.

c) **Soil texture**: Finer the texture, higher is the N content.

d) **Depth in the profile**: Surface soil has more N content than deeper soils.

### 10.4.1.2. Forms of soil nitrogen in soil

The total nitrogen content of soils ranges from less than 0.02 % in sub soils to more than 2.5 % in peat soils. The N present in soil can generally be classed as inorganic (around 2 %) and organic (around 98 %).

1. **Inorganic nitrogen compounds**: The inorganic forms of soil nitrogen include ammonium ($NH_4^+$), nitrate ($NO_3^-$), nitrite ($NO_2^-$), nitric oxide (NO) nitrous oxide ($N_2O$) and elemental nitrogen. $NH_4^+$, $NO_3^-$ and $NO_2^-$ are important in soil fertility and represent 2 to 5 % of total nitrogen.
2. **Organic nitrogen compounds** : occur as consolidated amino acids or proteins, free amino acids, amino sugars and other unidentified compounds like materials that result from the reaction of $NH_4^+$ with lignin, polymerisation of quinones and nitrogen compounds, the condensation of sugars and amines. bound amino acids are to the extent of 20-40 %, amino sugars (hexosamines) 5 to 10 %, purine and pyrimidine derivatives 1 % or less.

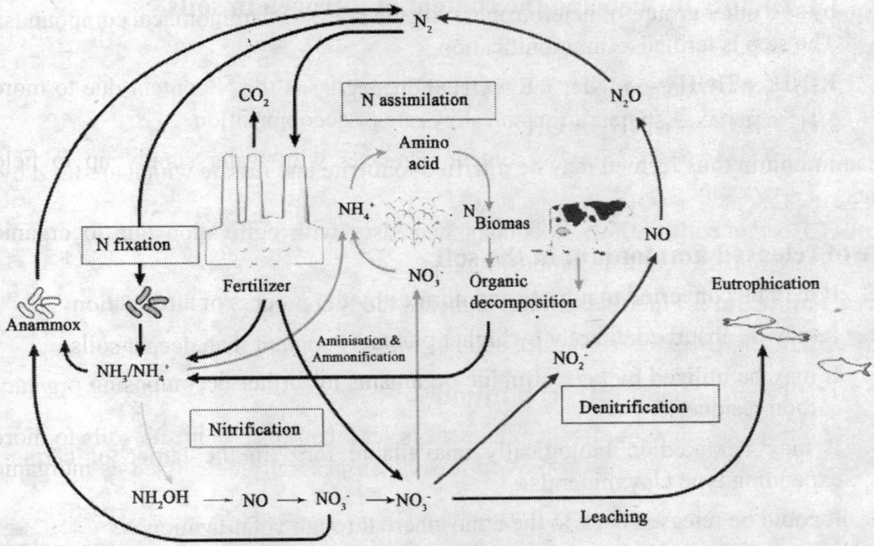


**Fig. 8:** Nitrogen cycle in the soil and environment (adopted from Srivastava and Singh, 1996)

## 10.4.1.3. Nitrogen transformations in soil

Plants absorb most of the N in the $NH_4^+$ and $NO_3^-$ forms. Nitrate is the dominant source as its concentration is higher than $NH_4^+$ and it is free to move to the roots. Potatoes, sugarbeet, pine apple, prefer both the forms; tomatoes, celery, bush beans, prefer $NO_3^-$, rice and blue berries prefer $NH_4^+$. $NO_3^-$ -N uptake is usually high and is favoured by low pH conditions. $NH_4^+$ -N is less subjected to losses by leaching and denitrification. $NH_4^+$ uptake is best at neutral pH values. When the plants are supplied with $NH_4^+$-N, it leads to acidity in the soil.

## 10.4.1.1. Mineralization of N compounds

N mineralization is simply the conversion of organic nitrogen to mineral form ($NH_4^+$, $NO_3^-$, and $NO_2^-$). When organic residues having a c: N ratio wider than 30 are added to the soil, immobilisation of nitrogen takes place. If c:N ratio is narrow i.e., less than 20 (for legume residues), mineralisation is the result. It takes place essentially by three steps.

1. Aminisation 2. Ammonification 3. Nitrification.

1. **Aminisation**: Heterotrophic soil microbes, mostly, bacteria like *Pseudomonas* and *Bacillus* are believed to dominate in the break down of proteins in neutral and alkaline soils. Under acidic conditions fungi prevail. In this step hydrolytic decomposition of proteins and release of amines and amino acids takes place.

   Proteins $\longrightarrow$ R-$NH_2$ + $cO_2$ + Energy + other products.

2. **Ammonification** : The amines and amino acids so released are further utilized by still other groups of heterotrophs with the release of ammoniacal compounds. The step is termed as ammonification.

$$R\text{-}NH_2 + HOH \longrightarrow NH_3 + R - OH + \text{Energy.}$$

$$NH_3 + H_2O \longrightarrow NH_4^+ + OH^-$$

The ammonium thus formed may be nitrified to nitrite and nitrate which are used by plants.

## Fate of released ammonium in the soil

1. It may be converted to nitrites and nitrates by the process of nitrification.
2. It may be absorbed directly by higher plants.
3. It may be utilized by heterotrophic organisms in further decomposing organic carbon residues.
4. It may be fixed in biologically unavailable form in the lattice of certain expanding type clay minerals.
5. It could be released back to the atmosphere through volatilization

## Ammonium fixation

In agricultural soils, 5 – 20 % of the total nitrogen is found as fixed ammonium ion, with an average of 10 %. When ammonium containing or ammonium forming fertilizers are added to the soil, most of ammonium gets adsorbed on the cation exchange sites, but a part of it is susceptible to strong adsorption in the interlayer spaces of some 2:1 layer silicate clay minerals with an expanding lattice. The order of fixation is as follows: vermiculite > fine grained mica > smectites. The ammonium ion has an ionic diameter 2.96 °a, which is close to the 2.8 °a diameter of inter lattice spaces in 2:1 type clay minerals. Fixed ammonium is not easily available for plants and microbes. It comes out by a replacement of $NH_4^+$ for inter layer cations in the expanded lattices of clay minerals. The fixed ammonium can be replaced by cations that expand the lattice like $ca^{2+}$, $Mg^{2+}$, $Na^+$.

The presence of $K^+$ (ionic diameter of 2.68 °a) will often restrict $NH_4^+$ fixation since this ion can also fill fixation sites. consequently it has been suggested that K fertilization prior to $NH_4^+$ application is a practical way of reducing $NH_4^+$ fixation.

Ammonium can also be chemically fixed or bound by the organic matter. The reactions take place most readily in the presence of oxygen and at high soil pH. In organic soils with high fixing capacity, the reaction could result in a serious loss of available nitrogen and would dictate the use of fertilizers other than those that supply $NH_4^+$, however, the fixed $NH_4^+$ will be slowly released by mineralization.

## Factors affecting ammonium fixation

i. **Type and amount of clay**: $NH_4^+$ fixation increases with increase inthe content particularly 2:1 type of clay minerals like vermiculite, fine grained mica and smectite.

ii. **Moisture content of the soil**: The moisture content and temperature of the soil will affect the fixation of $NH_4^+$. Freezing and drying increases the fixation. alternate cycles of wetting and drying; freezing and thawing are believed to contribute to the stability of recently fixed $NH_4^+$

iii. **Amount of $K^+$** : The presence of $K^+$ will restrict NH4 $^+$ fixation since $K^+$ also fills the fixation sites.

iv. **Depth of the soil**: Fixation of $NH_4^+$ is generally higher in sub soil than surface soil due to higher clay content and lower rate of nitrification.

v. **Organic matter content** : Higher the organic matter content more is the $NH_4^+$ fixation.

vi. **Population of nitrifying bacteria** : Nitrifying bacteria oxidize $NH_4^+$ to nitrite and nitrates thus reducing the chances for $NH_4^+$ fixation.

**3. Nitrification** : The biological oxidation of $NH^+$ released by the process of ammonification to nitrate is known as nitrification. This process is carried out by nitrifying bacteria referred to as nitrifiers. It is a two step process in which $NH^+$ is first converted to nitrite ($NO^-$) and then to nitrate ($NO^-$). conversion to nitrite is brought about largely by a group of obligate autotrophic bacteria known as *Nitrosomonas* as follows:

$$2\,NH_4^+ + 3O_2 \longrightarrow 2\,NO_2^- + 2\,H_2O + 4H^+$$

The conversion from nitrite to nitrate is affected by *Nitrobacter* as follows :

$$2\,NO_2^- + O \longrightarrow 2\,NO_3^-$$

## Factors affecting nitrification

i) **Supply of the ammonium ion** : because the substrate for the nitrifying bacteria is the ammonium ion, a supply of this ion is the first requirement fornitrification.

ii) **Population of nitrifying organisms** : Under similar conditions of temperature, moisture and added ammonia, the nitrification is greatly influenced by population of nitrifying bacteria)

iii) **Soil reaction** : Nitrification takes place between pH of 5.5 to 10.0, with an optimum around 8.5.

iv) **Soil aeration** : The nitrifying bacteria are anaerobes and hence require sufficient oxygen supply.

v) **Soil moisture** : The rate at which nitrification proceeds in a soil is governed to a marked extent by the water content being retarded by both very low or very high moisture content.

vi) **Temperature** :Very low nearer freezing and increases rapidly up to 35°c.

vii) **C : N ratio** : when organic residues with wide c : N ratio are added, general purpose decay organisms are dominant and nitrifiers become inactive.

viii) **Pesticides** : Nitrifying organisms are quite sensitive to some pesticides.

### Fate of nitrate nitrogen

Nitrate supplied in commercial fertilizers or produced by nitrification of ammonium may be :

a) Absorbed by the plants.

b) Utilised by soil microorganisms.

c) Lost through excess water by leaching because of its higobility.

d) Escapes from soil in volatile condition through nitrate reduction referred to as denitrification.

### Leaching losses of nitrate nitrogen

Leaching loss of nitrogen from the soil is a loss from the soil-plant system and causes lower efficiency of applied nitrogen. It may vary from 2 to 50 % of applied N. Of the total leaching loss, more than 90 % is in the form of $NO^-$ as mostly nitrate ions are present in soil solution because of the reason that the anion exchange capacity of the soil is much less than cation exchange capacity. Therefore, its movement is related to the movement of water in the soil. The magnitude of leaching loss of nitrogen depends on soil texture, hydrological condition, water table depth, permeability and hydraulic conductivity of soil. Leaching loss can be reduced by increasing water holding capacity through organic matter addition, split application of fertilizer N, controlled irrigation and by using nitrification inhibitors like N – serve, slow release N fertilizers like neem coated urea, sulphur coated urea)

### 10.4.1.3.2. Denitrification

Denitrification is a process limited to anoxic soils in which bacterial reduction of $NO^-$ and $NO^-$ takes place leading to the release of NO, N O and N gases. When soils become water logged, oxygen is excluded and anaerobic decomposition takes place. Some anaerobic organisms have the ability to obtain their oxygen from nitrates and nitrites with the accompanying release of nitrous oxide and nitrogen. The most probable biochemical pathway leading to these losses is indicated as

$$2\,NO_3^- \underset{-2(O)}{\longrightarrow} 2\,NO_2^- \underset{-2(O)}{\longrightarrow} 2\,NO \underset{-2(O)}{\longrightarrow} N_2O \underset{-O}{\longrightarrow} N_2$$

Facultative anaerobic bacteria belonging to the genera *Pseudomonas*, *Bacillus denitrificans* and *Paracoccus* are responsible for this. autotrophs like *Thiobacillus denitrificans* also brings about denitrification.

**Nitrification inhibitors**: Several products have been developed with the purpose of slowing the release and/or nitrification of applied N to synchronise the supply of N with the crop demand and consequently decrease nitrogen losses via leaching and / or denitrification. These chemicals restrict the growth of *Nitrosomonas* and keep nitrogen in $NH_4^+$ form.

**Natural nitrification inhibitors** Neem cake (*Azadiracta indica*) Karanj cake (*Pongamia glabra*) Neem oil and nimin

### Chemical nitrification inhibitors

- N - serve or nitrapyrin (2 chloro – 6 tri chloromethyl pyridine) dcd (dicyandiamide)
- aM (2 amino, 4 chloromethyl pyrimidine)
- aTc (4-amino 1,2, 4 triazole) Sulphathiozole
- Thiourea Hydroquinone
- calcium carbide ($cac_2$)
- Phenyl phosphorodiamidate (PPd)

However, in India the use of such chemicals to retard nitrification is limited due to the high costs involved)

### 10.4.1.3.3. Nitrogen immobilization

Immobilisation of nitrogen is the reverse of mineralisation and it occurs when large quantities of low nitrogen crop residues with wide c:N ratio (>30:1) such as coconut coir or cereal straw begin decomposing in soil, the high amounts of carbohydrates in such residues cause the microbial population to build up quickly. as new cells are formed, nitrogen and other essential elements from soil are used to build protoplasm. This leads to a decrease in the levels of inorganic nitrogen for crops. a shortage of nitrogen can be avoided in such situations by supplying enough fertilizer nitrogen to compensate for immobilization and to meet the crop requirements. This lockup of inorganic N of soil is temporary and slowly will be released after mineralization.

### 10.4.1.4. Biological Nitrogen Fixation

**Nitrogen fixation** is a process by which molecular nitrogen in the air is converted into ammonia ($NH_3$) or related nitrogenous compounds in soil. [1] atmospheric nitrogen is molecular dinitrogen, a relatively nonreactive molecule that is metabolically useless to all but a few microorganisms. biological nitrogen fixation converts $N_2$ into ammonia, which is metabolized by most organisms. bacteria are capable of reducing nitrogen to ammonia with the help of an enzyme called nitrogenase. The process is known as biological nitrogen fixation (bNF). They contribute 140-170 MT N/yr whereas N input through fertilizer is only 65 MT N/yr.

**Different groups of $N_2$ fixing microbes in soil** : There are three main groups of $N_2$ fixers as given below :

**I. Symbiotic nitrogen fixers:** Symbiotic nitrogen fixers reduce nitrogen in association with plants by forming some specialized structures in plants.

**a) Legume – rhizobium symbiosis**

Some plants of leguminosae family form a symbiotic association with bacteria with the genus *Rhizobium* which fix atmospheric nitrogen. *Rhizobia* infect the root moving to the root cortex through an infection thread which results in the formation of a tiny outgrowth called root nodule. *Sesbania rostrata* forms nodules on the stem as well. This association is host specific) The *Rhizobia* turn to bacteroids which have nitrogenase enzyme embedded into leghaemoglobin. The symbiotic association of the host and the bacterium is mutually beneficial to both organisms. Rhizobium has got host specificity i.e. specific species of bacterium infects the roots of a particular group of legumes (Table 5). This host specificity of different *Rhizobium* speces or the group of legume crops nodulated by single *Rhizobium* sp. is referred to as 'cross inoculation group'.

**Table 5:** Host specificity of different *Rhizobium* speces

| *Rhizobium* sp. | Plants nodulated |
|---|---|
| *R. meliloti* | *Medicago sativa* (alfalfa), *Melilotus* sp. (sweet clover), *Medicago* sp., *Trigonella foenum grakum* (fenugreek) |
| *R. trifolii* | *Trifolium* sp. (red clover, white clover) |
| *R. leguminosarum* | *Pisum sativum* (garden and field peas), *Vicia faba* (broad bean), *Lens esculenta* (lentils), *Lathyrus* sp. |
| *R. phaseoli* | *Phaseolus vulgaris*, *P. coccineus* |
| *R. lupini* | *Lupinus* sp., *Ornithopus sativus* |
| *R. japonicum* | *Glycene max* (soybean) |
| *Rhizobium* sp. | *Vigna ungiculata* (cow pea), *V. radiata* (mung bean), *Arachis hypogeae* (groundnut), *Cajanus cajan* (redgram) |
| *Rhizobium* sp. | *Cicer arietinum* (chickpea), *Sesbania sp.* |

## b) Symbiosis with non legumes

i. Non legume – *Frankia* symbiosis: The member actinomycetes, the *Frankia* forms symbiosis with forest trees belonging to the families *Casuarinaceae* and *Myricaceae*.

ii. azolla – anabaena symbiosis : azolla is a fresh water fern found floating on the surface of water. all the species of *Azolla* have an algal symbiont called *Anabaena azollae* in a specialized cavity in the upper leaf surface. In rice fields the symbiosis can fix $N_2$ upto 30-40 kg $Nha^{-1}$.

**II. Non specific associative $N_2$ fixers:** Rhizobacteria like *Azospirillum*, *Acetobacter*, *Azotobacter*, *Flavobacterium*, *Pseudomonas* etc) These bacteria are capable of using exudates of roots as the source of energy.

**III. Free living nitrogen fixers:** Many soil bacteria are capable of reducing molecular $N_2$ to $NH^+$ without forming any association with plants. They supply about 10-15 kg N $ha^{-1}$. They also function as plant growth promoting rhizobacteria (PGPR).

$$N_2 + 6\,H^+ + 6\,e^- \xrightarrow{\text{Nitrogenase}} 2\,NH_3$$

ammonia is the end product of the nitrogenase catalysed reaction. The intermediate products are as follows

$$\underset{\text{Nitrogen}}{N2} \xrightarrow{+2H+} \underset{\text{diimide}}{[NH=NH]} \xrightarrow{+2H+} \underset{\text{Hydrazine}}{[NH2=NH2]} \xrightarrow{+2H+} \underset{\text{ammonia}}{2\,NH3}$$

The requirements of biological nitrogen fixation are i)an active nitrogenase enzyme system; ii) $Mg^{2+}$ and a continuous supply of aTP, iii) a strong reducing agent i.e., suitable $e^-$ donor and iv) Low oxygen tension.

### 10.4.1.5. Nitrogen balance sheet in soil

**A. Gains of N by soil** : acquisition of N by soil can happen through four ways.

i. **N fixation by symbiotic bacteria** : The bacteria which live in symbiotic relationship with the host plant (legume) and fix N in the root nodules belong to the genus *Rhizobium*. They derive the energy needed for this purpose from the carbohydrates of the host plant.

ii. **Azofication** : Heterotrophic free living bacteria *Azotobacter and Clostridium* fix N referred to as azofication. These organisms derive energy from the decomposing organic matter.

iii. **Additions through rain** : on an average it is a believed that 5 kg N/ha is added to the soil every year.

iv. **Additions through manures and fertilizers** : The greater part of N gain by the soil is acquired through crop residues, organic manures, green manures and N fertilizers.

**B. Losses of N from the soil** : The removal of N from the soil may be through

i. **Removal by crops and microorganism** : This can not be considered as a loss in real sense, but a loss to the soil. crop removal is the major loss of nitrogen to the soil

ii. **Leaching** : Nitrogen can be lost due to leaching either as $NO_3^-$ or $NH_4^+$. More than 90 % in the form of $NO_3^-$ because aEc of soil is much lesser than cEc. The magnitude of leaching loss depends on soil texture, soil moisture status, water table depth, permeability and hydraulic conductivity of soil.

iii. **Gaseous loss of nitrogen** : the gaseous loss of N as $NH_3$ by volatilization or as NO, $N_2O$ and $N_2$ through denitrification.

a) Ammonia volatilization : Losses of N through $NH_3$ ranges from 5 to 35 %. The Loss is less in flooded soil. Increase in pH to alkaline side results in volatilization

$(NH_4)_2 SO_4 + ca^{2+} + 2 OH^- à 2 NH_3 + 2 H_2O + caSO_4$.

b) Denitrification : When the redox potential of the soil falls below + 250 mV in soil due to submergence, the microbes like *Thiobacillus denitrificans* utilize the oxygen from N compounds. These losses range from 10-40 %.

c) Chemodenitrification : Non microbial processes by which nitrogen may be reduced in soil to gaseous form is known as chemodenitrification.

iv. **Immobilization losses** : conversion of inorganic nitrogen to organic forms by soil microorganisms is called immobilization.

v. **Ammonium fixation by clay mineral :** $NH_4^+$ is adsorbed strongly in the inter layer spaces of 2:1 clay minerals.

vi. **Loss of N by soil erosion and runoff**. Nitrogen loss by runoff may be as high as 70 %.

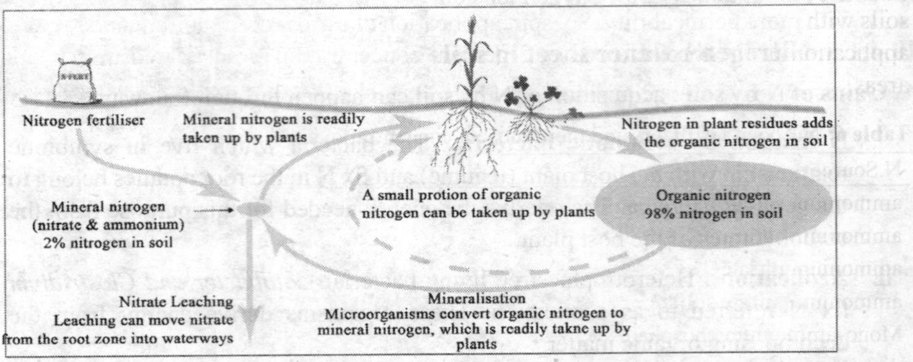


**Fig. 9:** Nitrogen balance sheet in soil: Nitrogen balance = (amount of fertilizer N added to soil + N added through crop residues, irrigation and rain water, manure) – (crop uptake of N + N loss through various processes + N remaining in the soil)

### 10.4.1.6 Functions of nitrogen in plants

In N sufficient plants, its concentration varies from 1 to 5 %. cell cytoplasm and organelles contain N in combination with c, H, O, P and S.

- It's an essential component of amino acids, proteins, nucleic acids, porphyrins, flavins, purines and pyrimidine nucleotides, flavin nucleotides, enzymes, coenzymes and alkaloids.
- N containing chlorophyll fixes atmospheric $cO_2$ through photosynthesis.
- Being a constituent of RNa and dNa, N is responsible for transfer of genetic code.
- Improves the quality of leafy vegetables and fodders.
- Improves the quality by increasing protein content.

### 10.4.1.7. Deficiency symptoms of nitrogen in plants

Plants having less than 1% nitrogen are usually regarded as deficient in N.

- Due to high mobility of N in plants, its deficiency symptoms first appear on the older leaves in the form of light green to pale yellow coloration due to proteolysis).
- Stunted growth is the manifestation.
- In grasses, the lower leaves usually fire or turn brown beginning at the leaf tip, and progressing along the midrib in the form of inverted 'V' shape.
- Reduction in flowering and crop yields and lower protein content are associated with N deficiency.

### 9.4.1.8. Corrective measures of nitrogen deficiency

Nitrogen in the form of $NO_3^-$ is more prone for leaching especially on light textured soils with more permeability. So, split application of nitrogen is recommended) Foliar application in the form of urea @ 2 percent concentration is advocated in dry land areas.

**Table 6:** Nitrogen fertilizers and their nutrient concentrations

| N Source | % N |
|---|---|
| ammonium sulfate | 21 |
| ammonium chloride | 25 |
| ammonium nitrate | 33 |
| ammonium nitrate sulfate | 30 |
| Mono-ammonium phosphate | 11 |
| diammonium phosphate (daP) | 18 |
| calcium nitrate | 15 |
| Urea | 46 |
| Urea ammonium nitrate | 28 |

### 10.4.1.9. Toxicity symptoms of nitrogen in plants

Under conditions of large nitrogen availability, succulence of the plant increases; taller plants and heavier heads cause lodging. a nitrogen rich, luxuriant crop is more susceptible to insect pest and disease attacks.

### 10.4.2. Phosphorus

Phosphorus (P), is an essential nutrient for all living organisms, is a vital component of the substances that are building blocks of genes and chromosomes. Its reserves in soil are found mostly as sedimentary formations and unrefined ore known as phosphate rock. countries like china, Morocco, Jordan, Egypt, Western Sahara, South africa and USa are the major phosphate rock share holders. In India, Jhamarkotra (Rajasthan) has reserves of 13.5 Mt of high grade phosphate rock (30% $P_2O_5$). The total P content in Indian soils ranges from 100 to 200 ppm. The total P content is generally highest in soils developed from granite and gneiss followed by shales. The total P consists of inorganic P and organic P.

### 10.4.2.1. Forms of Phosphorous in soil

### 10.4.2.1.1. Inorganic Phosphorus

The inorganic phosphorus content of soils is frequently higher than that of organic phosphorus (excepting organic soils) accounting for 54 to 84%. The inorganic P can be further divided into different pools.

a) **Soil solution Phosphorus** : Phosphorus is absorbed by plants largely as primary and secondary orthophosphate ions ($H_2PO_4^-$, $HPO_4^{2-}$). The amount of each form is dependent on soil pH. Very little P is present in soil solution, concentration being less than 0.1 ppm.

$$H_2PO_4^- \underset{H^+}{\overset{OH^-}{\rightleftharpoons}} H_2O + HPO_4^{2-} \underset{H^+}{\overset{OH^-}{\rightleftharpoons}} H_2O + PO_4^{3-}$$

Strong acidic Solution — Strong alkaline solution

Soil solution P — Labile soil P — Non labile P

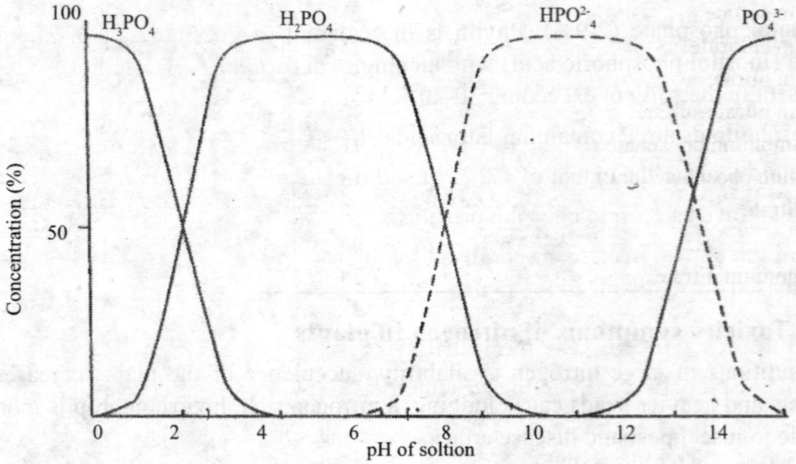


| | | |
|---|---|---|
| Readily available P<br>In soil solution $H_2PO4$ - and $HPO_4$ - ions released from other forms of P or added P | Slowly available P<br>(labile) Fe, al and Mn Phosphates in acid soils and ca, Mg phosphates in alkaline soils that are freshly formed | Very slowly avail P<br>(non labile) precipitate of Fe, al, Mn phosphates aged and well crystallized Stable org. P compounds. |

**Fig. 10:** Relation between solution pH and the relative concentration of three soluble forms of phosphorous.in the pH range common to soils, $H_2PO_4^-$ ions predominate

b) Labile soil phosphorus is the readily available portion of the quantity factor and it has a high dissociation rate, permitting rapid replenishment of solution phosphorus. depletion of labile P causes non labile P to be labile again, but at a very slow rate. Phosphorus bound to aluminium (al-P), iron (Fe-P) and calcium (ca-P) constitutes the major active labile form of inorganic P. The Fe-P and al-P constitute 1-25 % of total P in soils. The ca-P constitutes 40 - 50 % or even more of total P in neutral to alkaline and calcareous soils.

c) Non labile Phosphorous: The relatively less active non labile forms are the occluded and reductant soluble phosphates. calcium Phosphate compounds found in soil are., Flourapatite- 3 $ca_3$ $(PO_4)_2$ $caF_2$; carbonate apatite- 3 $ca_3$ $(PO_4)_2$ $cacO_3$; Hydroxy apatite-3$ca_3(PO_4)_2$ $ca(OH)_2$; Oxy apatite-3 $ca_3$ $(PO_4)_2caO$; Tricalcium phosphate- 3 $ca_3$ $(PO_4)_2$; Octa calcium phosphate- $ca_8$ $H_2$ $(PO_4)_5$.5 $H_2O$; dicalcium phosphate- $caHPO_4.2H_2O$; Monocalcium phosphate- ca $(H_2Po_4)_2$. Of all the above, the readily available forms for plant growth are mono and dicalcium phosphates. The apatite minerals are the most insoluble and unavailable compounds. Iron (Fe) and aluminium (al) containing P compounds are, Strengite- $FePO_4.2H_2O$; Vivianite- $Fe_3\ (PO_4)_2\ 8\ H_2O$; Variscite- al $PO_4.2H_2O$; Wavellite- $al_3(OH)_3(PO_4)_2.5\ H_2O$.

### 10.4.2.1.2. Organic phosphorus

In Indian soils, the organic P ranges from 0 to 2160 ppm. It constitutes 20 to 90 % of total soil P. The principal organic P compounds in soil are

a) Inositol phosphate (2-9%): Phytin is the calcium magnesium salt of phytic acid (Inositol phosphoric acid) with an empirical formula $(cH)_6\ (H_2PO_4)_6$. It is present in the soil not exceeding 30-40 %

b) Phospholipids are P containing fatty acids (1-2 %)

c) Nucleic acids to the extent of 1-2 % of soil organic P

d) Unidentified esters and phospho proteins.

P present in the above forms is mineralised by microorganisms and the phosphates that are released are available to the plants.

### 10.4.2.2. Factors affecting the availability of inorganic phosphorus in soils

**i). Soil pH** : Soil pH severely influences the reactions of P with different ions

$$H_2PO_4^- \underset{H^+}{\overset{OH^-}{\rightleftharpoons}} H_2O + HPO_4^{2-} \underset{H^+}{\overset{OH^-}{\rightleftharpoons}} H_2O + PO_4^{3-}$$

Very acid Solution — Very alkaline solution

Monovalent forms are preferred by plants than divalent or trivalent. a soil pH of 6.5 – 7.5 is considered to be the best for optimum availability of phosphorus. In strongly acid soils, the concentration of Fe and al ions greatly exceeds, forming insoluble compounds with H2PO4- rendering it unavailable. the orthophosphate $(H_2PO_4^-)$ also forms complex insoluble hydrous oxides as follows :

$$al{<}\begin{matrix}OH\\OH\\OH\end{matrix} + \underset{(soluble)}{H_2PO_4^-} \longrightarrow al{<}\begin{matrix}OH\\OH\\H_2PO_4\end{matrix} + OH^- \quad \text{(Insoluble)}$$

at high pH, $H_2PO^-$ may form insoluble tri-calcium phosphate by reacting with $cacO_3$.

$$\underset{\text{Monocalcium P}}{ca(H_2PO_4)_2} + cacO_3 \longrightarrow \underset{\text{di calcium P}}{2\ caHPO_4} + cO_2 + H_2O$$

$$6\ \underset{\text{di calcium P}}{caHPO_4} + 2\ cacO_3 + 3H_2O \longrightarrow \underset{\text{Octa calcium P}}{ca_8H_2(PO_4)_5.5H_2O} + 2cO_2$$

$$\underset{\text{Octa calcium P}}{ca_8H_2(PO_4)_6.\ 5\ H_2O} + cacO_3 \longrightarrow \underset{\text{Tri calcium P}}{3\ ca_3(PO_4)_2} + cO_2 + 6H_2O$$

ii) **Soluble Fe, Al and Mn in soils:** Fe, al and Mn react with soluble phosphate ions and form precipitations of respective phosphates.

$Fe^{2+} + 2.H_2PO_4 \longrightarrow Fe(H_2PO_4)_2$

iii) **Presence of Fe, Al and Mn containing minerals in soil:** the minerals containing Fe, al and Mn such as bauxite, goethite, gibbsite, pyrolusite, manganite would release soluble Fe, al and Mn in soil on weathering. These soluble cations reaction with phosphate ion and form insoluble phosphates of respective metal cations.

iv) **Available calcium and calcium minerals in soil:** calcium ion ($ca^{2+}$) react with soluble phosptate ion and converts it to insoluble calcium phosphate.

$ca^{2+} + 2.H_2PO_4 \longrightarrow ca(H_2PO_4)_2$

v) **Amount and decomposition of organic matter in soil:** Organic matter decomposition forms humus which chelates the metal cations and thus keep the soluble phosphate ion in the soil.

vi) **Activities of microorganisms in soil:** Micro organisms solubilise the non-lebile phosphate forms to soluble phosphorous.

### 10.4.2.3. Phosphorus fixation

The fixation of P is a chemical reaction that makes it unavailable to crops. another term often used in this sense is 'P retention'. Phosphate fixation can be the removal of phosphate from solution which reduces the amount that plant roots can absorb) Thomas Way (1850) demonstrated the P fixation. Soil pH has a deep effect upon P fixation. In neutral to alkaline soils (pH 7.0 and above), the presence of lime causes the precipitation of phosphorus. In acid soils (pH < 6.5), iron and aluminium react with phosphorous to form highly insoluble compounds. These reactions can be represented by the chemical equations. Three mechanisms are involved depending on pH.

1. Phosphorus fixation in neutral to alkaline soils

   $ca(H_2 PO_4)_2 + 2\ ca\ cO_3 \longrightarrow ca_3 (PO_4)_2 + 2\ cO_2 + 2\ H_2O$

   Monocalcium phosphate — Tricalcium phosphate

2. Phosphorus fixation in acid soils

   $al^{3+} + H_2PO_4^{\ -} + 2\ H_2O \longrightarrow 2H^+ + al_2(OH)H_2PO_4$

   dihydrogen P (Soluble) — aluminium Hydroxy P (insoluble)

   Since the products of phosphate fixation are practically insoluble, only minute amounts of phosphorus are immediately available for plant growth.

3. Phosphorus fixation in neutral soils

   P is fixed in the crystal lattices of 1:1 type of clay minerals like the Kaolinite having high aEc by anion exchange phenomenon.

## Factors affecting P fixation in soils

1. **Clay contest** : The main seat of P fixation in soils rests on the clay fraction. The more the surface area exposed with a given type of clay, the greater is the tendency to retain phosphorus.
2. **Type of clay** : Phosphorus is retained to a greater extent by 1:1 than by 2:1 clays. Soils high in kaolinitic clay will fix larger quantities of added P due to more aEc.
3. **Hydrous metal oxides of iron and aluminium** : In soils with significant amounts of iron and aluminium oxides, the P fixing capacity is larger.
4. **Amorphous colloids** : In andisols, due to the presence of large quantities of amorphous colloids like allophanes with large negative charges which are balanced by aluminium cations, the phosphate becomes fixed to these aluminium ions.
5. **Calcium carbonate** : calcium carbonate precipitates phosphates to form insoluble apatites. The activity of P will be lower in those soils that have a high $ca^{2+}$ activity.
6. **Soil pH**: adsorption of P by iron and aluminium oxides declines with increasing pH.
7. **Organic matter** : High organic matter content decreases P fixation.
8. **Temperature** : The rise in temperature generally enhances P fixation.
9. **Over liming** : Over liming increases P fixation by forming more insoluble ca-P compounds.

## Ways to reduce P fixation

i. **Organic matter addition to soil:** Incorporation of organic residues into soil influences the reactions of phosphates and its availability to plants. Organic matter counteracts the effect of $cacO_3$. It also reduces P sorption by soil. addition of organic matter also increases P concentration in soil solution through mineralization of organic P and solubilisation of native soil P compounds. This effect is more pronounced at higher moisture level. during decomposition of organic matter various organic acids are produced which solubilize the phosphates and other P bearing minerals and thereby lower the P fixation. The reason is

   a) by formation of phospho - humic complexes that are easily assimilated by plants.

   b) Through the mobilisation of the phosphate by humate ion by anion exchange

   c) by forming a coating on sesquioxide particles by humus as a protective cover which otherwise fix P.

ii. **Placement of fertilizers in soil** : The finer the soil texture, the greater is the retention of added P due to increased speed of reaction with surface area of contact. If finely divided fertilizer P is added to a soil by applying through broadcast the phosphate is exposed to a greater amount of surface, hence more fixation takes place. While band placement reduces the surface of contact between the soil and fertilizer with consequent reduction in the amount of fixation.

iii. **Mycorrhiza inoculation**: certain mycorrhizal fungi known as VaM colonize plant roots and through this association help the plant to absorb more P.

iv. **Addition of phosphorous solubilising microorganisms**: P solubilising bacteria like *Bacillus megatherium* and *Pseudomonas striatus*; or fungal species namely *Aspergillus awamori* are reported to enhance the availability of fixed P and reported to add 30 – 35 kg $P_2O_5$ $ha^{-1}$ under ideal conditions.

v. **Liming of acid soils**: about one- third of Indian soils are acidic in reaction. The beneficial effect of liming of these acid soils is due to the release of fixed P that was held with iron, aluminium oxides &hydroxides.

### 10.4.2.4. Quantity– Intensity relationships of Phosphorous (Schofield,1955)

**Intensity factor,** which indicates the soil solution phosphate and is measured by phosphate potential. **Quantity factor,** which indicates the total quantity of P in the soil. This pool supplies P to the soil solution pool (intensity) when P is taken up by plants.

In a limited range of P concentration where Q/I relationship is linear, the slope of the curve is given by Q/I which is referred to as phosphate buffering capacity or simply capacity, which indicates the resistance of soil to a depletion in quantity of P. For a given value of intensity, the quantity values are higher for fine textured soils than coarse textured soils. availability of P to plants is controlled by the mutual interplay of the three factors viz., quantity, intensity and Pbc. Hence, in clay soils depletion is less than sands, hence can supply P to the crops for longer period)

### 10.4.2.5. Functions of phosphorus in plants

Phosphorus is the second most abundant mineral in human body making up for more than 20 % of body minerals. It is largely absorbed by the plant roots as dihydrogen orthophosphate ion ($H_2PO_4^-$). In normal P sufficient plants, P content varies from 0.1 to 0.4 per cent by weight.

- Phosphorous is a constituent of nucleoproteins, phytins and phospholipids.
- It isessential constituent of number of enzymes -important in energy transfer.
- Phosphorous is essential for cell division and development.
- Phosphorous stimulates root development and growth.
- It is responsible for early establishment of seedlings.
- Strengthens the straw and decreases lodging.
- It brings about early maturity in plants.
- counteracts the effect of excessive nitrogen.

- Phosphorous Increases grain to straw ratio.
- Improves the quality of food grain.
- Increases rhizobial activity, increases the formation of root nodules thus helping in more N - fixation.

### 10.4.2.6. Deficiency symptoms of phosphorous in plants

In general, plants having less than 0.1 % phosphorus are designated as P-deficient. because of its faster mobility in plants, P gets translocated from older tissues to the meristematic tissue. Therefore, deficiency symptoms of P first appear on the older leaves. Phosphorous deficiency in plant results in

- Production of dark green color leaves.
- Severe restriction of root growth.
- Thin, erect and spindly plants with sparse and restricted foliage.
- Suppressed lateral bud production.
- bluish green foliage, and under continued deficiency the older leaves become bronzed or develop reddish purple tip or leaf margins.

### 10.4.2.7. Toxicity symptoms Phosphorous in plants

Excess P indirectly affects plant growth by reducing Fe, Mn and Zn uptake; thus, potentially causing deficiency symptoms of these nutrients to occur. Zn deficiency is most common when excess P present in soil.

### 10.4.2.8. Correction measures of deficiency

Generally, P is applied as a basal application by band placement. The following are the phosphatic fertilizers.

a) Single Super Phosphate (SSP) contains 16 to 22% $P_2O_5$

b) Rock phosphate contains 30-38 % $P_2O_5$

c) Triple Super Phosphate contains 46 % $P_2O_5$

d) diammonium phosphate contains 46 % $P_2O_5$

e) Monoammonium phosphate contains 52 % $P_2O_5$

f) Basic slag contains 10-20 % $P_2O_5$

## 10.4.3. Potassium

Potassium is one of the three major plant nutrient elements. The gap between removal of K and its application to crops is widening. Potassium content in the earth's crust averages about 1.9 % with the range of 0.5 – 2.5 %. In tropical soils the total content of K is quite low because of their origin, high rainfall and high temperature.

### 10.4.3.1. Sources of potassium in soil

The K in soil originates from the disintegration and decomposition of rocks containing potassium bearing minerals. The minerals considered original source of K are potassium feldspars, orthoclase and microcline ($KalSi_3O_8$), muscovite $Kal_3Si_3O_{10}(OH)_2$, biotite $K(MgFe)_3alSi_3O_{10}(OH)_2$. Potassium is also found in the soil as secondary silicate clays like illite, vermiculite and chlorite.

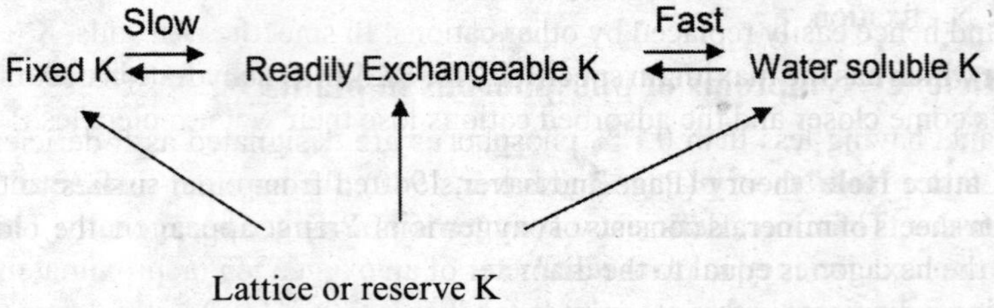


**Fig. 11:** Forms of potassium in soils

1. **Water soluble K** : Potassium is present as soluble cation in soil solution. It is in equilibrium with exchangeable K.
2. **Exchangeable K** : This fraction constitutes the K adsorbed on soil clay complex and replaceable with neutral salts. In mineral soils, it is less than 1% of the total K.
3. **Fixed potassium** : The part of the added K which is firmly bound by the soil and not immediately replaceable with neutral salts is said to be fixed K. The fixed K is bound between the basal planes of micaceous minerals.
4. **Lattice potassium** : This form constitutes the major part of the total K in mineral soils and present in primary minerals such as muscovite, biotite and potash feldspars. It accounts for 92 – 98 % total K.

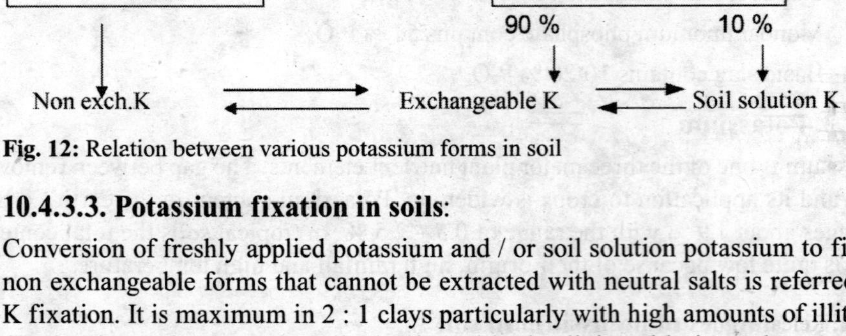


**Fig. 12:** Relation between various potassium forms in soil

## 10.4.3.3. Potassium fixation in soils:

Conversion of freshly applied potassium and / or soil solution potassium to fixed or non exchangeable forms that cannot be extracted with neutral salts is referred to as K fixation. It is maximum in 2 : 1 clays particularly with high amounts of illite. The fixation is nearly absent in soils dominated by kaolinite, chlorite and unweathered micas; slight in montmorillonite and substantial in illite and high in vermiculite dominated soils. For the clay minerals like illite, vermiculite and weathered mica three different adsorption sites can be distinguished) These sites are at the planar surfaces (planar position), at the edges of layers (e positions) and in interlayer spaces (*i*-positions). The binding of $K^+$ with organic colloids and kaolinite is at 'p' position which is weak and hence easily replaced by other cations. In smectite rich soils, $K^+$ is held at *i* position which has the maximum specificity for $K^+$. When dehydration occurs, the lattice sheets come closer and the adsorbed cations lose their water molecules.

According to 'Lattice Hole' theory (Page and baver, 1940), the exposed surface and surfaces between sheets of minerals consists of oxygen ions arranged hexagonally. The opening within the hexagon is equal to the diameter of an oxygen ion (approximately 2.8°a). Ions having a diameter in this magnitude (eg. For $K^+$ it is 2.66°a) will fit snugly into the lattice holes and such ions will be held very tightly as they come in contact with the negative electrical charges within the crystal. However, ions like $NH^+$ (dia) 2.86°a) has nearly the same ionic radius as the $K^+$ and is subject to similar fixation by 2 : 1 clays.

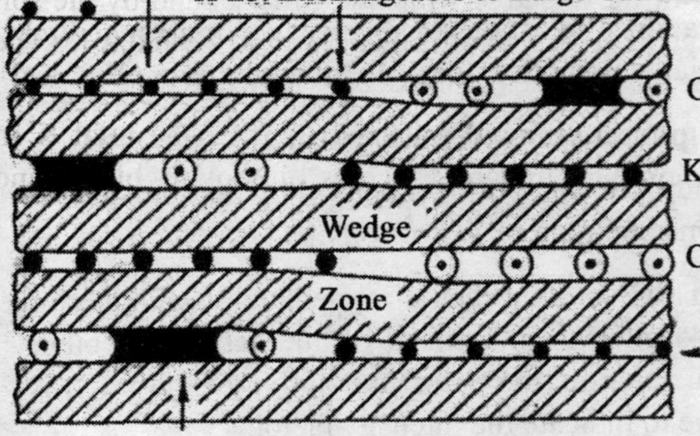


**Fig. 13:** Model of an expandable layer silicate with interlayers, wedge zone, p⁻, e⁻ and i⁻ positions

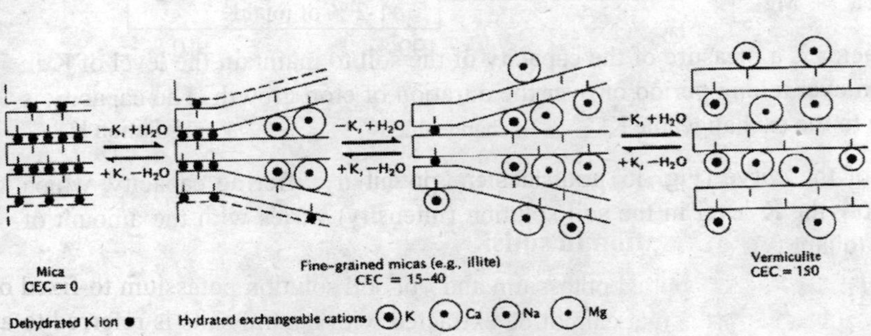


**Fig. 14:** Release of K ions from Mica mineral

## Factors affecting potassium fixation

1. **Clay minerals:** K fixation is controlled mainly by clay minerals, with both their quantity and nature being important. Illite, weathered mica, vermiculite take part in K fixation but 1:1 type of clays such as kaolinite do not fix potassium.
2. **Soil reaction:** K fixation capacity can be reduced by the presence of $al^{3+}$ and al-hydroxide cations.
3. **Cation Exchange Capacity (CEC):** Fine textured soils have a higher cEc and hold more exchangeable K.
4. **Concentrations of added K:** Increasing the concentration of $K^+$ in soils with high fixing capacity encourages greater fixation.
5. **Wetting and drying:** Soils with high exchangeable $K^+$ on drying will result in fixation due to air drying.
6. **Freezing and thawing:** of moist soils may release fixed K.
7. **Sub soil K and rooting depth:** Exchangeable K varies with depth.
8. **Soil temperature: Higher** temperature favors dehydration and contraction of the crystal lattice resulting in higher K fixation.
9. **Ca and Mg (exchangeable cations):** The size of $K^+$ and other ions that would replace K is important in K fixation. The cations of smaller size of the hydrated ions can easily enter into clay lattices and replace some of the fixed potassium.

### 10.4.3.4. Quantity and intensity parameters of potassium

The concept of intensity, quantity and buffering power proposed by beckette (1964) are useful in measuring the potassium supplying power of soils.

Intensity factor is a measure of K in soil solution that is immediately available for absorption by plants. Since this absorption is influenced by the activity of other cations like $ca^{2+}$ and $Mg^{2+}$ in the soil solution, the potassium activity ratio ($aR^k$) is used in place of K concentration alone to indicate the intensity factor.

$$AR^k = \frac{[K^+]}{\sqrt{[Ca^{2+} + Mg^{2+}]}}$$

Quantity factor is a measure of the capacity of the soil to maintain the level of K in soil solution over a long period or over the duration of crop growth. The capacity is mainly due to the exchangeable K.

The slope of the curve (Fig. 15) indicates the potential buffering capacity, which indicates how the K level in the soil solution (intensity) varies with the amount of labile form (quantity)

$$Pbc = \Delta Q / \Delta I$$

The wider the ratio, the more buffered is the soil. Soil having smectitic clay have more Pbc, than having kaolinite.

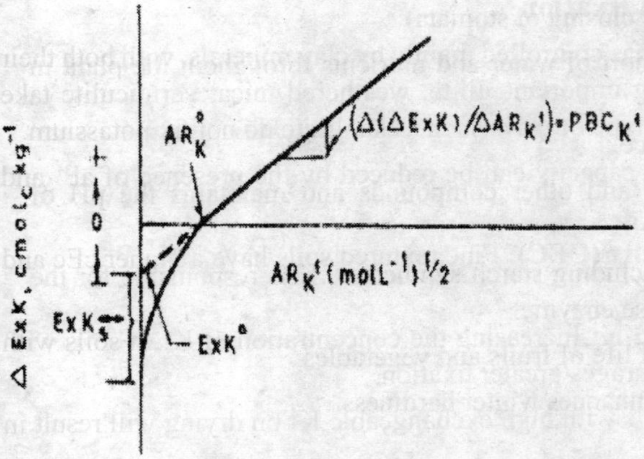


**Fig. 15:** A typical quantity/ intensity (Q/I) relations (adopted from Evangelou, 1994)

## 10.4.3.5. Luxury consumption of potassium

Some crops tend to absorb and accumulate potassium for in excess of their needs if it is present in sufficiently large quantities in the soil. This tendency is termed luxury consumption because the excess K absorbed does not increase crop yields to any appreciable extent. Wasteful luxury consumption occurs especially with forage crops.

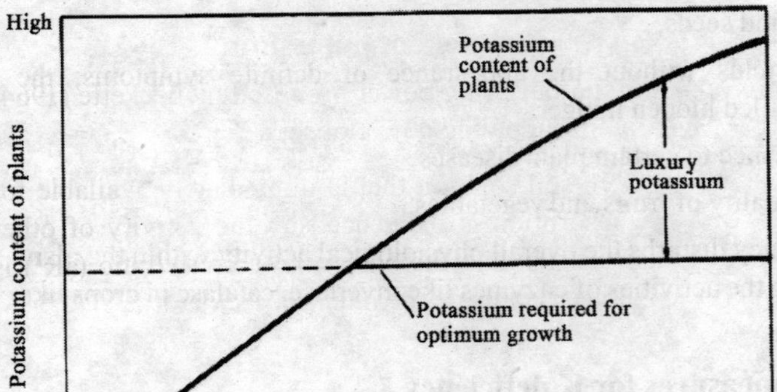


**Fig. 16:** Graphical representation of Luxury consumption of potassium

## 10.4.3.6. Functions of potassium in plants

Potassium is absorbed by the plant as $K^+$. Its concentration in healthy tissue varies from 1-5 %. Potassium is a unique element that plants can accumulate it in abundant amounts without exhibiting any toxicity symptoms. This behaviour is referred to as luxury consumption. Potassium plays a regulatory role in plant metabolism. Potassium is not a component of any organic compound in plants.

- It regulates the opening and closing of stomata)
- Plays a major role in transport of water and nutrients throughout the plant in xylem.
- It improves drought tolerance.
- Neutralizes organic anions and other compounds and maintains the pH of cytoplasm in the range of 7-8.
- Activates > 60 enzymes, including starch synthetase, K is responsible for the activation of nitrate reductase enzyme.
- Enhances crop quality, shelf life of fruits and vegetables.
- Reduces lodging of crops, enhances winter hardiness.
- Imparts disease resistance.

### 10.4.3.7. Deficiency symptoms of potassium in plants

- Potassium deficiency does not manifest immediately in the form of visible symptoms. First growth rate decreases and later deficiency symptoms appear. deficiency symptoms first develop on the older leaves.
- Chlorosis along the margins followed by scorching and browning of tips of older leaves which gradually progresses inwards giving burning appearance. Slow and stunted growth of the plant and crop lodging.
- Shrivelled fruits and seeds.
- Reduced crop yields without the appearance of definite symptoms; the phenomenon is called hidden hunger.
- Decrease in resistance to certain plant diseases
- Decrease in the quality of fruits and vegetables.
- Potassium deficiency disturbs the overall physiological activity within the plant system by altering the activities of enzymes like invertase, catalase in crops like sugarcane.

### 10.4.3.8. Correction measures for K deficiency

Potassium chloride or muriate of potash constitutes 99% of usage of K fertilizer and potassium sulphate accounts for 1%. Muriate of potash contains 60% $K_2O$ or 50% K and sulphate of potash contains 50 % $K_2O$ or 41.5% K. Some crops are sensitive to high amounts of potassium chloride. These include tobacco, grapes, cotton, fruit trees, sugarcane, potatoes, tomatoes, straw berries, onion and cucumber. Oil palm and coconut on the contrary appear to be chloride loving crops. Generally, potassic fertilizers are applied as basal dose, but for light textured soils, split application is advocated) In andhra Pradesh split application of N and K are recommended for light soils. Potassium nitrate containing 44% $K_2O$ (37% K) and 13% N is preferred for spraying on fruit trees and horticultural crops. Spraying of $KNO_3$ during winters (*rabi* season) to impart cold hardiness is also in practice by the farmers in andhra Pradesh.

## 10.5 Secondary nutrients

Secondary nutrients include calcium, magnesium and sulphur. These nutrients received little attention due to their inadvertent application through fertilizers and irrigation water. ammonium sulphate (100 kg) supplies 20 kg N and 24 kg S; While single super phosphate supplies 16 kg of P2O5, 21 kg of ca and 12 kg of S. Potassium sulphate supplies 52 kg K2O and 18 kg of S. These secondary nutrients though required in equal amounts as primary nutrients, but are added inadvertently along with primary nutrients. Lime which contains ca and Mg is applied to raise the pH and sulphur compounds are used to lower the pH of soil. Irrigation water may contain on an average 25 ppm of $SO_4^{=}$, 50 ppm of $ca^{++}$ and 5 ppm of $Mg^{++}$.

### 10.5.1. Calcium

### 10.5.1.1. Content and sources of calcium in soil

The calcium concentration of the earth's crust is about 3.20 %, it may be as low as 0.015 % in humid region to as high as 5 % in arid region. High calcium content in arid regions could be due to low rainfall and hence little leaching. The major sources of calcium are the weathered products of rocks and minerals containing calcium like the following sources. The mineral sources of calcium are given in the table 7.

**Table 7:** Mineral sources of calcium and their calcium concentration

| Mineral | chemical formula | Total ca (%) |
|---|---|---|
| calcite | $cacO_3$ | 40 |
| apatite | $ca_{10}(PO_4)_6$ | 41 |
| dolomite | $cacO_3.MgcO_3$ | 22 |
| Gypsum | $caSO_4.2H_2O$ | 23 |

Besides these secondary minerals, primary minerals like augite (18.7%), Hornblende (4.6%), anorthite (10.0 – 20.0 %) Epidote (22.0-25.0%) etc), are the sources of calcium in the soil.

### 10.5.1.2.Forms of calcium in soils

calcium is generally the dominant cation in normal cultivated soils. It is present in different forms. The calcareous soils have higher amount of ca than acid soils.

a) **Mineral particles** : calcium is mostly present as primary minerals such as basic plagioclase like anorthite, and basic rocks like basalt, gabbro. They all release ca on weathering.

b) **Calcium carbonate** : calcite mineral is very common. The nodular form of calcium carbonate and amorphous calcium carbonate exist.

c) **Simple salt :** calcium is present in many simple salts such as calcium chloride, calcium sulphate, calcium nitrate and calcium bicarbonate.

d) **Exchangeable calcium :** among all the cations on the exchange complex of soils $ca^{2+}$ is the most dominant cation.

### 10.5.1.3 The fate of released mineral calcium in soil

1. **Absorbed by organisms** : Legumes use much more calcium than grasses. Prolonged cropping with crops like alfalfa on sandy soils low in ca depletes ca resource.
2. **Losses due to leaching** : The loss of calcium in drainage water due to leaching by rain or irrigation is very large. The loss of $ca^{2+}$ increases with increase in rainfall or with increase in permeability of the soil. Light textured soil suffers greater loss of calcium than heavy soils. Leaching losses range between 84 to 224 kg $ha^{-1}$.
3. **Adsorbed by clay colloids** : If the activity of calcium in soil solution suddenly increases, there will be a shift in equilibrium with subsequent adsorption of some $ca^{2+}$ by exchange complex. conversely, if the activity of calcium in the soil solution phase decreases due to either leaching or crop removal, the adsorbed calcium releases into soil solution.
4. **Reprecipated calcium :** calcium is reprecipitated as secondary calcium compounds in arid climate. In arid region, due to low rainfall, ca in soil solution will be precipitated as least soluble calcium compounds.

### 10.5.1.4. Factors affecting the availability of Ca in soils

i. **Total calcium supply** : Total supply of calcium in very sandy acid soils with low cation exchange capacity can be too low to provide sufficient available calcium to crops.

ii. **Soil pH** : High $H^+$ activity occurring at low soil pH will impede calcium uptake due to high base unsaturation.

iii. **Cation exchange capacity and degree of saturation of exchange complex** : The absolute amount of calcium present is not that important but the relative amount of other cations or degree of calcium saturation is important. Montmorillonite requires a 70 % base saturation to release ca, while kaolinite requires only 40 % base saturation.

iv. **Type of soil colloid**. The type of clay influences the degree of calcium availability. a much higher degree of saturation is required for 2 : 1 clay minerals than 1:1 clay.

v. **Ratio of calcium to other cations in soil solution** : calcium availability is influenced by the ratio between calcium and other cations. calcium / total cation ratio of 0.10 to 0.15 is desirable. blossom end rot, a calcium deficiency disorder in tomatoes can be prevented by maintaining a ca / soil solution salt ratio in the range of 0.16 and 0.2.

### 10.5.1.5. Functions of calcium in plants

Calcium is absorbed by the plant as calcium ions ($ca^{2+}$), the sufficiency range is between 0.2 – 1.0 %.

1. Calcium is essential for the formation of cell wall and calcium pectate in the middle lamella of the cell wall which regulates the entry of only those nutrients which are not toxic to plants. In seeds, calcium is present as calcium phytate.
2. In root tip, calcium is very essential for the meristematic activity.
3. Provides a base for neutralisation of organic acids and other toxins (like al) produced in plants.
4. It plays a role in mitosis (cell division) and helps to maintain the chromosome structure.
5. Essential co-factor or an activator of a number of enzymes like hydrolases.
6. It activates phospholipase, argnine kinase, amylase and adenosine tri phosphatase (aTPase) enzymes.
7. Favours the assimilation of nitrogen into organic constituents especially proteins.

### 10.5.1.6. Deficiency symptoms of calcium in platns

Soils seldom become calcium deficient, as long as soil pH is maintained towards neutral range. deficiency of calcium is characterised by a reduction in meristematic tissue

- Though calcium is highly mobile in soil, in the plant system it is the immobile nutrient and hence the deficiency symptoms manifest at the growing tips of shoots and youngest leaves.
- Failure or desiccation of terminal bud development.
- No unfolding of new leaves in corn, whose tips are colorless and are covered with sticky gelatinous material which makes them adhere to one another.
- Chlorosis of young leaves followed by distortion of the growing points of the stem.
- In fruit trees, the death of growing points followed by die back.
- In guava, the old leaves are chlorotic with red brown spots.
- In apple, the discoloration of the fruit meat, the condition generally referred to as "bitter pit".
- In brassica, severe loss of color in young leaves, terminal bud leaves are hooked, leaves below become cup shaped) Old leaves collapse due to terminal bud disintegration.
- blossom end rot in tomato is due to ca deficiency

### 10.5.1.8. Correction measures of calcium deficiency

Calcium as a plant nutrient is more important in calcium deficient acid soils. The application of carbonate or sulphate salts of calcium @ 2 – 4 q $ha^{-1}$ in furrows could increase the yield by 48 %.

### 10.5.1.9.Balance sheet of calcium in Soil

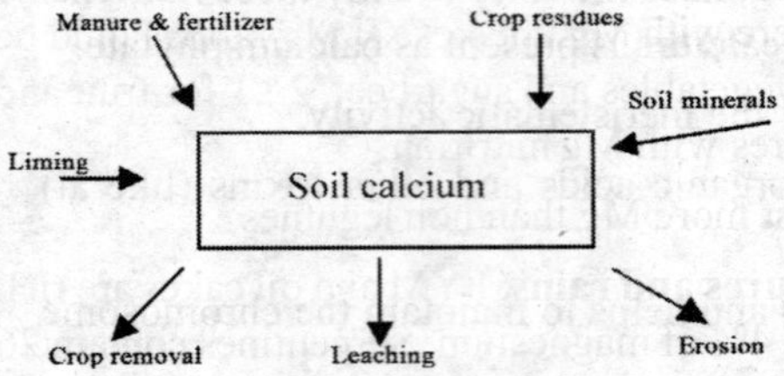


**Fig. 17:** Calcium balance sheet in soil

## 10.5.2. Magnesium

Magnesium is absorbed by plants from the soil solution as $Mg^{+2}$. It constitutes 1.93 % of the earth's crust ranging from 0.1 % in coarse sandy soils of humid region to 4 % in fine textured soils of arid and semi arid regions. Mg in soil originates from weathering of biotite, dolomite, hornblende, olivine, epsomite ($MgSO_4$. 7 $H_2O$) and serpentine. also found in secondary clay minerals like chlorite, illite, montmorillonite and vermiculite; excess Mg may be seen in soils developed from serpentine bed rock. Magnesium forms 12-80 % of exchangeable bases which is second to calcium (80-85%). If large quantities of magnesium are added to montmorillonite, it may convert to vermiculite.

### 10.5.2.1. Fate of released Mg in soil

1) Lost in the percolating water.
2) absorbed by living organisms.
3) adsorbed by surrounding clay colloids.
4) Reprecipitated as secondary mineral in arid regions.

### 10.5.2.2 Factors affecting Mg availability in soil

The soil Mg available to plants is in the exchangeable plus water soluble forms. The availability of Mg is influenced by

a) **Amount of exchangeable $Mg^{+2}$** : Soils are deficient in Mg when they contain less than 25 to 50 ppm of exchangeable $Mg^{+2}$.

b) **Soil pH** : Nutritional problems of $Mg^{+2}$ will arise in strongly acid soils with $al^{+3}$ saturation of 65 to 70 %. The availability of $Mg^{+2}$ can also be adversely affected by high $H^+$ activity in strongly acid soils.

c) **Degree of Mg saturation** : The more the degree of saturation the greater is its availability. When exchangeable magnesium is more than 45 %, it is toxic to plants. Magnesium availability is high when its saturation is 10 – 20 % on the exchange complex. Normally, the exchangeable magnesium accounts for 4 to 20 % of cEc. Magnesium saturation should not be less than 10 %.

d) **Nature of other exchangeable ions** : Plant deficiencies of Mg can occur in soils with wide ca / Mg ratio. Ideally it should not be greater than 7:1. High levels of exchangeable $K^+$ can interfere with Mg uptake. K / Mg ratio should be less than 5:1 for field crops, 3:1 for vegetables and sugarbeet, 2 : 1 for fruit and green house crops. $NH_4^+$ also interferes with Mg nutrition.

e) **Removal by crop** : Legumes exhaust more Mg than non legumes.

f) **Additions through fertilizers, manures and rains**: FYM and oil cakes are rich source of Mg. basic slag contains 3 – 4% of magnesium. Serpentines contain 26 %.

### 10.5.2.3. Functions of Mg in plants

The usual concentration of $Mg^{+2}$ in crops varies between 0.1 and 0.4 per cent. a large part of Mg is associated with organic anions like malate. Magnesium

i. Mg is the only mineral constituent of chlorophyll located at its centre.

ii. chlorophyll formation usually accounts for about 15 to 20 % of total Mg content of plants as Mg - porphyrin.

iii. Serves as a structural component of ribosomes. Mg activates the formation of polypeptide chains to form amino acids. about 70 % of Mg is associated with anions such as malate and citrate.

iv. Seeds contain Mg as salt of phytic acids. Mg is required for phosphate transfer from aTP (Phosphorylation) in carbohydrate metabolism.

v. Several enzymes (eg : *Ribulose carboxylose*) require $Mg^{+2}$ as cofactor.

vi. It promotes uptake and translocation of phosphorus and movement of sugars within the plants.

### 10.5.2.4. Deficiency symptoms of Mg in plants

$Mg^{+2}$ is a mobile element and is readily translocated from older to younger plant parts in the event of deficiency and hence deficiency symptoms are manifested in the older leaves. The magnesium deficient plants usually have less than 0.1% Mg. Magnesium deficiency is common in the plants grown on coarse textured acidic soils.

i. As a consequence of $Mg^{+2}$ deficiency, the proportion of protein nitrogen decreases and that of non protein nitrogen increases in plants.

ii. Shortage of $Mg^{+2}$ results in an interveinal chlorosis of the leaf in which only the veins remain green and the interveinal areas turn yellow with streaky or patchy appearance. In more advanced stages the leaf tissue becomes uniformly pale yellow, then brown and necrotic)

iii. Affected leaves turn small in final stage and curve upwards at the margins.

iv. In some vegetables, interveinal chlorosis with tints of red, orange and purple colors is observed)

v. Grass tetany : cattle consuming forages with low Mg may suffer from "Hypomagnesemia" (low level of blood Mg) commonly known as Grass tetany. This happens due to high levels of $NH_4^+$ - N and K application.

### 10.5.2.5 Correction measures of Mg deficiency

Use of dolomitic lime stone ca $Mg(cO_3)_2$ – 6.5%Mg; Magnesia @ MgO, 60 % (Mg); basic slag @ 3-4 % Mg; Epsomite ($MgSO_4$ $7H_2O$), 9.5% Mg; Kieserite ($MgSO_4$ $H_2O$), 17.4 % Mg at a basal dose of 30 – 50 kg $ha^{-1}$ is recommended)

### 10.5.3. Sulphur

The earth's crust contains about 0.06 % sulphur. It is mostly present as sulphides, sulfates and in organic combinations with c and N. More than 95% of total sulphur in soil is present in organic matter under temperate conditions. Under tropical and sub tropical conditions, the organic matter content in the soil is generally less than the temperate soils. The initial source of all sulphur is the sulphide contained in plutonic rocks. Through weathering the sulfides are released) In humid region organic sulphur, while in arid zone, the sulfates of calcium, Mg, Na and K predominate. another source of sulphur is the atmosphere. Owing to industrial activity or through combustion of fuel, sulphur dioxide is produced which is brought down by rain.

The continuous use of high analysis fertilizers, particularly those which are devoid of sulphur such as urea, triple super phosphate, ammonium nitrate and ammonium chloride may lead to sulphur deficiency in soils.

### 10.5.3.1. Forms of sulphur in soils

Sulphur is present in the soil in many different forms, both organic and inorganic) The inorganic forms are readily soluble sulphate, adsorbed sulphate, insoluble sulphate co-precipitated with calcium carbonate and reduced inorganic sulphur compounds.

1. **Easily soluble sulphate** : Sulphur is normally taken up by plants as the $SO^=$ ion. In sulphur deficient soils the amount of readily soluble $SO_4^=$ is frequently between 5 and 10 ppm. Sandy S deficient soils often contain less than 5 ppm. $SO_4^=$ when associated with monovalents like $Na^+$ are more prone for leaching. Medium losses when bound to divalent like $ca^{2+}$ or $Mg^{2+}$ and minimum losses when bound to $al^{3+}$ or $Fe^{3+}$.
2. **Adsorbed sulphate :** adsorbed sulphate is an important fraction in soils containing large amounts of hydrous oxides of iron and aluminium like in Oxisols and Ultisols. adsorbed sulphate can account for upto 1/3rd of total sulphur.
3. **Sulfate coprecipitated with calcium carbonate :** It is an important fraction in calcareous soils.
4. **Sulfides :** Under anaerobic conditions in water logged soils, there may be accumulation of $H_2S$ formed by the decay of organic matter. also, $SO^=$ present in the soil serves as an electron acceptor for sulphate reducing bacteria and it is

reduced to $H_2S$. This reduction takes place at a redox potential of less than - 150 mV and at a pH range of 6.5 to 8.5.

5. **Elemental sulphur** : It is an intermediate product formed during oxidation of sulfides by chemical processes.

6. **Organic sulhur** : Most of the sulphur in surface horizons of well drained agricultural soils of humid, temperate, semi arid and sub tropical regions is present in the organic form. It accounts for more than 90 %. average c : N:S ratio for calcarious soil 113 : 10 : 1.3 and for non calcareous soils it is 147 : 10 : 1.4. The N:S ratio of most soils is in the range of 6 to 8 in the form of amino acid like cystein, cystine and methionine.

## 10.5.3.2. Sulphur transformations in soil

The transformations of sulphur in soil are many and varied with changes often being cyclic) also, the element converts back and forth from inorganic and organic forms due to the presence of living organisms. When plant and animal residues are returned to the soil they are attacked by microorganisms thus releasing some of the sulphur as sulphate. Most of the sulphur, however, remains in organic form and eventually becomes part of the soil humus.

i. **Mineralisation** : Mineralisation of sulphur is the conversion of organic sulphur to inorganic $SO_4^{=}$ by microbes. It depends on the sulphur content of the decomposing material. Mineralisation of sulphur takes place at or below c/S weight ratio of approximately 200 : 1. Heterotrophic microorganisms bring about mineralization. Sulfatase enzyme plays an important role in mineralization.

$$R\text{-}O\text{-}SO_3^- + H_2O \rightleftharpoons ROH + HSO_4^-$$

Mineralisation is more in cultivated soils than fallow due to rhizosphere effect.

ii. **Immobilisation** : Sulfur is immobilized in soils in which the ratio of either c : N: S is too wide. at a c : S ratio of above 200 : 1, immobilisation or tie-up of $SO4^{=}$ with various organic forms is favored) Particularly, if the ratio is greater than 400: 1 i.e., when low sulfur containing organic materials are added to soils not plentifully supplied with inorganic sulphur immobilization of inorganic sulphur takes place. The energy rich material stimulates microbial growth and the inorganic sulphur ($SO4^{=}$) in the soil is locked in the microbial tissues. Only when the microbial activity subsides inorganic $SO4^{=}$ again appear in the soil solution.

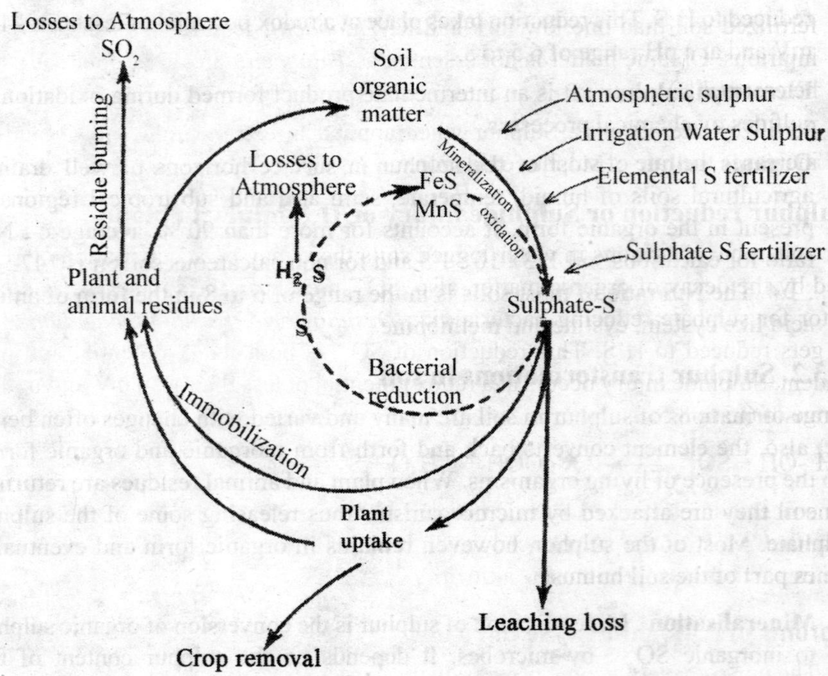


**Fig. 18:** Sulphur cycle

iii) **Sulphur oxidation** : Elemental sulfur, sulfides and several other inorganic sulfur compounds can be oxidized in the soil by microbial oxidation, though chemical non biological oxidation also takes place to a lesser extent. The rate of biological oxidation depends on three factors like i) the microbial population ii) characteristics of sulfur source, iii) Environmental conditions in the soil

$$cO_2 + S + H_2O_2 + 2\,H_2O \longrightarrow (c\,H_2O)n + SO_4^{=} + 2\,H^+$$

Two classes of bacteria are specially adapted for sulfur oxidation. These are chemolithotrophic sulfur bacteria which utilize the energy released from the oxidation. eg : *Thiobacillus thiooxidans*. *Thiobacillus thioparus*

## Factors affecting sulphur oxidation in soils

i. Soil microflora : The more the number of sulphur oxidizing bacteria like *Thiobacillus thiooxidans* the more is sulphur oxidation.

ii. Temperature : Temperature between 25° and 40°c will be close to ideal, the process is slow below 10°c.

iii. Soil pH : Optimum pH range is 6.5 to 7.2.

iv. Soil type and properties : S oxidation is not very dependent on soil texture, organic matter content etc) Rapid sulphur oxidation is observed in well fertilized soil than one low in P and K. *Thiobacilli* require $NH_4^+$ and $NO_3^-$ is injurious. Organic matter is not essential as *Thiobacilli* are autotrophic but not heterotrophic)

v. Fertilizer interactions : Sulphur when applied in combination with N and P increases the rate of sulphur oxidation.

## iv) Sulphur reduction or Sulphide injury or $H_2S$ injury to plants

Under anaerobic conditions in water logged soils there may be accumulation of $H_2S$ formed by the decay of organic matter. also, $SO_4^=$ present in the soil serves as an $e^-$ acceptor for sulphate reducing bacteria viz., *Desulfovibrio*, *Desulfatomaculum* and $SO4^=$ gets reduced to $H_2S$. This reduction of $SO_4^=$ is both redox potential and pH dependent. Sulphide injury occurs at a redox potential of less than -150 mV and in the pH range of 6.5 to 8.5.

$$2\ R\ cH_2\text{-}OH + SO_4^= \longrightarrow 2\ R\ cOOH + 2\ H_2O + S^=$$

In some rice paddies, high in organic matter and low in active metallic elements such as iron, the free $H_2S$ may be released which is harmful to rice roots. This $H_2S$ injury to rice in paddies is referred to as "akiochi".

## Symptoms of Akiochi disease in rice

- The upper leaves become progressively shorter.
- Lower leaves drying to a dirty yellow color with brown spots on the surface.
- Roots are pale white, mingled with black or rotten ones giving rotten egg smell and when roots are exposed to sun after washing they turn to reddish colour.
- Soil pH is around 6.0.
- bubbles will be coming out from the affected areas.

In normal submerged soils well supplied with Fe, the $H_2S$ liberated is almost removed from solution by reaction with $Fe^{2+}$ forming $FeS_2$ which forms protective coating on roots and prevent rotting.

$$Fe^{2+} + S^{-2} \longrightarrow FeS_2 \text{ (Iron sulphide)}$$

## Remedial measures for Achiochi disease in rice

- Use of ammonium sulphate or any S containing fertilizers should be avoided)
- application of material containing iron such as laterites or lake mud is recommended)
- Providing aeration to soil by draining the field)

## 10.5.3.3 Functions of sulphur in plants

- It is required for the synthesis of the S containing amino acids cystein, cystine and methionine and for protein synthesis.
- It activates certain proteolytic enzymes such as papainase and synthesis of papain.

- It is a constituent of certain vitamins viz., Thiamine and biotin, coenzymes and glutathione, acetyl coenzyme a (precursor for fatty acid synthesis), ferredoxin.
- It is present in the crops like onion, mustard, cabbage and cauliflower as polysulfides.
- It increases oil content of crops like flax, soybean, groundnut etc)
- disulfide linkages (-S-S - ) have been associated with the structure of proteins.
- Sulfhydryl (-SH) groups in plants are related to increased cold resistance.
- It is required for N fixation in legumes and is a part of nitrogenase enzyme system.

### 9.5.3.4. Deficiency symptoms of sulphur in plants

Sulphur content in plants ranges between 0.1 to 0.4 %. In view of the large field scale occurrences of sulphur deficiencies in India, it has been described as the fourth major nutrient after N, P and K. Plants suffering sulphur deficiency accumulate non protein nitrogen in the form of nitrate and amide. N:S ratio of plants is between 9 to12 : 1. as sulphur is immobile in the plant, its deficiency is manifested on young leaves.

i. The fading of normal green colour of the young meristem followed by chlorosis.
ii. Shoot growth is restricted)
iii. In *Brassica* plants, the lamina is restricted and the leaves show cupping owing to the curling of leaves.
iv. The older leaves become puckered inwardly with raised areas between veins.
v. The older leaves may develop orange or reddish tints and may be shed prematurely.
vi. The stem and leaf petiole may become brittle and may collapse.
vii. Reduced synthesis of proteins and oil.

### 9.5.3.5 Correction measures for S deficiency in plants

a) Application of elemental sulphur or gypsum particularly on alkaline soils.
b) Application of sulphur containing fertilizers like single super phosphate (12-14% S), Magnesium sulphate (30 % S), ammonium sulphate (24.2% S).
c) For correcting deficiencies of sulphur on the standing crop, foliar application of sulphate containing salts like Ferrous sulphate (32.8% S) and ferrous ammonium sulphate (16% S) etc)

## 10.6. Micronutrients

The term micronutrient or trace element is quite often used interchangeably though most of the scientists prefer to use the term micronutrients to denote the elements which are essential for the plant but are required in small amounts. They are constituents of enzymes and co-enzymes. The micronutrients essential for green plants are iron, copper, manganese, boron, zinc, molybdenum and chlorine.

**Sources of Micronutrients**: The main source of micronutrients in soil is the parent material . The micronutrient content of the soil entirely depends on the rocks from which the parent material is derived; soil forming processes, additions and removals. cobalt, Nickel and copper are preferentially attached to more basic rocks. Zinc has a greater tendency to be associated with sulphides. Shales contain boron (20 – 200 ppm), cobalt (10 – 50 ppm), zinc (20 – 100 ppm), copper (20 – 200 ppm) and molybdenum (1 – 5ppm).

**Table 8:** Sources of micronutrients and the forms of micronutrients in soil solution

| Major sources | Total content (ppm) | Forms is soil solutions |
|---|---|---|
| **Zinc** | 7 to 1000 | $Zn^{2+}$, $Zn(OH)^+$ |
| Sphalerite (ZnS) | | |
| Smithsonite $ZncO_3$ | | |
| **Copper** | 1.8 to 960 | $cu^{2+}$, $cu(OH)^+$ |
| chalcocite $cu_2S$ | | |
| covellite cuS | | |
| Malachite $cu_2(OH)_2cO_3$ | | |
| azurite $cu_3(OH)_2(cO_3)_2$ | | |
| **Iron** | 4000 to 27300 | $Fe^{2+}$, $F(OH)^+$, $Fe(OH)_2$ |
| Haematite $Fe_2O_3$ | | |
| Goethite FeOOH | | |
| Magnetite $Fe_3O_4$ | | |
| Pyrite $FeS_2$ | | |
| Olivine $(Mg\ Fe)_2SiO_4$ | | |
| **Manganese** | 37 to 115000 | $Mn^{2+}$ |
| Pyrolusite $MnO_2$ | | |
| Manganite MnOOH | | |
| **Boron** | 2.8 – 630 | $H_2bO_3^-$, $HbO_3^{2-}$, $bO_3^{3-}$ |
| borax $Na_2b_4O_7\ 10H_2O$ | | |
| Tourmaline $Na(Mg\ Fe)_3al_6(bO_3)_3Si_6O_{18}(OH)_4$ | | |
| **Molybdenum** | Traces to 12 | $MoO_4^{2-}$, $HMoO_4^-$ |
| Molybdenite $MoS_2$ | | |
| Wulfenite $PbMoO_4$ | | |
| Powellite $caMoO_4$ | | |
| **Chlorine** | 20 to 1000 | $cl^-$ |
| Muriate of potash (Kcl) | | |
| Sodium chloride (Nacl) | | |

**Inorganic forms of micronutrients:** Micronutrients present in soil clay minerals in various forms like oxides, sulphides, silicates, carbonates, chlorides etc (Table 9).

**Table 9:** Inorganic forms of micronutrients in soil

| Element | Major forms |
|---|---|
| Iron | Oxides, sulphides and silicates. |
| Manganese | Oxides, silicates and carbonates. |
| Zinc | Sulphides, carbonates and silicates |
| copper | Sulphides, hydroxy carbonates and Oxides. |
| boron | Sulphides, oxides and molybdates. |
| chlorine | chlorides |
| cobalt | Silicates |

**Pools of micronutrients in soils :** Viets (1962) postulated five distinct pools of micronutrients in soils.

**Pool A**: It is the soil solution pool and contains water soluble ions. It is usually small or almost nonexistent for copper and zinc and very small for iron and manganese in well aerated and natural soils. Poor aeration, low redox potential and low pH can markedly increase manganese and iron but not copper and zinc)

**Pool B** : This comprises of the micronutrient cations exchangeable by weak electrolytes like $NH_4^+$. It is small for cu and Zn.

**Pool C** : Includes the adsorbed, chelated or complexed ions exchangeable by other cations possessing high affinities for exchange sites or extractable with strong chelating agents, like EdTa, pyrophosphates, ammonium acetate and carbon tetra chloride and dithizone and other chelating compound which can extract such complexes.

**Pool D**: Micronutrient cations in secondary clay minerals and insoluble metal oxide. It depends upon the content of clay and on the amount of its saturation with micronutrients.

**Pool E**: comprises cations held in primary minerals. It is determined by the parent material and other geochemical considerations.

The micronutrient anions also occur in pools. For example chloride because of its negative adsorption, occurs mostly in pool a, but it also occur in pool E. borates present in pool E, as primary mineral tourmaline and in pool c probably as chelates with humus, but mostly it occurs in pool a in solution. Molybdates occur in all the pools, they can be replaced by oxalates, phosphates and hydroxyl ions. availability of molybdenum is more in the alkaline pH.

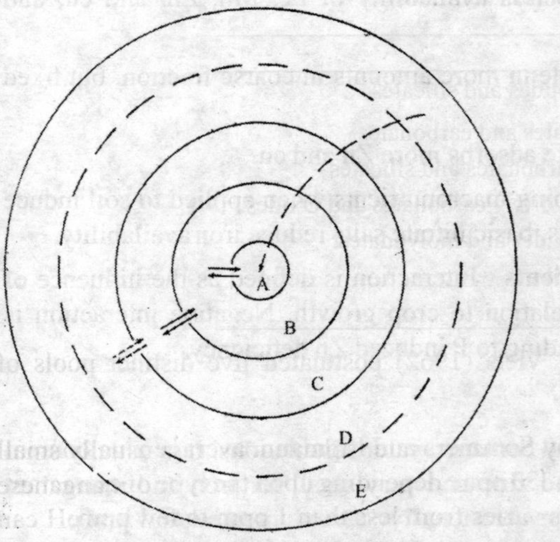


**Fig. 19:** The postulated pools of micronutrients in soil (Viets, 1962). a, Water soluble ions, b, cations exchangeable by weak electrolytes like $NH_4^+$; c, adsorbed, chelated, or complexed ions exchangeable by other cations possessing high affinities for exchange sites or extractable with stronger chelating agents; d, Micronutrient cationsin secondary clay minersls and insoluble metal oxides; E, cations held in primary minerals.

**Organic forms** : organic matter is an important secondary source of some trace element. They are held in complex combination with colloid complex.

## Predisposing factors for occurrence of micronutrient deficiencies

i) Parent material : The soils whose parent material originally contain low amounts of micronutrients are most likely to show micronutrient deficiency under leaching and intensive cropping.

ii) **Soil pH :** acid soil conditions are most conducive for Mo deficiency. Similarly, high pH soils such as sodic soils show Zn, Fe, Mn and cu deficiency. Overliming of acid soils also causes deficiency of micronutrient cations. In acid sulfate soils micronutrient toxicity limits crop growth.

iii) **Land leveling and shaping :** Land leveling and shaping removes the fertile top soil in which micronutrients are concentrated)

iv) **High yielding varieties** : HYV require more nutrients depleting the nutrients from soil causing their deficiencies.

v) **Imbalanced nutrient application** : High doses of macronutrient application induces micronutrient deficiencies.

vi) **Soil conditions** : Water logged condition reduce the availability of Zn and cu and increase availability of Fe and Mn.

vii) **Organic matter:** Organic compounds like humic and fulvic acid form organo - metallic complexes held as insoluble complexes. but some organic ligands can keep micro nutrient cations as soluble chelates and those are plant available.

viii) **$CaCO_3$:** More $caO_3$ leads to less availability of Fe, Mn, Zn, and cu; and increases Mo availability.

ix) **Soil texture :** boron is available in more amounts in coarse fraction, but fixed in fine textured soils.

x) **Clay content :** Montmorillonite adsorbs more Zn and cu.

xi) **Fertilizers :** Fertilizers containing macronutrients when applied to soil induce the deficiency of micronutrients. basic nitrate salts reduce iron availability.

xii) **Interactions with macronutrients** : Interaction is defined as the influence of an element upon another in relation to crop growth. Negative interaction is observed between P and Zn leading to P induced Zn deficiency.

## 10.6.1. Zinc

Essentiality of zinc was established by Sommers and Lipmann. average zinc content of mineral soils varies between 10 and 330 ppm. depending upon the type of extractant used, the available zinc in Indian soils varies from less than 1 ppm to few ppm.

### 10.6.1.1. Forms of zinc in soils

Zinc present in soil in the forms of a)Water soluble zinc, b)Exchangeable zinc, c) adsorbed zinc on surface of clay, organic matter, carbonatesand oxide minerals, d) Organically complexed Zn and f)Zinc substituted for $Mg^{2+}$ in the crystal lattices of clay minerals

### 10.6.1.2. Critical limits of Zn in soil and plants

The critical levels of Zn nutrient in soil differs with extractants used) different extracts had different critical levels (Table 10). In soil 0.6 ppm of dTPa extractable zinc is considered lower critical limit, but for black soil 0.8 ppm may be considered as critical limit.

**Table 10:** Critical levels of zinc in soil

| Type of extractant | critical limits of deficiency (ppm) |
|---|---|
| 0.005M dTPa + 0.1 M TEa + 0.01 M $cacl_2$ (pH 7.3) | 0.6 – 1.0 |
| 0.1 N Hcl | 1 – 5 |
| 1 N $NH_4Oac$ (pH 4.6) | 0.2 – 0.5 |
| 0.005 N Hcl | 1.0 |

**Critical Limits of Zn in most plants** : If zn content in plant is below 20 ppm-deficient, 20 - 150 adequate level ppm; above 400 ppm toxic level.

**Critical Limits of Zn in rice plant:** If Zn content is less than 5 ppm (is the critical limit) - very high response is obtained to the application of zinc, if it is 5 - 10 ppm – High response; 10 - 15 ppm – Possible response ; 15 - 20 ppm – No response

### 10.6.1.3. Factors affecting availability of zinc in soils

1. **Soil pH** : Zinc is generally more available in acid soils than in alkali ones. Mostly pH induced deficiencies of zinc occur in pH range of 6 to 8. The critical pH for availability of Zn is 5.5 to 6.5. In the alkaline range particularly above pH 7.85, zinc forms negatively charged ions called zincate ions, thus reducing the availability due to the formation of calcium zincate.
2. **Organic matter** : The presence of organic matter may promote the availability of zinc by complexing with zinc)
3. **$CaCO_3$** : Zinc deficiencies are more common in calcareous soils. Zinc was adsorbed on $cacO_3$ rendering it unavailable.
4. **Clay minerals** : Zinc adsorption on montmorionite clays from $Zncl_2$. Solution zinc occurs as $Zn^{2+}$, Zn $cl^+$ and Zn $(OH)_2$ ions.
5. **Phosphates**: Zn deficiency is very common in soils containing very high amounts of phosphates.

### 10.6.1.4. Functions of Zn in plants

- Zinc is a constituent of three enzymes viz., alcoholic dehydrogenase, carbonic anhydrase, superoxide dismutase (SOd).
- Zn is involved in the synthesis of indole acetic acid, metabolism of gibberellic acid and synthesis of RNa.
- because of preferential binding to sulphydryl group, zinc plays an important role in the stabilization and structural orientation of the membrane proteins.
- Zn influences translocation and transport of P in plants. Under Zn deficiency conditions excessive translocation of P occurs resulting in P toxicity.

### 10.6.1.5. Deficiency symptoms of Zn in plants

i. Zn deficiency symptoms show wide variation in different plant species. The common symptoms are chlorosis between the veins, reduction in the size of the young leaves, which are often clustered, bronzing, purple, violet, reddish brown or brown coloration of the foliage.

ii. **Khaira disease of rice:** The first symptom of zinc deficiency appear in 3 - 4 week old seedlings when the young leaves develop reddish brown pigmentation. The pigmentation appears first in the middle of the leaves, then intensifies and spreads over the entire lamina) The affected tissue becomes papery and necrotic and under conditions of severe deficiency, the entire mass of leaves collapses and further growth of the plant is arrested)

iii. **White bud of maize** : Soon after the emergence of seedlings, areas between the veins of old leaves become light yellow and develop white necrotic spots, which later develop dark brown necrotic areas that enlarge and coalesce, resulting in the necrosis (death of the entire leaf). Leaves that emerge and unroll subsequently appear yellow and white.Mottle leaf or Frenching of citrus

iv. Reduced leaf size and shortening of internodes in brinjal and mango called as little leaf.

### 10.6.1.6. Corrective measures of Zn deficiency in crop plants

- Soil application of zinc sulphate (21% Zn) @ 50 kg ha$^{-1}$ once for three crops or years is effective and economic to overcome its deficiency. Zinc sulphate need to be applied on the surface and mixed in the soil through light harrowing.
- In case of rice, zinc sulphate can be broadcast after final puddling before transplanting.
- In case of orchards, zinc sulphate can be applied in the basin mixing into the soil.
- When deficiency appears on standing crop, spraying of 0.2 % $ZnSO_4$ twice or thrice at weekly intervals or at 0.5 per cent concentration with lime.
- In case of alkali soils, the dose of zinc sulphate for soil application needs to be doubled to 100 kg ha$^{-1}$.
- Zinc is also available as zinc chelate (mostly Zn - EdTa form). In case of highly alkaline soils (pH>8.0) zinc chelate is better source than zinc sulphate.
- In calcareous soils, Zn- HEdTa (Hydroxy Ethylene diamine Tetra acetic acid) form is efficient for soil application.

### 10.6.2. Copper

copper concentration in earth's crust ranges from 55 to 70 ppm. Sulfides are the predominant minerals of copper in the earth's crust. copper deficiency is often the first nutritional disorder to appear in plants grown on newly reclaimed acid Histosols (organic soils) and this condition is referred to as 'reclamation disease'.

### 10.6.2.1. Forms of copper in soil

1. **Solution form of copper** : Ionic and complexed cu in the soil solution range from $10^{-8}$ to $10^{-6}$ (0.6 to 63 ppb). at soil pH below 6.9, it exists as $cu^{2+}$ and above 7.0 as cu $(OH)_2$ and cu $(OH)^+$

   Exchangeable & soluble (cu) ⟷ Specifically adsorbed cu ⟷ Organically bound cu

2. **Occluded and co-precipitated Cu** : cu is buried in various mineral structures or substituted for other ions or as an impurity in carbonate minerals or trapped in oxides of Fe and al.
3. Adsorbed cu: $cu^{2+}$ is adsorbed to layers silicate clays, organic matter and oxides of either iron or al.
4. Clay minerals and oxides: cu is held tightly with cu – O – al or cu – O – Fe surface bonds.
5. Organic matter: In many soils one fifth to one half of cu occurs in organically bound form. Organic acid molecules solubilise cu and make it available to

plants. Humic and fulvic acids contain multiple binding sites for cu, with $cOO^-$ playing a prominent role.

### 10.6.2.2. Critical limits of copper in soils and plants

critical limits of copper in plants, less than 4 ppm- deficient; 5- 10 ppm adequate; morethan 20 ppm toxic) The critical levels of cu in soils are given in the table11.

**Table:11:** Critical limits of copper in soils

| Type of extractant | critical level of deficiency (ppm) |
|---|---|
| 0.005M dTPa + 0.1 M TEa + 0.01 M $cacl_2$ (pH 7.3) | 0.2 – 0.5 |
| 0.1 N Hcl | 1 – 3 |
| 1 N NH4Oac (pH 4.8) | 0.2 |

### 10.6.2.3. Factors affecting availability and movement of copper

i. Texture : $cu^{+2}$ in soil solution is usually lower in excessively leached podzolic sands and calcareous sands

ii. Soil pH: Mobility of $cu^{+2}$ in soil solution increases with increasing pH due to, increased generation of pH dependent sites on colloids; reduced competition with $H^+$, and conversion of highly soluble $cu^{2+}$ to less soluble $cu(OH)^+$ and $cu(OH)_2$.

iii. Interrelationship with other elements: Excessive use of NPK fertilizers aggravate cu deficiency. acid forming fertilizer increase cu deficiency due to increased abundance of $al^{3+}$ in soil. High concentration of P, Fe, Zn and al depress absorption of cu.

iv. Incorporation of crop residue: Incorporation of crop residues reduce cu mobility due to

a) Chemical reactions of $cu^{2+}$ and organic compound and other substances originating from decomposing organic residues.

b) Competition for available $cu^{2+}$ by microbes

c) Inhibition of root development and ability to absorb $cu^{2+}$.

v) Plant factors: crops highly responsive to cu are carrot, lettuce, spinach, table beets, citrus and onions. crops tolerant to low cu are beans, peas, potatoes, lotus species.

### 10.6.2.4. Functions of copper in plants

Similar to other micronutrient cations, copper is absorbed by plant roots as $cu^{2+}$.

- Concentration of $cu^{2+}$ in copper sufficient plants is from 5 to 30 ppm and toxicity occurs between 20 and 100 ppm.
- $Cu^{2+}$ is a component of large number of proteins and enzymes like plastocyanin, SOd, diamine oxidase, polyphenol oxidase, ascorbate oxidase.

- Copper is important in imparting disease resistance.
- The enhances fertility of male flowers.

### 10.6.2.5. Deficiency symptoms of copper in plants

- Plants having a copper content of less than 5 ppm are regarded as cu deficient.
- Male flowers' sterility, delayed flowering and senescence are the most important effects of cu deficiency.
- Chlorosis of the younger shoot tissue, white tips, reclamation disease,necrosis, leaf distortion and die back are characteristics of cu deficiency. Necrosis of apical meristem results in elongation of shoot in cereals and auxiliary shoots in dicots.
- The shoot apex may cease to grow, resulting in the development of several auxiliary bunds.
- In cereals symptoms appear as bleaching and withering of young leaves.
- Exanthema and dieback in citrus which manifests as dark brown spots on the leaves, terminal twigs and fruits.
- Yellowish brown blotches on the leaf particularly in legumes.
- Tip drying and bluish green leaf tips are the symptoms in rice.
- Empty glumes in wheat
- Total amino nitrogen accumulates in plants

### 10.6.2.6. Correction of copper deficiency in soils and plants

Soil and foliar application are both effective. Soil application @ 1.0 – 5 kg /ha of $cuSO_4$. 5 $H_2O$ (24 % cu). Foliar application of $cuSO_4$ @ 0.2 % concentration. cu- EdTa contains 9-13 % cu.

### 10.6.2.7. Copper toxicity in plants

It occurs in acid soils and also due to the usage of sewage sludge, municipal composts, pig and poultry manures and mine waste and repeated use of copper containing pesticides like bordeaux mixture, $cuSO_4$. Toxicity symptoms include reduced shoot vigor, poorly developed discolored root system and leaf chlorosis resembling iron deficiency.

### 10.6.3. Iron

Iron comprises of about 5 % of the earth's crust and is the fourth most abundant element in the lithosphere. Normal concentration of total iron in soil varies from 0.7 to 5.5 %, average being 3.8 %. Haematite and goethite are the most common iron oxides in earth's crust. available Fe in Indian soils ranges between 0.09 and 225 ppm.

### 10.6.3.1. Forms of iron in soil

Iron is taken up by plants as $Fe^{2+}$. In soil solution, iron can be transported to root surface as iron chelates. The low solubility of compounds containing $Fe^{3+}$ severely limits availability and uptake of $Fe^{3+}$. diffusion and mass flow are found to be the two mechanisms responsible for the transfer of iron from soil to root surface. The organic substances capable of complexing or chelating iron originate as root exudates. The insoluble $Fe^{3+}$ form predominates in well drained soils, while $Fe^{2+}$ increases with a decrease in oxygen content or increase in water content of soil.

### 10.6.3.2. Critical limits of iron in soils and plants

In plants if the Fe is less than 50 ppm it is deficient; 50 ppm – 250 ppm adequate level. The critical levels of Fe in soil are given in table 12.

**Table 12:** Critical limits of iron in soils

| Extractant | Range |
|---|---|
| N $NH_4Oac$ (pH4.8) | 2.0 ppm |
| 0.005M dTPa + 0.1 M TEa + 0.01 M $cacl_2$ (pH 7.3) | 2.5 to 5.8 ppm |

### 10.6.3.3. Factors affecting availability and movement of iron

i. **Iron imbalance** : Imbalance of ions such as cu, Fe and Mn reduce the availability of iron.

ii. **Soil pH** : Iron deficiency is most often encountered in high pH and calcareous soils in arid regions. For each unit increase in pH the activities of $Fe^{3+}$ and $Fe^{2+}$ decrease by 1000 fold and 100 fold respectively.

iii. **Excessive water and poor aeration :** Flooding and submergence of soils improve iron availability. Under reduced conditions, sulfides are released which combine with $Fe^{2+}$ forming iron sulfide, otherwise if the soils are deficient in iron, the sulfides combine with $H^+$ forming $H_2S$ which causes 'akiochi'. The longer the period of submergence the greater is the amount of soluble plus exchangeable iron. Reduction of iron takes place at a redox potential of -130 mV.

iv. **Organic matter :** Iron reduction was greatly accelerated by the addition of organic matter.

v. **Interrelationship with other nutrients :** High phosphorus levels usually aggravate iron deficiency. $NO_3^-$ application lowers iron solubility. On the contrary, $NH_4^+$-N addition increases iron availability. Other metallic cationic micronutrients compete with iron for adsorption.

vi. **Plants factors :** The ability of plants to absorb and translocate iron appears to be genetically controlled) Roots of iron efficient plants alter their environment to improve the availability and uptake of iron.

### 10.6.3.4. Functions of iron in plants

Iron is taken up in $Fe^{2+}$ form by plants. Its concentration in the range of 100 – 500 ppm in mature leaf tissue is regarded as sufficient for crop production. Variable valency of iron assigns it a role in biological redox systems.

- Iron is a constituent of two groups of proteins viz., a) Heme proteins containing Fe – porphyrin complex. eg., peroxidase, leghaemoglobin.
- Fe-S proteins in which Fe is coordinated with thiol group Eg : Ferredoxin.
- It activates a number of enzymes including aminolevolinic acid synthetase and coproporphyrinogen oxidase.
- It plays an essential role in the nucleic acid metabolism.
- It is necessary for synthesis and maintenance of chlorophyll in plants.
- It is structural constituent of pigments in micro-organisms; the black pigment in *Aspergillus niger* contains iron)

## 10.6.3.5. Deficiency symptoms of Fe in plants

- The critical limit of iron in plant is 30 ppm, the sufficiency range being 50 and 250 ppm.
- It has been established that $Fe^{2+}$ content of the plant rather than total Fe content resolves Fe def.
- deficiency of Fe results in interveinal chlorosis appearing first on the younger leaves with leaf margins and veins remaining green. Plants having less than 50 ppm of Fe are usually classified as iron deficient.
- In later stage burning of the chlorotic leaves start from the tips and margin, spread inwards.
- The chlorotic leaves may become white and the leaf tissues devoid of chlorophyll die.
- Leaves with large necrotic areas fall off and twigs defoliate.
- In mild cases, mottled pattern may be seen with primary and secondary veins retaining their green color.
- In graminaceous crops, chlorosis consists of alternate strips with greenveins and yellow interveinal tissues.
- In case of barley, maize and jowar, leaves show reddish brown spots on leaves away from the base on margin.
- Under conditions of severe deficiency, growth cessation occurs with the whole plant turning necrotic)

## 10.6.3.6. Correction of Fe deficiency in soils and plants

In general soil application of iron salts such as ferrous sulfate is not practiced because of their rapid oxidation to much less soluble ferric iron. correction of Fe deficiency is generally done by foliar sprays.

## Fertilizer Sources of iron (Fe)

- $FeSO_4.7\,H_2O$ :19 % Fe
- Iron chelates NaFe EdTa :5-14 % Fe

- Fe – EddHa (Ethylene diamine dihydroxy acetic acid salt of Fe) for calcareous soils
- Fe dTPa (diethylene Triamine Penta acetate salt of Fe) for alkaline soils
- Fe HEdTa (Hydroxy Ethylene diamine Tetra acetate salt of Fe) for alkali soils
- Fe EdTa (Ethylene diamine Tetra acetate salt of Fe) for acid soils

**Iron chlorosis in paddy nurseries:** Iron chlorosis is common in rice nurseries due to the prevailing irrigated dry condition in the soil which oxidises $Fe^{2+}$ to $Fe^{3+}$ thus reducing its availability. It can be corrected by spraying 1.0% of Fe $SO_4$ or FaS (Ferrous ammonium Sulphate) mixed with 0. 1% citric acid, twice or thrice at 4 to 5 days intervals or till chlorosis disappears.

### 10.6.3.7. Iron toxicity

Toxic situations occur primarily on acid soils (< pH 5.0) and where excess soluble iron salts have been applied as foliar sprays or soil amendments. Injury due to high soil iron concentration is not common under neutral or high pH soil but under conditions that are saturated, poorly draining, compacted or poorly aerated) Iron toxicity also occurs where zinc levels are low.

### Symptoms of iron toxicity in plants

One of the most common symptoms is necrosis or death of the leaves. Necrotic spots will appear on the leaves of plants suffering from iron toxicity. Other symptoms include dark green foliage, stunted top growth and root growth, as well as leaf bronzing—especially on rice—which causes dark brown to purple spots on the foliage. The symptoms developed in rice grown in acid soils are browning of roots and bronzing of leaves.Reduced manganese uptake by plants.

### Management of iron toxicity in plants

- Management of iron toxicity is mainly through cultural practices. One of the most effective ways to avoid and to treat iron toxicity is by adding lime to the topsoil to raise the soil pH.
- Providing proper aeration to the soil for oxidation of ferrus iron ($Fe^{2+}$) to ferric ion ($Fe^{3+}$) form.

### 10.6.4. Manganese

Manganese concentration in the earth's crust averages 1000 ppm. It is an element that is widely distributed and at least traces of it are found in most rocks. Manganese when released through weathering of primary rocks will combine with $O_2$, $cO_3^{2-}$ and $SiO_2$ to form a number of secondary minerals including pyrolusite ($MnO_2$), manganite (MnOOH) and hausmannite ($Mn_3O_4$).

The total manganese in Indian soils ranges between 92 to 11500 ppm but majority of soils contain 300 to 1600 ppm of Mn. Forms of manganese are usually expressed as water soluble $Mn^{2+}$, exchangeable $Mn^{2+}$, water soluble and insoluble organically bound manganese, easily reducible manganese and manganese oxides. These forms are in equilibrium with one another.

The principal ion species in solution is $Mn^{2+}$ and its concentration decreases 100 fold for each unit increase in pH. concentration of $Mn^{2+}$ measured in the soil solution of acid and neutral soils have varied from less than 0.01 ppm to almost 13 ppm.

### 10.6.4.1. Factors affecting availability and movement of manganese

i. **Effect of pH and carbonates** : Soil pH will also greatly influence $Mn^{2+}$ availability and uptake. High pH favors the formation of less available complexes of Mn. activity of the soil microorganisms which oxidise soluble manganese to unavailable forms reaches a maximum near pH 7.0

ii. Excessive water and poor aeration: Soil submergence and water logging lower the redox potential and increase the amount of soluble $Mn^{2+}$ in soils.

iii. **Organic matter** : availability of Mn can be affected in several ways by the presence of organic materials. The low availability of manganese in basic soils high in organic matter is attributed to the formation of unavailable Mn compounds. It may also be held in unavailable organic complexes in peats or muck soils.

iv. **Imbalance of metal ions:** High levels of cu, Zn or Fe in the soil will impede manganese uptake by plants.

v. **Interrelationship with other nutrients:** acid forming N fertilizers like $(NH_4)_2$ $SO_4$, $NH_4cl$, $NH_4NO_3$ will enhance plant uptake of manganese. Neutral chloride containing salts like Nacl, Kcl and $cacl_2$ have increased the manganese concentration of plants.

vi. **Seasonal and climatic effects:** It is believed that wet winter weather favors the presence of $Mn^{2+}$, whereas warm dry summer encourages the formation of less available oxidized forms of manganese.

vii. **Soil microorganisms** : deficiencies of Mn is caused by soil micro organism oxidizing $Mn^{2+}$ to $Mn^{4+}$.

viii. **Plant factors** : among cereals, Oats are sensitive to Mn deficiency. Low reductive capacity at the root due to root exudates may be the factor restricting uptake and translocation of manganese.

### 10.6.4.2. Functions Mn in plants

Healthy Mn sufficient mature plants contain 20 to 300 ppm of Mn. It is a transitional metal, present in plants in $Mn^{2+}$ form. because of its variable valence, Mn plays an important role in the photosynthesis and detoxification of superoxide free radicals.

- Mn is an integral component of the water splitting enzyme associated with photosystem II. because of this role, Mn deficiency is associated with adverse effects on photosynthesis and $O_2$ evolution.

- It is a constituent of superoxide dismutase (Mn-SOd). Mn SOd, present in mitochondria protects cells against the deleterious effects of superoxide free radicals.
- Mn has a role in Tca cycle in oxidative and non oxidative decarboxylation reactions with Nicotine adenine diamide (Nad)

Malic dehydrogenase enzyme catalyzing the reaction.

Malate + $NadP^+$ ⟶ pyruvate + NadPH + $H^+$ + $cO_2$

Isocitrate dehydrogenase catalyzing the reaction.

Isocitrate + $NadP^+$ ⟶ Oxalosuccinate + NadPH

### 10.6.4.3. Deficiency symptoms of manganese in palts

Mn deficient plants contain less than 25 ppm Mn. deficiency symptoms of Mn are more severe on middle leaves than on the younger ones because Mn is preferentially translocated to the younger tissues. Interveinal chlorosis in dicotyledons is characterized by the appearance of chlorotic and necrotic spots in the interveinal areas. In monocotyledonous plants like cereals, Mn deficiency symptoms appear as greenish grey spots, fleeks and stripes more on the basal leaves (Grey speck). chlorotic leaf areas soon become necrotic and turn red, reddish brown or brown. Symptoms of Mn deficiency are popularly known as Grey speck in Oats, Speckled yellow in sugar beet, Marsh spot inpeas, Pahala blight in sugar cane and Frenching in tung grass

### 10.6.4.4. Critical limit of Mn in soil

**Table 13.** Critical levels of Mn in Soil

| Extractant | Critical limit |
|---|---|
| 0.005 M dTPa + 0.1 M TEa + 0.01 M $cacl_2$ | 2-4 ppm |
| 1 N $NH_4Oac$ | 3-4 ppm |
| 3 N $NH_4H_2PO_4$ + 0.1 N $H_3PO_4$ | 15-20 ppm |

### 10.6.4.5. Corrective measures of Mn deficiency

Soil application of Mn $SO_4$. $3H_2O$ (26-28 % Mn) @ 10-25 kg/ha and $MnSO_4$ $H_2O$ (30-32 % Mn) @ 10-25 kg/ha) $MnO_2$ contains 55-65 % of Mn EdTa - Mn supplies 5-12 % of Mn

### 10.6.5. Boron

The total boron content in Indian soils has been found to vary from 7 to 630 ppm. It is observed that more boron is concentrated in the surface soils than in the sub surface.

### 10.6.5.1. Factors affecting available boron content in soils

i. **Parent material** : The boron content of granite is 19.3, shales and lime stone 39.2, basalt 42.5 and of alluvium 42.3 ppm.

ii. **Texture** : boron content of the soil depends on the coarse sand fraction than on the clay content. arid zone soils have higher boron content.

iii. **Calcium carbonate** : calcareous soils are supposed to contain more total boron than non calcareous soils.

iv. **Effect of cultivation** : cultivated soils have more boron than their counter part virgin soils due to its accumulation through irrigation water.

v. **Irrigation water :** continuous use of irrigation water containing even a small amount of boron would build up the boron content of the soil.

vi. **Soil pH :** This element generally becomes less available to plants with increasing pH greater than 6.3 to 6.5. adsorption of boron by iron and aluminium hydroxides is maximum at pH of 8 to 9 and 7 respectively.

vii. **Organic matter :** It is one of the main source of boron in acid soil.

viii. **Soil moisture :** boron deficiency is often associated with dry weather and low soil moisture conditions.

ix. **Plant factors :** crops most sensitive for boron deficiency are sugar beet and celery. High boron requiring crops are apple, asparagus, broccoli, cabbage, cauliflower. boron needs of graminae are low.

Soils where boron deficiency is common are Soils low in boron such as those derived from acid igneous rocks; Highly leached acid soils like laterites and podsols; Light textured soils; Soils low in organic matter and alkaline soils with high amount of free $cacO_3$.

### 10.6.5.2. Functions of boron in plants

Boron is absorbed by plants as boric acid) However, it can also be absorbed in anionic form viz., dihydrogen borate ($H_2bO_3^-$), monohydrogen borate ($HbO_3^{2-}$) under acidic conditions and borate $bO_3^{3-}$ under high pH condition. Normal boron sufficiency in plants is between 10 and 200 ppm.

- Boron is neither a constituent of enzymes nor it activates any of the enzymes.
- Most important property of boron ($H_3bO_3$) is to form stable complexes with organic compounds with cis - diol configuration.
- It is responsible for cell wall formation and stabilization, lignification and xylem differentiation. as a consequence, b deficiency causes changes in chemical composition and ultra structure of cell wall, accumulation of toxic phenols, inhibition of lignin synthesis and a decrease in the production of indole acetic acid (Iaa). decrease in Iaa is responsible for the induction of ca deficiency.
- It imparts drought tolerance to the crops.
- Plays a role in pollen germination and pollen tube growth.
- It facilitates ion uptake by way of increasing the activities of plasma membrane bound $H^+$ - aTPase.
- It facilitates the transport of K in guard cells as well as stomatal opening.

### 10.6.5.3. Deficiency symptoms of boron in plants

Plants having b concentration in the order of 5 to 30 ppm are suspected to be boron deficient. critical deficiency range of b varies from 5 to 10 ppm in graminae plants and 20 to 70 ppm in dicotyledons.

- Boron deficiency symptoms are conspicuous on the terminal buds or the youngest leaves; which become discolored and may die under acute deficiency.
- Internodes become shorter and give appearance of a bush or rosette.
- Increased diameter of stem and petiole gives rise to the typical cracked stem of celery.
- Specific names given to b deficiency in different crops are Heart rot of sugar beet.
- Browning or hollow stem of cauliflower.
- Top sickness of tobacco.
- Internal cork of apple.

### 10.6.5.4. Boron containing fertilizers

Generally crops like alfalfa and other legumes, cruciferous plants such as cabbage, cauliflower and turnip and fruit trees such as apple, need boron fertilization.

i) Foliar aplication of borax ($Na_2\ b_4\ O_7$ 10 $H_2O$ which contains 10.6 % b) at 0.1 % concentration.

ii) Calmonite $ca_2b_4O_7$ 5 $H_2O$ is suitable for sandy soils due to slow solubility.

iii) Boric acid $H_3bO_3$ contains 17 % b

iv) Solubor : $Na_2b_40_7$. 5 $H_2O$ + $Na_2b_{10}0_6$ 10 $H_2O$ contains 20-21 %

### 10.6.6. Molybdenum

The essentiality of Mo was established by arnon and Stout (1939). The total Mo content of most soils varies from 0.2 to 5.0 ppm with a mean of 2 ppm.

**10.6.6.1 Forms of molybdenum in soils :** Molybdenum is generally present as Part of the crystal lattice of primary and secondary minerals; adsorbed as molybdate ($MoO_4^{=}$) on clay or exchange complex; Part of the organic matter and as water soluble molybdenum compounds. The available Mo content of Indian soils extractable with ammonium oxalate at pH 3.3 (Tamm's extractant) varies from traces to 1.65 ppm. The Mo content in alluvial soils is 1.0 to 5.5 ppm, in black cotton soils 0.6 to 11.80 ppm, in laterite soils1.34 to 2.0 ppm

### 10.6.6.2. Critical limits of molybdenum in soils and plants

Critical limits of Mo in soil: ammonium oxalate extractable molybdenum is 0.05 to 0.2 ppm. Soils with less than 0.15 ppm of available Mo respond to the application. critical limits of Mo in plants: less than 0.1 ppm – deficiency; 0.1 to 0.5 ppm – adequacy level.

### 10.6.6.3. Factors affecting the availability of molybdenum

i. **pH** : availability of Mo is low in acid soils, with increase in pH the availability increases upto 7.9.

ii. **Organic matter** : The role of organic matter in increasing the availability of Mo is due to the protection of molybdenum from anion adsorption.

iii. **Calcium carbonate** : Liming of acid soils increases Mo availability. Liming makes the native molybdenum available by increasing its availability.

iv. **Texture :** Mo availability increases with increase in clay content.

v. **Phosphate and sulphate fertilizers :** Phosphate and sulfate ions may compete with molybdate ions for exchange sites in the soil and plant.

vi. **Manganese :** Mn induces Mo deficiency because of antagonism between them due to changes in pH.

vii. **N and K fertilizers :** N fertilizers reduce the response to Mo in case of legumes due to reduced biological N fixation. However, a positive interaction exists between K and Mo.

viii. **Oxides of iron and aluminium :** availability of Mo may depend upon the molybdate adsorbed to free iron oxides.

### 10.6.6.4. Functions of Mo in plants

- Mo is a constituent of nitrate reductase.
- Activates several enzymes like catalase, peroxidase and polyphenol oxidase.
- Inhibits acid phosphatase.
- Required by N fixing organisms like azotobacter, clostridium, Nostoc, anabaena)
- Being a constituent of the Hydrogenase enzyme, helps in enzymatic transfer of $e^-$ for reduction of N≡ N bonds.
- deficiency of Mo results in accumulation of nitrates.
- Involved in carbohydrate metabolism and sugar formation.

### 10.6.6.5. Deficiency symptoms of Mo in plants

- In most plants with reticulate venation, the first effect of Mo deficiency appears as chlorotic mottling between the veins.
- Brassica crops are very susceptible to Mo deficiency, the symptoms developing in 3-4 week old plants.
- Tomato, lettuce, spinach, beet root and brassica speces especially cauliflower, broccoli and rape seed are very sensitive to restricted Mo supply.
- Legumes develop symptoms which resemble N deficiency.
- Grasses seem to have low Mo requirement.
- Citrus plants develop 'yellow spot'.
- Cauliflower exhibits whiptail (leaves get twisted elongated)

- Cabbage shows cupping, veins become purple, leaves become necrotic and malformed along the margins.
- In tomato, Mo deficiency showed chlorotic mottling,old leaves roll inwards along leaf margin.
- In rice, slight chlorosis between the vein in the middle of the upperand middle leaves and extending across the width.

### 10.6.6.6. Correction measures of Mo defficiency

- Application of 400 -500 gm of sodium molybdate ($Na_2MoO_4$) along with super phosphate to the soil.
- Mo deficiency can be corrected by spraying 0.1 % sodium molybdate.
- Liming of acid soils helps in correcting Mo deficiency.
- Seed treatment with Mo as sodium molybdate is very effective method of correcting the deficiency.

### 10.6.6.7. Toxicity of Mo in plants

- Plants develop brilliant tints of golden yellow or blue due to formation of granular complex between Mo and tannins.
- In potato, tomato and flax, Mo toxicity manifests as reddish or golden yellow color of the shoots.

### 10.6.7. Chlorine

Chlorine has recently been established as one of the essential nutrients for plant growth. Wilting is considered the most general symptom of chlorine deficiency. It is present primarily in igneous and metamorphic rocks. In the soil, $cl^-$ occurs as Nacl, ca $cl_2$ and Mg $cl_2$. The quantity of $cl^-$ in soil solution may range from 0.5 ppm or less to over 6000 ppm. chloride is highly mobile in soils. chloride ion will accumulate in ground water which will move into the root zone by capillarity.

### 10.6.7.1. Functions of Chlorine in plants

- Chlorine has been shown to be involved in the oxygen evolution in photosynthetic reactions.
- Claimed to be involved in photophosphorylation.
- In microorganisms, chlorine containing metabolically active compounds like chlorometabolites were found)
- In higher plants, chlorides largely accumulate in free ionic form in cell vacuoles.

### 10.6.7.2. Deficiency symptoms of Chlorine in plats

- Common symptoms are wilted appearance of the foliage and stuffy roots with laterals showing branching.
- Tomato leaves show chlorotic mottling, bronzing and tissue necrosis.

### 10.6.7.3. Toxicity symptoms of Cl in plants

Excess chloride will increase the osmotic pressure of soil water and lower the water availability to crops. Some plants are sensitive to chloride and develop leaf burn symptoms when chloride concentration reaches about 0.5%.leaves of tobacco and tomatoes thicken and begin to roll due to excess chloride. The other toxicity symptoms include reduction in the number and size of leaves, burning or firing and scorching of the leaf tips or margins, bronzing, premature yellowing and abscission of leaves. Sometimes growth reduction may occur without any leaf symptom.

High concentration of $cl^-$ in soil is due to high amount of chloride in irrigation water, inadequate leaching of salts from the root zone of the plant, inadequate permeability and drainage of the land and high water table and capillary movement of $cl^-$ into the root zone.

### The crops have been classified into different grades of tolerance to $Cl^-$

i. Low tolerance to $cl^-$ (< 20 me $L^{-1}$) eg : Peach, avocado, Lemon, beans, apple

ii. Medium tolerance (20-25 me $L^{-1}$), eg : Orange, apricot.

iii. High tolerance (> 25 me $L^{-1}$), eg : Tomato, beet

Chloride is beneficial for some salt loving plants like beetroot, spinach, cabbage etc) There is a negative interaction between chloride and nitrate ($NO_3^-$) ions in plants.

## 10.7. Soil fertility evaluation

Soil fertility concerned with the inherent capacity of soil to provide nutrients in adequate amounts and in proper balance for the growth of specified plants when other factors such as light, moisture, temperature and the physical condition of the soil are favourable. Soil fertility is an aspect of the soil plant relationship viz., plant growth with reference to plant nutrients available in soil.

Justus Von Liebig 1840 propounded the 'law of Restitution' which states that in order to maintain soil fertility nutrients removed from the soil by crops must be restored by the application of manures and fertilizers. The assessment of nutrient supplying capacity of the soil is soil fertility evaluation. It necessitates understanding of certain major concepts having definite bearing on soil fertility. The law of minimum was put forward by Von Liebig which envisages that if a soil contains optimum / adequate amounts of all but one nutrient element, crop growth is regulated by that single nutrient.

**Approaches for soil fertility evaluation** : The wide variety of diagnostic techniques used so far can be broadly grouped into i)Soil analysis, ii)Plant analysis, iii) biological methods and iv) Visual symptoms of nutrient deficiency or toxicity. Soil testing and plant analysis are useful tools for making recommendations for application of fertilizers to crops. Soil testing gives a measure of the availability of nutrients to crops, plant analysis indicates the actual removal of the nutrients from the soil.

### 10.7.1. Soil analysis

It is set of various chemical processes that determine the amount of available plant nutrients in the soil.

#### 10.7.1.1. Objectives of soil analysis

- Grouping soils into classes relative to the levels of nutrients for suggesting fertilizer practices.
- Predicting the probability of getting profitable responses.
- Helping to evaluate soil productivity.
- Determining specific soil conditions like alkali, salinity and acidity which limits crop yields.

**Available nutrients** : Plants draw their nutrients from air, water and soil. The bulk of mineral nutrients come from soil. Soil available form of nutrient is that fraction which is distributed in different discrete chemical forms, which often exist in a state of dynamic equilibrium and constitute the pool from which plants draw it. Soil available form of a nutrient is also that fraction whose variation in amount is responsible for significant changes in yield and responses. The nutrient available to biological organisms is termed as bioavailable nutrient.

#### 10.7.1.2. Chemical methods for estimating nutrients

Soil testing includes measurement of available N, P, K, S, micronutrient, lime and gypsum requirement, besides measuring pH, Ec and calcium carbonate, texture by bouyoucos hydrometer method)

**Table 14:** The different extractants for the available nutrients

| Nutrient | Extractant |
|---|---|
| available N | 0.32% $KMnO_4$ in alkine medium |
| available P | 0.5 M NaH $cO_3$, pH 8.5 Olsen's extractant |
| | 0.03 N $NH_4F$ + 0.025 NHcl bray's No.1 extratant |
| available K | Neutral normal ammonium acetate |
| available S | 0.15 % $cacl_2$ |
| available Zn, Fe, cu, Mn | 0.005 M dTPa, pH 7.3 diethylene Triamine penta acetate |
| Gypsum requirement | Schoonover method |
| Lime requirement | Shoemaker et al. method |

**Table 18:** Rating limits of soil test values

| Nutrient | Low | Medium | High |
|---|---|---|---|
| Organic carbon (%) | below 0.5 | 0.5 – 0.75 | above 0.75 |
| avail. N (kg $ha^{-1}$) | below 280 | 280 – 560 | above 560 |
| avail. P (kg $ha^{-1}$) | below 10 | 10 – 24.6 | above 24.6 |
| avail. K (kg $ha^{-1}$) | below 108 | 108 – 280 | above 280 |
| avail. S (ppm) | deficient < 10 ppm | Sufficient > 10 ppm | |
| dTPa Zn | deficient < 0.6 ppm | Sufficient > 0.6 ppm | |

## 10.7.2. Plant analysis

Although plant analysis is an indirect evaluation of soil, it is a valuable supplement to soil testing. Plant analysis is useful in confirming nutrient deficiencies, toxicities or imbalances, identifying hidden hunger, evaluating fertiliser programme and determining the availability of elements. Sometimes adequate nutrients may be present in the soil, but because of other problems like soil moisture and inadequate amounts of some other nutrients, the plant availability of the nutrient in question may be constrained) Plant analysis is based on the fact that the amount of a given element in plant is an identification of the supply of that particular nutrient and as such is directly related to the quantity in the soil.

For most diagnostic purposes, plant analyses are interpreted on the basis of critical value approach, which uses tissue nutrient concentration calibrated to coincide 90% or 95% of the maximum yield, below which the plants are considered to be deficient and above that value sufficient. Two general types of plant analysis are in use.

1. The tissue test which is customarily made on fresh tissue in the field)
2. Total analysis performed in the laboratory with precise analytical techniques.

### a) Rapid Tissue tests

Rapid tests for the determination of nutrient elements in the plant sap of fresh tissue. In these tests, the sap from ruptured cells is tested for unassimilated nitrogen, phosphorus and potassium. They are semi quantitative tests mainly intended for verifying or predicting deficiencies of N, P or K. The results are read as low, medium or high. Through the proper application of tissue testing it is possible to anticipate or forecast certain production problems which still in the field) The concentration of the nutrients in the cell sap is usually a good indication of howwell the plant is supplied with nutrients at the time of testing.

i. **Plant Part to be Selected**: In general the conductive tissue of the latest mature leaf is used for testing.

ii. **Time of Testing**: The most critical stage of growth for tissue testing is at the time of bloom or from bloom to early fruiting stage. Nitrates are usually higher in the morning than in the afternoon if the supply is short.

Test for nitrates by diphenylamine: dark blue colour-sufficient N; light blue coloure slightily dificianet N; No colour-highly deficiental.

Test for Phosphates by ammonium molybdate + Stannous oxalate: dark blue-sufficient P; bluish, green-slightly deficient P; No colour-highly deficient P; slightly deficient P; No colour-highly deficient P.

Test for potassium by Sodium cobaltinitrate: High turbidity-sufficient K; slight turbidity moderatate K deficiency; No turbidity-K deficient

## b) Total Plant Analysis

Total analysis is performed on the whole plant / plant parts. Precise analytical techniques are used for measurement of the various elements after the plant material is dried, ground and ashed and used for estimating total nutrient content.

**Table 16:** Average concentrations of Nutrients in plants

| Plant Nutrient | average concentration* |
|---|---|
| H | 6.0% |
| O | 45.0% |
| c | 45.0% |
| N | 1.5% |
| K | 1.0% |
| ca | 0.5% |
| Mg | 0.2% |
| P | 0.1% |
| S | 0.1% |
| cl | 100 ppm (0.01%) |
| Fe | 100 ppm |
| b | 20 ppm |
| Mn | 50 ppm |
| Zn | 20 ppm |
| cu | 6 ppm |
| Mo | 0.1 ppm |

*Concentration expressed by weight on a dry matter basis.

### 10.7.2.1. Critical nutrient concentration

Critical Nutrient concentration is the level of a nutrient below which crop yield, quality or performance is unsatisfactory. However it is difficult to choose a specific concentration. For crops such as sugarbeet excessive concentration of N seriously affects the quality. So, cNc is maximum rather than a minimum consequently it is more realistic to use the critical nutrient range (cNR) which is defined as the range of nutrient concentration at a specified growth stage above which the crop is amply supplied and below which the crop is deficient.

### 10.7.2.2. Diagnosis and Recommendation Integrated System (DRIS)

This concept was proposed by beaufils (1973) which considers nutrient concentration ratios rather than individual elemental concentration for interpreting plant tissue composition. The dRIS approach measures the relative balance between nutrients by means of index values with negative values indicating insufficiencies and vice versa) dRIS reveals not only the limiting nutrient but also the order in which the nutrients are likely to become limiting. It is a comprehensive system which identifies all the nutritional factors limiting crop production and in doing so increases the chances of obtaining high crop yields by improving the fertilizer recommendations. Index values which measure how far particular nutrients in the leaf or plant is deviating from the optimum are used in the calibration to classify yield factors in the order of limiting importance.

To develop a dRIS for a given crop the following requirements must be met. all factors suspected of having an effect on crop yield must be defined)

1. The relationship between these factors and yield must be defined)
2. Calibrated norms must be established)
3. Recommendations suited to particular set of conditions and based on correct and judicious use of these norms must be continuously refined)

## Establishment of DRIS norms

Large number of sites is selected at random in order to represent the whole production area) at each site plants and soil samples are taken for all essential element analyses. The entire population of observation is divided into two sub populations (high and low yielders) on the basis of vigour, quality and yield) Each element in the plant is expressed in as many ways as possible. For eg: Nutrient ratios N/P, N/K or products NxP, NxK etc) Each form of expression which significantly discriminates between high and low yielding sub populations is retained as a useful diagnostic parameter. The mean values of all the sites for each of these forms of expression then constitute the diagnostic norms.

NPK requirement of the crop is diagnosed using dRIS chart. The chart is constructed of three axes for N/P, N/K and K/P represented with mean values of the sub populations of the high yielder. The concentric circle can be considered asconfidence limits. Horizontal arrows (→) in the inner circle indicate the balance between nutrients. diagonal arrows indicate(↗↘) a tendency to imbalance. The inner being set at ± 15% and outer at the mean ± 30% for each expression. Vertical (↓↑) arrows representing nutrient imbalance. The arrow notation can be replaced by dRIS indices.

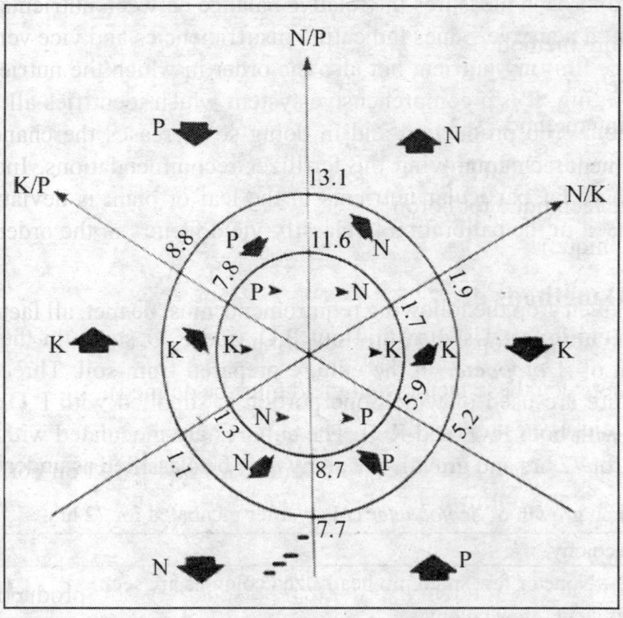


**Fig. 21:** DRIS chart for obtaining the qualitative order of requirements for N-P-K in corn. Means of significant expressions (values in origin of chart) are N/P 10.04, N/K 1.49 and K/P 6.74 (Sumner, 1978)

## Advantages of dRIS

1. The importance of nutritional balance is taken into account.
2. The norms for the elemental content can be universally applied)
3. diagnosis can be made over wide ranges of stages.
4. The nutrients limiting the yield either through excess or insufficiency can be readily identified)

## 10.7.3. Biological methods of soil fertility evaluation

For calibrating crop response, besides chemical soil test values other procedures are also available.

## Microbiological methods

1. Sackett and Stewart technique (*Azotobacter* test for $P_2O_5$ and $K_2O$)
2. Mehlich technique for available $K_2O$
3. Mehlich *Cunninghamella* plaque method for phosphorus
4. The Mulder's *Aspergillus niger* test for copper and magnesium

## Potculture methods

5. Mitscherlich pot culture method
6. The Jenny pot culture test
7. The Neubauer seedling method
8. The Stanford and dement technique
9. Sunflower pot culture technique for boron
10. A – value (tracer technique)

## 10.7.3.1. Microbiological methods are

**1) Sackett and Stewart technique** : Used to find out $P_2O_5$ and $K_2O$ status in the soil judged by colonization of *Azotobacter* in the culture prepared from soil. Three containers having soil culture are used of which one portion is supplied with $P_2O_5$ another with $K_2O$ and rest with both $P_2O_5$ and $K_2O$. The cultures are inoculated with *Azotobacter* and incubated for 72 hrs and growth of colony may be classified as under.

**Table 17:** Classification based on growth of *Azotobacter* colony after incubated for 72 hrs

| class | Growth of the colony |
|---|---|
| class I | Very deficient – None or few small pin head sized colonies are seen. |
| class II | Moderately deficient – few colonies |
| class III | Slightly deficient – The colonies on unfertilized cultures are equal in number and development. |
| class IV | Not deficient – colonies on both fertilized and unfertilized plaques are equal in number and development. |

**2) Mehlich technique for available $K_2O$** : A small amount of soil is taken in conical flasks in which appropriate nutrient solution is added and then it isinoculated with *Aspergillus niger* and incubated for four days. Weight of mycelial pad and its $K_2O$ content are taken into account.

**Table 18:** Critical limits for available K by using the *Aspergillus niger* method

| Weight of 4 pads (g) | $K_2O$ absorbed by *A niger* per 100 g soil (mg) | degree of potassium deficiency |
|---|---|---|
| <1.4 | <15 | Very deficient |
| 1.4 to 2.0 | 15 to 20 | Moderate to slight deficiency |
| 2.0 | >20 | Not deficient |

**3) Mehlich's *cunninghamella* plaque method for P** : *Cunninghamella* is sensitive for $P_2O_5$ status. The soil is mixed with nutrient solution and paste is prepared which is spread in clay dish. Then inoculated with cunninghamella and allowed to incubate for 4-5 days. The diameter of the mycelial growth is considered as an index for P status.

**Table 19:** P – Deficiency and mycelial growth

| diameter of colonies (mm) | degree of P deficiency |
|---|---|
| <10 | Very deficient |
| 11-15 | Moderately deficient |
| 16-21 | Slightly deficient |
| >22 | Not deficient |

**4) Mulder's *aspergillus niger*** test **for Cu and Mg** : color of the mycelia and spores give an indication of either deficiency or sufficiency of cu and Mg. For comparison, known standards are prepared as follows and their colors are compared with those on the unknown soil.

**Table 20:** Ranges for cu and Mg in Mulder's test

| cu in µg /g of air –dry soil | deficiency degree | mg in µg/3 g of air – dry soil |
|---|---|---|
| < 0.4 | Very deficient | < 50 |
| 1 – 1.5 | Slightly deficient | 50 - 100 |
| >2.0 | Not deficient | >100 |

### 10.7.3.2. Pot culture methods

Besides plant analysis there are some biological tests which may be used to evaluate soil fertility.

**1) The Mitscherlich pot culture method** : In this method pots containing 2.72 kg soil are taken for growing oats as test crop. N, P and K are added in different combinations in these pots [$N_o$ - one pot, $P_o$ - three pots (NK), $K_o$- three pots (NP) and NPK - three pots) ]. The crop is grown till maturity and percentage increase in yield is calculated by using Mitcherlich tables from rotation of given quantity of fertilizers over native fertility status (control).

**2) The Jenny's pot culture test** : Smaller pots consisting of 1.81 kg soil are used for growing lettuce (*Lactuca sativa* longifolia) as test crops for 6 weeks. Following treatments are used in four replications.

| | |
|---|---|
| Control | NoPoKo |
| Full fertilizer | N150 P150 K100 |
| No nitrogen | N0 P150 K 100 |
| No phosphorus | N150 P0 K100 |
| No potash | N150 P150 K0 |

**Table 21:** The percentage values are categorized as deficiency, probable deficiency and uncertain deficiency

| Jenny's values | % yield | | |
|---|---|---|---|
| | definite deficiency | Probable deficiency | Uncertain deficiency |
| N | 20 | 20-50 | 51-70 |
| P | 20 | 20-50 | 51-65 |
| K | 70 | 70-75 | 76-80 |
| S | 66 | 66-76 | 77-83 |

**3) The Neubauer's seedling method:** In this technique, 100 seedlings of rye or oats are made to feed exhaustively on 100 g of soil mixed with 50 g of sand for 17 days in dishes of 11 cm and 7 cm depth. a blank without any soil also is taken. The total $P_2O_5$ and $K_2O$ uptake is calculated and the blank value is deducted to obtain root soluble $P_2O_5$ and $K_2O$ in 100 g of air dry soil. These values are designated as Neaubauer's numbers and expressed as mg/100 g of dry soil. The following Neaubauer limit values are used to determine the deficiency.

**Table 22:** Neaubauer limit values mg/100 g soil

| Nutrient | barley | Oats | Rye | Wheat | Turnip | Potato | Sugarbeet |
|---|---|---|---|---|---|---|---|
| $P_2O_5$ | 6 | 6 | 5 | 5 | 7 | 6 | 6 |
| $K_2O$ | 24 | 21 | 27 | 20 | 39 | 37 | 25 |

**4) The Stanford and Dement technique** : Round waxed cardboard cartons of about 100 g capacity with bottom removed which are nested in similar containers having intact bottom filled with 680 g of sand) The seeds of the test crop are sown about 1.25 cm deep. after growing the seedlings for 2 to 3 weeks, a carton containing the plants are nested in second carton holding 200 g of soil or soil mixed with fertilizers. The plant roots enter the second carton where these plants are allowed to feed for 3 to 5 days. Four plants of maize and 30 plants of wheat are maintained for the study. after 5 days the plant samples are taken to determine the nutrient content.

**5) Sunflower pot culture technique for Boron** : In this method 500 g soil is taken in small pot and 5 sunflower seedlings are allowed to grow. The soil is fertilized with a solution containing all the nutrients except b and deficiency of b is noticed and ranked)

**Table 23:** Identifying boron deficiency

| class | days after which b deficiency is noticed |
|---|---|
| Marked deficiency | < 28 |
| Moderate deficiency | 28 – 36 |
| Little or no deficiency | > 36 |

**6) A value** : By using radioactive isotopes, it has now become possible to calculate the available nutrients in the soil. Fried and dean (1952) defined a-value as that amount of nutrients in soil which behave in a similar way as the applied fertilizer nutrient doses. This can be calculated by the formula)

$$A = B\frac{(1-y)}{y}$$

Where

a = Available soil nutrient

b = Amount of fertilizer nutrient applied)

Y = The fraction of the nutrient derived from fertilizer contained in the plant.

### 10.7.4. Visual symptoms of nutrient deficiency or toxicity

Chemical analysis of plants may indicate the presence of more than 90 elements but 16 of them have been established to be essential for their successful growth and development; as per the criteria of essentiality. a constant balanced supply of these nutrients is essential for normal plant growth. any imbalance among them leads to the emergence of nutritional disorders owing to their deficiencies or toxicities when an essential nutrient is in extremely short / excess supply, the plant suffers from its deficiency which is manifested in the form of specific sign termed as deficiency / toxicity symptom of the nutrient.

Visual deficiency symptoms are generally characteristic enough to permit easy identification of the deficiency of a nutrient as these appear on particular plant part at specific growth stage. The mobility of nutrients with in a plant differs markedly. Nutrients like N, P and K are readily translocated from old to young leaves under stress condition and are termed as mobile nutrients within the plant and they show up their deficiencies initially on the old leaves. The nutrients such as calcium, sulphur, boron and iron which are not retranslocated are called immobile nutrients and their deficiency symptoms first appear on young leaves. Mobility of other nutrients is intermediate. For example, shortening of internodes due to Zn deficiency results primarily from impaired auxin metabolism. The visual identification of nutrient deficiencies or toxicities is considered as a simple and inexpensive diagnostic tool in plant nutrition as it does not involve the use of any analytical equipment.

Certain plants are very sensitive to deficiency of a specific plant nutrient and they produce specific symptoms which are different from other deficiency symptoms. Thus the deficiency of that element can easily be detected)

**Table 24:** Indicator plants for plant nutrients

| Element | deficiency indicator plant |
|---|---|
| N | cauliflower, cabbage |
| P | Rape seed |
| K | Potato |
| ca | cauliflower, cabbage |
| Mg | Potato |
| Fe | cauliflower, cabbage, Potato |
| Na | Sugar beet |
| Mn | Sugarbeet, Oats, Potato |
| b | Sunflower |

### Limitations of visual deficiency symptoms

- Confident diagnosis by this approach requires much experience as the symptoms of some nutrient deficiencies are difficult to differentiate.
- By the time the deficiency / toxicity symptoms appear, the crop has undergone marked set back and the ameliorative measures taken at that time may not produce optimum yields. The appearance of deficiency symptoms is an extreme

limit of nutrient deficiency but even if the symptoms are not manifest, reduction in the yield of crop may occur. This condition has been termed as hidden hunger.

## 10.8. Nutrient recommendations

### 10.8.1. Soil test based fertilizer recommendations or fertility gradient approach

This soil test based fertilizer recommendations approach was put forth by Ramamoorthy (1971). The underlying principle of this approach is to get as a large variation as possible in the soil fertility levels in one and the same field in a given locality where an elaborate field experiment is conducted so that the real relationship between the yield and the level of soil fertility could be evaluated without interference from other factors affecting the yield)

A large field of certain amount of natural variation in fertility is chosen. It is divided into four strips which are treated with four different doses of N, P and K viz., control ($N_0P_0K_0$), half of the normal dose ($N_{1/2}P_{1/2}K_{1/2}$), normal dose ($N_1P_1K_1$) and double normal dose ($N_2P_2K_2$) of N, P and K fertilizers that are usually applied to the high yielding varieties of the crops in the country. Exhaustive crop like sorghum, maize, wheat or rice are grown on this field to deplete the soil fertility level in the strips treated with fertilizers. an experiment is then laid out on this field for the purpose of Soil Test crop Response (STcR) correlation with 21 treatments having different N, P and K randomly arranged in each strip in such a way that there is an untreated check plot on either side of the treated plot.

The soil test crop response correlation work is then carried out taking the yield of treatments and the average of the two controls (check) plots into consideration.

### 10.8.2. Critical soil test level approach

The critical soil test level, concept advanced by cate and Nelson (1965) is the level of the nutrient below which a reasonably satisfactory economic response should be expected from the application of that particular nutrient and above which the probability of such response is low. In order to apply this concept, the soil is collected from each field are analysed, field experiments are conducted with application of graded doses of nutrient and response curve is fitted) a scattered diagram of percentage yield on Y-axis and soil test value on X-axis is then plotted) It is divided into four quadrants positioning the lines in such a way that the number of points in the upper right and lower left quadrants is maximum. The point where the vertical line crosses the X-axis is defined as the critical soil test value.

**Critical levels of nutrients in soils** critical value may be defined as the nutrient concentration below which the plants are expected to respond to the application of that nutrient.

**Table 25:** Critical level of micro nutrients in soils

| Micronutrient | Indices | Critical level (ppm) | |
|---|---|---|---|
| | | Range | Mean |
| b | Hot water soluble | 0.1 – 2.0 | 0.7 |
| cu | Mehlich No.1 | 0.1 – 10.0 | 3.0 |
| | dTPa + ca cl$_2$ (pH 7.3) | 0.12 – 2.5 | 0.8 |
| | ab-dTPa (pH 7.6) | | 1.8 |
| Fe | dTPa + ca cl$_2$ (pH 7.3) | 2.4 – 5.0 | 4.0 |
| | ab-dTPa (pH 7.6) | | 4.0 |
| Mn | Mehlich No.1 | 4.0 – 8.0 | 7.0 |
| | dTPa + ca cl$_2$ (pH 7.3) | 1.0 – 2.0 | 1.4 |
| | 0.03 M $H_3PO_4$ | 0-20.0 | 10.0 |
| | ab-dTPa (pH 7.6) | | 1.8 |
| Mo | (NH4)$_2$c$_2$O$_4$ (pH 3.3) | 0.04 – 0.2 | --- |
| Zn | 0.1 N Hcl | 2.0 – 10.0 | 5.0 |
| | Mehlich No.1 | 0.5 – 3.0 | 1.1 |
| | dTPa + ca cl$_2$ (pH 7.3) | 0.25 – 2.0 | 0.8 |
| | ab-dTPa (pH 7.6) | | 1.5 |

## 10.8.3. Use of empirical equations

Krishnamoorthy et al., (1963) worked out empirical relationships between the nature of the soil and the extractable P. They found that the capacity to fix added phosphates is largely a function of the type and amount of clay present in the soil.

- In the case of typical black cotton soil consisting of montmorillonite and beidellite type of lay minerals, the extractability of added phosphates is given by the formula
  Percentage extractable P = 100 – clay percentage
- In the case of red and alluvial soils where illite dominates, the relationship appears to be

Percentage extractable P = 100 – 2 (clay percentage)

## 10.8.4. Soil test crop response (STCR) method

After introduction of high yielding varieties and hybrid crops, the need for systematic soil test crop response research in different soil agro-climatic regions become evident. IcaR established the aIcRP on STcR in 1967 and the STcR concept was developed by Ramamoorthy, in 1987. STcR provides the relationship between a soil test value and crop yield)

The soil test values are needed to be correlated with actual crop response obtained under field conditions. Separate calibration charts are needed for each crop and soil. Fertility gradient and regression approach and targeted yield concepts were evolved) This is also called as "rationalized fertilizer prescription approach" in which inherent soil fertility and yield level of the crop are taken in to account while recommending the fertilizer doses.

Objective of STcR include 1)to prescribe fertilizer doses for a given crop based on soil test values to achieve the "Targeted yields" in a specific soil agro-climatic region under irrigation or protective irrigation conditions by using mathematical equations for different crops and different soil agro-climatic zones separately. 2)This takes in to consideration-the efficiency of utilization of soil and added fertilizer nutrient by the crops and its nutrient requirements for a "desired yield level".

Concept of STcR: STcR approach is aiming at obtaining a basis for precise quantitative adjustment of fertilizer doses under varying soil test values and response conditions of the farmers and for targeted levels of crop production. These are tested in follow up verification by field trials to back up soil testing laboratories for their advisory purpose under specific soil, crop, and agro climatic conditions. The fertilizers are recommended based on the following criterias.

- Fertilizer recommendations based on regression analysis approach
- Recommendations for certain % of maximum yield

STcR methodology takes in to account the three factors;

1. Nutrient requirement (NR) in kg/ quintal of the produce
2. Percentage contribution from soil available nutrients (SE)
3. Percentage contribution from added fertilizers towards making effective fertilizer prescriptions for specific yields.

i) Nutrient Requirement (NR) = Total nutrient uptake (grain+straw) kg/ha (in kg for producing 1 Quintals of grain)/Yield of grain (q/ha)

ii) Fertilizer Efficiency (FE in %) =

$$\frac{\text{Total nutrient uptake (kg/ha)} - \text{soil test value in treated Plot (kg/ha) X soil nutrient efficiency (\%) X 100}}{\text{Fertilizer dose applied (kg/ha)}}$$

iii) Soil Efficiency (SE in %) = $\dfrac{\text{Total nutrient uptake in control kg/ha X 100}}{\text{Soil test value in control kg/ha}}$

With the help of above parameters, adjustment equations have been developed for a number of crops in various soils.

E.g.: For Rice crop.

a) Fertilizer N = 4.39 T -0.6723 Soil N

b) Fertilizer P2O5 = 2.83 T – 6.110 Soil P

c) Fertilizer K2O = 1.41 T – 0.329 Soil K

Where, T= Targeted yield of rice

**Advantages of STCR approach**

1) Efficient and profitable site specific fertilizer recommendation for increased crop production and for maintenance of soil fertility.
2) Aims to provide balanced, efficient and profitable nutrient application rates for pre- set yield targets giving due consideration to basic fertility status of soil

**Targeted yield concept**: These are soil test based recommendations but given for different yield goals and not for a single optimum yield level. a large variety of fertilizer prescription have been made available by putting soil test values in to certain mathematical equations and finding out the amounts of nutrients needed for a given yield target.

## 10.8.5. Soil Test Calibration and interpretation method

It is essential that the results of soil tests be calibrated against crop responses from applications of the plant nutrients in question. This information is obtained from field and greenhouse fertility experiments conducted over a wide range of soils. Yield responses from rates of applied nutrients can then be related to the quantity of available nutrients in the soil.

The results of long-term soil test calibration studies on different soil types are then utilized to establish recommended amounts of plant nutrients to apply to a particular crop at a given soil test level.

In this example to the right, more than 85 percent of the fields testing very low in a particular plant nutrient may give a profitable yield response to the added nutrient. at the very high level, there is only a 15 percent probability of a profitable yield increase to the added nutrient. These values are arbitrary, but they illustrate the idea of expectation of response.

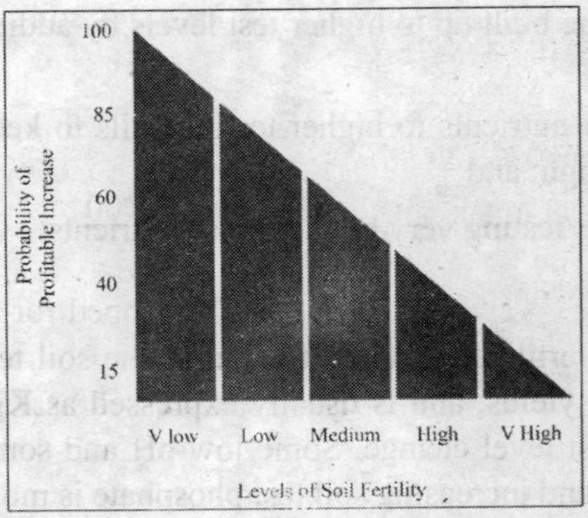


**Fig. 22:** Probanbility of profitable increase to soil fertility levels

The tools of site-specific precision management now allow farmers to manage more homogeneous areas within fields. Some of those areas have much higher yield potentials than the database with which most of today's soil tests were calibrated)

When interpreting soil test results, several issues should be kept in mind:

- The chances of getting a profitable response to fertilization are much greater on a soil that tests low in a given nutrient than on one that tests high.
- This does not rule out the possibility of a profitable response from nutrient application at a high level of fertility or lack of a profitable response on soils of low fertility.
- Soil tests are better at predicting the probability of a profitable response to nutrient application than predicting the actual quantity of nutrient that will be needed in any one year.
- Interpretation of soil test results and recommendations often becomes a matter of how to improve the fertility status of soils testing less than optimum. How much will be needed to change the soil from low to medium or high in that element? What will be the most economical level at which to maintain the nutrient status of the soil?
- With top-level management practices, yields increase and the probability of a response at any given soil test likewise increases.
- Wise use of soil testing incorporates a long-term approach to fertility management, in which site-specific soil test target levels are established for each field and nutrient management plans developed to reach and maintain the target levels.

## Philosophy of soil test fertilizer recommendations

The goal of soil testing is to help farmers achieve economical optimum yields while protecting the environment. The basic philosophy of soil test fertilizer recommendations is:

- Base them on soil test results
- Recommend that lower-testing soils be built up to higher test levels by adding extra fertilizer;
- Apply maintenance amounts of plant nutrients to higher-testing soils to keep them there and to keep productivity high; and
- Do not apply specific nutrients to soils testing very high in these nutrients.

## Soil test buffer potential

Soil test buffer potential is the quantity of fertilizer required to change the soil test level to the level supporting optimum crop yields, and is usually expressed as Kgs of $P_2O_5$ or $K_2O$ required per ppm of soil test level change. Some low-pH and some high-pH soils fix applied phosphate readily, and increasing soil test phosphate is more costly, decreasing the optimum soil test level. Soil test phosphate and potassium levels are usually easier to change in coarse soils than on medium or fine-textured soils, except with very sandy soils, where potassium leaching becomes significant.

Recommendations when soil test value is high:

- Sufficiency: add necessary rates of deficient nutrients so yields are not limited in present crop.
- Build-Maintenance: add enough of needed nutrient(s) to supply present crop need, and gradually increase soil supply to non-limiting level. Replace crop harvest–removed nutrients to keep plant nutrient levels at non-limiting levels.

Failure to apply any of these nutrients will result in soil test depletion. also, under some conditions, crops will respond profitably to a nutrient even with a high test. For example, on early-planted corn, the addition of N, P and K as a row application may produce response on soils testing high.

### 10.8.6. Integrated Nutrient Management (INM)

Integrated nutrient management(INM) is one of the agronomic practices aiming at the usage of the harmonious properties of both sources by making a combination that can be used for decreasing the enormous use of chemical fertilizers and accreting a balance between fertilizer inputs and crop nutrient requirement options, which can maintain the soil fertility, restore the soil health and continuous supply of plant with nutrient requirements to obtain an optimum level of yield production, maximize the profitability, and subsequently reduce the environmental pollution. In brief, Integrated Nutrient Management refers to the maintenance of soil fertility and of plant nutrient supply at an optimum level for sustaining the desired productivity through optimization of the benefits from all possible sources of organic, inorganic and biological components in an integrated manner.

The continuous recycling of nutrients into and out of the soil is known as the nutrient cycle. The cycle involves complex biological and chemical interactions, some of which are not yet fully understood) a simplified version of this cycle of plant growth is shown in Figure 1. The simplified cycle has two parts: "inputs" that add plant nutrients to the soil and "outputs" that export them from the soil largely in the form of agricultural products. Important input sources include inorganic fertilizers; organic fertilizers such as manure, plant residues, and cover crops; nitrogen generated by leguminous plan and atmospheric nitrogen deposition. Nutrients are exported from the field through harvested crops and crop residues, as well as through leaching, atmospheric volatilization, and erosion. The difference between the volume of inputs and outputs constitutes the nutrient balance. Positive nutrient balances in the soils (occurring when nutrient additions to the soil are greater than the nutrients removed from the soil) could indicate that farming systems are inefficient and, in the extreme, that they may be polluting the environment.

Negative balances could well indicate that soils are being mined and that farming systems are unsustainable over the long term. In the latter instance, nutrients have to be replenished to maintain agricultural output and soil fertility into the future. The supply of nutrients from various sources of organic, inorganic, biological constitute integrated nutrient management.

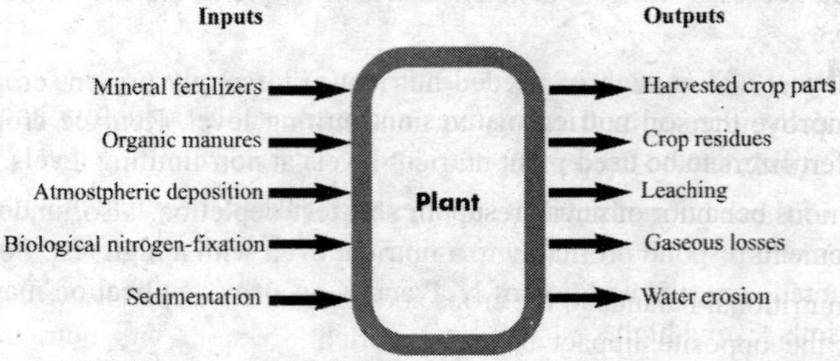


**Fig. 22:** The plant nutrient balance system

**Concepts of INM:** Integrated nutrient management is a practice that combines old and modern techniques of fertilizer use and nutrient management. The idea of INM depends on a number of factors, including harmony in nutrient properties, a balance between crop nutrient demands, what sort of nutrient, in general, is available in soil and in the farmer's hand, information and skills about the most suitable nutrient can be harmonized in combination, and which materials can be safely used that lead to increase nutrient-use effciency. additionally, it is a method and a way of disposing organic wastes safely and also an effective method of recycling wastes into goodquality compost.

The main objective of the integrated nutrient management is to maintain economic yield for a long period with little effect on native soil fertility and environmental pollution, making some changes in farmer's awareness toward the eco-friendly technique (organic farming system) for producing healthy food free from contaminants and insuring satisfactory economic returns.

## Components of INM

a) **Soil Source of nutrients:** Mobilizing unavailable nutrients and to use appropriate crop varieties, cultural practices and cropping system.

b) **Mineral Fertilizer sources :** Super granules, coated urea, direct use of locally available rock PO4 in acid soils, Single Super Phosphate (SSP), MOP and micronutrient fertilizers.

c) **Organic Sources :**by products of farming and allied industries. FYM, droppings, crop waste, residues, sewage, sludge, industrial waste.

d) **Biological Sources :** Microbial inoculants substitute like nitrogen fixing (Rhizobium, azatobacter, azospirellum, azolla- anabena complex (bGa)), phosphorous augmenting microbial inoculations (bacillus, Pseudomonas, Mocorrhizae).

e) **Crop residues**: Incorporation of crop residues in the soil will improve soil physical health besides plant nutrient supply.

f) **Residual soil fertility**: The unutilized fertilsiers applied for main crop will become available for second crop. It is evident in Rice-pulse and Rice- wheat rotations.

## Advantages of INM

- Systems can improve the soil nutrient natives and increase the solubility and availability of fertilizers to be used
- Use the harmonious behavior of nutrient supplies and making them match with the crop requirements
- INM offer the nutritional balance to the crops and lessen the aggressive effects resulting from the opposite impact between nutrient fractions and nutrient imbalance
- Advance and sustain the physiochemical and biological functions of soil properties
- Reduce the rate of soil degeneration, water, and ecosystem by enhancing carbon confiscation and decreasing nutrient losses to ground and surface water forms and/or to environment pollution
- Minimize higher total costs of production and raise the farmer's returns (increasing profitability)
- Improve the resistance to both biotic and abiotic stresses
- An effective method of agricultural practices to ensure healthy food, covering population food demands alongside with many soil and environmental impacts, especially in countries with rapid growth in population
- Additional benefits can also be gained; it does not only save the total costs at the satisfactory level with an increase in crop production but also can be easily practiced by farmers; therefore, it is considered one of the most promising techniques in line with the future needs
- INM can have positive effects on the susceptibility or plant resistance against many types of biotic and abiotic stresses.
- Following INM will enable to explore a larger volume of soil in order to access water and nutrients; additionally, improved root development enables the plant to absorb water from deeper soil layers and then reflect an increase in the ability of crops toward drought resistance
- Changes in awareness of farmers toward the climate changes from season to season, which have greater ecological impacts in order to produce safe food rather than achieving higher yield aiming at attaining higher profit.

## Constraints of INM

- due to the diffculties in obtaining FYM and biofertilizers, some poor farmers have some trouble in obtaining organic manure as well as lack of knowledge, poor advisory services, and skills of recycling organic wastes to produce high-quality compost.

- Biotic or abiotic stresses
- Poor and improper soil tillage practices
- Lack of needful equipment at farmer level fro timely operations.
- Absence of effective extension services and NGOs to support farmers and direct their attention to the importance of using integrated nutrient management and its benefits in conserving soil properties, soil nutrient balance, environmental impacts, and their role in increasing their profitability is a constraint.

### 10.8.7. Site specific nutrient management (SSNM)

The SSNM helps in improving NUE as it provides an approach for feeding crops like rice, maize, wheat, etc) with nutrients as and when needed) The major benefit for farmers from improved nutrient management strategy is an increase in the profitability. The SSNM eliminates the wastage of fertilizers by preventing excessive rates of fertilization and by avoiding fertilizer application when the crop does not require nutrient inputs. It also ensures that N, P, and K are applied in the ratio required by the intended crop.

**Principles of SSNM** : SSNM is an approach to feeding crops with nutrients as and when needed) The application and management of nutrients are dynamically adjusted to crop needs of the location and season.

The SSNM approach aims to increase farmers profit through

- Increased yield of crops per unit of applied fertilizer.
- Higher rice yields, and
- Reduced disease and insect damage.

Important Features of SSNM

1. Optimal use of existing indigenous nutrient sources such as crop residues and measures.
2. Application of N, P and K fertilizers is adjusted to the location and season specific needs of the crop.
   - Use of the leaf color chart ensures that nitrogen is applied at the right time and in the amount needed by the crop which prevents wastage of fertilizer.
   - Use of nitrogen omission plots to determine the P & K fertilizers required to meet the crop needs. This ensures that P and K are applied in the ratio required by the rice crop.
3. Local randomization for application of Zn, S and micronutrients are followed)
4. Selection of most economic combination of available fertilizer sources.
5. Integration with other integrated crop management (IcM) practices such as the use of quality seeds, optimum plant density, integrated pest management and good water management.

**Approaches of SSNM:** The relatively new approach of nutrient recommendations is mainly based on the indigenous nutrient supply from the soil and nutrient demand of the crop for achieving targeted yield) The SSNM recommendations could be evolved on the basis of (a)solely plant analysis or (b)soil cum plant analysis.

**Plant analysis based SSNM:** It is considered that the nutrient status of the crop is the best indicator of soil nutrient supplies as well as nutrient demand of the crops. Thus, the approach is built around plant analysis. There are five key steps (given below) for developing field specific fertilizer NPK recommendations for rice, through the basic principles remain the same for other crops as well.

**Selection of the Yield goal**: a yield goal exceeding 70–80% of the variety-specific potential yield (Ymax) has to be chosen. Ymax is defined as the maximum possible grain yield limited only by climatic conditions of the site, where there are no other factors limiting crop growth. The logic behind selection of the yield goal to the extent of 70–80% of the Ymax is that the internal NUEs decrease at very high yield levels near Ymax. crop growth models (eg. dSSaT) can be used to work out Ymax of crop variety under a particular climatic conditions.

**Assessment of Crop nutrient requirement**: The nutrient uptake requirements of a crop depend both on yield goal and Ymax. In SSNM, nutrient requirements are estimated with the help of quantitative evaluation of fertility of tropical soils (QUEFTS) models. The nutrient requirements for a particular yield goal of a crop variety may be smaller in a high yielding season than in a low yielding one.

**Estimation of indigenous nutrient supplies**: Indigenous nutrient supply (INS) is defined as the total amount of a particular nutrient that is available to the crop from the soil during the cropping cycle, when other nutrients are non-limiting. The INS is derived from soil incorporated crop residues, water and atmospheric deposition. It is estimated by measuring plant nutrient uptake in an omission plot embedded in the farmers' field, wherein all other nutrients except the one (N, P or K) in question, are applied in sufficient amounts.

**Computation of fertilizer nutrient rates**: Field specific fertilizer N, P or K recommendations are calculated on the basis of above steps (1–3) and the expected fertilizer recovery efficiency (RE, Kg of fertilizer nutrient taken up by the crop per Kg of the applied nutrient). Studies indicated fertilizer recovery efficiency values of 40–60% for Nitrogen, 15–20% for Phosphorous and 40–50% for Potassium in rice under normal growing conditions, when the nutrients are applied as water soluble fertilizer sources.

**Dynamic adjustment of Nitrogen rates**: Whereas fertilizer P and K, as compute above, are applied basally i.e., at the time of sowing/planting, the N rates and application schedules can be further adjusted as per the crop demand using chlorophyll meter (popularly known as SPad) or leaf color chart (Lcc).

## Soil and Plant analysis based SSNM

In this case, nutrient availability in the soil, plant nutrient demands for a higher target yield (not less than 80% of Ymax), and RE of applied nutrients are considered for developing fertilizer use schedule to achieve maximum economic yield of a crop variety. In order to ascertain desired crop growth, not limited by apparent or hidden huger of nutrients, soil is analyzed for all macro and micronutrients well before sowing/ planting. Total nutrient requirement for the targeted yield and RE are estimated with the help of documented information available for similar crop growing environments. Field-specific fertilizer rates are then suggested to meet the nutrient demand of the crop (variety) without depleting soil reserves. These soil–test crop response-based recommendations are now in practice to achieve desired yield targets in many field crops. Thus, recent studies with intensive cropping systems have shown that fertilizer recommendations with above approach offer greater economic gains as compared with NPK fertilizer schedule conventionally prescribed by soil testing laboratories.

## 10.9. Nutrient use efficiency

Nutrient use efficiency (NUE) may be defined as yield per unit input. In agriculture, this is usually related to the input of fertiliser, whereas in scientific literature, the NUE is often expressed as fresh weight or yield per content of nutrient. Improvement of NUE is an essential pre-requisite for expansion of crop production in marginal lands with low nutrient availability. NUE not only depends on the ability to efficiently take up the nutrient from the soil, but also on the transport, storage, mobilization, usage within the plant, and even on the environment.

Nutrient use efficiency is defined as the extent to which the nutrients and management practices interact to give a specified yield level.

$$\text{NUE (\%)} = \frac{\text{Yield with applied nutrient} - \text{Yield without applied nutrient}}{\text{amount of nutrient applied}} \times 100$$

### 10.9.1. Factors affecting Nutrient use efficiency (NUF)

a) **Soil factors** : The most important factors are soil physical conditions, soil fertility and soil reaction. On soils with poor physical condition, plant growth will be poor due to impeded drainage, restricted aeration and unfavourable soil temperature due to which the nutrients will not be used efficiently. coarse textured soils are usually poorer in available nutrients than fine textured soils. On such soils nitrogen and potassic fertilizers should be more frequently applied than in fine textured soils. The higher the fertility status of the soil, the lower is the response. Soil reaction is an important consideration in the selection of right type of P fertilizers. The higher the organic matter status, the more is the nutrient use efficiency.

b) **Climatic factors**: include temperature, rainfall and its distribution, evaporation, length of day and growing season. Rate of nitrification is slower in cooler climate than in warmer climate, hence more amounts of fertilizers should be added in cool climate. Higher amount of fertilizers are required in high rainfall region due to leaching to obtain an expected yield potential. In arid regions, soil moisture is a limiting factor to get higher nutrient use efficiency. The higher the light intensity, the better is the nutrient use efficiency.

c) **Crop factors** : cEc of plant roots influences the fertilizer responsiveness of the crop. a large ramifying root system of the plant absorbs nutrients more efficiently. The time of application of fertilizer should match the pattern of nutrient uptake to increase nutrient use efficiency. For legumes N fertilizer may be reduced as they can fix atmospheric N to increase N use efficiency.

d) **Agronomic factors** : Include selection of fertilizer responsive crops and varieties, timely sowing, proper spacing, proper dose, time and method of fertilizer application to increase the yield and thereby increasing NUE.

## 10.9.2. Enhancement of nutrient use efficiency

Use efficiency of any nutrient can be increased by achieving potential yields of crops by optimizing the factors of crop production.

1. Selection of suitable crops and varieties, which are input responsive recommended for the region.
2. Sowing or planting the crops at optimum time.
3. Maintaining optimum plant population.
4. Use of organic manures and biofertilizers to supplement nutrients and also to bring ideal conditions for crop growth.
5. Inclusion of legumes in the cropping system as intercrop.
6. The crops should be irrigated at least to save life at critical growth stages.
7. Fertilizer scheduling must be based on soil test values to prevent nutrient deficiencies or luxury consumption.
8. P and K fertilizer and part of N fertilizer should be applied as basal and N in splits doses; for light textured soils K also should be applied in splits.
9. band placement of fertilizers preferable to prevent losses. (Especially P to reduce fixation).
10. Under moisture stress condition, foliar application of urea at 2% concentration is effective.
11. Micronutrient deficiencies should be corrected instantly.
12. $ZnSO_4$ should be applied as package once in two seasons @ 25-50 kg/ha)
13. Problem soils must be ameliorated by taking reclamation measures.

NUE of individual nutrients also can be increased by following the above management practices along with some specific measures as follows

## Nitrogen use efficiency

- Split application of nitrogenous fertilizers to prevent losses due to leaching.
- The use of slow release nitrogenous fertilizer like urea formaldehyde, sulphur coated urea, neem coated urea etc)
- Use of nitrification inhibitor (Eg: N-serve) to retard the conversion of $NH_4^+$-N to $NO_3^-$-N to prevent leaching and make it available to crops for quite longer period)
- by the integration of inorganic N with organic sources the soil physical condition can be optimized besides adding nutrients to the soil.

## Phosphorous use efficiency

Can be increase by decreasing P fixation and balanced application of the nutrients. P fixation can be reduced by judicious application oforganic manures, application of P fertilizer by plâcement, inoculation (either seed or soil) with phosphorus solubilising bacteria like *Pseudomonas*, *Bacillus megathrium* var. *phosphaticum*.

## Potassium use efficiency

Can be enhanced by preventing leaching loss either by split application on light soils, applying organic manure and balanced application of nutrients.

## Sulphur use efficiency

Sulphur in soil solution is present as $SO_4^{=}$ and more prone for leaching losses. The losses can be prevented by applying organic manures to improve water holding capacity of the soil and it also acts as a source of S. S oxidation can be facilitated by providing oxidized conditions in the soil.

## Iron use efficiency

Most available form of iron is $Fe^{2+}$. all the measures which govern the soil reaction will influence Fe availability. Fe availability is more in acidic soil pH.

- application of organic manures including greenmanuring improve the use efficiency of iron by
  i. Acidifying the rhizosphere due to the release of organic acids
  ii. Supplementing with iron after decomposition
  iii. Act as substrate for heterotrophic bacteria that can reduce ferric to ferrous form (eg., *Bacillus, Clostridium and Klebsiella* etc)).
  iv. The microbes also produce chelating ligands called as 'siderophores' that can form complex with $Fe^{3+}$, which can be absorbed into the plant.
- Reclamation of alkali soils
- If deficiency appears on standing crop foliar application of Fe.

## Zinc use efficiency

- Zn fertilizer should not be applied with phosphatic fertilizers.
- Maintaining the soil pH between 5.5 – 6.5
- Applying organic manures.

## Review questions

define soil fertility and how it differs from soil productivity.

Write the criteria of essentiality of plant nutrients and classify plant nutrients based on quantities required by plants and biochemical factions in plants.

List out various mechanisms of movement of plant nutrients from soils to plant roots and explain diffusion and lundigardhs hypothesis.

Sraw a neat sketch of nitrogen cycle. Explain the processes of nitrification and denitrification.

What is ammonium fixation in soil ? write factors affecting ammonium fixation in soil

What is bNF? discuss different groups of nitrogen fixing microbes in soil.

Mention three each of important Nitrogen function and deficiency symptoms in plants

Describe different forms of phosphorous in soil.

What is phosphorous fixation? Explain factors affecting P fixation in soil.

Write functions and deficiency symptoms of potassium in plants.

Explain sulphur transformation in soil with suitable chemical equations.

What is akiochi disease? Write its symptoms and management.

Write functions and deficiency symptoms of Mg in platns.

Describe Viet's pools of micronutrients in soil

Wrote the causative factors for micronutrient disorders in soils

Warite the symptoms of boron in plants.

How you correct deficiency symptoms of Zn and Fe in soils.

## Objective questions

1. Zn is
   a) Primarynutrient
   b) Secondary nutrient
   c) Macronutrient
   d) Micronutrient
2. The following plant nutrient is important for root development of plant
   a) Nitrogen
   b) Phosphorous
   c) Potassium
   d) Sulphur

3. balanced Nutrient supply capacity of soil is
   a) Soil sustainability
   b) Soil Productivity
   c) Soil fertility
   d) Soil profitability
4. Essentiality criteria for plant nutrients was given by
   a) Beckette
   b) Schofied
   c) Arnon
   d) Springle
5. Mineralization take place when c:N ratio is
   a) < 20:1
   b) > 30:1
   c) > 60:1
   d) > 90:1
6. Rhizobium is
   a) Protozoa
   b) bacteria
   c) Fungi
   d) Virus
7. Urea is
   a) K-Fertiliser
   b) P-Fertiliser
   c) N-Fertiliser
   d) complex fertilsier
8. The Following element is constituent of chromosomes
   a) Mn
   b) Na
   c) K
   d) Mg
9. Total Number of essential elements for plants
   a) 27
   b) 10
   c) 17
   d) 35
10. Gypsum is
    a) Pesticide
    b) Manure
    c) Soil amendment
    d) Fungicide
11. Kaolinite is the following type of clay minerals
    a) 2:1 type
    b) 1:1 type
    c) 1:2:1 type
    d) 1:3 type
12. Phosphorous containing mineral is
    a) Sphelerite
    b) Calcite
    c) Epsomite
    d) Apatite
13. pH of acid soil is
    a) > 8.5
    b) 6.5 – 7.5
    c) 7.5 – 8.5
    d) < 6.5
14. Mycorrhizae is an association between
    a) Fungi and higher plant roots
    b) Fungi and bacteria
    c) Fungi and actinomyceties
    d) Fungi and virus

15. The Enzyme responsible for bNF is
   a) Kinase  b) Nitrogenase
   c) dehydrogenase  d) Urease
16. The plant nutrient Phosphorus absorbed by plants through
   a) Mass flow  b) diffusion
   c) Ion exchange  d) $cO_2$ theory
17. Labile phosphate is
   a) Fixed P  b) Exchangeable P
   c) Soil solution P  d) Apatite P
18. boron containing primary mineral is
   a) Borax  b) Feldspar
   c Boric acid  d) Tourmaline
19. The following form of Zn fertilizer is recommended for alkaline soils
   a) Chelated form  b) Oxide form
   c) Carbonate form  d) Hydroxide form
20. The % of sulphur in Single Super Phosphote (SSP) fertilizer is
   a) 6  b) 11
   c) 46  d) 61
21. Sulphohedral groups is related to increased
   a) Disease resistance  b) Drought resistance
   c) Cold resistance  d) All three
22. The amendment suitable for reclamation of calcareous- alkali soil is
   a) Gypsum  b) Pyrites
   c) Lime  d) Feldspar
23. Rhizobium strain suitable for soybean crop.
   a) Rhizobiummeliloti  b) Rhizobium japonicum
   c) Rhizobiumsolaneserum  d) Rhizobium phaseoli
24. akhiochi disease is seen in the following crop
   a) Rice  b) Wheat
   c) Maize  d) cotton
25. The author of reference book " The nature and properties of soil" is
   a) Danial Hillel  b) Schofield
   c) Liebig  d) Nyle.c. brady
26. "Law of minimum" was proposed by
   a) Lindsay and Norvel  b) Rutherford
   c) Stefen Hakings  d) Justus Von Liebig

27. Muriate of Postash contains

a) 60% $K_2O$ b) 50% $K_2O$

c) 46% $K_2O$ d) 18% $K_2O$

28. World Soils day is celebrated on

a) 5th Janunary b) 5th may

c) 5th September d) 5th december

29 Rapid tissue test for determining nitrates is

a) Olsens reagent b) Sodium cobalt nitrate

c) Brays reagent d) Diphenylamine

30. Mulders aspergillus niger test is for evaluating following nutrient(s)

a) N b) $K_2O$

c) cu &Mg d) $P_2O_5$

31. If the concentration of hydrogen ion solution is $5x10^{-4}$ Mol, then the pH of the solution is

a) 4.301 b) 3.401

c) 1.340 d) 0.431

32. The concentration of $cO_2$ in soil air is

a) 0.33% b) 0.03%

c) 3.33% d) 0.003%

33. The range of SaR for $S_3$ water is

a) 0-10 b) 10-18

c) 18-26 d) More than 26

34. Ninety percent of gaseous exchange between soil and atmosphere occur by

a) diffusion process b) Mass flow process

c) conduction process d) Lundigardh process

35. Working principle of Spectrophotometer is based on

a) Fouriers law b) Ficks law

c) beer and Lambert law d) Stokes law

36. The tendency of a system to reduce or oxidize, expressed as negative and positive potentials respectively is measured as

a) Pressure potential b) gravitational potential

c) Redox potential d) b & c

37. If the available soil potassium is 200 kg $ha^{-1}$, it is rated as

a) Low b) Medium

c) High d) Very high

38. The average concentration of $N_2$ in air is

a) 78%
b) 21%
c) 0.03%
d) 7.8%

39. The extractant used for the estimation of available phosphorous in acid soils is

a) Olsen reagent
b) Brays & Kurtz reagent
c) dTPa
d) EdTa

40. On an average the % of solar radiation reaches the earth

a) 50
b) 60
c) 75
d) 85

41. The fraction of incident radiation that is reflected by the land surface is termed as

a) Diffraction
b) Swath
c) Albedo
d) Refraction

42. Soil microbial activity and decomposition of organic matter is restricted at the temperature of

a) below 10 °C
b) 20 °C
c) 30 °C
d) 40 °C

43. Soil microbial activity and decomposition of organic matter is ceases at the temperature of

a) 20 °C
b) 10 °C
c) below 5 °C
d) 30 °C

44. The % of solar radiation reaches the earth is only 50 because

a) interaction with clouds
b) interaction with dust particles
c) interaction with ozone
d) a&b

45. The Ec of the saline soil is 400 μmhos/cm, it is equivalent to

a) 4000 $dSm^{-1}$
b) 400 $dSm^{-1}$
c) 0.4 $dSm^{-1}$
d) 4.0 $dSm^{-1}$

46. Irrigation water with Ec of 2.1 $dSm^{-1}$ and SaR of 12 is assignated to the following class

a) c2S2
b) c3S2
c) c4S1
d) c2S3

47. The index leaf to be sampled in case of rice plant is

a) 1st leaf from the bottom
b) 2nd leaf from the top
c) 3rd leaf from the top
d) 4th leaf from bottom

48. If the soil organic carbon value is 0.55%, its rating is

a) Low
b) Medium
c) High
d) Very high

49. The average concentration of Nitrogen in plant is
 a) 5.0% b) 0.5%
 c) 0.1% d) 1.5%
50. The cation that brings about deflocculation of soil colloids is
 a) calcium b) Sodium
 c) Magnesium d) aluminium
51. The most favourable structural type for plant growth is
 a) Columnar b) Platy
 c) Crumby d) Prismatic
52. The mineral that is most resistant to weathering is
 a) Micas b) Feldspars
 c) Olivines d) Quartz
53. The constraint associated with black clay soil is
 a) High permeability b) Good drainage
 c) Narrow workable range moisture d) Low draft power
54. black plastic mulch
 a) reduces out going temperature b) increases outing temperature
 c) green house effect d) thermal insulators
55. The extractant used for soil available sulphur is
 a) 0.15 % $cacl_2$ b) 1.5 % $cacl_{2\backslash}$
 c) 0.15 % $Mgcl_2$ d) 1.5 % $Mgcl_2$
56. The method employed for estimating Lime requirement of acid soil is method
 a) Woodruff method b) Shoemaker method
 c) Shoonover method d) a or b
57. Soil available Zn is low when its concentration in soil is
 a) < 1.6 ppm b) >0.6 ppm
 c) < 0.6 ppm d) > 1.6 ppm
58. boron toxicity in soil when b concentration is
 a) >1.0ppm b) >4.0ppm
 c) >2ppm d) >0.4ppm
59. The range of available soil sulphur status said to be medium is
 a) 5- 10pm b) 15- 30ppm
 c) 30- 45ppm d) 10- 15ppm
60. When soil test value of phosphorous is low, the probable crop response for added P fertilizers is
 a) Low b) Medium
 c) High d) a and c

61. The scientist suggested the term pH is

a) Ramamoorthy
b) Scofield
c) Sorenson
d) Liebig

62. The value of ionic product of water is

a) $10^{7}$
b) $10^{-14}$
c) $10^{14}$
d) $10^{-7}$

63. The ultimate source of nitrogen used by the plants is

a) Urea
b) Atmospheric inert gaseous nitrogen
c) Anhydrous ammonia
d) Ammonium sulphate

64. The nitrogen content of ploughed layer of majority of cultivated soils is

a) 0.02 – 0.04 %
b) 0.10 – 0.40 %
c) 0. 2 – 0. 4 %
d) 2.0 – 4.0 %

65. In general Indian soils have very low content because

a) Low N reserves
b) Low organic matter accumulation due to tropical climate
c) Low rainfall
d) Low nitrogen fixation ins soil

66. Relatively more nitrogen content found in

a) Sandy soils
b) Red soils
c) Black clay soils
d) Arid soils

67. The N present in soil is

a) 88 %inorganic and 12
b) 98 %inorganic and 2 %organic
c) 12 %inorganic and 88 %organic
d) 2 %inorganic and 98 %organic %organic

68. Plants absorb the N in

a) $NO_3^-$ form
b) Both $NH_4^+$ and $NO_3^-$ forms
c) $NH_4^+$ form
d) $N_2$ form

69. The nitrogen form less subjected to losses by leaching and denitrification is

a) $NH_4^+$ -N
b) $NO_3$-N
c) $N_2$ form
d) amide N

70. When organic residues having a c: N ratio wider than 30 are added to the soil

a) nitrogen volatalisation takes place
b) mineralisation of nitrogen takes place
c) nitrogen fixation takes place
d) Immobilisation of nitrogen takes place

71. Nitrogen mineralization occur in soil by
    a) Nitrification, aminisation and ammonifiation in sequence
    b) Ammonification, aminisation and Nitrification in sequence
    c) Aminisation, ammonification and Nitrification in sequence
    d) Ammonification, nitrification and aminisation in sequence
72. In the process of aminisation the microbes involved are
    a) Saprophytes b) Autotrophs
    c) A&B d) Heterotrophs
73. The fate of ammonium in soils is
    a) It may be converted to nitrites and nitrates by the process of nitrification.
    b) It may be absorbed directly by higher plants.
    c) It may be fixed in biologically unavailable form in the lattice of certain expanding type clay minerals.
    d) All the three
74. In agricultural soil, of the total % N found as fixed ammonium ion is
    a) 40% b) 20%
    c) 30% d) 10%
75. The order of ammonium fixation by minerals is as follows
    a) Vermiculite > smectites > illite b) Vermiculite > illite > smectites
    c) Illite > smectites >vermiculite d) Illite >vermiculite > smectites
76. The presence of $K^+$ (ionic diameter of 2.68 ⁰a) will often restrict $NH_4^+$ fixation in soils due to
    a) the $K^+$ ion can also fill fixation sites
    b) the $K^+$ ion repels the $NH_4^+$
    c) the $K^+$ ion increases base saturation
    d) the $K^+$ ion decreases solubility of $NH_4^+$ ion
77. Practical and cost effective way of reducing $NH_4^+$ fixation is
    a) Zn fertilization prior to $NH_4^+$ application
    b) P fertilization prior to $NH_4^+$ application
    c) Mo fertilization prior to $NH_4^+$ application
    d) K fertilization prior to $NH_4^+$ application
78. $NH_4^+$ fixation in soil increases with increase in
    a) Allophate clay minerals b) 1:1 type of clay minerals
    c) 2:1:1 type of clay minerals d) 2:1 type of clay minerals

79. Fixation of $NH_4^+$ is generally higher

a) Surface soil
b) Subsurface soil
c) Organic soil
d) Sandy soil

80. Nitrification in soil is

a) Chemical reduction
b) Chemical oxidation
c) Biological reduction
d) Biological oxidagtion

81. In soil nitrification process is

a) Conversion of ammonia to nitrate and nitrate to nitrite in sequence
b) Conversion of nitrite to nitrate and ammonia to nitrite in sequence
c) Conversion of ammonia to nitrite and nitrite to nitrate in sequence
d) Conversion of nitrite to ammonia and ammonia to nitrate in sequence.

82. Nitrification process in is brought out by

a) Obligate heterotrophic bacteria
b) Facultative autotrophic bacteria
c) Obligate autotrophic bacteria
d) Facultative heterotrophic bacteria

83. Optimum soil pH for the nitrification process in soils is

a) 7.0
b) 8.5
c) 5.0
d) 9.5

84. Nitrification process in soil increases up a soil temperature of

a) up to 45 $^0$c
b) up to 25 $^0$c
c) up to 35 $^0$c
d) up to 15 $^0$c

85. In soil nitrification process influenced by

a) Soil aeration
b) Soil temperature
c) Soil moisture
d) All the three

86. Leaching losses of N from soil occur mostly in the form of

a) Ammoniacal
b) Nitrite
c) Nitrate
c) Nitric oxide

87. Leaching loss of nitrogen from soil can be reduced by

a) Organic matter addition
b) Split application of fertilizer N
c) Using nitrification inhibitors
d) All the three

88. Slow release N fertilizers like neem coated urea, sulphur coated urea are useful in

a) Reducing the ammonium fixation in soil
b) Increasing the nitrification process in soil

c) Reducing leaching losses of N from soil

d) Increasing the nitrogen – plant root contact

89. denitrification occur in

a) Oxic soils b) Anoxic soils

d) Aerobic rice soils d) Organic soils

90. The process of bacterial reduction of $NO_3^-$ and $NO_2^-$ to NO, N2O and N2 gases is

a) Denitrification b) Nitrification

c) Volatalisation d) Gleyzation

91. denitrification process in soil brought about by

a) Pseudomonas b) Bacillus denitrificans

c) Paracoccus d) All the three

92. These chemicals restrict the growth of Nitrosomonas and keep nitrogen in $NH_4^+$ form are referred as

a) Nitrifiers b) Nitrification inhibitors

c) Ammonia inhibitors d) Denitrifiers

93. Natural nitrification inhibitor is

a) Nitrapyrin b) N – serve

c) Neem cake d) Sulphathiozole

94. In India the use of Nitrapyrin to retard nitrification is limited due

a) Unawareness of farmers b) Cause soil and water pollution

c) Difficult to application d) High cost

95. biological nitrogen fixation (bNF) is

a) Hydrolysis b) Hydration

c) Reduction d) Oxidation

96. The enzyme required for bNF is

a) Amylase b) Nitroginase

c) Kianase d) Phosphatase

97. Sesbania rostrata forms nodules on

a) Roots b) Leaves

c) Stem d) Fruits

98. In the process of biological nitrogen fixation the Rhizobia turn to bacteroids which have nitrogenase enzyme embedded into

a) Chlorophyll b) Protein

c) Leghaemoglobin d) Carbohydrate

99. This host specificity of different Rhizobium spices or the group of legume crops nodulated by single Rhizobium sp. is referred to as

a) Square inoculation groups b) Line inoculation group
c) Cross inoculation group d) Radial 38.inoculation groups

100. Nodulation and nitrogen fixation in clover plant occur by

a) Rhizobium phaseoli b) Rhizobium trifolii
c) Rhizobium japonicum d) Rhizobium leguminosarum

101. The member actinomycetes involve in the symbiotic N fixation in non-legume plants is

a) Azatobacter c) Nostoc
c) Frankia d) Clostridium

102. On an average the rain add nitrogen to soil every year about

a) 20 kg N/ha b) 10 kg N/ha
c) 15 kg N/ha d) 5 kg N/ha

103. Heterotrophic free living bacteria azotobacter and clostridium fix N in soil referred to as

a) Azofication b) Clostification
c) Nitrification d) Ammonification

104. Nitrogen can be lost due to leaching as

a) $NH_3$ b) $NH_4^+$
c) $NO_3^-$ d) a&b

105. The magnitude of leaching loss of N from soil depends on

a) Soil colour b) Soil texture
c) Soil consistence d) Soil odour

106. the ranges nitrogen losses through $NH_3$ volatilization

a) 2 to 10 % b) 30 to 65 %
c) 5 to 35 % d) 50 to 55 %

107. Losses of nitrogen through denitrification ranges from

a) 2 to 10 % b) 5 to 35 %
c) 10 to 40 % d) 50 to 55 %

108. $NH_4^+$ is adsorbed strongly in the inter layer spaces of

a) 2 : 1 clay minerals b) 1 : 1 clay minerals
c) 2 : 1 : 1 clay minerals d) Sesquioxide clay minerals

109. The element, essential component of amino acids, proteins, nucleic acids, porphyrins, flavins, purines and pyrimidine nucleotides, flavin nucleotides, enzymes, coenzymes and alkaloids is

a) Zinc
b) Nitrogen
c) Molybdenum
d) Potassium

110. The element responsible for transfer of genetic code

a) Molybdenum
b) Zinc
c) Nitrogen
d) Potassium

111. Nitrogen deficiency occur in plants when they contain

a) Less than 4.0 % nitrogen
b) Less than 2.0 % nitrogen
c) Less than 3.0 % nitrogen
d) Less than 1.0 % nitrogen

112. deficiency symptoms of nitrogen first appear on the older leaves in the form of light green to pale yellow coloration due to

a) Etiolation
b) Albinism
c) Bleaching
d) Proteolysis

113. Reduction in flowering and crop yields and lower protein content are usually associated with

a) B deficiency
b) Zn deficiency
c) N deficiency
d) K deficiency

114. Nitrogen toxicity in plants causes

a) Succulence
b) Lodging
c) Susceptible to insects pests and diseases
d) All the three

115. The plant nutrient is a vital component of the substances that are building blocks of genes and chromosomes

a) Phosphorous
b) Nitrogen
c) Potassium
d) Calcium

116. balanced Nutrient supply capacity of soil is

a) Soil elasticity
b) Soil Productivity
c) Soil sustainability
d) Soil fertility

117. Soil Samples is collected from field in

a) Near bunds
b) In random location
c) Under tree shades
d) Circles

118. Gypsum is

a) Manure
b) Amendment
c) Conditioner
d) Fungicide

119. Chlorite minerals is

a) :1 type b) 1:2 type

c) 2: 1:1 type d) 1:3 type

120. According to Sorensen, pH is negative logarithm of

a) $OH^{+}$ ion concentration b) $ca^{++}$ion concentration

c) $Mg^{++}$ ion concentration d) $H^{+}$ ion concentration

121. Buffering capacity is high for following soil

a) Sandy soil b) Loamy soils

c) Clay soils d) All the three

122. Liming practiced for

a) Acid soil b) Saline soil

c) Alkali soil d) Saline alkali soil

123. The concept of functional nutrient was given by

a) Arnon b) Liebig

c) Helmond d) Nicholas

124. Silicon is essential to

a) Ferns b) Polysiphonia

c) Sunflower d) Diatoms

125. Plants absorbs 'Zn' in the form of

a) Cation b) Anion

c) Both d) None

126. 'Calcium' in plants is

a) Mobile c) Partly mobile

d) Immobile d) All these three

127. The chemical formula for calcite or lime is

a) $caSO_4$ b) $cacO_3$

c) $ca(NO_3)_2$ d) $cacl_2$

128. The base saturation require for release sufficient ca in montmorillonite dominant soil is

a) 40% b) 50%

c) 60% d) 70%

129. Ca present in middle lamella of cell wall as

a) Calcium phytate b) Calcium pectate

c) Calcium sulphate d) All three

130. Sulphohedralgroups is related to increased ion

a) Disease resistance
b) Drought resistance
c) Cold resistance
d) All three

131. When organic manure with very wide c:N ration is added to soil, it result in

a) Mineralization
b) Immobilization
c) Both
d) Humification

132. The amendment suitable for reclamation of calcareous alkali soil is

a) Gypsum
b) Pyrites
c) Lime
d) Feldspar

133. The microorganism reported to solubilise phosphorus and reduce P- fixation is

a) Rhizobium
b) Bacillus
c) Thiobacillus
d) Frankia

134. ppm transfer

a) Parts per mole
b) Patrs per million
c) Parts per milli equivalent
d) Parts per meter

135. An example for autotropic denitrifying bacteria is

a) Pseudomonas florescence
b) Thiobacillusdenitrificaus
c) Pseudomonas
d) Paracocusdentrificaus

136. Rhizobium strain suitable for pea crop.

a) *Rhizobium meliloti*
b) *Rhizobium japonicum*
c) *Rhizobium solaneserum*
d) *Rhizobium phaseoli*

137. Nutrients move from soil to plant against concentration gradient in

a) Contact exchange
b) Mass flow
c) Diffusion
d) Carbon-dioxide theory

138. Critical level for dTPa extractable Zn in red soil is

a) 0.8 ppm
b) 2.0 ppm
c) 4.0 ppm
d) 0.6 ppm

139. The following element causes dispersion of soil aggregates

a) ca
b) Mg
c) K
d) Na

140. The structural shape of alkali sub soil is

a) Granular
b) Blocky
c) Plate
d) Columnar

141. The average Potassium contact of soil is
a) 1.9% b) 2.9%
c) 3.9% d) 4.9%

142. Available potassium in soil is
a) Water soluble -K b) Exchangeable -K
c) Inter layer - K d) Both a & b

143. Optimum soil pH for Nitrification process is
a) 5.5 b) 7.5
c) 6.5 d) 8.5

144. Zinc deficiency in rice crop is known as
a) White bud b) Speakled yellow
c) Little d) Khaira

145. Indicator is used in Wet digestion method of organic carbon stimation
a) Mixed indicator b) Methyl orange
c) Muroxide d) Diphenyl amine

146. The author of reference book " The nature and properties of soil" is
a) Danial Hillel b) Schofield
c) Nyle.C. brady d) Tisdale

147. "Law of minimum" was proposed by
a) Lindsay and Norvel b) Rutherford
c) StefenHakings d) Liebig

148. Tamato crop prefer the following form of nitrogen soil
a) Nitrite b) Nitrate
c) amoniacal d) amide

149. The nitrogen content in Urea fertilizer is
a) 64% b) 46%
c) 26% d) 16%

150. The Phosphorous content of di-ammonium phosphate (daP) is
a) 16% b) 64%
c) 46% d) 26%

151. Potassium deficiency is observed in
a) Younger leaves b) Olderleaves
c) Middle d) Terminal

152. Bloosm- end rot is caused by
a) Excessive Ca b) Excessive Mg
c) Deficient Ca d) Deficient Mg

153. Potassium held in following sites in soil clays are more available to plants
a) P-sites b) E-sites
c) I-sites d) None of them
154. The range of soil pH for moderately acidic soil
a) 4.5- 5.5 b) 5.5-6.5
c) 6.5- 7.5 d) 7.5 -8.5
155. Calcium deficiency on plant is first seen in
a) Middle leaves b) Older leaves
c) New leaves d) None of them
156. Manganese (Mn) containing mineral
a) Pyrite b) Pyralusite
c) both a&b d) Siderite
157. Iron (Fe) containing mineral
a) Manganite b) Magnetite
c) Both a&b d) Magnasite
158. borates mostly occur in Viets pool of
a) Pool E b) Pool c
c) Pool a d) Pool d
159. Micro nutrient cations exchangeable by weak electrolyses are present in
a) Pool a b) Pool b
c) Pool c d) Pool d
160. Mo deficiency mostly observed in
a) acid soils b) alkali soils
c) Saline soils d) calcareous soils
161. Essentiality of Zn was established by
a) Sprengal and Viets b) Rutherford
c) Schofield d) Sommers
162. When plant sample contains 80 ppm Zn, it is rated as
a) Deficient b) Adequate
c) Toxic d) None of them
163. The critical soil pH for Zn availability is
a) 5.0 to 5.5 b) 6.5 to 7.5
c) 6.0 to 7.0 d) 5.5 to 6.5
164. The diameter of $NH_4^+$ ion is
a) 2.66 b) 2.86
c) 2.90 d) 2.96

165. The clay mineral which do not fix K is

a) Illite
b) Vermiculite
c) Kaolinite
d) Montmorrilonite

166. Freezing and thawing process of soil result in

a) Leaching of solution K
b) Fixation of solution K
c) Absorption of solution K
d) Release of fixed K

167. Increase in temperature favours

a) Increase in K fixation
b) Decrease K fixation
c) Leaching of K
d) No change in K

168. Opening and closing of Stomata in plants is regulated by the following nutrient

a) Zn
b) P
c) N
d) K

169. Luxury consumption is associated with

a) N
b) P
c) K
d) ca

170. Tip burning is a symptom of

a) P deficiency
b) K deficiency
c) S deficiency
d) Mg deficiency

171. Muriate of Postash contains

a) 46% $K_2O$
b) 50% $K_2O$
c) 60% $K_2O$
d) 18% $K_2O$

172. Frankia is an example of

a) Symbiotic-legume association
b) Non-Symbiotic association
c) Symbiotic Non-legume association
d) Free living

173. Crop plants became susceptible to pests and diseases when the soil is

a) N deficiency
b) N toxicity
c) P deficiency
d) P toxicity

174. The nutrient element that play important role in energy transfer in plants is

a) ca
b) K
c) P
d) Mg

175. al-containing phosphorous mineral compound in soil is

a) Vivianite
b) Strengite
c) Variscite
d) Wevillite

176. Phosphorous reserves in soil are found mostly as
a) Residual formations b) Sedimentary formations
c) Glacial formation d) Colluvial formation

177. The total P content is generally highest in soils developed from
a) Shales b) Basalt
c) Granite and gneiss d) Marble

178. The range of total P content in Indian soils is
a) 100 to 200 ppm b) 200 to 300ppm
c) 300 to 400ppm d) 400 to 500 ppm

179. Strengite is
a) Al containing P mineral b) Iron containing P mineral
c) Mn containing P mineral d) Copper containing P mineral

180 Very slowly avail P (non labile) forms in soils are
a) Precipitate of Fe, al, Mn phosphates
b) Aged andwell crystallized Stable org. P compounds
c) Freshly formed Fe, al and Mn Phosphates in acid soils
d) s&b

181. The concentration of phosphorous in soil solution is
a) Less than 0.1 ppm b) 0.1 to 1.0 ppm
c) 1.0 to 2.0 ppm d) 2.0 to 5.0ppm

182. In general, the concentration of Inositol phosphate in soils is
a) 0.1 to 1.0% b) 1-2 %
c) 2-9% d) 20 to 90 %

183. The range soil pH considered to be the best for optimum availability of phosphorus is
a) 4.5 – 5.5 b) 5.5 – 6.5
c) 6.5 – 7.5 d) 7.5 – 8.5

184. At high soil pH, $H_2PO_4^-$ may form insoluble tri-calcium phosphate by reacting with
a) $caCO_3$ b) $caSO_42H_2O$
c) $MgSO_47H_2O$ d) $MnSO_4$

185 The product of the chemical reaction $ca_8H_2(PO_4)_5$. 5 $H_2O$ + $caCO_3$ à
a) Octa calcium phosphate b) Monocalcium phosphate
c) Dicalcium phosphate d) Tricalcium phosphate

186. Phosphorous is fixed in the crystal lattices of 1:1 type of clay minerals like the Kaolinite having high aEc occur in

a) Neutral soils
b) Acid soils
c) Alkali soils
d) Calcareous soils

187. Phosphorous fixation in soils aggravated by

a) Amorphous clays
a) 1:1 type clays
c) Sesqui oxide clays
d) All the three

188. Phosphorous fixation is more in

a) Andisols
b) Inceptisols
c) Entisols
d) Gelisols

189. Phosphorous fixation in soil is controlled by

a) Placement of P fertilizers
b) Addition of Organic matter
c) Application Mycorrhizae
d) All the three

190. Incorporation of organic matter in to soil increases phosphorous availability due to

a) By formation of phospho - humic complexes that are easily assimilated by plants.
b) Through the mobilisation of the phosphate by humate ion by anion exchange
c) By forming a coating on sesquioxide particles by humus as a protective cover which otherwise fix P.
d) All the three

191. Phosphorous solubilising microorganisms are

a) *Bacillus megatherium*
b) *Pseudomonas striatus*
c) *Aspergillus awamori*
d) All the three

192. The nutrients show localized deficiencies in soils, crops plants require them in moderate amounts, and they also added to soil inadvertently through carriers of the primary nutrients

a) Micronutrients
b) Primary nutrients
c) Secondary nutrients
d) Scarce nutrients

193. Micronutrient elements are often been called as

a) Primary elements
b) Secondary elements
c) A&B
d) Trace elements

194. The nutrient immobile in plant

a) Calcium
b) Boron
c) Sulphur
d) a&b

195. Partially mobile nutrient in plant is
   a) Fe  b) P
   c) N  d) ca

196. The % of calcium in Single Super Phosphote (SSP) fertilizer is
   a) 16  b) 21
   c) 46  d) 61

197 **Match the following**

| | | |
|---|---|---|
| Hidden hunger | ( ) | 1. K deficiency |
| Achiochi | ( ) | 2. Sulphur toxicity |
| Chlorosis in rice nurssaries | ( ) | 3. Iron deficiency |
| White bud of Maize | ( ) | 4. Zn deficiency |

a) 1342  b) 1243  c) 1432  d) 1234

| | | |
|---|---|---|
| 198. Essentiality of N | ( ) | 1. Ramoorthy |
| Law of diffusion | ( ) | 2. Becket |
| Q/I of Potassium | ( ) | 3. Theodore de Sassure |
| Fertility equations | ( ) | 4. Fick |

a) 3421  b) 3124  c) 1324  d) 1432

| | | |
|---|---|---|
| 199. Strengite | ( ) | 1. $Fe_2O_3$ |
| Haematite | ( ) | 2. $PbMoO_4$ |
| Wulfinite | ( ) | 3. $cacO_3$ $MgcO_3$ |
| dolomite | ( ) | 4. $FePO_4$ $2H_2O$ |

a) 3124  b) 3142  c) 4231  d) 4123

200.

| | | |
|---|---|---|
| Saline Soil | ( ) | 1. Gypsum |
| Alkali soil | ( ) | 2. Pyrite |
| Calcarious alkali soil | ( ) | 3. Leaching |
| Acid Soil | ( ) | 4. Lime |

a) 3124  b) 3421  c) 3142  d) 3412

201.

| N Source | | % N |
|---|---|---|
| Ammonium sulfate | ( ) | 1) 21 |
| Ammonium chloride | ( ) | 2)25 |
| Ammonium nitrate | ( ) | 3)33 |

| | | |
|---|---|---|
| Mono-ammonium phosphate | ( ) | 4)11 |
| Urea | ( ) | 5)46 |

Ans:1, 2, 3,4,5

202. Assertion (a) and Reasoning ® type questions

Assertion(a): Calcium deficiency seen in terminal bud of plants

Reason ( R ): Calcium is immobile in plants

a) Only A is True

b) Only B is True

c) Both A & R are true, R adequately explains A

d) Both A & R are true, R does not adequately explains A

203. Assertion (a):Gypsum is recommended for non-calcareous sodic soils is an amendment Reason ( R ): Gypsum is less soluble product

a) Only A is True

b) Only B is True

c) Both A & R are true, R adequately explains A

d) Both A & R are true, R does not adequately explains A

204. Assertion (a): Rhizobium japonicum fix atmospheric nitrogen in Soybean crop.

Reason ( R): Rhizobium japonicum is non-specific strain in legume plants

a) Only A is True

b) Only B is True

c) Both A & R are true, R adequately explains A

d) Both A & R are true, R does not adequately explains A

205. Assertion (a): the clay mineral vermiculite has high cation exchange capacity

Reason (R ): Isomorphous substitution is high in vermiculite clay mineral

a) Only A is True

b) Only B is True

c) Both A & R are true, R adequately explains A

d) Both A & R are true, R does not adequately explains A

206. Assertion (a):Urea is most widely used nitrogen complex fertilizer

Reason (R ): Urea has highest nitrogen concentration among N fertilizers

a) Only A is True

b) Only B is True

c) Both A & R are true, R adequately explains A

d) Both A & R are true, R does not adequately explains A

**Answer key**

| | | | | | | | | | | | | | | | | | | | | | |
|---|---|---|---|---|---|---|---|---|---|---|---|---|---|---|---|---|---|---|---|---|---|
| **1** | d | **21** | d | **41** | c | **61** | c | **81** | c | **101** | c | **121** | c | **131** | b | **141** | a | **151** | b | **161** | d |
| **2** | b | **22** | b | **42** | a | **62** | c | **82** | c | **102** | d | **122** | a | **132** | b | **142** | d | **152** | c | **162** | b |
| **3** | c | **23** | b | **43** | c | **63** | b | **83** | b | **103** | a | **123** | d | **133** | b | **143** | d | **153** | a | **163** | d |
| **4** | c | **24** | a | **44** | d | **64** | a | **84** | c | **104** | c | **124** | d | **134** | b | **144** | d | **154** | b | **164** | b |
| **5** | a | **25** | d | **45** | c | **65** | b | **85** | d | **105** | b | **125** | b | **135** | a | **145** | d | **155** | c | **165** | c |
| **6** | b | **26** | d | **46** | b | **66** | c | **86** | c | **106** | c | **126** | d | **136** | d | **146** | c | **156** | b | **166** | d |
| **7** | c | **27** | a | **47** | c | **67** | d | **87** | d | **107** | c | **127** | b | **137** | c | **147** | d | **157** | b | **167** | a |
| **8** | d | **28** | d | **48** | b | **68** | b | **88** | c | **108** | a | **128** | d | **138** | d | **148** | c | **158** | c | **168** | d |
| **9** | c | **29** | d | **49** | d | **69** | a | **89** | b | **109** | b | **129** | a | **139** | d | **149** | b | **159** | b | **169** | c |
| **10** | c | **30** | c | **50** | b | **70** | d | **90** | a | **110** | c | **130** | d | **140** | d | **150** | c | **160** | a | **170** | b |
| **11** | b | **31** | b | **51** | c | **71** | c | **91** | d | **111** | d | **121** | c | **131** | b | **141** | a | **151** | b | **161** | d |
| **12** | d | **32** | a | **52** | d | **72** | d | **92** | b | **112** | d | **122** | a | **132** | b | **142** | d | **152** | c | **162** | b |
| **13** | d | **33** | c | **53** | c | **73** | d | **93** | c | **113** | c | **123** | d | **133** | b | **143** | d | **153** | a | **163** | d |
| **14** | a | **34** | a | **54** | a | **74** | d | **94** | d | **114** | d | **124** | d | **134** | b | **144** | d | **154** | b | **164** | b |
| **15** | b | **35** | c | **55** | a | **75** | b | **95** | c | **115** | a | **125** | b | **135** | a | **145** | d | **155** | c | **165** | c |
| **16** | b | **36** | c | **56** | d | **76** | a | **96** | b | **116** | d | **126** | d | **136** | d | **146** | c | **156** | b | **166** | d |
| **17** | b | **37** | b | **57** | c | **77** | d | **97** | c | **117** | b | **127** | b | **137** | c | **147** | d | **157** | b | **167** | a |
| **18** | d | **38** | a | **58** | b | **78** | d | **98** | c | **118** | b | **128** | d | **138** | d | **148** | c | **158** | c | **168** | d |
| **19** | a | **39** | b | **59** | d | **79** | b | **99** | c | **119** | c | **129** | a | **139** | d | **149** | b | **159** | b | **169** | c |
| **20** | b | **40** | a | **60** | c | **80** | d | **100** | b | **120** | d | **130** | d | **140** | d | **150** | c | **160** | a | **170** | b |

| | | | | | | | | | | | | | | | | | | | | | |
|---|---|---|---|---|---|---|---|---|---|---|---|---|---|---|---|---|---|---|---|---|---|
| 171 | c | 175 | c | 179 | b | 183 | c | 187 | d | 191 | d | 195 | a | 199 | d | 201 | a | 205 | c | | |
| 172 | c | 176 | b | 180 | d | 184 | a | 188 | a | 192 | c | 196 | b | 200 | a | 202 | c | 206 | b | | |
| 173 | b | 177 | c | 181 | a | 185 | d | 189 | d | 193 | d | 197 | d | 199 | d | 203 | d | 205 | c | | |
| 174 | c | 178 | a | 182 | c | 186 | a | 190 | d | 194 | d | 198 | a | 200 | a | 204 | a | 206 | b | | |

# 11

# Manures and Fertilizers

The word “Manure” is originated from the French word “Manoeuvrer” which refers to “work with soil”. The word Manure is also originate from Latin word “Manu” (hand) and operate (to work). Manures are the substances which are organic in nature, capable of supplying plant nutrients in available form, bulky in nature having low analytical value and having no definite composition and most of them are obtained from animal and plant waste products. Formerly, the word in use was “Dung ”in English (at present fertilizer). It is derived from the old German word “Tung” meaning storage pit covered with manure for protection of soil against cold. From Tung “Tungen’ is derived (to cover) and then dung (fertilizer) as the cover. The corresponding english word “Fertilizer” which is now accepted internationally in view of the old English word Dung was derived from Latin word “Fertil”. A fertilizer can be defined as a mined or manufactured material containing one or more essential plant nutrients in potentially available forms in commercially valuable amounts. Manuring is the process of increasing the productive capacity of land by adding plant foods to the soil in different forms.

Manures obtained from plant, animal and human origin while fertilizers are synthetically manufactured. Manures are Organic in nature while fertilizers are inorganic in nature. Manures are slow acting and fertilizers are quick acting. Manures have low analytical value, fertilizers have high analytical value. Manures donot have definite chemical composition but fertilsiers have definite chemical composition. Manures on application to soil improve soil physical properties, fertilizers do not improve soil physical properties. Manures supply almost all macro and micro plant nutrients, while that of fertilsiers supply very few plant nutrients. Manures are bulky in nature, fertilizers are non-bulky in nature

**Classification of Manures and Fertilizers**: The detailed classification of manures and fertilizers are given in table1 and figure1.

**Table 1:** Classification of manures

| Manures | |
|---|---|
| A) Bulky organic manures (Slow acting with large quantity of organic matter) | B)Concentrated organic manures (Quick acting with small quantity of organic matter) |
| 1.**Farm Yard Manure (FYM)**<br>i. Cattle manure<br>ii. Sheep manure<br>iii. Poultry manure<br>iv. Pig manure<br>v. Biogas slurry<br>2. **Composts**<br>Rural or village compost<br>Urban compost<br>Vermicompost<br>3.**Green manures**<br>i. Green manuring in-situ<br>ii.Green leaf manure<br>4.**Agro-based Industry wastes**<br>i)Press mud cake<br>ii)Cotton dust<br>5.**Human habitation waste**<br>i) Night soil<br>6.**Municipal wastes**<br>i) Municipal rubbish<br>ii) Street sweepings<br>iii) Market wastes | 1. **Edible oil cakes** (suitable for cattle and poultry feeding )<br>i) Mustard cake<br>ii) Ground nut cake<br>iii) Sesame cake<br>iv) Linseed cake<br>2.**Non –edible oil cakes**(suitable for crop fertilization)<br>a) Plant origin<br>i. Castor cake<br>ii. Neem cake<br>iii. Sunflower cake<br>iv. Mahua cake<br>v. Karanj cake<br>b) Animal origin<br>Bone meal<br>Blood meal<br>Horn meal<br>Meat meal (Tankage)<br>Fish meal<br>Bird product- Guano(excreta and dead bodies of sea birds) |

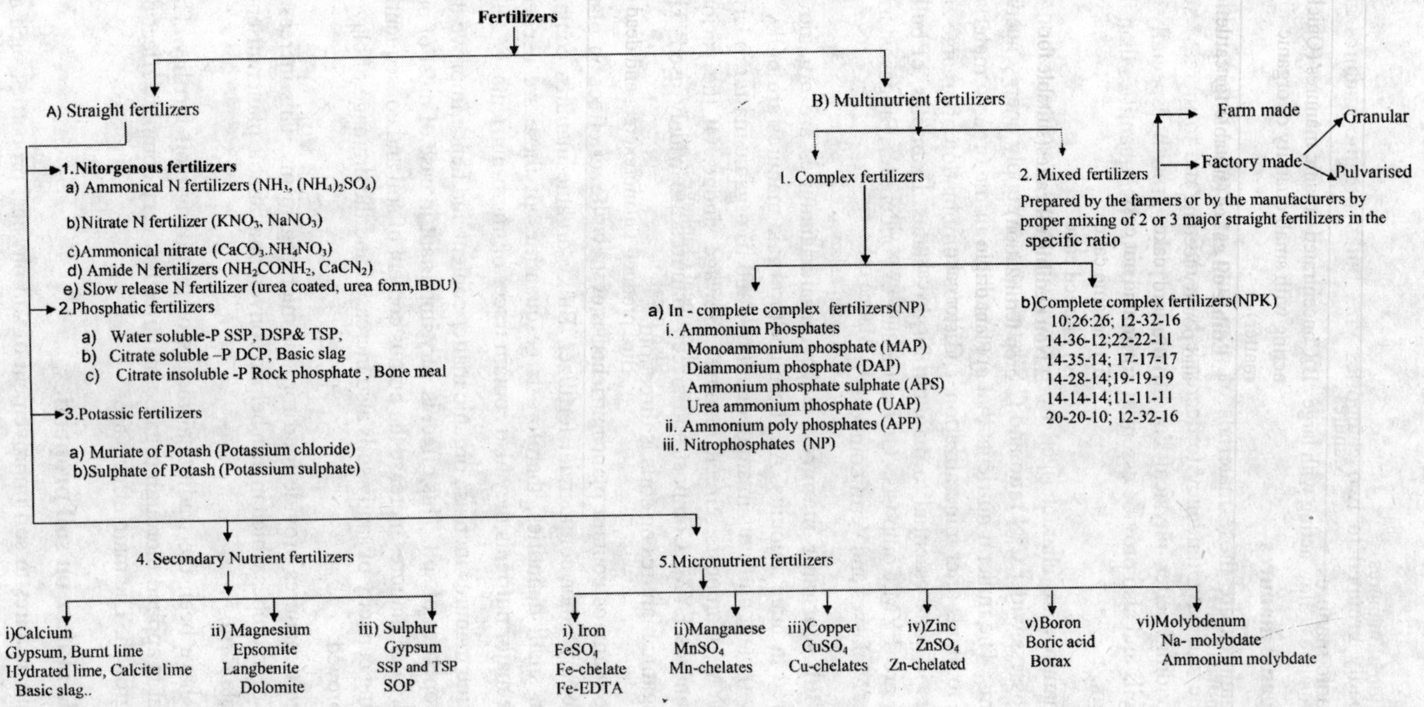


Fig. 1: Classification of Fertilizers

## 11.1. Manures

Manures are broadly grouped to two categories viz., bulky organic manures and concentrated organic manures.

### 11.1.1. Bulky organic manures

Bulky organic manures are those materials of plant and /or animal origin, which when added to the soil have tendency to decrease bulk density and to increase soil volume,thus providing better physical conditions for plant growth especially in coarse textured soils and also provide essential nutrients in smaller quantities than the chemical fertilizers.

Bulky organic manures are characterized by relatively poorer in concentration of plant nutrients; possess wider C:N ratio and C:S ratios and so supply energy needed for microorganisms; The mineral nutrients that are available in the organic materials become available to plants after mineralization; Judicious combination of organic and inorganic manures is quite essential to maintain fertility status. Examples, of bulky organic manures are FYM,Composts, Green manures, Pouderette, Biogas slurry, Sewage or sludge, Molasses and Vermicompost

**Importance of Bulky organic manures:** Bulky organic manures are organic in nature and required in huge amounts. All these bulky orgaic manures are bulky in nature and supply i) plant nutrients in small quantities and ii) organic matter in large quantities. Eg. FYM, Composts, Green manures, Sewage sludge. Of the various bulky organic manures FYM, Compost and Green manures are widely used. The effect of bulky organic manures on soils is three fold.

a) Since these manures contain plant nutrients, they have direct effect on plant growth like any other commercial fertilizer. Bulky organic manures contain nutrients in small quantities, therefore large quantities of these are needed. Besides the major nutrients,they also contain traces of micro-nutrients

b) Bulky organic manures increase organic matter content and hence improve the physical properties of soil. This effect is very important in case of most of our arable land. Such manures increase the humus content of soil and consequently water holding capacity of sandy soils also increased and the drainage of clayey soils is improved.

c) Bulky organic manures provide food for soil microorganisms. This increases the activity of microbes which in turn help to convert unavailable plant nutrients into available form.

Organic manures have been the traditional means of improving soil fertility. The organic matter that is applied thorough organic manures has very complex effect on soil and on plant growth. The main effects are

## Effect of organic matter on soil properties

Addition of organic manures to soil improvement of soil physical properties such as improvement of soil structure, improvement of water holding capacity, improvement of soil aeration and reduce of soil loss through erosion. Organic manures also improve soil chemical properties such as supply of essential plant nutrients in balanced ratio makes slow release of nutrients and has high residual value. Organic manures improvement of biological activity such as soil stimulation of soil fauna and flora.

## i. Farm Yard Manure (Cattle manure)

**Definition:** The physical composition of cattle manure is called Farm Yard Manure (FYM), which consists of dung and urine of cattle and the litter, a bedding material like hay, straw used for cattle. Cattle manure is slow acting, bulky organic and however is a low analysis fertilizer, obtained from dung and urine of farm animals mixed with litter and other miscellaneous farm wastes.

## Constituents of FYM

a) **Dung:** The dung is a solid excreta voided by farm animals, which represents the undigested and non-digestible portion of the animal feed. Besides, it is also admixed with digestive juices, tissues of alimentary canal, microbial cells etc. The percentage of moisture varies from 70-80 per cent.

b) **Urine:** The liquid excreta of farm animals, which is composed essentially of the byproducts of nitrogen metabolism of farm animals.It also consists of 90 per cent water with small amounts of urea, hippuric acid, sulphates of sodium and potassium and also chlorides and phosphates of Ca, Mg, K and Na etc.,

c) **Litter:** Litter is a variable mixture consisting of straw, leaves, stems used as bedding material for cattle during their rest and eventually absorbs the urine and adds carbonaceous matter in the manure to facilitate decomposition.

**Quality and composition of FYM:** The quality of manure and chemical composition in particular is highly variable as the following factors affect the product.

i) Kind animal: The quality of manure depends on the class of manure.viz., cattle,horse manure.With in the same class,quality varies according to the kind of animal, such as milch cattle, dry cattle, work cattle, breeding bulls etc., By and large the dung and urine from animals, which assimilates less (little ) for their maintenance and production will provide better quality manure.

ii) Age and condition of animal: Growing animals,milch cattle,pregnant or carrying cattle utilize much of the ingredients in the feeds for building up their growing bodies, milk production and for the development of the embryo[calf]. Old or adult animals kept on light work or no work utilizes little from feeds and as such, most of nitrogen is voided through urine and dung. Eventually, the adult old cattle provide better manure.

iii) Quality and quantity of feed consumed: Nutritious and protein rich feeds like oil cakes enriches the nitrogen content to the resulting manure than the bulky feeds like straw and green grass. Animals fed on concentrated feeds yield better quality manure.

iv) Kind of litter used: The quality of manure depends to a considerable extent on the nature of litter used. Remnants of leguminous hays (*Bhusa* ) give richer manure than usual straws.

v) Collection of manure: The method adopted for collection of dung, urine and litter primarily decide the quality of manure as the loss of nutrients particularly nitrogen occurs from the time urine and dung are voided by cattle. The quality of manure depends upon the methods of collection viz., Byre, Lose box and Dry earth systems.

a) Byre system: Cattle are stalled in a shed with a non absorbent floor provided with necessary slope towards the urine drains. The urine that flows into the drains is collected into a covered tank. From where it is periodically removed and sprinkled on the manure stored in a covered pit. The urine which is an important component of FYM can be properly stored (conserved) in this system. The perfect cleanliness and hygienic conditions of the stalls as well as cattle can be maintained in this system.

b) Dry earth system: The floor of the cattle shed is well rammed and compacted. Layers of fine sand, red earth of loamy soil are spread as an absorbent for urine. The wet portions are properly covered with dry layers or any of the above materials and once a week the surface layer is removed and dumped in the manure pit. Available saw dust, paddy husk, groundnut shell, paddy winnowed dust would serve the purpose very well compared to the earth absorbents. This system is popular and extensively adopted in rural parts being cheep, convenient and practicable under the existing rural conditions in India.

vi) Storage of Manure: Method of storage of manure influences the quality of manure to a large extent. During storage the manure undergoes fermentative changes, decomposition which leads to losing its original structure and shape. There are three methods of storage viz., pit method and heap method and covered pit method.

a) **Pit method (Below ground level):** In this method, the manure is stored in a pit with non –absorbent bottom and sides. The pit is provided with a bund at the rim of the pit to prevent the surface run-off of waters during rainy season.The dimensions of the pit can be variable depending on the quantity of dung,urine and litter produced on the farm per day.The losses also occur in this method due to exposure to sun and rain,but it is relatively a better method than the heap method.

b) **Heap method (Above the ground level)**: This is the most common method adopted in Indian villages. Manure is heaped on the ground preferably under the shade of a tree. Ideal procedure is to dump the dung first and to cover it with litter soaked urine.This is further covered with a layer of litter /ash / earth to prevent the loss of moisture and to avoid direct exposure to sun.It is also desirable to put up a small bund around the base of the heap to protect against surface run-off washing out the manurial ingredients.It is beneficial to cover the exposed portion of the heap with Palmyra leaves or any other available material. The maximum losses of nutrients occur in this method of storage, resulting in poor quality manure. Direct exposure to the vagaries of climate such as sunshine and rainfall causes looseness and dryness of manure, which hasten the losses of nutrients and rapid oxidation of organic matter.

c) **Covered pit method:** Of all the methods described, it is the best method. In this method, the bottom and sides of the pit are made non-absorbent by granite stone lining. The pit is also provided with a bund of 1½ feet height to prevent surface flow of water (Rain water) and a suitable cover by way of roofing with locally available materials like Palmyra or phoenix leaves etc., organic matter and 8nutrient losses can be effectively controlled in this method of storage in order to obtain better quality manure [FYM: 0.68 % N; 0.5% P; 1 % K].

**d) Trench method of preparing FYM:** This method has been recommended by C.N.Acharya. The manure preparation should be carried in trenches of suitable size,say 20-25 feet length, 5-6 feet breadth and 3-3.5 feet depth.All available dry litter and refuse from the farm and the houses should be heaped up near the cattle shed and portions of litter mixed with earth if available should be spread in the shed in the evening @2.26 kg per animal for the absorption of urine.The litter should be localized in the areas where urine generally drops and soaks into the ground. Every morning the urine soaked litter and dung should be well mixed and then taken into the manure trench.A section of 3 feet length of the trench from one end should be taken up for filling with daily collection of refuse from cattle shed. When the trench is filled to the height of 1.5 to 2.0 feet above ground level, the top is made dome shaped and plastered with cow dung mixed with soil. The manure becomes ready for about three months. By this time the next 3 feet length of the trench being filled up. Generally 2 such trenches would be needed for three to four cattle. It is possible to prepare by this process 250-300 cubic feet of manure (3 to 5 tones or 10-12 cart loads) per animal. The FYM should be enriched by addition of super phosphate @ 30-40 kg per trench before application to fields. The content of nitrogen is at least 0.7 to 0.8 % N on the fresh weight basis or 1.4 to 1.6 % N on dry weight basis.

**Losses of Nutrients from FYM during collection and storage:** There are two types of losses of FYM, viz., Losses during handling and Losses during preparation and storage.

**A. Losses during handling:** FYM consists of two original components such as i) dung (solid portion) and ii) urine (liquid portion) which are subjected to different losses in different ways. Both the components contains N, $P_2O_5$ and $K_2O$. Approximately, half of nitrogen and potash in dung and other half in urine.Whereas nearly all the phosphorus (96%) in the solid portion. To conserve N, $P_2O_5$ and $K_2O$, it is most essential that both the parts of cattle manure are properly handled and stored

i) Losses of dung: Dung is a valuable manure. Still larger portion is dried in dung cakes and burnt as fuel.Besides this large portion of cattle excrements is dropped outside the cattle shed, when the animals are grazing on the uncultivated lands.

This can be used for preparation of FYM.

ii) Losses of urine: Urine contains nitrogen and potassium in large portions. But there is no good method of preserving the urine in our country. In our country most of the cattle sheds have un cemented or *kachha* floor and the urine gets soaked in the soils of *kachha* floor of the cattle shed and the large quantities of nitrogen are thus lost through the formation of gaseous ammonia as follows.

$CO(NH)_2 + 2H_2O \rightarrow (NH_4)_2CO_3$

Urea in urine — Ammonium carbonate

$(NH_4)_2CO_3 + 2H_2O \rightarrow 2NH_4OH + H_2CO_3$ (Carbonic acid)

$NH_4OH \rightarrow NH_3\uparrow + H_2O$

Ammonium hydroxideAmmonia

**B. Losses during preparation and storage:** Cow dung and other farm wastes are collected daily and these are accumulated in manure pit in open space for months together.The manure remain exposed to sun and rain during this period. Due to this effect the nutrients are lost in following ways.

i) **By leaching:** Nutrients of manures are water soluble and these are liable to get washed by rain water.The leaching loss of nutrients will vary with the surface exposed,the intensity of rain fall and the slope of the surface on which manure is heaped.The leaching loss may be prevented by erecting a roof over the pit.

ii) **By volatilization:** During storage, the urine and dung are decomposed and considerable amount of ammonia is produced.The ammonia combines with carbonic acid to form ammonium carbonate and bicarbonate, which are rather unstable and gaseous ammonia may be readily liberated and passes into atmosphere as indicated in the following equations.

Urea and other nitrogenous compounds in urine and dung by microbial decomposition liberates $NH_3$(Ammonia)

$2NH_3 + H_2CO_3 \rightarrow (NH_4)_2CO_3$

$(NH_4)_2CO_3 + 2H_2O \rightarrow 2NH_4OH + H_2CO_3$

$NH_4OH \rightarrow NH_3\uparrow + H_2O$

**Ways to minimize the losses from FYM during handling**

a) Adopt trench method as suggested by C.N. Acharya for handling of dung and urine

b) Use of Gobar gas plant: 50 % of dung is made dung cakes and burnt as fuel for cooking. The use of cow dung in gas plant produces a combustible gas, methane used as fuel gas which is an improved method of handling FYM.

c) Adopting covered method of storing FYM: Nutrients losses can be effectively controlled by this method

d) Adoption of Byre system in collection of FYM

e) Proper field management of FYM: During spreading of FYM in the field in small heaps leads to loss of nutrients from it.It is advisable to spread the FYM before ploughing.

f) Use of chemical preservatives: Chemical preservatives are added to FYM to decrease nitrogen losses. To be most effective, the preservatives are applied in the cattle shed to permit direct contact with the liquid portion of excreta or urine. This has to be done because the loss of nitrogen from urine starts immediately.

Commonly used preservatives are 1) Gypsum 2) Super phosphate.It is recommended that 450 g to 900 g of super phosphate should be applied per day per animal in the cattle shed.Super phosphate should be applied in places where animal pass urine. The reaction of gypsum with ammonium carbonate (intermediate product from decomposition of urea present in urine) is

$$(NH_4)_2\,CO_3 + CaSO_4 \rightarrow Ca\,CO_3 + (NH_4)_2\,SO_4$$

With this reaction ammonium carbonate is converted to $(NH_4)_2$ $SO_4$.As long as manure is in moist, no loss of ammonia will occur but if dried the chemical reaction is reversed and loss of ammonia may occur. Use of gypsum also prevents the bad smell caused by production of ammonia in the cattle sheds. As such in Indian conditions use of gypsum to decrease N loss does not offer practical solution. Super phosphate has been extensively used as a manure preservative. Since ordinary superphosphate contains up to 50to 60 per cent gypsum besides mono calcium phosphate. The reaction with ammonium carbonate is given below

$$2CaSO_4 + \underset{(MCP)}{Ca\,(H_2PO_4)} + 2(NH_4)_2\,CO_3 \rightarrow \underset{(TCP)}{Ca_3\,(PO_4)_2} + 2(NH_4)_2SO_4 + 2H_2O + 2CO_2$$

In this reaction, tricalcium phosphate is formed which does not react with ammonia sulphate when manure becomes dry as such there is no loss of ammonia.

**Straw**: Straw is sometimes left on the field as waste product from the harvest. It is useful as organic fertilizer especially on farms without live stock. Although straw contains few nutrients, it yields decomposable organic matter (Eg. Cellulose).Fertilization with straw thus supplies energy [1 kg straw = 12 Kilo Joules energy], but mainly improves the structure of medium and heavy soils.The nutrient contents of straw vary but are small in any case viz., 0.5 % N; 0.1 % P; 1 % K and dry matter of 85 %. The C: N ratio is very large about 100:1. This may cause inhibition of decomposition because the microbes lack in nitrogen, [Compensating supplies of N about 1% of the quantity of straw] is thus required.

**II. Compost:** Dry and fresh plant tissues and also animal wastes can be introduced directly into or on the soil. They are then decomposed at varying speeds,but have only slight and sometimes detrimental effects and bring about scarcely any substantial soil improvement. Process known since antiquity make it possible to convert such wastes into useful agents for soil improvement i.e., compost. The word compost is derived from the Latin word "Componere" to mean put together. Compost is a product of decomposition of plant and animal wastes with various additives. The compost had the largest variation of all organic material ranging from neglected garbage dumps to carefully composted and treated substances with high fertility.

**Composting:** Composting is a process of converting organic matter in to manure in a short time by accelerating fermentation process under controlled conditions is called composting. The organic material that is usually handled for composting are waste vegetative material such as spoiled straw, peanut hulls, saw dust, dried leaves etc., stubbles, chilli plants, cotton stalks, tobacco stems, weeds, municipal rubbish etc., according to availability.

## Basic rules are important for the production of good quality compost

i. The purpose of composting is to convert organic matter into growth promoting substances, for sustained soil improvement and crop production.

ii. The organic matter is partially decomposed and converted by microbes. These microbes require proper growth conditions, for their activity.i.e moisture content: 50% and 50% aeration of total pore space of the composting material. This is achieved through stacking and occasional turning over. Microbes also need sufficient nitrogen for synthesizing their body cells [the optimum C:N ratio of the composting material is 20:1 to 30:1]

iii. Soil microorganisms constitute sufficiently to the decomposition of organic matter through their continuous activities. The majority of these soil animals provide optimal conditions in their digestive track for their synthesis of valuable permanent humus and stable soil crumbs. A typical compost earthworm is *Eisenia foetida*.

iv. Certain additives accelerate the conversion and improve the final product. The materials such as lime, earth, gypsum, rock phosphate act as effective additives.The addition of nitrogen (0.1 to 1 %) is important in case of large C:N ratio of the composting material. Addition of lime (0.3 to 0.5 %), if sufficient lime is not present.The preparation of compost takes 2-3 months. The composition of compost varies with in wide limits.

Composition of compost: The following average values are rough estimates viz., Dry matter - 30-50%; organic matter 10- 15%; nitrogen- 0.3%; phosphorous 0.1% and potassium 0.3%.

**Methods of composting:** The process of composting was first initiated in England during the period of First World War (1914 1918). The various systems of composting are ADCO process (Agricultural Development Company); Activated compost process; Indore process; Bangalore process; Coimbatore process; Rain -water composting; Rural compost; Urban compost; Mechanical compost and Vermicompost.

**1. ADCO process:** Agricultural Development Company was initiated, [A private concern operating at Harpenden, England] developed by Hutchinson, H.B and Richards, E.H. during 1914-1918, at Rothamsted Experimental Station, England.

Materials needed: Straw and other wastes -Basic raw material) and Ammonium sulphate/Ammonium phosphate /Super phosphate/Muriate of potash Ground limestone/ urea - Starters

**Procedure:** The basic raw material straw is spread in layers and sprinkled over with a solution of ammonium sulphate. Then powdered lime stone is applied as broadcast. Then another straw layer is put on. The piling of the layer is continued

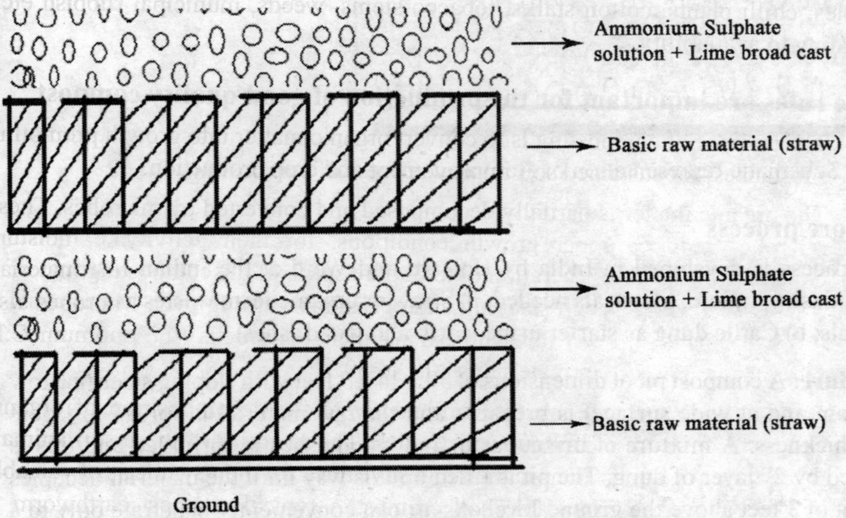


**Fig 2.** Schematic representation of ADCO process of composting

**2. Activated compost process:** This method was developed by Fowler and Ridge in 1922 at Indian Institute of Science, Bangalore. Materials needed are Basic raw materials (Straw and farm wastes) and Starters such as a) Cow dung b) Urine c) Night soil d) Sewage and sludge Procedure:

In this process the basic raw material for composting straw and other farm wastes is treated with mixture of cattle dung and urine as decoction. So that every portion of mass comes in contact with the inoculants(dung + urine ) and fermentation takes place evenly. On piling up in a heap of 3 feet or 4 feet height and turning over form time to time, keeping moist with dung and urine decoction, very high temperatures attained. When the temperatures begin to drop at the end of one week. The volume of the material gets reduced. further quantity of the basic material is added onto the heap. About 25 % of the new materials should be added at one time and thoroughly mixed with starters (dung +urine decoction) at intervals as before. If properly carried out, the compost will be ready in 5-6 weeks. Night soil and sewage and sludge are also used as starters in this method.

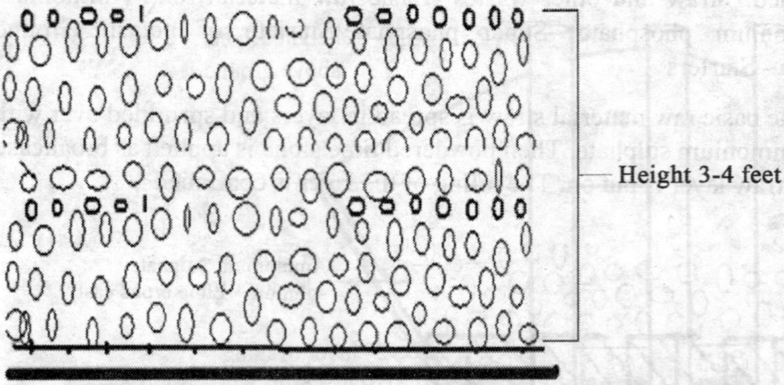


**Fig. 3:** Schematic representation of Activated process of composting

### 3. Indore process

This process is developed in India by Howard and Ward at the Indian Institute of plant Industry, Indore. Materials needed: a) Straw or organic farm wastes as basic raw materials; b) Cattle dung as starter urine, earth and wood ashes)

**Procedure:** A compost pit of dimensions of 30 x 14 x 3 feet with sloping sides (narrow at bottom and at wide surface) is prepared and the raw material is spread in layers of 3" thickness. A mixture of urine, earth, and wood ashes is sprinkled and this is followed by 2" layer of dung. The pit is filled up this way until the material occupies a height of 3 feet above the ground level. As air can conveniently penetrate only to a depth of 1.5 to 2.0 feet extra aeration has to be provided, which is done by means of artificial vents (holes) of 4" diameter pipe for every 4 feet length of the pit. The pit is watered twice a day i.e., morning and evening with rose can. The material is turning over 3 times, i.e., First at the end of the first fort- night; second at the end of the second fort night; Third when the material is two months old in the process of composting.

Observations during composting:

i) After 10 days of composting the following things happens: Synthesis of humus begins i.e., development of fungi and the height of the material is reduced by half. Check anaerobic decomposition, as indicated by the foul smell and fly breeding. If there is an anaerobic decomposition,turn over material for proper aeration. If insufficient fermentation, hasten by watering the material.

ii) At the end of two months: Fungal activity is completes, Materials become dark, Now the bacterial activity takes place.

Stock the material on the ground after 2 months, so that 25 % of additional nitrogen will be fixed from atmosphere. Compost is ready by 3-4 months. In this process one cattle pair produced 50-60 cartloads per year.

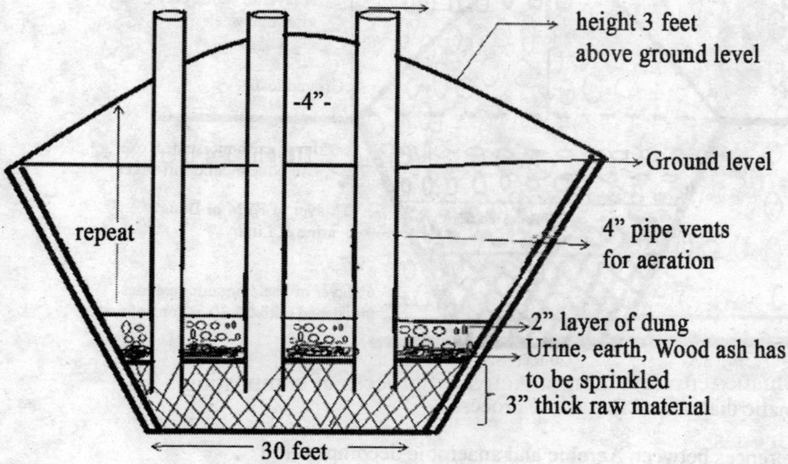


**Fig. 4:** Schematic representation of Indore process of composting

**4. Bangalore process [Aerobic and anaerobic process]:** This process of composting was developed by Dr.C.N.Acharya in the year 1949. Basic raw material used: Any organic material like FYM; Starters or inoculants: dung + urine or litter mixture (un-decomposed); 3. Additives: Bone meal or oil cakes, wood ash

**Procedure** [Pit size: 20 x 4 x 3 feet]: The basic raw material is spread in a pit of 20 x 4 x 3 feet dimensions to a depth of 6" layer, moistened with 20-30 gallons of water if the material is dry. Over this FYM or preferably a mixture of dung, urine and litter (un-decomposed) from the cattle shed is placed as a layer of 2" thickness. It is again covered on the top with a layer of earth to a thickness of 6". It is beneficial to mix the earth with bone meal or oil cakes, wood ash etc., to improve manurial value of the compost. The piling of layers is continued till the heap raises above the ground level to a height of 2 feet. Then the heap is kept open for one week to facilitate aerobic decomposition. Later the heap is plastered with a layer of moist clay for anaerobic fermentation to occur. Fissures, or cleavages (cracks) that occur in the clay layer, have to be sealed off periodically. The compost will be ready in 4-5 months period starting from the day of preparation. This process is called as aerobic and anaerobic decomposition of compost.

In this process the basic raw material is not so well decomposed as in the other methods But organic matter and N contents are well conserved. The number of turnings is reduced. The out turn of the compost is process is relatively greater and cheapest.

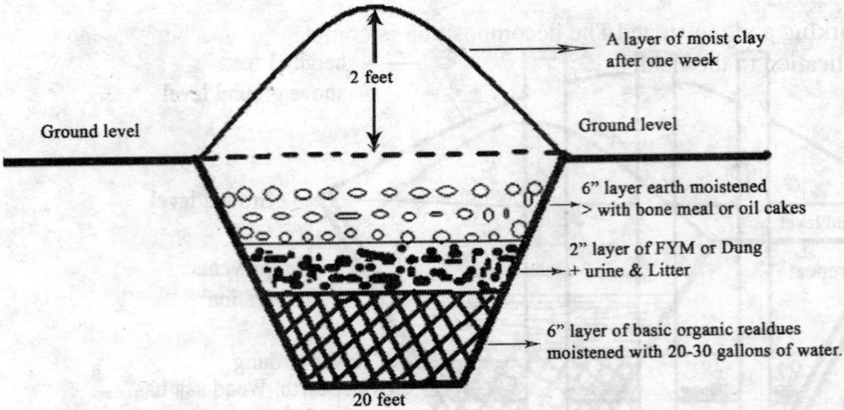


**Fig. 5:** Schematic diagram of Bangalore process

**Table 2:** Differences between Aerobic and anaerobic decomposition

| S.No. | Aerobic decomposition | Anaerobic decomposition |
|---|---|---|
| 1 | This takes place in the presence of oxygen | This takes place in the absence of oxygen |
| 2 | $CO_2$ is produced | Methane gas is produced |
| 3 | High water content is required | Less water content is required |
| 4 | Organic matter around 50-60 % conserved | Organic matter conserved is around 75% |
| 5 | Around 25-50% N is lost | 20% nitrogen is lost conserved and more N is |
| 6 | Temperature built up is more. Hence, pathogens weed seeds are killed | Pathogens are only killed due to antibioses. |
| 7 | More labour is required turnings and watering | Less labour is required |
| 8 | Time taken for decomposition is 3-4 months | Time taken is 5-6 months |

**The coimbatore process:** The basic raw material are, raw organic matter; starters: Powdered bone meal and cattle dung and water emulsion prepared by mixing gung in water @ 5-10 kg dung in 5 to 10 liters water.

**Procedure**: The basic raw material loosely spread [Pit: 12 x 6 x 3 feet] to a depth of 9" and water is sprinkled till the entire material is moist. Then about one kg of powdered bone meal is broadcasted uniformly above the layer and above this an emulsion of 5-10 kg of fresh cattle dung in 5-10 liters of water is applied.Repeat this process until a heap 2 feet above ground level is formed. Then the entire exposed surface area of heap is plastered with mud to facilitate semi-aerobic fermentation which would takes place for above 4-6 weeks depending upon the nature of the raw material. After 4-6 weeks, the mud plaster is removed to permit aerobic fermentation. If the heap has sunk unevenly which is a sign of defective fermentation, the material is re-heaped after forking and moistened.The decomposition is complete in 3-4 months and is fit for application to the field.

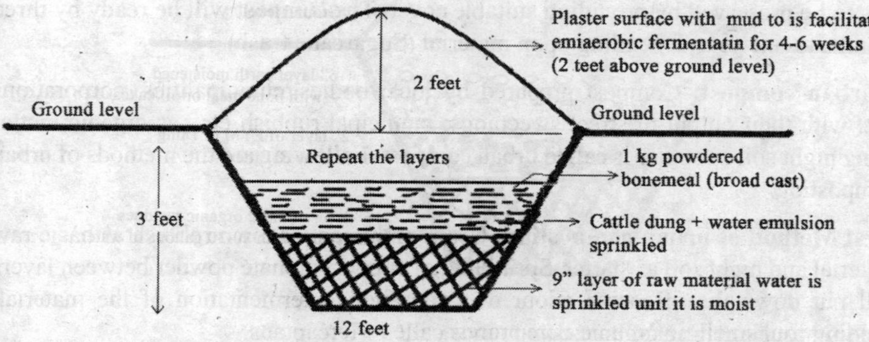


**Fig. 6:** Schematic representation of Coimbatore process of composting

**6. Rain watered compost:** In dry areas where it is difficult to obtain water for watering, the composting can be done with the aid of rain fall. The compost heap is built up as usual before the rains set in. The turnings are given during the rainy period at the end of rains the material will be ready for application. About 400 mm rain fall received in 3-4 months is considered sufficient.

**7. Rural compost:** Compost prepared using farm organic wastes and fresh dung and / or urine soaked earth as starter is called as Rural Compost. This is within the reach of farmers, as they can individually prepare the compost in their farm sheds.Composting can be done in a pit size of 3 x 6 x 3 feet size located on an elevated place. Even length and width are adjustable depending upon the raw material available. Keep the sides of the pit slanting.When a number of pits are dug, keep 12 feet gap between the pits for facilitating turnings. Locate the compost pits away from the civilians areas.

Basic raw material: i) Green succulent, non-woody and non-fibrous material (12"thick); ii)Dry materials like sugarcane trash, stubbles etc., (3" thick); ii)Starter: Dung and water; iv)Neutralizing agents for organic acids: Bone meal is preferred to ash as it provides Ca and also phosphate, which are desirable constituents in the final product.

**Procedure:** The organic material is spread in layers. The thickness depends upon compostability of materials. Green succulent non-woody materials can be laid in 12" layer while tough woody materials like sugarcane trash and stubbles are spread in 3" layer. Bone meal or ash is sprinkled in a thin layer over the material to provide the base for neutralizing organic acids produced during the decomposition. Dung and water is then sprinkled over the layer. Dung functions as starter while the water provides the moisture. The layers are thus built up to a height of one foot above the ground level. It is desirable to cover the surface of the heap with a layer of earth. Periodically watering is done.

**Precautions to be taken:** Avoid over watering, over watering creates anaerobic decomposition. Turnings are necessary to hasten the aerobic decomposition. After turning the compost must be re-heaped and covered. Use the compost immediately. It is to be preserved by providing suitable cover. The compost will be ready by three months to one year depending upon material (Sugarcane trash)

**8.Urban compost:** Compost prepared by local bodies,municipalities, corporations etc., with their output of street sweepings, municipal rubbish etc., as organic matter using night soil as starter is called urban compost. Following are the methods of urban composting.

**First Method of urban composting**: Using municipal or town rubbish as basic raw material and night soil as starter. Sprinkling of copper sulphate powder between layers will put down the offensive odour due putrefactive fermentation of the material, yielding foul smelling organic compounds called Mercaptans.

**Second Method of composting**: Night soil and rubbish are mixed in 1:4 ratio. Copper sulphate is added between layers at the rate of one kilogram for every 400 kg of the mixture and the mixture is racked up to heaps of convenient size.

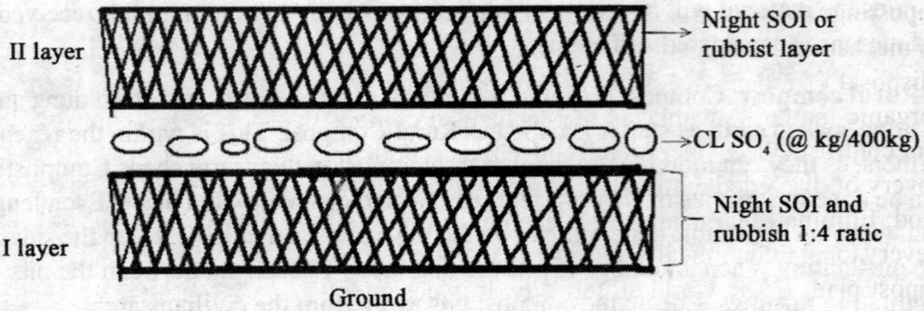


**Fig 7:** Urban composting Method II

The heaps are moisned and racked up for every fort night and rebuilt.During the period of two months, the manure will be ready for use. The product is dry and powdery, free from foul smell.

**Third Method of composting**: Town refuse is heaped into a pit of dimensions of 8 x 4 x 4 feet. A groove measuring 6x1 x 1 feet is made at the top in the centre and is filled up with night soil and covered over with refuse. Heap is moistened daily with sullage water and racked up once a week. For six weeks by which the decomposition process would be complete and the product would be ready for use as manure.

**Fourth Method of composting**: The layer of waste material of 6" thick and night soil 2 " thick are altered; the final top layers will be the waste material.The heap is covered by 9" thick layer of soil.The material will be ready in about five months.

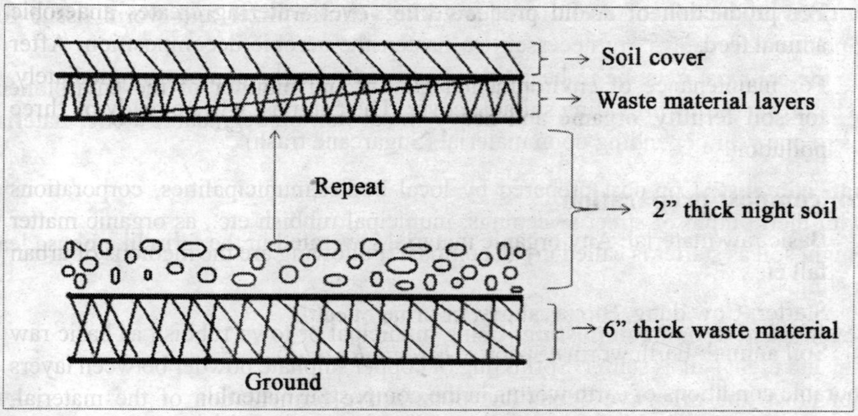


**Fig. 8:** Urban composting method IV

**Mechanical Compost**: In big cities like Delhi, Bombay, Kolkata, while disposing the urban wastes, the municipal administration innumerable problems such as cost, land, labour for composting operations involving human contact with filthy and obnoxious materials with a result mechanical can be composting is undertaken.mechanical composting undertaken at central location and insmall compact area. It is therefore only mechanized composting which may prove effective in not only serving as a means of disposal of urban wastes but also of providing the country with large quantities of organic matter. Advantages of Mechanical compostingare sanitary control with odour control device; Operational convenience, both in dry and wet seasons; Efficient recovery of discarded materials like metal,glass etc.; High quality manure in a short period; Eliminates large anaerobic decomposition; Reduces preparation cost @ 10% for every tone; Economically, cheaper than chemical fertilizers. Israel built up largest compost plant in the world at Tel-Aviv. It is designed to cope up with all domestic refuse from TEL-AVIV i.e., 500 tonnes in one shift of 8 hours. Second biggest plant is in Bangkok. In this method one tonne compost contains 13 kg N, 11 kg $P_2O_5$ and11 kg $K_2O$, which is equivalent to 65 kg ammonium sulphate, 66 kg single super phosphate and 22 kg muriate of potash.

## Vermicompost

An emerging technology for recycling of crop residues and other organic solid wastes is the utilization of earthworm technology to convert them into vermicompost. Vermicomposting is a method of making compost, with the use of earthworms, which generally live, in soil eat biomass and excrete it in digested form. This compost is generally called vermicompost or Wormicompost. Vermiculture means scientific method of breeding and rising earthworms in controlled conditions. Vermitechnology is the combination of vermiculture and vermicomposting. Thus, earthworms can be used in the following areas.

1. For development of arable soils, turnover of soil, break down of plant organic matter aeration and drainage

2. For production of useful products like vermifertilizer and worm tissue for animal feed.
3. For maintenance of environmental quality and monitor of the environment for soil fertility, organic and heavy metal non-biodegradable toxic material pollution.

## Vermicompost preparation

1. Basic raw material: Any organic material generated in the farm like bhusa,leaf fall etc.,
2. Starter: Cow dung, Biogas slurry, or urine of cattle
3. Soil animal: Earth worms(Species: *Eisenia foetida* )

Favourable conditions of earth worms in the composting material:

- pH: Range between 6.5 and 7.5
- Moisture: 60-70 % of the moisture below and above range moderately of worms taking place
- Aeration: 50 % aeration from the total pore space
- Temperature: Range between 18 $^0$C to 35 $^0$C

**Procedure:** It is mostly prepared in either pit or heap method. The dimensions either heap or pit are 10 x 4 x 2 feet. The length and width can be increased or decreased depending on the availability of material but not the depth because the earth worms' activity is confined to the 2 feet depth only.

1$^{st}$ layer: bedding material of 1" thick with soft leaves

2$^{nd}$ layer: 9" thick organic residue layer finely chaffed material

3$^{rd}$ layer: Dung + water equal mixture of 2" layer.

At the day of 24$^{th}$, 4000 worms are introduced in to the pit [1m$^2$ =2000 worms ] without disturbing the pit by regular watering the entire raw material will be turned into the vermicompost in the form of worm excreta. The turnover of the compost is 75 % [the total material accommodated in the pit is 1000 kg; The out turn will be 750 kg.

**Harvesting of the vermicompost from the pit:** Stop watering before one week of harvest. All the worms spread across the pit come in close and penetrate each other in the form of ball in 2 or 3 locations. Heap the compost by removing the balls and place them in a bucket, then the material is sieved in 2 mm sieve, the material passed through the sieve is called as vermicompost which is stored in a polythene bags [Note: Vermicomposting is done under thatched roof to protect worms against rain and sun]

**Table 3:** Nutrient composition of vermicompost (adopted from Kale (1983) )

| S. No. | Nutrient | Content |
|---|---|---|
| 1 | Organic carbon | 9.15 to 17.98 % |
| 2 | Total nitrogen | 1.5 to 2.10 % |
| 3 | Total phosphorus | 1.0 to 1.50 % |
| 4 | Total potassium | 0.60 % |
| 5 | Ca and Mg | 22.00 to 70.00 m.e / 100 g |
| 6 | Available S | 128 to 548 ppm |
| 7 | Copper | 100 ppm |
| 8 | Iron | 1800 ppm |
| 9 | Zinc | 50 ppm |

Besides the above nutrients the vermicompost also contains following enzymes Enzymes: Protease, Lipase, Amylase, Cellulose. The conversion rates: 1000 earth worms may convert 5 kg waste material per day 1000 worms weighs about a kilogram.

## Advantages of composting over direct application or farm wastes

1. There will be no immobilization in compost because of narrow C:N ratio
2. Application is easy, because the compost is humified and have a structure of crumb and granular.
3. It is hygienic, pathogens and weeds seeds are destroyed.

Advantages of direct application: No loss of nutrients and it improves physical properties better than compost on soil application.

## 11.1.2. Green manuring

Green manuring can be defined as the growth of a crop for the specific purpose of incorporating it into soil while green, or soon after maturity with a view to mproving the soil and benefiting subsequent crops or Practice of ploughing or turning in to the soil un decomposed green plant tissues for the purpose of improving physical condition as well as fertility of the soil

## Objectives of green manuring

i) Increasing organic matter content of soil.
ii) Maintain and improve soil structure.
iii) Reduce the loss of nutrients, perticularly nitrogen.
iv) Provide a source of nitrogen for the next growing crop.
v) Reduce the soil loss by erosion.

**Types of green manuring:** The practice of green manuring is adopted in various ways in different states of India to suit soil and climatic conditions. Broadly speaking, the following two types of green manuringcan are differentiated.

**Classification of green manuring:** classified into two groups as i)Green manure *in situ* and ii)Green leaf manuring

**I. Green manuring *in situ*:** In this system, green manure crops are grown and buried in the same field which is to be green-manured, either as a pure crop or as intercrop with the main crop. This is most common green manure crops grown under this system are sunnhemp(***Crotalaria juncea***), daincha (***Sesabania aculeata***), Pillipesera (***Phaseolus trilobus***) and guar (***Cyamopsis tetragonoloba***). The green manure crops are mostly legumes, which are fast growing and yield substantial succulent vegetation. There is little or no preparatory cultivation. Sowing is effected by broad cast adopting a heavy seed rate.Green manuring can be safely adopted for irrigated and irrigated dry crops viz., rice, sugarcane, tuber crops,vegetables and orchards. In case of dry crops it is unsafe because of limiting moisture. But when rains are sufficient and evenly distributed green manuring could be followed even under rain fed conditions when the rain fall is above 900 mm (Application of lime is suggested to neutralize organic acids that are formed during decomposition (Bone meal preferred). The following are some of green manure crops.

## Green manure *in situ* crops

**Sunnhemp (*Crotalaria juncea*):** Sunhemp is a unique crop possessing,fiber, fodder and green manurial value with nutrient composition of 2.3 % N, 0.2 % P and 1.4 %K. It can be raised beneficially for irrigated dry conditions. Under high rain fall conditions it is grown in dry lands. Grown in medium fertile soils. Seed rate is 45 kg $ha^{-1}$. Green matter yield 9-17 tonnes $ha^{-1}$.

**Daincha (*Sesbania aculeata*) and (*Sesbania speicosa*):** Daincha plants are erect growing deep rooted crops and useful to open soil and improve drainage in heavy soils. Nutrient composition (%) [3.5N, 0.3P and 1.0K]. These crops are grown on heavy soils. They are non fodder crops and non palatable. They correct sodic soils specially *S. spaciosa*as it is less woody and less fibrous, which gives heavy foliage and easily decomposable. Seed rate required is 30 kg $ha^{-1}$. Green fresh matter yield 5 tonnes $ha^{-1}$. Seeds require scarification for easy germination (Scarification means lightly pounding with sand).

**Indigo (*Indigofera tinctoria*):** Slow growing, deep rooted drought resistant crop. It is not relished by cattle. It can be grown in fruit gardens and plantations during non–monsoon season. Recommended seed rate is 20 kg $ha^{-1}$. Green matter yield is 5 tonnes $ha^{-1}$.

**Wild indigo (*Tephrosia purpurea*):** It is suited for hard coarse gravelly textured soil and poor soils. It is used as a green leaf manure also. Self grown crop when sown once. Suitable for unirrigated orchards like mango, sapota. Nutrient composition (%) crop: 1.8N, 0.1 P and 0.3 K; leaf: 3.2 N, 0.1 P and 1.2K.

**Pillipesara ( *Phaseolus trilobus*):** Regular green manure, minor pulse crop and fodder crop ( triple purpose crop). Popular green manure crop for black and alluvial soils. It has good ratooning capacity. The crop could be incorporated in to the soil after two cuttings for fodder. Yield: 3-5 tonnes $ha^{-1}$. Recommended sed rate: 35 kg $ha^{-1}$. Chemical composition (%): 0.3 N, 0.1 P and 0.3K.

**Horse gram (*Dolichus biflorus*):** It is suitable as green manure crop for poor and hard soils. It can also withstands drought. Seed rate is 35 kg $ha^{-1}$ and yield a green matter of 3.5 tonnes $ha^{-1}$.

**Benefits of green manuring (*in situ*)**: Green manure crops can be chosen to suit the soil, season, water facility and cropping pattern. Reduces expenditure on collection and transportation of green leaf. It is easy to incorporate the green manure crop in right time. It reduces the loss of nitrogen from the soil.

**Limitations of green manure crops (*in situ*):** There must be sufficient time available for growing the green manure crop, nearly 2-3 months Extra expenditure has to be incurred for growing green manure crop. Some of the green manure crops are of fodder value, they are liable for cattle tress pass. They are susceptible for pests and diseases as such they may harbour them as alternate hosts. Need timely rainfall or irrigation etc., for growing. Seeds may not be available in time.

## Criteria for selection of green manure crop

1. Capacity to fix atmospheric N in good amounts in symbiosis with micro organisms
2. Plethorcity of vegetative growth (heavy foliage)
3. Succulent vegetation with limited fibrous material
4. Deep root system to open the soil-for recycling of nutrients
5. Short duration with maximum and faster vegetative growth.

**II. Green leaf manuring:** Green leaf manuring refers to turning into the soil green leaves and tender twigs collected from shrubs and trees grown on bunds,waste lands and near by forest areas. The common shrubs and trees used are Glyricidia, Sesbania speciosa, Karanj (Pongamia pinnata ) etc., Plants used as a source of green leaf manure are *Aeschynomene aspera, Azolla filiculoidea, Azolla pinnata, Calotropis gigantean, Cassia auriculata, Cassia siamea, Cyamopsis tetragonoloba, Ipomea carnea, Glyricidia maculate, Leucaena leucocephala, Pongamia glabra, Sesbania rostrata, Sesbania cannabina, Tephrosia purpurea, Tinctonia diversiflora.*

Advantages of green leaf manuring: All the quantity of green leaf applied is entirely an addition to soil –neither the moisture nor nutrients are utilized from the soil. There is no fear of spread of pests and diseases. It can be adopted at any time irrespective of the season

Advantages of leguminous crops as green/greenleaf manures: Legumes have the capacity to fix nitrogen and generally decompose more easily than others. Legumes come up better in poor soils where non –leguminous may fail to put satisfactory growth.

Limitations of green leaf manuring: The green leaf is not available every where except in forest regions and waste lands. Green leaf which ever is available has to be used without choice. Green leaf may not be available sufficient quantity in all seasons. Extra expenditure on collection and transport has to be incurred

Organic materials act as binding agents: Organic materials acts as binding materials for holding soil particles as aggregates. There are there groups of binding agents as identified viz.,

1. Transient binding agents: These are plant and microbial products, rapidly decomposed, having more polysaccharides. They are produced by plant roots as mucigel or by bacteria as extra cellular components
2. Temporary binding agents: Fine plant root and fungal hyphae, mycorrhizal hyphae.They accumulate in soil over a period of time and persist for months or even years
3. Persistent binding agents: Microbially resistant humic polymers derived from decomposition of plant residues, bacterial cells and fungal hyphae associated with Fe and Al in soils.

**III. Penning**: Keeping the boven animals (cattle and sheep) in the fallow land after the harvest of the last crop, throughout day /and or night provided with suitable food and shelter.

Types of penning: There are two types of penning viz., cattel penning and sheep penning

**a. Cattle penning:** Keeping the cattle in the fallow land after harvest of the last crop, throughout day and night by providing them suitable food and shelter is called cattle penning. It is an ideal practice of collection and storage of dung and urine directly in the field. Urine is absorbed by soil while dung and litter are incorporated in situ. Losses in storage and collection are avoided by penning. Cattle are shifted from one field to other field once in 2-3 days for covering the maximum area. The dung and urine are uniformly spread and ploughed in it. The period congenial for cattle penning varies from 3-6 months i.e., form January –June or March to May depending on the extent of non cropping period. Cattle penning is largely done for seedbeds of rice, root crops, sugarcane, vegetable etc., which require very heavy manuring. It is one of the oldest practice followed in India.About 800-1000 cattle per acre are penned which would be equivalent to 10-15 tonnes of cattle manure per acre [Composition (%): 0.5 N, 0.25P and 0.5K].

**b. Sheep penning:** It is a popular practice followed in drier regions of Andhra Pradesh. Flocks of sheep and goats are penned in the fields during nights and allowed to graze during day time.The flocks are frequently disturbed during nights to increase droppings as they tend to void droppings when disturbed.The sheep penning is generally adopted for cash crops like vegetables, tuber crops, chillies and sugarcane. Penning about 2000 heads per acre is the common practice. Sheep and goat manure are relatively lower in moisture content and higher in nutrient value than cattle manure. Average nutrient composition (%) of sheep and goat manure 1.93 N, 0.6 P and 1.90 K. Some of the other organic (bulky) manures are obtained from the solid excretions from piggery, poultry and Human excreta (Night soil)

**IV. Pig manure**: It is collected in dry state stored and extensively used whenever it is available for rice, banana, and vegetables.The percent nutrient composition is 3.7 N,1.4 P and 0.3 K.

**V) Poultry manure**: It has been become popular consequent to the rapid growth of poultry industry. It is used for extensive cropping such as rice, sugarcane and chillies. The per cent nutrient composition of poultry manure is 0.9 N,0.8P and 0.5K.

**VI) Night soil:** Human excrements both of solid and liquid material put together constitute Night soil. It is a rich source of N and P than cattle dung.

**Table 4:** Nutrient value of human excreta

| S.No. | Nutrient | Human excreta nutrients supply(kg /person /year) | Relative values in terms of cattle dung (kg /cow /year ) |
|---|---|---|---|
| 1 | N | 4.7 | 2.9 |
| 2 | $P_2O_5$ | 1.1 | 0.8 |
| 3 | $K_2O$ | 1.0 | 2.3 |

## 11.1.3. Concentrated Organic Manures

Like bulky organic manures, these organic manures contain plant nutrients and considerable amount of organic matter. They have direct effect on plant growth. Nitrogen content is higher in concentrated organic manures and varied from 2.5 per cent in mahua cake to 7.9 per cent in decorticated safflower cake. In addition to nitrogen all the oil cakes contain $P_2O_5$ (0.8 to 2.9 %) and $K_2O$ (1.2 to 2.2%) since they contain nitrogen in relatively large quantities,they are usually classified as organic nitrogenous manures. By virtue of this high nitrogen content they enrich the nitrogen status of soil.

i) **Bone meal:** Bone meal is a white to whitish material produced by treating the bones obtained in abattoirs (Slaughter houses).The bones are dried, crushed, degreased and cleaned to obtain bone grist.Finely ground,it serves as an organic N- P fertilizer. Bone meal is a P-fertilizer of organic origin i.e., it contains $Ca_3(PO_4)_2$. It has 1.0 to 2.0 per cent N and 10-13 per cent P.In general young bones contain less P and more nitrogen than older bones. Bone meal has some residual effect. Rice and other cereals and other orchard crops respond well to its application. It is particularly useful for soils high in Fe and Al content and applied along green manures with advantage prior to sowing or planting. It can be used for crops rather indiscriminately without fear of salt damage (burning ) unlike chemical fertilizers. Deamination is a process of removal of proteins from the grist which yields deaminated bone meal. It is also referred as steamed bone meal

ii) **Horn meal:** Horn powder, horn grist or horn chips can be obtained depending on the degree of crushing and collectively termed as horn meal.This is a slow acting fertilizer of Nitrogen containing 14 per cent N. About 3-4 kg horn and hoof material can be obtained from each animal.Horn and hoof meal manufactured in India annually to the extent of 14,000 tones provided horns and hoofs of all the dead animals are collected and processed. Horn and bone meal mixed fertilizers yield organic N-P depending on their composition (N from Horns and P from Bones).

iii) **Blood meal/ Blood powder:** Blood is collected from abattoirs (slaughter houses) dried and ground. Two types of dried blood are manufactured. i)Red product: It is obtained by drying the blood with super heated steam and hot air. ii)Black product: It is obtained by evaporating the liquid blood over sand bath subjected to higher temperatures.It is evenly causes the loss of N and causes Charring. Both red and black products are spread on the concrete floor covered with a net and allowed to sundry. This powder is used as manure

Blood meal absorbs moisture on exposure to atmosphere. Of all the protein organic manures, dried blood has the highest availability of N and given a rating of 80 (i.e.,) it is 80 per cent as efficient as the inorganic N fertilizer in providing the nitrogen to the crop. An adult bovine (cow, buffalo /ox) gives 15 kg where as sheep or goat yields 1.5 kg of dried blood. About 30-40 kg of dried blood is obtained from 100 kg fresh blood. The principal component of blood is N.Which is 10-14 % in slow acting form.

iv) **Meat meal:** It is also referred as Tankage obtained from rejected carcasses (meat products) mostly meat and waste products such as leather scrap, feather,wool etc. These materials are cooked in steel tanks under pressure of 2 to 7 kg / $cm^2$ for 30-90 minutes. Addition of sulphuric acid (0.5%) facilitates hydrolysis at low temperature. It is dark brown and fluffy material. It contains 7 %N, 1 to 5 % P and 3 to 10 % K. Tankage has the rating of 60 per cent.

v) **Fish meal:** The non-edible fish, fish carcasses and offels (parts of butchered animal) are used to prepare fish meal.Such material are crushed or powdered after drying.The oil is generally removed before the meal is ground and facilitate easy decomposition.It is quick acting fertilizer suitable for all crops on all soils. It contains 7 to 8% N,2 to 3 % $P_2O_5$ and 0.2 to 0.5 % $K_2O$.

vi) **Guano:** The name Guano is originated in Peru, from the word "Hauno" to mean manure. Guano is a product of sea bird (Pelican, Gannets, and Albatrosses) excrement covered over long periods and occurring in natural deposits. These birds live on islands with no rain or vegetation along with the pacific ocean coasts of PERU and CHILE and feed on abundant fish in the sea. These deposits may also have the excrements of turtles and seals together with the remains of dead bodies of birds, bone, feather,sand and gravel etc., guano deposits are up to 60 meters thick, however only central layer has a higher N-content. Guano has a colour varying from grey to dark brown, physical characters vary with the age, nature of deposit and amount of foreign material. It contains 8 to 15 % N and 2 to 3 % P. The chemical constituents are mainly ammonium oxalate and ammonium phosphate as well as calcium phosphates. There are important admixtures besides 2-4 % K. Raw guano is some times processed into guano fertilizer by acid decomposition with sulphuric acid. This is called peru guano:

6+12+2 (N+ $P_2O_5$+ $K_2O$). Guano also occurs elsewhere as "cane fertilizer" produced by bats.

**Oil cakes:** After removal of oil from seeds, the residue is made in to cakes. Oil cakes are used as organic fertilizers as they are rich source of organic nitrogen in protein form. In addition to N, small amounts of P, K and micronutrients. Oil cakes are classified into two groups viz., i) Edible oil cakes: Suitable for cattle and poultry feeding and also as a manure/fertilizer but not economical Eg. Groundnut, Gingelly cakes etc., ii) Non –Edible oil cakes: Suitable for crop fertilization. Eg. Castor cake, neem cake etc. Edible oil cakes serve as fertilizers, but their use is restricted due to economic reasons. Composition of oil cakes are variable. Oil cakes are quick acting organic manures. The decomposability increases with decrease in oil content.They nitrify in about 30-45 days on addition to the soil. The rate of decomposition can be hastened by grinding the oil cakes into fine powder and thorough mixing with the soil.

1) Castor cake: it is also called as castor pomace. It is the ground residue of beans from which oil has been extracted.it is poisonous to animals and used only as fertilizer.it is a by- product in the manufacture of castor oil. It contains 5 to 6 % N, 1 % $P_2O_5$ and 1.0 % $K_2O$. It has got a rating of 75 per cent.
2) Neem cake: Neem cake is prepared by crushing the neem seed (with shells) in expellers and oil is separated. Neem tree is regarded as a "village dispensary" by virtue of its medicinal and antiseptic value. Neem cake is useful for cash crops mainly due to insect repellant or insecticidal properties owing to the presence of residual bitter and sulphur. Comparatively it contains higher N (table5). It cannot be used as a cattle feed due to its bitter taste. Neem cake has also been used as coating material over urea super granules as the former is reported to improve the fertilizer efficiency of soil applied urea.

**Table 5:** Chemical composition of Neem cake

| S.No. | Constituent | Content (%) |
|---|---|---|
| 1 | Organic matter | 84.5 |
| 2 | Moisture | 9.9 |
| 3 | Carbohydrates | 17.5 |
| 4 | Protein | 36.2 |
| 5 | Fiber | 11.7 |
| 6 | Oil | 18.2 |
| 7 | Ash | 6.8 |
| 8 | N | 5.8 |
| 9 | P | 0.46 |
| 10 | K | 1.12 |

## 11.2. Fertilizers

These chemical fertilizers are also called as commercial fertilizers. A commercial fertilizer is defined as a material containing at least one of the primary nutrients in assemble or available form to plants in known amounts.

### 11.2.1. Classification of fertilizers

Fertilizers are broadly classified to two categoties viz., straight fertilsiers, complex fertilsiers and mixed fertilsiers (Table )

#### 11.2.1.1. Straight Fertilizers

Straight fertilizers are chemical substances that contain only one nutrient element in absorbable form by plants. The straight fertilizers include 1) Nitrogenous, 2) Phoshphatic and 3) Potassic fertilizers.

A) **Nitrogenous Fertilizers:** Nitrogenous fertilizers are chemical substances that contain the nutrient element nitrogen in absorbable form by plants chiefly as ammonium ($NH_4^+$) or nitrate ($NO_3^-$) or which yield these from after conversion.

### Origin and reserves of Nitrogen

i) Nitrogen reserves on the earth are abundant. The earth's atmosphere consists of nearly 80 percent of nitrogen by volume.

ii) Deposits of natural nitrates present in vast areas in CHILE and PERU of pacific coast of South America even up to 30 feet.The nitrogen containing salt of these deposits are called chile salt peter.

iii) Bulk of the nitrogenous fertilizers are produced synthetically from atmospheric N via ammonia synthesis.Though plants are surrounded by atmospheric nitrogen ($N_2$).They can not split this highly stable molecule and convert it into utilizable forms under normal pressure and temperature conditions. But this splitting (splitting of $N_2$) can be done only by certain microorganisms in the presence of specific enzymes.

iv) Production of N fertilizers from atmospheric nitrogen requires energy i.e., for producing 1 kg Nitrogen in the form of fertilizer it requires 40,000 Kilo Joules of energy.

**Production of nitrogen fertilizers:** Nitrogen fertilizers are produced by binding atmospheric nitrogen [$N_2$] by any one of the following process. Ammonia synthesis by Haber –Bosch process, Calcium cyanamide synthesis, Nitrate synthesis by arc-gap process. Nitrogen fertilizers are produced essentially via ammonia synthesis with ammonia being further processed in to various nitrogenous fertilizers. The other two processes could not economically compete with the Haber and Bosch process.

**Table 6:** Classification of chemical nitrogen fertilizers

| S.No. | Chemical form of N | Fertilizer | Chemical formula | N (%) |
|---|---|---|---|---|
| 1. | Ammonium [$NH^+$] | a)Gaseous ammonia | $NH_3$ | 82 |
| | | b)Ammonia water | $NH_3$,$NH_4$ OH | 24 |
| | | c) Ammonium sulphate | $(NH_4)_2$ $SO_4$ | 21 |
| 2. | Nitrate [$NO_3^-$] | a) Calcium nitrate | $Ca(NO_3)_2$ | 16 |
| | | b)Sodium nitrate | Na $NO_3$ | 16 |

| S.No. | Chemical form of N | Fertilizer | Chemical formula | N (%) |
|---|---|---|---|---|
| | . | c)Ammonium sulphate nitrate | $(NH_4)_2SO_4+NH_4NO_3$ | 26 |
| 3. | Ammonium + Nitrate $[NH^+] + [NO_3^-]$ | a) Ammonium nitrate | $NH_4\ NO_3$ | 35 |
| | | b)Calcium ammonium nitrate | $NH_4\ NO_3 + CaCO_3$ | 25 |
| 4. | Amide $[-NH_2]$ | a) Urea | $CO(NH_2)_2$ | 46 |
| | | b)Calcium cyanamide | $CaCN_2$ | 22 |

**Note:** For the sake of convenience,the list of chemical fertilizers also include urea, an organic compound which yields mineral N after conversion in the soil.

## Composition and properties of major nitrogenous fertilizers

### a) Ammonia

### Physical properties of ammonia

1. Ammonia is a colour less gas with a pungent odour,when concentrated,it is toxic to humans and plants. But it is a plant nutrient and in dilute form is harmless to humans
2. One kg of ammonia contains 0.82 kg of nitrogen at 10 $^0$C the gas exerts a pressure of 34.10 kg /inch $^2$ while the pressure at 38 0C is 89.50 kg / inch$^2$.It is stored in steel or plastic tanks that can with stand the pressure of 120.4 kg / inch $^2$.
3. Ammonia is handled on liquid under pressure and is released as a gas through an ammonia resistant hose into the soil. Copper and brass fittings reacts with ammonia and should not be used.
4. Ammonia contains 82 per cent nitrogen by weight.99.5 per cent nitrogen fertilizers are made from ammonia. It is the cheapest N-fertilizer to manufacture.
5. Ammonia is soluble in water up to 30% by weight.This gives a low pressure solution containing 24 per cent nitrogen called Aqua ammonia.
6. Ammonia is weak alkali. Ammonia and aqueous ammonia both being alkali (basic),the immediate effect of addition to soil is to raise the soil pH to above 9.0 in the zone of application.
7. Ammonia reacts with clay, organic matter and soil moisture. In warm aerated soil, microorganisms convert the applied ammonia to soluble nitrates after few days. The nitrate ions thus formed combine with $Ca^{2+}$, $Mg^{2+}$ and $K^+$ and moves towards the growing root or percolating water.Thus,the final effect of ammonia is to reduce the basicity of the soil and to increase concurrently the soil acidity. One kg nitrogen applied as ammonia will require 1.8 kg of lime for neutralization.

**b) Ammonium Sulphate $[(NH_4)_2\ SO_4\ ]$:** It is the oldest synthetic fertilizer.The production of Ammonium sulphate is related largely to the production of steel, a coke consuming process.

## Physical properties

1. It is white crystalline salt, but commercial product has light yellow to grey colour with free flowing character.
2. Thiocyanates when present are toxic to plants.
3. No problem in handling and storage,if it contains some powdered material, it cakes (stored in polythene bags )
4. Bulk density of $(NH_4)_2$ $SO_4$ is 876.60 kg /m$^3$

## Chemical properties

1. Ammonium sulphate fertilizer contains 20.6 per cent N and 23.45per cent sulphur (It is a acid producing fertilizer)
2. Soluble in water at ordinary temperatures.Solubility at 0 $^0$C is 70.60g/100 grams of water and at 100 $^0$C is 103.80 grams /100 grams of water.
3. It has free acidity (pH =5.0) of 0.025 per cent by weight.One kg of N applied as Ammonium sulphate fertilizer required 5.1 kg of lime for neutralization

### c) Calcium Ammonium Nitrate (CAN)

**Physical properties**

i. CAN is an easy flowing granular material (size 1 to 4 mm )
ii. Addition of calcium carbonate (lime powder) during the manufacture of CAN improve handling character and reduced the explosive and hygroscopic nature of ammonium nitrate.

## Chemical properties

1. CAN contains 25 per cent nitrogen with equal quantity of each NH4 $^+$ and NO3 $^-$ forms of nitrogen(12.5% and 12.5%)
2. It is a neutral fertilizer and leaves neither acidic nor basic residues on soil application
3. It contains 8.1per cent of calcium and 0.5percent by weight of calcium nitrate
4. It is readily soluble in water Flow diagram:

### d) Urea [Carbamide ($NH_2CO$ $NH_2$)]

Urea or carbamide as it is sometimes called "non-ionic" nitrogen compound used as a fertilizerfor crops and also as protein supplement in the feed of ruminants (farm animals). It is the most important N-Fertilizer constituting nearly 91 per cent of the total production of nitrogen in India. The major reason for its rapid growth is its very high nutrient content (46% N).

## Physical properties

1. White organic compound with low bulk density i.e., 0.7 kg /L
2. It is a solid fertilizer usually granulated to (1 to 2 mm)
3. Having specific gravity of 1.335

## Chemical properties

1. Synthetic protein, organic compound, richest source of N (46%) in amide form among solid N fertilizers
2. Soluble in water and solubility is 100 g /100 g of water at 20°C
3. Biuret content is about 1.5 per cent by weight
4. It is identical to urea found in animal urine.

**Reaction occur when urea is applied to soil:** Conversion of urea in to ammoniacal and nitrate form is complete in about a week. As such it is advisable to apply urea three to four days before sowing of any field crop. Loss of nitrogen by leaching is less. Urea is less acidic compared to ammonia sulphate. Since application of 100 kg urea leaves acidity, which requires 80 kg of $CaCO_3$ to neutralize it.Plants are capable of absorbing most of the nutrients through their leaves. Being very soluble and highly concentrated urea is most suitable fertilizer for spraying in the form of weak solution (3 to 6 %). Urea is easily hydrolyses to ammonia soon after its application to the soil in the presence of urease enzyme by micro organisms and is converted to ammonium carbonate and then by microbial oxidation into ammonium and nitrate and these forms are absorbed by plant. Some of the ammonia and nitrate formed is lost due to leaching, volatilization and denitrification and these losses are responsible for the low recovery of urea by rice which is generally 40-50 per cent or even less.

$$NH_2CONH_2 + 2H_2O \rightleftarrows (NH_4)_2CO_3$$

$$(NH_4)_2CO_3 + 3O_2 \rightleftarrows 2HNO_2 + 3H_2O + CO_2$$

$$2HNO_2 + O_2 \rightleftarrows 2HNO_3$$

Urea being converted in to ammonium carbonate, the following reaction occur in the soil

$$Ca^{++}\ \text{Clay colloid} + (NH_4)_2CO_3 \longrightarrow NH_4^{+}\ \text{Clay colloid} + CaCO_3$$

Calcium carbonate is in soluble in water and therefore its loss is minimum.

### B) Phosphate fertilizers

Phosphate fertilizers are chemical substances that contain the nutrient element phosphorus in the form of absorbable phosphate ions (anions) or that yield such phosphate anions after conversion

**Origin and reserves**: The raw material of P-fertilizers are essentially rock phosphates from phosphate deposits, phosphate ores and other P compounds. The deposits of phosphate rock exists in nature because of the low solubility of the compounds present. The composition of rock phosphates consists of various apatites (Calcium phosphate) which are partly as magmatic and partly as organogenic origin. Weathering and decomposition of primary minerals constitute magmatic origin. Bones, Teeth of animals, Guano deposits results in the formation of apatites constitute organogenic origin of P fertilsiers.

**Deposits of rock phosphate:** Large deposits of rock phosphates occurs in i) North Africa (Morocco, Algeria, Tunisia) in the form of organic crystalline phosphate, especially in a soft earth, finely crystalline form known as gafsa phosphates. ii) USA (eg.Florida apatite) in the form of pebbles. iii)USSR in the form of hard earth, coarsely crystalline kola apatite. In India phosphate rock deposits have been found in udaipur in Rajasthan and Mussoorie in Uttaranchal. Some deposits occur in Singbhum district(Bihar), Jhabhua district (M.P), Visakhapatnam district (AP), Tiruchanapalli (TN).

**Production of phosphate fertilizers:** Mineral P-fertilizers are obtained by chemical treatment or fine grinding of phosphates found in nature.Soft earth rock phosphates requires grinding where as hard earth raw material require chemical treatment. This is accompanied by Thermal methods 2) Chemical methods using $H_2SO_4$ or other acids 3) Combination of above two methods [1 and 2 methods].

**Classification of phosphate fertilizers:** Phosphatic fertilizers were classified by two ways.They are 1)Classification of phosphatic fertilizers based on relative solubility of phosphate, 2)Classification of phosphatic fertilizers based on the form in which orthophosphoric acid or phosphoric acid is combined with calcium.

**1) Classification of phosphate fertilizers based on relative solubility of phosphate**: Based on the relative solubility, the phosphatic fertilizers are classified into following three types viz., A)Water soluble phosphorus fertilizers, B)Water insoluble but citrate soluble phosphorus fertilizers, C)Water and citrate insoluble phosphorus fertilizers.

A) **Water soluble phosphorus fertilizers**: Phosphorus in these fertilizers is present in water soluble form. P is present in the form of monocalcium phosphate Ca $(H_2PO_4)_2$. This form of P is generally regarded as the most readily available to plants.

a) Single super phosphate (16-18% $P_2O_5$)
b) Double super phosphate (32 % $P_2O_5$)
c) Triple super phosphate (46-48% $P_2O_5$)
d) Ammonium phosphate ( 20% N and 20% $P_2O_5$).

These fertilizers are suitable for neutral to alkaline soils and should be applied at the time of sowing. Immediately after application, phosphorus gets converted into insoluble dicalcium phosphate. Hence the P fertilizers containing water soluble P should be applied in granulated form rather than powdered form. Contact between soil and fertilizers should be reduced. Hence pocketing of fertilizers is beneficial than broad casting. Under acidic conditions, water soluble phosphoric acid gets converted into un available iron aluminium phosphates.

B) **Water insoluble but citrate soluble phosphorus fertilizers**: Phosphorus present in these fertilizers is soluble in 2% citric acid or neutral normal ammonium acetate solution. P is present as dicalcium phosphate $Ca_2H_2(PO_4)_2$ / $CaHPO_4$

1. Basic slag (14 to 18% $P_2O_5$)
2. Dicalcium phosphate (34-39 % $P_2O_5$)

3. Raw and steamed bone meal (part of the $P_2O_5$ soluble in citric acid ) – suitable for acid soils and lateritic soils.

The fertilizers of this group are particularly suitable for the acidic soils, because with low pH citrate soluble phosphoric acid gets converted into monocalcium phosphate or water soluble phosphate, and there is less chances of phosphate getting fixed as iron and aluminium phosphate. Contact between soil and fertilizer should be more to solubilize the citrate soluble P present in the fertilizer. Hence they should be applied as broad casting to increase contact with soil.

C)**Water and citrate insoluble phosphorus fertilizers**: Phosphorus present in the fertilizer is not soluble both in water and citrate solution containing insoluble phosphoric acid or tri calcium Phosphate ($Ca_3(PO_4)_2$. Examples, a)Rock phosphate (20 to 40 % $P_2O_5$), b) Raw bone meal (20 to 25% $P_2O_5$), c)Steamed bone meal (22% $P_2O_5$). These fertilizers very well suited for acidic soils or organic soils, which require large quantities of phosphatic fertilizers to raise the soil fertility.They should be applied as broad casting to facilitate intimate contact with soil. They should be applied one month before taking up the crop so that insoluble-P gets solubilized by the time of sowing of crop.

**2) Classification of phosphatic fertilizers based on the form in which orthophosphoric acid or phosphoric acid is combined with calcium.**

The phosphatic fertilizers can be classified broadly into three groups, depending on the form in which orthophosphoric acid or phosphoric acid is combined with calcium. They are: 1. Ortho phosphates Eg. MAP, DAP, UAP,SSP, Nitro phosphates, 2. Polyphosphates Eg. Ammonium poly phosphate, Potassium poly phosphate, 3. Metaphosphates Eg. Ammonium meta phosphate, Potassium meta phosphate

**Available phosphorus:** The sum of the water soluble and citrate soluble phosphorus values is taken as an estimate of the fraction of the total P which is available to plants.

**Total phosphorus:** The total P is the sum of the available and citrate insoluble fractions. It is determined by treating a sample of the original material with strong acids and analyzing the solution for phosphates.

**1. Rock Phosphate:** Rock phosphate is mined from clay deposits that contain phosphorous purified and either directly used as or as raw material for manifacture of P-fertilizers

## 1. Physical and chemical properties

1. The fine powder of phosphate rock is known as float to mean un acidulated ground rock phosphate
2. The mineral in phosphate rock is apatite, has the general formula [$Ca_{10}(PO_4)_6$ ($F,Cl_2(OH)_2,CO_3$]
3. Contain about 10-16 % P and varying amounts of lime (7-10% Ca $CO_3$) and, silica fluoride [3-4% F] iron and aluminum as impurities.

4. It has no practically water soluble phosphates ($10^{-7}$ molar P), but less than 30% of P is soluble in 2.0% citric acid solution.
5. Waste pond phosphates contain 7-9 % P out of which 1 to 2 % phosphate rock is soluble in 1 N neutral ammonium acetate.

**2.Normal or Ordinary or Single Super Phosphate [16%P]:** Super phosphate is a term used in reference to phosphates of which is in a form readily available to plants. Single super phosphate is the oldest artificially produced fertilizer and its manufacture dates back to 1842, when J.B Lawes prepared for the first time in England by treating Rock phosphate with sulphuric acid.

## Physical properties of SSP

1) SSP is in granular form has bilk density 961.10 kg $m^{-3}$ and is easy to handle
2) SSP is also available in powder form, it is not free flowing and being slightly hygroscopic has a tendency to cake. Hydration of monocalcium phosphate may be the cause for hardening SSP.It has grey colour and an acidic odour.
3) Free acid in the SSP, will usually rot the jute fiber bags and hence the fertilizer has to be stored in polythene lined gunny bags or polyethylene bags.

## Chemical properties of SSP

1) SSP has 2/5 Mono calcium phosphate and 3/5 Gypsum by weight
2) SSP manufactured in India consists of two grades viz., Garde I: 16% $P_2O_5$ % or (7%P) by weight of water soluble $P_2O_5$ and Grade II: 14% $P_2O_5$ or (6%P) by weight of water soluble $P_2O_5$.
3) SSP also contains 21.0% calcium, 12% sulphur and traces of micronutrients. Obviously it contains more sulphur (12%) than phosphorus (6 to 7%).
4) SSP has a pH of about 3.0

**3.Triple Super Phosphate (TSP):** Its high plant nutrient content and its capacity to react with $NH_3$ in the production of multinutrient fertilizers. The 'prefix' 'Triple' was first used when SSP contained 16 % $P_2O_5$ and TSP contained three times as much as 48% $P_2O_5$.

## Physical properties of TSP

1) TSP in powdery form is not free flowing, has a tendency to form lumps on storage. However, the granulated product has excellent handling and storage characters and is free flowing.
2) It has bulk density 800-881 kg $m^{-3}$.
3) TSP is to be packed in polyethylene lined jute bags or multi wall paper bags to prevent rotting due to free phosphoric acid.

## Chemical properties of TSP

1) The main phosphate compound present in TSP is monocalcium sulphate.
2) It contains 46% total $P_2O_5$ by weight and 36.8% minimum by weight of water soluble $P_2O_5$.

3) It has 3% free phosphoric acid.

4) TSP also contains 12 to 16% of calcium and 1.0 to 2.0 % of sulphur.

**4) Basic slag [$(CaO)_5$ $P_2O_5$ $SiO_2$]:** Basic slag is a by product of steel industry. It also called as "Thomas slag" named after the inventor of the production process,the English metallurgist, **Thomas (1877).** Millions of tonnes of "slag" have been used as a source of P in European agriculture.

## Physical properties of basic slag

1) Heavy,dark brown powder, ground and pass through 100 mesh sieve.
2) Has relatively higher density than any other fertilizer material.
3) It is likely to solidify when exposed to moisture.

## Chemical properties of basic slag

1) The basic slag is a double silicate of phosphate and lime
2) The slag obtained from Indian steel industry contain only 3-8 % $P_2O_5$.Being low in $P_2O_5$ having highest viscosity, it is not popularly used in Indian agriculture.
3) Being low in $P_2O_5$ it is currently enriched with rock phosphate and marketed as pelophos which on analysis contains 11.0 % citrate soluble and 5.0% water soluble $P_2O_5$.
4) It is alkaline in reaction and has neutralizing effect equal to 70% $CaCO_3$ equivalent.

## C). Potassic fertilizers:

All the fertilizers potassium viz., KCl and $K_2SO_4$ used in India is imported as on today. Recently Geological Survey of India has reported the occurrence of K-deposits in Leh (Jammu and Kashmir) and Bikanir (Rajasthan)which await commercial exploitation. Potassium schoenite a double salt of $K_2SO_4$ and Mg $SO_4$ obtain from salt bitterns is being produced on a pilot scale at Bhavanagar (Gujarat).

**Table 7:** Principal potassium containing minerals

| S.No. | Mineral | Chemical formula | Content of K (%) |
|---|---|---|---|
| 1 | Carnallite | $KCl\ MgCl_2\ 6H_2O$ | 14.1 |
| 2 | Kainite | $KCl\ Mg\ SO_4\ 3H_2O$ | 15.7 |
| 3 | Langbeinite | $K_2SO_4\ 2\ Mg\ SO_4$ | 18.8 |
| 4 | Nitre | $KNO_3$ | 38.6 |
| 5 | Polyhalite | $K_2SO_4MgSO_4 2CaSO_4\ 2H_2O$ | 12.9 |
| 6 | Sylvite | KCl | 52.4 |
| 7 | Sylvinite | KCl NaCl | 29.4 |

**1. Potassium Chloride (KCl) [MOP]:** Potassium chloride is popularly known as muriate of potash [MOP]. The term muriate is derived from muriatic acid, a common name for hydrochloric acid(HCl). It is the most important K-fertilizer used directly or in conjunction with P and N fertilizers.

## Physical properties of MOP

1) MOP in pure form is white crystalline salt. However, colour ranges from white to red based on the impurities present in K minerals and methods of refinement.
2) It has a solubility of 37 grams per 100 grams of water at 30 ⁰C.
3) The crystalline MOP is not very hygroscopic,flows freely and does not cake.
4) Has specific gravity is 1.98.

## Chemical properties of MOP

1) MOP contains 58% by weight of $K_2O$ and about 47 % chloride
2) MOP is neutral in reaction and does not produce acidity and alkalinity on soil application.

**2. Potassium Sulphate ( $K_2SO_4$) [SOP]:** The entire potassium sulphate fertilizer used in India is imported and henceit is more expensive than MOP.

## Physical properties of SOP

1. Potassium sulphate [SOP] is a white crystalline salt, less hygroscopic as compared to MOP and free flowing.
2. It has specific gravity of 2.66 and has a solubility of 13 g/100 grams of water at 30 ⁰C.

## Chemical properties of SOP

1. SOP contains 48 % $K_2$ O and 18 % S by weight.
2. Like MOP, it is also neutral salt providing neither acidity nor alkalinity on soil application.
3. The SOP is widely desirable. The chloride content of KCl effects the burning quality of tobacco.Never use KCl (MOP) on tobacco crop.

## Mode of action of N, P and K fertilizers in soils or reactions of fertilizers in soils

- Nitrogenous fertilizers
- Phosphorus fertilizers
- Potassic fertilizers.

### I. Mode of action of Nitrogenous fertilizers

1. Urea (Refer fate of applied urea in soil)
2. Nitrate ($NO_3^-$) N fertilizers when applied to soil
   i. Nitrate fertilizers raise the pH of the soil.These are superior in highly acid soil
   ii. Nitrate fertilizes are immediately effective and as such are specially suited for top dressing
   iii. These are also suitable for moderately to slightly acidic soils

iv. Loss through leaching of Nitrate from soil are unavoidable but can be limited by precise N-fertilization

3. Ammoniacal N fertilizers [$NH_4^+$]

i) It is soil acidifier. It is suitable to neutral to very slightly alkaline soil.

ii) Application of $NH_4^+$ Fertilizers to alkaline soils (pH =7.5) resulting in losses of gaseous ammonia.For this reason there are generally inferior in alkaline soils.

iii) These are moderately quick in action. It becomes more mobile in the soil only after conversion to $NO_3$

iv) Soil acidification through some N-fertilizers *viz.*, Ammonium sulphate, urea, brings about changes in soil reaction. It is advantageous in soils with high pH,since contain trace elements are better mobilized through acidification.

v) $NH_3$ has fungicidal action against normal fungi in the soil.

In general N fertilizers are very rapid in action in soils.Salt damage to crops occur due to heavy application of N fertilizers, which effects the absorption of water and mineral nutrients. Fertilizers can act effectively only when soil has optimum structure and soil reaction.

## II. Mode of action of Phosphatic fertilizers

i) Water soluble P-fertilizers lead to better P-absorption in the neutral soils, whereas water insoluble P fertilizers are often superior in acid soils

ii) Fertilizers like SSP improves the soil structure because of its gypsum component. It also have a slight soil acidifying action.

## III. Mode of action of Potassic fertilizers

i) Water soluble and as such act rapidly

ii) In the soil K-fertilizers first enters the soil solution, but most of it directly absorbed on the complex and stored as loosely bound K and easily available to plants.

## 11.2.1.2. Complex Fertilizers

**Classification of complex fertilizers:** The complex fertilizers being produced in India can be classified into A)Incomplete complex fertilizes and B) Complete complex fertilizers.

**A) Incomplete complex fertilizes:** Contain only two major nutrients such as N and $P_2O_5$ in chemical combination. They are of two types

i. Ammonium phosphates: May be present in the combination of mono or diammonium salt or mixture of two.Most of these formulations are in granular form. eg. MAP 11:46, DAP18:46, and UAP-28:28:0

ii. Nitrophosphates eg. 20:20:0, 24:24:0

## 1. Mono Ammonium Phosphate [MAP]

**Physical properties**

- Free flowing, non hygroscopic and white to dark grey granular material with excellent storage without caking
- Has specific gravity of 1.8

**Chemical properties**

- MAP contains 11%N and 52% $P_2O_5$ out of which 44% $P_2O_5$ is in water soluble form.
- 10 per cent solution of MAP has a pH of 3.5 to 4.0. Thus it is acid forming fertilizer requires 86 kg of $CaCO_3$ to neutralize acidity developed due to 100 kg of MAP on soil application.

**2. Di Ammonium Phosphate [DAP]:** DAP is a popular fertilizer in India.

### Physical and chemical properties:

- Free flowing and non –caking fertilizer
- It contains 18% Nitrogen and 46% $P_2O_5$ out of which 41 per cent $P_2O_5$ is in water soluble form.
- 10% of DAP solution has a pH of 7.8

**3. Urea Ammonium Phosphate [UAP]:** Urea ammonium phosphate (UAP) is essentially made in order to enrich the nitrogen content of ammonium phosphates. This is popularly called as *Uramphos.* It was first manufactured in India that in Andhra Pradesh at Viskhapatnam in fertilizer Coramandal Ltd.,

### Physical properties

1. Granular, free flowing, good handling and storage properties.
2. Longer exposure to high humidity results in caking.

**Chemical properties**

1. UAP is manufactured in India consists of two grades viz., 28:28: 0 and 24:24: 0 containing N and $P_2O_5$, respectively.
2. UAP of grade 28:28: 0 contains 28% N of which is 9% is ammonical N for and the remaining N is amide form [$NH_2$].It also contain 28 % $P_2O_5$ of which 25.2% is in water soluble form. For grade 24:24: 0,7.5% Nitrogen in ammonical form and 16.5% N in amide form and.Of the 24% $P_2O_5$, 20.4% is water soluble form.

**Nitro Phosphates**

Nitro phosphates are also called as nitric phosphates.They are the fertilizers derived from the digestion of rock phosphate with nitric acid by ammoniation of the resultant slurry.

## Physical properties

1. Nitrophoshates are granular,free flowing fertilizers with good storage quality, provided it is packed in multi wall paper bags
2. Dicalcium phosphate present in nitro phosphates is soluble in normal ammonium citrate solution, but not in water, by using more sulphuric acid in the process, greater portion of water soluble phosphate such as mono ammonium phosphates is obtained.

## Chemical properties

1. Nitro phosphates of grade 20:20 contains 20%N out of which 10.6% is in $NH4^{+}$ form while 9.4 % $NO3^{-}$ form.As for phosphate it has 20% P2O5 in ammonia citrate soluble form out which 5.4 % is water soluble form.

**B) Complete complex fertilizers:** Contains all the three major nutrients in chemical combination.E.g. 17:17:17 and 19:19: 19.

**11.2.1.3. Mixed fertilizer:** Mixed fertilizers are simply physical mixture of solid fertilizers materials containing 2 or more major plant nutrients.Each individual particle of this blend represents only one of the ingredients of the mixture. Bulk blending was started in 1947. It has considerable potential in developing countries that cannot afford plant (factory) facility to manufacture complex fertilizers. Blending provides a convenient and economical means of mixing dry fertilizer materials to produce specified ratios and grades of varying nutrient percentage depending on the needs of the crops. Based on the method of preparation of the BBF is grouped as Farm Mixtures and Mechanical Mixtures

**I. Farm Mixtures:** The manually blended fertilizers are called "farm mixtures". Implying that they can be prepared by the farmers themselves based on the crop requirements.

## Types of fertilizer mixtures

i. **Open formula fertilizers mixtures:** The formula of such fertilizers in terms of kind and quantity of ingredients mixed are disclosed by the manufacturer. Knowing the formula,the farmer or extension worker is able to judge the type and quantity of the ingredients and suitability for specific soils and crops and is able to determine roughly the quantity of filler or other nutrient material contained in any fertilizer.

ii. **Closed formula fertilizer mixtures:** The ingredients or straight fertilizers used in such mixtures are not disclosed.So these mixtures have the disadvantage that the farmer cannot know the type and quantity of the nutrient in them.

## Advantages of mixed fertilizers over straight fertilizers

1. Less labour is required to apply mixtures than to apply its various components separately.This is an important factor in areas where farm labour is scarce and expensive

2. If proper mixture used to suit particular soil type and crop,the use of fertilizer mixture leads to balanced manuring
3. The residual acidity of fertilizers can be effectively controlled by the usage of proper quantity of liming material in the mixtures.
4. Micronutrients, which are applied in small amounts to soil, can be incorporated in fertilizer mixtures.This facilitates uniform soil application of plant nutrients required in small amounts.
5. Mixtures have better physical condition and are more easily applied than many straight fertilizers.This advantage is most marked when mixtures are prepared in granular form.

## Disadvantages of fertilizer mixtures

- Their use does not permit application of individual nutrients which may well suit the needs of the crop at specific times.
- The unit cost of plant nutrients in the mixtures is usually high than these straight fertilizers, because labour is required to prepare mixtures
- Farmers often use mixtures without careful study of their needs leads to economical loss.

## Model calculations formulation of fertilizer mixtures

Model question1.How do you prepare one tonne of the fertilizer mixtures of the fertilizer grade 10:6: 4[N, $P_2O_5$, $K_2O$] by using the following fertilizers and the percent nutrient content against each. Ammonium sulphate 20.5 N. Single Super Phosphate 16 $P_2O_5$. Muriate of potash 60 $K_2O$

Ans: The following are the quantties of individual fertilizers required for the mixture 10:6:4

$$\text{Qunatity of fertilizer required} = \frac{\%\text{nutrient desired}}{\%\text{ nutrient content of the fertilizer}} \times 1000$$

Thus

$N = \frac{10 \times 1000}{20.5} = 487.80$ kg Ammonium sulphate

$P_2O_5 = \frac{6 \times 1000}{16} = 375.00$ kg super phosphate

$K_2O = \frac{4 \times 1000}{60} = 66.70$ kg Muriate of potash

Sum of the quantities of fertilizers = 929.50 kg

[487.80+375+66.70]

Balance to be made good by addition of = 70.50 kg

the filler to obtain one tonne of the mixture

[1000-929.50= 70.50]

--------------

1000.00 kg

--------------

Model question.2:Calculate the quantities of different fertilizers required to prepare one tonne of the bulk blended fertilizers of the grade 6-4-3.The fertilizers available, and their nutrient content are given below. Neem cake- 5%N, Urea 46%N, Basic slag 18% P2O5, TSP 45%P2O5, SOP 50%K2O.

The fertilizer mixed should have the following specified ratios; Neem cake to urea: 1:2 and Basic slag to TSP: 1:3

Total N required for the mixtures : 6 N

N required through neem cake $\frac{6 \times 1}{3} = 2$ N

Urea

N required through urea $\frac{6 \times 2}{3} = 4$ N

Quantity of neem cake required = 2 x 1000/5 = 400 kg

Quantity of urea required= 4 x 1000/46 = 87 kg

Total $P_2O_5$ required for the mixture = 4 $P_2O_5$

So through basic slag: 4 x(1/4) = 1 $P_2O_5$

Through TSP = 4 x (3/4) = 3 $P_2O_5$

Quantity of basic slag required: 1 x 1000/18 = 55.5 kg Quantity of

TSP required: 3 x (1000/45) = 66.70 kg

Quantity of SOP required: 3 x (1000/50) = 60 kg

Sum of the quantities [400+87 +55.5 + +66.67 +60 ] = 669.20

Filler material required: [1000-669.20] = 330.80

Note the sum of three parts of the mixture calculated exceeds 1000 kg of mixture of the composition can not be prepared from the fertilizers in question

## 11.2.2. Unit Value of Fertilizers

The term unit value which refers to the cost per unit of the plant nutrient contain in a given fertilizer. The calculation of unit value of different fertilizers that would supply same kind of plant nutrient will enable the farmer to choose the cheapest fertilizer in terms of cost. This concept was first adopted in England,

## Model question 1

Nitrogen fertilizer available in the market are mentioned below along with their nutrient content and cost per tonne.Calculate the unit cost of nitrogen in each case to find out the cheapest N-fertilizer. Note:Ammonium sulphate (20.5%N) cost Rs.8000/ tone; Ammonium chloride (25%N) cost Rs.7000/ tone; Calcium ammonium nitrate (25%N) cost Rs.8000/ tone; urea(46%N) cost Rs. 6000/tone.

$$\text{Unit value} = \frac{\text{Cost of one tonne of fertilizer}}{\text{Concentration of plant nutrients per tonne of fertilizers}}$$

| | |
|---|---|
| Ammonium sulphate | 8000/ 205 = 39.02 |
| Ammonium chloride | 7000/250 = 28.00 |
| CAN | 6000/250 = 20.00 |
| Urea | 6000/460 = 13.00 |

Among the above nitrogen sources, urea will be the cheapest fertilizer based on unit value. In case of multi-nutrient fertilizers, difficulty in the unit value calculations arises as all the plant nutrients contained in the complex fertilizer should be assessed in such cases, there is a convenient method where in the unit values are calculated from the prices of standard fertilizers viz., 1)Ammonium sulphate 2) Single super phosphate 3) Muriate of potash.

## Model question 2

Find out which among the complex fertilizers and mixed fertilizer if prepared is economical when a complex fertilizer of graded 20:20:20 is purchased at a cost of Rs.5500 /- per tonne. Given unit value of mixed fertilizer

a) ammonium sulphate: Rs. 11.00 per kg per kg of N
b) SSP : Rs 11.50 per kg of P2O5
c) S.O.P : Rs 3.50 kg

| | | |
|---|---|---|
| 20 N @of Rs 11.00 a unit | =11 x 20 | = RS.220.00 |
| 20kg $P_2O_5$@ Rs.11.50/kg | = 11.50 x 20 | =RS.230.00 |
| 20 kg $K_2O_5$ @ Rs.3.50/kg | = 3.50 x20 | =RS. 70.00 |
| Cost of 100 kg of mixed fertilizer | | =Rs. 520.00 |
| Cost of one tone of mixed fertilizer is | | =Rs. 5,200.00 |
| complex fertilizer per tone is=Rs.5,500.00 | | |

So at normal rates the cost of mixed (prepared) fertilizer should be Rs.5,200/- per tonne where as the cost of complex fertilizer price offered at market per tonne is Rs.5500/- so an amount of Rs.300/- higher price than mixed fertilizer (prepared).

**Model question 3**

A rice crop requires 100 kg N, 50kg $P_2O_5$ and 50 kg $K_2O$ per hectare. Calculate the amount of urea, SSP, MOP required for application to one hectare.The contents given [46 % N in urea, 16 % $P_2O_5$ SSP and 60 % $K_2O$].

$$\text{Amount of fertilizer required} = \frac{\text{Nutrient desired}}{\text{\% nutrient content in fertilizer}} \times 100$$

Urea required = 100/46 x 100 = 217.40 kg SSP required

= 50/60 x 100 = 312.50 kg

MOP required = 50/60 x100 = 83.40 kg

**Model question 4**

Calculate the amount of DAP,Urea and MOP required to a ground nut crop of one hectare to supply 40 kg N,40 kg $P_2O_5$ and 50 kg $K_2O$.

Contents: {DAP- 18 % N and 46 % $P_2O_5$, Urea – 46 % N and MOP-60 % $K_2O$}

Quantity of DAP required t supply 40 kg $P_2O_5$ = 40/46 x100 = 87 kg

Note: incidentally it also supplies nitrogen also from 87kg of DAP

Therefore 87 kg DAP supplies = 18/100 x87 = 15.70 kg of N

The amount of N to supplied through urea is 40-15.70 = 24.30 kg N

So urea required to be supply 24.30 kg N is = 24.3/46 x 100 = 53 kg urea

MOP required to supply 50 kg $K_2O$ = 50/60 x 100 = 83 kg of MOP.

So total quantity individual fertilizers required to supply 40 kg N, 40 kg $P_2O_5$ and 50 kg $K_2O$ are: 87 kg DAP, 53 kg Urea and 83 kg MOP.

### 11.2.3. Soil amendments

Soil amendments are substances that influence the plant growth favorably by producing in the soil one or more of the following beneficial effects. Changing the reaction,that is making the soil less acidic or less alkaline; Changing the plant nutrients in the soil from unavailable to available forms; Improving the physical conditions of the soil and Counteracting the effects of injurious substances. Soil amendments usually contain plant nutrients also. Agricultural liming materials, for example, supply calcium and, sometimes magnesium as nutrient element.

**Types of soil amendments:** Mainly there are three types of soil amendments

1. Substances for correcting acidic soil
2. Substances for correcting alkaline soils ; and
3. Soil aggregating agents or soil conditioner s to stabilize soil aggregates and to form granular structure.

## 11.2.4. Secondary Nutrient Fertilizers [Ca, Mg and S]

**i) Calcium Fertilizers:** Nearly all multi nutrient liquid fertilizer formulations and more than 50 per cent multi nutrient solid fertilizers are almost devoid of calcium. Multi nutrient fertilizers use, warrants the use of calcium. The source of calcium with basic chemical formula, its content and solubility are mentioned in the following table.

**Table 8:** Basic mineral and fertilizer sources calcium

| S.N | Source | Formula | Ca (%) | Water solubility g /100 g at 25 °C |
|---|---|---|---|---|
| 1 | Burnt lime | $CaO$ | 70 | 0.12 |
| 2 | Hydrated lime | $Ca(OH)_2$ | 50 | 0.16 |
| 3 | Calcite lime | $CaCO_3$ | 36 | 0.01 |
| 4 | Dolamitic lime | $CaCO_3$ $Mg CO_3$ | 17 | >0.1 |
| 5 | Basic slag | $[CaO]_5$ $P_2O_5$ $SiO_2$ | 29 | >0.1 |
| 6 | Gypsum | $CaSO_4$ $2H_2O$ | 22 | 0.24 |
| 7 | Calcium nitrate | $Ca(NO_3)_2 2H_2O$ | 20 | 100 |
| 8 | Single Super phosphate | $Ca(H_2PO_4)_2$ $CaSO_4$ | 20 | 1.0 |
| 9 | Triple super phosphate | $Ca(H_2PO_4)_2$ | 13 | 1.80 |
| 10 | Rock phosphate | $Ca_5(PO_4)_3F$ | 33 | 0.002 |
| 11 | Calcium chloride | $CaCl_2$ | 36 | 100.00 |

Of the different sources, the Gypsum has been gaining importance for crops such as legumes. Global reserves of Ca are considerably large since whole mountain ranges consist of lime stone. Calcium nick named as root developer, which is slightly mobile in plants. Calcium deficiency in plants is rarely caused by shortage of available reserves in the soil, except in acidic soils.

**ii) Magnesium Fertilizers:** In multi nutrient fertilizers, finely ground dolamitic limestone is used as filler and it is a incidental supplier of Mg.

### Causes for Mg deficiency

i. Soils subjected to intensive NPK fertilization, reveal Mg deficiency
ii. Due to increased crop removal
iii. Due to leaching losses especially in acid soils
iv. Reduced supply of Mg through multi nutrient fertilizer
v. Antagonistic decrease in Mg uptake due to large K supply

**Table 9:** Important mineral Sources of Magnesium:

| S. No | Source | Formula | Mg (%) | Water solubility g/100 grams at 25 °C |
|---|---|---|---|---|
| 1 | Magnesium oxide | $MgO$ | 45.0 | $6.2 \times 10^{-5}$ |
| 2 | Dolamite | $CaCO_3$ $MgCO_3$ | 12.0 | 0.032 |
| 3 | Kiserite | $MgSO_4$ $H_2O$ | 18.2 | 68.40 |
| 4 | Langbeinite | $K_2SO_4$ $2MgSO_4$ | 11.2 | 100.0 |
| 5 | Magnesium sulphate | $MgSO_4$ $7H_2O$ | 10.5 | 91.10 |

**Sulphur fertilizers:** Plants take sulphur in the form of sulphate [$SO^{2-}$] ion. Sulphur fertilizers predominantely contain sulphate, some of which are easily soluble and some are slightly soluble. Gypsum [$CaSO_4$ $2H_2O$] is a calcium sulphate because its slight solubility in water which is slow acting. Elemental sulphur is also an important sulphur fertilizer with strong acidifying action.It can be used either directly or as an additive to other solid fertilizers eg sulphur coated urea.The following table gives the basic chemical data of sulphur sources.

**Table 10:** Important mineral and fertilizer sources of Sulphur

| S.No | Source | Formula | S (%) |
|---|---|---|---|
| 1 | Ammonium sulphate | $(NH4)_2$ $SO_4$ | 23 |
| 2 | Potassium sulphate | $K_2SO_4$ | 18 |
| 3 | Magnesium sulphate | $MgSO_4$ | 13 |
| 4 | Super phosphate | $Ca(H_2PO_4)_2$ $CaSO_4$ | 12 |
| 5 | Gypsum | $CaSO_4$ $2H_2O$ | 18 |
| 6 | Aluminium sulphate | $Al_2(SO_4)_3 18H_2O$ | 14 |

The sulphur requirements of plants are approximately 2/3 of their phosphorus requirements and are provided from various sources such as air, rain water,soil and fertilizer.Fields near Industrial zones are supplied with 10-30 kg $ha^{-1}$ per year from $SO_2$ waste gases.

## 11.2.5. Micronutrient Fertilizers

Higher green plants are known to require seven micronutrients viz., Fe, Mn, Cu, B, Mo, Cl etc. But this number may have to be increased in future.

### The need for the micronutrient fertilization has been increasing due to the following reasons

1. **Change in the plant types:** Change in the plant varieties from traditional to high yielding varieties (HYV). HYV have a capacity to remove more nutrients both major and minor. HYV have low mobilizing capacity of micro nutrients from the soils hence they are to be applied through external application.
2. **Changes in soil conditions**: Change in soil condition from acid to alkaline(Increased soil reaction) and aeration cause greater immobilization of most micronutrients except molybdenum (Fe, Zn, Cu and Mn). High acidic nature of the soil induces both calcium and magnesium deficiency.
3. **Changes in fertilization:** Using high doses of NPK in the form of complex fertilizers, induces deficiencies of secondary and micro nutrients.
4. **Antagonistic action:** It is due in part to excessive fertilization with NPK Eg. Higher K content in soils effects the uptake of Fe, Mn, and B. Excess P leads to deficiencies of Zn, Fe, Cu and increases Mo availability. More the sulphate present in the soils, lesser the availability of Mo.

5. **Changes in overall growing conditions:** Intensive cultivation leads to micro nutrient deficiencies. Non availability and application of organic manure. Under/ over limed condition.
6. **Parent material:** Most of the micronutrients originates from parent material and influences the availability or deficiency of micronutrients to plants
7. **Land levelling and shaping:** Most of the micronutrients are concentrated on the surface soil except Molybdenum. Levelling of land and deep tillage operations leads to deficiencies.
6. **Low Si/Mg ratio:** Leads to fixation of Zn in soil.
7. **Calcium carbonate:** More the $CaCO_3$ in soils lesser the availability of Fe, Cu, Mn, Zn.
8. **Soil texture:** Boron availability is more in coarse textured soils and in finer textured soils it gets fixed, unavailable to plants.
9. **Secondary clay minerals:** Montmorillonite clay adsorb more of Zn, Cu and leads to deficiency or unavailability to plants.
10. **Soil moisture:** Dry conditions of soil fix more of Boran and is released under wet conditions. Presence of more moisture reduces the availability of Mn.
11. **Interaction with macronutrients:** Heavy nitrogen interferes with availability of Cu, Mn, Zn and leads to deficiencies in plant. Liming reduces the availability of Mn, Zn, Fe, Cu.
12. **Seasonal variation:** Nutrient deficiencies are more during cold season. Out break of B deficiency is common in dry seasons. Summer drought aggravate Fe chlorosis. High and low soil temperatures induces Zn deficiency in soils having low Zn.

**Table 11:** Fertilizer sources of micronutrients

| S.No. | Micronutrients | Formula | Content (%) |
|---|---|---|---|
| **A** | **Sources of iron** | | |
| | 1. Ferrous sulphate | $FeSO_4\ 7H_2O$ | 20 |
| | 2. Fe-chelate | Fe-EDTA | 5 |
| | 3. Fe-Chelate | FeEDHA | 6 |
| **B** | **Sources of** manganese | | |
| | 1. Manganous sulphate | $Mn\ SO_4\ 4\ H_2O$ | 24 |
| | 2.Manganous sulphate(Monohydrate) | $Mn\ SO_4\ H_2O$ | 32 |
| | 3. Mn -chelate | Mn-EDTA | 13 |
| **C** | **Sources of zinc** | | |
| | 1. Zinc sulphate | $ZnSO_4\ 7H_2O$ | 23 |
| | 2. Zinc sulphate (Monohydrate) | $ZnSO_4\ H_2O$ | 36 |
| | 3. Zn-chelate | Zn-EDTA | 14 |
| **D** | **Sources of copper** | | |
| | 1. Copper sulphate | $Cu\ SO_4\ 5H_2O$ | 25 |
| | 2. Copper sulphate (Monohydrate ) | $Cu\ SO_4\ H_2O$ | 36 |

| S.No. | Micronutrients | Formula | Content (%) |
|---|---|---|---|
| E | **Sources of born** | | |
| | 1. Borax (Na-tetra borate ) | $Na_2B_4O_7$ $10H_2O$ | 11 |
| | 2. Borax anhdrous | $Na_2$ $B_4O_7$ | 22 |
| | 3. Boric acid | $H_3BO_3$ | 18 |
| F | **Sources of molybdenum** | | |
| | 1 Sodium molybdate | Na2$MoO_4$ $2H_2O$ | 40 |
| | 2. Ammonium molybdate | $(NH4)_6MoO_{24}$ | 54 |
| | 3. Molydenum trioxide | $MoO_3$ | 66 |
| | 4. Calcium molybdate | $CaMoO_4$ | 48 |

## 11.3. Quality Control of Fertilizers

Laws and regulations governing the manufacture and sale of fertilizers are imperative in order to check. Spurious standards and adultrated fertilizers entering in to the market. To ensure quantity of nutrients and quality of carriers present in the fertilizer. To ensure quantity of nutrients and quality of carriers present in the fertilizer. To eliminate black marketers off the market.

The total N,$P_2O_5$ and $K_2O$ must be guaranteed in terms of percentage of each of these nutrients in a given fertilizer.The oxide expression for P and K are actually in accurate and confusing as they are based on early practices with which chemists determined the elements by ignition and weighing the oxides.The current practice is to express the elemental concentration of these nutrients By considering the above said objectives, the government of India passed The Fertilizer Control Order [FCO] on 28th of April, 1957 in exercise of the power conferred by the section III of the essential commodities act of 1955. This order is intended to regulate the manufacture, distribution and supply of the fertilizers in India at a control cost.This has been effective from May 18th 1957.It is revised in 1985 with effect from 25-9-1985. The Government of India has delegated to powers to state Governments to implement the order. The Government of India (GOI) also passed the Fertilizer Movement Order (FMO) on 31st December,1960 in order to regulate the interstate movement of fertilizers and the export of fertilizers which came into force with effect from 1-1-1961

### 11.3.1. Fertilizer Control Order [FCO] regulations

1. All the fertilizer manufacturer should obtain licence from the Commissioner of Agriculture, state Government concerned for the manufacture of fertilizer and mixed fertilizers
2. The fertilizer dealers should on renewable basis, register their dealership with the Assistant Director of the Agriculture (ADA) Regular of the division concerned in a state.
3. The terms and conditions of manufacture,distribution and sales imposed by the government should be followed.
4. Duties of inspecting officers and the dealers are specified.

5. Fertilizer specifications and kind of package are stated
6. Method of drawing fertilizer samples for analysis in the fertilizer testing laboratories is stated.
7. Powers are vested with the FCO enforcing officials to book the cases against the fraudulent manufacturers, distributors and dealers of fertilizers.

## 11.3.2. Specifications for important fertilizers [As per FCO, 1957]

**Urea:** Moisture per cent by weight 1.0 Maximum, Total nitrogen per cent by weight 44.0 minimum, Biuret per cent by weight 1.50 maximum, Particle size:In the form of granule the material shall pass through2.8mm and not less than 80% by weight shall be retained on 1mm

**2. SSP:** Moisture per cent by weight maximum 12%, Free phosphates as ($P_2O_5$) percent by weight 4.0, Water soluble phosphates (as $P_2O_5$) by weight maximum 16.0.

**3. MOP:** Moisture per cent by weight maximum 12%, Water soluble potassium per cent by weight minimum 60.

**4. DAP:** Moisture per cent by weight 1.0%. total nitrogen per cent by weight minimum 18%, total phosphates (as $P_2O_5$) per cent by weight minimum 46.0 Water soluble phosphates (as $P_2O_5$) per cent by weight 41.6

## 11.4. Biofertilisers

**Bio-fertilizers** or **bio-inoculants** are productions containing living organisms such as nitrogen fixers or phosphorus solubulizers which are useful for agricultural production. Biofertilizers play a significant role in improving soil fertility and plant growth. A)by fixing atmospheric nitrogen both in association with plants and without plants. B)solubulize insoluble soil phosphates into soluble phosphates there by increasing the availability of phosphorus. C)Secrete growth promoting substances and supplies to the inoculated plants.

**Classification of Biofertilisers:** Classified the bio fertilizers based on the nutrient provided, which is as follows.

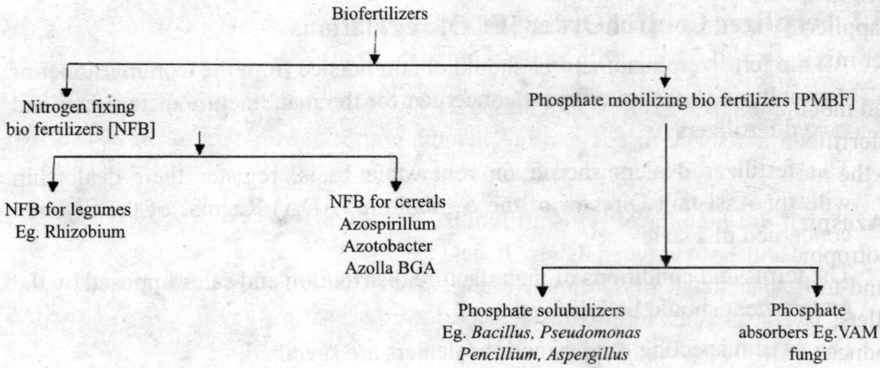

1) **Rhizobium inoculant**: Belongs to the family Rhizobiaceae and symbiotic in nature. *Rhizobium* establishes symbiotic relations with pulses and other legumes and can fix 20 to 70 kg N $ha^{-1}$ by which application of chemical nitrogen fertilizers could be reduced and supplemented nearly 30-50 per cent of nitrogen requirement of crop is met by nitrogen fixation in root nodules of legume crops. It is used for pulse legume crops like chick pea, red gram, chickpea, lentil, black gram etc., and oil seed legumes like soybean and groundnut and forage legumes like besrseem, lucerne and pillipesara. On average yield increase was 15-30% due to rhizobium inoculation. The rhizobia show host specificity to some extent. In other words rhizobia isolated from ground nut can not be used for red gram. Particular rhizobium species has the ability to form nodules only on roots of limited and related plant species. Again effective strains are to be selected for effective nodulation. The response of inoculation is governed by 1. Host plant 2. Strains effectiveness 3. Soil type and 4. Rainfall

Rhizobum biofertilizer is available commercially as carrier based (lignite or peat ) biofertilizers. Direct application of rhizobium to seed is the most common form of legume inoculation (seed coat). However, for groundnut,slurry of peat based inoculam in the seed furrows is recommended as the seeds are fragile. The recommended quantity of rhizobium fertilizer is 2-3 packets / ha (packet: 200 grams). Liquid refom rhizobium biofertilizer also being prepared and matured recommnded @ 750 to 1000 mL/ha. Liquid forms have more storage life.

**2. Azotobacter inoculants***:* Azotobacter belongs to the family Azotobacteriaceae, which is a chemoautotrophic in nature,free living, non-symbiotic in nature and is recommended for application to non-legume crops. It thrives well in many types of soils,with a minimal dose of phosphorus and carbon compounds. It grows best in neutral to alkaline soils.By the application of Azotobacter to various crops,the amount of recommended doses of nitrogenous fertilizers can be reduced by 10-20%. It produces plant growth promoting substances like vitamins of B group, Indole acetic acid,Gibberellic acid by which plants grow quickly and healthy. This biofertilizer is recommended for rice, wheat, millets and other cereals,vegetables, sunflower, mustard, tobacco sugarcane,cotton etc., and also fruit crops.

**Method of usage:** It is carrier based biofertilizer (lignite / peat). Azotobacter can be applied by seed inoculation, seedling inoculation, or by broad casting in the field after mixing with FYM or with well decomposed compost.

Seed inoculation: Seed is treated with Jaggery to make it sticky and then treated with biofertilizer and shade dried.Seedling inoculation: Make slurry of biofertilizer and dip the roots in it. The recommended dose of Azotobacter is 2-3 packets /ha.

**3. Azospirillum inoculatn:** Azospirillum belongs to the family Spirillaceae, chemo-autotrophic and associative in nature. It is an associative nitrogen fixing organism was found to be beneficial for sorghum, wheat, maize, barley, fodder grasses and minor millets. By the application of this fertilizer nitrogen saving is 20-30 kg /ha.It also produces growth regulating substances.

**Benefits of Azospirillum Application**: Azolla biofertilzer application results in increased mineral uptake by crop platns, increased water up take by plants, increased root development, increased vegetative growth and imparts the plant drought resistant. **Note**: Principle and mode of application is similar to that of Azotobacter.

**4. Azolla** [Botanical name is ***Azolla pinnata***]: It belongs to the family Azollaceae. Azolla is a water fern.It is considered as aquatic weed commonly found in floating in idle ponds, tanks shallow ditches and canals. Azolla is associated with rice field. A blue green algae (BGA) *Anabaena azollae* living in the epidermal cavity of lower side of the leaf of Azolla. The symbiotic association of *Azolla pinnata* and *Anabaena azollae* is termed as Azolla Anabaena complex.This algae fixes atmospheric nitrogen for Azolla and in exchange, the plant provides shelter and food to the algae. As this complex fixes atmospheric nitrogen, it has great potentiality for use, in agricultural field as biofertilizer and can be used as an alternative to nitrogen fertilizers. It is recommended as green manure because of its large biomass and high N content( 4-6%N) for sub merged rice fields, with in maximum temperature of 38°C. The yield increase is 15-20%. Fresh Azolla contains 90-95 water.

**Table 12:** Nutrient concentration of Decomposed Azolla.

| S.No. | Nutrient | Per cent (%) |
|---|---|---|
| 1 | Nitrogen | 4 to 6 |
| 2 | Phosphorus | 0.5 to 0.9 |
| 3 | Potassium | 2 to 6 |
| 4 | Calcium | 0.4 to 1 |
| 5 | Magnesium | 0.5 |
| 6 | Manganese | 0.11 to 0.16 |
| 7 | Iron | 0.06 to 0.16 |
| 8 | Total ash | 9 to 10 |

**Ideal conditions for growth and multiplication of Azolla:** Depth of water should be maintained at 8-10 cm in the field. Temperature in between 14 to 35 °C. Optimum for quick growth and multiplication is 20 to 28 °C. Suitable soil pH for azolla growth is 6.0 to 8.0. Single Super Phosphate should be applied at the rate of 20-25 kg $ha^{-1}$, should be applied. Furadon granules may be applied to control leaf eating insects and snails.

**Methods of growing Azolla:** Green manuring, Dual cropping, Compostingand Just growing.

**Green manuring:** When there is standing water in the field, Azolla is grown by spreading 200-500 kg /ha fresh Azolla.After 25 days of the whole field will be covered by Azolla, if condition is ideal. Green Azolla is incorporated by ploughing after dewatering in the field. This will add 30- 40 kg of N fertilizer /ha.

## Dual cropping

- Azolla nursery is prepared simultaneously with the paddy nursery.
- The area of Azolla nursery should be half that of paddy nursery (seed bed).
- The Azolla nursery should be kept free from other aquatic weeds and there should be 8-10 cm standing water in the Azolla nursery.
- When the paddy seeds are sown in the paddy nursery fresh Azolla is spread in the Azolla nursery @40-80 grams per square meter.
- After 25 days the Azolla will increase an extent of 60 times and super phosphate is applied @25 grams/Sq. meter/week.
- Within 25 days the paddy seedlings will be ready for transplanting. Before transplantation of paddy seedlings, there will be standing water in the main field.
- Azolla is taken out from the Azolla nursery and spread in the transplanted field @ 200- 500 kg /ha.
- Single super phosphate @25-50 kg /ha /week is applied for 3 weeks and water level is maintained 5-10 cm.
- After another 25 days, the whole field may be covered by Azolla. Azolla is incorporated after dewatering the field.This will add 30-40 kg N /ha besides another nutrients.

**Composting:** Azolla is grown in permanent Azolla nursery. The Azolla is collected every week from the nursery and kept in layers on the side of nursery. Within 10 days, the Azolla will decompose and the azolla compost can be used in any crop.

**Just gowing azolla:** Azolla when grown in one hectare of land throughout the year and ideal conditions are maintained it can fix 800 kg of nitrogen per hectare. Azolla is suitable for only flooded rice and fixes up to 40-80 kg N/ha. Symbiotically with *Anabaena azollae*.

**Advantages of azolla use:** It contains higher nitrogen content (4 to 6%). Release of nitrogen is easy. There is no leaching loss. Releases nutrient slowly. Weed interference is less when grown as dual crop

**5. Blue green algae or Cyanobacteria:** Blue green algae (BGA) is phototrophic in nature. These are photosynthetic, prokaryotic microorganisms and capable of fixing molecular nitrogen. Blue green algae are found to grown in a paddy filed and marshland. It is used as biofertilizer for irrigated rice crop. The rice field ecosystem provides an environment favorable for the growth of BGA with respect to their requirements viz., light, water and temperature. Nitrogen fixation by BGA have been estimated to be 25-30 kg N/ha. BGA has some species viz., Nostoc, Anabaena, Olocia, Calothrix,Tolipothix etc. They fix nitrogen in the presence of sunlight.They also contain growth promoting substances like Auxin, Indole acetic acid and Gibberellic acid.

Recommended dose of application: application of 10 kg /ha flakes of BGA. It is applied 10 days after the transplanting of rice crop.

**BGA production technology:** Mainly two methods are followed, viz., Trough method and, Pit method.

**Trough method:** Take 3 m x 2 m x 20 cm size (L x B x h) galvanized iron sheet trough or to the convenient size. Add about 8 to 10 kg of soil and well mixed with 200 grams of SSP. The soil pH should be about neutral. Add water to the trough 5-15 cm (height) to protect from insects, and add Carbofuran 3.0 G at the rate of 25 grams / tray. Sprinkle the algal culture on the surface of the standing water and keeping it in air (open) and exposed to sunlight. In hot summer months, the growth of algae will be rapid in about 7 to 10 days and form a thick mat, where algal growth becomes sufficiently thick stop watering and allow the water to evaporate completely,the dry algae cracks in to flakes. Collect the dry algal flakes from the trough and store in bags for the use in the field. One trough gives 1.5 to 2.0 kg mats.

**Pit method:** Adopted by number of farmers in south India. Demark the area in the field for algal production. Flood the area with waters to a depth of about 12.5 cm. Apply SSP @12.0 kg /40 m$^2$ area Apply the algal culture. In clayey soils good growth of algae takes place in about two weeks in clear sunny weather while in loamy soils it takes about 3-4 weeks. Once algal growing mats are formed, they are allowed to dry up in the field and dried algal flakes are thin collected and stored in bags. BGA benefit the rice crop up to 25-30 kg N /ha. apply BGA @10-15 kg /ha after transplanting in rice.

**6. Mycorrhizae:** Mycorrhizae means symbiotic association of fungi and roots of higher plants. It is also known as vesicular arbuscular mycorrhizae (VAM). These are obligate symbiotic fungi. They are ubiqutous in nature and with very few exceptions found on a wide range of plant species. They form vesicles and arbuscules hence called of vesicular arbuscular mycorrhizae (VAM). VAM fungi infects and spreads inside the plant root by providing an aseptate mycelium. Vesicles which are sac like structure provided by these fungi are known to store phosphorus as phospholipids. Arbuscules formed inside the cortical cells are attributed to help in the exchange of nutrients. Common genera that form VAM associations are Glomus, Gigaspora, Enterospora etc.

**Benefits of VAM:** VAM fungi enhancing the nutrient uptake of P, Zn and S. Enhances the water uptake under drought conditions. It enhances resistance to root damage by biotic and abiotic stress. It is recommended for forest trees, forage grasses, maize, millets, sorghum and barely and also leguminous crops. Mycorrhizae can substitute up to 560 kg phosphorous fertilizer per hectare for citrus.

**Constraints in usage of VAM fungi as biofertilizer:** Farmers acceptance has been far from satisfaction in use of VAM fungi as biofertilsier. The reasons are, it is difficult to demonstrate striking increase in yield due to biofertilizers unlike chemical fertilizers. VAM strains fail to establish if the soil moisture in the surface is not sufficient. Fungicide treatment to seed is a problem in crops like groundnut when we go with VAM because VAM is susceptible to seed treatning funcicides. Availability of quality of inoculants is another problem. Seed inoculation is not suitable when seeds are sown on dry soil conditions.

## 7. Phosphate solubilizing microorganisms

Phosphorus is present in soils in both organic and inorganic forms and most of which is insoluble and unavailable to plants. Several soil bacteria particularly those belonging to the genera Pseudomonas and Bacillus and fungi belonging to the genera Pencillium and Aspergillus possess the ability to bring insoluble phosphates in the soil to the soluble forms by secreting organic acids such as acetic acid,propeonic acid etc., These acids lowers the pH and bring about the dissolution of the bound phosphate forms.

## Objective questions

1. The word "Manoeuvrer" which refers to "work with soil" is derived from
   a) Latin b) Greek
   c) French d) Arabic word
2. The word "Tung" derived from ----- language, meaning storage pit covered with manure for protection of soil against cold
   a) Arabic word b) Greek word
   c) Latinword d) German
3. The organic substances with low analytical value, bulky in nature, slow acting and imprive soil physical condition are termed as
   a) Conditioners b) fertilsiers
   c) amendmants d) Manures
4. The inorganic substances with high analytical value, quck acting and doesnot affect soil physical condition are termed as
   a) Manures b) fertilsiers
   c) amendmants d) Conditioners
5. The process of increasing the productive capacity of land by adding plant foods to the soil in different forms is
   a) Manures b) Fertilisers
   c) husbandry d) crop production
6. Bulky organic manure is
   a) Neem cake b) Sesame cake
   c) Horn meal d) Press mud cake
7. concentrated organic manure is
   a) Press mud cake b) Vermicompost
   c) Horn meal d) Night soil
8. Guano is
   a) Concentrated organic manure b) Bulky organic manure
   c) Soil amendment d) Soil conditioner

9. Guano is

a) A product of reptiles
b) A product of animals
c) A product of birds
d) A product of plants

10. Ammonium sulphate is

a) Multinutriet fertilsier
b) Straight fertilsier
c) Bulk blended fertilsier
d) Filler

11. The nitrogen in urea is

a) Nitrite form
b) Ammoniacal form
c) Nitrate form
d) Amide forms

12. Basic slag is

a) Benzene soluble P fertiliser
b) Water soluble P fertilsier
c) Citrate-insoluble P fertilsier
d) Citrate soluble P fertilizer

13. Citrateinsoluble P fertilizer is

a) SSP
b) Rock phosphate
c) Basicslag
d) Dicalcium phosphate

14. MOP fertilsier is

a) $KMnO_4$
b) $K_2SO_4$
c) KCl
d) $KNO_3$

15. Urea ammonium phosphate (UAP) is

a) Complete complex fertilser
b) Incomplete complex fertilsier
c) Liquid fertilsier
d) Bulkblended fertilsier

16. Liquid fertilsier is

a) MAP
b) UAP
c) APP
d) DAP

17. EDTA is

a) Chelated form
b) Dispersed form
b) Liquid form
d) Ionized form

18. The substances that improve soil structure, supply essential plant nutrients and stimulate soil microbes are

a) Amendments
b) Fertilsiers
c) Conditioners
d) Manures

19. The materials on addition to the soil have tendency to decrease bulk density and to increase soil volume, thus providing better physical conditions for plant growth especially in coarse textured soils and also provide essential nutrients in smaller quantities are termed as

a) Fillers
b) Concentrated organic manures
c) Soil conditioners
d) Bulky organic manures

20. The materials possess wider C:N ratio and C:S ratios are referred to as

a) Bulky organic manures
b) Concentrated organic manures
c) Soil conditioners
d) Fillers

21. Quality and composition of FYM depends on

a) Kind of animal
b) Age and condition of animal
c). Quality and quantity of feed consumed
d) All three

22. Better quality manure provided by

a) Adult old cattle
b) Growing cattel
c) Milching cattle
d) Pregnant or carrying cattle

23. The popular method of cattle manure collection which is extensively adopted in rural parts being cheep, convenient and practicable under the existing rural conditions in India.

a) Byre method
b) Dry earth method
c) Loose box method
d) Wet earth method

24. the system of manure collection which maintain perfect cleanliness and hygienic conditions of the stalls as well as cattle is

a) Loose box method
b) Dry earth method
c) Byre method
d) Wet earth method

25. Maximum losses of nutrients occur in this method of manure storage, resulting in poor quality manure.

a) Staking method
b) Pit method
c) Covered pit method
d) Heap method

26. The method of manure storage in which nutrient losses can be effectively controlled in order to obtain better quality manure is

a) Covered pit method
b) Pit method
c) Heap method
d) Staking method

27. C N Acharya is associated with

a) Pit method
b) Covered pit method
c) Trench method
d) Heap method

28. FYM consists of
    a) Dung  b) Unine
    c) Litter  d) A&B
29. All the phosphorous is present in
    a) Phosphorous not present in cattle manure
    b) Liquid portion of cattle manure
    c) Both solid and liquid portions of cattle manure
    d) Solid portion of cattle manure
30. Cattle Urine contains
    a) Large portions of secondary nutrients
    b) Large portions of Phosphorous
    c) Large portions of micro nutrients
    d) Large portions of nitrogen and potassium
31. To minimize losses nutrients from FYM
    a) Adopt trench method  b) Adopt gobagas plant
    c) Use chemical preservatives like gypsum
32. Extensively used chemical preservatives to minimize nutrient losses from FYM are
    a) Dolomite  b) Gypsum
    c) Lime  d) SSP
33. A product of decomposition of plant and animal wastes is
    a) Conditioner  b) Amendment
    c) Compost  d) Filler
34. The process of converting organic matter in to manure in a short time by accelerating fermentation process under controlled conditions is
    a) Fermenting  b) Composting
    c) Gleying  d) Disposing
35. The purpose of composting is
    a) Convert organic matter into growth promoting substances
    b) Disposing waste to maintain hygine
    c) Converting waste to wealth
    d) all the three
36. For proper composting and production of good quality compost
    a) 100% moisture content and 0 % aeration of total pore space of the composting material is required

b) 75 % moisture content and 25 % aeration of total pore space of the composting material is required

c) 25 % moisture content and 75 % aeration of total pore space of the composting material is required

d) 50 % moisture content and 50 % aeration of total pore space of the composting material is required

37. A typical compost earthworm is

a) Eisenia foetida  b) Udrilus euginea

c) Peritima postuma  d) B&D

38. General composition of compost is

a) Dry matter 20-30%, Organic matter 20-30%, N 0.8%, P 0.5%, K 0.8%

b) Dry matter 30-50%, Organic matter 10-15%, N 0.3%, P 0.1%, K 0.3%

c) Dry matter 10-20%, Organic matter 20-30%, N 0.5%, P 0.5%, K 0.5%

d) Dry matter 30-50%, Organic matter 10-15%, N 1.0%, P 0.5%, K 1.0%

39. Rothamsted Experimental Station is associated with

a) Indore process  b) ADCO process of composting

c) Bangalore process  d) Coimbatore process

40. Activated compost process was developed by

a) Howard and Ward  b) Hutchinson and Richards

c) Fowler and Ridge  d) C.N.Acharya

41. Bangalore method of composting is

a) Aerobic process  b) Anaerobic process

c) Aerobic and Anaerobic process  d) Neither A, nor B

42. Indore process of composting was developed by

a) Howard and Ward  a) Fowler and Ridge

b) Hutchinson and Richards  d) C.N.Acharya

43. The starter used in Indore process of composting is

a) Ammonium sulphate  b) Cattle dung

c) Night soil  d) Powdered bone meal

44. The starter used in ADCO process of composting is

a) Powdered bone meal  b) Cattle dung

c) Night soil  d) Ammonium sulphate

45. The composting process in which 25 % of additional free nitrogen will be fixed from atmosphere is

a) Indore method of composting
b) ADCO method of composting
c) Bangalore method of composting
d) Coimbatore method of composting

46. Additives like Bone meal and oil cakes are use in

a) Coimbatore method of composting.
b) Indore method of composting
c) ADCO method of composting
d) Bangalore method of composting

47. Compost prepared by local bodies, municipalities, corporations etc., with their out put of street sweepings, municipal rubbish is

a) Vermicomposting
b) Rural composting
c) Farm composting
d) Urban composting

48. During urban composting ____ is sprinkled between layers of waste of material to put down the offensive odour due putrefactive fermentation of the urban waste during decomposition giving foul smelling organic compounds called Mercaptans.

a) Borax powder
b) Ammonium sulphate
c) Copper sulphate
d) red earth

49. During urban composting, foul smelling organic compounds produced due to putrefactive fermentation are feferred as

a) Conditioners
b) Enzymes
c) Hormones
d) Mercaptans

50. The composting method which prove effective in not only serving as a means of disposal of urban wastes providing the country with large quantities of organic matter

a) Urban composting
b) Mechanical composting
c) Rural composting
c) Vermicomposting

51. Largest mechanical compost plant in the world is located at

a) Sydney
b) Bangalore
c) Tel Aviv
d) Rome

52. The method of making compost, with the use of earthworms is

a) Coimbatore method
b) Rural composting
c) ADCO composting
d) Vermicomposting

53. The started used in vermicomposting is'

a) Cow dung
b) Biogas slurry
c) Urine of cattle
d) Any one of the three

54. The optimum pH for efficient vermicomposting is

a) 4.5 to 5.5 b) 7.5 to 8.5

c) 8.5 to 9.5 d) 6.5 to 7.5

55. Optimum moisture of the waste material in the beds for proper vermicomposting is

a) 60-70 % b) 20 to 30%

c) 30 to 40% d) 10- 20%

56. The range of temperature where in satisfactory vermicomposting process occur is

a) 10 ⁰C to 20 ⁰C b) 20 ⁰C to 45 ⁰C

c) 18 ⁰C to 35 ⁰C d) 5 ⁰C to 15 ⁰C

57. In vermicomposting, the worms are applied to the beds of waste material the rate of

a) 1000 worms/$m^2$ b) 2000 worms/$m^2$

c) 3000 worms/$m^2$ d) 4000 worms/$m^2$

58. The turn over of the compost is in the vermicompost process is

a) 15% b) 100%

c) 75% d) 125%

59. The organic carbon content of vermin compost is

a) 25 to 35% b) 2 to 5%

c) 10 to 18% d) 30 to 40%

60. Total nitrogen content in Vermicompost is

a) 0.1 to 0.2% b) 1.0 to 2.0%

c) 10 to 20% c) 8 to 10%

61. In general one kg of earthworms (isenia fotida) approximately counts in number

a) 1000 b) 2500

b) 3500 d) 5000

62. On an average 1000 earth worms may convert roughly ------ kg waste material per day

a) 1 kg b) 5 kg

c) 10kg d) 20kg

63. Relatively more nitrogen is conserved in

a) Aerobic decomposition b) Anaerobic decomposition

c) Both A &B d) Neither A nor B

64. Relatively Less time is required for compost preperations is

a) Aerobic decomposition b) Anaerobic decomposition

c) both A &B d) neither A nor B

65. Growing of a crop for the specific purpose of incorporating it into soil while green, or soon after maturity with a view to improving the soil and benefiting subsequent crops is referred to as

a) Brown manuring
b) Green leaf manuring
c) Green manuring
d) Penning

66. The following is not objective of green manuring

a) Reduce soil loss
b) Improve soil structure
c) Improve soil texture
d) Reduce nutrient loss

67. The following is not a green manuring crop

a) Sunhemp
b) Guar
c) Green gram
d) Pillipesara

68. Green manuring is possible under rain fed conditions when the annual rain fall is

a) Above 300 mm
b) Above 700 mm
c) Above 900 mm
d) Above 500 mm

69. Highest green matter is produced from

a) Pillipesara
b) Daincha
c) Sunhemp
d) Horsegram

70. The following is not used as criteria for selecting green manure crop

a) Succulent vegetation with limited fibrous material
B. Deep root system to open the soil-for recycling of nutrients
C. Short duration with maximum and faster vegetative growth.
d) high fertilizer requirement

71. Green manure crop

a) Fix atmospheric nitrogen
b) Decompose quickly in soils
c) Come up better in poor soil
d) All the three

72. The soil binding agents as identified by Tisdell and Oades (1981) for stable aggregate formation are

a) Transient binding agents such as plant secretions like mucigel or bacterial secretions as extra cellular components
b) Temporary binding agents such as fine plant root and fungal hyphae
c) Persistent binding agents such as microbially resistant humic polymers
d) all the three

73. Keeping the boven animals (cattle and sheep) in the fallow land after the harvest of the last crop, throughout day /and or night provided with suitable food and shelter is referred to as

a) Manuring b) Penning

c) Guano d) Poudrette

74. The penning is generally adopted for cash crops like vegetables, tuber crops, chillies and sugarcane is

a) Camel penning b) Cattle penning

c) Sheep penning d) Poulty penning

75. The nitrogen content is high in

a) Night soil b) Cattle manure

c) Poultry manure d) Pig manure

76. The manure product obtained from night soil without any admixture of other organic waste materials is

a) Guano b) Poudrette

c) Vermicompost d) Tankage

77. Central Public Health and Engineering Research Institute (CPHERI) is located at

a) Bhopal b) Nagpur

d) Delhi c) Hyderabad

78. A full scale night soils digester exists in

a) Nagapur b) Bengalore

c) Coimbatore d) Ernakulam

79. The composition of gas(%) generated fro night soil digester is

a) 65% $CH_4$, 34% $CO_2$ and 1% others on volume basis

b) 45% $CH_4$, 34% $CO_2$ and 21% others on volume basis

c) 65% $CH_4$, 14% $CO_2$ and 21% others on volume basis

d) 25% $CH_4$, 34% $CO_2$ and 41% others on volume basis

80. The used up water from towns and cities collected though a drainage system is referred to as

a) Sullage b) Sludge

c) Sewage d) Sewerage

81. The process of removal of proteins from the grist is

a) Desulphonation b) Volatilization

c) Denitrification d) Deamination

82. The following is called as organic Phosphorpus fertilizer
  a) Meat meal  b) Blood meal
  c) Bone meal  d) Tankage
83. Relatively high Nitrogen present in
  a) Fish meal  b) Blood meal
  c) Meat meal  d) Horn meal
84. Relatively high potassium present in
  a) Horn meal  b) Blood meal
  c) Meat meal  d) Fish meal
85. The phosphorous concentration in bone meal is
  a) 5%  b) 12%
  c) 46%  d) 18%
86. The concentrated organic manure particularly useful for soils high in Fe and Al content is
  a) Fishmeal  b) Meat meal
  c) Bone meal  d) Horn meal
87. Relatively high phosphorous present in
  A Blood meal)  b) Bone meal
  c) Meat meal  d) Tankage
88. The concentrated organic manure rated as 80 per cent as efficient as the inorganic N fertilizer in providing the nitrogen to the crop is
  a) Fish meal  b) Bone meal
  c) Horn meal  d) Blood meal
89. The manure obtained from rejected carcasses (meat products) mostly meat and waste products such as leather scrap, feather,wool etc.
  a) Guano  b) Tankage
  c) Poudrette  d) Tungen
90. The manural product of sea bird (Pelican, Gannets, and Albatrosses) excrement covered over long periods and occurring in natural deposits is
  a) Tungen  b) Poudrette
  c) Tankage  d) Guano
91. The chief chemical constituents of Guano are
  a) Ammonium oxalate  b) Ammonium phosphate
  c) Calcium phosphates  d) All three

92. The following is not present in Guano
a) Ammonium acetate
b) Ammonium oxalate
c) Ammonium phosphate
d) Calcium phosphates

93. The guano formed by bats is also referred as
a) Farm fertilsiers
b) Cane fertilsier
c) Residue fertilsier
d) Indirect fertiliser

94. Oil cakes are used as organic fertilizers as they are rich source
a) Potassium
b) Phosphorous
c) Micronutrients
d) Nitrogen

95. Neem cake is useful for cash crops mainly due to
a) Insect repellent
b) Manure value
c) Rodent control
d) Consumed

96. The nitrogen content of neem cake is
a) 18%
b) 14%
d) 6%
c) 46%

97. For setting up a biogas plant
a) 5 cattle are required
b) 10 cattle are required
c) 15 cattle are required
d) 20 cattle are required

98. The manure obtained from biogas plants has nitrogen content of
a) 0.5%
b) 5%
c) 3.0%
d) 1.5%

99. The thermal efficiency of cow dung bunt for fuel through biogas plant is
a) 21%
b) 11%
c) 60%
d) 85%

100. Biogas slurry is
a) Concentrated organic manure
b) Bulky organic manure
c) Additive manure
d) A&B

101. Ammonium sulphate is
a) Micronutrient fertilizer
b) Complex fertilizer
c) Straight fertilizer
d) Liquid fertilizer

102. Chile salt peter is
a) Magnesium nitrate
b) Ammonium nitrate
c) Potassium nitrate
d) Sodium nitrtate

103. To produce 1 kg Nitrogen in the form of fertilizer it requires

a) 20,000 Kilo Joules of energy
b) 40,000 Kilo Joules of energy
c) 10,000 Kilo Joules of energy
d) 2000 Kilo Joules of energy

104. Nitrogen fertilizers are produced by binding atmospheric nitrogen by

a) Ammonia synthesis by Haber –Bosch process
b) Calcium cyanamide synthesis
c) Nitrate synthesis by arc-gap process
d) Any one of the given three processes

105. The nitrogen content in gaseous ammonia on weight basis is

a) 82%
b) 46%
c) 20.6%
d) 25%

106. Neutral fertilizer is

a) Urea
b) Sodium nitrate
c) Ammonium sulphate
d) CAN

107. The following fertilizer has the nitrogen in amide form

a) Urea
b) calcium cyanamide
c) Ammonium sulphate
d) A&B

108. Haber –Bosch process is associated with

a) Urea synthesis
b) Ammonia synthesis
b) Nitrate synthesis
d) Protein sysntesis

109. In the Haber –Bosch process the reaction between molecular nitrogen and hydrogen takes place at

a) 200 $^0$C
b) 550 $^0$C
c) 350 $^0$C
d) 700 $^0$C

110. Ammonia sysnthesis is

a) Endothermic process
b) Exothermic process
d) Oxidation process
d) Hydration process

111. Ammonia sysnthesis is

a) Hydration process
b) Oxidation processes
c) Reduction process
d) Carbonation process

112. A major portion of ammonia produced in the world use ---------- as hygrogen source

a) Coal
b) Naphtha
d) Fuel oil
d) Natural gas

113. A major portion of ammonia produced in India use ---------- as hygrogen source

a) Fuel oil
b) Natural gas
d) Naphtha
d) Coal

114. The Fertilizer plant is based on coal gasification to produce hydrogen

a) Nangal (Punjab)
b) Visakhapatnam (Andhra Pradesh)
c) Ramagundam (Telangana)
d) Rurkela (Uttarakhand)

115. The cokeoven gas consists of

a) 26% $H_2$
b) 38% $H_2$
c) 75% $H_2$
d) 52% $H_2$

116. During systhesis of ammonia the nitrogen and hydrogen gases are allowed to react at the mole ratio of

a) 3:1
b) 1:3
c) 1:4
d) 4:1

117. Gaseous ammonia is stored in steel or plastic tanks that can with stand the pressure of

a) 120 kg / inch$^2$
b) 80 kg / inch$^2$
c) 60 kg / inch$^2$
d) 180 kg / inch$^2$

118. The cheapest N-fertilizer to manufacture is

a) Urea
b) Gaseous ammonia
c) Ammonium chloride
d) Ammonium sulphate

119. Ammonia is

a) Neutral in reaction
b) Acidic in reaction
c) Alkali in reaction
d) Hygroscopic

120. One kg nitrogen applied as ammonia will require ---- kg of lime for neutralization

a) 1.0
b) 0.9
c) 1.2
d) 1.8

121. The oldest synthetic fertilizer is

a) DAP
b) Urea
c) CAN
d) Ammonium sulphate

121. The N content of aqua ammonia is

a) 50%
b) 30%
c) 40%
d) 20%

122. Byproduct process of Ammonium sulphate synthesis is associated with

a) Paper industry
b) Iron & steel industry
c) Leather industry
d) Pharmaceutical industry

123. The fertilizer ammonium sulphate is manufactured by

a) By product Process b) Gypsum process

c) Neutralization process d) All three

124. The advantage of Gypsum process or leuna process is

a) In countries where sulphur supplies are meager and natural resources of gypsum are present ammonium sulphate can be produced without purchasing sulphur from abroad.

b) The by product $CaCO_3$ could be utilized for the manufacture of Calcium Ammonium Nitrate (CAN)

c) It has high N content compared to any nitrogen fertilizer

d) A& B

125. The sulphur content of Ammonium sulphate fertilizer is is

a) 8% b) 12%

c) 33% d) 23%

126. One kg of N applied as Ammonium sulphate fertilizer required ------of lime for neutralization

a) 5.1 kg b) 3.8kg

c) 8.6kg d) 10.5kg

127. Bulk density of following fertilizer is relatively high

a) Urea b) Ammonium sulphate

c) SSP d) CAN

128. Limestone or Dolomite used as raw material in manufacture of

a) Liquor ammonia b) Urea

c) Ammonium sulphate d) CAN

129. CAN fertilizer granules are coated with

a) Beef tallow b) soap stone

c) Acetone d) Ether

130. Calcium carbonate powder is added during the manufacture of CAN for the purpose of

a) Improve handling character and reduced the explosive

b) Reducing hygroscopic nature of ammonium nitrate

c) Increasing N content

d) A & B

131. The CAN fertilizer contain $NH_4^+$ and $NO_3^-$ forms of nitrogen in the ratio of

a) 2:1 b) 1:1

c) 3:1 d) 4:1

132. The Ca content of CAN fertilizer is

a) 8% b) 12%

c) 23% d) 25%

133. The fertilizer urea is used as

a) Fertilizer for crops

b) Protein supplement in the feed of ruminants

c) Amendment for reclamation f problemsoils

d) A&B

134. The most important N-Fertilizer constituting nearly 91 per cent of the total production of nitrogen in India is

a) Ammonium sulphate b) Urea

c) Calcium ammonium nitrate d) Aqueous ammonia

135. The scientist first prepared urea in the year 1828

a) John Helmont b) Leibig

d) Fedric Wholer d) Thomson Way

136. When concentrated urea solution is exposed to elevated temperatures i.e., greater than 100 °C during evaporation in prilling process

a) Uric acid is formed b) Biuret is formed

d) Urea ammonium sulphate is formed d) N volatilization occur

137. Synthetic protein, organic compound, richest source of N (46%) in amide form among solid N fertilizers

a) Gaseous ammonia b) Urea

c) Salt peter d) Aqua ammonia

138. Solubilityof Urea is

a) 30 g /100 g of water at $20^0$C b) 50 g /100 g of water at $20^0$C

c) 75 g /100 g of water at $20^0$C d) 100 g /100 g of water at $20^0$C

139. Acceptable limit of Biuret in Urea fertilizer is

a) 2.5 per cent by weight b) 1.5 per cent by weight

c) 4.0 per cent by weight d) 3.5 per cent by weight

140. In soil the conversion of urea in to ammoniacal and nitrate form is complete in about

a) A week. b) Two weeks.

c) Three weeks. d) Four weeks.

141. $CaCO_3$ required to neutralize the acidity caused by application of 100 kg urea

a) 40 kg b) 60 kg

c) 110kg d) 80 kg

142. most suitable N fertilizer for foliar spraying spraying is

a) Urea b) Ammonium sulphate

c) Gaseous ammonia d) Ammonium nitrate

143. In general the recovery of urea by rice crop is

a) 40 per cent b) 80 per cent

c) 20 per cent d) 15 per cent

144. Urea is more popular nitrogenous fertilizer in India because

a) High nitrogen content 44 to 46 per cent

b) Good physical condition

c) Suitable for foliar application

d) All the three

145. The deposits of phosphate rock exists in nature because of

a) Low solubility of the compounds present

b) High solubility of the compounds present

d) High specific gravity of the material

d) Low specific gravity of the material

146. The apatite minerals (Calcium phosphate) are

a) Magmatic origin b) Organogenic origin

c) Man made d) A and B

147. An example for organic phosphates are

a) Bones and Teeth of animals b) Guano deposits

c) Poudrette d) A and B

148. finely crystalline gafsa phosphates found in

a) North Africa b) USA

c) USSR d) Udaipur

149. The apatites found in USSR si

a) Kola phosphate b) Pebble phosphate

d) Gafsa phosphates d) B & C

150. In India phosphate rock deposits have been found in

a) Udaipur in Rajasthan b) Mussoorie in Uttarakhand

c) Singbhum district d) All three places

151. An example for Water soluble phosphorus fertilizers is

a) SSP b) Bone meal

d) Basic slag d) Rock phosphate

152. The Phosphorous content in monoammonium phosphate is

a) 8% b) 52%

c) 18% d) 46%

153. The phosphorous content in basic slag is

a) 25% b) 46%

c) 20% d) 8%

154. The first full scale ammonia plant was established in 1913 by FRITZ HABER and CARP BOSCH at OPPAU, Germany for

a) Dow chemicals b) Bayer

c) Badische, Anilin and Soda Fabric (BASF) d) Monsanto

155. Match the following

| | | |
|---|---|---|
| Single super phosphate | (a) | a.16% P2O5 |
| Double super phosphate | (b) | b. 32 % P2O5 |
| Triple super phosphate | (c) | c. 46% P2O5 |
| MAP | (d) | d. 52% P2O5 |

156. The following phosphatic fertilsier is recommended to apply by broadcasting and completely mix in soil

a) TSP b) SSP

c) Rock phosphate d) DAP

157. The following phosphatic fertilsier is recommended to apply by placement method

a) Basic slag b) Rock phosphate

c) Dicalcium phosphate d) SSP

158. The phosphatic fertilsize belong to ortho phosphates group is

a) Potassium metaphosphate b) Ammonium poly phosphate

c) Single Super Phosphate d) Ammonium metaphosphate

159. The phosphatic fertilsize belong to poly phosphates group is

a) SSP b) APP

c) UAP d) MAP

160. The phosphatic fertilsize belong to meta phosphates group is

a) UAP b) SSP

c) Potassium metaphosphate d) MAP

161. The grain size of rock phosphate fertilizer is

a) 0.14 mm sieve b) 0.20mm size

c) 0.25mm size d) 0.5mm size

162. The 58m of phosphorous present in root phosphat is
a) Mdnocalcium phosphate
b) Dicalcium phosphate
c) Pi Calcium phosphate
d) Tetracalcium phosphate

163. Waste pond phosphate containing soft phosphates and clay impurities is marketed under trade names
a) Mineral colloids
b) Calphos
c) Phos-cal-oids
d) All three

164. The mineral in phosphate rock is
a) Dolomite
b) Calcite
c) Apatite
d) Strengite

165. Waste pond phosphates contain
a) 2- 3%P
b) 7-9 % P
c) 10-14%P
d) 16-18%P

166. Direct application of rock phosphates is recommended to
a) Neutral soils
b) Strong alkaline soils
c) Strongly acid soils
d) Calcareous soils

167. Rock phosphate is recommended for
a) Long duration commercial crops
b) Short duration leafy vegetable crops
c) Short duration millet crops
d) Short duration fodder crops

168. Efficiency of rock phosphate is improved by
a) Addition of un-decomposed organic matter
b) Application P-solubilising bacteria
c) Acid forming N fertilizers
d) All the three

169. Single super phosphate is also known as
a) Normal super phosphate
b) Ordinary super phosphate
c) Abnormal super phosphate
d) A&B

170. The byproduct of Den process and continuous rock acidulation process is
a) Lime
b) Gypsum
c) DCP
d) MCP

171. Cause for hardening SSP may be
a) Hydration of monocalcium phosphate
b) Hydrolysis of monocalcium phosphate

c) Oxidation of monocalcium phosphate
d) Reduction of monocalcium phosphate

172. During manufacture of single super phosphate it is stored for ____ weeks to cure and attain the desired physical condition.
a) 1 – 2 weeks
b) 4 – 8 weeks
c) 6 – 10 weeks
d) 2 – 6 weeks

173. The synthesis reaction of SSP produces
a) Toxic methyl isocynate gas
b) Toxic hydro-fluoric acid gas
c) Carbon dioxide gas
d) Methane gas

174. The fertilizer SSP is available in
a) Granular form
b) Powder form
c) Pellets form
d) A&B

175. The reason for hardening of SSP when exposed to atmosphere
a) Hydration of gypsum
b) Hydration of monocalcium phosphate
c) Expulsion of toxic hydro fluoric acid gas
d) Binding of Monocalcium phosphate and calcium sulphate

176. The SSP fertilizer is not stored in jute bags and has to be stored in polythene lined gunny bags or polyethylene bags due to
a) Reduction of P content when stored in jute bags
b) Weight loss of SSP occur when stored in jute bags
c) Free acid in the SSP rot the jute fiber bags
d) Risk of pollution is high when stored in jute bags

177. The fertilsier SSP has
a) 2/5 Mono calcium phosphate and 3/5 Gypsum by weight
b) 3/5 Mono calcium phosphate and 2/5 Gypsum by weight
c) 1/3 Mono calcium phosphate and 2/3 Gypsum by weight
d) 2/3 Mono calcium phosphate and 1/3 Gypsum by weight

178. the calcium content of SSP is
a) 24%
b) 12%
c) 16%
d) 21.0%

179. The Sulphur content of SSP is
a) 21%
b) 12%
c) 16%
d) 24%

180. The pH of single super phosphate fertilizer is about

a) 7.5 b) 7.0

c) 5.5 d) 3.0

181. In the manufacture of single super phosphate the rock phosphate is treated with

a) Suphuric acid b) Orthophosphoric acid

c) Superphosphoric acid d) Metaphosphoric acid

182. In the manufacture of triple super phosphate the acid rock: ratio (expressed as eh mole ratio of total $P_2O_5$ to CaO in the acidulate) mixed is

a) 0.95 to 0.91 b) 0.91 to 0.95

d) 0.81 to 0.85 d) 0.85 to 0.81

183. The phosphatic fertilizer has higher bulk density is

a) SSP b) TSP

c) Basic slag d) Bone meal

184. Basic slag obtained from Indian steel industry contain about

a) 9-14% $P_2O_5$ b) 3-8 % $P_2O_5$

c) 12-16% $P_2O_5$ d) 16-20% $P_2O_5$

185. The $P_2O_5$ enriched basic slag is marketed as

a) Mineral colloids b) Waste pond phosphate

c) Colloidal phosphate d) Pelophos

186. The $CaCO_3$ equivalence of basic slag is

a) 35% b) 50%

c) 70% d) 85%

187. Crude potassium salts are

a) Kola apatites b) Natural sea water minerals

b) Gafsa apatites d) Sedimentary rocks

188. Potassium content of earth crust is

a) 0.11% b) 5.0%

c) 2.40 % d) 7.5%

189. Phosphorous content of earth crust is

a) 7.5% b) 2.40 %

c) 5.0% d) 0.11%

190. Large deposits of potassium salts are present in

a) South Africa b) Noth Africa

c) South east Asia d) North America

191. The mineral Langbeinite is

a) KCl NaCl
b) KCl $MgCl_2$ $6H_2O$
c) $K_2SO_4$ $2MgSO_4$
d) KCl $MgSO_4$ $3H_2O$

192. Highest K containing mineral is

a) Sylvite
b) Langbeinite
c) Carnallite
d) Kainite

193. The following is not K bearing mineral

a) KCl NaCl
b) KCl $MgCl_2$ $6H_2O$
c) KCl
d) KCl $MgSO_4$ $3H_2O$

194. Find the correct statement

a) With a rise in temperature the solubility of KCl increases rapidly whereas the solubility of NaCl varies very slightly.
b) With a rise in temperature the solubility of NaCl increases rapidly whereas the solubility of KCl varies very slightly.
c) With a rise in temperature the solubility of both KCl and NaCl increases rapidly
d) With a rise in temperature the solubility of both KCl and NaCl increases very slightly

195. in the manufacture of MOP fertilizer the raw material used is

a) Langbeinite ore
b) Sylvite ore
c) Syvinite ore
d) Carnallite ore

196. In the floatation process of MOP manufacture, the reagents used for sylvinite flotation which selectively film the sylvinite particle is

a) Starch or mannogalactan gums
b) A mixture of primary aliphatic amine salts derived from beef tallow
c) Polysaccharides
d) Plant based lignins

197. The specific gravity of MOP is

a) Specific gravity is 1.50
b) Specific gravity is 1.26
c) Specific gravity is 1.18
d) Specific gravity is 1.98

b) the MOP fertilizer is

a) Neutral in reaction
b) Acidic in reaction
c) Alkaline in reaction
d) None of the three

199. The chloride content of MOP is

a) 58%
b) 47 %
c) 50%
d) 60%

200. In the Mannheim furnace process of SOP manufacture sulphuric acid is reacted with

a) KCl NaCl
b) KCl MgCl2 6H2O
c) KCl
d) KCl MgSO4 3H2O

201. Mannheim furnace process of SOP manufacture is

a) Exothermic process
b) Endothermic process
c) Both A&B
c) Neither A nor B

202. The sulphur content in SOP fertilizer is

a) 13%
b) 46%
c) 18%
d) 50%

203. On addition of nitrate fertilizers to soil

a) No influence on soil pH
b) Soil pH decreases
c) Soil pH increases
d) None of the three

204. Soil application of SSP fertilizer improves the soil structure because

a) Its P component
b) Its inherent acidity
c) Its gypsum component
d) Its conditioning effect

205. The fertilizers that contain the three plant nutrient elements N P K is designated as

a) Straight fertilizers
b) Completefertilizers
c) Primary fertilizers
d) Secondary fertilizers

206. Each fertilizer granule of complex fertilizer contain

a) Two or more major plant nutrients in a specified ratio
b) One major plant nutrients in a specified concentration
c) each fertilizer granule one major plant nutrients in a disproportionate concentration
d) Two or more major plant nutrients in disproportionate concentration

207. The fertilizer containing two or more plant nutrients is

a) Additive fertilizers
b) Complex fertilizers
c) Mixed fertilizers
d) All the three

208. The principle techniques involved in the manufacture of complex fertilizers

a) Granulation, acidulation and ammonifiation in the order
b) Acidulation, Granulation and Ammonification in the order
c) Ammonification, Acidulation and Granulation in the order
d) Acidulation, Ammonification and Granulation in order

209. The technique is primarily intended to improve handling and storage properties of fertilizers is

a) Gasification b) Acidulation

c) Ammonification d) Granulation

210. Complex fertilizers are

a) Low analysis fertilizers b) High analysis fertilizers

c) A&B d) Neither A nor B

211. The fertilizer UAP is

a) Incomplete complex fertilizer b) Complete complex fertilizer

c) Straight fertilizer d) Organic manure

212. The following fertilizer has highest N content

a) APP b) DAP

c) UAP d) MAP

213. Water soluble $P_2O_5$ content of MAP is

a) 52% b) 44%

c) 11% d) 46%

214. To neutralize acidity developed due to 100 kg of MAP on soil application it require

a) 100 kg of $CaCO_3$ b) 86 kg of $CaCO_3$

c) 120 kg of $CaCO_3$ d) 50kg of $CaCO_3$

215. Water soluble P2O5 content of DAP is

a) 52% b) 41%

c) 44% d) 46%

216. Raw materials required for manufacture of Ammnium phosphate sulphate is

a) Ammonia b) Phosphoric acid

c) Sulphuric acid d) All the three

217. Ammonium polyphosphate is

a) Liquid fertilizer b) Solid fertilizer

c) Gaseous fertilizer d) None of the three

218. The viscosity of APP fertilizer at 24 °C is

a) 120 centipoise b) 40 centipoise

c) 100 centipoise d) 80 centipoise

219. Ammonium polyphosphate fertilizer has

a) 46% P2O5 b) 16% P2O5

c) 65% P2O5 d) 52%P2O5

220. UAP fertilizer was first manufactured in India by
   a) Coramandal fertilizers Ltd Viskhapatnam AP
   b) National Fertilisers Ltd, Nangal
   c) Madras fertilizers, Chennai
   d) Spic fertilizers, Udaypur

221. The first Liquid Nano Nitrogen fertilsier in India manufactured by
   a) Spic fertilizers  b) Coramandal fertilizers Ltd
   c) IFFCO  d) National fertilizers Ltd

222. The amide form of nitrogen in UAP (28-28-0) is
   a) 9%  b) 21%
   c) 14%  d) 17%

223. Nitrophosphate fertilizers are manufactured by
   a) PEC process  b) ODDA process
   c) Manhem process  d) A&B

224. PEC process of nitrophosphate manufacture is also known as
   a) Den process  b) Multi acid process
   c) Continuous acidulation process  d) Manhem process

225. In manufacture of nitrophosphates, greater portion of water soluble phosphate such as mono ammonium phosphates is obtained by using
   a) More sulphuric acid  b) More nitric acid
   c) More rock phosphate  d) All the three

226. In the complex fertilizer grade 14:35:14, the water soluble P2O5 is
   a) 14%  b) 6%
   c) 29%  d) 16%

227. Chemical problems associated with bulk blended fertilizers is
   a) Reduction  b) Oxidation
   c) Hydration  d) Carbonation

228. An example of fillers use in mixed fertilizers is
   a) Dolomite  b) Well ground corn cobs
   c) Diatomaceous earth  d) Hydrated silica

229. An example of conditioners use in mixed fertilizers is
   a) Sand  b) Dolomite
   c) Lime stone  d) Diatomaceous earth

230. The cost per unit of the plant nutrient contain in a given fertilizer is referred as
   a) Grade  a) Unit value
   c) Dose  d) Unit cost

231. The concept of unit cost of fertilizer will help the farmers to
   a) Shoose the cheapest fertilizer in terms of cost
   b) Choose high analytic fertiliser
   c) Choose right form of nutrient
   d) Choose required number of nutrients

232. The substances that influence the plant growth favorably by Changing the reaction or Changing the plant nutrients in the soil from unavailable to available forms or improving thc physical conditions are termed as
   a) Fillers b) Fertilizers
   c) Amendments d) Conditioners

233. Agricultural lime contains
   a) P b) S
   c) Ca d) K

234. Kiserite is
   a) Magnesium mineral b) Calcium mineral
   c) Phosphorous mineral d) Potassium mineral

235. Iron content in ferrus sulphate ($FeSO_4$ $7H_2O$) is
   a) 15% b) 20%
   c) 11% d) 25%

236. Zinc content in Zinc sulphate ($ZnSO_4$ $7H_2O$) is
   a) 11% b) 15%
   c) 23% d) 25%

237. Boron content in borax (Na2B4O7 10H2O) is
   a) 23% b) 11%
   c) 15% d) 25%

238. Molybdenum content in ammonium molybdate is
   a) 23% b) 54%
   c) 15% d) 41%

239. Government of India passed The fertilizer control order [FCO] in the year
   a) 1985 b) 1961
   c) 1957 d) 1947

240. The Government of India (GOI) passed the Fertilizer Movement Order (FMO) in the year
   a) 1960 b) 1970
   c) 1985 d) 1957

241. In Indore method of composting one cattle pair produced ------ cartloads per year

a) 20-30 cartloads b) 30-40 cartloads
c) 40-50 cartloads d) 50-60 cartloads

242. In the following method of composting 25 % of additional nitrogen will be fixed from atmosphere

a) ADCO method b) Coimbatore method
c) Indore method d) Bangalore method

243. Additives used in Indore method of composting is

a) Bone meal b) Oil cakes
c) Wood ash d) Any one of three

244. The gas is produced during anaerobic decomposition of Bangalore process of composting

a) Carbon dioxide b) Methane
d) Oxygen d) Sulphur dioxide

## Answer Key

| | | | | | | | | | | | | | | | | | | | | | | | |
|---|---|---|---|---|---|---|---|---|---|---|---|---|---|---|---|---|---|---|---|---|---|---|---|
| 1 | c | 16 | c | 31 | a | 46 | d | 61 | a | 76 | b | 91 | d | 106 | d | 121 | d | 136 | b | 151 | a | 166 | c |
| 2 | d | 17 | a | 32 | d | 47 | d | 62 | b | 77 | b | 92 | a | 107 | d | 122 | b | 137 | b | 152 | b | 167 | a |
| 3 | d | 18 | d | 33 | c | 48 | c | 63 | b | 78 | d | 93 | b | 108 | b | 123 | d | 138 | d | 153 | d | 168 | d |
| 4 | b | 19 | d | 34 | b | 49 | d | 64 | a | 79 | a | 94 | d | 109 | b | 124 | d | 139 | b | 154 | c | 169 | d |
| 5 | a | 20 | a | 35 | d | 50 | b | 65 | c | 80 | c | 95 | d | 110 | b | 125 | d | 140 | a | 155 | a | 170 | b |
| 6 | d | 21 | d | 36 | d | 51 | c | 66 | c | 81 | d | 96 | c | 111 | c | 126 | a | 141 | d | 156 | c | 171 | a |
| 7 | c | 22 | a | 37 | a | 52 | d | 67 | c | 82 | b | 97 | a | 112 | d | 127 | c | 142 | a | 157 | d | 172 | d |
| 8 | a | 23 | b | 38 | b | 53 | d | 68 | c | 83 | d | 98 | d | 113 | d | 128 | d | 143 | a | 158 | c | 173 | b |
| 9 | c | 24 | c | 39 | b | 54 | d | 69 | c | 84 | c | 99 | c | 114 | c | 129 | b | 144 | d | 159 | b | 174 | d |
| 10 | b | 25 | d | 40 | c | 55 | a | 70 | d | 85 | b | 100 | b | 115 | d | 130 | d | 145 | a | 160 | c | 175 | b |
| 11 | d | 26 | a | 41 | c | 56 | c | 71 | d | 86 | c | 101 | c | 116 | b | 131 | b | 146 | d | 161 | a | 176 | c |
| 12 | d | 27 | c | 42 | a | 57 | b | 72 | d | 87 | b | 102 | d | 117 | a | 132 | a | 147 | d | 162 | c | 177 | a |
| 13 | b | 28 | d | 43 | b | 58 | c | 73 | b | 88 | d | 103 | b | 118 | b | 133 | d | 148 | a | 163 | d | 178 | d |
| 14 | d | 29 | d | 44 | d | 59 | c | 74 | c | 89 | b | 104 | d | 119 | c | 134 | b | 149 | d | 164 | c | 179 | b |
| 15 | b | 30 | d | 45 | a | 60 | b | 75 | d | 90 | d | 105 | a | 120 | d | 135 | d | 150 | d | 165 | b | 180 | d |
| 181 | a | 196 | b | 201 | a | 206 | a | 211 | a | 216 | d | 221 | c | 226 | c | 231 | a | 236 | c | 241 | d | | |
| 182 | b | 197 | d | 202 | c | 207 | b | 212 | c | 217 | a | 222 | b | 227 | c | 232 | c | 237 | b | 242 | d | | |
| 183 | c | 198 | a | 203 | c | 208 | d | 213 | b | 218 | d | 223 | d | 228 | a | 233 | b | 238 | b | 243 | d | | |
| 184 | b | 199 | b | 204 | c | 209 | d | 214 | b | 219 | c | 224 | b | 229 | d | 234 | a | 239 | c | 244 | b | | |
| 185 | d | 200 | c | 205 | b | 210 | b | 215 | b | 220 | a | 225 | a | 230 | b | 235 | b | 240 | a | | | | |

# 12

# Soil Health

Soil is a dynamic, vital living natural resource. Soil health is the condition of the soil and its potential to sustain biological functions, maintain environmental quality, and promote plant and animal health. It is qualitative trait of soil. Healthy soil is the foundation for profitable, productive, and environment friendly agricultural systems. Deterioration in soil health implies loss of the vital functions of soil: (i) providing physical support, water and essential nutrients required for growth of terrestrial plants; (ii) regulation of the flow of water in the environment and (iii) elimination of the harmful effects of contaminants by means of physical, chemical and biological processes, i.e., environmental buffer or filter. The health of soil determine agricultural sustainability and environmental quality, which jointly determine plant, animal and human health.

## Healthy soils perform multiple functions

Healthy soil performs multiple functions: (i) providing physical support to terrestrial plants, (ii) supplying fundamental resources viz., water, nutrients and oxygen required for terrestrial primary production, (iii) providing habitat to a variety of soil organisms, with taxonomic identity and functions of several organisms still unknown/lesser known to the scientific and wider community, (iii) regulating hydrological and mineral/nutrient cycling, with significant impacts on global climate, (iv) detoxification of organic and inorganic substances, leading to purification of water resource and (v) resisting erosion. A given soil function is achieved through several and a given mechanism/process may contribute to several functions. Thus, litterdecomposition and mineralization contribute to detoxification as well as nutrient supply/agricultural production functions of soil. The overall assessment about whether a soil is good or bad depends on the objective of assessment and the net outcome of different soil processes and functions in given conditions. For example in intensive production in marginal soils, one may get high crop productivity but with low quality food, contamination of water and infected food products. once we used to consider a situations of low agricultural productivity is the main problem but ignore functions of soil other than production function viz., capacity of soil to supply water, nutrients and oxygen and to reduce crop yield losses due to pests and diseases. However, in the present widespread scenario of multiple problems (including increased levels of greenhouse gases in the atmosphere, soil erosion and land degradation, production of infected crop products, depletion and pollution of water resources, and depletion of bio- diversity), there is a need of addressing multiple functions of soil in an objective manner. The concept of soil health is essentially an elaboration of the concept of soil productivity/fertility to deal with the multiple and complex problems faced by the world today. This perspective of optimizing multiple functions makes soil health an integral dimension of agro-ecosystem health and sustainable development.

The soil functions can be weighted according to the relative importance of each function in fulfilling the management goals based on expert opinions. Regulation of each function is determined by a large number of soil attributes and a single attribute or a statistical/ mathematical derivative of several attributes (in the form of an index) can be viewed as an indicator of one or more soil functions if a systematic relationship exists between the attribute(s) or its derivative with the soil functions. As a single measurable soil attribute is unlikely to be correlated with soil function(s) and measurement of 'all' soil attributes is not practical, one needs to draw attention on indicators soil health and how they interact with each other, and thus, values of soil is affected by one or more of these selected parameters.

The soil, which interacts with landscape features and plant cover, is a key element in regulating and partitioning water flow through the environment. The biological, chemical, and physical processes that occur in soils buffer environmental changes in air quality, water quality, and global climate. Soil degradation through erosion, compaction, loss of biological activity, acidification, salinization, or other processes can degrade soil health. These processes affecct soil health by changing the soil attributes, such as nutrient status, organic and labile carbon content (organic carbon is the total amount of carbon held in the organic matter in the soil while labile carbon is that fraction of organic carbon that is most readily decomposable by soil microorganisms), soil texture, available water-holding capacity (the amount of water that can be held in the soil and made available to plants), soil structure, maximum rooting depth, and soil pH (a measure of the acidity or alkalinity). Some changes in these soil attributes can be reversed by external inputs. Nutrient losses, for example, can be replaced by adding fertilizers. Other changes such as loss of the soil depth available for rooting because of soil erosion, degradation of soil structure because of subsoil compaction are much more difficult to reverse.

## 12.1. Concepts ofsoil health

A healthy soil provides many functions that support plant growth, including nutrient cycling, biological control of plant pests, and regulation of water and air supply. These functions are influenced by the interrelated physical, chemical, and biological properties of soil, many of which are sensitive to soil management practices.

### a) Nutrient cycling

Nutrient cycling refers to the many pathways through which nutrients are added to soil, removed from soil, and changed within the soil. Nutrients are found in two basic forms in the soil: organic and inorganic (mineral) forms. Organic forms of nutrients contain carbon in the structure of the molecule, while inorganic forms do not contain carbon in the structure of the molecule. Nutrients are stored in several pools within the soil: as inorganic forms in soil particles, as organic forms in soil organic matter, as inorganic forms on cation exchange sites, and as organic and inorganic forms dissolved in the water surrounding soil particles, known as the soil solution.

Soil organic matter is a storehouse of several plant nutrients, including nitrogen, phosphorus, and sulfur. Every one percentage point of organic matter in the top 6 inches of soil contains about 450kgs of nitrogen, 105kgs of phosphorus, and 75kgs of sulfur per acre. However, most nutrients in organic matter are not directly available to plants. To be used by plants, nutrients in organic matter must be converted to inorganic forms through decomposition and mineralization by soil organisms.

Soil organisms form a food web that decomposes organic matter and releases nutrients in the process. At the base of the food web are bacteria and fungi, which obtain energy by decomposing soil organic matter directly. Protozoa and some nematodes are organisms that graze on bacteria and fungi, releasing nitrogen that can then be utilized by plants (fig 1). Soil organisms are affected by soil pH as well. Earthworms and bacteria prefer a near-neutral soil pH. Fungi do well at most soil pH levels, so in acidic soils, fungi tend to dominate the soil microbial community. Soil pH also influences the cation exchange capacity supplied by organic matter. As soil pH increases, the cation exchange sites on soil organic matter will also increase.

## b) Biological control of pest organisms

Biological control is the term for reduction of pest organisms by natural enemies, which include predators, parasites, and pathogens (disease-causing organisms). Healthy agricultural soil typically include a wide range of predators, parasites, and pathogens that contribute to the suppression of agricultural pests. Spiders, and ground (carabid) beetles are important ground-dwelling natural enemies of insect pests. Ground beetles play a major role in agroecosystems by contributing to the mortality of insects, weed seeds, and slugs.

## c) Microbes and soil critters

Microbes are microorganisms like bacteria, fungi and protozoa, and are essential components of healthy soil. Microbes decompose dead plants and animals, manure, and pesticides; protect water quality; increase the ability of soil to retain water; and make critical nutrients bioavailable to living plants. This is especially important when it comes to nitrogen fixing-bacteria. Nitrogen is a nutrient required for plant growth but frequently limited in soil. Some microbes even have a symbiotic relationship with plants where they inhabit plant roots and feed on the root's sugars and carbon, and in return, the microbe (in this case, rhizobia) provide nitrogen to the plant. Certain fungi have similar symbiotic relationships, where they infect and feed off of the plant roots but in return, help the plant better absorb water and nutrients. In addition to supporting plant life, microbes are also a primary producer in the food chain, providing the basic nutrition for soil critters (invertebrates) like worms (nematodes) and beetles (arthropods). These animals further help decompose organic matter and add nutrients to the soil, while also providing a food source for larger animals like birds and mammals. .

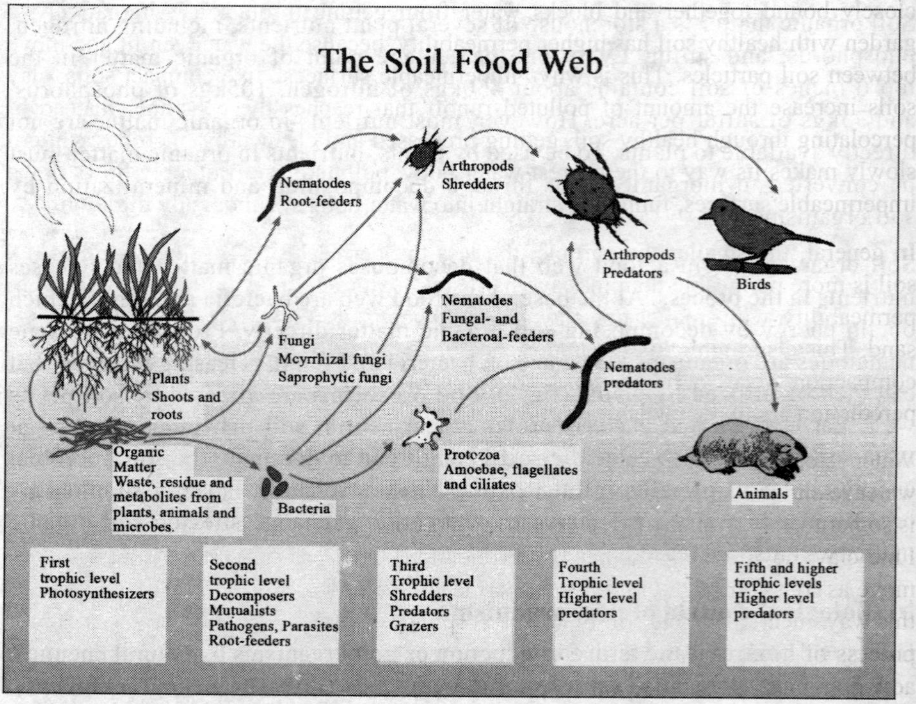


**Fig. 1:** Soil Food Web (*from USDA's Natural Resource Conservation Service)*

As mentioned, microbes feed on organic matter (like roots, twigs, and leaves) but they also need water and oxygen to survive. Fortunately, microbes and other soil critters like earthworms and centipedes help create air pockets in the soil, providing room for oxygen. If there is no oxygen, the soil becomes anaerobic and can promote "bad bacteria" that result in plant diseases. This can happen from soil compaction and/or over watering.

Life in the soil includes all of the bacteria, protozoa, nematodes and fungi, the food they eat, the excretions they make, and the root systems they sustain. Living microbes most quickly can be incorporated into soil by adding really good quality compost. Plants attract microbes to their roots by feeding them carbon. Bacteria and fungi hold the soil together with microscopic glues and binders. The microbes consume organic matter and are then consumed themselves by larger creatures (worms, ants, slugs, centipedes, insect larvae, etc.) In turn, these creatures are consumed by creatures further up the food chain. Carbon and other nutrients are cycled through these many life forms, creating healthy, living, well-structured soil, no matter what the soil type.

## d) Soil permeability and Water retention

Soil permeability is the ease of percolation of water or liquid through a soil. For instance, a massive compacted soil has extremely low permeability, because the particles are closely bound together and blocks water from getting through it. Alternatively, a garden with healthy soil has higher permeability, because the water can fit and move between soil particles. This is why "impermeable surfaces" like compact soils, clay soils increase the amount of polluted runoff that reaches the coastline. Instead of percolating through healthy soil, getting processed and cleaned by microbes while it slowly makes its way to the nearest water body, polluted runoff quickly flows down impermeable surfaces, funneling straight into water bodies without any filtration

In general, the smaller the particle, the less the permeability. So though almost all soil is more permeable than massive hard soil, texture of soil can impact the level of permeability, with smaller fine clay generally being less permeable than larger coarse sand. This also highlights the importance of reducing soil compaction, as too much compaction forces particles closer together, blocking air pockets, preventing water percolation and reducing soil oxygen content.

Water retention is the ability of a soil to hold water, with larger particles having less water retention. For instance, a sandy soil generally has low water retention because it is so permeable. Water infiltrates a sandy substance and then percolates right through. Fine clay, however, has higher water retention. Once water gets in, it is unable to move as freely through the soil (low permeability) so water is better held in place. In this way, medium to fine soil types act as sponges, absorbing and holding water. The process of holding water is due to a more technical mechanism known as capillary action.

The best soil for gardening and crop agriculture is one that balances water permeability and retention, which is generally found with loam (equal parts sand, silt, and clay). This mix provides a balance where water is able to infiltrate into the soil but not move too quickly through it. The water is held between the soil particles, and becomes "available soil water" to plant roots. If too much water is applied and held, the soil can become saturated, meaning all air pockets fill with water, and the system becomes anaerobic. Not only do anaerobic environments suffocate the good microbes, but they also create an environment where certain disease causing anaerobic bacteria like *Sclerotium* spp / *Phytophthera* spp can flourish and harm plants and root systems. Alternatively, if there is not enough water in the soil, plant roots are unable to access water, the permanent wilting point is met, and plants will not survive.

## e) Carbon sequestration and storage

Plants provide a natural method of removing carbon from the atmosphere through photosynthesis, and then storing or fixing much of that carbon in their tissue, roots and soil. This process is known as carbon sequestration. Over time, when the plant dies or gets consumed by animals and returned to the soil as litter, it helps create soil organic matter (SOM), also known as compost, which is extremely nutrient rich and loaded with fixed carbon. Healthy soils mean stronger carbon reservoirs and healthier plants, which will continue to take carbon out of the atmosphere, build SOM, and promote growth of new plants, providing a feedback loop that could help bring the system back into balance.

Approximately 75% of the terrestrial carbon is actually in the soil, and in its healthy state can store up to 11 gigatons of $CO_2$ every year. Some estimate that "soil sequesters four times more $CO_2$ than all the world's plants including forests". However, when we lose healthy soils by developing over natural areas, or by conducting conventional agricultural practices like clear cutting, tilling, and over-grazing; we reduce the ability to use this reserve as a natural storage of carbon. We also release the carbon that's previously been captured back into the atmosphere. To prevent the severe impacts from exacerbated climate change, we can't just rely on healthy soils to offset our impact. We need to stop adding additional carbon into the atmosphere as soon as possible, but to help alleviate the damage we have already done to our planet, and to sequester some of the carbon that we have already added to the atmosphere, we need to cultivate and protect healthy soils.

## f) Compost and Mulch

Compost helps kick start the transition from dirt to healthy soil, providing water, oxygen, and essential microbes. Once the plants grow, their own leaf litter and root systems can act as self-made mulch and compost. Compost itself is pre-decomposed organic matter through the use of carefully selected and fed microorganisms and worms. While compost is added and mixed within the soil itself, mulch is more of a topping, placed over the soil to help maintain soil moisture, prevent weed growth, and mitigate effects of temperature change on root systems. It is recommended to use organic materials for mulch, like untreated wood chips and leaf litter, as they will decompose overtime and add nutrients back into the soil.

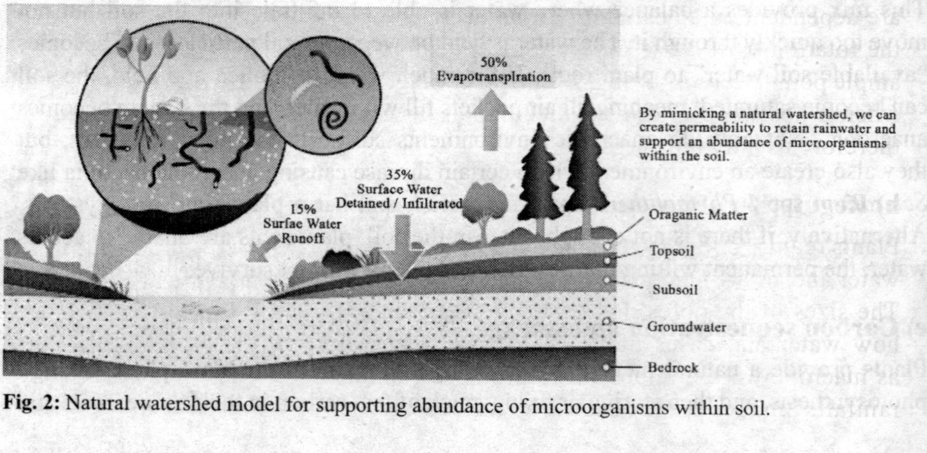


**Fig. 2:** Natural watershed model for supporting abundance of microorganisms within soil.

## g) Arbuscular Mycorrhizal fungi

Arbuscular mycorrhizal fungi are beneficial soil organisms that contribute to many aspects of soil health. Mycorrhizal fungi form a symbiotic association with plant roots. Symbiosis is a close association between different species. This association provides the fungus with relatively constant and direct access to carbohydrates supplied by the plant. In return, the plant benefits from the ability of the fungus to grow out into the soil, creating a threadlike network of fungal biomass known as hyphae or mycelium, thus effectively increasing root volume. Mycorrhizal fungi are dependent on the host plant for an energy source and cannot survive for long periods of time without a plant host. Approximately 80 percent of land plants form the symbiotic relationship with mycorrhizal fungi. A few notable crops and weeds that are nonmycorrhizal include brassicas (broccoli, cabbage, radish, canola, etc.) and chenopods (spinach, chard, lambsquarters, etc.).

Mycorrhizal fungi are especially effective in helping plants acquire phosphorus, a nutrient that is highly immobile in the soil. Because of the low mobility, when plant roots extract phosphorus from the soil, a phosphorus depletion zone develops around the root. Mycorrhizal fungi act as an extension of the plant root system, acquiring phosphorus from non-depleted zones and transporting it to the root. The external hyphae of mycorrhizal fungi also improve soil aggregation by exuding a gluelike compound called glomalin. Glomalin helps soil particles stick together in aggregates that resist erosion and maintain soil porosity.

Mycorrhizal symbioses increase a plant's stress tolerance. The network of fungal hyphae around the roots can block infection of the plant roots by plant pathogens. Mycorrhizal fungi can also suppress plant pathogens by enhancing plant nutrition, increasing root toughness, changing the chemical composition of the plant tissues, alleviating abiotic stress, and changing the microbial community on roots.

Tillage disrupts the network of delicate fungal strands, reducing populations. High levels of phosphorus in the soil also suppress mycorrhizal populations because plants are less likely to support the symbiosis. Since mycorrhizal fungi are dependent on a host plant for an energy source, if a nonhost crop is grown in the rotation, will cause populations to decline over time. Most native soils have ample populations of living mycorrhizal fungi or dormant spores that will awaken when a host crop is grown. Inoculation of field soil with mycorrhizal fungi is therefore usually unnecessary.

## h) Regulation of air and water in soil

Plants require both oxygen and water in the root zone for optimum growth. In soil, water and air are held in the pore space between soil particles and soil aggregates. The sizes of the pores that occur between and within soil aggregates determine how water and gases move in and are held by the soil. Larger pores, known as macropores, are important to promote good aeration and rapid infiltration of rainfall. Smaller pores, known as micropores, are important for absorbing and holding water. Macropores are often visible to the naked eye, while micropores between and within microaggregates are not. To maintain both adequate aeration and water supply for optimum plant growth, it is necessary to have both macro- and micropores in the soil.

Pores in the soil are formed when soil particles clump together into a hierarchy of aggregates. Soil organisms play an important role in developing soil aggregates and improving aggregate stability. Clay, organic matter, root hairs, organic compounds from bacteria and fungi, and fungal hyphae help "glue" soil aggregates together. Aggregate stability refers to the ability of soil aggregates to hold together against the erosive forces of water. Good aggregate stability will help maintain macropores in the soil, reduce surface crusting, promote aeration and reduce rainfall runoff, and reduce soil erosion. Aggregates also help conserve soil organic matter, as particles of organic matter that reside within aggregates are physically protected against microbial consumption.

Many large soil organisms are capable of moving soil and creating macropores in the soil. These include such organisms as ants, dung beetles, and earthworms. Earthworms are probably the best known soil organism that contributes to the development and maintenance of soil structure. The burrowing activity of earthworms benefits soil health through increased nutrient availability, better drainage, and a more stable soil structure.

### i) Soil organic matter (SOM)

Soil organic matter plays an important role in integrating many aspects of soil health. It serves as food and energy source for soil microorganisms. Organic matter increases water and nutrient retention in soil, it imparts buffering capacity to soil, it increases cation exchange capacity of soil, adsorb pollutants. Soil organic matter can be divided into labile and stable pools, each of which has different characteristics and functions in the soil. In agricultural soils, organic matter can range from 1 to 8 percent depending on climate, soil type, and soil management practices.

The labile pool of organic matter, which accounts for 5–20 percent of the total pool of soil organic matter, includes the living biomass of soil organisms and plant roots, fine particles of organic detritus, and relatively simple organic compounds such as polysaccharides, organic acids, and other compounds that are synthesized by microbial activity. Labile organic matter is readily decomposed by microbes and is the principal energy source that fuels the soil food web. It is the principal reservoir of organic nitrogen that can be readily mineralized and made available for plant use.

Polysaccharides in labile organic matter also enhance aggregate stability. When microbial consumption of labile organic matter is greater than the input of fresh organic matter into the soil, labile organic matter levels will decline. Excessive tillage of the soil can speed the decline of labile organic matter by oxygenating the soil, which increases microbial activity, and by exposing organic matter that had been protected within soil aggregates.

The stable pool of organic matter, which accounts for 60–95 percent of the total pool of soil organic matter, consists of organic compounds that are relatively resistant to decomposition because of either their chemical structure, their adsorption to clay particles, or their protection within microaggregates. Stable organic matter contributes cation exchange capacity and water-holding capacity to soil. The pool of stable organic matter is increased or depleted slowly as only a small portion of the labile organic matter that cycles through the food web is stabilized into forms that are resistant to decomposition.

The quantity of organic matter in a given soil is the result of a balance between organic matter inputs, such as crop residues, manure, and compost, and the rate of organic matter decomposition. Organic matter inputs can be influenced by crop management, such as the use of cover crops, crop rotations, and residue management, as well as soil management, such as using organic forms of nutrients like compost and manure. The quantity of labile organic matter generally responds to changes in management practices more quickly than the quantity of stable soil organic matter, so changes in labile organic matter levels can serve as a leading indicator of long-term trends in total organic matter levels.

## 12.2. Threats of soil health

### a) Indiscriminate use of inorganic fertilizers

Conventional crop agriculture relies on soil additives instead of investing the time and effort into nourishing naturally nutrient rich soils. To get the nutrients needed for successful plant growth, nitrogen and phosphorus rich fertilizers are added to the soil indiscriminately. Because fertilizers are generally affordable, they are frequently over-applied, resulting in runoff with high concentrations of nitrogen and phosphorus. If this nutrient rich runoff reaches water bodies without being filtered through healthy soil (for example, is directed to a storm drain or channelized river heading straight to the coastline) the excess nutrients cause "eutrophication" and algal growth. These algae blooms can result in "dead zones", or areas of water with no dissolved oxygen that are unable to sustain aquatic life. Recent studies have also found that fertilizers introduce salts into the environment, causing freshwater environments to become saltier, a phenomenon known as "Freshwater salinization syndrome". This "syndrome" changes important ecological parameters, including the pH, and releases toxic metals and harmful nitrogen compounds from streambeds and soils. In addition to becoming an aquatic pollutant, nitrogen rich fertilizers can also contribute to atmospheric pollution (releasing nitrous oxide) and soil acidification. To help prevent the over-use of nitrogen rich fertilizers, developing affordable, rapid soil testing methods is helpful for farmers to use when determining compost or fertilizer needs. Adopting organic farming practices and soil test based fertilsier management is one viable opportunity to check over use of fertilizers and chemical amendments.

This soil test based fertilizer recommendations approach was put forth by Ramamoorthy (1971). The underlying principle of this approach is to get as a large variation as possible in the soil fertility levels in one and the same field in a given locality where an elaborate field experiment is conducted so that the real relationship between the yield and the level of soil fertility could be evaluated without interference from other factors affecting the yield.

A large field of certain amount of natural variation in fertility is chosen. It is divided into four strips which are treated with four different doses of N, P and K viz., control ($N_0P_0K_0$), half of the normal dose ($N_{1/2}P_{1/2}K_{1/2}$), normal dose ($N_1P_1K_1$) and double normal dose ($N_2P_2K_2$) of N, P and K fertilizers that are usually applied to the high yielding varieties of the crops in the country. Exhaustive crop like sorghum, maize, wheat or rice are grown on this field to deplete the soil fertility level in the strips treated with fertilizers. An experiment is then laid out on this field for the purpose of Soil Test Crop Response (STCR) correlation with 21 treatments having different N, P and K randomly arranged in each strip in such a way that there is an untreated check plot on either side of the treated plot. The soil test crop response correlation work is then carried out taking the yield of treatments and the average of the two controls (check) plots into consideration.

### Philosophy of soil test fertilizer recommendations

The goal of soil testing is to help farmers achieve economical optimum yields while protecting the environment. The basic philosophy of soil test fertilizer recommendations is:

- Base them on soil test results
- Recommend that lower-testing soils be built up to higher test levels by adding extra fertilizer;
- Apply maintenance amounts of plant nutrients to higher-testing soils to keep them there and to keep productivity high; and
- Do not apply specific nutrients to soils testing very high in these nutrients.

## b) Non-native plants and invasive species

Conventional agriculture and landscaping generally use exotic, non-native plants. This is especially common for fruit and vegetable crops, as well as ornamental greenery. For instance, soy beans are native to East Asia, yet they are grown all over the Midwest of the United States. Because soybean plants developed to thrive with the conditions, pests, nutrients, and climate of East Asia, soybeans require a lot of habitat manipulation to successfully grow in the Midwestern United States. This includes irrigation to make up for different soil types and climate, application of pesticides to prevent damage or competition from Midwestern bugs, plants, and critters that they naturally wouldn't be exposed to, and addition of fertilizers to make up for the difference in soil nutrient levels. To note, fertilizers are also used to increase a crop's production, be it quantity or size of fruit/vegetable.

### c) Pesticides and Herbicides

By their very nature, pesticides and herbicides are toxic, as they are meant to kill unwanted plants and animals. However, there are certain herbicides that are not toxic to mammals (including humans), so be sure to read the ingredients carefully. Unfortunately, like fertilizers, pesticides and herbicides rarely stay where they are applied, and can enter the watershed and environment through runoff and airborne transport. Even when applied carefully, certain pesticides are able to penetrate the skin of various fruits and vegetables, exposing humans and wildlife to toxic consumption, even after rinsing. Pesticide consumption and exposure has been linked a wide-variety of human and wildlife health issues

### d) Soil pollution

Soil pollution is change in physical, chemical and biological conditions of the soil through man's intervention resulting in soil degradation in soil quality. Soil pollutants have an adverse effect on the physical chemical and biological properties of the soil and reduce its productivity. Pesticides, fertilizers, organic manure, chemicals, radioactive wastes, industrial effluents, municipal wastes, discarded food, clothes, leather goods, plastics, paper, bottles, and tins /cans - all contribute towards causing soil pollution. Chemicals like iron lead mercury, copper, zinc, cadmium, aluminium, cyanides, acids and alkalies etc. are present in industrial wastes and reach the soil either directly with water or indirectly through air. (Eg. through acid rain). The improper and continuous use of herbicides, pesticides and fungicides to protect the crops from pests, fungi etc. alter the basic composition of the soils and make the soil toxic for plant growth. Organic insecticides like DDT, Aldrin, BHC etc. are used against soil borne pests. They accumulate in the soil as they degrade very slowly by soil and water bacteria. Consequently, they have a very deleterious effect on the plant growth reducing the yield and size of fruit. The degradation products of pollutants may be absorbed by the plants from where they reach the animals and man through the food chains. Radioactive wastes from mining and nuclear processes may reach the soil via water or as 'fall-out'. From the soil they reach the plants and then into the grazing animals (livestock) from where ultimately reach man through milk and meat etc. resulting in retarded and abnormal growth of man. Human and animal excreta without proper processing used as organic manure to promote crop yield, pollute the soil by contaminating the soil and vegetable crops with the pathogens that may be present in the excreta.

Soil pollution can lead to water pollution if toxic chemicals leach into groundwater, or if contaminated runoff reaches streams, lakes, or oceans. Soil also naturally contributes to air pollution by releasing volatile compounds into the atmosphere. Nitrogen escapes through ammonia volatilization and denitrification. The decomposition of organic materials in soil can release sulfur dioxide and other sulfur compounds, causing acid rain. Heavy metals and other Potentially toxic elements are the most serious soil pollutants in sewage. Sewage sludge contains heavy metals and, if applied repeatedly or in large amounts to soil may accumulate heavy metals in soil and consequently become unable to support plant life. Therefore, ntensification of agricultural production by practices of irrigation (causes Salinisation), excessive fertilisers, pesticides, insecticides etc. have created the problems of soil pollution.

The rapid industrialization of agriculture, expansion of the chemical industry, and the need to generate cheap forms of energy has caused the continuous release of manmade organic chemicals into natural ecosystems. Consequently, the atmosphere, bodies of water, and many soil environments have become polluted by a large variety of toxic compounds. Many of these compounds at high concentrations or following prolonged exposure have the potential to produce adverse effects in humans and other organisms. These include the danger of acute toxicity, mutagenesis (genetic changes), carcinogenesis, and teratogenesis (birth defects) for humans and other organisms. Some of these manmade toxic compounds are also resistant to physical, chemical, or biological degradation and thus represent an environmental burden of considerable magnitude

## 12.3. Soil health assessment

Given the large number of soil characteristics that can be measured as indices of the soil health, a minimum data set (MDS), i.e. a set of specific soil measurements considered as the basic requirement for assessing the soil health needs to be selected. Some attributes for any indicators of MDS must be (i) compatible with basic ecosystem processes in soil as well as physical or chemical indicators of soil health, (ii) sensitive to management in acceptable time frames, (iii) easy to assess or measure, (iv) composed of robust methodology with standardized sampling techniques, (v) cost-effective, and (vi) relevant to human development goals, food security, agricultural production, sustainability and economic efficiency.

Ideal soil health Indicators should correlate well with ecosystem processes, Perform the functions necessary for its intended use, integrate soil physical, chemical, and biological properties & processes, be accessible to many users, be sensitive to management & climate, be interpretable.

**Soil health indicators**: Some of the commonest soil health indicators are

**Aggregation**: Aggregate stability is a measure of a soil's ability to remain aggregated despite stresses from tillage, raindrop impact, etc.

**Soil Reaction (pH):** Soil pH is an important indicator because it regulates nutrient availability/toxicities.

**Salinity and Sodicity:** Salinity is an excess of soluble salts in the soil. Sodicity is an excess of sodium in sodic soils.

**Organic Matter:** Organic matter comprises all carbon-based constituents of the soil, including living organisms, decaying plant material, and humus. The amount of organic matter in the soil is an important indicator because it is the storehouse for the energy and nutrients used by plants and soil organisms.

**Microbial activity:** in the rhizosphere can also affect nutrient availability to the plant through complex of interactions. The effects of microbial interactions on nutrient availability have practical implications for improving the efficiency of use of fertilisers. The physical and chemical components of soil are important, but the organisms that live in the soil ensure that it remains fertile and productive in the long term.

**Biological Diversity-**is the number of species in the soil or plant community, either at a particular time or over time. Most soils, except for highly disturbed or young soils, have inherent biological diversity. Diversity is an important indicator because biological composition of the soil can affect nutrient and pest cycles, disease suppression and pollutant degradation, as well as development of soil structure. Soils have redundancy for organisms responsible for organic matter decomposition and nutrient cycling. Measuring soil health provides site-specific interpretations for indicator results.

**Categories of soil health indicator:** Typical soil tests only look chemical indicators. Soil health attempts to integrate all three types of indicators. These categories do not neatly align with the various soil functions. Broadly indicator's categories are:

- **Physical indicators of soil health:** These indicators provide information about soil hydrologic characteristics such as water entry and retention which influences availability to plants. Some indicators are related to nutrient availability by their influence on rooting volume and aeration status.
- **Chemical indicators of soil health:** These indicators can give information about the equilibrium between soil solution and exchange sites, plant health, nutritional requirements of plant, levels of soil contaminants and their availability for uptake by plants.
- **Biological indicators of soil health:** These indicators can tell about the organisms that form the soil food web, which are responsible for decomposition of organic matter and nutrient cycling. individuals and species, that perform similar jobs or niches, can indicate a soil's ability to function or bounce back after disturbance (resistance and resilience). The set of indicators used to determine soil health is also called a minimum data set. The table below shows the relationship between indicator type and soil function.

**Combining indicators scores:** Once scored, indicators can be combined in a variety of ways, such as additive, weighted or multiplicative indexes.

## Parameters of evaluating soil health

**i)Minimum database method:** When the dataset is few or less selecting only five soil attributes, namely acidity (pH), OM, phosphorus, potassium, and EC can be combined to construct an index to represent the soil quality and health. The SHI constructed using the formula

SHI = (DpH + DOM + DP + DK + DEC)/5

where DpH = 1, if pH > 6.5 and otherwise 0

DOM = 1, if OM > 2% and otherwise 0,

DP = 1, if P > 20 kg/ha and otherwise 0,

DK = 1, if K > 80 kg/ha and otherwise 0, and

DEC = 1, if EC < 2.0 and otherwise 0,

D = constant.

SHI is bounded between 0 to 1, and higher the SHI, better is the health of the soil. Based on the SHI values, the soil health scale can be rated as very good (>0.7), good (0.6– 0.7), average (0.5–0.6), and poor (0.4–0.5).

### ii) Biochemical soil fertility index method

To compare the effect of a varied organic and mineral fertilization on the soil bioactivity in many year fertilization experiments (LTFE) it can be useful to involve indices which would define the relationship between the activity and the soil fertility. Such indices prefer the use of enzymatic activity, contrary to defining the count of selected groups of microorganisms, since defining the activity of enzymes is more convenient and easier to carry out serial analyses. Based on the obtained values of the enzymatic activity and the content of Ntotal and Corg Biochemical Soil Fertility Index (B) is calculated as:

**B = Corg + Ntotal + DHA + Fal + Prot + Amyl**

where: Corg – OC%

Ntotal – total nitrogen content, %

DHA – dehydrogenases activity, $cm^3\ H_2 \cdot kg^{-1} \cdot 24h^{-1}$

Fal – alkaline phosphatase activity, mmol $PNP \cdot kg^{-1} \cdot h^{-1}$

Prot – proteases activity, mmol $N\text{-}NH4^{+} \cdot kg^{-1} \cdot h^{-1}$

Amyl – amylases activity, mg of decomposed $starch \cdot h^{-1}$

In order to classify the respective soil samples to evaluate soil fertility, the values of the biochemical fertility index can be divided into the following ranges: 3 to 4 – low fertility, 4 to 5 – average fertility, 5 to 6 – high fertility, 6 to 7 – very high fertility.

**iii)Biological index of fertility (BIF)**: Several indices have been proposed to assess soil health which are mostly microbial in nature. They use a weighted average to calculate the BIF.

BIF = (DA+K x CA)/2

Where DA represent dehydrogenase and CA catalase activityy, and K is a proportionality coefficient.

**iv) Enzyme activity (EA)** as biological index based on 5 different enzymes given by the expression: EA= 0.2(TPF + Catalase/10 + Phenol/40 + amino-N/2+ amylase/20)

**v) Sustainable soil health index (SSHI):** It is usually based on the area of triangle in which nutrient index, microbial index and crop index of soil represents the three vertices of a triangle. Nutrient index can be calculated from the analysis of different soil parameters. Microbial index can be calculated as soil microbial and biochemical activities and crop index by measuring of crop yield parameters. Different indicators can be assessed in order to compare the effect of different sources of nutrients such as green manure, fym, and chemical fertilizer on cropping systems/patterns. The nutrient index can be calculated by following the different chemical parameters. In calculating NI each parameter is divided by the respective threshold value (arithmetic mean value of treatments for a parameter in field experiment). Thus each parameter has five calculated threshold value and each field experiment is being considered as a separate systems with difference equilibrium with reference to nutrient status, microbial activity and productivity. The sustainable soil health index value of the treatment (Iji)) calculated by dividing the value by the respective threshold of a parameter is given as:

**Iji= Aij/Thj**

Where Iji is the index value for ith treatment corresponding to jth parameter in an experiment,

Aji is the actual measured value for ith treatment and jth aparameter in an experiment and

Thj is the threshold for jth parameter.

**vi) Microbial index of soil health (MIi):** Microbial index can be calculated as the nutrient index, the measured microbial parameters as Microbial Biomass Carbon (MBC), Microbial Biomass Nitrogen (MBN), Potential Microbial Nitrogen (PMN), Soil Respiration (SR), Dehydrogenase activity (DA). The microbial index for each treatment can be calculated as an average of index values of all the parameters in each experiment.

$$MIj = 1/8 \sum_{j=1}^{8} Iij$$

where MIi is the microbial index for ith treatment and i is the number of parameters considered in deriving microbial index.

**vii) Soil health evaluation by Larson and Pierce, (1991) :** In this model they have proposed more parameters for evaluating soil health parameters. It is suggested that changes in indicator values reflect the combined effects of land use. If the changes in health parameters are positive and higher, then it is of better soil health. A brief methodology is as follows:

a) **Weights of the indicator**-The contribution of each indicator towards soil health is different and can be indicated by weighing coefficient. Several ways are there to assign weighing coefficient. This may include existing soil condition, cropping pattern, agro-climatic condition besides others. The sum of all weights is normalized to 100%. The indicators are subdivided based on their marks into Class I, Class II, ClassIII and Class IV. The Class I is the most suitable for plant growth, Class II is suitable for plant growth with slight limitations, Class III with more serious limitations than Class II and Class IV with severe limitation to plant growth. Marks of 4,3,2,1 are given to each class from I to IV.

b) **Quantitative evaluation of changes in soil health**- The concept of Relative Soil health index (RSHI) useful for quantitative evaluation of soil health. The equation is given as

RSHI= (SHI/SHIm) x 100

Where, SHI = Soil health index

SHIm = maximum value of SHI.

The maximum value of SHI for soil is 400 and the minimum value is 100.

SHI=$\sum$ Wi Ii

Where, Wi= Weights of the indicators; Ii= the marks of the indicators classes

Then SHI of each indicator can be calculated separately by multiplying weights of indicators and marks allotted to each class. An optimized soil in any region will have a normalized RSHI score of 100, but actual soil readings will have a value of less than 100 which indicate their deviation from the optimum range. According to the RSHI values, soils can be classified into 5 classes from best to worst, represented as I, II, III, IV and V respectively.

**Table 1:** Classification of soil classes based on RSHI values

| Soil health Classes | RSHI value |
|---|---|
| I | 90-100 |
| II | 80-90 |
| III | 70-80 |
| IV | 60-70 |
| V | <60 |

By comparing RSHI values, soil health in different regions can be compared even if they are evaluated with different evaluation systems, weightings and classes. Similarly, the ΔRSHI will quantify changes in soil health in a comparable way between two places. Changes in soil health (ΔRSHI) can be further grouped as

**Tabel 2:** Grouping based on change in soil quality

| Change Classes | ΔRSHI |
|---|---|
| Great change | >10 |
| Moderate increase | 10 to 5 |
| Slight increase | 5 to 0 |
| Slight decrease | 0 to 5 |
| Moderate decrease | -5 to- 10 |
| Great decrease | <-10 |

ΔRSHI can also be used to quantify small changes during a specified time period. The RSHI and ΔRSHI can provide useful tools for evaluating spatial or temporal changes in soil health. With this indexing, it is possible to quantify the degree and direction of soil health changes as affected by different land use and management and intrinsic quality of the original soil.

### viii) Rice soil health index

Rice cultivation by flooding is unique system followed for to facilitate rice roots to access plant nutrients, for effective weed control and to reduce percolation loss of water, has many beneficial effects of general amelioration of chemical fertility, increased availability of nutrients like N, P, K, Ca, Mn, Fe, Si, etc, preferential accumulation of organic matter and biological N fixation, reduced toxicity of Al and Mn, while availability of sulfur (S) may be reduced due sulfate reduction. Flooding soil is a great pH neutralizer. In problem soils this neutralizes acidity and alkalinity thereby influencing favorably to an extent in the release and availability of plant nutrients. On the contrary, the electro- chemical changes upon flooding could also lead to release of certain nutrients into toxic levels (Fe, $H_2S$ toxicity), and deficiency of micronutrients like Zn, Cu etc while the physical changes could constrain establishment and nutrition of non rice crops in the cropping system. Some indicators are crop specific. For instance, the optimum score for soil pH depends on what crop is being grown. Most indicators are site-specific, including interpretation based on inherent soil properties.

### Scoring and indexing system for RSHI

RSHI = $\sum$ (Sp + Sc+ Sf+ Sb)/18

RSHI = $\sum$ [(Sp1+ Sp2) x Scores) + $\sum$ (Sc1+ Sc2 + Sc3 x Scores) + (Sf1+Sf2…Sfn12 x Scores) + (Sb1 x Scores)]/18

RSHI = (Two physical parameters x Their scores) + (Three chemical parameters x Their scores) + (Twelve fertility parameters x Their scores) + (One Biological parameters x its score) whole divided by 18 makes a five way soil health indexing and scoring system

Hypothesis: A rating of 100 per cent expresses most favourable conditions and lower per cent indicates less favourable conditions.

The Rice Soil Health Index

0.8 –1.0 is assigned Grade 1 – Excellent

0.60–0.79 is assigned Grade 2 –Very good

0.40–0.59 is assigned Grade 3 – Good

0.20–0.39 is assigned Grade 4 – Poor

0.10–0.19 is assigned Grade 5 – Very poor

## 12.4. Soil health management

By understanding how soil processes (decomposition of organic matter, mineralization, immobilization, exchange of air, water filtration, retention and movement etc.) that support plant growth and regulate environmental quality are affected by management practices, it is possible to design a crop and soil management system that improves and maintains soil health over time.

1. Reduce inversion tillage and soil traffic: Excessive tillage is harmful to soil health in the following ways.
   - Excessive tillage increases oxygen in the soil, stimulating microbial activity, and results in the quick decomposition of organic matter.
   - Tillage also disrupts soil aggregates, exposing particles of organic matter that had been physically protected within aggregates to microbial consumption. If additions of organic matter are not sufficient to counteract the losses from decomposition, organic matter levels will decline over time, reducing soil health.
   - Inversion tillage also reduces the soil coverage provided by crop residues, leaving soil more exposed to erosion (see tillage sidebar).
   - Tillage can also disrupt the hyphal network of mycorrhizal fungi, which can lead to their decline over time.
   - Most inversion and noninversion tillage methods compact the subsoil, creating a plow pan, which restricts root growth and access to water and nutrients in the subsoil. Excessive wheel and foot traffic can compact the surface soil, reducing macroporosity and impeding root growth (see compaction sidebar).
   - Compaction and removal of surface residue may contribute to reduction in soil moisture and living space for soil-dwelling organisms.
   - Natural control of pest insects in soil may be enhanced in conservation tillage systems. Diversity and abundance of arthropod predators associated with the soil surface can be greater under conservation tillage management in comparison to conventional inversion tillage.
   - Beneficial insects associated with the soil are more likely to survive in fields where noninversion (e.g., chisel plowed) tillage is used.

   However, Some tillage is still a necessary practice in certain production systems, especially organic systems that do not use herbicides for weed control. When tillage is used, it is necessary to offset the increased rate of organic matter decomposition with addition of more organic matter through crop residues, manure, and compost.

2. **Increase Organic Matter Inputs:** To maintain or increase soil organic matter levels, inputs of organic matter must meet or exceed the losses of organic matter due to decomposition. Healthy crops can be a valuable source of organic matter, and crop residues should be returned to the soil to the extent possible.

Incorporation of cover crops or perennial crops and judicious additions of animal and green manure and compost can also be used to increase or maintain soil organic matter.

3. **Use Cover Crops:** Cover crops contribute numerous benefits to soil health.
   - Cover crops keep the soil covered during the summer/winter and other periods of time when crops are not growing, reducing the risk of erosion.
   - The biomass produced by cover crops is usually returned to the soil, enhancing organic matter levels.
   - Cover crops with taproots can create macropores and alleviate compaction.
   - Fibrous-rooted cover crops can promote aggregation and stabilize the soil.
   - Species of cover crops that host mycorrhizal fungi can sustain and increase the population of these beneficial fungi.
   - Legume cover crops can add nitrogen to the soil through nitrogen fixation.
   - Cover crops can retain nitrate and other nutrients that are susceptible to leaching losses.
4. **Reduce Pesticide use and provide habitat for beneficial organisms**: Beneficial insects that contribute to biological control or pest organisms can be harmed by the application of broad spectrum insecticides. Farmscaping is a whole-farm, ecological approach to increase and manage biodiversity with the goal of increasing the presence of beneficial organisms. Farmscaping methods include the use of insectary plants, hedge rows, cover crops, and water reservoirs to attract and support populations of beneficial organisms such as insects, spiders, amphibians, reptiles, bats, and birds that parasitize or prey on insect pests. Farmscapes placed in contours between fields, steep ditches, or places that are easily eroded give stability to the soil. Farmscaping can also be used as a filter strip to prevent water runoff and soil erosion. Plants used in farmscapes contribute to healthy soil by adding organic matter, the base of the soil food web.

   Using diverse nutrient sources can help maintain soil health. Manure and compost add organic matter as well as an array of nutrients, but using just compost or manure to meet the nitrogen needs of the crop every year can result in excessive phosphorus levels in the soil. Combining modest manure or compost additions to meet phosphorus needs with additional nitrogen inputs from legume cover or forage crops in a crop rotation can help balance both nitrogen and phosphorus inputs.
5. **Practice crop rotation**: Diverse crop rotations will help break up soilborne pest and disease life cycles, improving crop health. Rotations can also assist in managing weeds. By growing diverse crops in time and space, pests that thrive within a certain crop are not given a chance to build their populations over time. Rotating crops can also help reduce nutrient excesses.

6. **Nutrient management:** Carefully planning the timing, application method, and quantity of manure/ compost, and other fertilizers will allow to meet crop nutrient demands and minimize nutrient excesses. Healthy and vigorous plants that grow quickly are better able to withstand pest damage. However, over fertilization to crops can increase pest problems. Increasing soluble nitrogen levels in plants can decrease their resistance to pests, resulting in higher pest density and crop damage. Maintaining a soil pH appropriate for the crop to be grown will improve nutrient availability and reduce toxicity. Maintaining adequate calcium levels will help earthworms thrive and improve soil aggregation.

## a) Nitrogen (N) Management

- Nitrate nitrogen is susceptible to leaching losses because the negative charge of the molecule is not held by cation exchange sites of soil particles. Leaching occurs mainly in the fall, winter, and early spring.
- Nitrogen in urea-containing fertilizers and manure is susceptible to volatilization losses as ammonia gas when not incorporated into the soil.
- Nitrate nitrogen can be lost to the atmosphere through conversion into nitrous oxide and nitric oxide gases by microorganisms in warm, poorly aerated soil.
- Nitrogen losses can be minimized with appropriate timing and application of fertilizers and manures and by using cover crops to limit leaching losses in the winter.
- Split application of nitrogenous fertilizers to prevent losses due to leaching.
- The use of slow release nitrogenous fertilizer like urea formaldehyde, sulphur coated urea, Neem coated urea etc.
- Use of nitrification inhibitor (Eg: N-serve) to retard the conversion of $NH_4^+$-N to $NO_3^-$ -N for preventing leaching and make it available to crops for quite longer period.
- By the integration of inorganic N with organic sources.
- The soil physical condition can be optimized besides adding nutrients to the soil.

## b) Phosphorus (P) Management

- Phosphorus is tightly bound to soil particles and does not easily diffuse through the soil.
- Mycorrhizal fungi can assist plant roots in P acquisition in low-P soils.
- Adding organic matter can mask the P binding sites on soil particles, increasing P availability.
- Phosphorus can accumulate to excessively high levels when P inputs in manure and fertilizer exceed P removal by crops; this can occur in soil that receives annual manure applications at rates to supply crop nitrogen needs.

- Erosion can transport soil particles with high levels of P into waterways where P can become a pollutant.
- Environmental P pollution can be limited by reducing erosion and maintaining soil P levels in the optimum range of 30–50 ppm Mehlich 3 Phosphate.
- P fixation can be reduced by judicious application of organic manures,
- Application of P fertilizer by placement reduces fixation and P availability to crops
- Inoculation (either seedor soil) with phosphorus solubilising bacteria like *Pseudomonas, Bacillus megathrium var. phosphaticum.*

**c) Potassium management**

- Preventing leaching loss of K by split application on light soils,
- Applying organic manure helps efficient management of K in soil.
- Balanced application of nutrients ensure proper K availability to crops.

**d) Sulphur management** : Sulphur in soil solution is present as $SO_4^{=}$ and more prone for leaching losses. The losses $SO_4^{=}$ can be prevented by

1) Applying organic manures to improve water holding capacity of the soil and it also acts as a source of S.
2) Sulphur oxidation can be facilitated by providing oxidized conditions in the soil.

**e) Iron (Fe) management** : Most available form of iron is $Fe_2^{+}$. All the measures which govern the soil reaction will influence Fe availability. Application of organic manures including greenmanuring improve the use efficiency of iron by

1) Acidifying the rhizosphere due to the release of organic acids
2) Supplementing with iron after decomposition
3) Act as substrate for heterotrophic bacteria that can reduce ferric to ferrous form (eg., *Bacillus, Clostridium and Klebsiella etc.).*
4) The microbes also produce chelating ligands called as 'siderophores' that can form complex with $Fe^{3+}$, which can be absorbed into the plant.
5) Reclamation of alkali soils
6) If deficiency appears on standing crop foliar application of Fe.

**f) Manganese management**

- Using acid forming nitrogen fertilizers
- Compacting loose soils
- Preventing excessive drying
- Supply of easily decomposed organic matter ,which on conversion creates reducing conditions and thus Mn released .It is a practical means of Mn supply than to add Mn fertilizer .About ½ to 1.0 kg of Mn /ha/year is removed by crops in general

### g) Copper management

Copper is highly immobile in soil and as such needs thorough mixing with top soil. Copper removal by crops in general varies 30-100 g/ha /year. Copper deficiency can be overcome by Copper sulphate [Blue Vitriol] is the oldest water soluble Cu-fertilizer. It can be applied as soil dressing or foliar nutrient. However; its acidic side effects are likely to cause leaf scorch on foliar application. Hence less caustic agents like green copper $Cu_2Cl(OH)_3$ or Cu-chelates are safe to use

### h) Zn management

Zn fertilizer should not be applied with phosphatic fertilizers. Maintaining the soil pH between 5.5 – 6.5 by applying organic manures. Improvement of natural zinc (Soil zinc) can be done by the application of acid forming N -fertilizers to combat Zn-deficiency .Crop removal of Zn varies from 100-400 g/ha /year

**i) Boron management:** Boron removal by crops is about 50 g/ha/year. Deficiency of B can be met by Borax (Na –tetra borate) a historically famous water soluble boron fertilizer. It can be applied to soil or foliage

## 12.5. Soil Quality

Soil quality is a measure of the capacity of a soil to perform necessary functions. Soil functions include providing nutrients and water to plants, filtering and cleaning water, regulating temperatures, recycling and storing nutrients, and providing habitats for organisms. Soil quality and soil health are vital for sustainable agro-ecosystem management and survival on planet Earth. In order to evaluate soil quality, we need to use a set of indicators/ factors affecting soil quality. These factors/indicators may be qualitative or quantitative.

**Table 3:** Key soil indicators for soil quality assessment

| Selected indicator | Rationale for selection |
|---|---|
| Organic matter | Defines soil fertility and soil structure, pesticide and water retention, and use in process models |
| Topsoil-depth | Estimate rooting volume for crop production and erosion |
| Aggregation | Soil structure, erosion resistance, crop emergence an early indicator of soil management effect |
| Texture | Retention and transport of water and chemicals, modeling use |
| Bulk density | Plant root penetration, porosity, adjust analysis to volumetric basis |
| Infiltration | Runoff, leaching and erosion potential |
| pH | Nutrient availability, pesticide absorption and mobility, process models |
| Electrical conductivity | Defines crop growth, soil structure, water infiltration; presently lacking in most process models |
| Suspected pollutants | Plant quality, and human and animal health |
| Soil respiration | Biological activity, process modeling; estimate of biomass activity, early warning of management effect on organic matter |

| Selected indicator | Rationale for selection |
|---|---|
| Forms of N | Availability of crops, leaching potential, mineralization/ immobilization rates, process modeling |
| Extractable N, P and K | Capacity to support plant growth, environmental quality indicator |

**Table 4:** Interrelationship of soil quality indicators

| Selected indicator | Other soil quality indicators affecting the selected indicator |
|---|---|
| Aggregation | Organic matter, microbial (especially, fungal) activity, texture |
| Infiltration | Organic matter, aggregation, electrical conductivity, ex-changeable sodium percentage (ESP) |
| Bulk density | Organic matter, aggregation, topsoil-depth, ESP, biological activity |
| Microbial biomass | Organic matter, aggregation, bulk density, pH, texture, ESP and/or respiration |
| Available nutrients | Organic matter, pH, topsoil-depth, texture, microbial parameters (mineralization and immobilization rates) |

## 12.5.1. Importance of soil quality

Soils quality has important impacts on agricultural productivity, water quality, and the global climate. Soils make it possible for plants to grow by mediating the biological, chemical, and physical processes that supply plants with nutrients, water, and other elements. Microorganisms in soils transform nutrients into forms that can be used by growing plants. Soils are the storehouses for water and nutrients. Plants draw on these stores as needed to produce roots, stems, leaves, and, eventually, food and fiber for human consumption. The biological, chemical, and physical processes soils make possible are a fundamental resource on which the productivities of agricultural and natural ecosystems depend.

The soil, which interacts with landscape features and plant cover, is a key element in regulating and partitioning water flow through the environment. The biological, chemical, and physical processes that occur in soils buffer environmental changes in air quality, water quality, and global climate. Soil degradation through erosion, compaction. loss of biological activity, acidification, salinization, or other processes can reduce soil quality. These processes reduce soil quality by changing the soil attributes, such as nutrient status, organic and labile carbon content (organic carbon is the total amount of carbon held in the organic matter in the soil; labile carbon is that fraction of organic carbon that is most readily decomposable by soil microorganisms), texture, available water-holding capacity (the amount of water that can be held in the soil and made available to plants), structure, maximum rooting depth, and pH (a measure of the acidity or alkalinity). Some changes in these soil attributes can be reversed by external inputs. Nutrient losses, for example, can be replaced by adding fertilizers. Other changes such as loss of the soil depth available for rooting because of soil erosion or degradation of soil structure because of subsoil compaction are much more difficult to reverse.

## 12.5.2. Soil quality indices

Soil quality indices are decision tools that effectively combine a variety of information for multi-objective decision making. Selection of soil quality indicators or synthetic indices is guided by the goal of ecosystem management. If achieving sustainability is the goal of agroecosystem management, a soil quality index will constitute one component within a nested agroecosystem sustainability hierarchy.

Soil quality indexing involves three steps: (i) selection of soil properties/indicators constituting the minimum data set (ii) transformation of indicator scores enabling quantification of all indicators to a common measurement scale and (iii) combining the indicator scores into the index (Figures 4). Selection of soil properties/indicators of soil quality and their statistical/mathematical treatment to derive a composite index vary a lot.

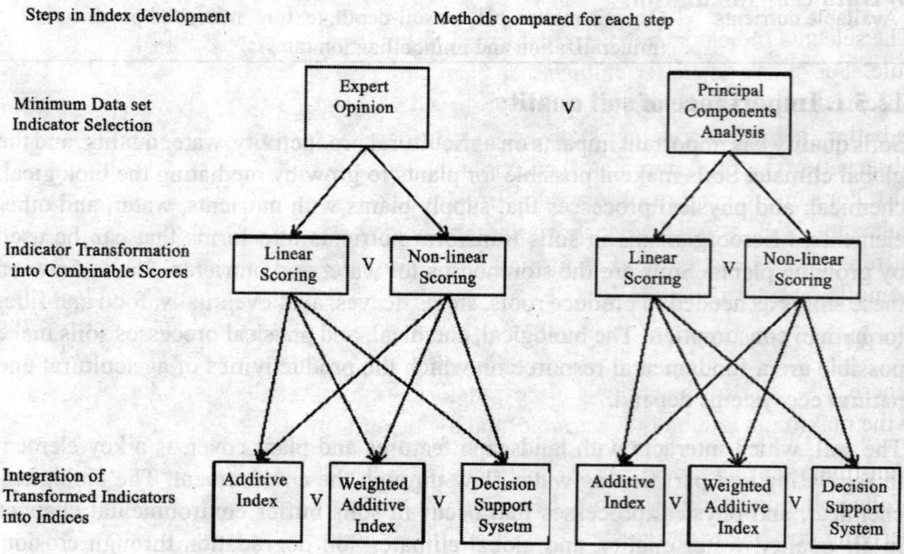


**Fig. 4:** Flow diagram depicting the three steps of index creation and the alternative methods for each step compared in this study

## a) Data compression

Principal Component Analysis (PCA) is a data compression technique designed for data that are in the form of continuous measurements, though it has been also been applied to other kind of data such as presence/ absence of an element or measurements in the form of discrete variables. Ordination, a collective term for multivariate techniques that arrange sites along axes on the basis of soil properties can help to show whether important environmental variables have been over- looked. Ordination is like a linear regression model, but with the major difference that the explanatory variables here are theoretical variable and not known environ- mental variables. Principal Components (PCs) for a data set are defined as linear combinations of the variables that account for maximum variance within the set by describing vectors of closest fit to the n observations in p-dimensional feature space, subject to being orthogonal to one another. The PCA output gives as many PCs as the input variables but it is assumed that PCs receiving high eigenvalues (setting a threshold, e.g., eigenvalues > 1) or those explaining variation in the data exceeding a limit (e.g., > 5% of the variability) are 'important' and not the others.. Contribution of a variable to a particular PC is represented by a weight or factor loading. Only the highly weighted variables are retained from each PC and highly weighted factor loadings identified based on thresholds such as those variables with absolute values within 10% of the highest factor loading or > 0.4. When more than one factor is retained under a single PC, multivariate correlation coefficients are employed to determine if variables could be considered redundant and if the variables are correlated, that with the highest value is chosen for multi dimensional scaling (MDS).

**b) Data transformation**

The selected indicators can be transformed following a linear or a non-linear scoring rule. For 'more is better' indicators, each observation is divided by the highest observed value such that the highest observed value received a score of 1. For 'less is better' indicators, the lowest observed value (in the numerator) is divided by each observation (in the denominator) such that the lowest observed value receives a score of 1. For some indicators, observations are scored as 'higher is better' up to a threshold value and as 'lower is better' above the threshold (Leibig et al. 2001). The values of different variables can be transformed to a common range, between 0.1 to 1.0 with homothetic transformation:

$$y = 0.1 + (x-b)/(a-b) * 0.9$$

where, y = value of the variable after transformation, x = the variable to transform, a = the maximum value of the variable, and b = the minimum value of the variable

Non-linear scoring functions are constructed based on literature review and consensus of the collaborating researchers. Masto et al. (2007) used the following equation for deriving non-linear scores:

$$\text{Non-linear score } (y) = 1/1+e^{-b\,(x-a)}$$

where, x = soil property value, a = the baseline or value of the soil property (the scoring function equals 0.5 and equals the midpoint between the upper threshold value and the lower threshold value), and b = slope.

The upper threshold value is the soil property value for which the score equals 1 and which corresponds to the most favourable level. The lower threshold value is the soil property value where the score equals 0 and which corresponds to an unacceptable level. Baselines are generally regarded as the minimum target values.

**c) Data integration**

There are basically two ways of integrating indicators to derive one soil quality index – by summing the scores from MDS indicators and by summing MDS variables after weighting them by considering the % variation explained by a PC, standardized to unity, as the weight for variable(s) chosen under a given PC.

The quality of a soil is determined by a combination of physical, chemical, and biological properties such as texture, water-holding capacity, porosity, organic matter content, and depth. Since these attributes differ among soils, soils differ in their quality. Some soils, because of their texture or depth, for example, are inherently more productive because they can store and make available larger amounts of water and nutrients to plants. Similarly, some soils, because of their organic matter content, are able to immobilize or degrade larger amounts of potential pollutants.

Soil management can either improve or degrade soil quality. Erosion, compaction, salinization, sodification, acidification, and pollution with toxic chemicals can and do degrade soil quality. Increasing soil protection by crop residues and plants; adding organic matter to the soil through crop rotations, manures, or crop residues; and careful management of fertilizers, pesticides, tillage equipment, and other elements of the farming system can improve soil quality.

### 12.5.3. Monitor changes in soil quality

A system to monitor changes in soil quality will require the following:

- Identification of the soil attributes that can serve as indicators of change in soil quality,
- Standard field and laboratory methodologies that can be used to measure changes in indicators of soil quality,
- A coordinated monitoring program that can quantify changes in soil quality indicators, and
- A coordinated research program designed to support, test, and confirm models that can be used to predict the impact of management practices on soil quality.

### 12.5.4. Indicators of soil quality

The quality of a soil is a composite of its physical, chemical, and biological properties. Indicators of soil quality are needed that relate to all three functions soils perform in natural and agroecosystems: (1) promote plant growth, (2) protect watersheds by partitioning and regulating precipitation, and (3) prevent air and water pollution by buffering agricultural chemicals, organic wastes, industrial chemicals, and other potential pollutants.

There are many soil properties that may serve as indicators of soil quality. The minimum data set, however, need only include those indicators that are most generally applicable to soils in varying climates and landscapes. Additional indicators could be added to the minimum data set to address properties that are particularly important in certain types of soils or regions. Table 4 presents a list of indicators that may be most useful for a minimum data set.

**Table 4:** Indicators of change in Soil quality and their relationship to soil functions

| Soil Quality Indicator | Soil Functions Promote plant growth | Regulate water Flow in soil | Buffer environmental changes |
|---|---|---|---|
| Nutrient availability | Direct | Indirect | Direct |
| Soil organic carbon | Indirect | Indirect | Direct |
| Labile carbon | Indirect | Direct | Direct |
| Texture | Direct | Direct | Direct |
| Water-holding capacity | Direct | Direct | Indirect |
| Soil structure | Direct | Direct | Indirect |
| Maximum rooting | Direct | Indirect | Indirect |
| depth Salinity | Direct | Direct | Indirect |
| Acidity/alkalinity | Direct | Direct | Indirect |

a) **Nutrient Availability:** Nutrient availability is an important soil attribute for plant productivity and water quality and is significantly altered by soil management practices. Nutrient availability can be estimated by extracting nutrients from the different components in the soil with chemicals and measuring the nutrient content in the extract. Nitrogen, phosphorus, and potassium are the major nutrients in the soil that are measured by extraction.

b) **Soil organic carbon :** Soil organic carbon or soil organic matter is perhaps single most important indicator of soil quality and productivity. Depletion of soil organic carbon is followed by depletion of plant nutrients, deterioration of soil structure, diminished soil workability, and lower water-holding capacity of the soil. The amount of organic carbon in the soil affects permeability, water retention, and hydraulic conductivity, which all determine the way rainfall is portioned and potential pollutants transported..It also alters the efficacies and fates of applied pesticides. Depletion of soil organic carbon and erosion are interrelated, since a decrease in organic carbon increases the susceptibility of a soil to erosion, thereby increasing the rate of depletion of soil organic carbon.

Because of its importance and its susceptibility to change by soil erosion, organic carbon should be included in the minimum data set and monitored periodically. Total organic carbon in the soil can be affected by management and has been shown to be directly related to the amount of organic matter added to the soil through crop residues, manures, or other sources. The total organic matter in the soil may change slowly, however; even changes restricted to the few millimeters of surface soil can have substantial effects on infiltration, aeration, and erosion. Organic residues and soil organic matter at the interface between the soil and atmosphere are extremely influential in partitioning water and influencing surface soil structure.

c) **Labile carbon:** Although total organic carbon provides important information, it is the labile carbon fraction that is most active in the soil. The amount of labile carbon is most directly related to important biological processes in the soil including rates of mineralization of nutrients, the generation of soil structure, and the attenuation of potential pollutants. Simple chemical procedures to assess the labile carbon fraction are available.

d) **Soil texture:** The soil's texture (particle-size distribution) in the surface soil layer may be altered as a result of the selective removal of fine particles during the erosion process, as a result of mixing subsoil into the surface layer during tillage (as cumulative erosion reduces the thickness of the surface soil layer) and as a result of deposition of eroded sediments on the soil surface. Changing the surface soil texture can have important effects on crop productivity, for example, by reducing the amount of nutrients or water the soil can hold or by restricting the growth of plant roots. Texture also influences the partitioning of rainfall and the flow of water and potential pollutants through the soil.

e) **Water-Holding Capacity:** An important attribute of a soil is its ability to store and release available water to plants. Plant-available water capacities are a required input for nearly all crop simulation models. The plant-available water capacities should be determined to the depth of rooting, and temporal changes in plant-available water capacities those that are either natural or induced by management or erosion should be determined in the surface layers.

f) **Soil Structure:** The term *soil structure*, as broadly has three components. The first is structural form, which refers to the geometry of the soil pore space (porosity, pore size distribution, and pore continuity). The second is aggregate stability, which refers to the size distribution and resistance of aggregates to degradation. The third is structural resiliency, which refers to the ability of the soil structure to re-form once it has been degraded. Measurements of structural form include bulk density, macroporosity (porosity (pores >60 μm), and saturated hydraulic conductivity.

Determination of compact soil layers that impede root growth are important for determination of effective soil rooting volume. Either bulk density or penetration resistance measurements (interpreted with respect to water contents) can be used to identify root-impeding layers. Compaction by wheeled traffic has direct and sometimes irreversible effects on soil structure. Texture, organic matter, and labile carbon are also related to structure, and soil management that results in changes in these soil attributes will also affect soil structure.

g) **Rooting depth:** Soil thickness has been related to crop productivity, particularly in mine reclamation studies. Soils in which the rooting depth is limited by the presence of a physical or chemical constraint are generally less productive. As limiting layers are moved closer to the soil surface as erosion removes the topsoil, crop productivity generally declines. Rooting depth varies by crop species, and the limits for various species were well defined. Maximum rooting depth should be determined at the time of physiological maturity of the crop species under study. Management of the soil can have important effects on rooting depth. Erosion reduces rooting depth by removing layers of surface soil, and compaction reduces rooting depth by creating layers in the soil that are impenetrable by crop roots.

h) **Acidity and Alkalinity:** Soil pH is a measure from which many general interpretations about the chemical properties of a soil can be made. The acidity, neutrality, or alkalinity of a soil suggests the solubilities of various compounds in the soil, the relative bonding of ions to exchange sites, and the activities of microorganisms. A pH of less than 4 indicates the presence of free acids, generally from oxidation of sulfides; a pH of less than 5.5 indicates the likely occurrence of exchangeable aluminum; and a pH from 7.8 to 8.2 indicates the presence of calcium carbonate. Acidity and alkalinity can be readily managed, in many soils, by careful management of fertilizer and lime applications.

i) **Water Regulation:** Soil quality and the changes in soil quality that occur with soil management can be expected to affect natural resource models. For example, consider the Water Erosion Prediction Project (WEPP) model in which soil quality is assumed to affect both the water infiltration and the soil erosion portions of the model.

Infiltration in the WEPP model is quantified with the Green-Ampt infiltration model (a simple mathematical equation used to estimate how much and how quickly water soaks into the soil). The soil capillary potential is assumed to be proportional to bulk density and soil texture (sand, clay, and porosity). The saturated hydraulic conductivity is determined by the amount of coarse fragments, the amount of soil cover, whether the soil is frozen, and whether crusting occurs. Thus, for example, an increase in the amount of clay decreases hydraulic conductivity and a decrease in bulk density increases water infiltration. A baseline bulk density is assumed to be proportional to the amount of sand, clay, and organic matter plus the cation- exchange capacity.

Erosion in the WEPP model is predicted by interrill and rill soil erodibility terms. For croplands, the interrill erodibility term is assumed to be a function of the soil's texture and the magnesium and aluminum concentrations. Rill erodibility is estimated from the soil organic matter, cation-exchange capacity, sodium absorption ratio, aluminum concentration, and soil texture (very fine sand, clay, and sand). The critical shear (the amount of water-induced shear that initiates sediment movement) is also quantified from soil properties such as the amounts of clay and sand, the specific surface area of the soil, and a sodium absorption ratio.

On rangelands, interrill erodibility is estimated from the amount of sand, silt, and organic matter and estimates of the soil's water-holding capacity. Rill erodibility is estimated from the amount of clay, organic matter, bulk density, and root mass in the soil. Critical shear is assumed to be proportional to the amount of sand, organic matter, and bulk density.

In general, improvements in such soil quality attributes as organic matter and bulk density can be expected to increase infiltration, reduce runoff, and decrease soil erosion (with the exception of interrill erodibility).

j) **Buffering Capacity:** Soil quality attributes are needed to make such estimations as the desirability of soils for use in waste management. For example, in

Minnesota soils, texture, pH, total organic carbon content, and cation-exchange capacity are used as indicators of the suitability of applying processed sewage sludge to the land and the application amount. Reliable measures or estimates are required for each mapping unit in the proposed area of application. In this example, texture is used as an indicator of the susceptibility of soils to leaching of chemicals, pH is an indicator of the solubility of heavy metals and their ease of uptake by plants, and total organic carbon content and cation-exchange capacity are indicators of the capacity of the soil to absorb chemicals

k) **Pedotransfer Functions:** Prediction of the direction in which soil attributes are changing can often be made without direct measurement of the specific attribute. For example, a simple regression equation could predict organic matter changes in several soils on the basis of the amount of crop residue that was added to the soil. Important hydraulic properties of soils including water retention, hydraulic conductivity, and water-holding capacity can be estimated from data on texture and organic matter content. It is also suggested the development of pedotransfer functions that can be used to evaluate changes in soil quality from a minimum data set of indicators.

Pedotransfer functions could dramatically increase the utility of a minimum data set of soil quality indicators. Estimations of changes in many important soil attributes could be simulated from measures of relatively few indicators of soil quality.

### 12.5.5. Assessment of soil quality

The selection of soil attributes for use in a minimum data set, however, depends not only on our understanding of the relation of those attributes to soil quality, but also on the utility of the attributes for use in sampling programs and in pedotransfer functions to provide more quantitative estimates of change in soil quality. Sampling can be expensive. It is important that the attributes selected for a minimum data set be as few as possible. Testing and empirical evaluation of proposed indicators will be required to identify those most suited for use based on the ease and accuracy with which they can be sampled, the degree of spatial and temporal variability of the attribute, their utility as parameters in pedotransfer functions, and their applicability to a wide range of soils, climates, and landscapes. The following discussion summarizes some of the experience with soil attributes in quantitative assessments of soil quality.

a) **Soil organic carbon and carbon management index:** Soil organic matter serves as a primary indicator of soil quality and health for both scientists and farmers. The loss of SOC following conversion of natural ecosystems to agro ecosystems occurs at rates much faster than the rates of recovery following abandonment of agricultural land use. Chemically labile carbon fractions include a variety of organic substances, e.g., water soluble carbon (carbon extracted in distilled water, 1:5 solid : liquid, shaken for 2 hrs), water soluble carbohydrates (carbohydrates in above solution). Labile fractions, microbial biomass, dehydrogenase activity and ATP levels may be highly correlated. In general, bacteria contribute more in terms of decomposition of labile/soluble components of residues and fungi of the resistant (lignocellulose) component. Microbial biomass consists of both dormant and metabolically active organisms and has been considered as an integrative indicator of microbial significance of soils. Soil organic matter is the primary source and temporary sink for plant nutrients and soil organic carbon in agroecosystems has been considered as the best surrogate for soil quality.

i) **Amount of soil carbon (Cm)** : IPCC (1997) laid out calculating soil carbon amounts following a land use change was:

$$Cm = Cn \times B \times T \times I$$

where,

Cm = the amount of soil carbon some time after land use change;

Cn = the amount of soil carbon under the original native vegetation;

B = base factor, with values varying from 0.5 to 1.1 depending on environmental factors and the type of agricultural activities following the transition and the lowest values referring to long-term cultivated aquic soils or degraded land in the tropics and the highest values to improved pasture and rice paddies;

T = tillage factor which takes on higher values (1.1) for no tillage and lower values for full tillage (0.9-1.0);

I = input factor accounting for different levels of input from different residue management systems varying between 0.8 for shortened fallow under shifting cultivation to 1.2 for high input systems, such as those receiving regular fertilizer additions.

ii) **Carbon management index (CMI): It is** a multiplicative function of Carbon Pool Index (CPI) and Lability Index (LI) as an indicator of the rate of change of soil organic matter in response to land management changes, relative to a more stable reference soil:

Carbon Pool Index (CPI) = Total C of a given land use/ Total C of the reference land use

Lability Index (LI) =

[Labile carbon content of a given land use/Non-labile carbon content of a given land use] x [Labile carbon content of the reference land use/Non-labile carbon content of the reference land use]

Carbon Management Index (CMI) = CPI × LI × 100

b) **Enzymes as indicators of organic matter quality and microbial activity:** Soil enzyme assays generally provide a measure of the potential activity, i.e., that encoded in the genotype, but this will rarely be ever expressed. Three enzymes viz., phospho- monoesterase, chitinase and phenol oxidase, as a group reflect relative importance of bacteria and fungi, as well as the nature of organic matter complex, Phosphomonoesterase (acid phosphatase) activity is often correlated with microbial biomass, fungal hyphal length and nitrogen mineralization. Chitinase is a bacterial enzyme which converts chitin, a substance intermediate in its resistance to microbial metabolism produced by fungi and arthropods, into carbohydrates and inorganic nitrogen. Phenol oxidase is produced primarily by white rot fungi, and is specific for highly recalcitrant organic matter, such as lignin-

c) **Soil microbiological degradation index (SMDI):** Computation of this index involves : (i) selection of appropriate parameters, e.g., total organic carbon, water soluble carbon, water soluble carbohydrates, microbial biomass carbon, respiration, ATP, dehydrogenase, urease, protease, phosphatase and beta-glucosidase acitivity. (ii) transformation and weighting of values and (iii) combining the scores into an index. Factor analysis can be used to identify the most important parameters. As absolute values of some parameters are bigger than those of others, the values of the selected parameters are normalized. The SMDI is the sum of the normalized and weighted values of the most important parameters.

d) **General indicator of soil quality (GISQ):** Soil organisms and biotic parameters (e.g., abundance, diversity, food web structure, or community stability) meet most of the desired criteria of soil quality indicators . Use of faunal groups as indicators for soil quality needs a choice of organisms, that (a) form a dominant group and occur in all soil types, (b) have high abundance and high biodiversity and (c) play an important role in soil functioning, e.g., food webs. General indicator of soil quality (GISQ) is based on estimation of around 50 soil properties related to macrofauna, chemical fertility, physical state, organic matter fractions and soil morphology. The computational procedure involved four steps: (i) PCA analysis of the variables allowing testing of the significance of their variation among land use types; (ii) identification of the variables that best differentiate the sites according to the soil quality; (iii) creation of sub-indicators as (a)soil physical quality, (b)chemical fertility, (c)organic matter, (d)morphology and (e)soil macrofauna with values ranging from 0.1 to 1.0; (iv) combination of all five sub-indicators into a general one. This indicator allows the evaluation of soil quality and facilitates identification of problem areas through the individual values of each subindicator.

e) **Soil fertility, land quality and farm level environment indicators:** Land quality indicators represent generic directives for the functional role of land, indicating condition and capacity of land, including its soil, weather and biological properties, for purposes of production, conservation and environmental management. A soil test is a chemical method for estimating the nutrient supplying capacity of a soil and has an edge for biological methods of evaluating soil fertility in that it can be done rapidly and before the crop is planted. Soil quality thus could be viewed as a component of land quality and the most useful soil or land quality index is the one that is able to provide early warning of adverse trends and to identify problem areas. Classical soil fertility rating is a function of the crop response to added nutrients and fertilizers recommendations are primarily based on expected financial returns from the crop from applied nutrients rather than an integrated consideration of the costs and benefits of the outcomes of fertilizer addition, e.g., of environmental cost associated with leaching and volatilization of added fertilizers.

## Review questions

What is soil health? Discuss the concepts of soil health.

How soil organic matter contributing to soil health?

What are threats to soil health due to human interventions?

List the indicators of soil health.

Explain soil health indicators

How you mange Soil health?

What is oil quality? Differentiate between soil health and soil quality.

Discuss soil quality assessment by taking any three parameters of soil quality assessment.

## Objective questions

1. The condition of the soil and its potential to sustain biological functions, maintain environmental quality, and promote plant and animal health is referred to as
   a) Soil resilience b) Soil quality
   c) Soil health d) soil adaptability
2. Soil functions as
   a) supplying fundamental resources viz., water, nutrients and oxygen required for terrestrial primary production,
   b) providing habitat to a variety of soil organisms
   c) regulating hydrological and mineral/nutrient cycling
   d) all the three
3. Litter decomposition in soil contribute to
   a) detoxification b) nutrient supply
   c) base saturation d) A&B
4. The terrestrial carbon present in soil is approximately
   a) 35% b) 25%
   c) 50% d) 75%

5. It is estimated that estimate that carbon sequestration power of soil is
   a) Two times more CO2 than all the world's plants including forests
   b) Four times more CO2 than all the world's plants including forests
   c) Three times more CO2 than all the world's plants including forests
   d) Six times more CO2 than all the world's plants including forests.
6. Soil health is degrades by
   a) Tilling b) Overgrazing
   c) clear cutting d) all the three
7. Mycorrhizal fungi are especially effective in helping plants acquire
   a) Phosphorus b) Nitrogen
   c) Calcium d) Potassium
8. Mycorrhizal fungi
   a) Act as an extension of the plant root system, acquiring phosphorus from non-depleted zones and transporting it to the plant root.
   b) Hyphae of mycorrhizal fungi also improve soil aggregation by exuding a gluelike compound called glomalin.
   c) Symbioses of Myicorrhiza increase a plant's stress tolerance
   d) All the three
9. Tillage degrade soil health by
   a) Exposes labile organic matter which will decay fast
   b) Disrupts the network of delicate fungal strands, reducing populations
   c) Alters soil texture
   d) A&B
10. Soil organisms are capable of moving soil and creating macropores are
   a) Fungi b) Bacteria
   c) Earthworms d) Actinomycetes
11. Soil organic matter
   a) Serves as food and energy source for soil microorganisms.
   b) increases water and nutrient retention in soil
   c) Imparts buffering capacity to soil
   d) All three
12. The labile pool of organic matter general accounts for
   a) 5–20 percent of the total soil organic matter
   b) 40-50 percent of the total soil organic matter
   c) 60-70 percent of total soil organic matter
   d) 1-10 percent of the total soil organic matter

13. The fraction of organic matter readily decomposed by microbes and is the principal energy source that fuels the soil food web is
    a) Total b) Exchangeable
    c) Residual d) Labile
14. Stable organic matter in soil accounts for
    a) 1-10 percent of the total soil organic matter
    b) 5–20 percent of total soil organic matter
    c) 40-50 percent of total soil organic matter
    d) 60–95 percent of total soil organic matter
15. In evaluation of soil health index by minimum dataset method, the following parameter(s) is used
    a) soil texture b) Soil organic matter
    c) Soil N d) Soil Zn
16. According to Minimum database method of soil health evaluation the scale of soil health index 'very good' is
    a) >0.7 b) 0.6-0.7
    c) 0.5-0.6 d) 0.4-0.5
17. If the rice soil health index score is 0.5, it is regarded as
    a) Poor b) Very good
    c) Excellent d) Good
18. Boron removal by crops is about
    a) 50 g / ha /year b) 100 g / ha /year
    c) 150 g / ha /year d) 200 g / ha /year
19. The microorganism that transforms ferric iron to ferrous iron form is
    a) *Bacillus* b) *Clostridium*
    c) *Klebsiella* d) all the three
20. The whole-farm, ecological approach to increase and manage biodiversity with the goal of increasing the presence of beneficial organisms is
    a) Land scaping b) Farm scaping
    c) Land reclamation d) Farm management

**Answer key**

| | | | | | | | |
|---|---|---|---|---|---|---|---|
| **1** | c | **6** | d | **11** | d | **16** | a |
| **2** | d | **7** | a | **12** | a | **17** | d |
| **3** | d | **8** | d | **13** | d | **18** | a |
| **4** | d | **9** | d | **14** | d | **19** | d |
| **5** | b | **10** | c | **15** | b | **20** | b |

# 13

# Soil Pollution

Soil pollution is defined as presence of xenobiotics (e.g. chemical compounds, radioactive elements) that alters the soil properties viz., both chemical, physical and biological properties of soil. Soil pollution, including heavy metals, may be of natural origin, like volcanic eruptions, atmospheric lightning, animal excrements or ore leaching. Nevertheless, human activity and mostly chemical industry, mining and metallurgy, as well as municipal waste management and traffic emissions are the main source of environmental pollution. Chemical compounds, entering the ecosystem as a result of different human activities, may accumulate in soil and water environments. Therefore, soil may be regarded as a long-term reservoir of pollutants, from which these compounds may be introduced to food chains or groundwater. Inappropriate and careless disposal of municipal and industrial waste often results in soil and air pollution. The pollution includes point sources such as emission, effluents and solid discharge from industry, vehicle exhaustion and metal smelting or mining, as well as nonpoint sources (e.g. the use of pesticides or excessive use of fertilizers). Each of the sources have their own damaging effects on soil plant, animal and human health, but those that add heavy metals to soils are of serious concern due to the persistence of these elements in the soil environment. They cannot be destroyed, but are only transformed from one form to another. These builtup persistent toxic compounds, chemicals, salts, radioactive materials, and disease-causing agents in soils, adversely effects plant growth, microbial activity, nutrient availability in soil.

## 13.1. Causes of soil pollution

### 13.1.1. Natural causes (Geogenic)

Some pollutants are naturally accumulated in soils. This can occur due to the differential deposition of soil by the atmosphere. Another manner in which natural soil pollution can occur is, via the transportation of soil pollutants with precipitation of water. An example of natural soil pollution is the accumulation of compounds containing the perchlorate anion ($ClO_4^-$) in dry and arid ecosystems. Some contaminants can be naturally produced in the soil under the effect of certain environmental conditions like thunder storms. For example, perchlorates can be formed in soils containing chlorine and certain metals during a thunderstorm. Weathering of rocks and minerals which have heavy metals, erosion and volcanic activities as well as forest fires also contribute for geogenic causes of soil pollution.

### 13.1.2. Anthropogenic causes (manmade)

Most cases of soil pollution are anthropogenic in nature. A variety of human activities can lead to the contamination of soil.

- Demolition of old buildings can involve the contamination of nearby soil with asbestos.
- Usage of lead-based paint during construction activities can also pollute the soil with hazardous concentrations of lead.
- Spillage of petrol and diesel during transportation can contaminate soils with the hydrocarbons found in petroleum.
- Activities associated with metal casting factories often cause the dispersion of metallic contaminants into the nearby soils.
- Underground mining activities can cause the contamination of land with heavy metals.
- Improper disposal of highly toxic industrial/chemical waste can severely pollute the soil. For example, the disposal of toxic wastes in landfills can result in the seepage of the waste into the soil. This waste can go on to pollute groundwater as well.
- Chemical pesticides contain several hazardous substances. Excessive and inefficient use of chemical pesticides can result in severe soil pollution.
- Unscientific disposal of municipal sewage can also contaminate soil. These wastes also contain several carcinogenic substances and heavy metals.
- Other forms of waste that can pollute soil include nuclear waste, e-waste, coal ash, etc.

## 13.2. Sources of soil pollution

### 13.2.1. Agricultural practices

Use of agrochemicals viz., inorganic fertilizers, pesticides, weedicides in excessive quantities causes soil pollution.

#### a) Excess use of inorganic fertilizers

Excessive use of inorganic nitrogen fertilizers leads to acidification of soil and contaminate the agricultural soil. Excessive use of inorganic fertilizes causes deleterious impactes not only on soil but also on environment.

#### Nitrogen fertilizers

Nitrate pollution of groundwater from fertilizer N is caused by leaching. The consequences are two main alleged health hazards are blue baby disease of young babies and cancer due to nitrate ingestion in food and water. World Health Organization (WHO) recommends that drinking water should not more than 10mg $NO_3$-N $L^{-1}$. The nitrate is converted into nitrite in the intestine and then absorbed in the blood stream. Young babies cannot detoxify this nitrite, which combine with hemoglobin to form inactive form methaemoglobin thus reducing the capacity of blood to carry oxygen. When the conversion of nitrate to nitrite exceeds 10% blood is incapable of carrying oxygen and clinical symptoms such as gray or blue skin develop known as "methaemoglobinaemia" or "blue baby syndrome".

**Ammonia Volatilization:** Volatilization of $NH_3$ is not only a major loss of N but also a cause of environmental pollution. From the atmosphere $NH_3$, is washed out by clouds and redeposit on the terrestrial ecosystem In the atmosphere it is oxidized to $N_2O$, which acts as a greenhouse gas and is responsible for the destruction of ozone layer.

**Acid Rain:** The effect of acid rain on ecosystems is evident in temperate region. Nitrogenous fertilizers contribute substantially towards emissions of ammonia, one of the agents causing acid rain. A high atmospheric concentration of ammonia can result in acidification of land and water surfaces, because plant damage and plant bio-diversity reduction in natural systems.

**Nitrogen Oxides (NO and $NO_2$):** Nitrogen oxides are not GHGs. Nitrogen fertilizer input accounts for only 0.5 per cent of NO emissions. Both nitric oxide (NO) and nitrogen dioxide ($NO_2$) react in sunlight with volatile organic compounds to form tropospheric ozone ($O_3$). Ozone is toxic to crops, even at low concentrations, and detrimental to the health of sensitive individuals.

**Nitrous Oxide ($N_2O$):** Nitrous oxide has a greenhouse effect and is considered to be detrimental to the ozone layer. According to experts of the Intergovernmental Panel on Climate Change (IPCc), $N_2O$ is responsible for 7.5 per cent of the calculated greenhouse effect caused by human activity. The concentration in the atmosphere is increasing at a rate of about 0.2 per cent per year. Although nitrogen fertilizers can be a direct or indirect source, they account for only 0.8 per cent of the $N_2O$ emissions.

## Phosphorous fertilizers

Minaral rock phosphate is used in manufacture of SSP and phosphoric acid is used for manufacturing DAP and complex fertilizers. Presently, there are 12 phosphoric acid plants in India having a combined installed capacity of about 2.0 Mt. per annum as $P_2O_5$. The phosphate fertilizer plant produces large amount of waste in the form of liquid and solid effluents. Phosphogypsum is generated as a by-product of the phosphoric acid based fertilizer industry and is well recognized as a source of sulphur (16%) and calcium (21%), chemically present in the form of $CaSO_4.2H_2O$. It also contains some amount of phosphorus (0.2–1.2% $P_2O_5$), almost half of which is present in water soluble form.

**Environmental threat from Phosphogypsum:** One of the main concerns on the use of phosphogypsum is the higher amounts of the radionuclide. During acidifying process on rock phosphate to produce phosphoric acid, about 80–90%

of $^{226}Ra$ (due to similarity with Ca in chemical properties) is concentrated in phosphogypsum while nearly 86% of U and 70% of Th end up in the phosphoric acid. $Radium^{226}$ in phosphogypsum further decays to produce Radon gas ($Rn^{222}$) emitting carcinogenic radioactive alpha particles exceeding 370 Bq $kg^{-1}$ of radioactivity. The $^{238}U$ levels in phosphogypsum sample were higher than those compared worldwide. Phosphogypsum also contains fluoride (0.5–1.5%), which may leach down and contaminate groundwater. It contains heavy metals (cadmium, chromium, lead etc.), which if applied in agricultural land, may enter into food chain and damage human and animal health.

**Eutrophication:** One of the problem associated with excess fertilizer use is the eutrophication of surface water causing several diseases. Arable soils leak considerable amounts of nitrate, phosphate, potassium and other nutrients mainly through run-off and erosion, which enrich the water body in terms of nutrients leading to luxurious growth of algae and other organisms and resultant eutrophication problems in water fresh bodies.

## b) Indiscriminate use of pesticides

Pesticides are synthetic toxic chemicals that kill different types of pests and insects causing damage to agriculture but it has many ecological repercussions. Pesticides are generally insoluble in water and non-biodegradable consequently persist longer periods in soils (Table1). Therefore, these chemicals will not gradually decompose and keep on accumulating in the soil. Hence, the concentration of these chemicals will increase when they transfer from lower to higher tropic level via the food chain. Hence, it will cause many metabolic and physiological disorders in humans. The harmful effect of DDT and other chemicals led to the introduction of less persistent organic and more-biodegradable substance such as carbamates and organophosphates. However, these chemicals act as harmful toxins for nerves, hence they are more dangerous to humans. It led to pesticides related to the death of field workers in some agricultural fields.

**Table 1:** Persistence time for some selected pesticides in soil

| S.No | Pesticide | Persistence time |
|---|---|---|
| 1 | BHC | 11 yrs |
| 2 | DDT | 10 yrs |
| 3 | 2,4,D | 2- 8 weeks |
| 4 | Aldrin | 9 yrs |
| 5 | Diuron | 16 months |
| 6 | Atraxine | 18 months |
| 7 | Simazine | 17 months |
| 8 | Chlordane | 12 yrs |
| 9 | 2,3,6- Trichlorobenzene(TBa) | 2- 5 yrs |

## c) Weedicides

Initially the industries began production of weedicides like sodium arsenite ($Na_3AsO_3$), sodium chlorate ($NaClO_3$), etc. Herbicides can decompose in a span of few months. Even though they are not as harmful as organo-chlorides but most of the herbicides are toxic. They are known to cause birth defects. Furthermore, research suggests that spraying weedicides causes more insect attack and diseases of plants in comparison to manual weeding. One thing to note here is all the above factors occupy just a small portion of the causes. Majority of the causes is related to manufacturing activities in chemical and industrial processes of weedicides associated with large scale release of effluents and toxic gases in to soil and atmosphere.

### d) Methane (CH4) emission from paddy fields and livestock

Methane is a GHG. Within agriculture, $CH_4$ is emitted mostly by ruminant digestive process and from livestock wastes. Rice paddy fields are also a major source of $CH_4$ that is formed by the anaerobic decomposition of organic matter. The addition of readily decomposable organic matter significantly increases $CH_4$ emissions. The impact of mineral fertilizers on $CH_4$ emissions is not clear, but seems minor.

### e) Inferior irrigation practices

Poor irrigation methods increase the soil salinity. Moreover, excess watering, improper maintenance of canals and irrigation channels, lack of crop rotation and intensive farming gradually decreases the quality of soil over time and cause degradation of land.

### f) Use of sewage sludge in agriculture

The sewage sludge is generated after treatment of effluents from cities and towns using different technologies. Sewage treatment and consequent production of sewage sludge compost provides a win-win situation for both urban society and farming community as it allows beneficial recycling of wastes. Although estimates of volume of sewage being treated are available, quantitative information on sludge generation from sewage treatment plants (STPs) in India is not available. The CPCB evaluated performance of 152 sewage treatment plants located in 15 states across the country. The total treatment capacity of these plants was reported to be 4716 MLD (CPCB 2013). However actual performance of these plants was only 66% of the total capacity (3126 MLd) as revealed from various studies. Moreover BOD and COD of treated effluents from 49 and 7 STPs respectively exceeded permissible levels for discharge into the water bodies, indicating inefficiency of several STPs in the country. Average sewage effluent treatment capacity of some important states followed the order: Tamil Nadu with 16.9% rank top position followed by Uttar Pradesh (16.4%), Andhra Pradesh (15%), Punjab (14%), West Bengal (10%), Haryana (7%), Maharashtra (6%), Gujarat (4.9%), Madhya Pradesh (3.6%), Bihar (3.4%) in the order. The maximum efficiency of BOD removal (90%) was observed in the STPs designed on Trickling filter and Sequential Batch Reactor (SBR) technologies.

Composition and characteristics of sewage sludge varies widely depending on UWW quality and its treatment processes. Organic and inorganic pollutants (non-essential trace metals, toxic metals), salt ions, microorganisms and pathogens, beneficial plant macro- and micro-nutrients and some organic compounds are widely found in sewage sludge. Composts prepared from municipal sewage sludge have proved useful in agriculture because of presence of higher amount of organic carbon and plant nutrients.

### 13.2.2. Industrial pollution

The incorrect way of chemical waste disposal from different types of industries can cause contamination of soil. Human activities like this have led to acidification of soil and contamination due to the disposal of industrial waste, heavy metals, toxic chemicals, dumping oil and fuel, etc.

#### a) Thermal power plants

In general, the major problems associated with the application of flyash for crop production and dumping includes high electrical conductivity, unfavourable physical properties associated with pozzolanic nature of the material, boron toxicity and accumulation heavy metal particularly chromium and arsenic in soil. The most frequently cited cause of these effects is heavy metals and boron toxicity. In some cases, ash is shown to induce P deficiency, salt injury, pozzolanic reactions etc.

#### b) Tannery Industries

In India, leather and leather products are processed from about 2334 major production centers. The main process with adverse environmental consequences in these industries is tanning of raw hides by using chromium salts. These industries are concentrated in a few pockets of Tamil Nadu, West Bengal and Uttar Pradesh (having 88.5% of the total tanneries in the country). Tannery waste contains both organic (chlorinated phenols) and inorganic (chromium) pollutants. Other pollutants of concern with the tanning industry include azodyes, cadmium compounds, cobalt, copper, antimony, barium, lead, selenium, mercury, zinc, arsenic, polychlorinated biphyenls (PCb), nickel, formaldehyde resins and pesticides residues. All these pollutants that released into the environment due to the tanning process lead to deterioration of air, water and soil quality. There is a significantly higher prevalence of morbidity among tannery industry workers, mostly from respiratory diseases because of chromium exposure.

#### c) Pesticide Industries

India is the fourth largest producer of agro-chemicals globally after US, Japan and China. However the per hectare consumption of agrochemicals in India is very low (only 0.58 kg) when compared to US (4.5 kg $ha^{-1}$) and Japan (11 kg $ha^{-1}$). Among the crops, paddy consumes maximum share of pesticide (around 28%), followed by cotton (20%) in India. There are about 150 industrial units technically involved in the manufacturing of pesticides and more than 500 units engaged in preparation of formulations in India. India is the largest producer of pesticides in Asia and ranks 12th in the world in terms of pesticide use. Pesticides are not only used for agricultural purposes, but also in industries (for wood and carpet, wood preservation, etc.), public health (malaria, dengue fever and cholera) and domestic purposes such as fly killer, ant killer, repellants. Rodenticides, fungicides, etc.

The waste water of pesticide industries contains pollutants like halomethanes, cyanides, haloethers, phenols, heavy metals, pesticides, etc. However, the BOD, COD and TSS contents are very low. Many of the chemicals used are persistent in nature and severely affects soil ecology and find its way in food chain through plant food materials and affects the human health.

### d) Fertilizer Industries

There are about 93 fertilizer plants producing single super phosphate in India with an installed capacity of 7.7 Mt. per year, which are located in AP (9), Assam (1), Bihar (1), Chhatisgarh (2), Gujrat (7), Haryana (1), Karnataka (2), MP (16), Maharashtra (17), Rajasthan (22), TN (3), UP (7), WB (5). Besides these 12 fertilizer plants are manufacturing di-ammonium phosphates with an installed capacity of 8.3 Mt. per year. Phosphoric acid is mainly used for manufacturing DAP and complex fertilizers. Presently, there are 12 phosphoric acid plants having a combined installed capacity of about 2.0 Mt. per annum as $P_2O_5$. The phosphate fertilizer plant produces large amount of waste in the form of liquid and solid effluents. Phosphogypsum is generated as a by-product of the phosphoric acid based fertilizer industry and is well recognized as a source of sulphur (16%) and calcium (21%), chemically present in the form of $CaSO_4.2H_2O$. It also contains some amount of phosphorus (0.2–1.2% $P_2O_5$), almost half of which is present in water soluble form.

One of the main concerns on the use of phosphogypsum is the higher amounts of the radionuclide. During acidifying process on rock phosphate to produce phosphoric acid, about 80–90% of $^{226}Ra$ (due to similarity with Ca in chemical properties) is concentrated in phosphogypsum while nearly 86% of U and 70% of Th end up in the phosphoric acid. $Radium^{226}$ in phosphogypsum further decays to produce Radon gas ($Rn^{222}$) emitting carcinogenic radioactive alpha particles exceeding 370 Bq $kg^{-1}$ of radioactivity. The $^{238}U$ levels in phosphogypsum sample were higher than those compared worldwide. Phosphogypsum also contains fluoride (0.5–1.5%), which may leach down and contaminate groundwater. It contains heavy metals (cadmium, chromium, lead etc.), which if applied in agricultural land, may enter into food chain and damage human and animal health.

### e) Iron and Steel industry

After China and Japan, India is the world's third-largest producer of crude steel and has been a major contributor to India's manufacturing output. There are about 11 major integrated steel industries operating in the country in the states of Jharkhand, Odisha, Chattisgarh, West Bengal, Tamil Nadu, Karnataka, Andhra Pradesh and Maharastra. Besides these, several mini steel plants (about 650) having electric arc furnaces are using steel scrap and sponge iron. Most of these industries are located nearby coal mining area. Slag generated from iron and steel making industries have been found to possess lime supplying potential in acidic soils. Slag materials have been found to contain high amount of Ca (about 30–55% CaO) and Si (12–35% $SiO_2$) which are essential/ beneficial elements for plants. Acid soils of India are deficient in these two elements, requiring their frequent application in crop production.

The main problem concerning the utilization of steel slags in agriculture consists of the possible leaching of heavy metals. Among heavy metals, steel making slags contain Vanadium (V) and chromium (Cr) in significantly high concentration. In Sweden, death of cattle was reported from vanadium poisoning which occurred due to grazing from steel slag amended land.

## f) Aluminium industry

India is richly endowed with bauxite resources, which is the basic raw material for aluminium production. Alumina and aluminium processing creates a range of waste products, the most significant being bauxite residues (also known as red mud). Red mud is a by-product of the Bayer process, the principal means of refining bauxite to produce alumina using NaOH. Heavy metal concentration of bauxite residues from alumina refinery are presented in table 2.

**Table 2:** Hazardous elemental concentrations in bauxite residues from alumina refinery (adopted from *Wang and Liu* 2012)

| Trace elements | Concentration (mg $kg^{-1}$) |
|---|---|
| As | 267.3 |
| Pb | 56.6 |
| Hg | 67.3 |
| Cd | 27.1 |
| Cr | 537.8 |
| Ni | 984.9 |
| Ba | 212.0 |
| Zn | 103.2 |
| Cu | 78.2 |
| Mn | 187.5 |

Depending on quality bauxite, approximately 1.5 metric tons of bauxite residues is generated with every metric tonne of alumina produced. Red mud is composed of a mixture of solid and mud of metallic oxides. Iron oxides comprise up to 60% of the mass of the red mud. The mud is highly basic with a pH ranging from 10 to 13. In addition to iron, the other dominant components include silica, residual alumina, and oxides of Ti, Ca and Na. Besides these, red mud also contains high amount of several hazardous trace elements As, Pb, Hg, Cd, Cr, Ni, and these are present in concentrations several times higher than commonly found in unpolluted soils.

### g) Copper Refineries and Smelters

The copper industry in India accounts for 3% of the world copper market at about 500 thousand tonnes by size in terms of annual consumption of refined copper. The country copper production accounts for only 0.2% of the world's global production. Because of the large waste volumes associated with copper production, the processing facilities are usually located near the mines.

All aspects of copper production, from leaching to milling, smelting, refining and electro twinning have potential impacts on surface and groundwater quality. The waste from copper industry contain toxic heavy metals mainly As, Pb, and Cd.

### h) Zinc and Lead Smelting industry

Zinc and lead smelters India are either located near zinc ore mines (in Rajasthan) or near sea ports (as in Vishakhapatnam and Alwaye, primarily dependent on imported mineral concentrates). Major zinc and lead smelters and refineries in India are located in Rajasthan state (Chanderiya, Debari and Dariba) with total metal production of 8,80,000 per year.

The major air pollutants emitted from the smelting processes are particulate matter, and $SO_2$, along with gaseous forms several metals like As, Hg etc. in trace amount. Emitted particulate matters contain oxides of Zn, Pb, As, Hg, Cd, Cu, Sb. All these air pollutantsupon settling can contaminate water, soil and food crops. Wastewater effluent from the industry contains dissolved and suspended solids, heavy metals, and oil and grease having adverse effects on soil and water bodies. The smelting process typically produces less than 3 tons of solid waste (slag and sludge) per ton of lead/zinc produced. Slag material from the industry may contain 0.5–0.7% lead & zinc along with many other toxic heavy metals, particularly Cd.

### i) Paper Industry

There are about 759 paper mills in India having a total installed capacity of 12.7 Mt. and was anticipated to produce around 14 Mt. paper, paper board and newsprint per year by the year 2016 according to Govt. of India report. Paper & pulp industries also use significant amount sodium hydroxide, sodium carbonates, sodium sulfide & bi-sulfites, elemental chlorine and hypochlorite, calcium oxide, hydrochloric acid, etc. during manufacturing of paper and generate larger quantities of effluents containing organic and inorganic salts and toxic pollutants.

### j) Molasses based Distillery industry

India is the second largest producer of sugar and fourth largest producer of alcohol in the world. Molasses is used as a raw material in the distillery industries where alcohol is produced by fermentation processes. The effluent generated from distilleries is called raw spent-wash (RSW). Depending upon the technology and quality of molasses, the spent wash generation rate can vary from 7.0 to 15.0

liters per liter of alcohol produced. Spent wash generated in the distilleries is subjected to anaerobic digestion (biomethanation) in many industries for methane recovery and effluent generated after this process is called post-methanation effluent (PME). The chemical analysis of RSW and PME clearly showed their high manurial potential (table 4). Even application of 1 ha-cm effluent to the soil can meet the N, P, K and S requirements of most of the agricultural crops.

Distilleries are classified as "Red Category" among the 17 categories of highly polluting industries by CPCB. Disposal of spent wash, a distillery waste, has long been a problem for the industry. Seepage of effluent from storage lagoon pond to contaminate groundwater has been reported at several places which resulted increase in salinity and coloration making it unsuitable for drinking purposes. Release of distillery effluent may cause rise in salinity, biological oxygen demand, chemical oxygen demand in surface water bodies deteriorating water quality for aquatic life. Although the effluent may facilitate better crop yields at the beginning due to its richness in nutrients (N, P, K, Ca, S, micronutrients), its continuous or indiscriminate/heavy use may cause accumulation of salts endangering the soil productivity and quality.

**Table 3:** Chemical composition of raw spent wash (RSW) from distillery industry. *Source:* Adopted fro.n Biswas et al. (2011)

| Parameter | RSW Composition |
|---|---|
| pH | 4.2 |
| EC (dS $m^{-1}$) | 25.3 |
| Total solid (g $L^{-1}$) | 98.0 |
| Organic C (g $L^{-1}$) | 37.5 |
| Total N (g $L^{-1}$) | 2.02 |
| Ammoniacal-N (g $L^{-1}$) | 1.25 |
| Nitrate-N (g $L^{-1}$) | Trace |
| Total P (g $L^{-1}$) | 0.24 |
| Organic P (g $L^{-1}$) | 0.14 |
| Total K (g $L^{-1}$) | 13.0 |
| Water soluble K (g $L^{-1}$) | 7.9 |
| Total S (g $L^{-1}$) | 2.82 |
| Sulphate-S (g $L^{-1}$) | 1.31 |
| Chloride (g $L^{-1}$) | 4.05 |
| BOD (mg $L^{-1}$) | 36,000 |
| COD (mg $L^{-1}$) | 1,13,240 |

**k) Leather/ tannery industry**

There are about 2000 tannery industry existing in Indiatry, of which about 70%

are present in TN and West Bengal. Mineral tanning usually uses chromium in the form of basic chromium sulfate and produces stretchable leather which is excellent for use in handbags and garments. The leather processing requires large quantities of chemicals like sodium chloride, chromium sulphate, calcium and ammonium salts and organic dyes. Wastes and chemicals produced during washing, dehairing and tanning of the leather are released to nearby soils and water causing pollution.

Effects of Tannery Industry Effluents: The effluents generated from tannery and textile industries are characterized by high value of BOD, COD, Na and dissolved solids (Table 8). The tannery effluents are also ranked as the highest polluting waste amongst all industrial waters. Use of tannery effluents in crop land increases several plant nutrient contents like N, P and K levels in soil. However it has adverse effect on soil physical properties as well as builds up heavy metals, particularly Cr in soil.

## l) Textile Industry

The organized textile sector has more than 3400 textile mills both in small scale industry (SSI) and the non-SSI sectors, producing about 2500 million $m^2$ of cloth apart from 2500 million kg of man-made fibres & filament yarns. Most of the medium to large cotton textile industries are located in Maharastra, Gujarat and Tamil Nadu; while jute mills and woolen industries are concentrated in West Bengal and Punjab respectively. The textile industry uses high volumes of water throughout its operations, from the washing of fibres to bleaching, dyeing and washing of finished products. On average, approximately 200 L of water are required to produce 1 kg of textiles.

Effects of Textile Industry Effluents: Colour, high contents of suspended and dissolved solids, residual chlorine, toxic metals (like Cr, As, Cu) and organic pollutants (mainly dyes) are the major pollutants present in textile mill effluent. All of these have adverse effect on soil physical properties (due to salts), plant growth (due to residual chlorine and Na, sulphate), microbial biodiversity (due to dye compounds and heavy metals) and food quality (due to salts and toxic elements) when used as irrigation.

## m) Industrial clusters

Katedan industrial area of Hyderabad is hosting more than 300 industries which are involved in dyeing, edible oil production, battery manufacturing, metal plating, chemicals production etc. Dumping of the waste materials around this area is commonly observed and is a major cause of soil pollution which occurs due to spreading hazardous material through rainwater and wind. (Bhupal Raj et al. 2009) reported the prevalence of very high concentrations of Pb, Cr, Ni, Zn, As and Cd throughout the industrial area. Hazardous metals like As and Cr, Cu, Pb and Zn had also been observed in some of the residential area nearby industrial complex. The wastes and effluents generated from different industrial activities in and around this area are discharged into the ponds without adequate treatment. It resulted in contamination of groundwater in and around Katedan area with salts and several heavy metals (Cd, Cr, Ni and Cu). All these heavy metals exceeded the permissible limits in the groundwater, except Pb. Continuous use of the contami- nated groundwater for agricultural production increased the contents of Pb, Ni and Cr in the soil (Bhupal Raj et al. 2009).

Coimbatore industrial area is the 2nd largest industrial area in Tamil Nadu. The industrial area is having about 500 textile industry, 200 electroplating industry,100 foundries and 300 dyeing industries. All the industrial effluent/ sewage finds its way to Ukkadam river, which is source of irrigation in nearby area. Heavy metal contents in the city sewage water were quite high and varied widely with season. ICAR-Indian Institute of Soil Science, Bhopal investigated changes in soil properties in agricultural land nearby different industrial clusters (Table 3); namely electroplating industry, textile industry, dye industry and city sewage irrigated areas (Saha et al. (2013)). Groundwater near industrial area has developed salinity due to contamination mainly with salts of $Na^+$ and $Cl^-$; magnitude of contamination was more near textile and dye industries. Sulphate contamination was the highest in the ground- water near electroplating industries.

Soils of agricultural land near textile and dye industries have developed severe salinity (EC > 6 dS $m^{-1}$) and slight alkalinity (pH >8.0). Soils of agricultural land near industrial areas contained 47–178 mg Ni $kg^{-1}$, 47–214 mg Pb $kg^{-1}$, 0.5–4.2 mg Cd $kg^{-1}$ and 43–241 mg Cr $kg^{-1}$ (Table 4). Most of the soils had all the heavy metals more than the safe concentration limits determined by Saha et al. (2013), which indicate that these soils may pose threat to the environment. Nickel and Pb concentrations were high in soils near the electroplating and sewage from industrial area; Cd concentration was higher in soils irrigated with mixed effluents, and sewage; Cr concentration was higher in soils irrigated with textile, dye and sewage effluent.

**Table 4:** Total heavy metal contents (mg $kg^{-1}$) in soils of agricultural land nearby different industrial clusters at Coimbatore (modified from Saha et al. 2013)

| Source of contamination | Ni | | Pb | | Cd | | Cr | |
|---|---|---|---|---|---|---|---|---|
| | Range | Median | Range | Median | Range | Median | Range | Median |
| Sewage | 78–140 | 107 | 116–214 | 158 | 1.2–4.2 | 2.17 | 75–176 | 124 |
| Electroplating | 84–178 | 122 | 74–167 | 136 | 1.1–2.7 | 1.63 | 48–152 | 109 |
| Textile | 47–126 | 72 | 52–159 | 114 | 0.6–3.1 | 1.28 | 86–241 | 168 |
| Dye | 62–111 | 89 | 58–147 | 82 | 0.5–2.4 | 1.34 | 79–187 | 127 |
| Mixed effluents | 58–132 | 96 | 47–171 | 92 | 1.3–3.8 | 2.39 | 43–158 | 96 |

### 13.2.3. Heavy metal pollution

Natural sources of heavy metals are, i.weathering of rocks and minerals, ii.erosion and volcanic activities, iii. forest fires and biogenic sources and iv. particles released by vegetation. Anthropogenic sources of heavy metals are presented in table 5.

**Table 5:** Anthropogenic sources of heavy metal contamination to soils

| Heavy metal | Anthropogenic sources of pollution |
|---|---|
| Arsenic (As) | Pesticides, wood preservatives, biosolids, ore mining and smelting |
| Cadmium (Cd) | Paints and pigments, plastic stabilizers, electroplating, phosphate fertilizers |
| Chromium (Cr) | Tanneries, steel industries, fly ash |
| Copper (Cu) | Pesticides, fertilizers, bio-solids ore mining and smelting |
| Mercury (Hg) | Au-Hg mining, coal combustion, medical waste |
| Nickel(Ni) | Effluents, kitchen appliances, surgical instruments, automobile batteries |
| Lead (Pb) | Aerial emission from combustion of lead fuel, batteries waste, insecticides and herbicides |

## Heavy metal pollution by agricultural practice

Consumption of chemical fertilizers and organic manure bear a direct relationship with food grain production. The total nutrient consumption (N, $P_2O_5$, $K_2O$) was about 24 million tonnes during 2013–2014, with about 141 kg fertilizer per ha (Annual Report 2014-'15, Department of Fertilizers, Govt. of India). Excessive use of fertilizers and pesticides, antibiotics and hormones in livestock and irrigating farms with contaminated wastewater are agricultural factors affecting soil pollution. Many of these agricultural inputs have been reported to be contaminated with heavy metals (Table 6).

**Table 6:** Agricultural sources of trace element Pollution in soils (mg $kg^{-1}$)

| Element | Sewage sludge | Phosphate fertilizers | Limestone | Nitrogenous fertilizers | Municipal solid waste compost | Manures | Pesticides |
|---|---|---|---|---|---|---|---|
| As | 2–26 | 2–1200 | 0.1–24 | 2–120 | - | 3–150 | 22–60 |
| Cd | 2–1500 | 0.1–170 | 0.04–0.1 | 0.05–8.5 | Tr.–8.4 | 0.3–0.8 | - |
| Cr | 20–40,600 | 66–245 | 10–15 | 3–19 | 14–401 | 5.2–55 | - |
| Cu | 50–3300 | 1–300 | 2–125 | 1–15 | 25–865 | 2–60 | 12–50 |
| Hg | 0.1–55 | 0.01–1.2 | 0.05 | 0.3–3 | - | 0.09–26 | 0.8–42 |
| Mn | 60–3900 | 40–2000 | 40–1200 | - | - | 30–550 | - |
| Ni | 16–5300 | 7–38 | 10–20 | 7–38 | 8.6–190 | 7.8–30 | - |
| Pb | 50–3000 | 7–225 | 20–1250 | 2–1450 | 11–647 | 6.6–15 | 60 |
| Zn | 700–49,000 | 50–1450 | 10–450 | 1–42 | 82–946 | 15–250 | 1.3–25 |

Adopted from Kabata-Pendias (2000) and Saha et al. (2010)

### 13.2.4 Greenhouse gases

Greenhouse gases (GHGs) are atmospheric compounds that store energy, thus influencing the climate. Each of the GHGs has a different global warming potential that takes into account the effectiveness of each gas in trapping heat radiation and its longevity in the atmosphere. For instance, one kilogram of methane ($CH_4$) is estimated to have the same warming potential as 21 kilograms of carbon dioxide ($CO_2$), and one kilogram of nitrous oxide ($N_2O$) has an equivalent impact to approximately 310 kilograms of $CO_2$. According to the Organization for Economic Cooperation and Development (OECd), $CO_2$, $CH_4$ and $N_2O$ emissions in agriculture at global level are estimated to account for 14 per cent of the total emission of GHGs. However, estimates of both absorption and emission of these three gases are subject to significant uncertainties.

## 13.3 Solid wastes pollution

Disposal of plastics, cans, and other solid waste falls into the category of soil pollution. Disposal of electrical goods such as batteries causes an adverse effect on the soil due to the presence of harmful chemicals. For instance, lithium present in batteries can cause leaching of soil.

## Nutrient potential of sewage sludge

Sewage sludge contains considerable amount of organic carbon and plant nutrients that are essential for crop production (Table 7). Therefore, application of sewage-sludge in agricultural land has shown to enhance crop yield. Urban sewage sludge compost has the potential to correct soil acidity due their high contents of bases and organic matter. Sludges generally contain higher amount of salts as compared to soil and therefore, their application may result in development/ enhancement of soil salinity particularly in low rainfall area.

**Table 7:** Typical characters and composition of sewage sludge: (Jayanta K. Saha et al 2017)

| Parameters | Range |
|---|---|
| Bulk density (g $mL^{-1}$) | 0.33–0.8 |
| Organic matter (%) | 7–80 |
| N (%) | 0.48–3.23 |
| P (%) | 0.08–2.22 |
| K (%) | 0067–1.44 |
| S (%) | 0.4–0.88 |
| Ca (%) | 0.37–4.9 |
| Mg (%) | 0.07–1.1 |
| As (μg $g^{-1}$) | 1.1–87 |
| Pb (μg $g^{-1}$) | 22–335 |
| Cd (μg $g^{-1}$) | 0.5–16 |
| Cr (μg $g^{-1}$) | 14.7–1290 |
| Cu (μg $g^{-1}$) | 150–484 |

| Parameters | Range |
|---|---|
| Mn ($\mu g\ g^{-1}$) | 32–605 |
| Ni ($\mu g\ g^{-1}$) | 5–96 |
| Hg ($\mu g\ g^{-1}$) | Tr.–6.5 |
| Zn ($\mu g\ g^{-1}$) | 261–1490 |

## Heavy metals in sewage sludge

Although several benefits can be harvested from use of sewage sludge compost in crop production, this byproduct from sewage treatment plants are screened and monitored by government organization before permitting for its safe use insustainable agriculture production. During treatment process, toxic substances in waste water are concentrated into solid phase, making it riskier to be applied on agricultural lands. The risks due to the presence of undesirable substances such as heavy metals/trace elements, potentially toxic synthetic organic compounds, pathogenic organisms; and possible element imbalance in composts have become major apprehension for potential compost users including farmers. The heavy metals concentrations in sewage sludges are vary widely. The heavy metal contents in sludge are especially higher if it is contaminated with industrial effluent. Therefore their level depends largely on industrialization profile of a city. Continuous application of sludge compost in fields may buildup heavy metals in soils beyond the permissible levels for its safe use in agriculture. Composting of sewage sludge tends to diminish heavy metals availability and therefore threat of metals transfer to plants with composted sludge is generally lower as compared to other forms of stabilized sludge.

## 13.4 Mechanisms of interaction of waste with soil

Waste on addition to soil undergo degradation by various mechanisms viz., physical, chemical and biological processes. Physical mechanisms include filtration and dilution; chemical mechanisms viz., ion exchange, adsorption, precipitation and complexation. Biological mechanisms include organic waste decomposition and assimilation.

### a) Physical mechanisms

Soil filtering systems are used to deplete Biological Oxygen Demand (BOd), remove biostimulants such as phosphates and nitrates, provide long term storage of heavy metals, and deactivate pathogens and pesticides. Soils suitable for use as filtering systems have permeability slow enough to allow adequate time for purification of water percclating through the soil system. A balance of air, water, and nutritive substances at a favorable temperature is important to a healthy microbial population and an effective filtration system. For example, overloading the filtration system with wastewater that has high amounts of suspended solids causes clogging of soil pores and a reduction of soil hydraulic conductivity. Management and timing of wastewater application are essential to maintaining soil filter systems. Climate, suspended solids in the wastewater and cropping systems must be considered to maintain soil porosity and hydraulic conductivity.

The wastewater application rate should not exceed the waste decomposition rate, which is dependent on soil temperature and moisture content. Periods of wetting and drying increase microbial decomposition and byproduct uptake by the crop and decrease potential soil pore clogging. In areas where the temperature is warm for long periods, the application rates may be higher if crops or other means of using the by-products of waste decomposition are available. Tillage practices that maintain or improve soil tilth and reduce soil compaction and crusting should be included in the land application part of agricultural waste management systems. These practices help to maintain soil permeability, infiltration, and aeration, which enhances the biological decomposition processes.

### (b) Chemical mechanisms

Management for utilization of organic waste material must take into account the chemical reactions that occur between the soil and the waste components. These reactions are broadly grouped as ion exchange, adsorption, precipitation, and complexation. The mechanisms and rates of these reactions are dependent upon physical, chemical, and biological properties of the soil and organic waste material. Organic waste mineralization byproducts consist of macro- and micro-plant nutrients, soluble salts, gases, and heavy metals. These by-products dissolve and enter soil water solutions as precipitation or irrigation water infiltrates the soil surface and percolates through the soil profile. The dissolved by-products are subject to the interactions of ionic exchange, adsorption, precipitation, or complexation. These processes store and exchange the macro- and micro-plant nutrient by-products of organic waste mineralization. They also intercept and attenuate heavy metals, salts, and other detrimental mineralization by-products that can adversely affect plant growth and crop production.

**Ion exchange**: Ion exchange reactions involve both cations and anions (table 8). Ionic exchange and adsorption is the replacement or interchange of ions bonded electrostatically to exchange sites on soil particles and soil organic materials with similarly charged ions in the soil solution. This ionic interchange occurs with little or no alteration to exchanging ions. Cation exchange is the adsorption and exchange of non-metal and metal cations to negatively charged sites on soil particles and soil organic materials. Cation exchange capacity (CEc) is the measure of a soil's potential to exchange cations and is related to soil mineralogy, pH, and organic matter content. Anion exchange is the exchange and replacement of negatively charged ions to positively charged sites on soil particles. Anion exchange capacity is relatively low in most soils when compared to cation exchange; however, anion exchange is important because the anion exchange potential of the soil is related to its ability to retain and exchange nitrate nitrogen ($NO_3^-$–N), sulfate, chloride, boron, molybdenum, and phosphorus.

**Adsorption and precipitation**: Adsorption and precipitation are processes that remove an ion from the soil solution. Sorption occurs as ions attach to the solid soil surface through weak chemical and molecular bonds or as strong chemical bonds.

Precipitation is the deposition of soluble compounds in soil voids. It occurs when the amount of the dissolved compounds in the soil solution exceeds the solubility of those compounds.

**Complexation** : Complexation is the interaction of metals with soil organic matter and some oxides and carbonates, resulting in the formation of large, stable molecules. This process extracts phosphorus and heavy metals from the soil solution. These stable complexes act as sinks for phosphorus, heavy metals, and some soil micronutrients.

**Table 8:** Common exchangeable soil cations and anions

| Elements | Cations | Anions |
|---|---|---|
| Aluminum | $Al^{+3}$ | |
| Boron | | $BO_3^{-3}$ |
| Calcium | $Ca^{+2}$ | |
| Carbon | | $C0_3^{-2}$, $HCO_{3-}$ |
| Chlorine | | $Cl^-$ |
| Copper | $Cu^+$, $Cu^{+2}$ | |
| Hydrogen | $H^+$ | $OH^-$ |
| Iron | $Fe^{+2}$, $Fe^{+3}$ | |
| Magnesium | $Mg^{+2}$ | |
| Manganese | $Mn^{+2}$, $Mn^{+3}$ | |
| Molybdenum | | $MoO_4^{-2}$ |
| Nitrogen | $NH_4^+$ | $NO_2^-$, $NO_{3-}$ |
| Phosphorus | | $HPO_4^{-2}$, $H_2PO_{4-}$ |
| Potassium | $K^+$ | |
| Sulfur | | $S0_3^{-2}$, $SO_4^{-2}$ |
| Zinc | $Zn^{+2}$ | |

## (c) Biological mechanisms

**Organic waste decomposition**: It is the breakdown of complex organic molecules such as organic waste by soil organisms. Several factors affect biological degradation of various agricultural waste organics when the waste is applied to soil. These factors interact during the biological degradation process and can be partitioned into soil and organic factors. Soil factors that affect biological degradation are temperature, moisture, oxygen supply, pH, available nutrients (N, P, K, and micronutrients), porosity, permeability, microbial population, and bulk density. Organic factors are carbon to nitrogen ratio (C:N), lignin content, and BOD. The soil and organic factors interact and determine the environment for microbial growth and metabolism. The physical and chemical nature of this environment determines the specific types and numbers of soil micro-organisms available to decompose organic material.

The decomposition rate of organic waste is primarily controlled by the chemical and biological composition of the waste material, soil moisture and temperature (Figs. 1 & 2), and available oxygen supply. Rapid decomposition of organic wastes and mineralization of organic nitrogen and phosphorus by soil micro-organisms are dependent on an adequate supply of oxygen and soil moisture. High loading rates or high BOD waste may consume most of the available oxygen and create an anaerobic environment. This process can cause significant shifts in microbial populations, microbial metabolisms, and mineralization by-products. Under anaerobic conditions, the by-products may be toxic and can be in sufficient concentrations to inhibit seed germination and retard plant growth, even after aerobic conditions have been restored.

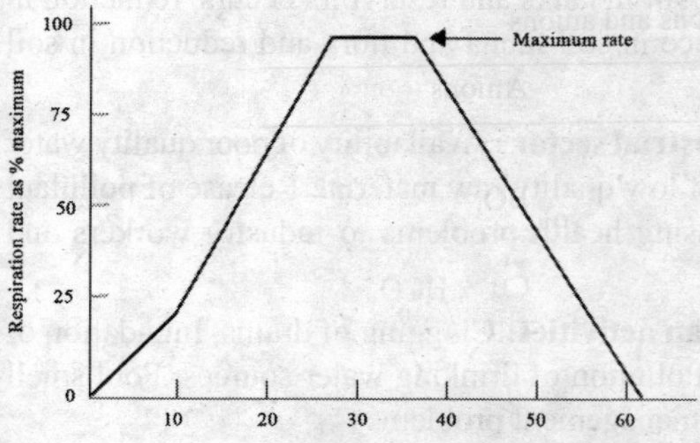


**Fig. 1:** Relation between microbial respiration rate and temperature

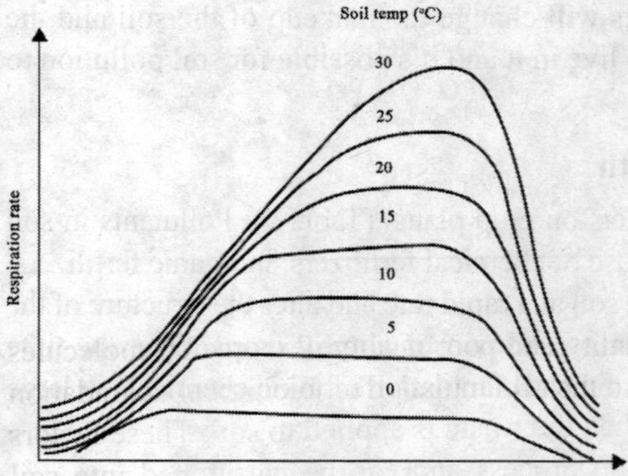


**Fig. 2:** Relation between microbial respiration rate and soil temperature and soil moisture

**Assimilation**: The microbial ability to absorb and transform metals is a promising aspect in respect of solving the pollution problems. The potential of numerous microbial metal transformations in treatment of environmental pollution may be employed and some processes are already in commercial operation. However, many processes are still at the laboratory scale and yet to be tested in a rigorous applied and/ or commercial context. Another mechanism that has the potential for the application in heavy metal-polluted sites is the production of siderophores by different microbial genera. Siderophores are the largest class of compounds that can bind and transport Fe. They are highly specific Fe(III) ligands and are excreted by a wide variety of fungi and bacteria to aid Fe assimilation.

## 13.5. Effects of Soil Pollution

It causes harmful effect on the soil and the environment at large. Broadly, contamination of soil will decrease the agricultural output of a land. Effects of soil pollution on agriculture, industry, urban and environmental activities are mentioned hereunder.

a. **Effect of soil pollution on agricultural sector:** Soil pollution reduced soil fertility, it reduces nitrogen fixation, increases soil erodibility, causes loss of soil and nutrients, deposition of silt in tanks and reservoirs occurs, reduction in crop growth and yield, imbalance in soil fauna and flora and reduction in soil microbial biodiversity.

b. **Effect of soil pollution on Industrial sector :** Availability of poor quality water for industrial use, availability of low quality raw material, Release of pollutant gases and radioactive rays causing health problems to industry workers and increased cost of production

c. **Effect of soil pollution on urban activities:** Clogging of drains, Inundation of areas, Public health problems, Pollution of drinking water sources, Foul smell and release of gases and Waste management problems

d. **Effect of soil pollution on Environment:** Emission of green house gases will cause climate change, soil becomes unavailable to grow food, loss of biodiversity, cause accelerated erosion, because a lack of plants on the soil will cause more erosion, the pollutants will change the makeup of the soil and the types of microorganisms that will live in it and it's possible for soil pollution to change whole ecosystems.

## Effects of pollutants on crop growth

Pollutant elements cause deleterious effects on crop plants (Table 9). Pollutants in soil decrease the quality of the crop. Regular use of chemical fertilizers, inorganic fertilizers, pesticides will decrease the fertility of the soil at a rapid rate and alter the structure of the soil. This will lead to decrease in soil quality and poor quality of crops. Over the time the soil will become less productive due to the accumulation of toxic chemicals in large quantity.

**Table 9:** Major impacts of different pollutants on plants.

| Pollutant | Its impact on plant |
|---|---|
| As | Red brown necrotic spots on old leaves, yellow browning of roots, growth reduction |
| Cd | Restricted growth, brown margin to leaves, chlorosis: chlorotic patches in older leaves and severe chlorosis in younger leaves at high level of toxicity, necrosis, curved leaves, brown stunted roots, reddish veins and petioles, reduction in growth, purple coloration |
| Cr | Affect seed emergence, stunted plant growth and decreased dry matter production |
| Pb | Dark green leaves, stunted foliage, increased amount of shoots |

| Pollutant | Its impact on plant |
|---|---|
| Ni | Chlorosis similar to Fe deficiency, necrosis, stunting, inhibition of root growth, decrease in leaf area |
| Hg | Severe stunting of seedlings and roots, chlorosis, browning of leaf tips, reduction in growth, tumor-like proliferations in deformed carrots |
| Cu | Chlorosis, yellow coloration, purple colouration of the lower side of the midrib, less branched roots, inhibition of root growth |
| Se | Interveinal chlorosis, black spots, bleaching and yellowing of young leaves, pink spots on root |

## Effects of soil pollution on Human health

Consuming food grown in contaminated soils will increase the exposure of humans to toxic and harmful chemicals (pollutants) thus increasing health threats to people living nearby and on the degraded land. Besides consuming agricultural products with high levels of contaminants, living, working or playing in the contaminated soil can lead to respiratory diseases, skin diseases, and other diseases. Impacts of some heavy metal pollutants on human health are furnished hereunder.

**Arsenic:** It is an analogue of phosphate and interferes with the ATP synthesis pathway. It is categorized as highly toxic metalloid, inorganic in nature and causes carcinogenic diseases to skin, lungs, liver and bladder. Smaller level exposure caused vomiting and nausea, it impairs the function of blood cells, i.e. red and white; abnormal the heart beat and very high level results death. Long term exposure lead to dark color of skin and soles; failure of kidney and liver.

**Cadmium:** It has carcinogenic properties. Inhalation of very high Cd concentration causes abnormal body biocycles, and leads to severely irritating stomach, vomiting and diarrhea. Long term contact with lesser Cd concentration causes damage to the kidneys and lungs, and makes the bones fragile. Disease caused by Cd poisoning in human has been termed as itai-itai disease which was detected in Japan in 1968. Disease is characterized by renal tubular dysfunction associated with atrophic kidney and osteomalacia (softening of the bones) and has been found among middle aged and elderly women with rela- tively frequent pregnancy. Survival rate of itai-itai disease patients is signifi- cantly lower.

**Chromium:** Hexavalent Cr(VI) is more toxic to human health. Generally Cr(VI) is reduced to Cr(III) in soil in presence of oxidizing com- pounds, which also reduces its carcinogenicity. Intake of the high amount of Cr compounds causes irritation of skin, ulcers in nose; and its longer exposure causes damage to body organs, i.e. kidney, blood circulatory system, nervous system and also leads to cancer.

**Lead:** It affects children more than adults. Higher concentration in children reduces the proper growth of body and mind. As a result it reduces intelligence; causes short memory, renal failure as well as poor development of cardiovascular organs. In adults, adverse effects like weakness in fingers, higher blood pressure and abnormal nervous system are observed. At high level of contamination, lead may cause death due to failure of multipleorgans. High level of Pb in human body reduces the fertility level of man and causes abnormal growth of the child during the pregnancy or miscarriage.

**Mercury:** As per the EPA guidelines, mercury has been described as a human carcinogen. Among the body organs, nervous system is more sensitive to all types of Hg compounds. Neurological syndrome caused by severe mercury poisoning is known as *minamata* disease. Symptoms include loss of physical control over bodily movements, numbness in the hands and feet, general muscle weakness, loss of peripheral vision, and damage to hearing and speech. In extreme cases, insanity, paralysis, coma, and death follow within weeks of the onset of symptoms. The disease was first detected in Japan in the year 1956. The toxicity can cause lasting damage to most of the body organs specially brain, kidneys, and developing fetuses. Long term exposure of lead results in complete failure of organs and may leads to death. Short term exposure causes irritation, diarrhea, abnormal blood pressure, vomiting and lung damage.

**Selenium:** It is beneficial for human and animal in trace amounts as it plays valuable role in proper functioning of thyroid gland. However at higher level of contamination, it causes toxic effect and imparts abnormality in organs. Upper tolerable intake level of Se is 400 µg/day (ATSDR 2001). Short period oral intact causes vomiting, nausea and diarrhea. The higher level intake of Se causes selenosis in human and animals, which is characterized by loss of hair, weak nail. Neurological abnormalities are also observed. It also causes stomach pain due to heavy intake of Se. Short term high level of Se exposure causes respiratory abnormalities, i.e. difficulty in respiration, bronchitis pain or coughing.

### Pollution Impact on soil ecosystem and biodiversity

Soil pollution can cause an imbalance of the ecosystem of the soil. The soil is an important habitat and is the house of different type of microorganisms, animals, reptiles, mammals, birds, and insects. Thus, soil pollution can negatively impact the lives of the living organisms and can result in the gradual death of many organisms. It can cause health threats to animals grazing in the contaminated soil or microorganisms residing in the soil. Therefore, it is very important to educate people around you the importance of environment if they are not aware. Prevention of soil erosion will help to cease soil pollution. Therefore, it is essential for industries, individuals and businesses to understand the importance of soil and prevent soil pollution and stop the devastation caused to plant and animal life.

## 13.6. Control of soil pollution

Different steps have been suggested to control soil pollution. However, prevention is better than cure, to help prevent soil erosion, we can limit excessive soil use in sensitive area. Reduce the use of chemical pesticides and fertilizers, recycle and reuse the farm wastes. In general we would need less fertilizer and fewer pesticides solid waste. One of the classifications divides the methods of pollution control are *in situ* and *ex situ* control technologies. *In situ* (in place) means that the polluted soil is treated in its original location, i.e. it remains at the site or in the subsurface. Such technologies remove the pollutant from soil without excavation or removal of the soil. In this case fixing agents are applied on the unexcavated soil. This technique's advantages may include low invasiveness, simplicity and rapidity. Moreover, it is fairly inexpensive and generates relatively low amount of waste. However, it is only a temporary solution. *Ex situ* means that the treated soil is removed or excavated from the site. It is applied in areas where heavily polluted soil must be removed from its place of origin and its storage is associated with high ecological risk.

### 13.6.1. Extraction and separation methods

In solvent extraction, the contaminated & oil is mixed with an extracting agent in general (an aqueous solution but preferably an organic solvent). Potential applications include the removal of metals such as cadmium, copper, zinc, nickel, chromium, arsenic, antimony and lead using a mineral solution, zinc lead, organo-metallic compounds and some cyanides using sodium hydroxide solution. Hydrocarbons and halogenated hydrocarbons can also be removed. Contamination is often preferentially present in the finer or coarser fraction of the soil or the organic components (ex. humus), contamination can therefore, be removed in some cases using a process which separates the soil into fractions on the basis of specific gravity or particle size or settling velocity.

### 13.6.2. Thermal methods

In thermal methods, there are two ways of heat treatment; removal of contaminants by evaporation either by direct heat transfer from heated air or an open flame or by indirect heat transfer, and destruction of the contaminants directly or indirectly at an appropriate temperature. The gas leaving the heating appliance must be treated to destroy or remove any

Contaminants or unwanted products of combustion. A related process is stream stripping in which steam is injected into soil to aid evaporation of relatively volatile contaminants which may be water soluble or insoluble.

### 13.6.3. Chemical methods

Treatment of the soil in suspension in a suitable liquid and without sludging is the two possible methods. In these, intimate, contact between soil and chemical is essential and should be frequently done so that the process of detoxification is complete.

### 13.6.4. Biological methods

Biological methods are getting more attentive in soil pollution control, particularly heavy metal pollution. These are two types 1. Phyto-remediation and 2. Microbial-remediation.

**1.Phyto-remediation** is one of the best techniques for treatment of heavy metal-polluted sites. It is an *in situ* strategy that uses vegetation and associated microbiota together with agronomic practices to aid in metal remediation. It is based on the use of special type of plants to decontaminate soil by inactivating metals in the rhizosphere or translocating them in the aerial parts. Some plants developed mechanisms to remove ions selectively from the soil to regulate the uptake and distribution of metals. Potentially useful phytoremediation technologies for heavy metal-polluted sites include *phytoextraction*, *phytostabilization* and *rhizofiltration*.

a) **Phytoextraction** uses hyperaccumulating plants to remove metals from soil by absorption into the roots and shoots of the plant. The aboveground shoots can be then harvested to remove metals from the site and subsequently stored as hazardous waste or employed for the recovery of metals. The ideal plant for phytoextraction should grow rapidly, produce high amount of biomass and be able to tolerate and accumulate high metal concentrations in shoots. Hyperaccumulating plants belong to the families of *Brassicaceae*, *Fabaceae*, *Euphorbiaceae*, *Asterraceae*, *Lamiaceae*, and *Scrophulariaceae*. Studies indicate that many *Brassica* species, such as *B. juncea*, *B. napus* or *B. rapa* exhibit enhanced accumulation of Zn and Cd.

b) **Phytostabilization** is based on the use of plants to limit the mobility and bioavailability of metals in soil. Plants used in this method are characterized by high tolerance of metals in surrounding soils together with their low accumulation. Phytostabilization can be carried out through the process of sorption, precipitation, complexation, or metal valence reduction. This technique is useful for the removal of Pb, As, Cd, Cr, Cu, and Zn. This process is advan tageous because in this case disposal of hazardous material/biomass is not required, and it is very effective when rapid immobilization is needed to preserve soils or ground and surface waters.

c) **Rhizofiltration** (or phytofiltration) removes metals from contaminated soil via absorption, concentration and precipitation by plant roots. This technique is used to remove pollutants from groundwater and aqueous-waste streams rather than for the remediation of polluted soils. Apart from the above described phytoremediation methods, some authors include also phytovolatization and phytodegradation.

d) **Phytovolatization** involves the use of plants to volatilize pollutants from their foliage such as Se and Hg.

e) Phytodegradation uses plants and associated microorganisms to degrade organic pollutants.

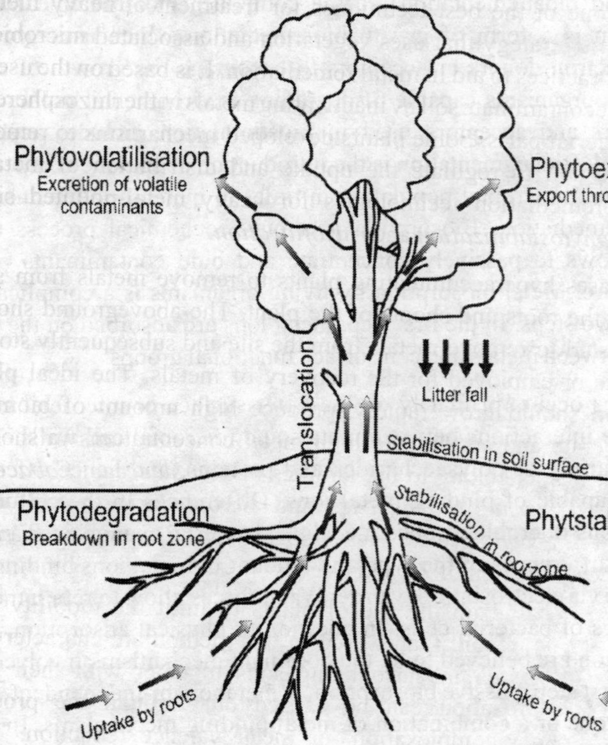


**Fig. 3:** Mechanisms of phytoremediation involved in purifying contaminated soils and physiological processes that occur in plants during phytoremediation.

**2. Microbial remediation:** Another approach for biological remediation of heavy metal-polluted soils includes the use of microorganisms to detoxify metals by valence transformation, extracellular chemical precipitation or volatilization etc. Bioleaching is the method that uses microorganisms to solubilize heavy metal pollutants either by direct bacterial processes, or as a result of interactions with metabolic products, or both. It can be used *in situ* or *ex situ* to help to remove the pollutants from soils. This process is based on mobilization of metal cations from insoluble ores by biological oxidation and complexation. This process was adapted from mining industry for the use in soil remediation and a general term covering both bioleaching and biooxidation techniques could be "biomining". This technique is mainly employed for copper, cobalt, nickel, zinc and uranium, which are extracted either from insoluble sulfides or (in the case of uranium) from oxides. The classical bioleaching bacteria belong to the genus *Acidithiobacillus* (*A. thiooxidans* and *A. ferrooxidans*), *Acidiphilium*, *Acidimicrobium*, *Ferromicrobium* or *Sulfobacillus*.

Another solution for soil bioremediation using microorganisms is to apply microbially mediated biochemical processes, such as oxidation/reduction or methylation reactions.

Often, biostimulation and bioaugmentation are the components of bioremediation strategies. Biostimulation is a form of in situ bioremediation which uses growth stimulation nutrients, electron donors or acceptors to encourage the growth of site-specific indigenous microorganisms capable of degrading environmental pollutants. Common electron donors and acceptors used in biostimulation include: acetate, sulfate, nitrate and ethanol. Bioaugmentation is the introduction of specific competent microorganisms to the local microbial population in order to increase the metabolic capacities needed for remediation. Biosorption is a physicochemical process that occurs naturally and allows to passively concentrate and bind contaminants onto the microbial cell structure. Metal biosorption by living organisms is a complicated process that consists of two steps. In the first step, metal ions are adsorbed on the cell surface by interactions between metals and cell surface functional groups.

Biosorption of metal ions occurs primarily on the outer surface of microbial cells and is the first step in the interactions between metals and microbial cell walls. The cell wall consists of a variety of polysaccharides and proteins, and hence offers a number of active sites capable of binding metal ions. Differences in the cell wall composition among various microbial groups, i.e. algae, bacteria, cyanobacteria and fungi, cause significant differences in the type and amount of metal ions binding to them. Physical adsorption via electrostatic or van der Waals forces allow to retain metal ions on the outer surfaces of bacterial cells. In addition to physical adsorption, ion exchange and complexation are believed to be the dominant mechanisms involved in metal biosorption. The first step, passive biosorption, is metabolism-independent and proceeds rapidly by any one or a combination of metal binding mechanisms. In the second step, due to active biosorption, metal ions penetrate the cell membrane and enter into the cells. This is, however, a slowly occurring process. Active mode is metabolism-dependent and related to metal transport and deposition. There are several microbial genera and species capable of metal biosorption. Fungi were found to be efficient biosorbent organisms, as their cells are characterized by a high percentage of cell wall material, which shows excellentmetal binding properties. *Aureobasidium pullulans*, *Cladosporium resinae*, *Aspergillus niger*, *Aspergillus versicolor* or *Rhizopus nigricans* are the fungal species proved to be effective in heavy metal biosorption . Numerous studies also identified several species of bacteria as efficient metal accumulating microorganisms. For instance, *Bacillus* spp. has been reported to have a high potential of metal sequestration and has been used in commercial biosorbent preparation. Other bacterial species capable of metal transformation include, among others: *Escherichia coli*, *Pseudomonas maltophilia*, *Shewanella putrefaciens*, *Pseudomonas aeruginosa*, *Enterobacter cloacae*.

Mechanisms involved in biochemical interactions between bacteria and metal ions involve specific enzymes that catalyze the oxidation, reduction, methylation, de-alkylation and precipitation reactions. Microorganisms transform a substantial number of metals and metalloids by reducing or oxidizing them directly to a lower or higher redox state. Additionally, indirect oxidation or reduction is an alternative for immobilization of toxic metals in the environment. Methylation is an important process involved in geochemical cycling of metals and the removal of metal pollutants from soils. Methylation processes derive the methyl group from methylocarbolamine ($CH_3B_{12}$) which is implicated in the methylation of multiple metals and metalloids, such as Pb, Sn, Pd, Pt, Au, Ti, As, Se and Te. Methylation of Hg, Sn and Pb can be mediated by a range of microbes, including *Clostridium spp.*, methanogens and sulfate-reducing bacteria under anaerobic conditions and principally by fungi (e.g. *Penicillium spp.* and *Alternaria spp.*) under aerobic conditions. Methyl groups are enzymatically transferred to metals and a given species may transform a number of different metals. Methyl-metal compounds are generally highly volatile and available to plants. Another mechanism that has the potential for the application in heavy metal-polluted sites is the production of siderophores by different microbial genera. Siderophores are the largest class of compounds that can bind and transport Fe. They are highly specific Fe(III) ligands and are excreted by a wide variety of fungi and bacteria to aid Fe assimilation.

### 13.6.5. Reducing chemical fertilizer and pesticide use

Applying bio-fertilizers and manures can reduce chemical fertilizer and pesticide use. Biological methods of pest control can also reduce the use of pesticides and thereby minimize soil pollution.

### 13.6.6. Reusing of materials such as plastics, metals, paper, cloth etc.

Materials such as glass containers, plastic bags, paper, cloth etc. can be reused at domestic levels rather than being disposed, reducing solid waste pollution.

### 13.6.7. Recycling and recovery of materials

This is a reasonable solution for reducing soil pollution. Materials such as organics, paper, some kinds of plastics and glass can and are being recycled. This decreases the volume of refuse and helps in the conservation of natural resources. For example, recovery of one tonne of paper can save 17 trees.

### 13.6.8. Reforestation

Control of land loss and soil erosion can be attempted through restoring forest and grass cover to check wastelands, soil erosion and floods. Crop rotation or mixed cropping can improve the fertility of the land.

### 13.6.9. Solid waste treatment

Proper methods should be adopted for management of solid waste disposal. Industrial wastes can be treated physically, chemically and biologically until they are less hazardous. Acidic and alkaline wastes should be first neutralized; the insoluble material if biodegradable should be allowed to degrade under controlled conditions before being disposed. As a last resort, new areas for storage of hazardous waste should be investigated such as deep well injection and more secure landfills. Burying the waste in locations situated away from residential areas is the simplest and most widely used technique of solid waste management.

## 13.6.10. Integrated crop management practices

- Quantify nutrient source such as soil reserve (residual), manure, crop residue
- Practice soil test based fertilizer management : pH, lime requirement, phosphorus, potassium (calcium and magnesium optional)
- Calibration of manure and fertilizer spreaders: tones or litres or kgs per acre
- Proper fertilization plan: manure application rate, supplemental fertilizer; utilize excess manure on alternative crops (hay crops); avoid applying large amounts of manure on fields with excessive P found using soil tests; do not over apply nitrogen from manure or fertilizer, and nitrogen soil test: side- or top-dressing supplemental nitrogen fertilizer
- Growing cover crop: to reduce soil loss and nitrate leaching, consider a legume based cover crop on vegetable farms and on distant fields on dairy farms where manure is not spread
- Proper planting plan: to ensure early harvest of crops to allow early cover crop planting on most erosion prone fields; and
- Adoption of minimum tillage: to reduce nutrient loss through soil erosion.
- Adopt integrated post management and integrated weed management

### Review questions

What is soil pollution? Explain causes of Soil pollution.

Explain the mechanisms of interactions of waste with soil.

Briefly discuss on Agricultural pollution.

Explain soil pollution caused by industrial activities.

Write briefly on phyto remediation of soil pollutants.

Explain microbially mediated biochemical processes for soil pollution control.

Mention the effects of soil pollution

Briefly discuss the effects of soil pollution on human health

### Objective questions

1. Soil pollution is due to presence of

   a) Probiotics　　b) Abiotics

   c) Xenobiotics　　d) Debiotics

2. Point source of soil pollution

   a) Effluents and solid discharge from industry

   b) Metal smelting or mining

   c) Fertiliser use in agriculture

   d) A&B

3. Geogenic soil pollution
   a) Perchlorates formed in soils containing chlorine and certain metals during a thunderstorm
   b) Demolition of old buildings can involve the contamination of nearby soil with asbestos
   c) Activities associated with metal casting factories
   d) Underground mining activities
4. Methaemoglobinaemia is associated with
   a) Borate  b) phosphate
   c) Sulphate  d) Nitrate
5. $NH_3$ in atmosphere converts to
   a) $NH_2$  b) $NH_4$
   c) $N_2O$  d) $NO_3$
6. Nitric oxide (NO) and nitrogen dioxide ($NO_2$) of atmosphere react in sunlight with volatile organ
   a) Oxygen ($O_2$)  b) Ozone ($O_3$)
   c) Elemental nitrogen ($N_2$)  d) Ammonia ($NH_3$)
7. Ozone is toxic when it present in
   a) Troposphere  b) Ozonosphere
   c) ionosphere  d) Stratosphere
8. IPCC is abbreviation for
   a) Intereuropian Panel for Change in Climate
   b) International Party for Climate Change
   c) Intercontinental Public Contact for Change
   d) Intergovernmental Panel on Climate Change
9. By-product generated in the phosphoric acid based fertilizer industry and is well recognized as a source of sulphur (16%) and calcium (21%) is
   a) Basic slag  b) Phosphogypsum
   c) Molases  d) Rockphosphate
10. Carcinogenic radioactive rays associated with phosphoric acid based fertilizer industry is
   a) Gamma rays  b) Beta Rays
   c) Alpha rays  d) Cosmic rays
11. Phosphatic fertilizers contain
   a) Mercury, Arsenic, boron  b) Cadmium, chromium, lead
   c) Fe, Mn, Cu  d) Li, Zn, Ti

12. The problem associated with enrichment of water bodies in terms of nutrients viz., nitrate, phosphate mainly through run-off and erosion is
   a) Geotrophism b) Eutrophication
   c) Peritropism d) Amalgumation
13. Less persistent organic and more-biodegradable substance is
   a) Chlordane b) Organochlorides
   c) Atrazin d) carbamates
14. Sodium arsenite ($Na_3AsO_3$) is
   a) Weedicide b) Insecticide
   d) Fungicide d) Acaricide
15. Major source of $CH_4$ emission from agricultural activities is
   a) Ruminant digestive process b) Rice paddy fields
   c) Use of Pesticides d) A&B
16. The maximum efficiency of BOD removal (90%) in sewage treatment plant is observed in
   a) Facultative lagoon
   b) The trickling filter solids contact process (TF-SCP)
   c) oxidation ditches
   d) Trickling filter and Sequential Batch Reactor (SBR) technology
17. Boron toxicity and accumulation heavy metal particularly chromium and arsenic in soil associated with
   a) Sugar industry b) Leather industries
   c) Thermal power plants(Flyash) d) Paper industry
18. The pollutants in Tannery waste
   a) chlorinated phenols and chromium
   b) Lead and Mercury
   c) Arsenic and Nickel
   d) Iron and copper
19. Production of agrochemicals globally in the following order
   a) US> Germany> China >Australia b) US> Japan> China > India
   c) Australia>China>India> US d) China> Cuba> Brazil> India
20. Per hectare consumption of pesticides in India
   a) 11 kg b) 4.5 kg
   c) 0.58 kg d) 5.8kg

21. In India crop wise pesticide consumption is in following order
   a) Cotton> Bengalgram> rice> Soybean
   b) Soybean> Cotton> Bengalgram> Rice
   c) Cotton> Bengalgram> rice> Soybean
   d) Rice >Cotton>Bengalgram>Soybean
22. Persistance of BHC in soils is
   a) 2-8 weeks b) 11 days
   c) 11 years d) 6years
23. Slag materials of iron and steel industries have been found to contain high amount of
   a) Al and Fe b) Ca and Si
   c) Cd&Hg d) As&Ni
24. The mineral resource for aluminium industry is
   a) Bauxite b) Hematite
   c) Magnetite d) Apatite
25. Hazardous heavy metals associated with Aluminium industry is
   a) Fe Cu b) Mn, Cr
   c) As, Pb d) Sb, Ni
26. The waste from copper industry contain toxic heavy metals pollutants
   a) Ni, Fe, Sb b) As, Pb, and Cd.
   c) Fe, Cu, Sr d) Mn, Cr, Sb
27. Leachate waste water from pulp and paper industry is known for containing heavy metals
   a) Fe, Cu, Mn b) Hg, As, Sb
   c) Mn, Cr, Sb d) **Cd, Ba, Cu**
28. CPCB classified Distilleries as
   a) Green category c) Yellow category
   c) Red category d) Blue category
29. The untreated spent wash from distillery industries is
   a) Neutral b) Acidic
   c) Basic d) Ultra basic
30. the heavy metal associated with tannery industries is
   a) Hg b) Cr
   c) Pb d) Cd

31. Tannery effluents are
   a) Basic b) Acidic
   d) Neutral d) None
32. highest polluting waste amongst all industrial waters
   a) Fertiliser industry effluents b) Distilleries effluents
   c) Tannery effluents d) paper industry effluents
33. Textile industry associated with
   a) Fe, Cu, Mn b) Cr, As, Cu
   c) Hg, Ba, Sb d) As, Pb, and Cd.
34. Match the following

| | | |
|---|---|---|
| Arsenic (As): | ( ) | 1. Pesticides, wood preservatives |
| Cadmium (Cd): | ( ) | 2. Paints and pigments |
| Chromium (Cr): | ( ) | 3. Tanneries, steel industries, fly ash |
| Copper (Cu): | ( ) | 4. Pesticides, fertilizers |
| Mercury (Hg): | ( ) | 5. Au-Hg mining, coal combustion, medical waste. |
| Nickel(Ni): | ( ) | 6. Effluents, kitchen appliances, surgical instruments, batteries. |
| Lead (Pb): | ( ) | 7.batteries waste, insecticides and herbicides |

   a) 1,2,3,4,5,6,7 b) 1,2, 5, 7, 3,6, 4
   c) 7,6,5,4,3,2,1 d) 4,5,6,7,1,2,3
35. Uses growth stimulation nutrients, electron donors or acceptors to encourage the growth of site-specific indigenous microorganisms capable of degrading soil pollutants is
   a) Bioenshrining b) Bioaugmentation
   c) Bioboosting d) Biostimulation
36. Introduction of specific competent microorganisms to the local microbial population in order to increase the metabolic capacities needed for remediation is known as
   a) Bioboosting b) Biostimulation
   c) Bioaugmentation d) Bioenshrining
37. the microbial ability to absorb and transform metals is a promising aspect in respect of solving the pollution problem is
   a) Microbial stimulation b) Microbial augmentation
   c) Microbial assimilation d) Microbila decomposition

38. One kilogram of methane ($CH_4$) is estimated to have the same warming potential of

a) 1.2 kg $CO_2$ b) 12 kg $CO_2$

c) 2.1 kg $CO_2$ d) 21 kg $CO_2$

39. One kilogram of nitrous oxide ($N_2O$) has an equivalent impact of

a) 210 kg $CO_2$ b) 310 kg $CO_2$

c) 31 kg $CO_2$ d) 21 kg $CO_2$

40. Microorganisms transform metals and mettalloids by

a) Reduction b) Oxidation

c) Cutroplication d) A and B

**Answer key**

| | | | | | | | | | | | | | | | |
|---|---|---|---|---|---|---|---|---|---|---|---|---|---|---|---|
| **1** | c | **6** | b | **11** | b | **16** | d | **21** | d | **26** | b | **31** | a | **36** | c |
| **2** | d | **7** | a | **12** | b | **17** | c | **22** | c | **27** | d | **32** | c | **37** | c |
| **3** | a | **8** | d | **13** | d | **18** | a | **23** | b | **28** | c | **33** | b | **38** | d |
| **4** | d | **9** | b | **14** | a | **19** | b | **24** | a | **29** | b | **34** | a | **39** | b |
| **5** | c | **10** | c | **15** | d | **20** | c | **25** | c | **30** | b | **35** | d | **40** | d |

# 14

# Soils of India

Geographic location of India is in Northern hemisphers and located between $8^{0}4'$ & $37^{0}6'$ N latitude and $64^{0}7'$ & $97^{0}25'$ E longitude. Extent of area is 329 M Ha or 329 x 104 Sq.Km or 3.29 MSq.Km. The physiography of India can be broadly grouped into triangular plateau, mountains, Indo-gangetic plains and islands. India is divided into twenty agro ecological regions and sixty five agro ecological sub regions as per NBSS and LUP. Agro ecological regions( AER) is referred to identification and demarcation of land units having comparable landform, soils, bioclimatic conditions and the length of growing period (LGP).

India has great heterogeneity expression of soil due to climatic variations. Some important soil groups are Alluvial soils: 75 mha; Black soils : 72mha; Red soils : 70 mha; Laterite soils: 25 mha; Coastal sands: 5-8 mha; Desert soils : 29 mha; Forest & hill soils: 75 mha; Salt Affected soils : 7 mha.

**There are 8 major group of soils in India:** 1)Alluvial soils 2) Black soils 3) Red soils 4) Laterite soils 5) Desert soils 6) saline sodic soils 7) Peat and marshy soils 8 ) Forest soils and 9) Coastal sands.

**Table 1:** Major soils of India, their extent and equivalents in Soil Taxonomy and FAO legend.

| S.No | Major soils | Area (M ha) | Equivalent taxonomic unit | |
|---|---|---|---|---|
| | | | USDA Soil Taxonomy | FAO Legend |
| 1 | Alluvial soils | 75 | Inceptisols, Entisols, Alfisols, Aridisoils | Cambisols, Fluvisols, Luvisols, Gleysols, Solonchalks,Solonetz |
| 2 | Black soils | 72 | Vertisols, Inceptisols, Entisols | Vertisols, Cambisols, Leptosols, Regosols |
| 3 | Red soils | 70 | Alfisols, Inceptisols, Entisols | Cambisols, Alfisols, Leptosols, Regosols |
| 4 | Laterite and Lateritic soils | 25 | Oxisols, Ultisols, Inceptisols | Cambisols, Alsols, Leptosols, Plinthosols |
| 5 | Saline and Alkali soils | 7 | Aridisols, Inceptisols, Alfisols, Entisols, Verisols | Solonchaks, Solonetzs, Cambisols, Luvisols, Arenosols, Vertisols |

| S.No | Major soils | Area (M ha) | Equivalent taxonomic unit | |
|---|---|---|---|---|
| | | | USDA Soil Taxonomy | FAO Legend |
| 6 | Desert soils | 29 | Aridisols, Entisols | Cambisols, Leptosols, Calcisols, Arenosols, Fluvisols |
| 7 | Forest and Hill soils | | Inceptisols, Alfisols, Mollisols, Ultisols, Entisols | Cambisols, Luvisols, Leptosols, Phaeozems |
| 8 | Peaty and Marshy soils | | Histosols, Inceptisols, Entisosls | Gleysols, Cambisols, Fluvisols |
| 9 | Coastal sands | | Entisols, Inceptisols | -- |

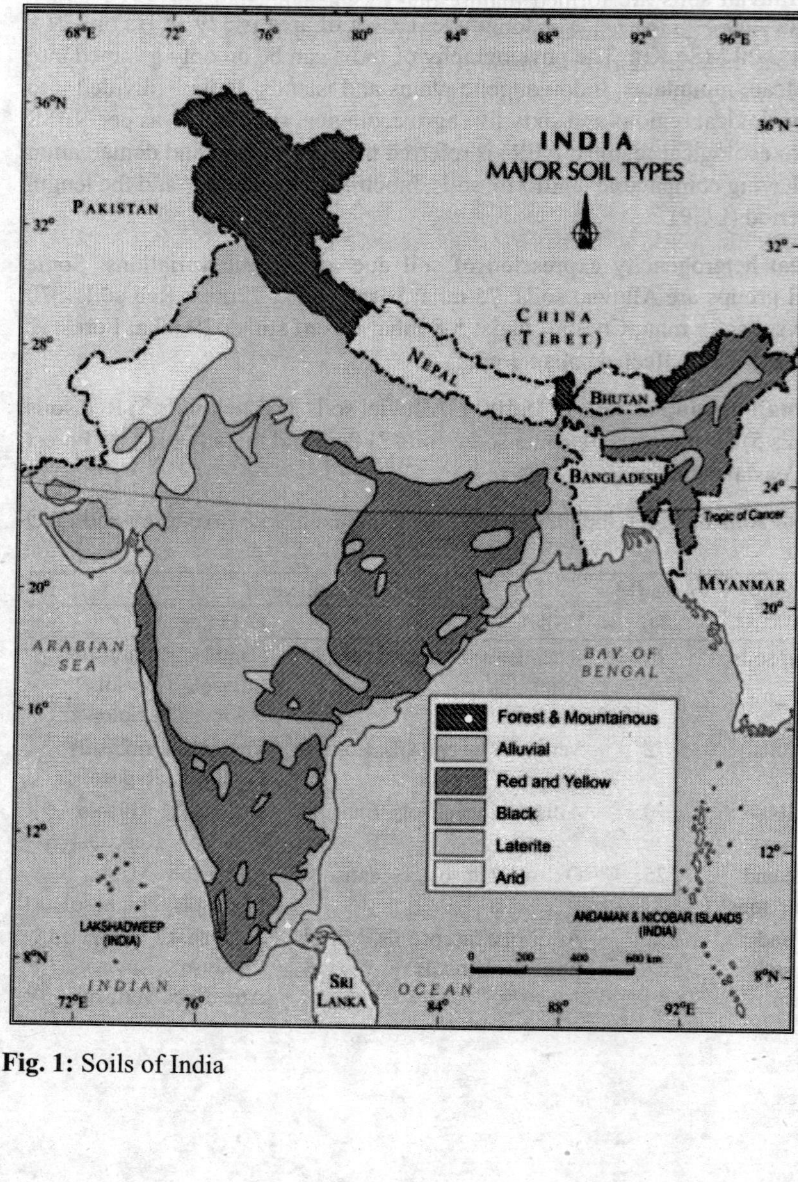


**Fig. 1:** Soils of India

## 14.1. Alluvial soils

The extent of alluvial soils in India is 75 mha. These soils are widespread in northern plains and river valleys. In peninsular-India, they are mostly found in deltas and estuaries etc. They are depositional soils i.e. transported and deposited by rivers, streams etc. They occur all along the Indo-Gangetic-Brahmaputra plains except in few places where the top layer is covered by desert sand. They also occur in deltas of the Mahanadi, the Godavari, the Krishna and the Cauvery, where they are called deltaic alluvium (coastal alluvium). Some alluvial soils are found in the Narmada, Tapi valleys and Northern parts of Gujarat.

**Genesis:** Alluvial soils are formed mainly due to silt deposited by Indo-Gangetic-Brahmaputra rivers. In coastal regions some alluvial deposits are formed due to wave action.. Parent material is river alluviums from erosion products of rivers, coastal alluviums from coastal sands; and deltaic alluviums from heterogeneous sediments.

Geologically, the alluvium of the Great plain of India is divided into newer or younger khadar and older bhangar soils. The bhabar belt is about 8-16 km wide running along the Shiwalik foothills. It is a porous, northern most stretch of Indo-Gangetic plain. Rivers descending from the Himalayas deposit their load along the foothills in the form of alluvial fans. These alluvial fans (often pebbly soils) have merged together to build up the bhabar belt. The porosity of bhabar is the most unique feature. The porosity is due to deposition of huge number of pebbles and rock debris across the alluvial fans. The streams disappear once they reach the bhabar region because of this porosity. Therefore, the area is marked by dry river courses except in the rainy season. The area is not suitable for agriculture and only big trees with large roots thrive in this belt.

The Bhangar is the older alluvium along the river beds forming terraces higher than the flood plain (about 30 metres above the flood level). It is of a more clayey composition and is generally dark colored. A few metres below the terrace of the bhangar are beds of lime nodules known as "Kankar". The Khadar is composed of newer alluvium and forms the flood plains along the river banks. The banks are flooded almost every year and a new layer of alluvium is deposited with every flood. This makes them the most fertile soils of Ganges. They are sandy clays and loams, more dry and leached, less calcareous and carbonaceous (less kankary). A new layer of alluvium is deposited by river flood almost every year.

**Salient features**

1) Texture is variable. Coarser near source and finer in proximity to the sea (delta).
2) Fluvial in nature, stratified which is reflected in their texture and OM distribution.
3) Colour depends on parent material.
4) Depth depends on geographic position,

5) Calcium carbonate content is variable
6) Mostly alkaline (aridic and ustic SMR) or acidic (udic SMR only) in reaction.
7) Profile development is variable A – C to A -B-C profiles.
8) Inherently rich in plant nutrients – fairly sufficient in P and K but deficient in N and organic matter.
9) These soils are porous because of its loamy (equal proportion of sand and clay) nature.
10) Porosity and texture provide good drainage and other conditions favorable for agriculture.
11) These soils are constantly replenished by the recurrent floods

**Potentials:** River alluviums and deltaic alluviums are best agricultural soils. Wheat, rice, sugarcane and other crops can be grown well

## Constraints

1) Stratification restricts leaching and drainage
2) Hydromorphic condition that promotes reduction and results in poor aeration for plant growth.
3) They can be rendered saline where evaporation exceeds rainfall (aridic).
4) The river alluviums and deltaic alluviums are becoming waterlogged and salt affected due to injudicious use of irrigation water.
5) Due to intensive use for cultivation, secondary problems are arising like deficiency of S and Zn.

## Management

- Judicious use of Irrigation Water and fertilizers.
- Providing proper drainage facilities where ever necessary.

## 14.2. Black soils

Black soil distributed in an extent of 72Mha. The parent material for most of the black soil are the volcanic rocks that were formed in the Deccan Plateau (Deccan and the Rajmahal trap). In Tamil Nadu, gneisses and schists form the parent material. The former are sufficiently deep while the later are generally shallow. These are the region of high temperature and low rainfall. It is, therefore, a soil group typical to the dry and hot regions of the Peninsula.

**Genesis:** Black soils formed generally on basic parent material like basalt, which is rich in lime, soda feldspars and clay. Climate: Semiarid-subhumid climate, and Lower/ plain topographic positions favour black soil formation.

## Soil Properties

- Black colour due to clay humus complex and/or titaniferous magnetite,
- Clay to Clay loam in texture (30-80% clay), highly argillaceous,

- Dominant clay mineral is smectite.
- Extremely hard on drying and very sticky and very plastic on wetting
- Because of swell-shrink nature, they develop deep wide cracks, gilgai relief and slickensides in lower depths due to churning / argillopedoturbation.
- $SiO_2/R_2O_3$ is 3 3 –3.5,
- Generally low in B.D. however High B.D is due to compaction in lower layers,
- High total pore space,
- Alkaline reaction pH (7.8 - 8.7/9.5),
- High CEC (35 – 55 cmol/kg),
- High in BS up to 95%, and high $CaCO_3$ (up to 40%),
- High Water holding capacity (150 - 250 mm/m)
- Generally low organic matter content.
- High water and nutrient holding capacity.
- Low hydraulic and permeability, hence ill drained and more runoff.

## Constraints of black soils

- Narrow workable moisture,
- Poor traficcability,
- High draft power (heavy soils),
- Low permeability,
- Poor drainage,
- Low HC,
- High EC and salinity in sub-soil.
- Micronutrient deficiency in calcareous black soils.

**Potentials**: Black soils are inherently fertile and suitable for dry farming. Under rainfed conditions cotton and millets, and under irrigation rice, sugarcane etc., can be grown profitably.

## Management

1. Adoption of pre-monsoon sowing, BBF method.
2. Deep ploughning in summer season to improve internal drainage during monsoon season.
3. Addition of organic matter/ green manuring to improve aggregation.
4. Needs to provide adequate drainage facilities for air permeability improvement.

## 14.3. Red soils

Redsoils distributed in India to an extentof 20-25% of the total geographical area of the Country, 70Mha. In Andhra Pradesh Red soils 65%, laterite soils 1%, black soils 25%, alluviasl 5% and coastal sands 3%. Morphologically thesese soils are also named as red loams, red earths, brown red and yellow soils. Brown soils are called parwa while red soils are called rakar.

**Genesis:** The name red is given to soils rich in sesquioxides that have developed of archean origin (granites, gneisses). Generally red soils are formed on well drained, stable higher landforms (higher topography). Under hot, semi-arid to humid sub-tropical climate, the weathering is moderately intense and leads to enhanced decalcification. The Fe and Al under oxidized conditions, form sesquioxides imparting red colour.

### Salient features of Red Soils

- Red to yellow in colour (hydration & coatings of Fe and Al oxides 30 – 40%),
- Loamy sand to clay texture, Shallow to deep in soil depth,
- Low to high clay content and dominant clay mineral is kaolinitic / illitic, low WHC & NHC
- Well to excessively drained as evident from their colour,
- CN ratio 10:1, pH 6 -7.5,
- CEC low (25 –40 cmol/kg of clay) and BS is low to medium,
- $SiO_2/R2O_3$ 2.5 –3.0, and Low nutrient status (O.C, N, P, K, lime)

**Constraints:** Shallow depth if formed on elevations, soil erosion, sub-surface hardening, sub-surface hard pans, surface crusting, rapid drying, excessive drainage or some times high runoff, high surface soil temperature in dry areas, deficient in OM, N, P and Ca. High in $Fe_2O_3$ & $Al_2O_3$, kaolinite, allophones. High P fixation hence low P availability, though K is sufficient,fast depletion of K is due to continuous cropping.

**Crusting and hardening of red sandy soils (chalkas)**

1. Proportion of clay to sand +silt ratio (3:7) – strength 36kg/cm2 on drying
2. Presence of free iron oxides which form irreversible bonds with clay fraction
3. Sealing pores by rain drops force – dispersion of soil particles.

**Management**

1. Tillage with spike roller, pulling thorny bush to break the crust.
2. Mulching of seed lines, sowing on the sides of the ridges
3. Incorporation of slowly decomposable organic residues – paddy husk, grounut pod shells, coir pith etc.

**Potentials:** Under good mangement practices, these soils are good for agricultural, horticultural and plantation crops.

## 14.4. Laterite and Lateritic soils

These soil distributed to an extent of 25 Mha in India. The term 'laterite' was coined by Buchanan in 1807 for highly ferruginous, vesicular, unstratified deposits of Malabar hills.

**Geetics**: Basic parent material, sub-tropical to tropical climate, higher topography under rain forest vegetation favour the formation of these soils. The siliceous matter is leached during weathering and sesquioxides are left behind.

## Soil properties

- These soils are in red colour (5YR or redder),
- Highly weathered hence very deep soils,
- High clay % (not due to migration but due to *in situ alteration),*
- *Massive* structure,
- Narrow $SiO_2/R_2O_3$ (<2) ratio ,
- Strongly acidic,
- LAC (<16cmol/kg clay),
- BS is low ( 40% ),
- Kaolinite dominant,
- Low in organic matter content.

**Table 2:** Differences between laterite and lateritic soils

| Property | Laterite soil | Lateritic soil |
|---|---|---|
| % base saturation | Less than 40 | More than 40 |
| $SiO_2/Al_2O_3$ ratio | Less than 2 | More than 2 |
| Dominant clay mineral | 1:1 type | 1:1 & 2:1 |
| Presence of Gibbsite mineral | Present | Absent |

**Constraints**: High permeability, low AWC, highly leached, Chemically degraded, low nutrient reserve, deficiency of P, K, Ca, Zn, B etc., Strongly acidic, high P fixation, Al & Mn toxicity.

**Management**: Generally used for plantation crops and shifting cultivation (should be 20 years or more), In low elevated areas - rice, banana coconut and high elevated areas - coffee, cocoa, rubber, tea are grown. Liming and fertilization (rock phosphates and silicates) is required

## 14.5. Forest and Hill (Mountain) soils

These soils are mainly found on the *hill slopes covered by forests*. These soils are heterogeneous in nature and their character changes with parent rocks, ground-configuration and climate. Consequently, they differ greatly even if they occur in close proximity to one another. In the Himalayan region, such soils are mainly found in valley basins, depressions.

Generally, it is the north facing slopes which support soil cover; the southern slopes being too precipitous and exposed to denudation to be covered with soil. Apart from the Himalayan region, the forest soils occur on Western and Eastern Ghats as well as in some parts of the Peninsular plateau.

The forest soils are very rich in humus but are deficient in potash, phosphorus and lime. Therefore, they require good deal of fertilizers for high yields. They are especially suitable for plantations of tea, coffee, spices and tropical fruits in Karnataka, Tamil Nadu and Kerala and wheat, maize, barley and temperate fruits in Jammu and Kashmir, Himachal Pradesh and Uttaranchal.

## 14.6. Arid and desert soils

A large part of the arid and semi-arid region in Rajasthan and adjoining areas of Punjab and Haryana lying between the Indus and the Aravalis, receiving less than 50 cm of annual rainfall, is affected by desert conditions. The Rann of Kuchchh in Gujarat is an extension of this desert. This area is covered by a mantle of sand which inhibits soil growth. These soils occupy an extent of 29 mha.

The desert soils consist of aeolian sand (90 to 95 per cent) and clay (5 to 10 per cent). Some of these soils contain high percentages of soluble salts, are alkaline with varying degree of calcium carbonate and are poor in organic matter.

The phosphate content of these soils is as high as in normal alluvial soils. Nitrogen is originally low but nitrate form of nitrogen is present some extent. Thus, the presence of phosphates and nitrates make them fertile soils wherever moisture is available. Therefore, great possibility of reclaiming these soils if proper irrigation facilities are available. The changes in the cropping pattern in the Indira Gandhi Canal Command Area are a living example of the utility of the desert soils.

## 14.7. Coastal sands

These soils occur all along coast 3 to 12 km away on both east and west coast. They are derived from marine sediments. These soils are deep to very deep. Ground water at as depth of 2 to 5 mts. Brackish ground water, Occasionally saline with neutral pH. Sandy texture with negligible clay content. Soil structure is single grain structure. The consistence is loose consistence. Permeability is rapid hence these soils are excessively drained. Low in water and nutrient holding capacity. Low organic matter, low in fertility, low in CEC and base saturation.

Owing to extreme sandy nature, these soils are able to withstand irrigation with high saline water – Doruvu technology can be followed. During cyclone periods these soils are inundated due to tidal overflow

## 14.8. Saline and Alkaline Soils

These soils are found in Andhra Pradesh, Karnataka, Bihar, Uttar Pradesh, Haryana, Punjab, Rajasthan and Maharashtra. There are salt-impregnated or alkaline soils occupying 70,000 sq km of area (7mha).

These soils are liable to saline and alkaline and are known by different names such as reh, kallar, usar, thur, rakar, karl and chopan. There are many undecomposed basic rock and mineral fragments which on weathering liberate sodium, magnesium and calcium salts and sulphurous acid.

In canal irrigated areas and in areas of high sub-soil water table, the injurious salts are transferred from below to the top soil by the capillary action as a result of evaporation in dry season. The accumulation of these salts makes the soil infertile and renders it unfit for agriculture. It has been estimated that about 1.25 million hectares of land in Uttar Pradesh and 1.21 million hectares in Punjab has been affected by usar. In Gujarat, the area round the Gulf of Khambhat is affected by the sea tides carrying salt-laden deposits. Vast areas comprising the estuaries of the Narmada, the Tapi, the Mahi and the Sabarmati have thus become infertile.

## 14.9. Peaty and Marshy soils

These soils occur in humid regions with accumulation of high organic matter. During monsoons the soils get submerged in water and the water receipts after the monsoon during which period rice is cultivated. Soils are black clay and highly acidic with pH of less then 3.5. Free aluminium and ferrous sulphate are present. The depressions formed by dried rivers and lakes in alluvial and coastal areas sometimes give rise to water logged soils and such soils are blue in colour due to the presence of ferrous iron. Peaty soils are found more in Kerala and marshy soils are found more in coastal tracks of Orissa, West Bengal and East coast of TamilNadu.

## Review question

1) List out major soils groups of India and write the characters and distribution of Black soils
2) Differentiate between red and laterite soils.
3) Differentiate between Laterite and Lateritic soils

## Objective questions

1. According to NBSS and LUP India is divided into -------- agro ecological sub regions

   a) 20 b) 65
   c) 35 d) 40

2. According to ICAR India is divided into -------- agro ecological regions

   a) 20 b) 65
   c) 35 d) 40

3. Dominant soils in India are

   a) Coastal sands b) Forest soils
   c) Laterite soils d) Alluvialsoils

4. Laterite and Lateritic soils are classified to

   a) Spodosols d) Histosols
   c) Oxisols d) Andisols

5. The constraint in black soils is
   a) Narrow workable range  b) Low water retention
   c) Water logging  d) A&C
6. Black cotton soils are classified to the Order of
   a) Spodosols  d) Vertisols
   c) Oxisols  d) Andisols
7. Bhangar is
   a) New Alluvium  b) Old alluvium
   c) Sedentary soils  d) Aeolian dunes
8. The dominant soils in Deccan plateau of India are
   a) Alluvial soils  b) Laterite soils
   c) Black soils  d) Acid sulphate soils
9. Acid sulphate soils are found in
   a) Andhra Pradesh  b) Himachal pradesh
   c) MadyaPradesh  d) Kerala
10. The approximate extent of acid soils in India is
   a) Mha  b) Lak ha
   c) 40,000 ha  d) 40 M ha
11. Desert soils of India can be classified to ......... soil order
   a) Alfisols  b) Gelisds
   b) Histosols  d) Aridisols
12. Deep black soils of central India are classified as
   a) Entisols  b) Mdlisols
   c) Vertisols  d) Alfisols
13. Most soils of Indo-gangetic and Brahmaputra Plains are
   a) Laterite soils  b) Black soils
   c) Red soils  d) Alluvial soils
14. Old alluvial soil are known as
   a) Khadar  b) Bhangar
   c) Multier  D Fluvial
15. Most of Black soils of India developed from
   a) Granite  b) Baralt
   b) Soundstone  d) Diorite

## Answer key

| | | | | | |
|---|---|---|---|---|---|
| **1** | b | **6** | b | **11** | d |
| **2** | a | **7** | b | **12** | c |
| **3** | d | **8** | c | **13** | d |
| **4** | c | **9** | d | **14** | b |
| **5** | d | **10** | c | **15** | b |

# 15

# Irrigation Water Quality

Water required for irrigation of cropped land is being degraded in terms of quantity and quality due to fast increasing demand for its use. Furthermore, the soil quality and crop productivity is linked to the quality of the water available for irrigation. In general, evaluation of irrigation water quality should focus on salt content, sodium concentration, the occurrence of nutrients and trace elements, alkalinity, acidity, and hardness of the water.

India receives 4000 billion cubic meters (BCM) of annually through rainfall with an average runoff generated is 1869 BCM. The annual replenishable groundwater resource of the country is 431 BCM and net groundwater available for utilization is 396 BCM. At present annual groundwater draft is 243 BCM out of which 221 BCM is used for irrigation and 22 BCM for domestic and industrial use. In India and many countries groundwater is the important source of irrigation as well as for drinking purpose too. Groundwater is extensively used for irrigation in India and the status of groundwater exploitation in India varies from 0.07 percent in the state of Arunachal Pradesh to 170 percent in Punjab. Other than availability it is the quality of the water that restricts its use for agriculture as well as domestic purpose. Due to injudicious and excessive use of fertilizers, discharge of untreated urban and industrial effluent in to land deteriorates the groundwater quality. Once the groundwater is contaminated, its quality cannot be restored by discontinuing the contaminants from the origin. It therefore becomes imperative to develop devices and means to prevent its contamination.

Conceptually, irrigation water quality refers to the characteristics of a water supply that will influence its suitability for a specific use, i.e. how well the quality meets the needs of the user. Water quality is defined by certain physical, chemical and biological characteristics. Irrigation water quality is being evaluated generally based on the chemical and physical characteristics. Irrigation and domestic uses have different quality needs and water supply is considered more acceptable (of better quality) if it produces better results or causes fewer problems.

### 15.1. Problems associated with poor quality irrigation water

Poor quality of water is one of the main factors turning good soil into saline or sodic. Several salts dissolved in it, as water is universal solvent. Irrigation with saline water adversely affects soil quality, crop growth and productivity. High subsoil water table, aridity, seepage from canals, poor drainage, back water flow, intrusion of sea water also leads to salinity and sodicity. Around 1.5 Mha areas are affected by poor quality water in India. The most affected state is Rajasthan. In world, over 50 million ha of soils are affected by poor water quality spread over 24 countries. The problems of poor quality irrigation water is discussed hereunder.

i) **Effect on water absorption of crop plants:** If excess soluble salts of irrigation water accumulated in crop root zone, crop has difficulty in extracting enough water. Root growth is also suppressed; increasing the difficulty of water uptake. Salinity stress in plants is often called physiological drought. Due to reduced uptake of water and other effects, yields are reduced. The reduction in yield due to salinity is more in warm climate than cool climate.

ii) **Reductions of soil permeability:** Soil permeability is reduced due to the deflocculation effect of sodium. If permeability is reduced, infiltration of water into and through the soil is reduced. Adequate root penetration is inhibited due to the presence of impermeable soil layer caused by $CaCO_3$ and high exchangeable sodium percentage (ESP). Crusting of seed bed, Water logging, reduced oxygen and nutrient supply to the crops are the problems due to high sodium content in irrigation water relative of Ca & Mg.

iii) **Toxicity Symptoms of B, Cl, Na, Bicarbonate appear on crop plants:** Excessive uptake of Boron, Chlorine, Sodium, Sulphate and Bicarbonate by plant creates toxicity problems. Vegetative growth decrease as osmotic pressure of the soil solution increases. Reduction in growth takes place even without any external toxic symptoms. Increase in salinity, salt injury appears. Thick cuticle, waxy bloom and deep blue-green colour of leaves are the symptoms of plants affected with salt injury. At high salt levels, leaf burn appears in barley, sorghum and field beans

iv) **Anatomical and Physiological disruption in crop plants:** Salinity reduces cell division, cell enlargement and protein synthesis. It affects the structure and integrity of plant membranes and causes mitochondria and chloroplast to swell. Sodium and chloride at toxic levels disrupt the structure of the protein molecules. High chloride content hinders the development of xylem tissue.

v) **Nutritional imbalance and toxicity effects in plants:** Higher level of certain ions affects the absorption of other nutrient elements. High concentration of sulphate reduces the uptake of calcium enhances the uptake of sodium. This process causes high level of sodium in plants, thus causing sodium toxicity. High concentration of Ca reduces the uptake of K. High concentration of Mg induces Ca deficiency in crop plants.

vi) **Effect on soil microbial activity:** Nitrite and Nitrate ($NO_2$ & $NO_3$) producing bacteria (nitrification bacteria) sensitive to high salt concentration than $NH_4$ producing bacteria. Azotobacter is resistance to salt concentration.

vii) **Other effects:** Excessive nitrogen in irrigation water causes excessive vegetative growth, lodging, delayed crop maturity. White and black deposit on soil surface is due to high salt content and sodium. Using salty irrigations water for sprinkler irrigation causes leaf burn in crop plants. Tilth of the soil will be poor due to high exchangeable sodium percentage (ESP). High exchgeable Na tends to make moist soil impermeable to air and water. Further, high ESP waters make the soil hard on drying and difficult takeup tillage and intercultural operations. The dense crusts formed on soil surface due to use of irrigation water with high exchangeable Na interfere with germination and emergence of seedlings. Soluble carbonates are in irrigation water applied to soil in absence/ low concentrations of Ca and Mg in soil, makes soil alkaline and unfavourable for crop plants. Excess sodium carbonate ($Na_2CO_3$) in irrigation water is toxic to crop plants. High concentration of sodium in irrigation water causes dispersion and swelling of soil which causes surface crusting, decrease in infiltration rates at the surface and reduces the hydraulic conductivity of the soil.

The groundwater contamination due to salinization is major issue which can be triggered by various processes, viz., intrusion of seawater, pollution by agrochemicals, geogenic contamination and salinization induced by irrigation (secondary salinisation). The ratio of sodium ions to calcium and magnesium ions called the sodium-adsorption ratio (SAR), is used to determine the sodium hazard for irrigation waters. Since, sodium hazard increases as SAR increases, therefore, the suitability of water for irrigation decreases. The effect of irrigation water on soil infiltration rates is dependent upon the interaction between the flocculating effects of specific conductance and the dispersion effects of sodium.

## 15.2. Assessment irrigation water quality

Irrigation water quality is determined according to the purpose for which it will be used. The suitability of irrigation water is mainly depends on the amounts and type of salts present in water. The main soluble constituents are calcium, magnesium, sodium as cations and chloride, sulphate, bicarbonate as anions. The other ions are present in minute quantities are boron, selenium, molybdenum and fluorine which are harmful to animals fed with plants grown in soils with excess concentration of these ions.

### i) Hydrogen ion activity or pH

The parameter pH is the negative logarithm of hydrogen ion activity. The pH scale ranges from 0 to 14. If pH is less than 7 it is considered as acidic in nature, greater than 7 is alkaline and pH of 7 is treated as neutral. Therefore, pH is a measure of acidity or alkalinity of water. The principal use of pH is quick evaluation of the possibility of water being normal or abnormal. The normal range of irrigation water is from pH 6.5 to 8.4. The pH of water changes with the production of hydrogen or hydroxyl ion in different chemical reactions. With the redox potential, temperature and pressure, the pH ascertains the compounds dissolved and precipitated in groundwater. In ground water the redox potential depends on the number of cations and anions in solution. Hence, the chemical concentration of anion with respect to pH may be inkling to its solubility. The pH of irrigation water furnishes critical information in many types of geochemical equilibrium or solubility calculations.

Acid water especially when the pH is below 6 has the potency to be corrosive to pipelines system laid in soil for carrying irrigation water. As the pH of the irrigation water increases above 8.2, the potential for sodium problems enhance. The higher pH of groundwater may be due to considerable sodium, calcium, magnesium, carbonate and bicarbonate concentration as carbonates and bicarbonates are hydroxyl generating ions.

**ii) Electrical conductivity**

The term "electrical conductivity" is synonymous with "specific electrical conductance." The standard unit for conductivity is milliMhos $cm^{-1}$ or microMhos $cm^{-1}$ but is also represented by $dSm^{-1}$. Electrical conductivity is used for indicating the total concentration of the ionised constituents of natural water. It is closely related to the sum of the cations (or anions) determined by chemical analysis and, therefore, correlates, well with the value for dissolved solids. Salinity alters accessibility of water to crops. Osmotic pressure of the soil water is increased by the excess salts thereby reduces absorption of water by plant roots, which results in a physiological drought condition. Even though the soil appears to have enough moisture, the plants may wilt because the roots do not absorb sufficient water to compensate for the water loss by transpiration. For diagnosis and classification, specific conductance is used to express the total concentration of soluble salts (salinity hazard) in irrigation water. Electrical conductivity has been universally accepted as a standard measure of salinity hazard of irrigation water. Electrical conductivity varies with concentration, degree of ionization of the constituents of water and temperature.

## Classification of irrigation water based on salinity hazard

C1 : Low salinity hazard: Electrical conductivity ranges from 0 to 250 micro mho $cm^{-1}$ (or) 0 to 0.25 dS $m^{-1}$. These waters are excellent quality. No specific management practices are suggested. These waters can be used safely for irrigation for most crops. .

C2 : Medium salinity waters. Electrical conductivity of water ranges from 250-750 micro mho $cm^{-1}$ or 0.25 – 0.75 dS $m^{-1}$, these are good quality water can be used with moderate leaching. Salt sensitive crops may get damage.

C3 : High salinity waters. Electrical conductivity of water ranges from 750 – 2250 micro mho $cm^{-1}$ (or) 0.75 – 2.25 dS $m^{-1}$. These are doubtful quality water and cannot be used on soil with restricted drainage. Crops with low salinity tolerance may get damaged. Provision of drainage and use of good quality irrigation water to leach excess salts is suggested.

C4 : Very high salinity hazard waters. Electrical conductivity of water ranges from > 2250 micro mho $cm^{-1}$ (or) > 2.25 dS $m^{-1}$. These irrigation waters are not suitable for irrigation. Use of this water requires salt tolerant crops with proper management practices.

### iii) Sodium Adsorption Ratio (SAR)

Sodium adsorption ratio (SAR) is the effective factor or parameter used for ascertaining the suitability of groundwater for irrigation purposes. Based on SAR values, irrigation water is classified into different classes which indicates that SAR value between 0-10, i.e., low sodium water poses almost no risk of exchangeable sodium, medium sodium water having SAR 10-18 can show considerable hazard, while on the contrary, high and very-high sodium water with SAR 18-26 and greater than 26, respectively are regarded as unfavourable as they can lead to detrimental levels of exchangeable sodium in soils. Sodium Adsorption Ratio is calculated by the equation,

$$SAR = \frac{Na^{+}}{\sqrt{\frac{Ca^{2+} + Mg^{2+}}{2}}}$$

where all ionic concentrations are expresses in meq $L^{-1}$. Ayers and Westcot reported that irrigation water having SAR values greater than 6-9 may induce permeability problem in swelling and shrinking types of soils. The SAR of soil solution is increased with the increase in SAR of irrigation water which eventually increases the exchangeable sodium of the soil. With elevated sodium adsorption ratio (SAR) values there is decrease in hydraulic conductivity and aggregate stability besides affecting clay dispersion, swelling of expandable clays, surface crusting and reduced tillage. The toxicity of $Na^{+}$ takes place with the accumulation of sodium in the tissues of plants thereby exceeding the tolerance limit of the crop. Moreover, excessive SAR levels can lead to soil crusting, poor seedling emergence, and poor aeration. It is evinced that SAR can specify the degree to which irrigation water tend to enter cation- exchange reactions in the soil.

### Classification based on the Sodicity hazard (SAR-Sodium Adsorption Ratio)

S1 : Low sodium hazard waters. Water can be used on all soils with little danger of development of normal level of exchangeable Na. SAR < 10.

S2 : Moderate sodium Hazard. SAR range from 10-18. These waters produce hazards with fine textured soils which have high CEC especially with low leaching conditions. Can be used on course textured soils with good permeability.

S3 : High sodium hazard waters. SAR range from 18-26. High Na water may produce harmful level of exchangeable Na in most soils. It requires special management practices such as good drainage, heavy leaching and organic matter additions as well as gypsum applicatin. Exchangeable Na should be replaced by the use of chemical amendments like gypsum.

S4 : Very high sodium hazard. SAR > 26; Very high Na water not suitable for irrigation this water can be used with the addition of gypsum or other amendments.

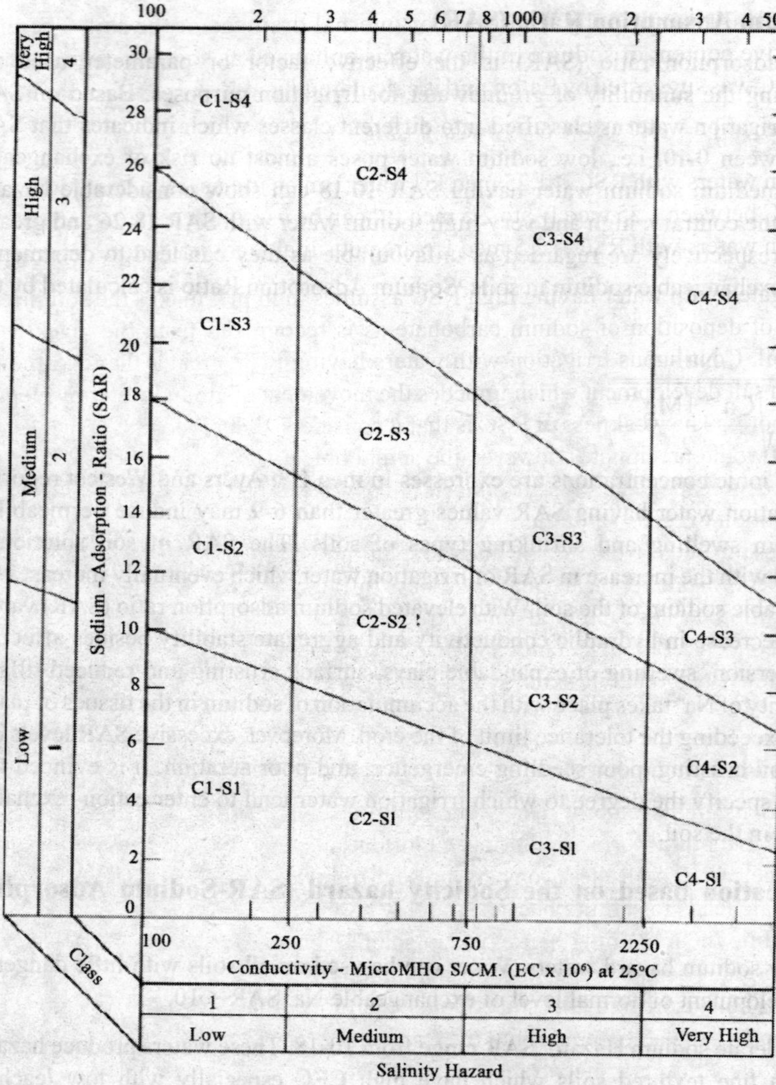


**Fig. 1:** Wilcox's graphic (USSL,1954) diagram for classification of irrigation waters

## iv) Residual Sodium Carbonate (RSc)

For agricultural purposes, residual sodium carbonate (RSc) is usually used to ascertain the dangerous effect of carbonate and bicarbonate on the quality of water. RSC of irrigationwater is influenced by pH, EC and SAR of the irrigation water. The continuous usage of water having high RSC will cause burning of plant leaves and reduces the yield of crops. Because of high concentration of bicarbonate in the water, the propensity for calcium and magnesium to precipitate increases as the water in the soil becomes more concentrated ensuing the decrease of permeability and finally leads to poor internal drainage of the soil. Therefore, the relative content of sodium in the water is enhanced as sodium carbonate. The term RSC was suggested by Eaton and is calculated by

$$\mathrm{RSC}\left(\mathrm{meql}^{-1}\right) = (\mathrm{HCO}_3^- + CO_3^{2-}) - (Ca^{2+} + Mg^{2+})$$

Irrigation waters with RSC < 1.25 meq $L^{-1}$ water can be used safely. Irrigation waters with RSC between 1.25 meq $L^{-1}$ to 2.5 meq $L^{-1}$ can be used with certain management. Irrigation waters with RSC > 2.5 meq $L^{-1}$ are not suitable for irrigation.

The irrigated with water having high RSC assumes high pH; makes the soil infertile because of deposition of sodium carbonate as is recognized from the black colour of the soil. Continuous irrigation with waters having RSC greater than 2.5 meq $L^{-1}$ results in salt development which impedes the movement of air and water by clogging the soil pores. The weakness of RSC is that it considers all bicarbonate in the water as though it would precipitate. However, the actual precipitation of bicarbonate depends upon the degree to which salts are concentrated in the root zone by evapotranspiration if the solution exceeds saturation index of calcite. For instance, if the evaporation of water does not occur from the soil and the solution stays at equilibrium or unsaturated with respect to calcite, bicarbonate will pass through the soil. A negative RSC is the best condition because the total concentration of carbonate and bicarbonate is lower than the concentration of calcium and magnesium combined which implies that there is no residual carbonate to react with sodium to enhance the sodium hazard in the soil.

## v) Chloride concentration

Owing to its high solubility in water chlorine exists as chloride ion and is the predominant natural form of chlorine. Chloride salts are more harmful than sulphate salts, since, when both these ions are present in high concentrations merely half of the sulphate leads to salinity because roughly half of the sulphates get precipitated as $CaSO_4$ while the remaining half exists in soluble form as Na $SO_4$ and $MgSO_4$ in the soil. Chloride is considered as the most common toxic ion in irrigation water. Since, chloride is not adsorbed by the soil colloids; therefore, it travels easily with soil water, is absorbed by the crop, moves into the transpiration stream, and accumulates in the leaves. If the chloride concentration in the leaves exceeds the tolerance of the crop, injury symptoms develop, such as leaf burn or drying of leaf tissue. The concentration of chloride in natural waters is somewhat low except the water is brackish or saline. In groundwater, the origin of chloride may be from diverse sources such as weathering, leaching of sedimentary rocks and soils, intrusion of saltwater, windblown salt in precipitation, domestic and industrial waste discharges, municipal effluents, etc. In groundwater the concentration of chloride ion upto 70 mg $L^{-1}$ (2 me $L^{-1}$) is considered safe and causes severe problem in the crops at concentration >350 mg $L^{-1}$ (10 me $L^{-1}$). The concentration of chloride in groundwater is best to study water-rock interaction.

**Table 1:** Chloride ion concentration and irrigation water quality

| Chloride concentration (me $L^{-1}$) | Water quality |
|---|---|
| < 4 | Excellent water |
| 4-7 | Moderately good water |
| 7-12 | Slightly usable |
| >12 | Not suitable for irrigation purpose |

### vi) Boron concentration

Boron is a micronutrient, therefore, essential for plant growth. However, slight excess of this element in the irrigation water or in soil solution can produce toxicity symptoms in certain crops. Boron is an inorganic essential nutrient that is dissolved in water and exists as a mixture of the $^{10}B$ (19.78%) and $^{11}B$ (80.22%) isotopes. The presence of naturally occurring boron in groundwater is primarily due to the leaching from rocks and soils containing borates and borosilicates. The possible sources of boron contamination in water resources are either natural (water rock interaction and seawater encroachment) or anthropogenic (sewage effluents and fertilizers). The concentration of boron in groundwater ranges widely from less than 3 to above100 mg/L throughout the world. McCarthy and Ellery proposed limits of boron concentration in irrigation water for semi-sensitive, semi-tolerant and tolerant crops as greater than 1.25, 2.5 and 3.75 ppm, respectively (table 2). The most sensitive crops can tolerate no more than 0.33 ppm of boron concentration.

**Table 2:** Boron concentration and irrigation water quality

| Boron toxicity/ hazard | Boron concentration (ppm) | | | Remarks |
|---|---|---|---|---|
| | Sensitive crops | Semi-tolerant crops | Tolerant crops | |
| Very low | < 0.33 | < 0.67 | < 1.0 | Can be used safely |
| Low | 0.33 – 0.67 | 0.67 – 1.33 | 1.00 – 2.00 | Can be used with management |
| Medium | 0.67 – 1.0 | 1.33 – 2.0 | 2.0 – 3.0 | Unsuitable for irrigation |
| High | 1.0 – 1.25 | 2.0 – 2.5 | 3.0 – 3.75 | Unsuitable for irrigation |
| Very high | > 1.25 | > 2.5 | > 3.75 | Unsuitable for irrigation |

### vii) Total dissolved solids (TDS)

The mineral components dissolved in water comprise dissolved solids. TDS are determined by evaporating the filtered sample of water to dryness, thereafter, weighing the residue. The concentration of dissolved solids in natural water is normally less than 500 mg/L, however, water with greater than 500 mg/L is unsuitable for drinking and many industrial uses. Therefore, the total concentration of dissolved minerals solids in water is a universal indication of the over-all suitability of water for various uses. The high TDS concentration in groundwater may be due to the occurrence of bicarbonates, carbonates, sulphates, chlorides and calcium. The methods which remove TDS are reverse osmosis, electrodialysis, exchange and solar distillation process. High value of TDS influences the taste, hardness, and corrosive property of the water.

### viii) Soluble sodium percentage (SSP, Na%)

This soluble sodium percentage (SSP) or percent sodium is a computed by the following equation,

$$Na\% = \frac{Na^+ + K^+}{Ca^{2+} + Mg^{2+} + Na^+ + K^+} \times 100$$

where the concentrations are in meq $L^{-1}$. High SSP value is the indication of soft water whereas a low value signifies hard water. Soluble sodium is important in categorizing irrigation water in terms of soil permeability to air and water i.e, ease with which a soil transmits water or air. When irrigation water contains higher concentration of $Na^+$, this ion tends to get adsorbed to the clay particles and displaces $Ca^{2+}$ and $Mg^{2+}$ ions. Exchangeable $Na^+$ ions are attracted weakly to soil colloids, thus, spread out to form a comparatively thick swarm of hydrated ions held in very loose outer-sphere complexes around the colloids, contrary to $Ca^{2+}$ which is strongly attracted to soil colloids by forming thin layer of hydrated ions. Therefore, highly sodium saturated colloids are held far apart that the forces of cohesion do not come into play to attract one colloid surface with another. Rather, the poorly balanced electronegativity of each colloidal surface repels other electronegative colloids and the soil becomes dispersed. So, the exchange process of $Na^+$ in water for $Ca^{2+}$ and $Mg^{2+}$ in soil decreases permeability consequently leads to soil with poor internal drainage. Irrigation water having soluble sodium percentage greater than 60 may lead to sodium accumulation and probably destruction of soil structure, infiltration and aeration. The higher Na in the groundwater may be due to long residence time of water, dissolution of minerals from lithological composition and addition of sodium containing chemical fertilizers with irrigation waters. Alkali (sodium) hazard is represented by sodium percentage, nevertheless, is not effective standard when compared to sodium adsorption ratio as percent sodium expresses percentage of sodium out of total cations in lieu of correlating sodium with calcium and magnesium only.

### ix) Lime deposition potential (LDP)

A special parameter, called as Lime deposition potential (LDP) is to indicate lime deposition when calcium or magnesium carbonates (lime) precipitate out of the irrigation water, which leave behind white residues or deposits. Increasing temperatures, increasing pH, loss of carbon dioxide in the water and evaporation are the factors which are responsible for lime deposition. Any of these environmental effects can initiate lime deposition on the equipment, vegetation and fruit as water passes through the irrigation system. The irrigation problems comprise plugging in micro- irrigation systems, precipitation of micronutrient fertilizers and SAR values. Water with high LDP usually results in the formation of a deposits of lime on fruits and leaves, which detract the value of crop.

### x) Nitrate concentration

During recent years, the pollution of groundwater by nitrates has been ascertained enormously across the globe. The concentration of nitrate greater than 45 mg $L^{-1}$ causes a disease in humans called as methemoglobinemia or blue baby syndrome. The possible origin of nitrate in agricultural areas include fertilizer, animal waste and mineralization of soil organic N (in plant residues, bacterial biomass and soil constituents). Due to intensive agriculture, large amounts of nitrogen fertilizers commonly urea, nitrate or ammonium compounds are applied to soils which result in higher concentration of nitrate in the areas of intensive arable production particularly Punjab and Haryana states.

**Table 3:** Nitrate concentration and irrigation water quality

| S.No | Nitrate ion concentration (m.e.$L^{-1}$) | Irrigation water quality |
|---|---|---|
| 1 | < 5 | Excellent quality |
| 2 | 5 to 30 | moderately injurious plants |
| 3 | >30 | Highly injurious to plants |

### xi) Magnesium Hazard (MH)

Broadly speaking, in groundwater alkaline earth metals are in state of equilibrium. Since, magnesium is an essential nutrient for plant growth. Excessive concentration also detrimental to the plant growth. The concentration of magnesium in water plays a pivotal role in deciding the quality of water for irrigation purposes, therefore, agricultural use. Magnesium hazard of water for irrigation is calculated by the formula.

$$MH = \frac{Mg^{2+}}{Ca^{2+} + Mg^{2+}} \times 100$$

Magnesium hazard less than 50 is considered suitable for irrigation whereas greater than 50 is insidious and unsuitable for irrigation as it decreases the yield of crops and makes soil alkaline. At the similar level of salinity and SAR, adsorption of sodium by soils and clay minerals is more at higher Mg: Ca ratios, since, the bonding energy of magnesium is lower than that of calcium, allowing more sodium adsorption and it occurs when the Mg: Ca ratio surpasses 4.

### xii) Permeability index (PI)

The long-term use of irrigation water had a profound effect on soil permeability as it is influenced by total dissolved salts, sodium content and bicarbonate content. Therefore, to integrate these three terms, Doneen empirically devised an equation referred to as 'Permeability Index (PI)' after carrying a series of investigations for which he used enormous number of irrigation water samples of variable ionic relationships and concentration.

$$PI = \frac{Na^{+} + \sqrt{HCO_3^{-}}}{Ca^{2+} + Mg^{2+} + Na^{+}} \times 100$$

Permeability index is a crucial parameter for assessing the suitability of irrigation water. From the soil quality viewpoint, in combination with subsurface structural features high permeability index would facilitate extensive contamination of groundwater. According to Doneen (1964), using the Permeability Index (PI), the irrigation water quality is classified into three different classes; more than 75% of the maximum permeability is Class I, suitable for irrigation, 25-75% of the maximum permeability is Class II, slightly suitable, and 25% of the maximum permeability is Class III, not suitable for irrigation.

### Xiii) Kelly's ratio

Kelly's Ratio was formulated by Kelly and is computed by dividing sodium ion concentration versus calcium and magnesium ion concentrations. The concentration of sodium calculated against calcium and magnesium ion is known as Kelly 1963. It is a significant index parameter to assess the suitability of groundwater for irrigation uses.

$$KR = \frac{Na^{+}}{(Ca^{2+} + Mg^{2+})}$$

where, concentrations of all ions were expressed in meq/L. Waters with a Kelly ratio value <1.0 are regarded suitable for irrigation, while those with higher than 1.0 are considered unsuitable.

### xiv) Hardness

Determination of water hardness is a utilitarian test to evaluate quality of water for domestic, agricultural and industrial uses. The hardness of water is generally caused by calcium and magnesium. However, total hardness of water can be classified into two types, i.e., temporary and permanent hardness. Total hardness is the summation of temporary and permanent hardness. The hardness of water suggests the nature of the geological formations with whichit has been in contact. The total hardness (as $CaCO_3$) of water samples can be calculated using the following equation,

$$[CaCO_3] = 2.5\,[Ca^{2+}] + 4.1\,[Mg^{2+}]$$

The water that contains <75 mg/ L $CaCO_3$ as soft, 75-150 mg/ L$CaCO_3$ as moderately hard, 150-300 mg/L $CaCO_3$ as hard and >300 mg/L $CaCO_3$ as very hard

### xv) Chloro-Alkaline Indices (CAI)

Understanding changes in chemical composition of groundwater during its travel through the soils geological strata is essential. To regulate the dissolution of unwanted elements in water is impractical during the subsurface movement but it is important to identify the different changes experienced by the water throughout the motion. Schoeller studied the ion exchange between the groundwater and its environment during the travel though the soil. The Chloro- alkaline indices (1 and 2) used in the base exchange are computed by the following equations:

I) Chloro-Alkaline Index 1

$$CAI\text{-}1 = \frac{[Cl\text{-}(Na+K)]}{Cl}$$

II) Chloro-Alkaline Index2

$$CAI\text{-}2 = \frac{[Cl\text{-}(Na+K)]}{(SO_4 + HCO_3 + CO_3 + NO_3)}$$

The exchange is identified as direct if there is ion exchange of Na and K from water with Mg and Ca in the rocks and the indices are negative. On the contrary, if the exchange is reverse then the exchange is known as indirect with positive indices and reported that 80% and 59% water samples in Cuddapah district of Andhra Pradesh showed positive ratio of chloro-alkaline indices in the post and pre-monsoon seasons, respectively, whereas 20% and 41% of the corresponding seasons exhibited negative ratios illustrating the type of base exchange.

### xvi) Saturation index (SI)

The degree of equilibrium between water and minerals is assessed by the saturation indices. The alterations in saturation sate are valuable in deciding various stages of hydrochemical evolution and assist in identifying the geochemical reactions that are crucial in checking chemistry of water. The saturation index of the mineral is calculated from the equation,

$$SI = \frac{\log (IAP)}{Kt}$$

where, IAP is the ion activity product of dissociated chemical species in solution, Kt is equilibrium solubility product for the chemical involved at the sample temperature. If water is precisely saturated with the dissolved mineral, SI is equal to zero. If the SI is below zero, the groundwater is considered as under saturated regarding any mineral thereby revealing the quality of water from a formation with inadequate quantity of mineral for solution or brief residence time. On the contrary, SI greater than zero indicates that groundwater is supersaturated regarding specific mineral phase and consequently unable to dissolve more of a mineral. Therefore, this index suggests groundwater discharging from an aquifer carrying sufficient amount of mineral with enough residence time to attain equilibrium. However, super saturation may also be developed by other causes like incongruent dissolution, common ion effect, evaporation, rapid increase in temperature and $CO_2$ exsolution.

Central Soil Salinity Research Institute located at Karnal, India gave classification of irrigation waters based on EC, SAR and RSC (Table 4). Three categories were given viz., Good quality waters, Saline waters and Alkali waters for which corresponding EC, SAR and RSC values were given under sub categories.

**Table 4:** Classification of irrigation water the basis of EC, SAR, and RSC, standards suggested by CSSRI, India

| Water quality | Electrical conductivity ($dSm^{-1}$) | SAR $(mmolL^{-1})^{1/2}$ | RSC (meq $L^{-1}$) |
|---|---|---|---|
| A. Good quality waters | <2 | <10 | <2.5 |
| B. Saline waters | | | |
| 1. Marginally saline | 2-4 | <10 | <2.5 |
| 2. Saline | >4 | <10 | <2.5 |
| 3. High SAR Saline | >4 | >10 | <2.5 |
| C. Alkali waters | | | |
| 1. Marginally alkali | <4 | <10 | 2.5-4.0 |
| 2. Alkali | <4 | <10 | >4.0 |
| 3. Highly alkali | Variable | >10 | >4.0 |

**Table 5:** Guidelines for interpretation of irrigation water quality (FAO)

| Potential Irrigation Problem | | Units | Degree of restriction on use | | |
|---|---|---|---|---|---|
| | | | None | Slight to Moderate | Severe |
| Salinity *(affects crop water availability)* | | | | | |
| | $EC_w$ (or) | dS/m | < 0.7 | 0.7 – 3.0 | > 3.0 |
| | TDS | mg/L | < 450 | 450 – 2000 | > 2000 |
| Infiltration *(affects infiltration rate of water into the soil (cm/h). Evaluate using $EC_w$ and SAR together)* | | | | | |
| SAR | = 0 – 3 | EC= | > 0.7 | 0.7 – 0.2 | < 0.2 |
| | = 3 – 6 | | > 1.2 | 1.2 – 0.3 | < 0.3 |
| | = 6 – 12 | | > 1.9 | 1.9 – 0.5 | < 0.5 |
| | = 12 – 20 | | > 2.9 | 2.9 – 1.3 | < 1.3 |
| | = 20 – 40 | | > 5.0 | 5.0 – 2.9 | < 2.9 |
| Specific ion toxicity *(affects sensitive crops)* | | | | | |
| | **Sodium (Na)** | | | | |
| | surface irrigation | SAR | < 3 | 3 – 9 | > 9 |
| | sprinkler irrigation | me/L | < 3 | > 3 | - |
| | **Chloride (Cl)** | | | | |
| | surface irrigation | me/L | < 4 | 4 – 10 | > 10 |
| | sprinkler irrigation | me/L | < 3 | > 3 | |
| | **Boron (b)** | mg/l | < 0.7 | 0.7 – 3.0 | > 3.0 |
| Miscellaneous effects *(affects susceptible crops)* | | | | | |
| | Nitrogen ($NO_3$ - N) | mg/L | < 5 | 5 – 30 | > 30 |
| | Bicarbonate ($HCO_3$) *(overhead sprinkling only)* | me/L | < 1.5 | 1.5 – 8.5 | > 8.5 |
| | pH | | Normal Range 6.5 – 8.0 | | |

## 15.3. Management of poor quality irrigation waters

The change in quality of irrigation water is technically not feasible. The effect of irrigation water on soil depends on the composition of water, the properties of soil and climatic conditions. The following measures should be adopted for improvement and management of poor quality water

### 1) Pre-treatment of water

If water contains excessive amounts of Na (>70%), bicarbonate (RSC > 5 me $L^{-1}$ or HCO3 > 8 me $L^{-1}$) and Mg (Mg/Ca ratio >4). It will induce sodicity. Hence water may either pre-treated with gypsum by suitable method or gypsum added directly to soil. For reducing 1 me $L^{-1}$of RSC water, about 169 kg of gypsum may be applied in the field before sowing.

### 2) Fertilizer management

Judicious use of fertilizer can increase crop yield in saline water irrigated areas. About 15% additional quantities of fertilizers (N, P) over the normal dose be applied. Ammonium sulfate or CAN is superior to urea due to less loss of N through gaseous loss and pH reducing tendency in root zone. DAP and super phosphate is better than rock phosphate. Response of fertilizers decreases as the salinity and SAR of water increases, so fertilizers should be added @ 1.25 -1.5 times the normal rate of their application and in split application to improve the yields. Application of zinc @ 20kg $ha^{-1}$ reduces the adverse effect of higher salinity and sodicity

### 3) Agronomic management

Salts tend to accumulate closer to the soil surface and tend to move downwards with reclamation and irrigation. Deep ploughing will loosen the sub soil and reduce salinity Mulching reduces salt accumulation on the surface. Planting technique like sowing on the slope of the ridge than sowing on flat land Pitcher or drip irrigation is good when water quality is poor since it keeps soil moist. Quantity of water need for irrigation is calculated based on tolerance limit of crops, degree of salt accumulation in soil and quality of water used.

For soils irrigated with saline water, slopping beds (Fig.2) are the best where the seedling can be safely established on the slope below the zone of salt accumulation. Under flood or sprinkler irrigation where water and salt transport is downward and away from the seedling, limited pre-planting leaching of the upper soil strata may take care of the germination and establishment inhibition. Under furrow and drip irrigation there is downward component of water and salt transport, but another component is lateral and upward in the spaces between furrows or laterals. With these methods the adjustment of the soil surface contour and seedling or planting position according to the expected salt distribution can limit significantly this damage.

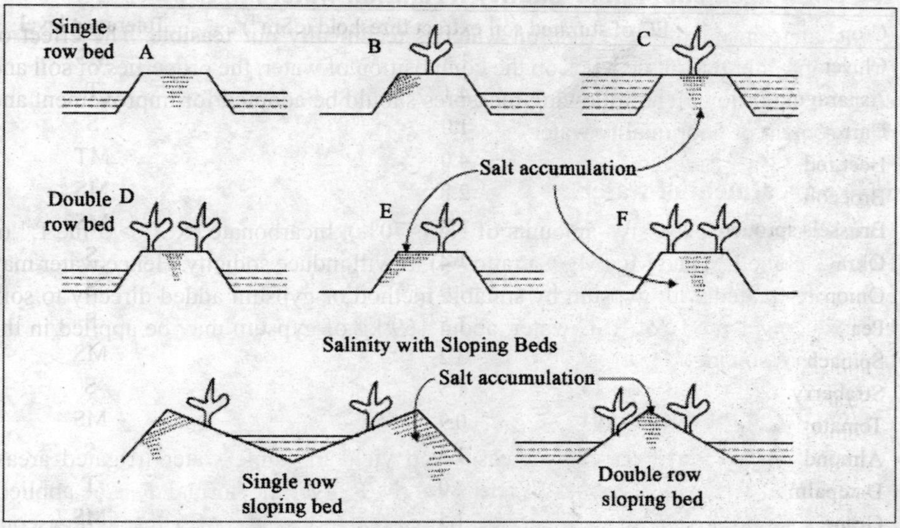


**Fig. 2:** Typical salt accumulation pattern in ridge and bed cross section in soils irrigated by furrow

## 4) Selecting salt tolerance of crops

Sensitive crops like wheat, barley, cow pea, pearl millet totally fail to grow under high saline water. There are certain crops which can grow under high salt stress since it has certain specific characters which offer resistance to grow in saline water. Among leguminous crop daincha is well known tolerance crop. High salt tolerant crops are rice, cotton, sugarbeet, tobacco, date palms, ber. Moderate salt tolerant crops are spinach, sugarcane, wheat, guava, pomegranate, tomato, sweet potato. Salt sensitive crops are beans, pea, grape, orange, apple, pear, carrot. FAO given threshold values for salt tolerance of field crops, vegetable crops and fruit crops (Tables 6)

**Table 6:** Salt tolerance threshold of field crops, vegetabele crops and fruit crops (FAO, 1935)

| Crop | EC of sturated soil extract threshold ($dSm^{-1}$) | Tolerance level |
|---|---|---|
| Barley | 8.0 | T |
| Bean | 1.0 | S |
| Braodbean | 1.6 | MS |
| Cotton | 7.7 | T |
| Maize | 1.7 | MS |
| Sorghum | 6.8 | MT |
| Soybean | 5.0 | MT |
| Sugarbeet | 7.0 | T |
| Wheat | 6.0 | MT |
| Alfalfa | 2.0 | MS |

| Crop | EC of sturated soil extract threshold (dSm$^{-1}$) | Tolerance level |
|---|---|---|
| Clover | 1.5 | MS |
| Asparagus | 4.1 | T |
| Carrot | 1.0 | S |
| Beet,red | 4.0 | MT |
| Broccoli | 2.8 | MS |
| Brussels sprouts | 1.8 | MS |
| Okra | 1.2 | S |
| Onion | 1.0 | S |
| Pea | 1.5 | S |
| Spinach | 3.2 | MS |
| Straberry | 1.5 | S |
| Tomato | 0.9 | MS |
| Almond | 1.5 | S |
| Datepalm | 4.0 | T |
| Grape | 1.5 | MS |
| Orange | 1.7 | S |
| Peach | 1.7 | S |

T: tolerant; S: sensitive; MS: moderately sensitive; MT: moderately tolerant

## 5) Crop rotation

Crops vary in their water requirement as well as in frequency with which they need water Crop with long duration of evapotranspiration will enhance much more salt accumulations compared to short duration crops. Crops like rice, alfalfa, and berseem require frequent helps reclamation irrigation. A best crop rotation not only provided good return but also help to control salinity of growing soil. Generally rice as first crop after burial of green manures followed by an alkali tolerant crops

## 6) Soil incorporation of organic materials

Organic materials (organic manures and organic wastes) could be incorporated in soil for improving physical properties and lowering the pH, EC and SAR of soil irrigated with saline water. For sodic water, use of FYM along with gypsum is more beneficial

## 7) Cyclic use with good quality water

Under situation of limiting quantity of good water, saline water can be used for irrigation in conjunction with good water. Pre sowing and first irrigation after germination should be done with good quality water

## 8) Treatment with amendments

In water containing excess sodium, gypsum can be used. On addition of gypsum, proportion of sodium to other ions is reduced.

## 9. Conjunctive use of saline and fresh water

The conjunctive use can be defined as the development and management of multiple water resources in a coordinated manner such that the total yield of the system over a period of years exceeds the sum of the yields of the individual components of the system resulting from an uncoordinated operation. The óbjective of conjunctive use implies not only the combined use of water resources of more than one type but also their exploitation through efficient management in technoeconomic terms by taking advantage of the interaction between them and the impact of one on the others. It refers to the integrated management of surface water and ground water and it requires:

(1) quantification of annual recharge and its spatial distribution to assess potential of conjunctive use,

(2) simulation of the ground water basin parameters to analyse the impacts of irrigation and development of the ground water on the changes in water levels in the aquifer, and

(3) identification of conjunctive use strategy that is most suitable for the given hydrologic, hydrogeologic, agroeconomic and hydrochemical conditions.

The conjunctive use planning methods include:

(1) Engineering considerations for feasible ground water operations based on simulation of ground water basin, and

(2) Resource allocations based on both simulation and mathematical programming approach.

In the conjunctive water use process, water balance is estimated considering rainfall, surface runoff, seepage from canals, drains and natural streams and irrigation water recycling. The water table fluctuations and its rise are determined. The optimum water yields that can be drawn from the wells or tubewells with different operation schedules are determined. The available quantity of surface water from canals, lakes or ponds is estimated. The water quality of surface water and ground water is evaluated. A matching cropping plan with the irrigation requirement is developed and the salt tolerances of crops are determined.

a) The conjunctive use planning must include principles involved in the two water systems considered independently, but must also include principles to guide the optimal development of the complementarity of the two systems. Conjunctive use is planned and practised with the following objectives:

b) Mitigating the effect of the shortage in canal water supplies often subject to steep variation in river flow during different periods in the year;

c) Increasing the dependability of existing water supplies;

d) Alleviating the problems of high water table and salinity resulting from introduction of canal irrigation;

e) Facilitating the use of poor quality water which cannot otherwise be used without appropriate dilution;

f) Storing water in ground water basins closer to the users, to ensure water supply to the users in case of interruption of surface water supply;

g) Minimizes drainage water disposal problem.

## 10. Alternating use of good and poor quality waters

Avoid the deterioration of the good water quality. This good quality water could be used at the time it should be most needed, for instance at the germination and seedling stages which are very sensitive to the salinity level of irrigation water as well as to satisfy the leaching requirements which requires water of relatively good quality.

With the plants which are sensitive to the salinity level in irrigation waters, satisfactory production could only be achieved with water of good quality through alternative application modes. The disadvantages appearing under mixing could be completely eliminated and offer a free-hand possibility in using the different water resources according to the prevailing conditions. The cyclic use of water of low and high salinity prevents the soil from becoming too saline while permitting, over a long period, the substitution of brackish water for a substantial fraction of the irrigation needs. Cyclic strategy provides a vast choice of the crops to be included in the crop rotation as compared with the blending technique where crop selection is limited to the tolerant ones.

## 11. Treatment of problem waters with amendments

Water of high RSC can be used for crops having low water requirement. The rain water should not be allowed to leave the field as runoff but should be allowed to percolate through the soil profile.

**Use of gypsum** : In water containing excess of sodium (when RSC > 4 me $L^{-1}$), gypsum can be successfully used. With the addition of gypsum, the adverse effects of sodium can be mitigated. Gypsum can be applied by

a) Keeping gypsum lumps or powdered gypsum in bags in small tanks constructed at the mouth of the tube well or water passage.

b) Application of gypsum @ 75 kg $ha^{-1}$ for each irrigation of 5 cm depth for 1 me $L^{-1}$ of RSC above 2.5 me $L^{-1}$ should be applied to the soil.

c) A saturated solution of gypsum may be added at a constant rate to the flowing water.

$$2\,NaHCO_3 + CaSO_4 \rightleftarrows Ca\,(HCO_3)_2 + Na_2\,SO_4$$

$$Ca\,(H\,CO_3)_2 \rightleftarrows CaCO_3 + H_2O + CO_2$$

$Na_2SO_4$ being soluble can be leached.

d) Water containing toxic amounts of boron can be reduced by mixing with gypsum. Sodium borate is converted to calcium borate which being soluble can be leached out of root zone.

**Use of ion exchange resins** : On a limited scale, water containing excess amounts of Na is passed through high CEC calcium and H-charcoal, which adsorbs Na.

$$Ca - charcoal + 2\,NaH\,CO_3 \overset{Na}{\rightleftarrows} Charcoal + Ca(H\,CO_3)_2$$

$$H - charcoal + 2\,NaH\,CO_3 \overset{Na}{\rightleftarrows} Na - charcoal + CO_2 + H_2O$$

Liberal application of organic manures specially growing of green manures like sunhemp and *dhaincha* and incorporating the same at the time of flowering in the soil also reduce the adverse effects of high RSC water.

Phosphorus should preferably be applied as SSP.

## 15.4. Use of saline water in agriculture

The change in quality of irrigation water is technically not feasible and not economically viable. The following measures should be adopted for the use of saline water.

1. Selection of salt tolerant crops and varieties.
2. Deep ploughing to break the hard pan of salts if any.
3. Furrow planting is best for saline conditions because the seed can be safely planted below the zone of high salt accumulation.
4. Poor germination, high mortality of young seedlings and poor tillering are common features when the crop is grown with poor quality water. So, higher seed rate and close spacing is advisable. An additional seed rate of 25% should be adopted.
5. Dhaincha as a green manuring crop improves physical properties of the soil.
6. Addition of organic manures to some extent mitigates the adverse effect of poor quality water.
7. Fertilizers should be applied 1.25 - 1.5 times the normal rate of their application.
8. Application of Zn @ 20 kg $ZnSO_4$ $ha^{-1}$ counteracts the negative effect of higher salinity and sodicity.
9. Split application of nitrogen to prevent N losses through volatilization and denitrification. Correction of nutrient deficiencies by foliar application of nutrients.
10. Dilution and cyclic use of good and saline waters. When good quality water is limited, it can be used as follows :
11. Pre-sowing and first irrigation should be with good quality water. Later saline water can be used.
    - Poor quality water can be mixed with good water.
    - Drip or pitcher irrigation is found suitable.
12. Use of mulches to reduce the requirement of water for evaporation. Use of mulches and intercultural operations reduce water requirement of crops, thus with saline water salinity develops at a relatively lesser intensity.

## Review questions

1) Write the importance of irrigation water quality,
2) Explain the Problems associated with poor quality irrigation water
3) Explain different criteria used for assessment water quality
4) Write the Classification of irrigation water the basis of EC, SAR, and RSC, standards suggested by CSSRI, India
5) Explain the Management of poor quality water
6) Write briefly on use of saline water in agriculture.

## Objective question

1. The following state registered highest ground water exploitation in India
   a) Andhrapradesh b) Arunachal pradesh
   c) Punjab c) Tamilnadu
2. The most affected state with poor quality waters in India is
   a) Rajasthan b) Andhrapradesh
   c) Arunachal pradesh d) Punjab
3. Salinity stress in plants is often called .
   a) Physiological drought b) Agronomic drought
   c) Genetic drought b) Economic drought
4. Soil permeability is reduced due to
   a) Flocculation effect of sesquioxides
   b) Deflocculation effect of calcium
   c) Flocculation effect of phosphorous
   d) Deflocculation effect of sodium
5. the problems due to high sodium content in irrigation water relative of Ca & Mg is
   a) Crusting of seed bed
   b) Water logging,
   c) reduced oxygen and nutrient supply to the crops
   d) All three
6. the symptom of salt injury in plants is
   a) Thick cuticle b) Waxy bloom
   c) Deep blue-green colour of leaves d) All three
7. In plants swelling mitochondria and chloroplast is the result of
   a) N deficiency b) Salt injury
   c) Water stress d) N toxicity

8. High chloride content in plants affects
   a) Development of xylem tissue
   b) Development of flowering
   c) Development of fruiting
   d) Cell division and elongation
9. high salt concentration in soil hinders the activity of
   a) $NH_4$ producing bacteria
   b) Azotobacter
   c) Nitrifying bacteria
   d) Rhizobium
10. Tilth of the soil will be poor due to
    a) High EC irrigation waters
    b) High ESP irrigation water
    c) High RSC irrigation water
    d) High chlorine irrigation water
11. Soil salinization induced by faulty irrigation practices is
    a) Primary selinization
    b) Secondary salinization
    c) Tertiary selinization
    d) Quaternary salinization
12. to determine the sodium hazard of irrigation waters, the criteria used is
    a) SAR
    b) RSC
    c) EC
    d) SSP
13. potency to be corrosive to pipelines system laid in soil for carrying irrigation water of reaction
    a) Acidic
    b) Basic
    c) Neutral
    d) Ultra basic
14. The units for EC is
    a) MilliMhos $cm^{-1}$
    b) MicroMhos $cm^{-1}$
    c) Desi Simons $m^{-1}$
    d) All the three
15. the standard measure of salinity hazard of irrigation water is
    a) TSS
    b) TDS
    c) EC
    d) % salts
16. Electrical conductivity of irrigation water influenced by
    a) Concentration salts
    b) Degree of ionization of the constituents of irrigation water
    c) Temperature
    d) All three
17. The salinity hazard of irrigation water when its EC record 0.20 dS $m^{-1}$
    a) Very high salinity hazard
    b) High salinity hazard
    c) Medium salinity hazard
    d) Low salinity hazard
18. One dS $m^{-1}$ is equals to
    a) 1.0 micro mhos $cm^{-1}$
    b) 10 micro mhos $cm^{-1}$
    c) 100 micro mhos $cm^{-1}$
    d) 1000 micro mhos $cm^{-1}$

19. Use of the following quality water requires salt tolerant crops with proper management practices
    a) C1 waters
    b) C2 waters
    c) C3 waters
    d) C4 waters
20. Irrigation water having SAR 10-18 grouped in to
    a) Low sodium hazard
    b) Moderate sodium hazard
    c) High sodium hazard
    d) Very high sodium hazard
21. Use of the following quality waters requires special management practices such as good drainage, heavy leaching and organic matter additions as well as gypsum application
    a) S1
    b) S2
    c) S3
    d) S4
22. To ascertain the dangerous effect of carbonate and bicarbonate on the quality of water the following criteria is used
    a) pH
    b) SAR
    c) RSC
    d) EC
23. Irrigation waters with RSC > 2.5 are
    a) Suitable for irrigation
    b) Moderately suitable for irrigation
    c) Not suitable for irrigation
    d) None of the three
24. Chloride is considered as the most common toxic ion in irrigation water is
    a) Chloride
    b) Sulphate
    c) Nitrate
    d) Phosphate
25. The chloride concentratinin irrigation water is 13 me $L^{-1}$
    a) Suitable for irrigation
    b) Moderately suitable for irrigation
    c) Not suitable for irrigation
    d) None of the above
26. Possible anthropogenic sources of boron contamination in water resources are
    a) Industrial activity\
    b) Mining activity
    c) Sewage effluents and fertilizers
    d) Pesticides

27. The most sensitive crops can tolerate a boron concentration of
   a) Less than 0.33 ppm  b) 0.33 to 0.67 ppm
   c) 0.67 to 1.0ppm  d) 1.0 to 2.0ppm
28. Irrigation water having soluble sodium percentage (SSP) of _____ may lead to sodium accumulation and probably destruction of soil structure, infiltration and aeration
   a) Less than 30  b) More than 30
   c) Less than 60  d) More than 60
29. The concentration of nitrate ------------ causes a disease in humans called as methemoglobinemia or blue baby syndrome
   a) Greater than 45 mg $L^{-1}$  b) 35 to 45 mg $L^{-1}$
   c) 20- 35 mg $L^{-1}$  d) 10 to 25 mg $L^{-1}$
30. Nitrate concentration in irriation water causes highly injurious to plants is
   a) Less than 5.0 m.e.$L^{-1}$  b) 5.0 to 30.0 m.e.$L^{-1}$
   c) Above 30.0 m.e.$L^{-1}$  d) Less than 10.0 m.e.$L^{-1}$
31. Magnesium hazard -------is considered suitable for irrigation
   a) Less then 70  b) Less than 50
   c) Less than 100  d) Less than 125
32. Irrigation water with less than 25% of maximum permeability index are
   a) Suitable for irrigation  b) Unsuitable for irrigation
   c) Moderately suitable for irrigation  d) None of the answers

## Answer Key

| | | | | | | | | | | | | | |
|---|---|---|---|---|---|---|---|---|---|---|---|---|---|
| **1** | c | **6** | d | **11** | b | **16** | d | **21** | c | **26** | c | **31** | b |
| **2** | a | **7** | b | **12** | a | **17** | d | **22** | c | **27** | a | **32** | b |
| **3** | a | **8** | a | **13** | a | **18** | d | **23** | c | **28** | d | | |
| **4** | d | **9** | c | **14** | d | **19** | d | **24** | a | **29** | a | | |
| **5** | d | **10** | b | **15** | c | **20** | b | **25** | c | **30** | c | | |

# References

# References


Abrol, I.P., Yadav J.S.P. and Massoud, F.I. 1988. Salt-affected soils and their management. FAO Soils Bulletin No. 39, Food and Agriculture Organization of the United Nations, Rome.

Anil R. Chinchmalatpure 2017. Reclamation and Management of Salt Affected Soils for Increasing Farm Productivity and Farmers' Income. ICAR-CSSRI/Bharuch/Technical Manual.

Antaryami Mishra 2007. Land Suitability Classification for Different Crops. Orissa Review. https://www.researchgate.net/publication/238101640Ayers R.S. and. Westcot D.W 1994 Water quality for agriculture. FAO irrigation and Drainage paper.

Balasubramanian A. 2015. The Wastelands in India. https://www.researchgate.net/publication/309786077

Bandyopadhyay, K. K., Misra, A. K., Ghosh, P. K. and Hati, K. M.,2010. Effect of integrated use of farmyard manure and chemical fertilizers on soil physical properties and productivity of soybean. Soil & Tillage Research., 110, 115–125

Bauder, T.A. Waskom, R.M. Sutherland P.L. and Davis J. G. 2007. Irrigation Water Quality Criteria. Colorado Water Institute; P.L. Sutherland, USDA/NRCS area resource conservationist. https://extension.colostate.edu/topic-areas/agriculture/irrigation-water-quality-criteria-0-506

Bhattacharyya, T. 2013. Soils of India: historical perspective, classification, and recent advances. Curr. Sci., 104, 1308– 1323.

Bhumbla, D.R. 1977. Alkali and saline soils of India, Paper presented at Indo-Hungarian Seminar on management of salt-affected soils, Feb. 7-12, Central Soil Salinity Research Institute, Kamal, India.

Bhupal Raj G, Singh MV, Patnaik MC, Khadke KM. 2009. Four decades of research on micro-and secondary- nutrients and pollutant elements in Andhra Pradesh. Research Bulletin. AICRP Micro- and Secondary-Nutrients and Pollutant Elements in Soils and Plants, IISS, Bhopal, pp 1–132

Brownmang Onwuka, 2018. Effects of soil temperature on some soil properties and plant growth. Advances in Plants & Agriculture Research 8(1):34-37

Bureau of Indina Standards 1986. Indian Standard Guidelies for Irrigation Water. Manakbhavan 9, Bahadur shah Jaffer Marg, New Delhi.

Chakravorty O. N., Sehgal J.L., Pal O.K. and Dev G.1992. Clay Mineralogy of Assam Soils Developed in Alluvium. Agropedology, 2: 1992 45-48

Chandran P, Ray S. K., T. Bhattacharyya, Srivastava P., Krishnan P., and Pal D. K. 2005. Lateritic soils of Kerala, India: their mineralogy, genesis, and taxonomy. Australian Journal of Soil Research, 43, 839–852

Chandran P., Ray D, S. K.,. Bhattacharyya T, Srivastava P., Krishnan P., and Pal D. K. 2005. Lateritic soils of Kerala, India: their mineralogy, genesis, and taxonomy. Australian Journal of Soil Research, 43, 839–852

Charles E. Kellogo 1961. Agriculture Handbook No. 210. Land Capability Classification. Soil Conservation Service, U.S. Department of Agriculture.

Charles White and Mary Barbercheck 2017. Managing Soil Health: Concepts and Practices. https://extension.psu.edu/managing-soil-health-concepts-and-practices

Chibuike G. U. and Obiora S. C. 2014. Heavy Metal Polluted Soils: Effect on Plants and Bioremediation MethodsApplied and Environmental Soil Science Volume 2014, Article ID 752708, 12 pages http://dx.doi.org/10.1155/2014/752708

CPCB (2013) Comprehensive Environmental Assessment of Industrial Clusters. Ecological Impact Assessment Series: EIAS/5/2009–2010. Central Pollution Control Board, Ministry of Environment and Forest, Government of India

Ghassemi, F., Jakeman, A.J., and Nix, N.A. 1995. Salinisation of land and water resources. Human causes, extent and management and case studies. CABI Publishing, Wallingford.

Goswami N. N. 1982. Distribution and characteristics of problem soils in India and their management for crop production. Proceedings of a symposium on tropical agriculture researches. International symposium on distribution, characteristics and utilization of problem soils Japan International Research Center for Agricultural Sciences. 25-39

Gowthami B and Gurumurthy P Effect of Biochar on Soil Physical, hydrological properties, growth, yield and quality of Rabi groundnut in red sandy loam soils of North coastal AP. Journal of Oilseed research. 36(3): 163-168

Gurbachan Singh, Bundela, D. S., Madhurama Sethi, Khajanchi Lal, and Kamra S. K. 2010. Remote Sensing and Geographic Information System for Appraisal of Salt-Affected Soils in India. American Society of Agronomy, Crop Science Society of America, and Soil Science Society of America. https://www.researchgate.net/publication/40847176

Gurumurthy P 2017. Laboratory manual of Soil Physical Analysis. Agricultural College, Naira, Acharya N.G Ranga Agricultural University, Andhra Pradesh, India.

Gurumurthy P 2018. Study of morphology, characters, genesis and classification of soils of Agricultural College farm, Naira. Journal of Research ANGRAU 46 (4) 39- 48.

Gurumurthy P and Yasmin C 2018. Laboratory manual of Analysis of problem soils and Irrigation water. Agricultural College, Naira, Acharya N.G Ranga Agricultural University, Andhra Pradesh, India.

Gurumurthy P, Balaguraviah D, Giridhar Krishna, Prabhuprasadini P and Sudhakar Rao D 2016. Soil Health Management – concepts and applications. Published by Agricultural college, Naira, ANGRAU on the occasion of World Soil Day.

Gurumurthy P. Nature, Properties and Management of Problem Soils. 2020. Satish Serial Publishing House, New Delhi.

Gurumurthy P., Santosh B., Yasmin C and Sudhakar Rao D 2018. Soil, plant, water and seed testing. Educreation publication, New Delhi.

Gurumurthy P., Srilatha T., Jagannadham J., Joga Rao P., Himabindu K.and Gowthami B. 2021. Soil Resources of North Coastal Andhra Pradesh: Physical and Fertility characters vis a vis management strategies for sustainable crop production. In.. National Web Conference on Strategies for buffering ecosystems for their sustainable services in the era of intensive agriculture and climate change held at Agricultural College, Rajmahendrvaram from – 26.03.2021.

Gurumurthy P., Srilatha T., Jagannadham J., Roy G.S., Sudhalkar Rao D.,JogaRao P., Nataraja K.C., Himabindu K.and Goutami N, Bhargava D.V. and Goutami B. 2021. Soil and water conservation through Watersheds and their impacts in Andhra Pradesh- An Overview. 2021. In 30th National Webconference on 'Soil and water management Technologies for climate Resilience, Agricultural and Environment sustainability' held at Bhubaneswar, organised by Soil Conservation Society of India, New Delhi from 14 – 16 December, 2021. Pp46

Gurumurthy P., Srilatha T., Jagannadham J., Himabindu K.and Gowthami, B. 2021. Soil related constraints in AndhraPradesh: Management options for intensification of Agriculture In. National web conference on "Sustainable Intensification and Agri Development in AP in the aftermath of COVID-19" held at Agril College,Naria from 02.03.2021 to 03.03.2021.

Gurumurthy, P. and Singa Rao,M 2006 Effect of tillage and irrigation regimes on Soil Physical properties and performance of Sunflower in Rice fallows, I. Soil Physical properties and Crop establishment. Journal of the Indian Society of Soil Science. 54 (1) 12- 17

Gurumurthy, P., Seshagiri Rao, M., Bhanuprasad, v., Pillai, R.N. and Veeraiah, K. 1996. The morphology and Taxonomy of red, black and associated soils of Giddalur mandal in Andhrapradesh. The Andhra Agricultural Journal. 43 (2-4): 93- 96.

Guy Fipps 2003. Irrigation Water Quality Standards and Salinity Management Strategies Texas A&M Agrilife Extension B-1667https://cdn-ext.agnet.tamu.edu/wp-content/uploads/2019/03/ EB-1667-irrigation-water-quality-standards-and-salinity-management-strategies-1.pdf.

Hamdy A. 2005. Saline irrigation management for a sustainable use in Hamdy A. ( ed.) , El Gamal F. ( ed.) , Lamaddalena N. ( ed.) , Bogliotti C. ( ed.) , Guelloubi R.( ed.)Non-conventional water use: WASAMED project. http://om.cih eam.org/article.ph pID PD F=8007 50

Hemlata Bhatt, Mohit Husain, Jagdeesh Prasad Rathore and Sah V.K. 2017. Bioremediation of problematic soils through Agroforestry practicesJournal of Pharmacognosy and Phytochemistry 6(5): 2044-2048

Indoria A.K., Sharma K. L,. Sammi Reddy K and Srinivasa Rao Ch. 2017. Role of soil physical properties in soil health management and crop productivity. Current Science. 112, (12), 2405-2414

Indoria A.K.,. Sharma K. L,. Sammi Reddy K and Srinivasa Rao Ch. 2016. Role of soil physical properties in soil health management and crop productivity in rainfed systems – II management technologies and crop productivity. Current Science.110 (3) 320-328.

Jayanta K. Saha, Rajendiran Selladurai, Vassanda Coumar M., Dotaniya M.L., Samaresh Kundu and Ashok K. Patra 2017. Soil Pollution - An Emerging Threat to Agriculture. Springer Nature publication, Singapore.

Kabata-Pendias A 2000. Trace element in soils and plants, Third edn. CRC Press, Baton Raton, p 432

Laryea K.B and Vittal K.P.R 1995. Degrades Soils. In 'Soils of Andhra Pradesh'- A Monograph. Ed by I.V.Subba Rao, Andhra Pradesh Agricultural University, Hyderabad, India. Pp.305-308.

Latha M R and Janki P 2015. Problem soils and their management http://agritech.tnau.ac.in/pdf/3.pdf

Mandal, S.C. 1976 Acid soils of India and their management. Bull. Indian Soc. Sci. 11, 191-197.

Mark H. Powell ed.1996. Nitrogen fixing trees for acid soils: a field manual, Published by Winrock International's Nitrogen Fixing Tree Association and the Tiawan Forestry Research Institute with financial support from the Council of Agriculture.

Michael C., Amacher, Katherine P. O'Neill, and Charles H. Perry 2007. Soil Vital Signs: A New Soil Quality Index (SQI) for Assessing Forest Soil Health. Natural Resources Research Center, 2150 Centre Avenue, Building A, Fort Collins, CO 80526

Minhas P. S and Sharma O.P 2003. Management of Soil Salinity and alkalinity Problems in Inda. In 'Crop Production in Saline Environments- Global and Integrative Perspectives' Ed. by Sham S. Goyal, Surinder K. Sharma and William Rains D. Food Products Press New York pp.181-230

MOEF 2000. Municipal solid wastes (management and handling) rules, 2000. Notification, dated 25th September, 2000. Ministry of Environment and Forests, New Delhi. Available via http:// www.moef.nic.in/legis/hsm/mswmhr.html. Accessed 14 Sept 2016

Muhammad Aqeel Ashraf, Mohd. Jamil Maah and Ismail Yusoff 2014. Soil Contamination, Risk Assessment and Remediation. In Environmental Risk Assessment of Soil Contamination. http://dx.doi.org/10.5772/57287.

Murthy, R.S. 1980. Acid sulphate soils of India. FAO World Soil Resources Report, 41, 24-29.

Nagarajarao, Y and Prabhu Prasadini 1995. Soil Physical Constraints. In 'Soils of Andhra Pradesh'- A Monograph. Ed by I.V.Subba Rao, Andhra Pradesh Agricultural University, Hyderabad, India. Pp.267-282.

Navjot Singh, Buta Singh Dhillon and Parminder Singh Sandhu 2014. Alleviation of soil physical constraints for improved crop production: A Review. Eco. Env. & Cons. 20 (1) : 31-38

Nayak, A.K., Chinchmalatpure, Anil R., Rao, G. Gururaja and Tyagi, N.K. 2004. Effect of electrolyte concentration, ESP and soil texture on the flocculation, hydraulic conductivity and bulk density of Vertisols. Extended Summaries, International conference on Sustainable management of Sodic lands, Lucknow, India. Pp 203-205.

NRSA 1996. National Remote Sensing Agency. Mapping of salt affected soils of India, 1:250000 map sheets, Legend, NRSA Hyderabad.

Pasricha, N.S., Bansal S.K. and Golakiya B.A. 2000. Balanced Nutrition of Groundnut and Other Field Crops Grown in Calcareous Soils of India. Proceedings of National Symposium. September 19-22, 2000, held at Junagadh, Gujarat Organized by Potash Research Institute of India, International Potash Institute and Gujarat Agricultural University

Phocaide A. 2007. Water quality for irrigation. In Handbook on Pressurized Irrigation Techniques. FAO, Rome. https://prime.erpnext.com/files/CHAPTER7-Waterqualityforirrigation.pdf

Rishikesh thakur and Somanath Sitharam 2007. Multipurpose tree species for bioremediation of problem soils http://www.jnkvv.org/PDF/20042020125523204200744_2.pdf

Roy PS, Dwivedi RS, Vijayan D 2015 Remote sensing applications. National Remote Sensing Centre (NRSc), Hyderabad. https://www.nrsc.gov.in/sites/default/files/pdf/ebooks/Chap_4_Soils.

Saha JK, Panwar N et al 2013a. Risk assessment of heavy metals in soil of a susceptible agro-ecological system amended with municipal solid waste compost. J Indian Soc Soil Sci 61:15–22

Shaon Kumar Das and Suvo Kumar Das 2015. Acid Sulphate Soil: Management Strategy for Soil Health and Productivity. Popular Kheti 3, (2) April-June 1-7.

Shashi A. Verma 2010 Degraded and Wastelands of India Status and Spatial Distribution. Published by Dr T P Trivedi, Project Director, Directorate of Information and Publications of Agriculture, Indian Council of Agricultural Research, Krishi Anusandhan Bhavan I, Pusa, New Delhi 110 012

Singa Rao M. and Padma Raju A. 1995. Soil air and thermal regimes. In 'Soils of Andhra Pradesh'- A Monograph. Ed by I.V.Subba Rao, Andhra Pradesh Agricultural University, Hyderabad, India. Pp.305- 308.

Soil Health Concepts part-1, Comprehensive Assessment of Soil Health - The Cornell Framework Manuahttp://www.css.cornell.edu/extension/soil-health/1concepts.pdf

Steffan J.J., Brevika, E.C. Burgessa, L.C. and Cerdà A.2018. The effect of soil on human health: an overview. Eur J Soil Sci. 2018 January ; 69(1): 159–171

Suganthi M., MuthukrishnanP., Chinnusamy C and SureshSubramonian B. 2017. Management of Soil Physical constraints- A Review. International Journal of Science, Environment and Technology, Vol. 6, No 2, 2017, 1215 – 1222

Swarajyalakshmi, G., Gurumurthy, P and Subbiah G.V. 2003. Soil Salinity in South Inida; Problems and solutions. In 'Crop Production in Saline Environments- Global and Integrative Perspectives' Ed. by Sham S. Goyal, Surinder K. Sharma and William Rains D. Food Products Press New York pp.247- 276

USDA Soil Conservation service 2010. Chapter 5 Role of Soils in Waste Management. Agricultural Waste Management Field Handbook.

USDA, NRCS Soil health, health soil for life https://www.nrcs.usda.gov/wps/portal/nrcs/main/soils/health/

Veerabhadra Rao K. and Srinivasa Reddy K. 1995. Acid Soils. In 'Soils of Andhra Pradesh'- A Monograph. Ed by I.V.Subba Rao, Andhra Pradesh Agricultural University, Hyderabad, India. Pp.335-344.

Wilcox L. V. 1955. Classification and Use of Irrigation Waters, United States Salinity Laboratory (USSL). Circular No. 969